THE OXFORD HANDBOOK OF

EEG FREQUENCY

OXFORD LIBRARY OF PSYCHOLOGY

The Oxford Handbook of Internet Psychology
Edited by Adam Joinson, Katelyn McKenna, Tom Postmes, and Ulf-Dietrich Reips

The Oxford Handbook of Job Loss and Job Search
Edited by Ute-Christine Klehe and Edwin van Hooft

The Oxford Handbook of Parenting and Moral Development
Edited by Deborah J. Laible, Gustavo Carlo, Laura M. Padilla Walker

The Oxford Handbook of Stigma, Discrimination, and Health
Edited by Brenda Major, John F. Dovidio, and Bruce G. Link

The Oxford Handbook of Deaf Studies in Learning and Cognition
Edited by Marc Marschark and Harry Knoors

The Oxford Handbook of 4E Cognition
Edited by Albert Newen, Leon De Bruin, and Shaun Gallagher

The Oxford Handbook of Attention
Edited by Anna C. Nobre and Sabine Kastner

The Oxford Handbook of Clinical Child and Adolescent Psychology
Edited by Thomas H. Ollendick, Susan W. White, and Bradley A. White

The Oxford Handbook of Group Creativity and Innovation
Edited by Paul B. Paulus and Bernard A. Nijstad

The Oxford Handbook of Organizational Citizenship Behavior
Edited by Philip M. Podsakoff, Scott B. MacKenzie, and Nathan P. Podsakoff

The Oxford Handbook of Digital Technologies and Mental Health
Edited by Marc N. Potenza, Kyle Faust, David Faust

The Oxford Handbook of Sexual and Gender Minority Mental Health
Edited by Esther D. Rothblum

The Oxford Handbook of Psychological Situations
Edited by John F. Rauthmann, Ryne Sherman, David C. Funder

The Oxford Handbook of Human Motivation, 2e
Edited by Richard Ryan

The Oxford Handbook of Integrative Health Science
Edited by Carol D. Ryff and Robert F. Krueger

The Oxford Handbook of Positive Psychology, 3e
Edited by C.R. Snyder, Shane J. Lopez, Lisa M. Edwards, Susana C. Marques

The Oxford Handbook of Dialectical Behaviour Therapy
Edited by Michaela A. Swales

The Oxford Handbook of Music and the Brain
Edited by Michael H. Thaut and Donald A. Hodges

The Oxford Handbook of Expertise
Edited by Paul Ward, Jan Maarten Schraagen, Julie Gore, and Emilie M. Roth

The Oxford Handbook of Singing
Edited by Graham F. Welch, David M. Howard, and John Nix

The Oxford Handbook of Evolutionary Psychology and Behavioral Endocrinology
Edited by Lisa L. M. Welling and Todd K. Shackelford

The Oxford Handbook of Autism and Co-Occurring Psychiatric Conditions
Edited by Susan W. White, Brenna B. Maddox, Carla A. Mazefsky

The Oxford Handbook of Adolescent Substance Abuse
Edited by Robert A. Zucker and Sandra A. Brown

THE OXFORD HANDBOOK OF

EEG FREQUENCY

Edited by

PHILIP GABLE, MATTHEW MILLER,
EDWARD BERNAT

OXFORD
UNIVERSITY PRESS

OXFORD
UNIVERSITY PRESS

Great Clarendon Street, Oxford, OX2 6DP,
United Kingdom

Oxford University Press is a department of the University of Oxford.
It furthers the University's objective of excellence in research, scholarship,
and education by publishing worldwide. Oxford is a registered trade mark of
Oxford University Press in the UK and in certain other countries

First Edition published in 2022
Impression: 1

Published in the United States of America by Oxford University Press
198 Madison Avenue, New York, NY 10016, United States of America

British Library Cataloguing in Publication Data
Data available

CIP data is on file at the Library of Congress

ISBN 978–0–19–289834–0

DOI: 10.1093/oxfordhb/9780192898340.001.0001

Printed and bound by
CPI Group (UK) Ltd, Croydon, CR0 4YY

Contents

PART II

PART III

PART IV

List of Contributors

John J. B. Allen, Distinguished Professor of Psychology, Department of Psychology, University of Arizona, Tucson, AZ

Selin Aviyente, Professor, Department of Electrical and Computer Engineering, Michigan State University, East Lansing, MI

Martha Ann Bell, University Distinguished Professor, Department of Psychology, Virginia Tech, Blacksburg, VA

Róbert Bódizs, Senior Research Fellow, Semmelweis University, Institute of Behavioural Sciences, Budapest, Hungary

Steven L. Bressler, Professor, Center for Complex Systems and Brain Sciences, Department of Psychology, Florida Atlantic University, Boca Raton, FL

James F. Cavanagh, Associate Professor, Department of Psychology, University of New Mexico, Albuquerque, NM

Michael X Cohen, Assistant Professor, Donders Centre for Neuroscience & Radboud University Medical Center, Nijmegen, Netherlands

Kimberly Cuevas, Associate Professor, Department of Psychological Sciences, University of Connecticut, Storrs, CT

Kyle J. Curham, Department of Psychology, University of Arizona, Tucson, AZ

Paige Ethridge, Department of Psychology, McGill University, Montreal, QC, Canada

Spencer Fix, Department of Psychology, University of Maryland College Park, College Park, MD

Dan Foti, Associate Professor, Department of Psychological Sciences, Purdue University, West Lafayette, IN

Philip A. Gable, Associate Professor of Psychology, Department of Psychological and Brain Sciences, University of Delaware, Newark, DE

Lilianne Gloe, Department of Psychology, College of Social Science, Michigan State University, East Lansing, MI

Benjamin Griffiths, School of Psychology, University of Birmingham, Edgbaston, Birmingham, UK

Simon Hanslmayr, Professor, Centre for Cognitive Neuroimaging, School for Neuroscience and Psychology, University of Glasgow, UK

Eddie Harmon-Jones, Professor, School of Psychological Science, University of New South Wales, Sydney, Australia

Jeremy Harper, Department of Psychiatry and Behavioral Sciences, University of Minnesota, Minneapolis, MN

Christoph S. Herrmann, Professor, Experimental Psychology Lab, Carl von Ossietzky University, Oldenburg, Germany

William G. Iacono, Professor, Department of Psychology, University of Minnesota, Minneapolis, MN

Florian H. Kasten, Experimental Psychology Lab, Carl von Ossietzky University, Oldenburg, Germany

Andreas Keil, Professor of Psychology, Department of Psychology and Center for the Study of Emotion & Attention, University of Florida, Gainesville, FL.

J. Leon Kenemans, Professor, Biopsychology and Psychopharmacology, Utrecht University, Utrecht, Netherlands

Alpár S. Lázár, Associate Professor, School of Health Sciences, University of East Anglia, Norwich, UK

Zsolt I. Lázár, Assistant Professor, Faculty of Physics, Babeş-Bolyai University, Cluj-Napoca, Romania

Courtney Louis, Department of Psychology, College of Social Science, Michigan State University, East Lansing, MI

Stephen M. Malone, Research Assistant Professor and Co-investigator, Minnesota Center for Twin & Family Research, University of Minnesota, Minneapolis, MN

Ali Mazaheri, Associate Professor, School of Psychology, University of Birmingham, Edgbaston, Birmingham, UK

Sebastian Michelmann, Princeton Neuroscience Institute, Princeton, NJ

Matthew W. Miller, Associate Professor, School of Kinesiology, Auburn University, Auburn, AL

Jason Moser, Professor, Department of Psychology College of Social Science, Psychology, Michigan State University, East Lansing, MI

Belel Ait Oumeziane, Department of Psychological Sciences, Purdue University, West Lafayette, IN

J. Matias Palva, Research Director, Department of Neuroscience and Biomedical Engineering, Aalto University, Helsinki

Satu Palva, Professor, Centre for Cognitive Neuroimaging, School of Psychology and Neuroscience, University of Glasgow, Glasgow

Taylor Popp, Carnegie Mellon University, Pittsburgh, PA

John E. Richards, Professor, Department of Psychology, University of South Carolina, Columbia, SC

Stefanie Russman Block, Department of Psychology, College of Social Science, Michigan State University, East Lansing, MI

Dennis J. L. G. Schutter, Professor, Experimental Biopsychology of Motivation and Emotion, Utrecht University, Utrecht, Netherlands

Kevin M. Spencer, Research Health Scientist, Veterans Affairs Boston Healthcare System; and Associate Professor of Psychiatry, Harvard Medical School, Boston, MA

Daniel Strüber, Adjunct Professor, Experimental Psychology Lab, Carl von Ossietzky University, Oldenburg, Germany

Nina Thigpen, Research Scientist at Google X, The Moonshot Factory, Mountain View, CA.

Bernadette C. M. Van Wijk, Marie Curie Research Fellow at the University of Amsterdam, Amsterdam, Netherlands

Bradley Voytek, Associate Professor, Department of Cognitive Science, UC San Diego, San Diego, CA

Anna Weinberg, Assistant Professor, Department of Psychology, McGill University, Montreal, QC, Canada

Wanze Xie, Assistant Professor, School of Psychological and Cognitive Sciences, Peking University, Beijing, China

Foreword

STEVEN J. LUCK

When Hans Berger conducted his pioneering human EEG recordings in the 1920s, his first major discovery was the *alpha rhythm*, a 10-Hz oscillation that grew larger when the subject's eyes were closed. A few years later, Lord Adrian (my intellectual great-great-grandfather) showed that the alpha rhythm also varied according to whether the subject was focusing intensely or daydreaming. Thus the study of EEG oscillations was born.

But this area of research underwent a protracted childhood, because the scientists of the mid-twentieth century could not easily see smaller oscillations amidst the chaotic twists and turns of the scalp EEG. To pull out specific neural processes from the complex and noisy EEG, they began to use signal averaging techniques that can isolate the brain potentials that are triggered by specific events such as the onset of a light (the *event-related potentials* or ERPs). However, these techniques assume that the phase of the signal is constant across trials, and the application of signal averaging to EEG data eliminates or hopelessly distorts oscillating activity. Indeed, for the first 30 years of my own research career, I viewed the alpha rhythm as a nuisance that should be suppressed lest it contaminate my precious ERP waveforms.

All of this began to change in the 1980s and 1990s, partly driven by high-profile studies of local field potential oscillations in animals and partly driven by the application of time-frequency analysis methods to human EEG recordings. The brain oscillations that were obscured by signal averaging could now be visualized and quantified. A new generation of researchers began studying human EEG oscillations and linking them with microelectrode data from animals and computational models of brain dynamics.

In science, the introduction of a new approach often leads to a burst of progress followed by the realization that things are not as simple as they seem. The new wave of oscillation research followed this common path, with the discovery of important new phenomena being accompanied by conceptual and methodological challenges. One such challenge—related to the famous Heisenberg Uncertainty Principle—is that precision in the time-domain is inversely related to precision in the frequency domain. The more precisely you determine the frequency of an oscillation, the less you can say about when the oscillation was present.

A second important challenge is the difficulty of distinguishing between bona fide oscillations and other kinds of neural events. Part of the genius of the Fourier transform is that *any* waveform—whether or not it contains oscillations—can be reconstructed by summing together a set of sinusoids. As a result, when you apply a method such as the Fourier transform to a time-domain signal such as the EEG, you will always see activity at some frequencies, whether or not the signal is actually oscillating. And what if the brain is oscillating, but not in a sinusoidal manner? This can lead to completely artifactual results, such as the apparent coupling of the amplitude of one frequency with the phase of another frequency.

A third key challenge is the interpretation of brain oscillations. Virtually any system containing multiple interconnected parts will oscillate when energy is introduced. Those oscillations can be essential to the function of the system, as when the sound of a single violin fills a concert hall or when a clock keeps perfect time. But oscillations can also be a sign of trouble, as when a bridge shakes violently following an earthquake or when a seizure spreads throughout the brain. The fact that the brain oscillates, and that the oscillations vary across states or tasks, does not mean that the oscillations themselves are playing a functional role in the brain's computations. They may be epiphenomenal. Or they may be fundamental. Distinguishing between these possibilities may require invasive recordings in animals or experimental manipulations of oscillations via brain stimulation.

The study of EEG oscillations has reached a key point in its development. The application of time-frequency methods to EEG data is now commonplace, aided by open source data analysis packages such as FieldTrip and EEGLAB. Labs that eschewed these methods for many years—including my own—are now examining oscillations alongside traditional event-related potentials. However, this success means that more researchers are using time-frequency analyses without understanding the challenges involved in properly quantifying and interpreting oscillatory activity. As a result, this edited volume has appeared at the perfect time.

The chapters in this volume will give scientists of all career stages the knowledge they need to understand how oscillations arise in the brain, how they can be accurately quantified, and how they can be appropriately interpreted. I encourage readers to think deeply about the many important issues that are raised in these chapters, especially with regard to the three challenges I have outlined. But if you think deeply and follow the "best practices" described in this volume, you will be able to see brain activity that would otherwise be invisible. And you may make important new discoveries about the human mind and brain.

Preface

Motivation

The time is ripe for a comprehensive book on the array of historical and cutting-edge frequency and time-frequency approaches to studying EEG/ERP because of the widespread interest in frequency research. There is a great need for a book organizing the diverse and important methods of EEG frequency analyses and interpreting the resultant measures.

One stream of research comes from traditional (band-based) frequency analyses. Likewise, understanding the cutting-edge frequency analyses which may not be familiar to many EEG researchers is increasingly important for investigators applying frequency analyses. However, there is a major need for a comprehensive handbook on analyses within this domain. Although research has been rapidly accumulating over the last decade, there has not been sufficient organization of research on the topic. We believe this comprehensive handbook is increasingly necessary to help delineate the boundaries of the area, the major scientific questions that need to be addressed, and the core theoretical frameworks that can guide future research and development.

Thus, a specific goal of this book is to bring together these various scientific perspectives and research approaches within a single reference volume that provides an integrated, cutting-edge overview of the current state of the field. This volume comprises contributions from leading researchers within various allied disciplines.

The use of electroencephalography (EEG) to study the human mind has seen tremendous growth across a vast array of disciplines due to increased ease of use and affordability of the technology. EEG is a non-invasive measure of electrical brain activity. Typically, researchers investigate the EEG signal using either time-domain (e.g., ERP) analyses or frequency analyses. Several books have examined practicalities of conducting ERP analyses and interpreting various ERP measures. However, a comprehensive book has yet to be developed organizing the numerous ways to process EEG frequency and interpreting frequency measures linked to cognitive, affective, and motor processes.

We (editors Philip, Matt, and Ed) felt a great need for a book organizing the diverse and fascinating methods of EEG frequency analyses and interpreting the resultant measures. Frequency analyses provide unique assessments of neural functioning, neural connectivity, and "resting" neural activity studied by EEG researchers. Further, frequency-domain measures are reliably associated with cognitive, affective, and motor processes of great interest to neuroscientists and psychological scientists. For example, asymmetrical activation of the frontal cortex as measured by the inverse of alpha-band activity is closely linked with motivation and emotion. In addition, analyses examining the synchrony of EEG frequencies recorded from different scalp locations allow researchers to examine brain connectivity without having to incur the costs of magnetic resonance imaging.

ORIGIN

As EEG frequency researchers, we wanted a resource that introduced the myriad ways in which EEG frequency analyses are being investigated. This volume began while the lead editor, Philip, was on sabbatical and seeking to begin a new chapter in his career. During that time, he visited with Matt Miller and the project quickly developed into a collaborative project. After developing the project more, Matt and Philip brought Ed Bernat on board.

As individual editors, we each had areas of expertise in EEG frequency research, but in developing this volume, we quickly discovered that each of our individual areas of expertise were far different from each other. Three editors were necessary to even try to cover the breadth of EEG frequency research being conducted. In addition, we did not want to create a handbook that focused on EEG frequency research specifically in emotion, cognition, or clinical applications. Instead, we wanted to provide a survey of the breadth of work being conducted with EEG frequency research. Try as we might, we also acknowledge this handbook will inevitably fail to cover everyone's interest across all topics. To that end, we hope to receive feedback from readers so that future editions of this handbook can be expanded to encompass the ever-growing field of EEG frequency research.

Together, the project has been a long labor, but well worth the time and effort to develop the resource. We have been especially excited to work with leading experts in the field as they develop chapters for the volume. We are excited for you as the reader to see what we have gotten to see throughout the editing process: the excitement and development of EEG frequency research across a wide range of fields and programs of research.

ORGANIZATION

To aid in reading the handbook, much thought and structuring has been given to the organization of the chapters. As a whole, the book provides a systematic summary of EEG frequency analyses and applications. Individual chapters give depth to each type of frequency analysis and interpretation of resultant measures. Chapters are organized into three sections.

The first section of the book is focused on basics of EEG frequency research and linking frequency analyses to other components of EEG research, such as event-related potential (ERP) components and the fundamentals of inference from EEG recording. For the second section, contributors focus on specific EEG frequency components that are commonly studied using traditional frequency bands of activity to study specific psychological processes related to cognition, motivation, and perception. The third section focuses on EEG frequency analyses in special populations and altered states. The fourth section of the handbook concludes with chapters focused on advancing methodology used in EEG frequency research.

The initial chapters in the first section describe methods for collecting EEG data for frequency analyses in humans as well as the basics of electrical activity and assumptions regarding the EEG signal. Following these chapters, contributors consider how the oscillations in the EEG signal may give rise to psychological phenomena, and how ERPs can be decomposed into time-frequency components. The chapters go on to consider the relationships between evoked (ERP-related) and induced EEG activity before shedding light upon frequency analyses of LFPs in non-human primates, which may inform frequency analyses of human EEG. The second section begins with the role of gamma oscillations in cognitive and sensory processing. Then, frontal midline theta, which is often linked to cognitive control, is addressed, followed by the role of alpha and beta oscillations in perception and memory. Next, research on asymmetries in frontal alpha oscillations are linked with motivation, followed by a description of the role of oscillatory activity in sensorimotor function. Moving into the third section of the book, the first chapter in this section focuses on changes in EEG frequency throughout infancy, childhood, adolescence, and early adulthood. Next, follows an examination of the ratio of theta to beta power in motivation and attentional control as they relate to normal and abnormal behavior. Then, the characteristics of the oscillations in persons with schizophrenia are described, followed by an examination of how frequency analyses clarify control processes in people with anxiety. The concluding section of the handbook begins with specialized frequency analyses for source localization and brain connectivity before concluding with chapters describing how to manipulate oscillatory activity with brain stimulation and how to parameterize neural field potential data to, for example, tease apart true oscillations from aperiodic signals.

LIMITATIONS OF THE CURRENT VOLUME

As with any book, the current volume is not a complete work on the topic of EEG frequency analyses. In addition, this is not the first book on EEG frequency research. Many excellent books have been published this topic. Here we mention what the current book does not include and refer the reader to additional resources available in other books.

One of the most inspiring resources for us as editors is Steven J. Luck and Emily S. Kappenman's *Oxford Handbook on Event-Related Potential Components*. This book has been used in our courses and labs, as well as by countless other EEG researchers. It focuses on the excellent work that has investigated the spectrum of ERP components derived from EEG research. We were inspired to build a similar handbook that would cover EEG frequency analyses. The current book does not address ERP research in much detail. The closest chapters addressing ERP research are Chapters 4 and 5. For those desiring a more comprehensive volume on ERP analyses, please see Luck and Kappenman.

Another extraordinary resource is Mike X. Cohen's *Analyzing Neural Time Series Data: Theory and Practice*. This book has served as the primary resource for many EEG researchers, including us, to learn how to conduct time-frequency analyses, and to teach our students how to perform the analyses. The book is particularly helpful in guiding the reader from the mathematical bases of frequency analyses to the implementation of these analyses in MATLAB. These analyses are referenced throughout our book and include fast Fourier transforms, complex Morlet wavelet convolution, intertrial phase clustering, surface Laplacian filtering, phase- and power-based connectivity measures, and cross-frequency coupling. Chapters in our book that address these "how-to" topics include Chapters 4, 5, 19, 20, 21, and 23.

Finally, Chapter 1 of this volume describes how to collect EEG data. For readers who which to have greater detail about implementing and collect EEG from human participants, we recommend Dickter and Kieffaber's *EEG Methods for the Psychological Sciences*. This book is an excellent resource for researchers beginning to implement EEG.

PART I

CHAPTER 1

INTRODUCTION

Methods for Collecting EEG Data for Frequency Analyses in Humans

PHILIP A. GABLE AND MATTHEW W. MILLER

1.1 Chapter Aims

This chapter aims to provide a structure for readers to understand the methodology behind collecting EEG (electroencephalography) presented in the subsequent chapters. It is important to first understand the research methods involved in recording EEG frequency before delving into more advanced frequency analyses and the interpretations. As researchers, we focus this first chapter on a brief introduction to the topic of EEG methodology and scientific practices. To begin, however, we feel it is important to lay down definitions for terms used throughout the book.

1.2 Definitions of EEG Frequency Research

EEG refers to the recording of electrical brain activity from the human scalp. It is one of the most common methods for measuring brain functioning in areas of mind, brain, and behavior science. EEG data contain rhythmic activity or waves that may reflect neural oscillations, or fluctuations in the excitability of populations of neurons (more on this later). These rhythmic fluctuations are typically described using two main descriptors. The first is frequency, which is the speed of the wave, and it is measured in hertz (Hz), which is the number of wave cycles per second. The second is power, which is the squared amplitude of the wave. The greater the power of an oscillation, the greater the energy of that oscillation. All of the chapters in this work discuss frequency with most referencing power. Sometimes researchers investigate the phase of the wave, which is the position of the wave measured in radians or degrees. Many of the chapters included here discuss phase.

The brain produces rhythms in multiple frequencies, which can be isolated from the raw EEG signal using multiple techniques described by Curhamn & Allen in Chapter 2, and Voytek in Chapter 23). Different psychological processes are linked to different frequencies, which are often grouped into bands. The most commonly studied bands include delta (1–4 Hz), theta (4–8 Hz), alpha (8–12 Hz), beta (13–30 Hz), and gamma (lower gamma 30–80 Hz; upper gamma 80–150 Hz). While these are not the only frequency bands, these are the bands most typically associated with processes of mind and behavior measured by EEG. Importantly, these bands are not defined without reason, but instead reflect biological changes at the cellular level (see Cohen, 2014 and Buzsaki, 2006 for reviews). However, these bands are not rigid and may vary depending on individual differences, such as brain development, structure, and chemistry. Chapters in the second section of this work note how different frequency bands are associated with cognitive, motivational, and sensorimotor processes.

These definitions are by no means complete. Individual chapters provide more precise definitions of terms used. With this initial framework, readers should be able to venture into subsequent chapters focusing in more detail on these definitions. Because EEG is a rather complex measure, we focus in more detail on the physiological basis and scientific methods used to record and process EEG.

1.3 Physiological Basis of EEG

EEG is measured because all nerve cells communicate using electrical signals, sending information throughout the brain and to the rest of the body. Within a neuron, an action potential is an electrical wave that runs from the axon hillock at the cell body to the axon terminals. At the axon terminals, the action potential causes neurotransmitter to be released. This neurotransmitter crosses the synaptic gap and binds to receptors on the membrane of the postsynaptic cell. Binding to the receptor causes voltage changes by activating ion channels or second messengers that either excite or inhibit the postsynaptic neuron. The summation of this voltage change in the membrane of the dendrites and cell body of the postsynaptic neuron is called a postsynaptic potential. Postsynaptic potentials tend to occur locally, rather than moving down the axon. This allows postsynaptic potentials to summate rather than to cancel, resulting in voltage changes that have larger amplitudes and can be recorded on the cortical surface or at the scalp.

When tens of thousands to millions of neurons are excited or inhibited at the same time, the voltage change outside the cell (extracellular potential) can be recorded at the scalp using EEG, which measures the sum of electrical activity from excitatory and inhibitory postsynaptic potentials over this collection of neurons. The activity can only be recorded on the scalp surface because tissue (cerebrospinal fluid, meninges, skull, and skin) between the neurons and scalp conducts the electrical signal. In addition, for electrical activity to be projected to the scalp, cellular alignment must be precisely arranged in parallel so that their effects cumulate to project the electrical activity to the scalp (see Curhamn & Allen, Chapter 2; Keil & Thigpen, Chapter 3). Neurons must be arranged so that the cluster of neurons all have dendrites at one pole and axons departing at the opposite pole. This arrangement is called an open field and occurs when neurons are organized in layers. The cortex, cerebellum, and parts of the thalamus tend to have this open field arrangement of neurons resulting from pyramidal cells.

1.4 A Dual Nature

Due to the electrical basis of the EEG signal, EEG has excellent ability to tell us *when* something is happening in the brain. This is called temporal resolution and is one of the greatest strengths of the EEG signal. The EEG signal measures neural activity at the accuracy of milliseconds, which allows for the ongoing measurement of psychological processes as they unfold (Luck, 2014).

However, the EEG signal is limited in its ability to measure *where* something is occurring. This is called spatial resolution and is one of the greatest weaknesses of the EEG signal. Depending on where the source of the EEG signal is generated, the orientation of the open field neurons might not be parallel to the scalp, thus generating EEG signals in multiple directions. In addition, resistors (e.g., the skull) in the tissue between the neurons and scalp can cause the EEG signal to spread out. Because of the volume conduction through the head, as well as the orientation of the pyramidal cells emitting the signal, the spatial location of the signal is difficult to ascertain. As Keil and Thigpen (Chapter 3) note, a difference in frequency power between two experimental conditions could be the result of a different number of neurons activated, the temporal order in which they were activated, or neurons with different orientations being activated. To address EEG's limited spatial resolution, cortical source localization techniques have been developed and are reviewed by Xie and Richards (Chapter 19). In sum, using an analogy from Steve Luck's ERP Boot Camp, the EEG signal is like being able to see every frame of a movie as it unfolds. However, because of the low spatial resolution, the movie appears a bit blurry.

1.5 EEG Equipment and Recording

The earliest method of EEG measurement was implemented by Hans Berger in the late 1920s. In his early experiments, he used two sponges soaked in saline connected to an amplifier (Berger, 1929). While the equipment and processing of EEG signal has advanced considerably since that time, the basic components remain similar. EEG electrodes are placed on or near the head, the signal from the electrode is transmitted to an amplifier, and the signal is digitized and recorded.

In psychological research labs, EEG is usually recorded from 32, 64, 128, or more electrodes. In other research (e.g., sleep, nonhuman) fewer electrodes (2–8 electrodes) is more typical. When larger numbers of electrodes are used, electrodes are mounted in an electrode cap or net. When fewer electrodes are used, electrodes may be positioned individually on the head using a bonding agent. Electrodes can either be wet electrodes or dry electrodes. Wet electrodes are made of silver, silver-chloride, or tin, and a conductive gel or liquid is placed inside or around the electrode. Dry electrode systems use electrodes coated in gold, silver, or nickel, and place electrodes directly on the scalp without a conductive medium. Wet electrodes generally have higher signal quality, but dry electrodes may be preferred when high impedance levels are tolerable, or when recording for long periods.

Electrode systems will have active or passive electrodes. Active electrodes contain a small pre-amplification unit directly attached to the conductive metal in the electrode. This allows the EEG signal picked up at the sensor to be immediately amplified before additional environmental noise can be introduced. Passive electrodes do not have amplification at the electrode, and instead carry the signal to an amplifier about a meter away. Compared to passive electrodes, active electrodes minimize noise introduced during signal transmission, tolerate high impedance recording, and reduce participant preparation time. Passive electrodes have a lower profile to benefit transcranial magnetic stimulation over the cap and can be used inside an MRI bore.

Electrode placement is predominantly based on the 10–20 system (Jasper, 1958). Electrodes are named using the first letter to refer to the brain region under the electrode from anterior to posterior (e.g., Fp—frontal pole, F—frontal, C—central, P—parietal, T—temporal, O—occipital). Numbers following the letter are used to indicate the lateral position of the electrodes. Ascending odd numbers indicate sites more lateral over the left hemisphere of the brain, whereas ascending even numbers indicate sites more lateral over the right hemisphere of the brain. The letter Z is used to designate medial sites. In addition to the recording electrodes, EEG also requires a ground electrode, which assists in reducing electrical noise, as well as a reference electrode placed on the head or face.

The raw EEG signal is usually filtered during recording. Signals below 0.1 Hz or above 200 Hz are removed because the frequency bands of interest fall within this range. A filter at 60 Hz (in North and South America) or 50 Hz (in Europe, Asia, and Africa) may also be used to further reduce electrical noise from alternating current.

1.6 Artifacts

The quality of the EEG is crucial to EEG frequency analysis. To best record EEG signal reflecting brain activity, researchers must remove signal that occurs because of anything other than neural activity. Signal that is not naturally present is called artifact. These can be biological (e.g., muscle movement) or nonbiological (e.g., electrical noise). Much artifact can by eliminated by taking preventative measures. Artifact that cannot be prevented should be removed.

Muscle artifact, or electromyography (EMG), is one of the most common types of artifacts and is usually high in frequency (100–500 Hz). Usually, this falls outside of the frequency range typically investigated by researchers. However, some muscle artifact may seep into lower frequencies. Researchers can reduce muscle artifact by instructing participants to limit their muscle movements. Muscle artifact that does occur can be removed through visual inspection, filtering, and automatic artifact detection algorithm. It should be noted that some muscle artifact may be related to the experiment (e.g., sensorimotor studies). In such cases, it may be beneficial to measure EMG at the site movement is expected (e.g., the hand), then control for it in analyses.

Eye movements are another common type of artifact. The eyeball is polarized which causes large artifact in the form of voltage changes resulting from moving the eyes. Like dealing with muscle artifact, eye movement artifact can be removed from signal using recordings near the eyes called electro-oculograms (EOG). One pair of EOG electrodes are placed above and below the eye, while another pair is placed just lateral to either eye on the temple. Eye blinks also create eye movement artifact. It is preferable to correct blink artifact using an artifact reduction algorithm based on regression, principal component analyses, or independent component analyses.

Artifacts occurring in the environment are the result of nonbiological factors. The most common sources of these artifacts are the result of external electrical noise coming from compact fluorescent lightbulbs, data hubs, or electrical junctions. Grounding will aid in reducing these sources of noise, as will electromagnetically shielded rooms.

1.7 Frequency Processing

Once an EEG signal is recorded, the raw data must go through several processing steps before it is in a format useable in analyses. The raw signal is collected in the time-domain but must be converted into a frequency-domain representation. One way this can be accomplished is in the form of a power spectrum, which collapses frequency data across time to map the frequencies present. A frequency analysis can be conducted over windows that are minutes, seconds, or milliseconds in length; these are called epochs. Epochs that are seconds or minutes can be analyzed for power spectra using a Fourier transform, which decomposes a signal into a series of sine and cosine functions of various frequencies. The function of each frequency begins with its own phase. A Fourier transformation assumes that the epoch repeats infinitely forward and infinitely backward in time. A process called windowing is used to prevent artifact created from the Fourier transform. However, windowing can also introduce artifact and data loss into the frequency analysis. Overlapping epochs is a way to prevent discontinuity, data loss near the ends of the epoch, and to help meet the assumptions of the Fourier transform in windowing.

One of the most common forms of signal frequency processing is to use a fast Fourier transform (FFT). An FFT provides the spectrum of frequency power for a period, which is often averaged across a range of frequencies comprising a band (e.g., theta). It also provides a spectrum of phase. The power spectrum reflects the energy of each frequency determined by the squared amplitude of the wave. The phase spectrum reflects the phase in radians or degrees of the sine or cosine wave at each interval (e.g., 1/T). Most frequency analyses focus exclusively on frequency power. However, there is increasing interest in examining the phase of frequencies (e.g., see Michelmann, Griffiths, & Hanslmayr, Chapter 10; Palva & Palva, Chapter 20). Crucially, the FFT has two limitations: it poorly depicts changes in the frequency spectrum over time, and it assumes the EEG data are stationary during the period to which the FFT is applied. To overcome these limitations, time-resolved frequency decomposition techniques are growing in popularity, particularly wavelet analyses (e.g., complex Morlet wavelets), which reveal changes in power at various frequencies with excellent temporal precision (see Aviyente, Chapter 4; Weinberg, Ethridge, Oumeziane, & Foti, Chapter 5).

1.8 Experimental Design

Regardless of the EEG recording and processing, researchers need to be considerate of experimental design. In any experiment, researchers should manipulate a single variable at a time to ensure internal reliability, but this procedure is challenging in EEG research. This follows because manipulating one variable between conditions may inadvertently change a second variable between conditions that affects EEG. For example, if an experimental condition attempts to manipulate participants' motivation while they are physically responding to stimuli, then EEG linked to motivation may change (see Harmon-Jones, Popp, & Gable, Chapter 11) but so may the vigor of their responses and EEG linked to sensorimotor function (see van Wijk, Chapter 12). Thus, it is crucial that researchers attempt to experimentally control for such factors in their experimental design or statistically account for them by collecting covariates, such as EMG to index the vigor of motor responses.

1.8.1 Reproducibility in Electrophysiology: Challenges and Recommendations

A scientific discipline benefits from reproducible results because they strengthen confidence that they reflect the way a system operates under certain conditions.[1]

[1] There have been special issues in the *International Journal of Psychophysiology* and *Psychophysiology* devoted to reproducibility and, relatedly, open science in cognitive electrophysiology, and readers are encouraged to read these special issues (Kappenman & Keil, 2017; Larson & Moser, 2017).

We use Goodman, Fanelli, and Ioannidis's (2016) definition of reproducible results as "obtaining the same results from the conduct of an independent study whose procedures are as closely matched to the original experiment as possible" (pp. 2–3). Although there are different definitions of what it means to obtain "the same results" (Open Science Collaboration, 2015), we hope that closely matched studies yield effects with confidence intervals that overlap substantially (for details on using confidence intervals in EEG studies, see Groppe [2017]), thus allowing us to make precise inferences about the true size of the effect being studied. Since reproducible results are crucial for a discipline, it is important to consider what can be done to obtain them. We discuss several research practices that increase the likelihood that an original result will be reproducible and that subsequent studies reproduce the original result, as well as challenges faced by cognitive electrophysiology researchers attempting to do so.

1.9 Pre-Register Specific Hypotheses

Each study should test specific hypotheses because results that confirm hypotheses are more likely to be true than results based on exploratory analyses (Ioannidis, 2005). Thus, it is crucial that researchers do not rewrite their hypotheses to fit with their results, a practice known as HARKing (hypothesizing after results are known), as it exaggerates the confidence the reader has that the results are true. One method to avoid HARKing is pre-registering hypotheses using the Open Science Framework (osf.io), aspredicted.org, or other repositories. However, formulating specific hypotheses for cognitive electrophysiology studies can be difficult, especially if researchers want to frame them statistically. For example, a researcher may be confident in predicting that an experimental condition will affect EEG activity, but they may struggle to define "EEG activity" as a dependent variable. In ERP studies, this is less of a concern because the dependent variables (ERP components) are well-characterized (Kappenman & Luck, 2012); however, there are fewer well-characterized time-frequency variables. Although researchers can be vague about defining their time-frequency variables, they should then use statistical analyses with strict corrections for multiple comparisons (see Cohen, 2014). This reduces the likelihood of making Type I errors, but consequently increases the likelihood of making Type II errors. (Researchers may also consider data-driven region-of-interest approaches; see Brooks et al., 2017). Therefore, time-frequency analyses will benefit from having well-defined dependent variables (Indeed, this was an initial impetus for this book!).

For example, if a researcher believes an experimental manipulation is likely to influence EEG activity related to cognitive control, they can have a clearly defined dependent variable of oscillatory activity in the theta frequency bandwidth measured from frontal midline electrodes over a certain time (Cavanagh & Cohen, Chapter 9). In a pre-registration, researchers should use a more precise definition of their dependent variable than "frontal midline theta". For example, they could specify that they will determine the wavelet of 4–8 Hz that exhibits the greatest peak power between 200 and 600 ms after stimulus onset at electrode Fz for each participant, and then compute the average power of this wavelet during the 200–600 ms time epoch at Fz. They could also introduce some flexibility into the specification of the dependent variable by noting that they will choose a different bandwidth, epoch, and/or electrode if the grand average time-frequency plot (averaged across all conditions) reveals an unexpected time-frequency and/or scalp distribution. In this example, the average across all conditions avoids biasing the analysis in favor of choosing a time-frequency window exhibiting differences between conditions. Besides reducing HARKing and facilitating the specification of dependent variables, pre-registration is crucial for holding researchers accountable to their research design, statistical analyses, and sample size; however, none of these positive features of pre-registration work if researchers deviate from their pre-registration without properly noting the deviation (Claesen et al., 2019).

1.10 Increase Power and Conduct A Priori Power Calculations

Another way that researchers can increase the likelihood that study results are reproducible is by increasing the power of their studies (Ioannidis, 2005). Alarmingly, Button and colleagues' (2013) analysis of neuroscience studies found their average power was very low (8–31%). There are two general ways that researchers can increase the power of their studies. First, researchers should attempt to maximize the effect they are studying and minimize its variance (i.e., increase the standardized effect size). This can be done by using strong experimental manipulations and reliable dependent variables, such as those discussed in this work. Additionally, researchers should increase the signal to noise ratio in their studies by optimizing the number of trials and collecting good EEG data (Cohen, 2017; Luck, 2014).[2] Further, when possible, researchers should use within-subjects designs, which is already the case in many cognitive electrophysiology studies. Second, researchers should collect larger samples, which is particularly important when testing between-subjects effects or within-between subject interaction effects. Cognitive electrophysiology studies can require a lot of time to collect and process data, so collecting more participants may seem burdensome, especially for researchers investigating small–medium effects. For example, a two-tailed

[2] A good way for researchers to benefit the field (and their citation count) may be for them to establish the number of trials required for different time-frequency variables in different paradigms, which has been done for ERP variables (e.g., Rietdijk et al., 2014). It is worth noting that simply adding more trials may not increase the signal to noise ratio, since participants may fidget more toward the end of long data collections, consequently increasing noise.

dependent *t*-test for an effect size of $dz = 0.35$, an alpha = .05, and power = .90 requires 88 participants, according to G*Power 3.1.9.4 (Faul et al., 2009). Indeed, it is likely that researchers will often find themselves studying small-medium effects. Specifically, when researchers conduct a priori power calculations to determine their sample sizes, they should assume that the effect sizes in the extant literature are inflated, due to publication bias by researchers and journals (i.e., only publishing significant results) (for a more detailed discussion on sample size calculations in EEG studies, see Larson & Carbine, 2017). Although it is difficult to collect and process large samples, it is crucial to the reproducibility of cognitive electrophysiology studies. To reduce the demands large sample sizes impose, researchers who mentor doctoral students, review for and sit on the editorial boards of journals, are involved in hiring decisions about faculty and post-doctoral researchers, and are involved in promotion and tenure decisions should reconsider expectations about the speed of science and the number of publications (for further discussion on these issues, see Bradley (2017) and Yeung [2019]). Also, if a researcher is concerned about allocating a lot of time to a study that may not yield significant results, they can conduct a sequential analysis where they pause data collection after a pre-specified sample has been collected and then determine whether to continue data collection based on if the incremental results are significant (given an adjusted alpha level) and if the incremental results suggest an effect size that is too small to be of interest (Lakens, 2014).

1.11 Make Methods, Materials, and Data Open

In addition to pre-registered hypotheses and adequately powered studies, researchers should also make their methods, materials, and data accessible[3]. In so doing, they will allow other researchers to "methodologically reproduce" the original study (Goodman et al., 2016), which should increase the likelihood of observing the same results. Researchers should make their stimuli and stimulus presentation scripts available and include specific details about instructions given to participants and the equipment used for the study. Further, researchers should also make signal and statistical processing scripts available and provide their data when possible so that other researchers can attempt to reproduce analyses or explore new ones. There are various ways to provide this information, including on the Open Science Framework and GitHub (github.com).

[3] There is a special issue in the *International Journal of Psychophysiology* devoted to open science in human electrophysiology, and readers are encouraged to read it (Clayson, Keil, & Larson, 2022).

1.12 Replicate and Expand

The recommendations were made to increase the likelihood that researchers' original results will be reproducible, but the recommendations also apply to researchers attempting to reproduce original results. Crucially, researchers attempting to reproduce original results should also attempt to replicate and expand an original finding (Cohen, 2017), preferably increasing the sample size by two and half times the original (Simonsohn, 2015). Specifically, cognitive electrophysiology will benefit if most studies include an attempt to reproduce an original result and then add a new result (e.g., by adding a new experimental condition). With results from replication attempts, more precise estimates about the direction and size of effects can be made. Of course, it is nearly impossible to reproduce a study methodologically. For example, different researchers may have different criteria for manually rejecting trials, and different independent component analyses may yield different components. However, researchers can still come quite close to a methodological reproduction, especially if methods and materials are available for them to use. A challenge for cognitive electrophysiologists is that they often employ different signal processing methods, such as subtracting or not subtracting the ERP from an epoch of EEG data prior to convolving the data with a wavelet. To this end, researchers should use the same methods as the original study, unless a different method is clearly superior. Of course, whether a different method is clearly superior is debatable; thus it is incumbent upon the researcher conducting the methodological reproduction to state their case in a compelling way.

1.13 Explore

Some of the most exciting and reproducible effects in cognitive electrophysiology have been discovered by accident (e.g., Kutas & Federmeier, 2011), meaning that they would not have occurred if researchers had only conducted confirmatory research testing a priori hypotheses. Thus, it is imperative that researchers conduct exploratory analyses in addition to confirmatory analyses. However, results from exploratory analyses should be clearly labeled as such to avoid misleading readers to having excessive confidence in the result. Ideally, then, a study will test pre-registered confirmatory hypotheses that replicate and expand an original result and conduct exploratory analyses. Compelling results from the exploratory analyses can then serve as a priori hypotheses in future confirmatory research. Finally, some of the most exciting exploratory research may come from analyzing old data in new ways. For example, Voytek (Chapter 23) proposes exciting new analytical methods that researchers can apply; his signal processing scripts are freely available (https://voyteklab.com/code), and researchers can use these scripts to analyze their old data or other openly available data.

References

Berger, H. (1929). Über das Elektrenkephalogramm des Menschen. *Archiv für Psychiatrie und Nervenkrankheiten*, *87*, 527–570. https://doi.org/10.1007/BF01797193.

Bradley, M. M. (2017). The science pendulum: from programmatic to incremental—and back? *Psychophysiology*, *54*, 6–11. doi: 10.1111/psyp.12608

Brooks, J. L., Zoumpoulaki, A., & Bowman, H. (2017). Data-driven region-of-interest selection without inflating Type I error rate. *Psychophysiology*, *54*, 100–113. doi: 10.1111/psyp.12682

Buzsáki, G. (2006). *Rhythms of the brain*. Oxford University Press.

Button, K. S., Ioannidis, J. P. A., Mokrysz, C., Nosek, B. A., Flint, J., Robinson, E. S. J., & Munafò, M. R. (2013). Power failure: Why small sample size undermines the reliability of neuroscience. *Nature Reviews Neuroscience*, *14*, 365–376. doi:10.1038/nrn3475

Claesen, A., Gomes, S., Tuerlinckx, F., & Vanpaemel, W. (2019). Comparing dream to reality: An assessment of adherence of the first generation of preregistered studies. *PsyArXiv* [online]. https://doi.org/10.1098/rsos.211037

Clayson, P. E., Keil, A., & Larson, M. J. (2022). Open science in human electrophysiology. *International Journal of Psychophysiology*, *174*, 43–46. https://doi.org/10.1016/j.ijpsycho.2022.02.002

Cohen, M. X. (2014). *Analyzing neural time series data: Theory and practice*. MIT Press.

Cohen, M. X. (2017). Rigor and replication in time-frequency analyses of cognitive electrophysiology data. *International Journal of Psychophysiology*, *111*, 80–87. doi: 10.1016/j.ijpsycho.2016.02.001

Faul, F., Erdfelder, E., Buchner, A., & Lang, A. (2009). Statistical power analyses using G*Power 3.1: Tests for correlation and regression analyses. *Behavior Research Methods*, *41*, 1149–1160. doi:10.3758/BRM.41.4.1149

Goodman, S. N., Fanelli, D., & Ioannidis, J. P. A. (2016). What does research reproducibility mean? *Science Translational Medicine*, *8*, 341ps12. doi: 10.1126/scitranslmed.aaf5027

Groppe, D. M. (2017). Combatting the scientific decline effect with confidence (intervals). *Psychophysiology*, *54*, 139–145. doi: 10.1111/psyp.12616

Ioannidis, J. P. A. (2005). Why most published research findings are false. *PLoS Medicine*, *2*, e124. doi: 10.1371/journal.pmed.0020124

Jasper, H. H. (1958). The ten-twenty electrode system of the International Federation. *Electroencephalography and Clinical Neurophysiology*, *10*, 371–375.

Kappenman, E. S., & Keil, A. (2017). Introduction to the special issue on recentering science: Replication, robustness, and reproducibility in psychophysiology. *Psychophysiology*, *54*, 3–5. doi: 10.1111/psyp.12787

Kappenman, E. S., & Luck, S. J. (Eds.) (2012). *The Oxford handbook of event-related potential components*. Oxford University Press.

Kutas, M. & Federmeier, K. D. (2011). Thirty years and counting: Finding meaning in the N400 component of the event-related brain potential (ERP). *Annual Review of Psychology*, *62*, 621–647. doi: 10.1146/annurev.psych.093008.131123

Lakens, D. (2014). Performing high-powered studies efficiently with sequential analyses. *European Journal of Social Psychology*, *44*, 701–710. doi: 10.10002/ejsp.2023

Larson, M. J. & Carbine, K. A. (2017). Sample size calculations in human electrophysiology (EEG and ERP) studies: A systematic review and recommendations for increased rigor. *International Journal of Psychophysiology*, *111*, 33–41. doi: 10.1016/j.ijpsycho.2016.06.015

Larson, M. J. & Moser, J. S. (2017). Rigor and replication: Toward improved best practices in human electrophysiology research. *International Journal of Psychophysiology*, *111*, 1–4. doi: 10.1016/j.ijpsycho.2016.12.001

Luck, S. J. (2014). *An introduction to the event-related potential technique* (2nd ed.). MIT Press.

Open Science Collaboration. (2015). Estimating the reproducibility of psychological science. *Science*, *349*, aac4716. doi: 10.1126/science.aac4716

Rietdijk, W. J., Franken, I. H., & Thurik, A. R. (2014). Internal consistency of event-related potentials associated with cognitive control: N2/P3 and ERN/Pe. *PLoS One*, *17*, e102672. doi: 10.1371/journal.pone.0102672

Simonsohn, U. (2015). Small telescopes: detectability and the evaluation of replication results. *Psychological Science*, *26*, 559–569. https://doi.org/10.1177/0956797614567341

Yeung, N. (2019). Forcing PhD students to publish is bad for science. *Nature Human Behaviour*, *3*, 1036. doi: 10.1038/s41562-019-0685-4

CHAPTER 2

LOGIC BEHIND EEG FREQUENCY ANALYSIS

Basic Electricity and Assumptions

KYLE J. CURHAM AND JOHN J. B. ALLEN

2.1 Introduction

THE brain is an electrochemical machine. Although nerve cells differ greatly in their size and morphology, all pass messages using electrical signals, sending information throughout the brain and to the rest of the body via the spinal cord. Within any neuron, an action potential is a wave of electrical activity that travels along the nerve membrane. Action potentials are triggered by the summation of input from other neurons using chemical neurotransmitters, which create voltage potential changes in the post-synaptic neuron. Surface-recorded EEG is blind to the activity of single neurons but can non-invasively measure the electrical activity resulting from summated excitatory and inhibitory post-synaptic potentials over millions of these signals in humans. EEG has excellent temporal resolution, but poor spatial resolution; it can tell us *when* something is happening in the brain, but not precisely *where* it is happening.

EEG gives an incomplete picture of the electrical activity occurring in the brain. The cortex is the outermost layer of the brain, and the primary generator of the electrical activity we measure with EEG. Our ability to detect cortical activity largely depends on the parallel arrangement of cortical pyramidal neurons. When millions of parallel neurons fire simultaneously, their electrical activity adds together to generate a signal large enough to detect at the scalp. However, when neural populations fire incoherently, or when they are not arranged in a parallel formation, the electrical activity does not add constructively, so there is no observed signal at the scalp. This chapter introduces basic concepts in electricity in signal processing, with some practical considerations for data collection and analysis, to help readers understand EEG measurement and interpretation.

2.2 Electricity—Voltage, Current, and Resistance

Every physical thing is made up of atoms, which in turn are made up of fundamental particles including protons, neutrons, and electrons. Protons have a positive charge, neutrons have a neutral charge, and electrons have a negative charge. Charges of the same sign repel, and opposite charges attract. A few simple experiments demonstrate the existence of electrical charge. For example, rubbing a balloon on a wool sweater makes the balloon negatively charged as electrons move from the wool to the balloon (Figure 2.1). The degree of attraction or repulsion between two point charges is proportional to the product of the charges divided by the inverse squared distance between them. Therefore, when the electron-rich balloon is brought into proximity of the electron-poor wool, or any neutral surface such as a piece of paper or the wall, the balloon will be attracted. If you subsequently rub a second balloon in the same way, the two balloons will repel each other since they are both negatively charged.

Most of the time, atoms have equal numbers of protons and electrons. However, atoms can become *ionized* when electrons are removed or added, resulting in a net charge. *Electricity* is the phenomenon that describes the behavior and movement of charge. In general, it doesn't matter whether it is the electrons or ions that are moving. The flow of charge can be accomplished either by the transfer of electrons from atom to atom, or, in the case of electrophysiology, by the diffusion of charged ions across cellular membranes.

The degree to which electrons are free to move from atom to atom varies by material type. For example, in metals, the outermost electrons are so loosely bound that they freely move in the space between atoms at room temperature. Because these unbound electrons are free to travel from atom to atom, they are called *free electrons*. The relative mobility of electrons within a material is known as electrical *conductivity*. Conductivity is determined by the types of atoms in a material, and how the atoms are linked together with one another. Materials with few or no free electrons are called *insulators*, and materials with many free electrons are called *conductors*. The directed motion of electrons is called electrical *current*. Just like water flowing through a pipe, electrons move within the empty space between atoms. Under normal conditions, the motion of free electrons in a conductor is random, with no particular direction or speed. However, electrons can be influenced to move in a coordinated fashion through a conductive material by supplying a *voltage*. Voltage is the "pressure" that pushes on free electrons to cause them to flow. The ability of a current to flow from one location to another depends on the *resistance*. In insulating materials, such as glass or rubber, electrons have little freedom to move from atom to atom. The less freedom electrons have to move from atom to atom, the greater the resistance to the flow of charge. A conductor's resistance generally increases as its length increases or its diameter decreases. It is again useful to refer to the water analogy: water can flow more easily through a short, wide pipe than a long, narrow pipe. The international system of units (SI) of current is called the *ampere*.

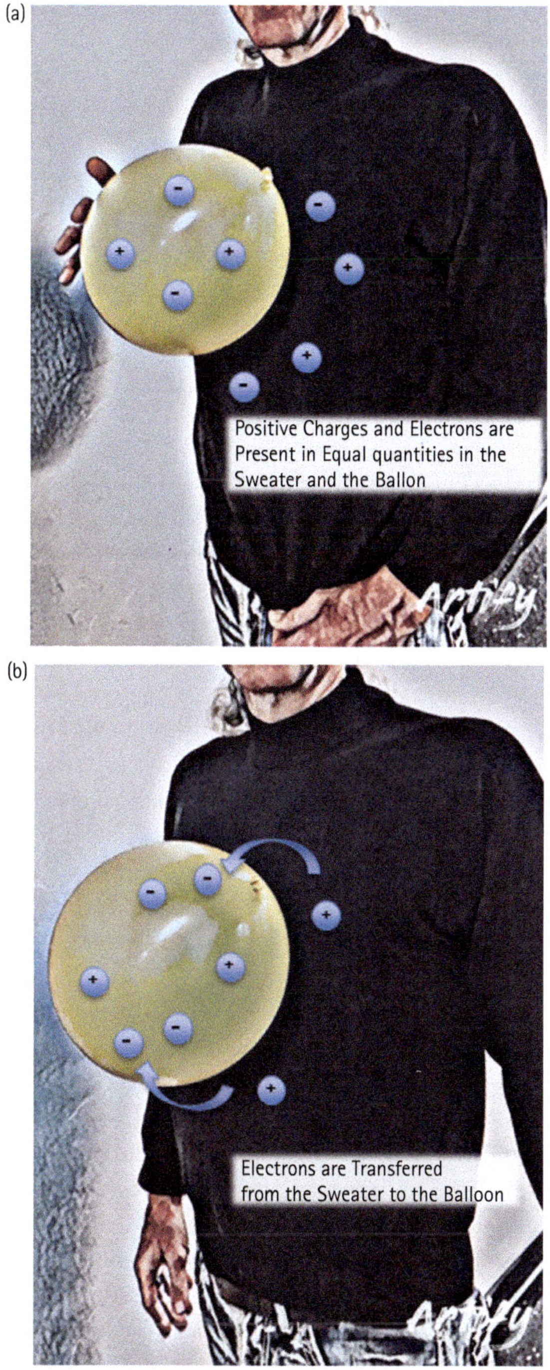

FIGURE 2.1 Before rubbing the balloon against the sweater (A), no net accumulation of electrons exists on the balloon. After rubbing the balloon on the sweater (B), the balloon has accumulated an excess of electrons, and the resultant negative charge of the balloon and positive charge of the sweater creates a force of attraction sufficient to keep the balloon from being pulled to the ground by gravity.

Figure credit: K. Ehrmann.

Table 2.1 Guide to electrical symbols and terminology

Symbol	Term	Definition	Unit
E	Voltage	Electromotive force	Volts (V)
I	Current	Rate of flow	Amperes (A)
R	Resistance	Opposition to current	Ohm (Ω)
C	Capacitance	Ratio of the change in charge to the change in voltage	Farad (F)
P	Power	Rate of work	Watt (w)
W	Energy	Ability to do work	Watt-second (Joule)

One ampere is defined as one coulomb of charge (or 6×10^{18} electrons) flowing past a given point in a conductor in one second. The *volt* is the unit of pressure, that is, the amount of *electromotive force* (EMF) required to push a current of one ampere through a conductor with a resistance of one ohm, or 1 volt/ampere.

When resistance is high, electrons tend to gather on one side of the insulating material and the positive ions tend to gather on the other, effectively storing *potential energy* in an *electric field*. At some point, the voltage across the material will exceed a threshold known as the dielectric constant, at which point current begins to flow. This tendency for high resistance to result in charge separation is known as *capacitance*. The amount of charge stored in the capacitor is directly proportional to the surface area of the dielectric (the electrical insulator polarized by the electric field) (Table 2.1).

2.3 Circuits

An electrical circuit consists of closed conductive paths between circuit elements. Elements may consist of resistors, capacitors, voltage sources, or current sources. Electrical components may be wired in series or parallel with each other (Figure 2.2). In many cases, circuits consist of some complex combination of series and parallel components. However, these circuits may often be represented by simpler equivalent circuits with identical electrical properties.

To gather some intuition for the flow of current in simple circuits, we refer back to the water analogy. The resistance increases as the pipe gets longer. This is equivalent to stringing together multiple resistors in series. The rule to combine resistors in series is additive:

$$R_{series} = R_1 + R_2 + \ldots + R_n$$

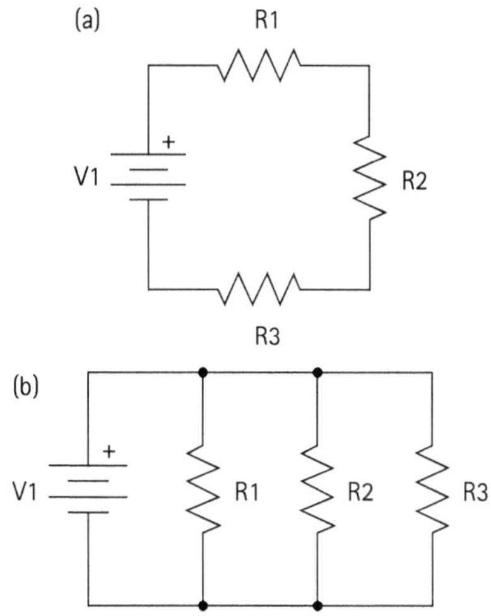

FIGURE 2.2 Simple series (left) and parallel (right) circuits. Circuit components labeled with R indicate resistors. Circuit components labeled with V indicate voltage sources.

Conversely, adding multiple resistors in parallel will decrease the overall resistance:

$$\frac{1}{R_{parallel}} = \frac{1}{R_1} + \frac{1}{R_2} + \ldots + \frac{1}{R_n}$$

As more "pipes" are added, the water has more paths to escape, decreasing the overall resistance to flow. Capacitors wired in series or parallel follow the same rules, but reversed:

$$\frac{1}{C_{series}} = \frac{1}{C_1} + \frac{1}{C_2} + \ldots + \frac{1}{C_n}$$

$$C_{parallel} = C_1 + C_2 + \ldots + C_n$$

Most circuits are some complex combination of series and parallel. We can approach these circuits one piece at a time, deriving a new equivalent circuit at each step (Figure 2.3).

The next few sections explore examples of equivalent circuit representations, and we use equivalent circuit representations to learn the voltages and currents at every point in the original complex circuit. We later show how models of neurons can be represented as a simple equivalent circuit of capacitors, resistors, and voltage sources.

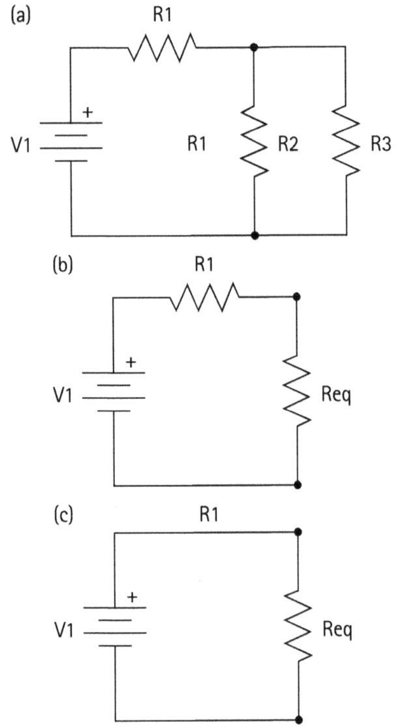

FIGURE 2.3 Reducing a complex circuit (left) to a simple equivalent circuit (right). In an intermediate step, we combine parallel resistors 2 and 3 (middle). Next, we combine resistor 1 with the equivalent resistor from the intermediate step.

2.4 DIRECT CURRENT

Circuits come in two basic flavors: direct current (DC) and alternating current (AC). The difference depends on whether the voltage and current change directions over time. DC circuits maintain currents flowing in a constant direction within a closed loop, whereas AC circuits have current that repeatedly reverses direction. A DC electric source feeds from one terminal to a set of circuit elements and then back to the other terminal, in a complete circuit. Figures 2.2 and 2.3 are both DC circuits due to their constant voltage power source. Note some resistors are connected in parallel, while others are connected in series. At each step, we can combine resistors according to the rules in Section 2.3 to derive a simpler equivalent circuit.

Some circuits may contain both resistors and capacitors. These are known as RC circuits. In a simple circuit with one resistor and one capacitor in series, the capacitor must discharge through the resistor. This discharge occurs at an exponential rate determined by the *RC time constant*. The time constant indicates the number of seconds for the capacitor to become 63.2% charged, or, equivalently, the time for current flow to have slowed by 63.2% from its starting value. This choice of time constant has an intuitive explanation: at any moment in time, the rate of change in voltage is equal to the voltage divided the time constant. For example, a 1 mF cap and a 1 kΩ resistor yields a time constant of one second. If the capacitor is charged to 5 volts, the voltage will fall at a rate of 5 V/s. If the capacitor is charged to 2 volts, the voltage will fall at a rate of 2 V/s.

2.5 Alternating Current

In contrast to DC signals, some sources of electricity produce AC, where voltages and currents periodically reverse direction and switch back and forth between positive and negative polarity. The electricity that comes from an American wall outlet is an example of AC. The current in North America is 120 VAC and changes direction 60 times per second. AC circuits can exhibit more interesting behaviors than DC circuits. For example, at low frequencies, a capacitor acts like an open circuit, so no current flows in the dielectric. However, when driven by an AC source, a capacitor will only accumulate a limited amount of charge before the potential difference changes polarity and the charge is returned to the source. The higher the frequency, the less charge will accumulate and the smaller the opposition to the current.

Both resistors and capacitors resist the flow of current when a voltage is applied. However, unlike in DC circuits, resistance may be frequency dependent. Frequency-dependent resistance is known as *impedance*, a complex-valued quantity that can be broken into two parts: magnitude (the ratio of the voltage amplitude to the current amplitude) and phase (quantifies how much the current lags the voltage). Alternatively, we can break impedance down into its real and imaginary parts. Like in DC circuits, the real part of the impedance acts like resistance, resisting the flow of electric current. The imaginary part is called *reactance*, and it quantifies the opposition to a *change* in the current of a capacitive circuit element. Ideal capacitors are purely reactive, that is, they have zero resistance, and the impedance of a resistor is purely real, or resistive. Note this implies the current in a capacitor always lags the voltage by 90°. Impedance devices add like resistors in a DC circuit. For a set of components in series, the total impedance is the sum of the component impedances. To obtain the impedance of parallel circuit components, the inverse total impedance is given by the sum of the inverses of the component impedances.

Using different combinations of resistors and capacitors, RC circuits can be used to attenuate some frequencies, while allowing others to propagate through the circuit unaffected. For example, wiring a resistor in series with a load, and a capacitor in parallel with the same load, significantly attenuates high-frequency signals. Conversely, wiring a resistor in parallel with the load, and the capacitor in series, attenuates low-frequency signals. In Figure 2.4, at low frequencies, the reactance of the capacitor will be very large compared to the resistance of the resistor. This means that the voltage across the capacitor will be much larger than the voltage across the resistor. At high frequencies the reverse is true: the voltage across the resistor is larger than across the capacitor. In other words, low frequencies pass to the output, and high frequencies are attenuated. This is known as a low-pass RC filter. Similarly, we can construct a high-pass filter by swapping the resistor and capacitor. The frequency cutoff for these *filters* is determined by the time-constant of the circuit, which is derived from the resistance and capacitance. The *cutoff frequency* for an RC circuit is:

$$f_c = \frac{1}{2\pi RC}$$

In the low-pass configuration, frequencies just above the cutoff are attenuated to half their original amplitude. Conversely, in the high-pass configuration, frequencies just below the cutoff are attenuated to half amplitude. The amount of attenuation increases as you move farther beyond the cutoff frequency. Figure 2.4 shows the amplitude roll-off as a function of frequency for a low-pass filter.

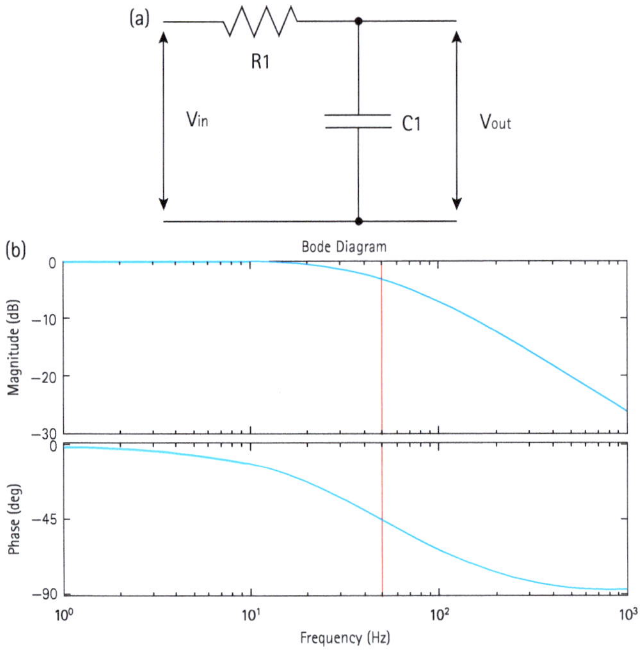

FIGURE 2.4 Low-pass RC filter (left). The parallel arm with the capacitor provides a low impedance path for high frequency signals. However, the capacitor saturates for low-frequency signals, providing a high-impedance path. Low-frequencies signals are thus preferentially observed at Vout. Frequency response of the low-pass RC filter (right). The signal is not appreciably attenuated below the cutoff frequency of 50 Hz (shown in red). At frequencies just above 50 Hz, the signal magnitude is cut in half. As frequency increases, the amount of attenuation increases.

2.6 Hodgkin–Huxley Model

We can use a simple circuit model to describe the electrical properties of neurons, including the initiation and propagation of action potentials. Action potentials are the result of the diffusion of sodium and potassium ions across neural membranes. The Hodgkin–Huxley model treats each component of a neuron as an electrical element in the circuit (Figure 2.5), where current is propagated by the movement of ions across cell membranes.

The cell membrane is represented by a capacitance (C_m). Cellular membranes are highly resistive, and act as a dielectric material due to their relatively impermeability. In the absence of special proteins called ion channels, ions like sodium and potassium are unable to diffuse across the membrane, effectively turning the membrane into the dielectric of a capacitor. As ionic currents add or subtract from the charge accumulating inside the neuron, ions line up along the cell membrane. The differing concentration of ions on either side of the membrane results in a net voltage potential, represented by voltage sources (E_n). To maintain these concentration gradients, neurons have active sodium-potassium pumps that exchange two sodium ions into the extracellular space for three potassium ions in the intracellular space.

Sodium and potassium ion channels are represented by electrical conductances (G_{NA}, G_K) that depend on both voltage and time. As the voltage potential increases, the permeability of the membrane is selectively modulated, that is, conductance is increased, for specific ion species. The flux of sodium or potassium ions across the membrane is represented by ionic currents (I_p). Since the membrane is not perfectly impermeable, leak channels are also included, represented by another conductance (G_L).

The Hodgkin–Huxley circuit model exhibits similar dynamics to real neurons. The amount of injected current controls the emergence of a stable limit cycle. For a sufficiently large input current, the circuit will exhibit repeating "action potentials" at a minimum firing rate. This means that either the neuron is not firing at all (corresponding to zero frequency) or is firing at the minimum firing rate. Increasing the injected current beyond the minimum threshold increases the firing rate of the neuron. When

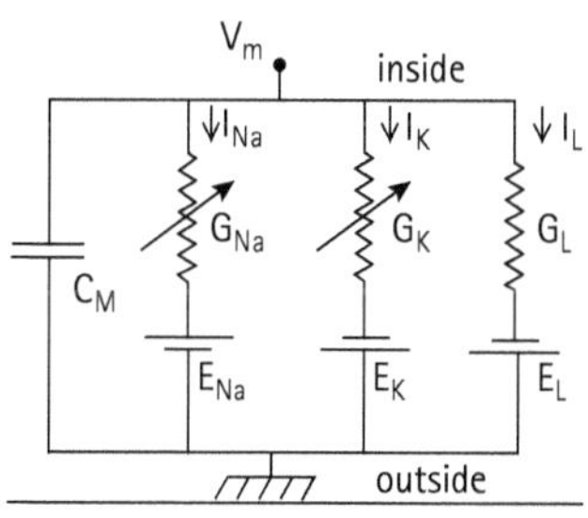

FIGURE 2.5 Equivalent circuit diagram for the Hodgkin–Huxley model.

the neuron fires, a series of channel activations occur to produce the action potential. As the membrane potential approaches threshold, sodium ion channels begin to rapidly open, depolarizing the membrane (i.e., discharging the capacitor). The influx of sodium changes the voltage gradient between the intracellular and extracellular space, increasing the membrane potential. Once the polarity of the potential changes direction (when enough sodium ions have crossed the membrane), sodium ion channels begin to deactivate (decreasing sodium conductance). As the sodium channels close, potassium channels begin to open (increasing potassium conductance), resulting in an efflux of potassium ions to the extracellular space, restoring the membrane potential to the resting state following a brief hyperpolarization.

2.7 Filtering

As discussed, simple RC circuits can selectively attenuate certain components or features of a signal, and not others. This is known as filtering. Filters come in a variety of forms (RC circuits, mechanical and optical filters, digital signal processing, etc.). Here we examine two of the most common digital filters used in electrophysiology. The response of a system to a brief input signal, or impulse, is called the *impulse response*. The impulse response of a *finite impulse response* (FIR) *filter* settles to zero in finite time. Given a finite sample of nonzero input values, an FIR filter will always yield a finite sample of nonzero output values. This contrasts with an infinite impulse response (IIR) filter, which does not settle in finite time. The RC filters seen in Section 2.6 are examples of IIR filters since the capacitors (or inductors) in the RC filter never completely relax following an impulse.

2.8 Analog-to-Digital Signal Conversion

Most FIR filters are implemented using digital signal processing. Our discussion of electricity thus far has dealt with *analog signals*, which are continuous in both time and in voltage. An economy of representation can be achieved by sampling discrete points in both the time and voltage domains, a process of creating a *digital signal*, which has a temporal resolution determined by the *sampling rate* and a voltage resolution determined by the resolution of the *analog-to-digital converter* (Figure 2.6). For example, a 16-bit converter will allow 10^{16} of 65,536 discrete voltage values, and a sampling rate of 1,000 Hz will allow one value every millisecond.

With sufficiently large sampling rates, a digital signal can closely approximate the analog signal it is attempting to represent (Figure 2.7). However, several considerations

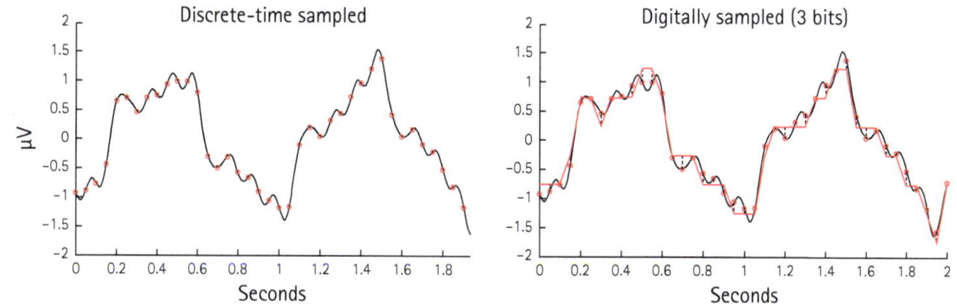

FIGURE 2.6 A signal sampled at 20 Hz. Discrete-time sampling (left panel) allows for continuous y-axis (μV) values, whereas digitally-sampled signals (right panel) must use a limited number of y-axis values. The three bit converter illustrated here (right panel) allows for $2^3 = 8$ distinct values, providing only a coarse approximation of the signal voltage. The right panel depicts the discrete sample value (red circle) and the 3-bit digital equivalent (red line), and the discrepancy (dashed vertical black lines).

are essential to ensure signal fidelity when digitizing an analog waveform. In order to recover all components of a periodic waveform, it is necessary to use a sampling rate at least twice the highest waveform frequency. This is known as the *Nyquist sampling rate*, and it determines whether or not aliasing will occur. Similarly, for a given sampling rate,

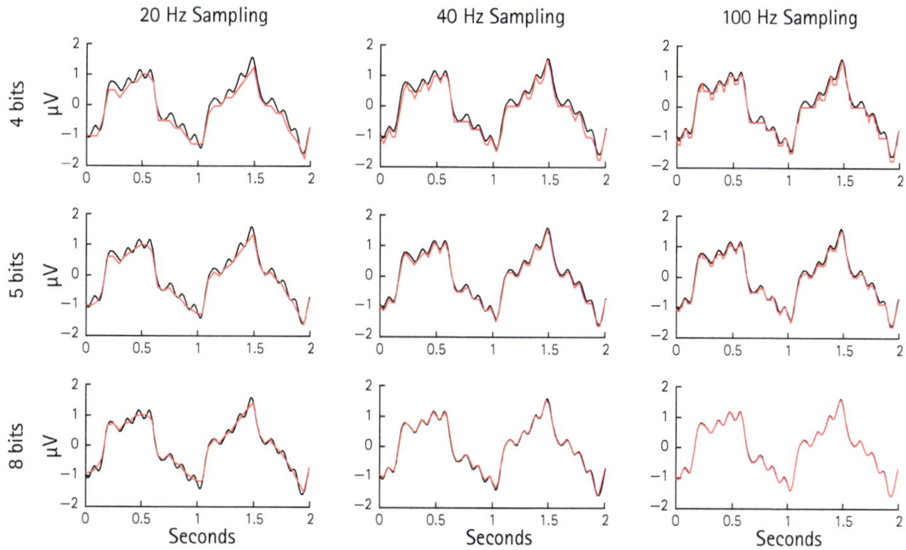

FIGURE 2.7 A comparison of a signal (black line) sampled (red line) at three sampling rates (20, 40, 100 Hz) and using three different converter resolutions (4-bit, 5-bit, and 8-bit) that allow for 16, 32, and 128 distinct μV values. Low bit-resolution was used here for illustrative purposes; commercial converters are typically 12-bit (4,096 values) or 16-bit (65,536 values).

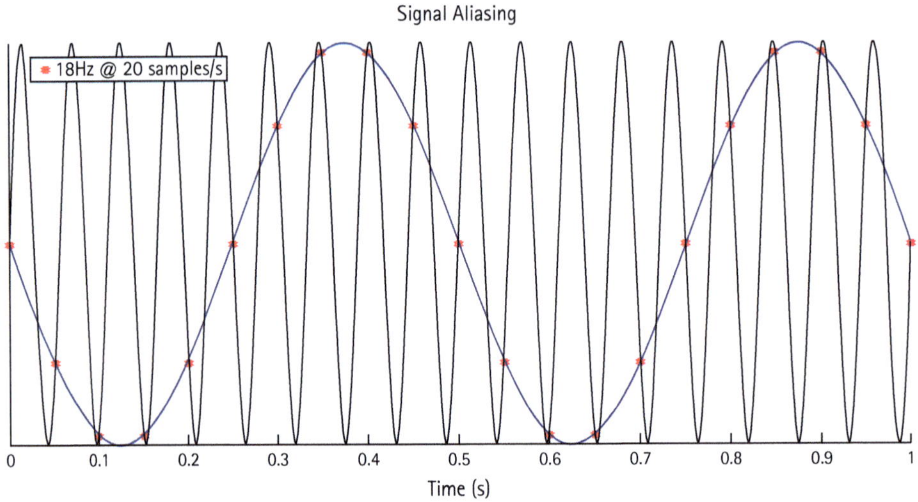

FIGURE 2.8 Signal aliasing due to insufficient sampling rate. Samples from an 18-Hz signal were sampled at a rate of 20 Hz. Identical samples are obtained for a 2-Hz sine wave sampled at 20 Hz. These digitized signals are indistinguishable from each other.

the highest frequency signal that can be represented is one-half the sampling rate, and this is known as the *Nyquist frequency*. *Aliasing* occurs when the set of samples obtained from the analog signal are indistinguishable from samples from a lower-frequency signal. For example, in Figure 2.8, an 18-Hz signal is sampled at 20 Hz, well below the Nyquist sampling rate. For this 20-Hz sampling rate, the Nyquist frequency is 10 Hz. Although the signal is in fact an 18-Hz signal, the sampled signal appears as a 2-Hz sine wave, so the signal is not well characterized. In general, signals that are x Hz above the Nyquist frequency will appear as a signal x Hz below the Nyquist frequency. Sampling above the Nyquist frequency will prevent aliasing but may still not characterize the signal well in the time domain. As a general guideline, it is recommended to sample at least 5× the highest frequency of interest to get a good signal.

2.9 Digital Filtering

FIR and IIR digital filters can be used to process digital signals. A simple moving average is an example of an FIR filter. If we take the average across the last five data samples and shift the five-sample window forward by one sample at each timestep, the result is a moving average window. We can denote the value of the filtered signal (x) at the nth timepoint using summation notation:

$$y(n)=\sum_{k=-2}^{2}\frac{1}{5}x(n-k)$$

We can take this one step further, using a different weight for each sample in the moving average window (h) of width M + 1:

$$y(n) = \sum_{k=-M/2}^{M/2} h(k)x(n-k)$$

This result is exactly the impulse response when the input signal x is an impulse, that is, one at the middle timestep and zero at all other timesteps. Note the filter output will clearly be zero outside the range of the window, demonstrating that this is in fact an FIR filter. Figure 2.9 shows the output of a moving average filter, given a noisy input signal.

The summation operation described is known as *convolution*. In general, convolution indicates the amount of overlap of one function or kernel (h) as it is shifted over another function (x). In these examples, the convolution is simple to compute. However,

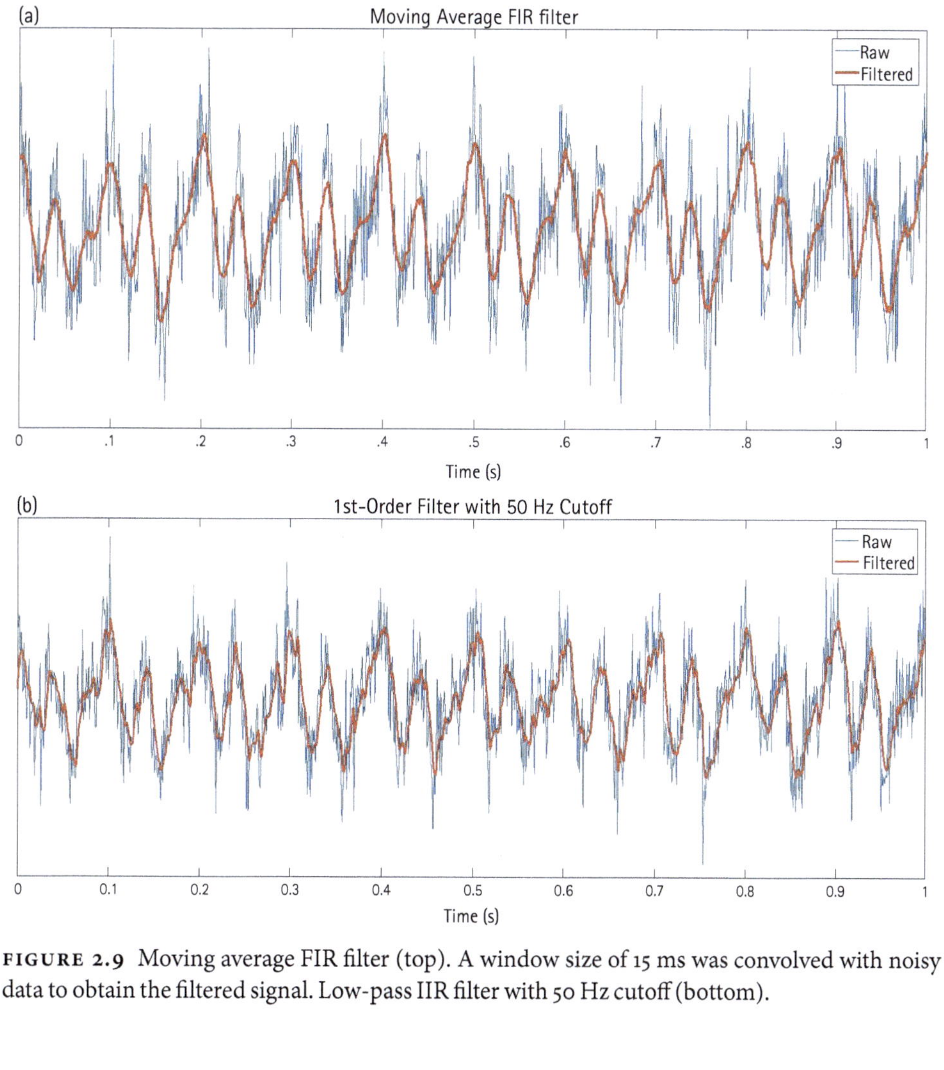

FIGURE 2.9 Moving average FIR filter (top). A window size of 15 ms was convolved with noisy data to obtain the filtered signal. Low-pass IIR filter with 50 Hz cutoff (bottom).

in practice, it can be computationally expensive. In the next section, we show how to efficiently evaluate convolutions using an advanced signal-processing technique called the Fourier transform.

2.10 Frequency Domain

In general, signals may have both AC and DC components. For example, you may see an AC signal that oscillates around a nonzero mean (a DC offset). Arbitrary complex signals may be approximated by the sum of two or more simpler signals. For example, we can synthesize a complex signal my summing together multiple sine waves of various frequencies. Fourier analysis is the reverse process—decomposing a signal into its constituent parts. The Fourier series approximates any complex periodic signal as a finite weighted sum of sine waves of various frequencies. The more sine waves included in the summation, the better the Fourier series can approximate the signal. In the limit that the number of frequencies included in the summation goes to infinity, the Fourier series converges to the Fourier transform. In this case, we can describe the signal as a continuous distribution, or spectrum, of frequencies, along with the phases at which each sine wave begins. The Fourier series can be applied to a wide array of mathematical, physical, and signal processing problems. Fourier analysis is now widely used across several domains, including audio, images, radar, sonar, X-ray crystallography, and more.

Rather than analyzing signals as a function of time, Fourier analysis allows us to study their properties as a function of frequency. Signals that are localized in the time domain have Fourier transforms that are spread out across the frequency domain, and vice versa. For example, the Fourier transform of a pure sine wave is a single point in the frequency domain (Figure 2.10, bottom). Points in the frequency domain may be characterized by properties such as power and phase, which are of interest for EEG analyses. The absolute value of a given frequency component of the Fourier series indicates the "amount" of that frequency present in the original signal. The squared absolute value is the signal power. The *power spectrum* describes how signal power varies as a function of frequency. EEG signals typically follow a $1/f$ trend, such that low frequencies have more power compared to high frequencies. However, the power spectrum may vary as a function of individual differences and task demands. Changes in power at a frequency may be due to alterations in the slope of the EEG frequency spectrum, or modulations in frequency-specific oscillatory activity (see Chapters 9, 10, and 23). Similarly, the *phase spectrum* of the signal can be extracted from the Fourier series. The signal phase indicates the amount of "shift" in each of the basis sine waves (for more info, see Chapter 7).

The Fourier transform is invertible. Given the power and phase spectrum, it is possible to reconstruct the original time-domain signal. Moreover, it is possible to compute the Fourier transform of a signal, perform mathematical operations in the frequency

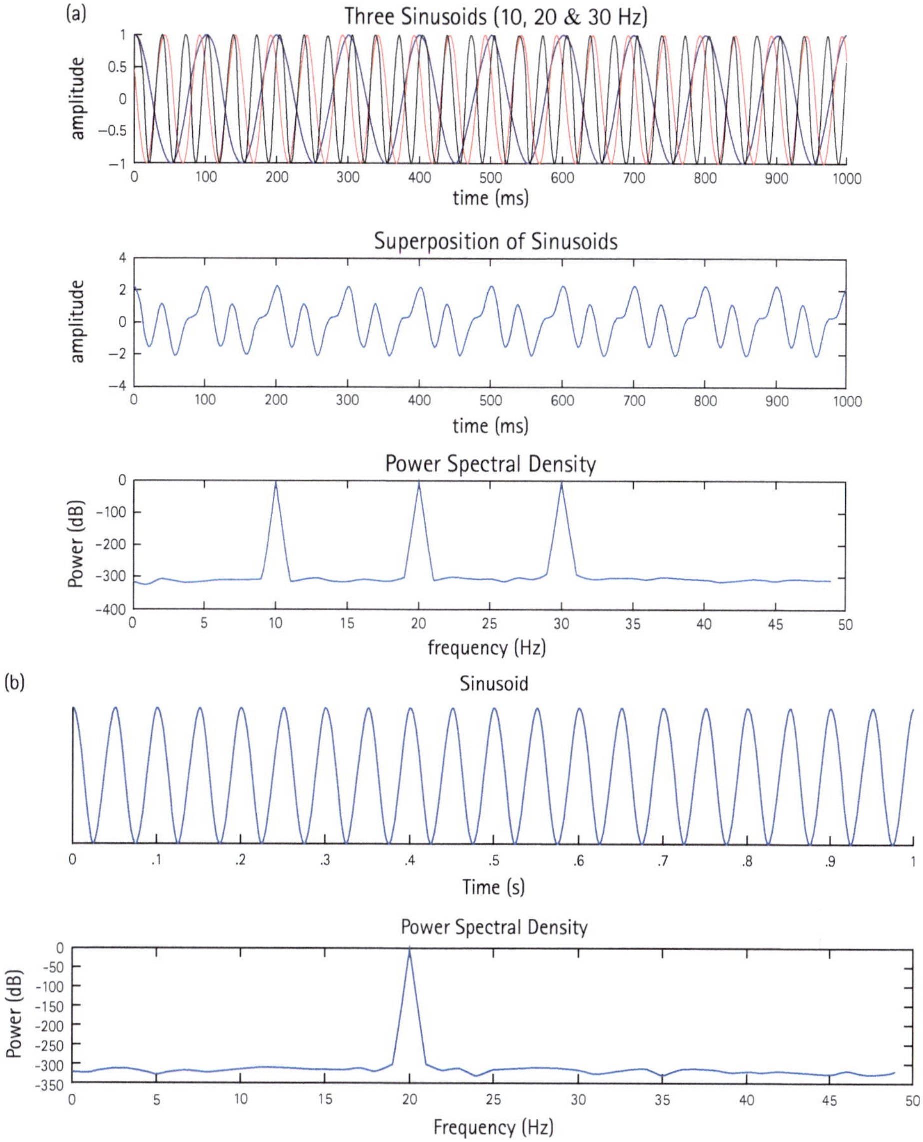

FIGURE 2.10 Constructing a complex signal from the superposition of sinusoids (top). The power spectrum of the signal show distinct peaks at the frequencies of the component sinusoids. A single sinusoid corresponds to a single peak in the power spectrum (bottom).

domain to alter the power and phase spectrum, and then compute the inverse Fourier transform to convert back to the time domain. For every mathematical operation in the time-domain, there is a corresponding equivalent operation in the frequency-domain. In some cases, operations may be easier to perform in one domain or another. This makes the Fourier transform very powerful. For example, convolution in the time-domain is equivalent to multiplication in the frequency domain.

To sum up: compute the Fourier transform of the time-domain signal and filter weights, multiply them in the frequency domain, and compute the inverse Fourier transform to obtain a filtered signal.

2.11 Windowing

The Fourier transform assumes the signal is periodic and includes sine waves of infinite period. However, EEG recording epochs are finite, spanning just seconds or minutes. Clearly low frequencies cannot be captured by data segments smaller than the period of the signal. For example, a segment of 250 ms will contain only a half-cycle of a 2-Hz sine wave. In contrast, a 50-Hz waveform would complete 10 full cycles within the allotted window. A general rule guideline to determine an adequate window size is to take 3–5 times the period of the lowest frequency of interest.

Window functions are often used to examine small segments of data from a longer signal to study transient events that have may have different spectral properties than other surrounding data segments. Windows are usually constructed such that they are zero-valued outside of the interval, symmetric around the middle of the interval, and tapering away to zero at the edge of the window. Several types of window functions are commonly used in EEG frequency analysis, including Hann, Hamming, and Gaussian windows. The duration of the window is application specific and is governed by requirements such time and frequency resolution.

However, we cannot perfectly resolve both time and frequency simultaneously. Time and frequency are *conjugate variables*, which are coupled such that knowledge of one makes knowledge of the other uncertain. Therefore, there is a tradeoff between time and frequency resolution, such that good time resolution comes at the expense of frequency resolution, and vice versa. As the window shrinks to zero width, short tones become clicks, with no discernable frequency. Clicks can be perfectly localized in time, but the frequency is undefined. As the window gets large compared to the signal length, it becomes impossible to localize a signal in time, but the frequency can be easily determined. Time-frequency analyses balance the time-frequency resolution tradeoff.

2.12 Time-Frequency Analysis

Frequency analyses are restricted to stationary signals. That is, the spectral characteristics do not change over time. However, EEG signals change as a function of state and task demands, which can vary over time. Time-frequency analyses are used to study the non-stationary characteristics of psychophysiological signals. Rather than analyzing signals in one domain—time *or* frequency—signal properties are localized to points or pixels in a two-dimensional time-frequency plane.

The simplest method to obtain a time-frequency representation of data is to compute the Fourier transform over a small sliding window that is chosen to be just wide enough to resolve the lowest frequency of interest. This approach yields local information about the signal, including frequency and phase content. This method is known as the short time Fourier transform (STFT) and can be used to obtain the power and phase of the signal at each point in the time-frequency plane.

The wavelet transform is slightly more advanced, considering the time-frequency trade-off that comes with windowing. This method convolves the signal with a set of complex gaussian-windowed sinusoids. When the similarity between the signal and wavelet is high, the convolution term is large. A family of wavelets of varying frequency and duration are constructed and convolved with the EEG signal to find points of peak similarity. The amplitude and phase of the signal can be obtained from the transform by computing the magnitude and angle of the complex wavelet coefficients. Unlike the STFT, the resolution of the wavelet transform has different properties at different frequencies. As frequency increases, the wavelets get shorter, effectively decreasing the window width. This leads to better time-resolution at the cost of worse frequency resolution (figure 11). In general, the wavelet transform has better time-resolution at high frequencies and better frequency resolution at low-frequencies. For a more in-depth look at the wavelet transform, see chapters 12 and 13in Cohen (2014).

Time-frequency decompositions provide a new way to identify psychophysiologically relevant components. While traditional time-domain methods could reveal components that occur at different timepoints, they cannot distinguish between components that occur at different frequencies at the same time-point. However, once the time-frequency decomposition is obtained, new analyses may be performed that were not possible with pure time-domain or

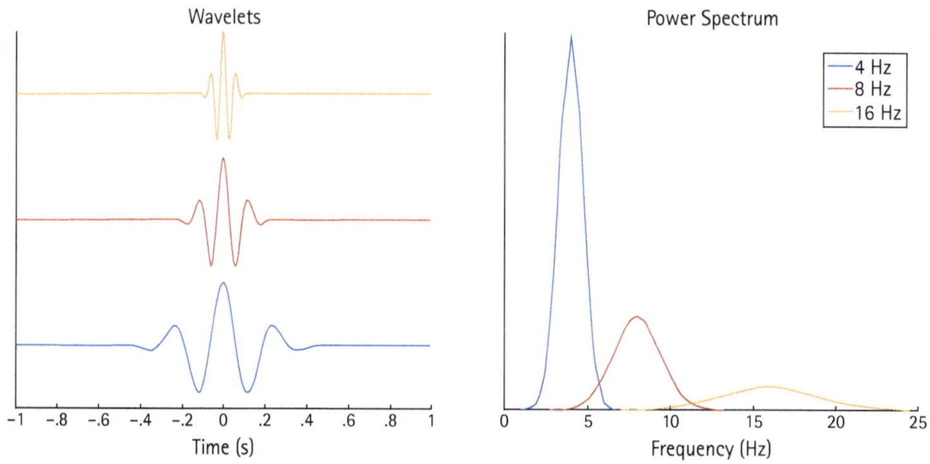

FIGURE 2.11 Wavelet time-frequency resolution trade-off. Higher-frequency wavelets (yellow) are more precise in time (left), but less precise in frequency (right). Conversely, low frequency wavelets (blue) have poor time resolution (left) but good frequency resolution (right).

frequency-domain approaches. For example, the phase consistency between two or more channels over time is often used to infer frequency-specific connectivity between regions. Inter-channel phase synchrony is evaluated by computing the covariance between the phases of two or more channels at a given frequency (see Chapter 20). Time-frequency analysis allows for the identification of event-related time-locked changes that are not phase-locked; such changes would not appear in traditional time-domain analyses using signal averaging to create event-related potentials (cf. Trujillo & Allen, 2007).

2.13 Practical Considerations for EEG

2.13.1 EEG Signal Artifacts

EEG is used to record electrical activity originating in the cerebral cortex, although it also detects electrical activity arising from sources other than the brain. EEG signals are highly sensitive and are easily influenced by electrical activity in the surrounding environment. Artifacts in EEG signals come in at least two distinct varieties: physiological and non-physiological noise. EEG signals generated by the body, but that are not cerebral in origin, are deemed physiological noise. Sources of physiological artifact include muscle movements, eye blinks, tongue movements, respiration sway, electrocardiographic activity, jaw clenching, etc. Conversely, non-physiological artifacts arise from sources outside the body, such as electronic noise, amplifier saturation, or loose electrodes. For example, thermal noise due to the random motion of electrons in resistive components, flicker due to irregularities in contact pins, and burst due to semiconductor impurities are common sources of noise. The most obvious electronic noise is flicker due to power-line interference, appearing at 60 Hz in North America and 50 Hz in much of the rest of the world. When the electrical impedance at the scalp-electrode interface is high (>5 kΩ), resulting in a poor connection, the electrode is more sensitive to these noise sources. To judge the acceptability of the scalp impedance for obtaining good quality signals, it is important to consider the input impedance of the amplifier. Contemporary high impedance amplifiers mitigate the susceptibility to electronic noise, as the fidelity of the observed EEG signal is directly related to the average impedance of the target and the reference electrode, and inversely related to the amplifier input impedance (Ferree et al., 2001). Thus some equipment can produce high quality signals with scalp impedances that may be higher than the often-used 5 kΩ standard.

There are multiple strategies to handle EEG recording artifacts: use of a ground electrode to remove noise that presents similarly across all electrodes, artifact rejection to remove electrodes or segments of data that are corrupted by noise, and artifact correction to separate signal from noise without sacrificing data.

2.13.2 Grounding

Ground electrodes are used for common mode rejection: parts of the signal shared across all electrodes are removed from the recording. A good ground connection is critical to reducing the impact of external noise sources. Impedances for all electrodes are compared to the ground electrode. Therefore, if the ground impedance is high, good impedances will not be possible on any other electrodes. However, electrical interference may persist even after ensuring good electrode impedances and secure grounding. Therefore, data rejection and correction are often used to handle residual artifacts during offline processing.

2.13.3 Artifact Rejection

Artifact removal may consist of rejecting entire segments of data, or individual electrodes from the recording. While the basic strategy is to identify data corrupted by artifact and exclude it from future analysis, it is not a trivial task. The determination of whether a signal is sufficiently corrupted by noise to justify removal is time consuming, subjective, and may vary by researcher. Moreover, significant training is required to learn to correctly identify different sources of noise. Judgements must also be made as to whether each artifact should be removed, even if it is correctly identified. In some cases, a filter may be used to attenuate electronic noise. For example, a 60-Hz artifact from power-lines may be removed from EEG recordings using a notch filter, which allows frequencies lower and higher than the specified frequency-band to pass unaffected, while significantly attenuating the artifact (e.g., between 55–65 Hz). However, filtering may not be an optimal solution for all artifacts. For example, blink timing is related to information processing (Stern et al., 1984), and blink rate is correlated to dopamine release (Taylor et al., 1999), although recent work has called this into question (Dang et al., 2017). However, many researchers remove data segments with eyeblinks due to the large artifact they produce in frontal EEG channels. If the phenomenon of interest is correlated to these physiological artifacts, it is possible to inadvertently remove valuable data and miss the psychophysiological phenomenon.

Sometimes artifacts may be localized to a single bad channel. In this case, remove only that channel from the recording and keep the remaining data. Many researchers examine the standard deviation of the signal to automatically detect bad electrodes. If it exceeds a threshold much larger than would be expected for a quality EEG signal, the channel is marked bad. If a channel is marked bad and it is required for a planned EEG analysis, the missing data can be interpolated. Interpolation uses information from the surrounding electrodes to guess what the signal would have been if there were a good electrode at that location. A popular choice is spherical spline interpolation (Perrin et al., 1989).

2.13.4 Artifact Correction

If possible, EEG researchers want to retain as much signal as possible and thus avoid removing noisy signals from data analyses. Several algorithms exist to attempt to separate signal from noise, thereby sparing valuable data. The three most common approaches are linear regression, independent component analysis (ICA), and principal component analysis (PCA).

Linear regression is by far the simplest method to remove ocular artifacts. Ocular artifacts typically have much higher amplitude than the EEG signals generated by neural sources. For the purpose of regression, the tiny EEG signals are assumed to be noise relative to the ocular artifacts. Blinks are assumed to affect electrodes in a linear combination, such that electrodes closer to ocular channels have a larger weight than electrodes farther from ocular channels. Once the optimal weights are calculated, a weighted composite of the ocular artifact is subtracted from each of the EEG electrodes (cf. Gratton et al., 1983), yielding EEG that should be (mostly) free of the artifact. However, it requires the ocular artifacts to be linearly dependent and normally distributed, which generally is not the case. This approach is fast and simple, but sometimes inaccurate, which introduces another source of residual artifact. Errors in the regression get "subtracted into" the recording.

ICA is a blind source separation technique, unmixing a multivariate signal into a set of additive components (Delorme et al., 2010). Unlike linear regression, independent components (ICs) are assumed to be non-Gaussian and statistically independent from each other. However, ICA can only separate linearly mixed sources, and even when the sources are not independent, ICA will return maximally independent components. In many cases, a subset of the ICs will capture artifacts such as ocular and motor components. Components are determined to be artifact or signal based on their similarity with the topography and time course of known artifacts. This can be determined based on experimenter ratings or algorithmically using spatial and temporal features of eye blinks, vertical and horizontal eye movements, and discontinuities in the EEG time series (Mognon et al., 2011). Artifactual components may also be identified automatically based on a supervised machine learning approach trained from expert ratings on a large corpus of EEG data (Winkler et al., 2014). ICA is more useful than linear regression because it can extract more than just ocular components, also capturing loose electrodes and muscle artifacts equally well. Once the bad components are identified, the signal can be reconstructed without the artifacts.

Other blind source separation techniques similar to ICA can be used in a similar manner. PCA is like ICA in that it separates the data into a set of linearly independent components. However, these components are additionally assumed to be mutually orthogonal (i.e., uncorrelated with each other).

2.13.5 Referencing

Voltage is always measured between two points. It does not make sense to talk about the potential at a particular EEG electrode without first defining a reference potential (i.e., an arbitrarily chosen "zero"). A common reference is usually chosen for all electrodes, to which all potentials recorded at each electrode are measured. The reference needs to be chosen carefully because the electrical activity under the reference site will be reflected in the activity at every other electrode. If the reference is not neutral, artifacts will be introduced due to the activity at the reference site. The reference electrodes should be placed on a presumed electrically neutral area. In many cases, researchers choose an electrode over the mastoid part of the temporal bone, or the left or right earlobe.

In reality, there are no electrically neutral points to choose as a reference. However, it is possible to construct a virtual reference that is more likely to be electrically neutral than any single EEG electrode. For example, choosing the left or right mastoid alone results in a systematic decrease of EEG amplitude in the electrodes which are closer to the reference site. The "linked" mastoids reference is obtained by using the average of both left and right mastoid electrodes as the reference level. This resolves the left/right asymmetry that would normally occur. Another popular choice is the average reference when the number of electrodes is large (typically >32). The average reference is obtained by subtracting the mean potential across all EEG electrodes from each electrode at each time-point. Theoretically, if we could obtain measurements sampling equally around a sphere containing the brain, then electric dipoles picked up by the electrodes would average out to zero, and the average reference would be a truly neutral point (Bertrand et al., 1985). However, this is a practical impossibility because the ventral surface is inaccessible for electrode placement. Still, by averaging the electrical activity across the entire scalp, the "responsibility" is distributed over all electrodes, rather than only one or two of them.

However, prominent deflections at any given scalp location can differ dramatically depending on the chosen reference scheme, affecting both the amplitude and latency of the recorded signal. Not all researchers use the same reference, leading to problems in the comparability between datasets. Fortunately, it is possible to convert from one reference to another with a simple mathematical transformation. Note the potential difference between two electrodes does not depend on the chosen reference. If your analysis is only concerned with comparisons between two electrodes, then the reference is irrelevant. A simple analogy is measuring height above sea level. A change in sea level does not change the shape of the landscape or the relative elevation between any two points. Thus, if there is a change in sea level, we can readjust to the new reference level by simply subtracting the change in elevation from every point. Similarly, we can transform between different references by subtracting the potential at the chosen reference site.

Other transformations can be used to obtain a reference-free representation of EEG signals. This is usually achieved using the scalp Laplacian, sometimes called the Current Source Density (CSD). Technically, the scalp Laplacian relates current generators within an electrical conductor to the second spatial derivative of the potential at each electrode. Approximately, the scalp-Laplacian at a given electrode is obtained subtracting the potentials of all neighboring sites weighted by their inverse distance. In other words, the scalp Laplacian measures how extreme the potential at one electrode is compared to the average of its neighboring electrodes. This intuitive description is complicated by the fact that scalp electrode montages do not form a flat grid of evenly spaced electrodes.

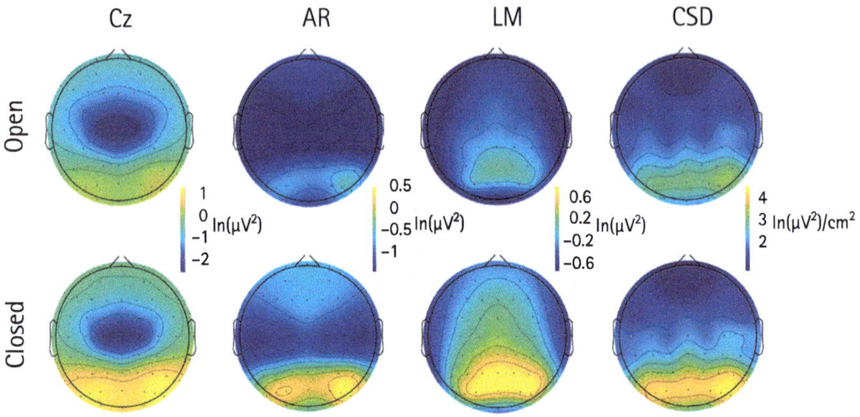

FIGURE 2.12 Topography of alpha power under eyes open (top) and eyes closed (bottom) conditions as a function of transformation (Cz, average (AR), or linked mastoid (LM) reference or current source density (CSD) transformation) from a sample of over 2400 resting recordings. Power values at each site represent natural-log transformed values; thus, negative numbers represent mean power values less than one. Each transformation is scaled independently, but within each transformation, eyes open and closed data are plotted on the same scale. Only the CSD transformation confines occipital alpha to occipital leads, whereas the other three montages show reflected alpha at frontal regions, visible most clearly by a comparison of frontal leads under eyes closed compared to eyes open recordings.

Figure modeled after Smith et al., 2017.

Instead, they conform to the nearly spherical geometry of the scalp. However, the Laplacian can still be efficiently computed using spherical spline interpolation (Delorme & Makeig, 2004). The scalp Laplacian has the added advantage of addressing volume conduction. Electrical fields originating from any particular neuronal structure will influence the electrical potential throughout the brain and surrounding physiological tissue. As a result, EEG electrodes detect a mixture of activity from across the whole brain, rather than directly under the recording site. Critically, the scalp Laplacian minimizes the influence of this effect since the Laplacian is close to zero when a signal is shared similarly across several electrodes (cf. Smith et al., 2017). Figure 2.12 shows a comparison across referencing schemes .

2.14 Conclusion

A basic understanding of electricity and signal processing is vital to EEG data collection and interpretation. Investigators need to report clearly how they acquired and analyzed the data, and consumers need to be able to evaluate the methods and claims critically. There are many steps to be taken in acquiring and processing EEG signals, from net application and impedance checking, to grounding and referencing, to artifact detection and correction, to data processing and analysis methods in the time and frequency domains. This chapter provided a basic introduction to these concepts, but more information can be obtained from several guidelines papers (Picton et al., 2000; Keil et al., 2014).

Glossary

Aliasing: When an analog signal is insufficiently sampled, the digitized signal may be indistinguishable from samples from a lower-frequency signal.
Ampere: The SI unit of electric current. It is defined as one coulomb of charge per second.
Analog signal: A continuous time-varying signal that may vary in frequency, amplitude, and phase.
Analog-to-digital converter: System that converts an analog signal into a digital signal.
Capacitance: Ratio of the change in an electric charge to the change in electric potential. A property that quantified a materials ability to store electric charge.
Conductivity: Property that quantifies how strongly a material conducts electric current.
Conductor: A material that allows the flow of electric current.
Conjugate variables: Pairs of variables that cannot be precisely estimated simultaneously. Knowledge of one variable necessitates uncertainty in the other variable. Time and frequency are examples of conjugate variables.
Convolution: A mathematical operation that indicates the amount of overlap of one function or kernel as it is shifted over another function.
Cutoff frequency: Frequency at which the signal power is reduced by half. The amount of attenuation increases as you move farther from the cutoff.
Current: The directed motion of electrons or electric charge.
Digital signal: A discrete sequence of finite values that represent a signal.
Electric field: The force per unit charge at each point in space.
Electricity: Phenomenon that describes the behavior and movement of electric charge.
Electromotive force (EMF): The rate at which energy is drawn from a 1 A current source, measured in volts.
Finite impulse response (FIR) filter: A filter whose impulse response settles to zero in finite time.
Free electrons: Any electron that is free to move under the influence of an electric or magnetic field.
Impedance: Measures the opposition to an electric current when a voltage is applied. Impedance is equal to resistance in a DC circuit.
Impulse response: The output of a circuit when presented with a brief input signal.
Insulator: A material that does not allow the flow of electric current.
Ionization: The acquisition or loss of electrons in an atom or a molecule, resulting in a net positive or negative charge.
Nyquist frequency: Half of the sampling rate of a digital signal. It is the highest frequency that can be accurately sampled for a given sampling rate.
Nyquist sampling rate: In order to adequately digitize an analog signal, it should be sampled at least twice the highest frequency of interest.

Phase spectrum: A measure of signal phase versus frequency.
Potential energy: The energy stored by an object due to electric charge.
Power spectrum: A measure of signal power versus frequency.
RC time constant: The time required to charge a capacitor to 63.2% of the applied DC voltage.
Reactance: Material property that measures the resistance to a change in current or voltage.
Resistance: Property that quantifies how strongly a material resists electric current.
Sampling rate: Rate at which discrete samples are acquired for a digital signal.
Volt: The amount of EMF required to push a current of one ampere through a conductor with a resistance of one ohm, or 1 volt/ampere.
Voltage: Difference in electric potential between two points.

References

Bertrand, O., Perrin, F., & Pernier, J. (1985). A theoretical justification of the average reference in topographic evoked potential studies. *Electroencephalography and Clinical Neurophysiology*, *62*(6), 462–464.

Cohen, M. X. (2014). *Analyzing neural time series data: theory and practice*. MIT press.

Dang, L. C., Samanez-Larkin, G. R., Castrellon, J. J., Newhouse, P. A., Zald, D. H., Perkins, S. F., & Ronald, L. (2017). Spontaneous eye blink rate (EBR) is uncorrelated with dopamine D2 receptor availability and unmodulated by dopamine agonism in healthy adults. *ENeuro*, *4*(5), 1–11.

Delorme, A. & Makeig, S. (2004). EEGLAB: An open source toolbox for analysis of single-trial EEG dynamics including independent component analysis. *Journal of Neuroscience Methods*, *134*(1), 9–21. https://doi.org/10.1016/j.jneumeth.2003.10.009

Delorme, A., Sejnowski, T. J., & Makeig, S. (2010). Enhanced detection of artifacts in EEG data using higher-order statistics and independent component analysis. *NeuroImage*, *34*(4), 1443–1449. https://doi.org/10.1016/j.neuroimage.2006.11.004.Enhanced

Ferree, T. C., Luu, P., Russell, G. S., & Tucker, D. M. (2001). Scalp electrode impedance, infection risk, and EEG data quality. *Clinical Neurophysiology*, *112*(3), 536–544.

Gratton, G., Coles, M. G., & Donchin, E. (1983). A new method for off-line removal of ocular artifact. *Electroencephalography and Clinical Neurophysiology*, *55*(4), 468–484.

Keil, A., Debener, S., Gratton, G., Junghöfer, M., Kappenman, E. S., Luck, S. J., Luu, P., Miller, G. A., & Yee, C. M. (2014). Committee report: Publication guidelines and recommendations for studies using electroencephalography and magnetoencephalography. *Psychophysiology*, *51*(1), 1–21.

Mognon, A., Jovicich, J., Bruzzone, L., & Buiatti, M. (2011). ADJUST: An automatic EEG artifact detector based on the joint use of spatial and temporal features. *Psychophysiology*, *48*(2), 229–240. https://doi.org/10.1111/j.1469-8986.2010.01061.x

Perrin, F., Pernier, J., Bertrand, O., & Echallier, J. F. (1989). Spherical splines for scalp potential and current density mapping. *Electroencephalography and Clinical Neurophysiology*, *72*(2), 184–187. https://doi.org/10.1016/0013-4694(89)90180-6

Picton, T. W., Bentin, S., Berg, P., Donchin, E., Hillyard, S. A., Johnson, R., … Taylor, M. J. (2000). Guidelines for using human event-related potentials to study cognition: recording standards and publication criteria. *Psychophysiology*, *37*(2), 127–152. http://www.ncbi.nlm.nih.gov/pubmed/10731765

Smith, E. E., Reznik, S. J., Stewart, J. L., & Allen, J. J. (2017). Assessing and conceptualizing frontal EEG asymmetry: An updated primer on recording, processing, analyzing, and interpreting frontal alpha asymmetry. *International Journal of Psychophysiology*, *111*, 98–114.

Stern, J. A., Boyer, D., & Schroeder, D. (1994). Blink rate: A possible measure of fatigue. *Human Factors*, *36*(2), 285–297. https://doi.org/10.1177/001872089403600209

Stern, J. A., Walrath, L. C., & Goldstein, R. (1984). The endogenous eyeblink. *Psychophysiology*, *21*(1), 22–33.

Trujillo, L. T., and Allen, J. B. (2007). Theta EEG dynamics of the error-related negativity. *Clinical Neurophysiology*, 118(3), 645–668

Taylor, J. R., Elsworth, J. D., Lawrence, M. S., Sladek, J. R. Jr, Roth, R. H., & Redmond, D. E. Jr. (1999). Spontaneous blink rates correlate with dopamine levels in the caudate nucleus of MPTP-treated monkeys. *Experimental Neurology*, *158*(1), 214–220.

Winkler, I., Brandl, S., Horn, F., Waldburger, E., Allefeld, C., & Tangermann, M. (2014). Robust artifactual independent component classification for BCI practitioners. *Journal of Neural Engineering* [online], *11*(3), 035013. https://doi.org/10.1088/1741-2560/11/3/035013

CHAPTER 3

FROM NEURAL OSCILLATIONS TO COGNITIVE PROCESSES

ANDREAS KEIL AND NINA THIGPEN

3.1 Oscillations in Complex Systems: An Overview

Cyclical rhythms are a hallmark property of natural systems, from the movement of the planets to the sleep and wake cycles of mammals, to neuronal membranes. Such rhythmic variation, if it occurs repeatedly across time, is called an *oscillation*. The back-and forth of tree branches on a windy day, ocean waves, and the rhythmic song of choir frogs are further examples for natural oscillations. As such, the notion of oscillations seems fairly straightforward: they occur in complex systems, in which many connected units interact, and they reflect activity perceived as recurrent, or rhythmic. However, a woman strolling on an ocean beach will soon realize that although waves break repeatedly and somewhat predictably, they also vary in terms of their exact timing, duration, size, and shape. This seems to be different from the orbit of our planet earth around the sun, or the vibrations of crystalline matter—rhythms that are regular and fixed enough for us to measure time on their basis. These latter are typically referred to as "periodic oscillations", which are a sequence of recurring events (often taking the shape of a wave when measured), each of which has identical, constant, duration. Although periodic oscillations are well defined in the language of mathematics (e.g., the sine function), the examples at the start of the chapter show that the concept of periodicity may not be as clear-cut in natural systems. For example, the duration of solar years is not constant when measured with precision, and even clock-setting crystal oscillators change their rhythm with temperature. Thus, periodicity may be a continuous dimension ranging from more periodic (the solar year) to less periodic (the tides) rhythms, rather than representing a qualitative (yes/no) feature. In fact, various natural systems are likely to oscillate in different fashion, and the same system may display oscillatory behavior at different levels of complexity when challenged in different ways.

Identifying how a complex system oscillates may go a long way towards its characterization. As a consequence, the natural sciences have developed an impressive toolbox for detecting, quantifying, and understanding oscillatory activity, and to separate it from non-oscillatory phenomena. This chapter discusses how these concepts are applied to fundamental problems in cognitive neuroscience, and ask what the significance is of oscillatory activity for human behavior and cognition? It also asks how we should define and conceptualize brain oscillations, and how they can be measured and interpreted. We address these questions in the light of current research, with a focus on electrophysiological recordings in human beings and experimental animals.

3.2 Brain Oscillations and Behavior—an Example

After over a decade of testing, in 1927 Hans Berger conducted the earliest in vivo recordings of electric fields from the human brain using his new invention, the electroencephalograph (Berger, 1929). Upon visual inspection of these first brain wave (electroencephalography, or EEG) recordings, Berger decided that the most salient feature of the EEG was its change in frequency content when the participant performed a task or was stimulated (often by being touched with a glass probe). Most prominently, Berger identified "first-order waves" oscillating at a rate of 10 cycles/second (i.e., 10 Hertz (Hz), today referred to as *alpha waves*) and faster "second-order waves" oscillating at about 20–30 Hz (today referred to as *beta waves*). Even without the ability to average across trials, and in the absence of digitally supported spectral analysis, Berger and his colleagues (i.e., his family; his wife Ursula von Bülow was his technical assistant and frequent test participant, along with himself, and his son, Klaus) made a striking observation: whenever a person was at rest, for example, with their eyes closed, first-order waves characterized by large-amplitude regular activity at 10 Hz dominated the EEG. However, touching the person with a glass probe, asking them to open their eyes, or just addressing them and asking a question led to a dramatic blocking of these alpha waves, which were replaced by faster and lower-amplitude EEG readings. This effect, referred to as "alpha blocking" has since been replicated thousands of times, and is still used today in clinical usage of EEG in neurology and psychiatry to assess the status of global brain electric states (Başar & Güntekin, 2012). Furthermore, this observation opened an avenue for experimental studies of EEG-behavior relations, in which participants are asked to respond to experimenter-defined task demands, and changes in the EEG are measured as a dependent variable (Klimesch et al., 2005). In addition to testing hypotheses about brain function, these variables possess practical value for testing hypotheses about behavioral and cognitive processes, their time course, and about specific characteristics of the participant or the environment, of interest in clinical and translational studies.

3.3 Quantifying and Categorizing Brain Oscillations

Oscillatory activity is typically described by quantifying its frequency, amplitude (power), and phase (Figure 3.1). Frequency refers to the rate of repetition of the oscillatory event (e.g., action potentials emitted by a neuron, or epileptic spikes in the EEG signal) and is measured in Hz, the number of events per second. Amplitude reflects the magnitude of the oscillatory signal at a given frequency, analogous to measuring the height of ocean waves. Power (e.g., microvolts squared) is often used instead of amplitude to describe the magnitude of an oscillation. This reflects the fact that the computation of spectral magnitude usually involves an integral across time, which may be thought of as an area under the curve at a given frequency, resulting in a squared unit of measurement. Thus, in many cases, depending on scaling and other normalization steps, power values given in published research may be roughly equivalent to the squared amplitude of the oscillatory signal.

The phase of an oscillation is a measure of its position in time, relative to a pre-defined cycle, or relative to a reference oscillation at the same frequency. Because oscillations are by definition cyclic and recurrent, phase is measured using circular metrics, typically in degrees or radians of arc. For example, a phase of 90 degrees (1/2 Pi) may indicate that the oscillation at this point in time is at its peak, going down, 180 degrees (Pi) may indicate that it is transitioning through zero, and 270 degrees may indicate that it is at its minimum, going up. In neuroscience studies, phase is often used to quantify the amount of temporal correspondence or similarity between two signals, measured

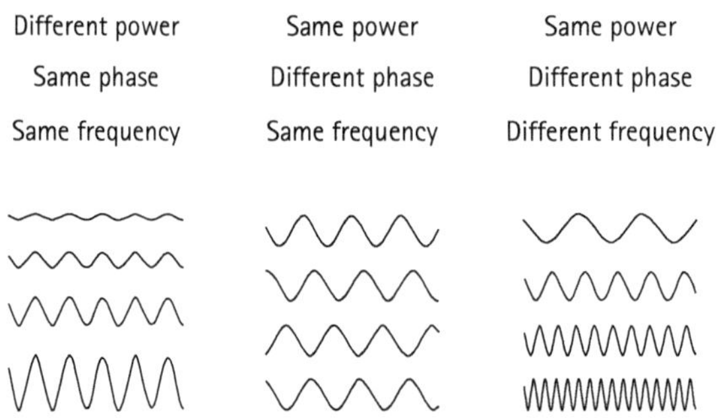

FIGURE 3.1 Illustration of oscillatory power, phase, and frequency. Oscillations are characterized these three properties and can independently vary along these three dimensions.

during different trials or at different locations (i.e., phase-locking, phase synchrony) [cross ref Chapter 20 Makeig and Palva]. We discuss phase-locking later in this chapter.

There are different ways for computing the oscillatory content of a signal, many of which are discussed in this volume [cross ref Chapter 23 Allen and Voytek]. Broadly, there are two main groups of methods for estimating spectral events in neural time series data. First, methods that result in a spectral representation, or frequency-domain representation. In these representations, frequency (in Hz) is plotted along the x-axis, and spectral power (or phase) along the y-axis. Power (and phase) values for each frequency in the spectrum are obtained by integrating the information across all time points entering the spectral analysis, which often is a variant of the well-known Fourier transform that uses sine and cosine functions as templates for quantifying the oscillatory content of a time-varying neural signal such as EEG or local field potentials. A second approach is used when researchers wish to retain information on the changes in oscillatory activity in their signal over time. Many methods exist for obtaining such a time-frequency representation, in which time is typically plotted on the x-axis, frequency along the y-axis, and a colormap or 3rd dimension in the figure is used to represent spectral power or amplitude for each time point and frequency. Often, this type of data analysis is used when the data set contains a substantial number of trials, locked to events (e.g., the onsets of a stimulus, the time of a motor action). Then, the time-frequency representations of each trial are compared and/or averaged. Figure 3.2 shows an example of the change in oscillatory power during a complex sequence of visual cues and behaviors as participants perform a working memory task. Wavelet analysis is a popular approach for obtaining time-frequency representations in studies of oscillatory brain activity, but many other methods exist (see Chapter 4). For a practical introduction, we recommend Cohen's *Analyzing Neural Time Series Data* (2014).

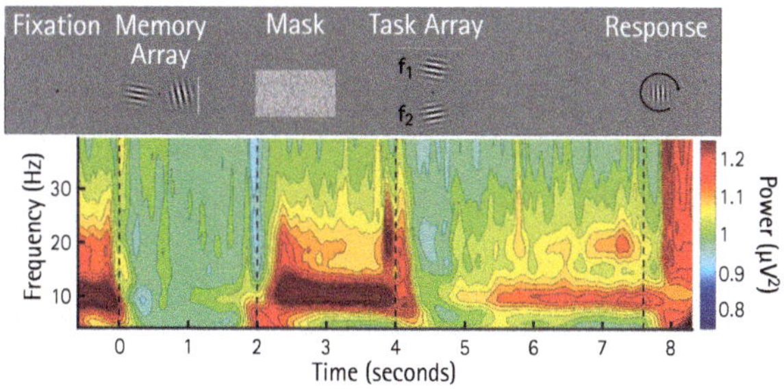

FIGURE 3.2 Oscillatory changes during a working memory task. Time-frequency decomposition of the EEG activity recorded at sensor Pz over a 9-s trial in which participants memorized the orientation of two gratings. They then viewed a mask, a task array, and finally were asked to indicate if the orientation of an item in the task array matched an item in the memory set. Each task element prompts specific time-frequency dynamics. For example, alpha-band activity (8–12Hz) is highest during the fixation period before the start of the trial, and during the retention interval when viewing the mask.

A useful approach for a classification of the brain's oscillatory activity is the widely-adopted nomenclature first introduced by Robert Galambos (1992). Considering brain oscillation across several species, Galambos distinguished:

1. spontaneous oscillations, which are not related to external stimuli;
2. evoked oscillations, which are elicited and precisely time-locked to the onset of an external stimulus;
3. emitted oscillations, which are time-locked to a stimulus that was expected but then did not occur; and
4. induced oscillations, which are prompted by a stimulus but are not time- and phase-locked to its onset.

Investigators new to the field are often curious about how different types of oscillations should be analyzed and interpreted in terms of hypotheses regarding a specific behavioral, cognitive, or neural process. To address these questions, it may be helpful to consider the different types of dependent variables that may be used in studies of oscillatory brain activity during cognitive processing.

3.4 Characterizing Brain Oscillations in the Context of Cognitive Tasks

The mathematical foundations of extracting power and phase from time domain neural signals as described earlier are straightforward and yield estimates of power at a given frequency, in a given time period, at a specific sensor. What is less clear is the functional and neurophysiological interpretation of differences in power, especially in population-level signals, such as local field potentials (LFPs), intra- and extracranial EEG, and MEG. These signals reflect the synchronous synaptic (dominantly post-synaptic) currents of a large number (at least several tens of thousands in the case of EEG/MEG) of neurons which need to be aligned favorably (in parallel) to create measurable extracranial electric fields (Olejniczak, 2006). Thus, a difference between two experimental conditions, say in EEG alpha-band power, may reflect differences in the number of neurons engaged in the respective cognitive process, but it may also represent a combination of differences in active population size, amount of dendritic engagement at each neuron, orientation of neurons involved, and temporal synchronization of the post-synaptic events. Neural mass power in a given frequency band is therefore difficult to map directly onto a physiological or cognitive concept, emphasizing the need for research that links measures of oscillatory brain activity to robust cognitive or behavioral processes, across different levels of observation (i.e., from single neurons to large populations).

In addition to measuring spectral power, researchers may use approaches that capitalize on additional spectral information, often the phase of the signal. Many algorithms exist for quantifying the similarity of the phase across different observations, such as experimental trials, time points, or channels. The resulting metrics are referred to as indices of phase-locking or phase coherency. These variables are often used to estimate the strength of oscillatory interactions across recordings sites (e.g., inter-site phase locking, often considered a proxy of connectivity), the stability of the temporal profile of oscillations across repeated trials (e.g., phase-locking value), or the interactions between different frequencies within or across recordings sites (e.g., phase-amplitude coupling (Chapter 20). Thus, a wide range of hypotheses and, increasingly, formal computational models of neurophysiological processes can be tested using these different variables, each of which may reflect entirely different facets of oscillatory brain activity. Later in this chapter we present examples for using these markers across a wide range of research questions in cognitive neuroscience.

3.5 What's in a Band? the Traditional Demarcation Of Oscillatory Events By Frequency

In the many decades since Berger's and many others' discoveries, brain oscillations have been traditionally divided into frequency bands in the 1–4 Hz range (delta), 4–8 Hz (theta), 8–12 Hz (alpha), 12–30 Hz (beta), and >30 Hz (gamma), with some variability in the demarcation of these bands. These frequency bands have been consistently observed at the level of spike trains, local field potentials, and EEG/MEG, making them apparent in multivariate analyses and meta-analyses across studies and even species (Lopes da Silva, 1991). The peak frequency in each of these bands is similar across bats, mice, rats, cats, dogs, horses, dolphins, macaques, and humans, despite the range of brain size and axon length across these species (Buzsáki et al., 2013). Given this phylogenetic stability, it has been suggested that the traditional frequency bands (and perhaps their logarithmic distance from one another) represent fundamental mechanisms underlying specific neurocomputations and behaviors (Steriade et al., 1990). Klimesch (1999) highlights the variability and task-dependency of these frequencies, however, with examples for gamma-like oscillations in the traditional beta band, and overlap between alpha and beta band oscillations, along with pronounced inter-individual differences. Similarly, limitations of assigning specific functional roles to oscillations in one of the traditional bands have been made apparent by phenomena in which oscillations continuously transition from one traditional band into another during the same task (Donner & Siegel, 2011).

Some of the stated confusion about these observations may be reflective of researchers' desire to equate oscillatory activity in one frequency range with a category of cognition of behavior, defined in the language of cognitive psychology, such as "alpha-blocking reflects attention", "gamma reflects feature binding", and similar notions. Such interpretations, which may be seen as examples for the problematic approach of reverse inference (Poldrack, 2011), have been considered less fruitful than, and have increasingly been replaced by, research aiming to characterize specific oscillatory correlates of a specific, quantifiable, behavior. With the advent of neurostimulation and neuromodulation techniques such as transcranial magnetic stimulation (TMS), direct current stimulation, neurofeedback, etc., recent research has also attempted to establish the causal role of oscillatory brain activity for specific cognitive and behavioral processes (Herrmann et al., 2016; Chapter 22 this volume). To understand how oscillatory activity relates to cognition, it is helpful to consider the rich neurophysiological literature based on work in experimental animals. This body of work has begun to outline neuromechanistic accounts for the emergence of brain oscillations and their relation to behavior. Importantly, these studies help to overcome the focus on one spatial scale (e.g. oscillations at membranes, in action potential, in circuits, or areas), and instead describe the rhythmic interplay between these levels of observation.

3.6 Oscillations at Different Levels of Observation: From Single Neurons to Neural Populations

Since the inception of brain electrophysiology, researchers measuring electrical potentials from individual neurons, or from groups of neurons, have noted the oscillatory character apparent in neural time series (Bernstein, 1868; Marrazzi & Lorente de No, 1944). Spatial scales of these observations are typically categorized into the microscale, mesoscale, and macroscale, where microscale refers to processes at single neurons, mesoscale to small functional units such as cortical columns, and macroscale to population responses of several tens of thousands of neurons (Nunez & Srinivasan, 2006). Recent work increasingly focused on interactions between oscillations across these levels of observation, for example, by linking the rate of action potentials (spikes) to oscillatory events at different spatial scales, and at different temporal rates (Lisman & Jensen, 2013; Schroeder & Lakatos, 2009). Although a comprehensive review of this literature is outside the scope of this chapter, a number of neurophysiological and computational principles of oscillatory processes within and across different scales are now widely accepted. These principles are discussed next.

At the core of many theories of brain oscillations is the observation that time series of action potentials (i.e., spike trains) tend to contain bursts of spikes. A burst represents a rapid sequence of action potentials, a group in time, which may vary in temporal rate, providing one potential source of oscillatory pace making. The frequency at which bursts occur (i.e., the inter-burst interval) has also been shown to relate to neurophysiological and behavioral processes, thus providing a second way for action potentials to temporally organize downstream neural—and ultimately behavioral—processes (Gütig, 2014). Such a temporal organization may be propagated in space if neurons act as coupled oscillators, facilitating specific phase relationships among the units of a functional network of connected neurons. Many such "coupled oscillator" models of neural oscillations exist, aiming to explain how temporal signatures are shared among populations of neurons (Moon et al., 2015; Naze et al., 2015). Recent empirical work as well as work in computational modeling has converged to suggest that such coupling involves not only action potentials but a range of mechanisms that also involve subthreshold and synaptic events, along with potential changes at glia cells (Buzsáki et al., 2012).

It is now well established that spike timing, that is, the timing of action potentials, is related to synaptic oscillations measured by local field potentials, likely both driving and being driven by these synaptic fields. This research has also demonstrated that subthreshold oscillations at membranes and post-synaptic potentials constrain the spiking rate of individual neurons and may also play a role in coordinating the firing among different neurons (Mazzoni et al., 2010). This is important because it demonstrates the interplay of all-or-nothing neuronal communication and post-synaptic events, opening avenues for research that identifies the convergent versus complementary roles of oscillations at different levels, within and across neurons, in behavioral and cognitive processes.

Research in human participants has increasingly made use of intracranial data obtained from patients under pre-surgical evaluation for neurological disorders, with implanted sensor arrays (Parvizi & Kastner, 2018). In these studies, one striking discrepancy arises between levels of observation regarding the frequency content of intracranial recordings vis-à-vis extracranial recordings. Intracranial recordings, for example, electrocorticogram (ECoG) data, tend to contain robust high-frequency oscillations (Osipova et al., 2008), which are small and less reliably observed in extracranial recordings such as EEG or MEG (Yuval-Greenberg et al., 2008). This salient difference between intracranial and scalp recordings may reflect the orders of magnitude difference in the summation of electrical potentials that comprise each signal. For example, the local field potential recorded from an intra-cranially sensor embedded in cortical tissue is thought to measure the extracellular electrical fluctuations of a few hundreds to thousands of neurons, while EEG data is thought to require synchrony across a few millimeters of cortex before any fluctuations are observed on the scalp (Nunez & Srinivasan, 2006). Based on simulations and in vitro studies, some authors suggest that postsynaptic changes at dendritic trees of 40,000 to 100,000 pyramidal cells are required to cause EEG changes in the 1–2 microvolt range—if the dendrites of these neurons are oriented in parallel, generating an open electric field, required for extracranial measurement. By contrast, ECoG and LFP data are thought to reflect activity in differentially oriented cells, including contributions from interneurons, pyramidal cells, and glial cells (Buzsáki et al., 2012). Thus, EEG oscillations reflect activity in a more specific subset of neurons than ECoG and LFP but integrated over a wider distribution of space.

Inter-scale oscillatory interactions also depend on the strength of the sensory input. Studies in macaque monkeys have shown that increasing the luminance contrast of a visual stimulus similarly affected spike trains, local field potential in intracranial electrodes over V1, and MEG signals, up to a contrast level of 60%. However, at levels above 60% luminance contrast, nonlinearities were observed: The MEG gamma continued to increase but the LFP gamma saturated, and in fact decreased with increasing contrast. Only the gamma frequency recorded at the macroscale (i.e., MEG, which continued to increase at luminance contrast levels above 60%) was correlated with perceptual performance (Hadjipapas et al., 2015). This is an example of how temporal summation across a wide spatial distribution (reflected in MEG recordings) reflects different activity from spikes or the LFP. Finally, it is worth noting that the relationship between spatial integration across neurons, measured through correlation of spike counts between neurons, and EEG power has also been shown to be nonlinear, such that higher spike-count correlations predict high and low EEG amplitude, while low spike-count correlations are observed during time windows where intermediate EEG amplitudes are observed (Snyder et al., 2015).

In summary, although systematic relations exist between neural oscillations at different levels of observation, it is not advisable to test hypotheses regarding one level of observation by measuring at another level. Inter-scale interactions however provide a promising area of research for understanding the neuromechanistic role of neural oscillations. They also represent an important source of constraints for computational models of brain function, in turn to be tested by multi-scale empirical studies. Researchers studying macroscopic oscillations in humans may generate more precise, and more mechanistic hypotheses regarding these measurements when considering research at the micro- and mesoscale. Thus, by manipulating the nature of the stimulus, for example, its intensity or temporal frequency, specific mesoscopic circuits can be challenged and experimentally isolated, dramatically increasing the neurophysiological specificity of extracranial signals such as EEG or MEG. This technique is increasingly used and further emphasizes the benefits of developing hypotheses for human research based on the animal model. However, as is the case with the majority of imaging modalities (Boynton, 2011), not all robust and neurophysiologically meaningful metrics of oscillatory brain activity are linearly related to the behavioral indices measured in the same task. This long-standing conundrum in cognitive neuroscience research may also be explained by the non-overlap of processes observed at different spatial scales, further highlighting the importance of multi-scale studies.

3.7 Brain Oscillations and their Role in Behavior: A Conceptual Overview

From its inception, the study of brain oscillations and their role in cognition has been multi-disciplinary, with a strong emphasis not only on physiological measurements, but also on mathematical analysis and modeling, as well as behavioral observations. The resulting literature can be daunting for its extent, scope, and methodological complexity. Therefore, readers may benefit from a brief overview of key concepts, providing scaffolding for integrating current findings, and context for making sense of the wide range of theoretical notions related to oscillatory brain dynamics.

3.7.1 Hebbian Cell Assemblies and Brain Oscillations

Adapting behavior based on experience is arguably one of the most essential functions of the human brain. Donald Hebb's (1949) classical theory in conjunction with theories of oscillatory brain activity are often used to account for this function. In these views, the brain's ability to form oscillatory networks comprising groups of neurons depends on a principle of association known as Hebb's second rule. This rule postulates that the synchronous activation of two cells or cell systems produces a facilitation of excitatory connections between them such that in the future activity of the one element produces a residual excitation of the other. A cell assembly refers to a distributed network of neurons that are bound together by the temporal synchronization of their sub-threshold membrane potentials and/or firing rates (Singer et al., 1990). Importantly, neuronal cell assemblies may be synchronized at local scales, separated in the millimeter range, but also at distances that span distinct cortical lobes (Pulvermuller & Fadiga, 2010). Theoretical and methodological advances in the neurosciences, including the discovery of long-term potentiation (Bliss & Lømo, 1973), whereby the responses of a post-synaptic cell are facilitated by oscillatory stimulation, have largely validated Hebb's notion, while also offering more detailed analyses of the physiological substrates that underlie the formation of oscillatory networks through experience (Nadel & Maurer, 2018).

Today, the ability of neurons to form networks based on coincidence of firing patterns can be explained at the molecular level (Andersen et al., 2017; Harris & Littleton, 2015), for example by reference to the unique properties of NMDA (N-Methyl-D-Aspartate) and AMPA (α-amino-3-hydroxy-5-methyl-4-isoxazolepropionic acid) glutamate receptors (Henley & Wilkinson, 2016). Additional physiological mechanisms for coincidence-based network formation have been elucidated, such as the dual sensitivity of L-type calcium channels to back-propagating action potentials and pre-synaptic excitatory currents (Nanou & Catterall, 2018). Importantly, neuronal cell assemblies may be synchronized at local scales, separated in the millimeter range, but also at distances that span distinct cortical lobes. Hebb's original suggestion (1949) held that reverberating activity within cell assemblies constituted a transient percept or memory trace that can become more consolidated with the passage of time. The concept of reverberation, by providing a metaphorical notion of how oscillatory phenomena arise, has stimulated a large body of work, which in turn has identified a plethora of mechanisms for network organization and oscillatory brain activity, beyond temporal synchrony (Morone et al., 2017).

3.7.2 Complex System Theory and Nonlinear Dynamics

With the advent of powerful digital computers, conceptual frameworks that had been elusive because of their computational complexity, became increasingly popular. One group of concepts with strong implications for the study of oscillatory brain activity drew from theories of complex systems ("chaos theory") and nonlinear dynamics (Elbert et al., 1994). Applied to brain oscillations, these approaches describe fluctuations among different elements, often in so-called non-equilibrium conditions, where some form of energy or environmental constraint affects a complex, macroscopic system (Fuchs et al., 2000). Research conducted using this framework demonstrates that dynamics in complex systems may be characterized by the interaction of some quantifiable and continuous variable, and the effect that it exerts on the collective system state. In one flavor of nonlinear system research, the collective system state is measured by one dominant parameter, the order parameter, taken to index the high-level formulation of the system (Haken et al., 1985), whereas variables that can change the nature of the system dynamics are referred to as the "control parameter". When the control parameter reaches a particular value, the macroscopic system undergoes a phase transition and acquires a qualitatively novel set of properties. Once it is established, the new high-level order parameter may "enslave" the lower components—a phenomenon demonstrated in everyday life when the clapping of a large audience determines the clapping frequency of individual members in the audience.

Metrics of nonlinear dynamics such as largest Lyapunov coefficients or the Grassberger–Procaccia dimension (Elbert et al., 1994) have been used to characterize neural time series, aiming to capture dynamics above and beyond the periodic, sine-wave-like, and stationary cycles that are assumed by linear methods such as the Fourier transform. In addition, approaches inspired by nonlinear systems research have made important contributions by emphasizing the role of oscillatory phase, and by providing tools for visualizing and analyzing nonlinear interactions among and between populations of neurons (Breakspear, 2017). Currently, some of these methods, such as phase space portraits, experience a renewed popularity and are applied in studies of motor preparation, perception, and awareness (Baria et al., 2017).

3.7.3 Spatio-Temporal Patterns and Travelling Waves

Although oscillatory patterns are most easily visualized as changes in magnitude over time, researchers have recognized for a long time that the spatial dimension of oscillatory activity also plays a crucial role in characterizing oscillatory processes. Walter Freeman, one of the first researchers to explore this topic in depth, established a mathematical model in which spatio-temporal patterns consisting of mesoscopic "phase cones"—moving two-dimensional Gaussian fields with systematically changing oscillatory phase—represent the nature of an odorant presented to the rabbit olfactory system (Freeman, 1991).

Current research has addressed this topic using refined mathematical tools and high-resolution electrophysiological recordings. As a result, traveling oscillatory waves of neural activity have been found at many different frequencies, across wide spectrum of brain areas. Spreading over cortical or subcortical tissue as time passes, these waves have been related to many different behaviors, ranging from sensory to motor processing (Muller et al,, 2018).

3.7.4 Oscillatory Hierarchies

Many prominent models of the functional role of oscillatory brain activity have considered interactions between oscillations at different temporal rates. A classical model of working memory proposed that hippocampal neurons encode memory content using timed gamma-frequency bursts, which in turn receive their temporal structure from the phase of low-frequency oscillations in the theta band (Lisman & Jensen, 2013). This account has received substantial support as well as extension and refinement based on computational and empirical studies.

In another widely recognized account, systematic relations across different frequencies are used by the brain to organize the interplay between behavioral and cognitive processes—the concept of "active sensing" emphasizes the active aspect of perceptual sampling (Schroeder et al., 2010). Examples for such active sensing include whisking and sniffing in rodents and saccadic sampling of visual scenes in primates. In all these cases, perception is tightly linked to the motor activity that causes the stimulation of sensory receptors as it initiates inflow of information through the sensory pathways. Each afferent volley of information meets ongoing brain states hypothesized to be partly determined by past experience and current goals, as well as the physiological properties of the tissue. In these hypothetical hierarchies, neural oscillations at lower frequencies (such as alpha or theta, 4–8 Hz as defined in adults) may provide temporal structure for higher-frequency phenomena, for example, beta- (13–30 Hz) or gamma-band (>30 Hz) oscillations. In a similar vein, studies in macaque monkey visual cortex have suggested that high-amplitude gamma oscillations more likely occur during specific phases of the alpha cycle (Bonnefond & Jensen, 2015). These findings have been taken to indicate that alpha phase represents the excitability of the neural tissue, possibly aligning excitability of sensory cortices with other events such as expectancy or memory-driven signals (Kizuk & Mathewson, 2016; Mathewson et al., 2011).

3.7.5 Synchrony and Oscillatory Communication

Most contemporary models of oscillatory brain activity emphasize the potential of rhythmic processes for organizing neural activity in time and space and thus for providing a mechanism for integrative coding across spatially separated units (neurons, columns, areas; Buzsáki & Draguhn, 2004). In earlier theories of oscillatory activity, temporal synchrony was seen as the crucial feature for coding information that belongs together but is distributed between different neurons or populations of neurons (von der Malsburg & Buhmann, 1992). This notion is readily aligned with the Hebbian perspective discussed earlier, adding to the appeal it has held for many decades. At the microscopic and mesoscopic level, synchronous firing in local circuits, especially in the gamma frequency range, has been proposed as a mechanism for cognitive processes as diverse as gestalt perception, predictive coding, motor preparation, and selective attention (Bauer et al., 2014; Keil et al., 2001).

Extending the concept of synchrony to long-range communication and integration, recent work focuses on the interaction between local and inter-area (long-range) synchrony, increasingly on oscillatory interactions across different frequencies (Chapter 20). For instance, oscillatory activity in the theta range (4–8 Hz) has been traditionally hypothesized as a potential mechanism for transmitting information between distant brain areas (Klimesch et al., 2005; Sarnthein et al., 1998). Empirical evidence in rodents, cats, and non-human primates supports this notion, with theta oscillations characterizing long-range signaling in large-scale networks, including the hippocampus, the amygdaloid, complex, and extensive cortical areas, during tasks that involve learning and memory (Popescu et al., 2009). Similar findings exist for alpha-band oscillations, discussed now as carriers of prediction signals from higher-order to sensory cortices. Importantly, the concept of synchrony in the context of these inter-area communications has been extended beyond zero-lag co-activation at different sites, and has instead identified inter-area interactions by the extent to which the phase lag between two recording sites is consistent over time, that is, so-called inter-site phase locking (Brovelli et al., 2004).

Together, local- and inter-area oscillations hold great promise for informing and constraining network-based models of human cognition. Because of their compatibility with findings obtained in the animal model, and the ability to relate neural activity across different levels of observation, indices of oscillatory brain activity also represent powerful dependent variables in cognitive neuroscience studies. The following example illustrate these benefits by selectively reviewing examples for such studies and address processes of perception, attention, learning, and memory.

3.8 Example 1: Alpha-Band Changes During Perception and Selective Attention

As discussed, oscillatory activity in the alpha frequency band is typically observed during stimulus-absent conditions such as resting or sitting with the eyes closed (Berger, 1929; Pfurtscheller, 1989). Many early studies established the robust relationship between alpha power changes and perceptual tasks, demonstrating that high-alpha states tend to be blocked following events such as the opening of the eyes, the presentation of a visual cue, or the direction of attention toward a salient visual stimulus (Adrian & Matthews, 1934; Klimesch, 1999; Pfurtscheller et al., 1996). Low levels of alpha power together with alpha phase also predict heightened performance in near-threshold detection tasks, where participants report weak sensory events (Mathewson et al., 2011; Weisz et al., 2014). This evidence supports a long-held notion that scalp-recorded alpha activity reflects a cortical state in which little sensory information is received (Pfurtscheller, 1992), mediated perhaps via a thalamic gating mechanism (Steriade et al., 1990).

Other findings are consistent with this notion.: Presenting a cue that directs attention to one hemifield prompts reduction of alpha power in the contralateral hemisphere (Foxe & Snyder, 2011). By contrast, alpha power over ipsilateral sensors (representing the ignored hemifield) remains stable (Thut et al., 2006), or increases (Kelly et al., 2006), taken to suggest suppression of irrelevant information. This interpretation is consistent with research examining attention to target items embedded in a rapid serial visual presentation stream of distractors. Accurate identification of rapidly presented targets amid non-targets is associated with higher pre-target alpha power (Petro & Keil, 2015). Thus, high alpha power may index behavioral states that benefit performance by reducing the processing of irrelevant sensory information (Klimesch et al., 2006). Importantly, these attention-related changes in alpha power are associated with faster and more accurate responses for the task, suggesting that they possess a functional role in the active selection of target stimulus features (Foxe & Snyder, 2011).

A further role of alpha oscillations during attention tasks is under consideration in the context of predictive coding. This has become a current topic in cognitive neuroscience. Increased alpha power and heightened inter-site phase-locking during target anticipation are proposed as a mechanism that optimizes the temporal organization of sensory processing and thus facilitates the sensory analysis of expected, task-relevant, visual stimuli (Samaha et al., 2015). Several of the different proposed functions of alpha-band oscillations during perception and attention are not mutually exclusive and have given rise to the idea that the alpha frequency is used by a wide range of neural processes, in the service of different cognitive and behavioral goals (Chapter 10).

3.9 Example 2: Learning and Memory

Work in the animal model abundantly demonstrates that associative learning induces Hebbian plasticity in widely distributed brain networks, both sub-cortical and cortical (Pape & Paré , 2010). These memory formation processes can be conceptualized as changes in the neural communication between nearby and distant brain loci, which dynamically sculpt neural signaling pathways at multiple levels of analysis. Recent findings suggest that neuronal oscillations are ideally suited to support the flexible formation of such network-level changes in neural architecture (Popescu et al., 2009), including those that support the acquisition and extinction of fear memories (Paré et al., 2002). In addition, computational models using simulated agents in artificial evolutionary environments show that network oscillations provide a fitness advantage, insofar as they help to promote rapid switches in perception and attention (Heerebout & Phaf, 2010). Oscillatory signals may therefore play an important role in the critical changes that occur when a previously innocuous stimulus acquires relevance for controlling behavior (Headley & Weinberger, 2011). At the level of neuronal populations, the associative principles that guide the formation of a newly acquired memory (e.g., the association between light and electric shock during classical fear conditioning) must effectively coordinate activity between neural representations of the conditioned stimuli (the light) and the systems that code biological value of the unconditioned event (the shock). Although conditioning-induced changes in neuroarchitecture and function occur on multiple spatio-temporal scales (Maren & Quirk, 2004), ranging from individual neurons to cortical sheets, and from minutes to days, the notion of a cell assembly (Hebb, 1949) provides the necessary conceptual framework for bridging across these multi-scale phenomena. Oscillatory synchronization between distributed neuronal assemblies in specific frequency bandwidths appears to represent a highly plausible substrate for synaptic plasticity transfer, that is, storing newly acquired memories as changes in synaptic weights (Paré et al., 2002). In particular, increased large-scale synchrony between subcortical and cortical networks may be crucial in producing the experience-dependent changes in the representation of fear-conditioned cues.

Another example is provided by a series of studies conducted by Walter Freeman and his colleagues (reviewed in Skarda & Freeman, 1987) in the olfactory system of the rabbit. The basic paradigm involves the placement of an 8×8 electrode grid onto the olfactory bulb in order to record electrocorticography (ECoG) signals associated with different odorant stimuli. To begin with, each odorant elicits a pattern of wave activity that appears as aperiodic noise with variability across trials. However, after the animal learns to associate an odor with a motivationally relevant outcome, for example, the delivery of food or aversive tactile stimulation, neural activity at the olfactory bulb undergoes a state transition whereby the odor comes to elicit a discriminant spatiotemporal pattern of amplitude across the electrode grid array. The learning-related establishment of a new global response pattern then acts to enslave the output of individual neurons (Freeman, 1994). Freeman interprets such selective changes in spatiotemporal amplitude as embodying the affective meaning of a stimulus for the animal, with meaning changing in accordance with the momentary relevance of a stimulus (e.g., an animal responds differently to an odorant associated with food once it is fed to satiety). A mechanism for stimulus selective changes in the output of olfactory neurons is provided by alterations of synaptic efficiency and neuropil structure, guided by Hebbian principles of association.

3.10 Example 3: Generating Brain Oscillations by Periodic Stimulation

Rhythmic modulation of stimulus contrast or luminance at a constant rate over a period of time evokes oscillatory field responses in the visual cortex at the same frequency as the modulation rate of the stimulus, often including higher harmonics—integer multiples of the stimulation frequency. The harmonic responses depend on the duty cycle, the stimulation method, and the complexity of the stimulus array (Norcia et al., 2015). In the visual domain, large-scale frequency-following responses evoked by periodic stimulation are referred to as steady-state visual evoked potentials (ssVEPs). These driven oscillations are best quantified in the frequency or the time-frequency-domain, where they can be reliably separated from noise and quantified as the spectral power in a narrow frequency range. Several studies have demonstrated that the flicker-evoked ssVEP is predominantly generated in the primary visual and to some extent in adjacent, higher order, cortices (Müller et al., 1997; Wieser & Keil, 2011). Interestingly, a number of studies have suggested different neural sources for the fundamental frequency and the higher harmonics, although these findings await replication and further interpretation (Kim et al., 2010). Generally, the ssVEP can easily be driven in lower-tier (retinotopic) visual cortices using high-contrast luminance modulation governed by a rapid square-wave (on-off) stimulation. For studies of higher-order cognitive processes, however, researchers may periodically modulate specific stimulus dimensions other than luminance or contrast, while holding these lower-level properties constant (or varying them randomly). Such an approach evokes ssVEPs in brain areas sensitive to the particular feature or stimulus dimension of interest (Giabbiconi et al., 2016; McTeague et al., 2015). For example, stimulation techniques have been used that isolate the ssVEP response to face identity generated in higher order visual areas such as the fusiform cortex (Rossion & Boremanse, 2011).

A related issue often discussed in the context of driven oscillations is the question to what extent ssVEPs can be regarded as a linear superposition of transient ERPs, or alternatively represent a nonlinear response that possesses properties beyond the linear combination of individual brain responses (Capilla et al., 2011; Regan, 1989). A similar current debate exists between the view that driven oscillations represent a temporal alignment or "entrainment" of oscillations that are already present spontaneously, and the alternative view that driven oscillations represent newly shaped waveforms, on top of ongoing oscillations (Keitel et al., 2019). Initial studies in the field have argued that observing higher harmonics when using harmonically simple (e.g., sinusoidal) modulation of luminance or contrast is considered evidence of nonlinearity, thus providing evidence for perspectives emphasizing resonance and entrainment (Regan, 1989). Subsequent work has examined the extent to which ssVEPs can be explained by properties of transient ERPs, and has observed that especially with square wave (on-off) modulation, ssVEPs may be modeled by transient ERP features with satisfactory accuracy (Capilla et al., 2011), providing evidence for a superposition perspective. The interpretation of these findings is not straightforward, however, because transient brain responses themselves represent a combination of individual trials that vary greatly in latency (phase) and amplitude, limiting their use for unambiguously explaining the generating mechanism underlying ssVEPs. In line with this notion, other studies have reported that the ssVEP amplitude is poorly predicted by single-trial power changes at the driving frequency, but is best predicted by locking of the single-trial phase with

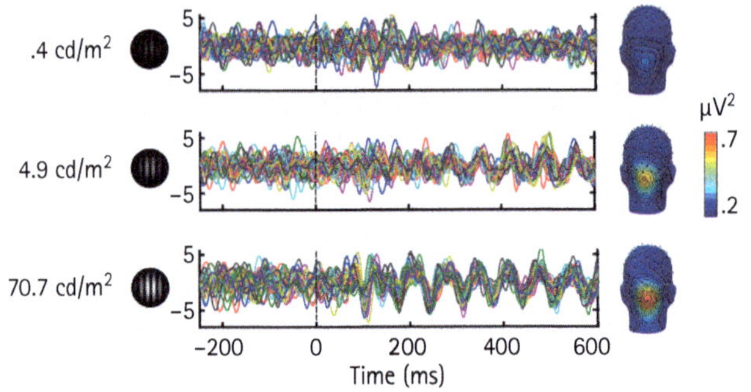

FIGURE 3.3 Effects of stimulus intensity on inter-trial phase locking of driven oscillations. Participants viewed sinusoidal gratings flickering at 15 Hz. Gratings varied in luminance, with 40 low-luminance trials (.4 cd/m^2), 40 medium-luminance trials (4.9 cd/m^2), and 40 high-luminance trials (70.7 cd/m^2). Each colored line represents one single trial from sensor Oz. Inter-trial phase locking increases with increasing luminance, whereas the overall magnitude of the signal within the single trials does not change. The increased phase-locking is associated with increases in 15-Hz power of the trial-averaged signal, as depicted in the topographies shown on the right.

Data from Thigpen et al., 2018.

the driving stimulus across trials (Moratti et al., 2007). Figure 3.3 shows an example of this phenomenon as it interacts with stimulus intensity, demonstrating increasing phase alignment of single trial EEG traces with increasing stimulus contrast. The debate of entrainment versus superposition perspectives is ongoing, and ssVEPs (because of their known frequency and pronounced signal) are well suited for examining competing hypotheses regarding the interaction of ongoing oscillations and sensory events.

3.11 Conclusions and Outlook: Elements of a Conceptual Framework of Oscillatory Brain Activity

This chapter aimed to illustrate how studies of oscillatory brain activity provide rich opportunity for testing hypotheses relating brain function to cognitive processes. Measuring oscillatory activity during cognitive task performance opens avenues into developing and testing neuromechanistic accounts of some of the most central building blocks of human behavior and experience. In conclusion, we discuss challenges for this field moving forward, and provide key concepts towards an integrative framework of oscillatory brain activity in the study of cognition.

3.11.1 Oscillations as Epiphenomena, and Other Challenges

A long-standing debate in the field centers around the question to what extent brain oscillations possess a primary functional role, as opposed to arising from other non-oscillatory processes as a by-product, or epiphenomenon (Buzsáki & Draguhn, 2004). For example, does a spike burst represent an oscillation, or a mere sequence of individual events? The question of functional relevance has been difficult to address because the required ground truth, for example, "there is an oscillation at 11.5 Hz", is often something the researcher wishes to establish, rather than something that can be used as a premise. Current research leverages causal manipulations such as electric micro-stimulation in experimental animals, synchronized to the phase of ongoing oscillations to determine if altering a robustly measured ongoing oscillation affects the animal's behavior in a systematic fashion (Tehovnik et al., 2006). Such data, challenging to obtain because it requires predicting the phase of the oscillatory signal over time, would present direct evidence supporting a causal role of oscillatory activity. Other approaches underway in several laboratories use similar causal manipulation, as well as driven oscillations, to address this question.

Additional challenges for researchers using metrics of oscillatory activity are rooted in the fact that highly complex data arrays that may contain sensors, conditions, time points, trials, etc., tend to get further inflated when adding a frequency dimension. This triggered attempts towards using multivariate methods, machine learning methods, or other approaches, all aiming at reducing the dimensionality of the data without losing information that is relevant to addressing the empirical question of interest]. The resulting methodological diversity has not only been enriching for the field, but also often prevented direct comparisons of results from different laboratories, and thus has been regarded as one obstacle towards reproducing and replicating findings. Ongoing efforts to improve the training of next-generation researchers are expected to address this issue, by adding computer science, digital signal processing, and advanced statistics to undergraduate and graduate curricula. Current trends of expressing hypotheses in terms of mathematical models that can be explicitly tested and gradually improved are crucial in this respect and are readily applied to indices of oscillatory brain activity. As with many other fields in the biomedical sciences, (re-)establishing a culture of knowing the history of the field, of systematic and programmatic research, and of applying rigorous methods is an obvious goal in research on brain oscillations. The following key concepts summarize areas of conceptual progress in the field and provide avenues for future research.

3.11.2 Oscillations as Drivers of Plasticity

Brain oscillations are widely considered reflections of network activity, but growing support is also evident for their role in mediating neuroplastic changes that result in the *formation* of networks, local and distributed. The high signal-to-noise ratio of many frequency-domain measures and the resulting potential for trial-by-trial analyses may be leveraged to quantify neural changes that occur over the course of an experimental session (McTeague et al., 2015). Such an approach complements trial averaging, which is still the dominant approach in cognitive neuroscience, and sheds light on adaptation dynamics and plasticity in response to the experimental situation, processes of high ecological validity and with substantial clinical relevance.

3.11.3 Oscillations as Pacemakers for Cognition and Behavior

A substantial body of research addresses the role of low-frequency oscillatory brain activity for embedding and organizing neural, cognitive, and motor processes. Cornerstone results from this line of research include that perceptual and motor performance varies in a rhythmic fashion: recurrent time windows exist, in which cognition and motor action can be optimally initiated and executed, interspersed with time windows of reduced performance (VanRullen, 2016). Relating these behavioral rhythms to brain oscillations holds promise for advancing models of the temporal organization of cognitive processing.

3.11.4 Oscillations as Carriers of Communication Signals

Several chapters in this volume address the potential of oscillatory activity in lower-frequency bands such as alpha or theta for integrating information in space, and for mediating the neural communication between units, local and distal (Chapters 9, 10, and 20). Establishing the functional role of oscillatory phenomena in neural communication, across levels of observation, will greatly benefit from emerging multimodal and multi-species methods, including new recording and stimulation techniques, well suited to cross-validate observations and combine measurement and stimulation approaches (Adesnik, 2018).

3.11.5 Oscillations as Organizing Principles for Regulating Complex System States

From a perspective of complex systems, a major function of brain oscillations lies in representing macroscopic states that emerge from systematic nonlinear interactions between units at the microscopic and mesoscopic level. Increasingly, computational models exist, which make quantitative predictions for how these different levels interact. For example, in-silico studies have leveraged the steady increase in computing capacity to offer alternative models of inter-scale nonlinear interactions, which may provide critical predictions to be tested in empirical work (Neymotin et al., 2013).

3.11.6 Oscillations as Substrate of Representations in Perception and Memory

Finally, oscillations have been long regarded as a candidate mechanism for coding percepts and memories. As discussed, oscillatory dynamics provide a unifying way for describing diverse mechanisms involved in these cognitive processes, including the spatio-temporal dynamics and plastic changes within and between neuronal ensembles. They offer a parsimonious account for integrating activity across distributed functional units without invoking the widely criticized top-down versus bottom-up dichotomy (Awh et al., 2012). Current developments in cognitive science and experimental psychology to increasingly rely on computational modeling will facilitate the integration of experimental observations from the level of brain oscillations to the level of task performance.

In summary, brain oscillations have opened exciting new avenues in the study of cognitive and behavioral processes. Beyond mere measurements of power in a specific frequency band, recent developments have provided the research community with tools that allow us to visualize the dynamic interplay among different frequencies, quantify the role of phase in coding information, and assess the rhythmic interactions between distal brain areas. Together with rapid theoretical and computational advances, these tools will continue to provide unique insights into brain systems dynamics and their relation to behavior and experience.

References

Adesnik, H. (2018). Cell type-specific optogenetic dissection of brain rhythms. *Trends in Neurosciences*, *41*(3), 122–124. https://doi.org/10.1016/j.tins.2018.01.001

Adrian, E. D. & Matthews, B. H. (1934). The Berger rhythm: Potential changes from the occipital lobes in man. *Brain*, *57*(4), 355–385.

Andersen, N., Krauth, N., & Nabavi, S. (2017). Hebbian plasticity in vivo: Relevance and induction. *Current Opinion in Neurobiology*, *45*, 188–192. https://doi.org/10.1016/j.conb.2017.06.001

Awh, E., Belopolsky, A. V., & Theeuwes, J. (2012). Top-down versus bottom-up attentional control: A failed theoretical dichotomy. *Trends in Cognitive Sciences*, *16*(8), 437–443. https://doi.org/10.1016/J.Tics.2012.06.010

Baria, A. T., Maniscalco, B., & He, B. J. (2017). Initial-state-dependent, robust, transient neural dynamics encode conscious visual perception. *PLOS Computational Biology*, *13*(11), e1005806. https://doi.org/10.1371/journal.pcbi.1005806

Başar, E. & Güntekin, B. (2012). A short review of alpha activity in cognitive processes and in cognitive impairment. *International Journal of Psychophysiology*, *86*(1), 25–38. https://doi.org/10.1016/j.ijpsycho.2012.07.001

Bauer, M., Stenner, M.-P., Friston, K. J., & Dolan, R. J. (2014). Attentional modulation of alpha/beta and gamma oscillations reflect functionally distinct processes. *Journal of Neuroscience, 34*(48), 16117–16125. https://doi.org/10.1523/JNEUROSCI.3474-13.2014

Berger, H. (1929). Über das Elektrenkephalogramm des Menschen. *Archiv Für Psychiatrie Und Nervenkrankheiten, 87*, 527–570.

Bernstein, J. (1868). Ueber den zeitlichen Verlauf der negativen Schwankung des Nervenstroms. *Archiv für die gesamte Physiologie des Menschen und der Tiere, 1*(1), 173–207. https://doi.org/10.1007/BF01640316

Bliss, T. V. P. & Lømo, T. (1973). Long-lasting potentiation of synaptic transmission in the dentate area of the anaesthetized rabbit following stimulation of the perforant path. *The Journal of Physiology, 232*(2), 331–356. https://doi.org/10.1113/jphysiol.1973.sp010273

Bonnefond, M. & Jensen, O. (2015). Gamma activity coupled to alpha phase as a mechanism for top-down controlled gating. *PLOS ONE, 10*(6), e0128667. https://doi.org/10.1371/journal.pone.0128667

Boynton, G. M. (2011). Spikes, BOLD, attention, and awareness: A comparison of electrophysiological and fMRI signals in V1. *Journal of Vision, 11*(5), 12–12. https://doi.org/10.1167/11.5.12

Breakspear, M. (2017). Dynamic models of large-scale brain activity. *Nature Neuroscience, 20*(3), 340–352. https://doi.org/10.1038/nn.4497

Brovelli, A., Ding, M., Ledberg, A., Chen, Y., Nakamura, R., & Bressler, S. L. (2004). Beta oscillations in a large-scale sensorimotor cortical network: directional influences revealed by Granger causality. *Proceedings of the National Academy of Sciences of the United States of America, 101*(26), 9849–9854. (15210971).

Buzsáki, G., Anastassiou, C. A., & Koch, C. (2012). The origin of extracellular fields and currents—EEG, ECoG, LFP and spikes. *Nature Reviews Neuroscience, 13*(6), 407–420. https://doi.org/10.1038/nrn3241

Buzsáki, G. & Draguhn, A. (2004). Neuronal oscillations in cortical networks. *Science, 304*(5679), 1926–1929. https://doi.org/10.1126/science.1099745

Buzsáki, G., Logothetis, N., & Singer, W. (2013). Scaling brain size, keeping timing: Evolutionary preservation of brain rhythms. *Neuron, 80*(3), 751–764. https://doi.org/10.1016/j.neuron.2013.10.002

Capilla, A., Pazo-Alvarez, P., Darriba, A., Campo, P., & Gross, J. (2011). Steady-state visual evoked potentials can be explained by temporal superposition of transient event-related responses. *PLoS One, 6*(1), e14543. https://doi.org/10.1371/journal.pone.0014543

Cohen, M. X. (2014). *Analyzing neural time series data: Theory and practice*. MIT Press.

Donner, T. H. & Siegel, M. (2011). A framework for local cortical oscillation patterns. *Trends in Cognitive Sciences, 15*(5), 191–199. https://doi.org/10.1016/j.tics.2011.03.007

Elbert, T., Ray, W. J., Kowalik, Z. J., Skinner, J. E., Graf, K. E., & Birbaumer, N. (1994). Chaos and physiology: Deterministic chaos in excitable cell assemblies. *Physiology Review, 74*(1), 1–47.

Foxe, J. J. & Snyder, A. C. (2011). The role of alpha-band brain oscillations as a sensory suppression mechanism during selective attention. *Frontiers in Psychology, 2*, 154. https://doi.org/10.3389/fpsyg.2011.00154

Freeman, W. J. (1991). The physiology of perception. *Scientific American, 264*, 78–87.

Freeman, W. J. (1994). Role of chaotic dynamics in neural plasticity. *Progress in Brain Research, 102*, 319–333.

Fuchs, A., Jirsa, V. K., & Kelso, J. A. (2000). Theory of the relation between human brain activity (MEG) and hand movements. *Neuroimage*, *11*(5 Pt 1), 359–369. https://doi.org/10.1006/nimg.1999.0532

Galambos, R. (1992). A comparison of certain gamma-band (40 Hz) brain rhythms in cat and man. In E. Basar & T. Bullock (Eds.), *Induced rhythms in the brain* (pp. 103–122). Springer.

Giabbiconi, C.-M., Jurilj, V., Gruber, T., & Vocks, S. (2016). Steady-state visually evoked potential correlates of human body perception. *Experimental Brain Research*, *234*(11),1–11. https://doi.org/10.1007/s00221-016-4711-8

Gütig, R. (2014). To spike, or when to spike? *Current Opinion in Neurobiology*, *25*, 134–139. https://doi.org/10.1016/j.conb.2014.01.004

Hadjipapas, A., Lowet, E., Roberts, M. J., Peter, A., & De Weerd, P. (2015). Parametric variation of gamma frequency and power with luminance contrast: A comparative study of human MEG and monkey LFP and spike responses. *NeuroImage*, *112*, 327–340. https://doi.org/10.1016/j.neuroimage.2015.02.062

Haken, H., Kelso, J. A., & Bunz, H. (1985). A theoretical model of phase transitions in human hand movements. *Biological Cybernetics*, *51*(5), 347–356.

Harris, K. P. & Littleton, J. T. (2015). Transmission, development, and plasticity of synapses. *Genetics*, *201*(2), 345–375. https://doi.org/10.1534/genetics.115.176529

Headley, D. B. & Weinberger, N. M. (2011). Gamma-band activation predicts both associative memory and cortical plasticity. *Journal of Neuroscience*, *31*(36), 12748–12758. https://doi.org/10.1523/JNEUROSCI.2528-11.2011

Hebb, D. (1949). *The organization of behavior: A neuropsychological theory*. Wiley.

Heerebout, B. T. & Phaf, R. H. (2010). Good vibrations switch attention: An affective function for network oscillations in evolutionary simulations. *Cognitive, Affective, & Behavioral Neuroscience*, *10*(2), 217–229. https://doi.org/10.3758/CABN.10.2.217

Henley, J. M. & Wilkinson, K. A. (2016). Synaptic AMPA receptor composition in development, plasticity and disease. *Nature Reviews Neuroscience*, *17*(6), 337–350. https://doi.org/10.1038/nrn.2016.37

Herrmann, C. S., Strüber, D., Helfrich, R. F., & Engel, A. K. (2016). EEG oscillations: From correlation to causality. *International Journal of Psychophysiology*, *103*, 12–21. https://doi.org/10.1016/j.ijpsycho.2015.02.003

Keil, A., Gruber, T., & Muller, M. M. (2001). Functional correlates of macroscopic high-frequency brain activity in the human visual system. *Neuroscience and Biobehavioral Reviews*, *25*(6), 527–534.

Keitel, C., Keitel, A., Benwell, C. S. Y., Daube, C., Thut, G., & Gross, J. (2019). Stimulus-driven brain rhythms within the alpha band: The attentional-modulation conundrum. *Journal of Neuroscience*, *39*(16), 3119–3129. https://doi.org/10.1523/JNEUROSCI.1633-18.2019

Kelly, S. P., Lalor, E. C., Reilly, R. B., & Foxe, J. J. (2006). Increases in alpha oscillatory power reflect an active retinotopic mechanism for distracter suppression during sustained visuospatial attention. *Journal of Neurophysiology*, *95*(6), 3844–3851. https://doi.org/10.1152/jn.01234.2005

Kim, Y.-J., Grabowecky, M., Paller, K. A., & Suzuki, S. (2010). Differential roles of frequency-following and frequency-doubling visual responses revealed by evoked neural harmonics. *Journal of Cognitive Neuroscience*, *23*(8), 1875–1886. https://doi.org/10.1162/jocn.2010.21536

Kizuk, S. A. D. & Mathewson, K. E. (2016). Power and phase of alpha oscillations reveal an interaction between spatial and temporal visual attention. *Journal of Cognitive Neuroscience*, 29(3), 480–494. https://doi.org/10.1162/jocn_a_01058

Klimesch, W. (1999). EEG alpha and theta oscillations reflect cognitive and memory performance: A review and analysis. *Brain Research Reviews*, 29(2), 169–195. https://doi.org/10.1016/S0165-0173(98)00056-3

Klimesch, W., Sauseng, P., & Hanslmayr, S. (2006). EEG alpha oscillations: The inhibition-timing hypothesis. *Brain Research Reviews*, 53(1), 63–88.

Klimesch, W., Schack, B., & Sauseng, P. (2005). The functional significance of theta and upper alpha oscillations. *Experimental Psychology*, 52(2), 99–108.

Lisman, J. E., & Jensen, O. (2013). The theta-gamma neural code. *Neuron*, 77(6), 1002–1016. https://doi.org/10.1016/j.neuron.2013.03.007

Lopes da Silva, F. (1991). Neural mechanisms underlying brain waves: from neural membranes to networks. *Electroencephalography and Clinical Neurophysiology*, 79(2), 81–93. https://doi.org/10.1016/0013-4694(91)90044-5

Maren, S. & Quirk, G. J. (2004). Neuronal signalling of fear memory. *Nature Reviews Neuroscience*, 5(11), 844–852. https://doi.org/10.1038/nrn1535

Marrazzi, A. S. & Lorente de No, R. (1944). Interaction of neighboring fibres in myelinated nerve. *Journal of Neurophysiology*, 7(2), 83–101. https://doi.org/10.1152/jn.1944.7.2.83

Mathewson, K. E., Lleras, A., Beck, D. M., Fabiani, M., Ro, T., & Gratton, G. (2011). Pulsed out of awareness: EEG alpha oscillations represent a pulsed-inhibition of ongoing cortical processing. *Frontiers in Psychology*, 2, 99. https://doi.org/10.3389/fpsyg.2011.00099

Mazzoni, A., Whittingstall, K., Brunel, N., Logothetis, N. K., & Panzeri, S. (2010). Understanding the relationships between spike rate and delta/gamma frequency bands of LFPs and EEGs using a local cortical network model. *NeuroImage*, 52(3), 956–972. https://doi.org/10.1016/j.neuroimage.2009.12.040

McTeague, L. M., Gruss, L. F., & Keil, A. (2015). Aversive learning shapes neuronal orientation tuning in human visual cortex. *Nature Communications*, 6, 7823. https://doi.org/10.1038/ncomms8823

Moon, J.-Y., Lee, U., Blain-Moraes, S., & Mashour, G. A. (2015). General relationship of global topology, local dynamics, and directionality in large-scale brain networks. *PLoS Computational Biology*, 11(4), e1004225. https://doi.org/10.1371/journal.pcbi.1004225

Moratti, S., Clementz, B. A., Gao, Y., Ortiz, T., & Keil, A. (2007). Neural mechanisms of evoked oscillations: stability and interaction with transient events. *Human Brain Mapping*, 28(12), 1318–1333. https://doi.org/10.1002/hbm.20342

Morone, F., Roth, K., Min, B., Stanley, H. E., & Makse, H. A. (2017). Model of brain activation predicts the neural collective influence map of the brain. *Proceedings of the National Academy of Sciences of the United States of America*, 114(15), 3849–3854. https://doi.org/10.1073/pnas.1620808114

Muller, L., Chavane, F., Reynolds, J., & Sejnowski, T. J. (2018). Cortical travelling waves: mechanisms and computational principles. *Nature Reviews Neuroscience*, 19(5), 255–268. https://doi.org/10.1038/nrn.2018.20

Müller, M. M., Teder, W., & Hillyard, S. A. (1997). Magnetoencephalographic recording of steady-state visual evoked cortical activity. *Brain Topography*, 9(3), 163–168.

Nadel, L. & Maurer, A. P. (2018). Recalling Lashley and reconsolidating Hebb. *Hippocampus*, 0(0), 1–18. https://doi.org/10.1002/hipo.23027

Nanou, E. & Catterall, W. A. (2018). Calcium channels, synaptic plasticity, and neuropsychiatric disease. *Neuron, 98*(3), 466–481. https://doi.org/10.1016/j.neuron.2018.03.017

Naze, S., Bernard, C., & Jirsa, V. (2015). Computational modeling of seizure dynamics using coupled neuronal networks: Factors shaping epileptiform activity. *PLoS Computational Biology, 11*(5), e1004209. https://doi.org/10.1371/journal.pcbi.1004209

Neymotin, S. A., Hilscher, M. M., Moulin, T. C., Skolnick, Y., Lazarewicz, M. T., & Lytton, W. W. (2013). I_h tunes theta/gamma oscillations and cross-frequency coupling in an *in silico* CA3 model. *PLoS One, 8*(10), e76285. https://doi.org/10.1371/journal.pone.0076285

Norcia, A. M., Appelbaum, L. G., Ales, J. M., Cottereau, B. R., & Rossion, B. (2015). The steady-state visual evoked potential in vision research: A review. *Journal of Vision, 15*(6), 4–4. https://doi.org/10.1167/15.6.4

Nunez, P. L., & Srinivasan, R. (2006). *Electric fields of the brain* (2nd ed.). Oxford University Press.

Olejniczak, P. (2006). Neurophysiologic basis of EEG. *Journal of Clinical Neurophysiology, 23*(3), 186–189. https://doi.org/10.1097/01.wnp.0000220079.61973.6c

Osipova, D., Hermes, D., & Jensen, O. (2008). Gamma power is phase-locked to posterior alpha activity. *PLoS One, 3*(12), e3990. https://doi.org/10.1371/journal.pone.0003990

Pape, H.-C., & Paré, D. (2010). Plastic synaptic networks of the amygdala for the acquisition, expression, and extinction of conditioned fear. *Physiological Reviews, 90*(2), 419–463. https://doi.org/10.1152/physrev.00037.2009

Paré, D., Collins, D. R., & Pelletier, J. G. (2002). Amygdala oscillations and the consolidation of emotional memories. *Trends in Cognitive Sciences, 6*(7), 306–314. https://doi.org/10.1016/S1364-6613(02)01924-1

Parvizi, J. & Kastner, S. (2018). Promises and limitations of human intracranial electroencephalography. *Nature Neuroscience, 21*(4), 474. https://doi.org/10.1038/s41593-018-0108-2

Petro, N. M. & Keil, A. (2015). Pre-target oscillatory brain activity and the attentional blink. *Experimental Brain Research, 233*(12), 3583–3595. https://doi.org/10.1007/s00221-015-4418-2

Pfurtscheller, G. (1989). Spatiotemporal analysis of alpha frequency components with the ERD technique. *Brain Topography, 2*(1–2), 3–8.

Pfurtscheller, G. (1992). Event-related synchronization (ERS): An electrophysiological correlate of cortical areas at rest. *Electroencephalography and Clinical Neurophysiology, 83*, 62–69.

Pfurtscheller, G., Stancak, A., Jr., & Neuper, C. (1996). Event-related synchronization (ERS) in the alpha band--an electrophysiological correlate of cortical idling: A review. *International Journal of Psychophysiology, 24*(1–2), 39–46.

Poldrack, R. A. (2011). Inferring mental states from neuroimaging data: From reverse inference to large-scale decoding. *Neuron, 72*(5), 692–697. https://doi.org/10.1016/j.neuron.2011.11.001

Popescu, A. T., Popa, D., & Paré, D. (2009). Coherent gamma oscillations couple the amygdala and striatum during learning. *Nature Neuroscience, 12*(6), 801–807. https://doi.org/10.1038/nn.2305

Pulvermuller, F., & Fadiga, L. (2010). Active perception: Sensorimotor circuits as a cortical basis for language. *Nature Reviews Neuroscience, 11*(5), 351–360. https://doi.org/10.1038/nrn2811

Regan, D. (1989). *Human brain electrophysiology: Evoked potentials and evoked magnetic fields in science and medicine*. Elsevier.

Rossion, B., & Boremanse, A. (2011). Robust sensitivity to facial identity in the right human occipito-temporal cortex as revealed by steady-state visual-evoked potentials. *Journal of Vision, 11*(2), 16. https://doi.org/10.1167/11.2.16

Samaha, J., Bauer, P., Cimaroli, S., & Postle, B. R. (2015). Top-down control of the phase of alpha-band oscillations as a mechanism for temporal prediction. *Proceedings of the National Academy of Sciences, 112*(27), 8439–8444. https://doi.org/10.1073/pnas.1503686112

Sarnthein, J., Petsche, H., Rappelsberger, P., Shaw, G. L., & von Stein, A. (1998). Synchronization between prefrontal and posterior association cortex during human working memory. *Proceeding of the National Academy of Sciences of the United States of America, 95*(12), 7092–7096.

Schroeder, C. E., & Lakatos, P. (2009). Low-frequency neuronal oscillations as instruments of sensory selection. *Trends in Neuroscience, 32*(1), 9–18. https://doi.org/10.1016/j.tins.2008.09.012

Schroeder, C. E., Wilson, D. A., Radman, T., Scharfman, H., & Lakatos, P. (2010). Dynamics of active sensing and perceptual selection. *Current Opinion in Neurobiology, 20*(2), 172–176. https://doi.org/10.1016/j.conb.2010.02.010

Singer, W., Gray, C., Engel, A., König, P., Artola, A., & Brocher, S. (1990). Formation of cortical cell assemblies. *Cold Spring Harbor Symposium on Quantitative Biology, 55*, 939–952.

Skarda, C. A., & Freeman, W. J. (1987). How brains make chaos in order to make sense of the world. *Behavioral and Brain Sciences, 10*(2), 161–173. https://doi.org/10.1017/S0140525X00047336

Snyder, A. C., Morais, M. J., Willis, C. M., & Smith, M. A. (2015). Global network influences on local functional connectivity. *Nature Neuroscience, 18*(5), 736–743. https://doi.org/10.1038/nn.3979

Steriade, M., Gloor, P., Llinás, R. R., Lopes da Silva, F. H., & Mesulam, M.-M. (1990). Basic mechanisms of cerebral rhythmic activities. *Electroencephalography and Clinical Neurophysiology, 76*(6), 481–508. https://doi.org/10.1016/0013-4694(90)90001-Z

Tehovnik, E. J., Tolias, A. S., Sultan, F., Slocum, W. M., & Logothetis, N. K. (2006). Direct and indirect activation of cortical neurons by electrical microstimulation. *Journal of Neurophysiology, 96*(2), 512–521. https://doi.org/10.1152/jn.00126.2006

Thigpen, N. N., Bradley, M. M., & Keil, A. (2018). Assessing the relationship between pupil diameter and visuocortical activity. *Journal of Vision, 18*(6), 7–7. https://doi.org/10.1167/18.6.7

Thut, G., Nietzel, A., Brandt, S. A., & Pascual-Leone, A. (2006). Alpha-band electroencephalographic activity over occipital cortex indexes visuospatial attention bias and predicts visual target detection. *Journal of Neuroscience, 26*(37), 9494–9502.

VanRullen, R. (2016). Perceptual cycles. *Trends in Cognitive Sciences, 20*(10), 723–735. https://doi.org/10.1016/j.tics.2016.07.006

von der Malsburg, C. & Buhmann, J. (1992). Sensory segmentation with coupled neural oscillators. *Biological Cybernetics, 67*(3), 233–242.

Weisz, N., Wühle, A., Monittola, G., Demarchi, G., Frey, J., Popov, T., & Braun, C. (2014). Prestimulus oscillatory power and connectivity patterns predispose conscious somatosensory perception. *Proceedings of the National Academy of Sciences, 111*(4), E417–E425. https://doi.org/10.1073/pnas.1317267111

Wieser, M. J. & Keil, A. (2011). Temporal trade-off effects in sustained attention: Dynamics in visual cortex predict the target detection performance during distraction. *Journal of Neuroscience, 31*(21), 7784–7790. https://doi.org/10.1523/JNEUROSCI.5632-10.2011

Yuval-Greenberg, S., Tomer, O., Keren, A. S., Nelken, I., & Deouell, L. Y. (2008). Transient induced gamma-band response in EEG as a manifestation of miniature saccades. *Neuron, 58*(3), 429–441.

CHAPTER 4

TIME-FREQUENCY DECOMPOSITION METHODS FOR EVENT-RELATED POTENTIAL ANALYSIS

SELIN AVIYENTE

4.1 Introduction

EEG reflects the electrical activity of a collection of neural populations in the brain. As such, it reflects the superposition of different simultaneously acting dynamical systems. When a large number of parallel-oriented cortical neurons receive the same repetitive synaptic input, their synchronous activity produces extracellular rhythmic field potentials. These electrical fields are volume conducted throughout the brain and recorded as EEG from the scalp (Nunez & Srinivasan, 2006). In recent years, it has been suggested that this synchronization optimizes relations between spike-mediated "top-down" and "bottom-up" communication, both within and between brain areas. This optimization might have particular importance during motivated anticipation of, and attention to, meaningful events and associations and in response to their anticipated consequences (Von Stein et al., 2000; Fries et al., 2001; Salinas & Sejnowski, 2001; Makeig et al., 2004). This new theory of cortical and scalp-recorded field dynamics requires new data analysis approaches for spatially distributed event-related EEG dynamics.

The standard analysis method to study EEG in an event-related fashion is to focus on event-related potentials (ERPs) by averaging. ERPs are positive or negative voltage deflections seen in the averages of EEG epochs time-locked to a class of repeated stimulus or response events. However, this approach assumes that ERPs are superimposed on ongoing background EEG "noise" with amplitude and phase distributions that are completely unrelated to the task. This view of ERPs has been challenged in the last two decades. First, time-frequency analysis of single-trial EEG signals shows that EEG is not solely random noise; rather, there are event-related changes in the magnitude and phase of EEG oscillations at specific frequencies (Makeig et al., 2004). The early work in this area focused on spectral analysis of EEG signals, which breaks down the oscillatory EEG waveforms into sinusoids at different frequencies, reflecting the role of different frequency bands in cognitive function. Second, ERPs themselves may represent transient phase resetting of ongoing EEG by experimental events, leading to transient time- and phase locking of frequency specific oscillations (Makeig et al., 2004; Makeig et al., 2002). Makeig and colleagues (2002) have introduced the event-related brain dynamics framework, which emphasizes the spectral decomposition of single-trial event-related EEG epochs in order to separately examine event-related changes in the magnitude and phase of oscillations at different frequencies. However, spectral analysis assumes stationarity of the signal, in spite of the fact that information processing in the brain is mostly reflected by fast dynamic changes in EEG. Even though the Fourier transform provides a complete representation of the time series, the resulting power spectrum is only interpretable for stationary signals; non-stationarities are encoded in the phase spectrum, which is typically impossible to interpret visually. Spectral analysis is only capable of reflecting the average or global frequency content, rather than illustrating the local activity. The non-stationarities in the EEG signal are the primary motivation for methods that can capture simultaneously the variation of signal energy across time and frequency.

Time-frequency analyses of EEG provide additional information about neural synchrony not apparent in ongoing EEG activity or in ERPs. They can tell us which frequencies have the most power at different time points and how their phase synchronizes across time and space. Thus, using time-frequency analyses, we can assess changes in power and synchronization of EEG within or between spatial locations across trials with respect to the onset of tasks. However, it is important to note that time-frequency analysis cannot by itself determine the cause of the changes in EEG power, that is, whether the change is due to changes in the magnitude of the oscillations or to changes in their degree of synchronization (Yeung, 2004; Roach & Mathalon, 2008).

The goal of this chapter is to give an overview of different time-frequency analysis tools for both quantifying the changes in EEG power across time and frequency and the changes in phase synchrony across time, frequency, and space. The different methods discussed in this chapter can be divided into three categories: linear, nonlinear, and data-driven or adaptive methods. The first category of methods focus on simple extension of Fourier transform to the non-stationary signal that is, short-time Fourier transform or sliding window approach, and the continuous wavelet transform. The second category of methods focus on Cohen's class of time-frequency distributions (TFDs) that compute the spectrum of the time-varying auto-correlation function of the signal, rather than the signal itself. In this manner, these distributions are highly nonlinear but offer high-resolution visualizations of EEG dynamics. The last category of methods stem from recent advances in signal processing that focus on sparse signal representation. The two methods reviewed under this category are matching pursuit (MP) and empirical mode decomposition (EMD).

4.2 Linear Time-Frequency Decomposition Methods

4.2.1 Short-Time Fourier Transform

The earliest method that allowed for the time-frequency representation of a signal's energy distribution was the short-time Fourier transform (STFT), also known as the windowed Fourier transform. It relies on an estimation of power spectra, through Fourier transform, for short time windows shifted along the time axis. The basic idea of STFT is to break up the signal into small time segments and Fourier analyze each time segment to determine the frequencies that existed in that segment. The totality of such spectra indicates how the spectrum is varying with time. STFT is a linear time-frequency transform that correlates the signal with a family of waveforms that are well-concentrated in time and frequency, compared to standard spectral analysis, which correlates the signal with eternal sinusoids resulting in good frequency localization at the expense of poor time localization. For a signal $x(t)$, its STFT can be expressed as:[1]

$$S(t,\omega)=\int x(\tau)g(\tau-t)e^{-j\omega\tau}d\tau, \tag{4.1}$$

where $g(\tau)$ is the window function. The localization of information provided by STFT depends on the time-frequency spread of the window g.

The choice of the length and shape of the window function are the two critical issues in applying STFT to EEG signals. The first problem is usually addressed by considering the existing trade-off between time and frequency resolution determined by the length of the window function. This trade-off, known formally as the uncertainty principle, states that as the length of the window increases in time, localization in time or time resolution goes down while the localization in frequency (frequency resolution) increases (Cohen, 1995). Similarly, if the window length decreases, time resolution increases at the expense of frequency resolution. In practice, the length of the window is usually selected by trying different window lengths in order to obtain an optimal trade-off. This optimal length is highly dependent on the underlying signal's properties, such as the rate with which it changes. The second issue of selecting the shape of the window has been addressed in a more rigorous way with the resolution of some common windows mathematically formulated (Mallat, 1999). For numerical applications, the window function is usually selected as a symmetric and real function. By choosing an appropriate window function, one can obtain localization around the time point of interest, t. For good time-frequency resolution, the ratio of the first sidelobes to the peak of the

[1] All integrals are from $-\infty$ to ∞ unless otherwise noted.

window function should be large, that is, the rate of decay the window function in the frequency domain, $G(\omega)$, should be fast. Some common window functions include the rectangular, Hamming, Hanning, and Gaussian window functions. Rectangular windows have the worst frequency resolution due to the sharp transition of the window function. Gaussian window achieves the best joint time and frequency localization as it meets the lower bound of the uncertainty principle. In STFT, the duration of the time window is fixed in contrast to the wavelet transform, which in turn provides uniform frequency resolution. STFT is a complex valued time-frequency representation, and an energy distribution can be obtained as $|S(t,\omega)|^2$, known as the spectrogram.

4.2.2 Continuous Wavelet Transform

Continuous wavelet transforms (CWT) describe a class of spectral decomposition methods that are conceptually related to the STFT. Unlike STFT, where the signals are decomposed into localized sinusoids, in the wavelet transform the signals are decomposed in terms of "small waves" of limited duration with an average value of zero. The wavelet transform of a signal $x(t)$ is given by:

$$W(t,s) = \int x(\tau)\frac{1}{\sqrt{s}}\psi^{*}\left(\frac{\tau - t}{s}\right)d\tau, \tag{4.2}$$

where ψ is the wavelet function and s is the scale parameter. Similar to STFT defined in Eq. 4.1, the wavelet transform finds the decomposition of a given signal in terms of time-frequency atoms. However, in this case, the time-frequency atoms are the time-shifted and scaled versions of the mother wavelet, ψ. For this reason, wavelet transform should be thought of as a time-scale decomposition rather than a TFD, as the signal is not decomposed in terms of sinusoids across different frequencies. One of the major differences between STFT and CWT is the time-frequency localization. In STFT, the frequency localization is uniform due to the fixed window size. On the other hand, for CWT the time resolution is variable, with shorter time windows for higher frequencies and longer time windows for lower frequencies. This variable time resolution closely matches the structural properties of ERP signals and has made CWT an attractive choice for time-frequency analysis of ERPs (Roach & Mathalon, 2008; Demiralp et al., 2001; Polikar et al., 2007). Using CWT, the ERP can be decomposed across several orthogonal functions, wavelets, with overlapping time courses at different scales.

While any number of wavelet functions can be used for ERP analysis, the wavelet must provide a biologically plausible fit to the signal being modeled. Some commonly used wavelets for EEG and ERP analysis are real wavelets, such as splines, and analytic wavelets, such as the Morlet wavelet and the Generalized Morse Wavelet (GMW). In particular, the analytic wavelets are useful to obtain both energy and phase information of the underlying oscillations. The Morlet wavelet is one that has been commonly used for EEG/ERP analysis. The Morlet wavelet is defined as a complex sinusoid tapered by a Gaussian given as:

$$\psi(t) = e^{j2\pi ft} e^{-\frac{t^2}{2\sigma^2}}, \tag{4.3}$$

where σ is the width of the Gaussian and is reciprocally related to the frequency in order to retain the wavelet's scaling properties. By this scaling, one obtains the same number of significant wavelet cycles at all frequencies. Thus, the Morlet wavelet transform has a different time and frequency resolution at each scale. Therefore, at high frequencies the temporal resolution of a wavelet is better than at low frequencies. However, the inverse is true for the frequency resolution of the wavelet transform. Convolution with the complex Morlet wavelet results in a complex-valued signal from which instantaneous power and phase can be extracted at each time point. Wavelet convolution can be conceptualized as a template matching or bandpass filtering. Convolutions with Morlet wavelets can be computed for multiple frequencies in order to yield a time-frequency representation. Since Morlet wavelet is a complex function, the corresponding wavelet transform is also complex, where the real part corresponds to the bandpass filtered signal and the imaginary part corresponds to Hilbert transform of the signal. There are several advantages of Morlet wavelets for EEG/ERP analysis (Cohen, 2018). One is that the Morlet wavelet is Gaussian shaped in the frequency domain and the absence of sharp edges minimizes ripple effects. Second, the results of Morlet convolution retain the temporal resolution of the original signal. Third, wavelet convolution is computationally efficient and can be implemented using the fast Fourier transform. The wavelet power spectrum, also known as the scalogram, can be obtained as $|W(t,s)|^2$.

4.3 Nonlinear Time-Frequency Analysis: Cohen's Class of Time-Frequency Distributions

For a signal, $x(t)$, a bilinear TFD, $C(t,\omega)$, from Cohen's class can be expressed as (Cohen, 1995):

$$C(t,\omega) = \iiint \phi(\theta,\tau)\, x\left(u+\frac{\tau}{2}\right) x^*\left(u-\frac{\tau}{2}\right) e^{j(\theta u - \theta t - \tau\omega)} du\, d\theta\, d\tau, \tag{4.4}$$

where $\phi(\theta,\tau)$ is the kernel function in the ambiguity domain (θ,τ). Unlike linear transforms, which first multiply the signal with a time-limited window function before computing its Fourier transform, Cohen's class of distributions compute the Fourier transform of the local autocorrelation function. In this manner, the signal acts like a window on itself. The kernel $\phi(\theta,\tau)$ acts like a filter on this local autocorrelation function and determines which parts to preserve. TFDs represent the energy distribution of a signal over time and frequency, simultaneously. Some of the most desired properties of TFDs are the energy preservation, the marginals, and the reduced interference. Some common TFDs used for EEG and ERP analysis include the Wigner distribution and its filtered versions (Tağluk et al., 2005; Abdulla & Wong, 2011).

4.3.1 Wigner Distribution

The Wigner distribution is defined as:

$$W(t,\omega)=\int x\left(t+\frac{\tau}{2}\right)x^{*}\left(t-\frac{\tau}{2}\right)e^{-j\omega\tau}d\tau, \tag{4.5}$$

where $x(t)$ is the signal and τ is the time lag variable. The Wigner distribution computes the Fourier transform of the local autocorrelation of the signal as the kernel $\phi(\theta,\tau)=1$. As the kernel acts as an all pass filter, Wigner distribution keeps the whole autocorrelation function before transforming it. This provides a high time-frequency resolution. In fact, it has been shown that the Wigner distribution provides the highest time-frequency resolution among all Cohen's class of TFDs (Cohen, 1995). It overcomes some of the shortcomings of the spectrogram as there is no dependency on the choice of the window. Moreover, the Wigner distribution provides accurate instantaneous frequency estimates and is an accurate energy distribution as it satisfies the marginals.

However, Wigner distribution is a bilinear function of signals. Therefore, Wigner distribution of multi-component signals or mono-component signals with curved time-frequency supports will be cluttered by spurious terms called cross-terms. The existence of cross-terms may decrease the interpretability of the TFD. ERPs, like other non-stationary signals, require an analysis technique that is free from cross-terms.

4.3.2 Reduced Interference Distributions (RIDs)

In order to reduce cross-terms that appear in the Wigner distribution, the past twenty years demonstrates a move towards using reduced interference distributions (RIDs), designed using kernel functions $|\phi(\theta,\tau)|\ll 1$ for $|\theta\tau|\gg 0$, which concentrate the energy across the auto-terms, and satisfy the energy preservation and the marginals (Jeong & Williams, 1992). The kernel function $\phi(\theta,\tau)$ is usually designed in the ambiguity plane as the auto-terms tend to be close to the origin and the cross-terms away from the origin of the ambiguity plane. This allows for interpretation of the kernel function as a low-pass filter that attempts to reject the cross-terms and leave the auto-terms unchanged. The shape of the filter can be adapted to the analysis of the desirable signal components depending on the nature of the signal. For RIDs, $\phi(\theta,\tau)=h(\theta\tau)$, where h is a monotonously decreasing function to ensure that the parameterization function is a low-pass filter in the ambiguity plane. Depending on the canonical form of h, the resultant RIDs have different cross-term rejection capabilities. Some commonly used RIDs include Choi–Williams (CW) (Jeong & Williams, 1992), B-distribution (Zhao et al., 1990), Born–Jordan, and Zhao–Atlas–Marks (Zhao et al., 1990).

Since these distributions will be implemented in discrete-time, discrete-time TFD of size $2N+1\times 2N+1$ can be defined as:

$$TFD(n,\omega;\psi)=\sum_{n_1=-N}^{N}\sum_{n_2=-N}^{N} x(n+n_1)x^*(n+n_2)\,\psi\left(-\frac{n_1+n_2}{2},n_1-n_2\right)e^{-j\omega(n_1-n_2)}, \tag{4.6}$$

where ψ is the discrete-time kernel in the time and time-lag domain. The most commonly used discrete-time kernel for EEG/ERP analysis is the binomial kernel, which is given by

$$\psi(n,m)=2^{-|m|}\begin{pmatrix}|m|\\ n+\dfrac{|m|}{2}\end{pmatrix} \text{ for } |n|\le\frac{|m|}{2}.$$

4.4 Data-Driven or Adaptive Time-Frequency Representations

Unlike the first two types of time-frequency transforms, the methods reviewed in this section do not directly result in time-frequency energy distributions, but rather result in a description of the ERP dynamics as a sum of a few numbers of distinct components. These components may be either data-driven, that is, derived directly from the signals, or selected from a large dictionary of atoms. Usually, these components are localized in time and frequency and thus can be thought of as a decomposition of the event-related EEG signals into a few distinct time-frequency elements. Thus, the resulting components can be visualized using any transform discussed in the previous sections to obtain a time-frequency spectrum. For this reason, these are sometimes considered as time-frequency analysis tools in the literature. By the nature of the algorithms used to derive these components, these methods are nonlinear. The first one of these methods, EMD, is data-driven and the resulting components usually represent the different frequency bands of EEG. The second method, MP, can be thought of as an extension of wavelet transform. However, unlike the wavelet functions, which usually form an orthonormal basis for the signal space, the time-frequency atoms used in MP form an overcomplete dictionary. This means that the set of functions used to express the signal is more than necessary. This overcompleteness results in sparse representation of the EEG/ERP signals. These atoms are pre-determined similar to the wavelet transform and can be selected by the user based on the characteristics of the signals. Some example atoms could be a combination of sinusoids and wavelets, Gabor functions, etc. This efficiency in representation of the signals with a few time-frequency atoms comes at the expense of computational complexity. Unlike wavelet transform, which is computationally efficient, MP usually relies on greedy algorithms (Mallat & Zhang, 1993; Tropp & Gilbert, 2007).

4.4.1 Empirical Mode Decomposition

EMD is a fully adaptive, data-driven approach that decomposes a signal into oscillations inherent to the data, referred to as intrinsic mode functions (IMFs) (Huang et al., 1998). Finding the IMFs is equivalent to finding the band-limited oscillations underlying the observed signal. Extracting the IMFs is similar to finding the harmonic components in Fourier analysis. However, the IMF is much more general than the harmonic component since it can be modulated both in amplitude and frequency, while a harmonic component has constant amplitude and frequency. The amplitude and frequency modulations are possible because the decomposition depends on the local characteristic time scale of the data. Because the decomposition depends on the local characteristic time scale, EMD is suitable for application to non-stationary signals such as ERPs. The EMD algorithm decomposes the signal $x(t)$ as $x(t)=\sum_{i=1}^{M} C_i(t)+r(t)$ where $C_i(t)$, $i=1,\ldots,M$ are the IMFs and $r(t)$ is the residue. The algorithm can be described as follows:

1. Let $\tilde{x}(t)=x(t)$.
2. Identify all local maxima and minima of $\tilde{x}(t)$.
3. Find two envelopes $e_{min}(t)$ and $e_{max}(t)$ that interpolate through the local minima and maxima, respectively.
4. Let $d(t)=\tilde{x}(t)-\frac{1}{2}\left(e_{min}(t)+e_{max}(t)\right)$ as the detail part of the signal.
5. Let $\tilde{x}(t)=d(t)$ and go to step 2 and repeat until $d(t)$ becomes an IMF. The two IMF criteria are: a) the number of extrema and the number of zero-crossings must either be equal or differ at most by one; b) At any point, the mean value of the envelope defined by the local maxima and the envelope defined by the local minima is zero.
6. Compute the residue $r(t)=x(t)-d(t)$ and go back to step 1 until the energy of the residue is below a threshold.

Although EMD has the ability to extract ERP components, it suffers from the mode mixing problem (Huang et al., 2003). That is, the different time-frequency components may not directly correspond to the different IMFs, which makes it difficult to determine the distinct ERP components. To overcome the problem of mode mixing, Wu and Huang (2009) recommend ensemble EMD (EEMD), a noise-assisted data-analysis method. The output of EEMD is a set of IMFs generated from ensemble means of trials by repeating EMD on the same signal with different sets of Gaussian noise. A further potential drawback of EMD has been put forth by Flandrin and colleagues (2004), who showed that EMD behaves as a dyadic filter bank. This poses the concern that EMD may naturally lead to such a decomposition in all data, which implies that important oscillations may not be identified if they do not adhere to a dyadic frequency relationship with one another.

Once the IMFs are obtained, Hilbert spectral analysis, also known as Hilbert–Huang Transform (HHT), can be used to evaluate the frequency content of each IMF. Hilbert spectral analysis provides the instantaneous frequency of each IMF. According to Huang and colleagues (2011), the instantaneous frequency could represent the nonlinear and non-stationary signals without resorting to the mathematical artifact of harmonics. Like measuring ERPs, averaging IMFs across trials provides event-related modes (ERMs). Based on the instantaneous frequency of ERMs, ERP components can be extracted by summing ERMs (Cong et al., 2009; Williams et al., 2011; Wu et al., 2012) or using an ERM within a frequency range.

4.4.2 Matching Pursuit

The MP algorithm, originally proposed by Mallat and Zhang (1993), relies on an adaptive approximation of a signal by means of waveforms chosen from a very large and redundant dictionary of functions. MP provides high-resolution signal analysis with good resolution in time-frequency space and allows for parametric description of both periodic and transient signal features. It is suitable for analysis of non-stationary signals and for the investigation of dynamic changes of brain activity. MP aims at obtaining a sparse linear representation of a signal, $x(t)$, in terms of functions, g_i (sometimes referred to as atoms), from an overcomplete dictionary, D, using an iterative search algorithm (Mallat & Zhang, 1993):

$$x(t)=\sum_{i=1}^{M} a_i g_i(t). \tag{4.7}$$

The problem of choosing M functions, which explain the largest proportion of energy of a given signal, is a computationally complex problem and is known to belong to the class of non-deterministic polynomial-time hard (NP-hard) problems. MP offers a tractable suboptimal solution, obtained by an iterative algorithm.

In the first step of the iterative procedure, we choose the element of the dictionary that gives the largest inner product with the signal, that is, $g_1 = \text{argmax}_{g_i \in D} \left|\left\langle x, g_i \right\rangle\right|$. This first element of the dictionary is subtracted from the signal to obtain the residue. The iterative procedure is repeated on the subsequent residual, $R^k x$. This procedure can be summarized as:

1. Define the 0th order residual as $R^0 x = x$.
2. For the k th order residual, $R^k x$, select the best atom such that the inner product between the residual and the atom is maximized

$$g_k = \text{argmax}_{g_i \in D} \left|\left\langle R^k x, g_i \right\rangle\right|. \tag{4.8}$$

3. Compute the residue $R^{k+1} x$ as

$$R^{k+1} x = R^k x - \left\langle R^k x, g_k \right\rangle g_k. \tag{4.9}$$

4. After M iterations, the following linear representation is obtained:

$$x = \sum_{k=1}^{M} \left\langle R^k x, g_k \right\rangle g_k + R^{M+1} x. \tag{4.10}$$

The procedure converges to x in the limit, that is, $x = \sum_{k=1}^{\infty} \left\langle R^k x, g_k \right\rangle g_k$ and preserves signal energy. From this representation, one can derive a TFD of a signal's energy by adding Wigner distributions of selected atoms.

The overcomplete dictionary, D, can be designed to fit the class of signals at hand. Two important requirements for a dictionary are its descriptive power, that is, its ability to represent the signals of interest with relatively few atoms (sparsity), and its interpretability, that is, that the parameters indexing the atoms convey information. Although overcomplete dictionaries do not provide uniqueness of decomposition, they have more descriptive power than more classical, orthogonal dictionaries, such as wavelets. Regarding interpretability, the choice of atoms and their parameters is motivated by the types of activities that will be encountered. The dictionary is furthermore supposed to depend continuously on the parameter space (even though for numerical reasons, this space will obviously be discretized). This constraint is imposed only for ease of presentation, but the approach could be generalized, at the cost of some added complexity. For analyzing EEG and ERP signals, previous work has shown that Gabor logons represent the signals with few coefficients (Durka & Blinowska, 2001; Brown et al., 1994; Aviyente, 2007) . Gabor logons also have the advantage of being the most concentrated signals on the time-frequency plane, achieving the lower bound of the time-bandwidth product. A dyadic Gabor dictionary also allows for computationally effective implementation and has been used widely in EEG analysis.

4.4.3 Multichannel Matching Pursuit

The principle of MP can easily be generalized to the simultaneous decomposition of multiple signals, $\mathbf{X} = \left(x^1, x^2, \ldots, x^r\right)$ of r signals into atoms from the same overcomplete dictionary, D. This approach is sometimes referred to as the multichannel MP (MMP) or simultaneous MP in the literature, since it is usually applied to multiple signals collected over multiple channels/sensors or trials (Gribonval et al., 2008; Lelic et al., 2011; Sieluzycki et al., 2008). Unlike the original MP, MMP represents every component x^l of $\mathbf{X}$ as a weighted sum of the same elements from the overcomplete dictionary and thus tries to achieve joint sparsity of a collection of signals for a given dictionary. The algorithm can be described as follows:

1. Define for each signal l the 0th order residual as $R^0 x^l = x^l$.
2. For the kth order residual, $R^k x^l$, select the best atom such that the total inner product between the atom and the residuals in each signal is maximized

$$\operatorname{argmax}_{g_i \in D} \sum_{l=1}^{r} \left| \left\langle R^k x^l, g_i \right\rangle \right|. \tag{4.11}$$

3. Compute the residue $R^{k+1} x^l$ for all signals:

$$R^{k+1} x^l = R^k x^l - \left\langle R^k x^l, g_k \right\rangle g_k. \tag{4.12}$$

4. After M iterations, the following linear representation is obtained for each signal:

$$x^l = \sum_{k=1}^{M} \left\langle R^k x^l, g_k \right\rangle g_k + R^{M+1} x^l. \tag{4.13}$$

Different implementations of MMP have been employed for ERP analysis, that is, evoked activity MP (EMP) and induced activity MP (IMP) (Bénar et al., 2009). In EMP, the atoms are determined to maximize the correlation with the average signal, and the amplitude is adapted to the individual trials. In this manner, the method accounts for amplitude variability across trials, but not for variability in the parameter space. IMP, on the other hand, maximizes the average energy across trials. Durka and colleagues

(2005) have also proposed an alternative approach for simultaneous time-frequency parametrization of multiple EEG recordings by applying the standard MP algorithm to the average of multichannel EEGs. This approach reduces the computational complexity by a factor of r at the expense of favoring data with equal phases in all of the channels.

4.5 Illustration of Different Time-Frequency Methods for ERP Analysis

This section illustrates the performance of some of the time-frequency analysis methods discussed earlier for an example ERP signal. In particular, we focus on the error-related negativity (ERN), a neurophysiological marker of performance monitoring. ERN is a negative deflection in the ERP that peaks within 100 ms of an erroneous response at frontocentral recording sites (Moser, 2017; Moran et al., 2015). The ERN is generally considered to be an index of cognitive control-related performance monitoring that is involved in coordinating optimal responding following mistakes. For the purpose of this illustration, we consider ERP obtained by averaging across trials recorded at the FCz electrode for one subject.

First, we compare the performance of linear and nonlinear time-frequency energy distributions, namely the continuous wavelet transform with the Morlet wavelet, and RID with the binomial kernel. From Figure 4.1, a typical ERN waveform with a negative potential occurring in the first 100 ms can be seen, along with different TFDs. From both

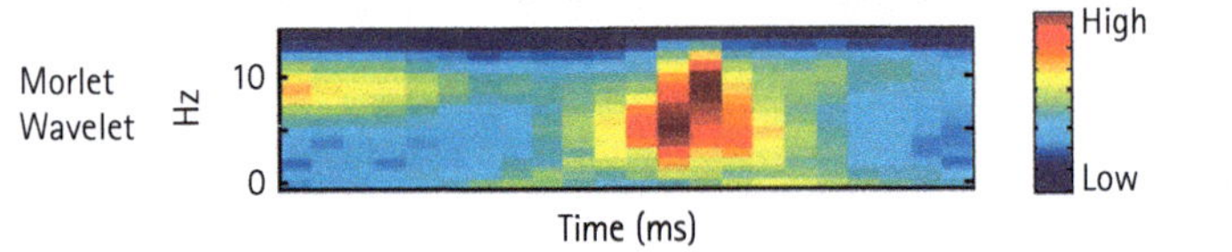

FIGURE 4.1 Time-frequency Analysis of ERN waveform. From top to bottom: ERN signal in the time domain, reduced interference distribution (binomial kernel), continuous wavelet transform with Morlet wavelet.

distributions, it is clear that there is high energy corresponding to ERN and P3e, P300 following the error response. The biggest difference between the different distributions is the time and frequency resolution of this component in the time-frequency plane. For RID, there are two distinct time-frequency components; one during ERN localized in the theta band (4–8Hz) and another distributed in time up till 400 ms in the delta (0–2 Hz) band. Morlet wavelet produces a distribution that captures the theta band activity without highlighting the delta band energy. It can be seen from these results that RID has the better time and frequency resolution, whereas the wavelet transform has poorer time and frequency localization. From this comparison, it can be concluded that RID provides better delineation of different ERP components. The results of the CWT may be improved by changing the wavelet type and wavelet scales. Unlike RID, which does not depend on the selection of any parameters, CWT is highly dependent on the user provided parameter values.

Next, we compare the performance of data-driven component extraction methods, EMD and MP. Figure 4.2 illustrates the first four IMFs, along with their Hilbert spectra. This figure shows that the first two IMFs correspond to the ERN time interval, whereas the third and fourth IMFs correspond to the P3e potential. The corresponding time-frequency energy distribution has a high energy concentration in the 100–200 ms and 300–400 ms time intervals within the delta frequency band. These components correspond to the P3e. However, the ERN component is not easily detected from this distribution as it has less energy than P3e.

In a similar manner, Figure 4.3 illustrates an MP spectrum obtained using a Gabor dictionary. The dictionary is constructed using Gabor functions, that is, Gaussian window functions shifted in time and modulated in frequency. The figure illustrates the time-frequency localization of the selected atoms. The figure also shows a high energy atom in the 100–200 ms time interval in the delta band. There are also Gabor functions with negative weights in the 0–100 ms time window around 10–12 Hz frequency band. This corresponds to ERN time window and alpha frequency band.

From these comparisons, it can be seen that RID is the best in terms of separating different ERP oscillations from each other with high resolution. The component-based methods work better in separating different oscillatory components, such as the IMFs extracted from EMD. However, EMD is not a true time-frequency energy distribution method, as it does not produce an actual spectrum like RID. The current method of visualizing IMFs in the time-frequency plane relies on the Hilbert transform of extracting the individual envelope and instantaneous frequency for each IMF. As the Hilbert transform is not a true time-frequency localization method, the IMFs can also be transformed to the time-frequency plane using different distributions to obtain higher-resolution visualizations. MP, on the other hand, is highly dependent on the selected dictionary atoms. The more suitable the dictionary is for the underlying signal, the sparser the resulting representation. In this particular example, MP wrongly detects alpha components for the ERN time window while missing the theta activity for the same time window. As EMD is data-driven and independent of priors, it performs better than MP for ERP analysis.

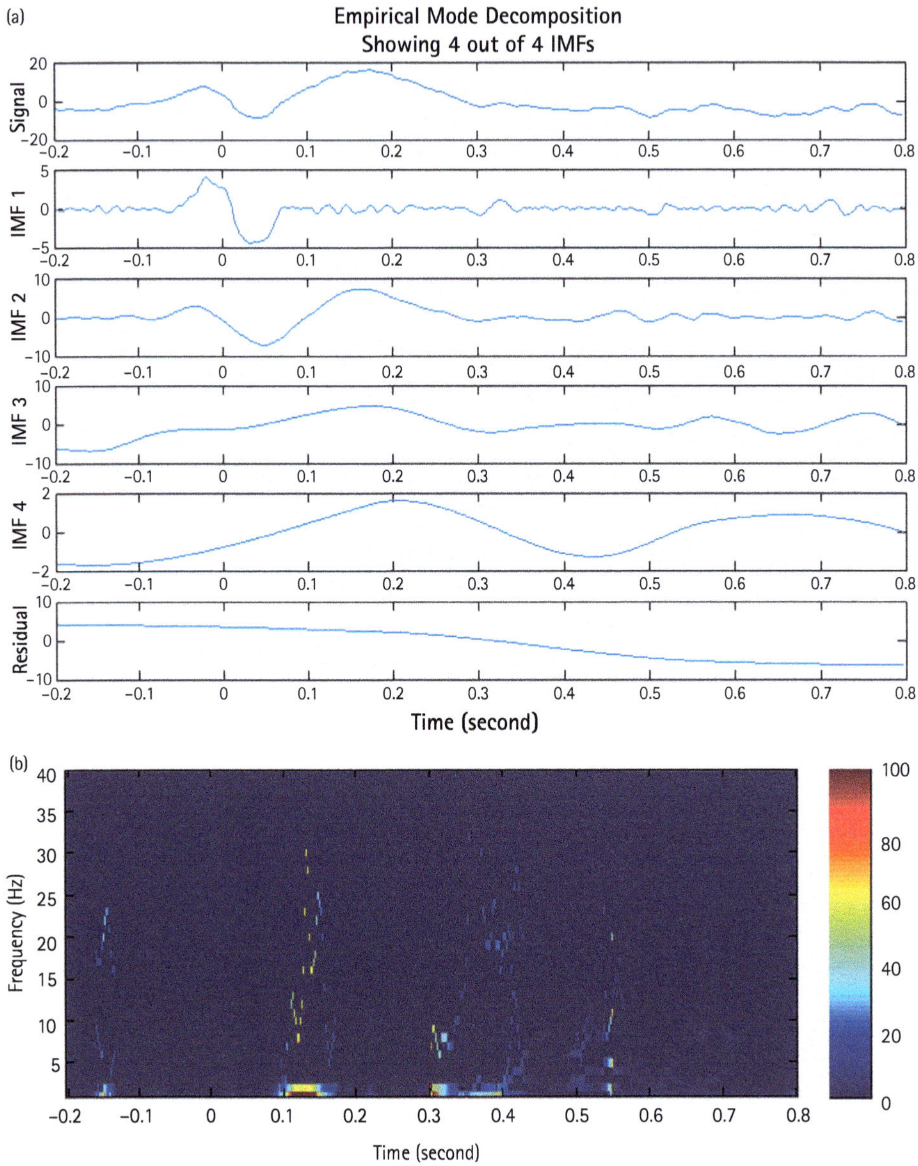

FIGURE 4.2 Empirical mode decomposition for ERP signal: (A) The first four intrinsic mode functions (IMFs) extracted from the ERP signal. (B) The time-frequency spectrum corresponding to the extracted IMFs.

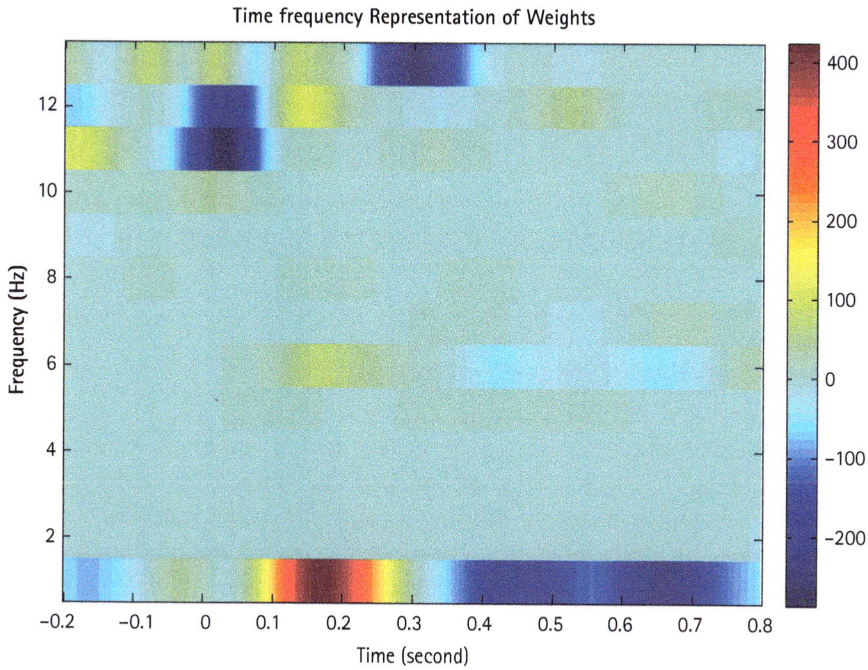

FIGURE 4.3 Time-frequency analysis using matching pursuit with a Gabor dictionary.

4.6 Time-Frequency-Based Phase Synchrony Analysis

Although time-frequency energy distribution is effective for studying the spectral content of EEG and ERP oscillations, most of the current approaches focus only on the magnitude spectrum of the oscillations and ignore the phase information. Recent research shows that ERPs result from event-related partial phase resetting of ongoing oscillatory activity along with transient increases in the magnitude of oscillations that are time-locked to the experimental events (Roach & Mathalon, 2008; Makeig et al., 2004). Therefore, it is important to characterize the change in phase information across time and frequency.

Phase synchrony quantifies the relation between the temporal structures of the signals, regardless of signal amplitude (Rosenblum et al., 2000; 2001). Two signals are said to be synchronous if their rhythms coincide. The amount of synchrony between two signals is usually quantified by first estimating the instantaneous phase of the individual signals around the frequency of interest. Traditionally, the instantaneous phase of a given signal is estimated first, and then the phase synchrony is quantified using statistical measures. Conventionally, the instantaneous phase is estimated using the Hilbert transform. However, the Hilbert transform does not provide any selectivity in frequency, the whole range of frequencies is considered to define the instantaneous phase. Therefore, if the signal is broadband, such as in the case of EEG signals, it is necessary to pre-filter it in the frequency band of interest before applying the Hilbert transform, in order to get a proper phase estimate. However, this method relies heavily on the proper selection of the bandpass filter and may suffer from estimation bias. In order to address the shortcomings of the Hilbert transform, different time-frequency transforms have been used to estimate instantaneous phase.

4.6.1 Wavelet-Based Phase Estimation

One common time-frequency based phase estimation approach employs CWT (Lachaux et al., 1999), in which the phase of the signals is extracted from the coefficients of their continuous wavelet transform using a complex wavelet function, such as the Morlet wavelet at the target frequency. These coefficients are the result of a convolution of the original signal with a time-shifted and frequency-modulated Gaussian function as:

$$W_x(t,f) = \int_{-\infty}^{\infty} x(u)\psi^*_{t,f}(u)\,du \tag{4.14}$$

where $\psi^*_{t,f}(u)$ represents the complex conjugate of the wavelet function. Using the Morlet wavelet $\psi_{t,f}(u) = \sqrt{f}e^{j2\pi f(u-t)}e^{-\frac{(u-t)^2}{2\sigma^2}}$, the phase spectrum of $x(t)$ can be evaluated as follows:

$$\Phi_x(t,f) = \arg\left[\frac{W_x(t,f)}{|W_x(t,f)|}\right]. \tag{4.15}$$

This method of computing time-frequency phase estimation has been shown to yield similar results with respect to the Hilbert transform with the main difference being the increased frequency resolution offered by CWT (Le Van Quyen et al., 2001). The main shortcoming of the CWT-based method is the non-uniform resolution across time and frequency, which results in biased phase estimates (Aviyente et al., 2011).

4.6.2 EMD-Based Phase Estimation

More recently, EMD- and RID-based phase synchrony estimates have been proposed. While HHT provides a time-frequency energy distribution, EMDs have also been used for quantifying phase synchrony between signal pairs. A problem with current synchrony detection methods such as the Hilbert transform is that they depend on *a priori* selection of bandpass filters. In response to this problem, EMD-based phase synchrony measures have been defined (Rutkowski et al., 2008; Looney et al., 2009; Sweeney-Reed & Nasuto, 2007; Mutlu & Aviyente, 2011). In most of these approaches, the IMFs for each time series were extracted individually and were compared individually against the IMFs from the other time series for computing phase synchrony. This approach has multiple shortcomings. First, the IMFs from the different time series do not necessarily correspond to the same frequency, thus making it hard to compute exact within-frequency phase synchronization across different EEG channels. Second, the different time series may end up having different numbers of IMFs, which makes it difficult to match the different IMFs for synchrony computation. Finally, Looney and colleages (2009) showed that univariate EMD is not robust under noise and may suffer from mode mixing, which refers to the phenomenon of different modes (or frequencies) existing in a single IMF due to noise or intermittent signal activity.

Recently, extensions of EMD to the multivariate case have been developed including Complex EMD (Tanaka & Mandic, 2007), Rotation Invariant EMD (RIEMD) (Altaf et al., 2007), and Bivariate EMD (BEMD) (Rilling et al., 2007). These complex extensions of EMD decompose data from different sources simultaneously. Looney and colleages (2009) showed that the IMFs obtained in this fashion are matched, not only in number, but also in frequency overcoming problems of uniqueness and mode mixing, and first suggested the idea of using bivariate EMD to compute phase synchrony between two signals and the BEMD was shown to perform better than univariate EMD for quantifying bivariate synchrony. However, this approach still has some shortcomings, as the frequency bands corresponding to IMFs from different bivariate pairs are not necessarily the same. As such, this method is mostly limited to bivariate synchrony analysis and is not as easily generalizable to the whole brain synchrony analysis. Moreover, the multivariate extensions of EMD have been shown to be computationally expensive for computing functional connectivity across the whole brain (Mutlu & Aviyente, 2011).

4.6.3 RID-Based Phase Estimation

In recent work, we have shown the effectiveness of RID-based phase synchrony estimates compared to Hilbert- and CWT-based estimates (Aviyente & Mutlu, 2011; Aviyente et al., 2011; Sponheim et al., 2011; Aviyente et al., 2017). As most members of Cohen's class of TFDs are real-valued, they do not preserve the phase information. For this reason, we introduce a complex TFD, the Rihaczek distribution, to extract time- and frequency-dependent phase estimates. Rihaczek introduced the complex energy distribution and gave a plausibility argument based on physical grounds (Rihaczek, 1968). For a signal, $x(t)$, Rihaczek distribution is expressed as $C(t,\omega)=\frac{1}{\sqrt{2\pi}}x(t)X^*(\omega)e^{-j\omega t}$, where $X^*(\omega)$ is the complex conjugate of the Fourier transform of the signal and $C(t,\omega)$ measures the complex energy of a signal around time, t and frequency, ω. The complex energy density function provides a fuller appreciation of the properties of phase-modulated signals that is not available with other TFDs. Rihaczek distribution is a bilinear, time- and frequency-shift covariant, complex-valued TFD belonging to Cohen's class. This distribution satisfies the marginals and preserves the energy of the signal. Rihaczek distribution provides both a time-varying energy spectrum as well as a phase spectrum, and thus is useful for estimating the phase synchrony between any two signals. One of the disadvantages of Rihaczek distribution, similar to other quadratic TFDs, is the existence of cross-terms for multicomponent signals. In order to get rid of these cross-terms, in previous work (Aviyente et al., 2011), we proposed to apply a kernel function such as the CW kernel to filter the cross-terms. The resulting distribution can be written as:

$$C(t,\omega) = \iint \underbrace{\exp\left(-\frac{(\theta\tau)^2}{\sigma}\right)}_{\text{CWkernel}} \underbrace{\exp\left(j\frac{\theta\tau}{2}\right)}_{\text{Rihaczekkernel}} A(\theta,\tau) e^{-j(\theta t+\tau\omega)} d\tau d\theta, \tag{4.16}$$

where $e^{j\frac{\theta\tau}{2}}$ is the kernel function for the Rihaczek distribution. This new distribution, which will be referred to as RID–Rihaczek, will have an equivalent time-frequency kernel $\phi(\theta,\tau) = e^{-\frac{(\theta\tau)^2}{\sigma}} e^{j\frac{\theta\tau}{2}}$. Since this kernel satisfies the constraints, $\phi(\theta,0) = \phi(0,\theta) = \phi(0,0) = 1$, the corresponding distribution will both satisfy the marginals and preserve the energy, as well as be a complex energy distribution at the same time. The value of σ can be adjusted to achieve a desired trade-off between resolution and the number of cross-terms retained. The phase estimates from RID-Rihaczek distribution can be obtained as $\phi(t,\omega) = arg\left[\frac{C(t,\omega)}{|C(t,\omega)|}\right]$. This phase estimate has been shown to be more robust to noise and has uniformly high resolution in time and frequency compared to wavelet-based phase estimates (Aviyente & Mutlu, 2011).

4.6.4 Different Measures of Phase Synchrony

Once the phase difference between two signals is estimated through a time-frequency-based method, it is important to quantify the amount of synchrony. The most common scenario for the assessment of phase synchrony entails the analysis of the synchronization between pairs of signals. In the case of noisy oscillations, the length of stable segments of the relative phase gets very short; further, the phase jumps occur in both directions, so the time series of the relative phase $\phi_{x,y}(t)$ looks like a biased random walk (unbiased only at the center of the synchronization region). Therefore, the direct analysis of the unwrapped phase differences $\phi_{x,y}(t)$ has been used seldomly. As a result, phase synchrony can only be detected in a statistical sense. Two different indices have been proposed to quantify the synchrony based on the relative phase difference, that is, $\phi_{x,y}(t)$ is wrapped into the interval $[0,2\pi)$. The first index uses an information-theoretic criterion to quantify synchronization. This measure studies the distribution of $\phi_{x,y}(t)$ by partitioning the interval $[0,2\pi)$ into L bins and comparing it with the distribution of the cyclic relative phase obtained from two series of independent phases. This comparison is carried out by estimating the Shannon entropy of both distributions, that is, that of the original phases, and that of the independent phases, and computing the normalized difference (Quiroga et al., 2002). The second metric, phase synchronization index, is also known as the mean phase coherence and computed as $\sqrt{<cos\left(\Phi_{xy}(t)\right)>^2 + <sin\left(\Phi_{xy}(t)\right)>^2} = \left|\frac{1}{N}\sum_{k=0}^{N-1} e^{j\Phi_{xy}(t_k)}\right|$. It is a measure of how the relative phase is distributed over the unit circle. If the two signals are phase synchronized, the relative phase will occupy a small portion of the circle and mean phase coherence is high. This measure is equal to 1 for the case of complete phase synchronization and tends to zero for independent oscillators.

These different measures of synchrony can be used to quantify three different types of synchronization from EEG/ERP signals, depending on the application. The major difference between the different implementations of phase synchrony is whether the consistency of phase differences is measured across time, trials, or channels. Phase Locking Value (PLV) quantifies the consistency of the phase differences across trials between two channels as follows: $PLV(t,\omega) = \frac{1}{N}\left|\sum_{k=1}^{N} \exp\left(j\phi_{x,y}^{k}(t,\omega)\right)\right|$, where k is the trial number and N is the number of trials. This measure is also known as the inter-channel phase synchrony (ICPS). It is commonly used to estimate phase-locking in experimental situations, common in neurocognitive studies, where a subject is presented with a sequence of similar stimuli. Single trial phase-locking value (S-PLV), on the other hand, allows us to measure the significance of synchronies in single trials, and does not depend on block repetition of events. The variability of phase-difference is not measured across trials, but across successive time-steps, around a target latency. Specifically, a smoothed or S-PLV is defined for each individual trial. Finally, inter-trial phase synchrony (ITPS) quantifies the consistency of phase values for a given frequency band at each point in time over trials, in one particular electrode. ITPC is defined as $ITPC(t,\omega) = \left|\frac{1}{N}\sum_{k=1}^{N} \exp\left(j\phi_k(t,\omega)\right)\right|$, where N is the number of trials and $\phi_k(t,\omega)$ is the phase of the kth trial for each time and each frequency point. ITPC thus reflects the extent to which oscillation phase values are consistent over trials at that point in time-frequency plane. Note that this measure of phase coherence does not differentiate between possible biophysical mechanisms underlying phase consistency, such as phase reset or phase "smearing". Rather, this measure simply indicates the statistical probability of increased phase consistency between trial and baseline epochs.

4.7 SUMMARY

This chapter reviewed some of the basic time-frequency methods for analyzing ERP dynamics. We first introduced the different transforms that have been used to analyze the transient ERP activity. These transforms can be divided into three categories: linear methods such as STFT and CWT, data-adaptive methods such as EMD, and nonlinear methods such as Cohen's class of distributions. Although the overall goal of all of these methods is to determine the dynamics of transient activity, the mathematical principles upon which they rely are quite different. These differences in mathematical formulation lead to different computational complexities. Therefore, it is important to understand how the different methods can be used for different applications and purposes. For example, if the goal is to visualize the energy distribution of the transient ERP activity in time and frequency simultaneously, then Cohen's class of distributions (such as the RID) performs the best. Even though this method is the most complex in terms of computational complexity, it offers very high time-frequency resolution, which in turn provides a way to delineate different ERP components from each other. However, if the goal is to obtain a compact representation of the ERP signals in terms of a few physiologically meaningful components, then data-adaptive methods like EMD and MP are more suitable and computationally efficient. This chapter also shows how the same mathematical transforms can be used to study ERP dynamics through both energy distributions and phase synchrony in the time-frequency domain. While the energy distributions focus on the univariate activity, that is, dynamics within a certain brain region or electrode, the phase synchrony quantifies the dynamics across brain regions. In this manner, it is possible to obtain a better understanding of ERP timing, synchronization across time, frequency, and space.

REFERENCES

Abdulla, W. & Wong, L. (2011). Neonatal EEG signal characteristics using time frequency analysis. *Physica A: Statistical Mechanics and its Applications*, *390*(6), 1096–1110.

Altaf, M. U. B., Gautama, T., Tanaka, T., & Mandic, D. P. (2007). Rotation invariant complex empirical mode decomposition. In *2007 IEEE international conference on acoustics, speech and signal processing*-ICASSP'07, vol. 3 (pp. III–1009). IEEE.

Aviyente, S. (2007). Compressed sensing framework for EEG compression. In *Proceedings of* IEEE/SP: 14th *workshop on statistical signal processing* (pp. 181–184). IEEE. Doi: 10.1109/SSP.2007.4301243.

Aviyente, S., Bernat, E. M., Evans, W. S., & Sponheim, S. R. (2011). A phase synchrony measure for quantifying dynamic functional integration in the brain. *Human Brain Mapping*, 32(1), 80–93.

Aviyente, S. & Mutlu, A. Y. (2011). A time-frequency-based approach to phase and phase synchrony estimation. *IEEE Transactions on Signal Processing*, 59(7), 3086–3098.

Aviyente, S., Tootell, A., & Bernat, E. M. (2017). Time-frequency phase-synchrony approaches with ERPs. *International Journal of Psychophysiology, 111*, 88–97.

Bénar, C. G., Papadopoulo, T., Torrésani, B., & Clerc, M. (2009). Consensus matching pursuit for multi-trial EEG signals. *Journal of Neuroscience Methods, 180*(1), 161–170.

Brown, M. L., Williams, W. J., & Hero, A. O. (1994). Non-orthogonal Gabor representation of biological signals. In Proceedings of ICASSP '94: IEEE *international conference on acoustics, speech and signal processing*, vol. 4 (pp. IV/305-IV/308). IEEE. doi: 10.1109/ICASSP.1994.389742.

Cohen, L. (1995). *Time-frequency analysis*, vol. 778. Prentice Hall.

Cohen, M. X. (2018). A better way to define and describe Morlet wavelets for time-frequency analysis. *bioRxiv* [online], 397182.

Cong, F., Sipola, T., Huttunen-Scott, T., Xu, X., Ristaniemi, T., & Lyytinen, H. (2009). Hilbert-Huang versus Morlet wavelet transformation on mismatch negativity of children in uninterrupted sound paradigm. *Nonlinear Biomedical Physics, 3*(1), 1.

Demiralp, T., Ademoglu, A., Istefanopulos, Y., Başar-Eroglu, C., & Başar, E. (2001). Wavelet analysis of oddball p300. *International Journal of Psychophysiology, 39*(2–3), 221–227.

Durka, P. J. & Blinowska, K. J. (2001). A unified time-frequency parametrization of EEGs. *IEEE Engineering in Medicine and Biology, 20*(5), 47–53.

Durka, P. J., Matysiak, A., Montes, E. M., Valdés Sosa, P., & Blinowska, K. J. (2005). Multichannel matching pursuit and EEG inverse solutions. *Journal of Neuroscience Methods, 148*(1), 49–59.

Flandrin, P., Rilling, G., & Goncalves, P. (2004). Empirical mode decomposition as a filter bank. *IEEE Signal Processing Letters, 11*(2), 112–114.

Fries, P., Reynolds, J. H., Rorie, A. E., & Desimone, R. (2001). Modulation of oscillatory neuronal synchronization by selective visual attention. *Science, 291*(5508), 1560–1563.

Gribonval, R., Rauhut, H., Schnass, K., & Vandergheynst, P. (2008). Atoms of all channels, unite! Average case analysis of multi-channel sparse recovery using greedy algorithms. *Journal of Fourier Analysis and Applications, 14*(5–6), 655–687.

Hassanpour, H., Mesbah, M., & Boashash, B. (2004). Time-frequency feature extraction of newborn EEG seizure using SVD-based techniques. *EURASIP Journal on Applied Signal Processing, 2004*, 2544–2554.

Huang, N. E., Shen, Z., Long, S. R., Wu, M. C., Shih, H. H., Zheng, Q., et al. (1998). The empirical mode decomposition and the Hilbert spectrum for nonlinear and non-stationary time series analysis. *Proceedings of the Royal Society of London. Series A: Mathematical, Physical and Engineering Sciences, 454*(1971), 903–995.

Huang, N. E., Wu, M-L., Qu, W., Long, S. R., & Shen, S. S. P. (2003). Applications of Hilbert–Huang transform to non-stationary financial time series analysis. *Applied Stochastic Models in Business and Industry, 19*(3), 245–268.

Huang, Y. X., Schmitt, F. G., Hermand, J-P., Gagne, Y., Lu, Z. M., Liu, Y. L., et al. (2011). Arbitrary-order Hilbert spectral analysis for time series possessing scaling statistics: Comparison study with detrended fluctuation analysis and wavelet leaders. *Physical Review E, 84*(1), 016208.

Jeong, J. & Williams, W. J. (1992). Kernel design for reduced interference distributions. *IEEE Transactions on Signal Processing, 40*(2), 402–412.

Lachaux, J-P., Rodriguez, E., Martinerie, J., & Varela, F. J. (1999). Measuring phase synchrony in brain signals. *Human Brain Mapping, 8*(4), 194–208.

Lelic, D., Gratkowski, M., Hennings, K., & Drewes, A. M. (2011). Multichannel matching pursuit validation and clustering—a simulation and empirical study. *Journal of Neuroscience Methods, 196*(1), 190–200.

Le Van Quyen, M., Foucher, J., Lachaux, J-P., Rodriguez, E., Lutz, A., Martinerie, J., & Varela, F. J. (2001). Comparison of Hilbert transform and wavelet methods for the analysis of neuronal synchrony. *Journal of Neuroscience Methods, 111*(2), 83–98.

Looney, D., Park, C., Kidmose, P., Ungstrup, M., & Mandic, D. P. (2009). Measuring phase synchrony using complex extensions of EMD. In *2009 IEEE/SP 15th workshop on statistical signal processing* (pp. 49–52). IEEE.

Makeig, S., Debener, S., Onton, J., & Delorme, A. (2004). Mining event-related brain dynamics. *Trends in Cognitive Sciences, 8*(5), 204–210,

Makeig, S., Westerfield, M., Jung, T-P., Enghoff, S., Townsend, J., Courchesne, E., & Sejnowski, T. J. (2002). Dynamic brain sources of visual evoked responses. *Science, 295*(5555), 690–694.

Mallat, S. (1999). *A wavelet tour of signal processing*. Elsevier.

Mallat, S. G. & Zhang, Z. (1993). Matching pursuits with time-frequency dictionaries. *IEEE Transactions on Signal Processing, 41*(12), 3397–3415.

Moran, T. P., Bernat, E. M., Aviyente, S., Schroder, H. S., & Moser, J. S. (2015). Sending mixed signals: Worry is associated with enhanced initial error processing but reduced call for subsequent cognitive control. *Social Cognitive and Affective Neuroscience, 10*(11), 1548–1556.

Moser, J. S. (2017). The nature of the relationship between anxiety and the error-related negativity across development. *Current Behavioral Neuroscience Reports, 4*(4), 309–321.

Mutlu, A. Y. & Aviyente, S. (2011). Multivariate empirical mode decomposition for quantifying multivariate phase synchronization. *EURASIP Journal on Advances in Signal Processing, 2011*(1), 615717.

Nunez, P. L. & Srinivasan, R. (2006). *Electric fields of the brain: The neurophysics of EEG*. Oxford University Press.

Polikar, R., Topalis, A., Green, D., Kounios, J., & Clark, C. M. (2007). Comparative multiresolution wavelet analysis of ERP spectral bands using an ensemble of classifiers approach for early diagnosis of Alzheimer's disease. *Computers in Biology and Medicine, 37*(4), 542–558.

Quiroga, R. Q., Kraskov, A., Kreuz, T., & Grassberger, P. (2002). Performance of different synchronization measures in real data: a case study on electroencephalographic signals. *Physical Review E, 65*(4), 041903.

Rihaczek, A. (1968). Signal energy distribution in time and frequency. *IEEE Transactions on Information Theory, 14*(3), 369–374.

Rilling, G., Flandrin, P., Gonçalves, P., & Lilly, J. M. (2007). Bivariate empirical mode decomposition. *IEEE signal processing letters, 14*(12), 936–939.

Roach, B. J. & Mathalon, D. H. (2008). Event-related EEG time-frequency analysis: An overview of measures and an analysis of early gamma band phase locking in schizophrenia. *Schizophrenia Bulletin, 34*(5), 907–926.

Rosenblum, M., Pikovsky, A., Kurths, J., Schäfer, C., & Tass, P. A. (2001). Phase synchronization: From theory to data analysis. In F. Moss & S. Gielen (Eds.), *Handbook of biological physics: Neuro-informatics and neural modelling*, vol. 4 (pp. 279–321). Elsevier.

Rosenblum, M., Tass, P., Kurths, J., Volkmann, J., Schnitzler, A., & Freund, H-J. (2000). Detection of phase locking from noisy data: Application to magnetoencephalography. In K. Lehnertz, C. E. Elger, J. Arnhold, & P. Grassberger (Eds.), *Proceedings of the workshop on chaos in brain* (pp. 34–51). World Scientific.

Rutkowski, T. M., Mandic, D. P., Cichocki, A., & Przybyszewski, A. W. (2008). EMD approach to multichannel EEG data-the amplitude and phase synchrony analysis technique. In D-S. Huang, D. C. Wunsch II, D. S. Levine, & K-H. Jo (Eds.), Proceedings of the international *conference on intelligent computing: Advanced intelligent computing theories and applications. With aspects of theoretical and methodological issues* (pp. 122–129). Springer.

Salinas, E. & Sejnowski, T. J. (2001). Correlated neuronal activity and the flow of neural information. *Nature Reviews Neuroscience, 2*(8), 539.

Sieluzycki, C., Konig, R., Matysiak, A., Kus, R., Ircha, D., & Durka, P. J. (2008). Single-trial evoked brain responses modeled by multivariate matching pursuit. *IEEE Transactions on Biomedical Engineering, 56*(1), 74–82.

Sponheim, S. R., McGuire, K. A., Kang, S. S., Davenport, N. D., Aviyente, S., Bernat, E. M., & Lim, K. O. Evidence of disrupted functional connectivity in the brain after combat-related blast injury. *Neuroimage, 54*, S21–S29, 2011.

Sweeney-Reed, C. M. & Nasuto, S. J. (2007). A novel approach to the detection of synchronisation in EEG based on empirical mode decomposition. *Journal of Computational Neuroscience, 23*(1), 79–111.

Tağluk, M. E., Çakmak, E. D., & Karakaş, S. (2005). Analysis of the time-varying energy of brain responses to an oddball paradigm using short-term smoothed Wigner–Ville distribution. *Journal of Neuroscience Methods, 143*(2), 197–208.

Tanaka, T. & Mandic, D. P. (2007). Complex empirical mode decomposition. *IEEE Signal Processing Letters, 14*(2), 101–104.

Tropp, J. A. & Gilbert, A. C. (2007). Signal recovery from random measurements via orthogonal matching pursuit. *IEEE Transactions on Information Theory, 53*(12), 4655–4666.

Von Stein, A., Chiang, C., & König, P. (2000). Top-down processing mediated by interareal synchronization. *Proceedings of the National Academy of Sciences, 97*(26), 14748–14753.

Williams, N., Nasuto, S. J., & Saddy, J. D. (2011). Evaluation of empirical mode decomposition for event-related potential analysis. *EURASIP Journal on Advances in Signal Processing, 2011*(1), 965237.

Wu, C-H., Lee, P-L., Shu, C-H., Yang, C-Y., Lo, M-T., Chang, C-Y., & Hsieh, J-C. (2012). Empirical mode decomposition-based approach for intertrial analysis of olfactory event-related potential features. *Chemosensory Perception, 5*(3–4), 280–291.

Wu, Z. & Huang, N. E. (2009). Ensemble empirical mode decomposition: A noise-assisted data analysis method. *Advances in Adaptive Data Analysis, 1*(1), 1–41.

Yeung, N., Bogacz, R., Holroyd, C. B., & Cohen, J. D. (2004). Detection of synchronized oscillations in the electroencephalogram: An evaluation of methods. *Psychophysiology, 41*(6), 822–832.

Zhao, Y. L., Atlas, E., & Marks, R. J. (1990). The use of cone-shaped kernels for generalized time-frequency representations of nonstationary signals. *IEEE Transactions on Acoustics, Speech, and Signal Processing, 38*(7), 1084–1091.

CHAPTER 5

TIME FREQUENCY ANALYSES IN EVENT-RELATED POTENTIAL METHODOLOGIES

ANNA WEINBERG, PAIGE ETHRIDGE, BELEL AIT OUMEZIANE, AND DAN FOTI

5.1 Introduction

Event-related potentials (ERPs) are powerful tools for measuring the dynamics of human brain activity, and they have been used for decades to measure sensory, cognitive, motor, and emotion-related processes—as well as individual differences in these processes—across the lifespan (Cohen, 2014; De Haan, 2013; Hajcak et al., 2012; Kappenman & Luck, 2016; Luck, 2014). ERPs are defined by voltage fluctuations in the ongoing electroencephalogram (EEG) that are time-locked to specific events. As we discuss further in the studies we describe, these events can be the onset of an external stimulus (e.g., a picture, tone, or feedback about performance), or the generation of motor response (e.g., a button press).

Typically, tasks are designed so that these events are repeated across multiple trials, which are then averaged within conditions, presumably canceling out substantial amounts of trial-level noise and yielding the "prototypical" waveform that is common across trials (Figure 5.1A shows an example waveform). Specific ERP components are then identified within these averaged waveforms, and they are quantified numerically as deviations from a pre-event baseline period. A component is defined "a set of voltage changes that are consistent with a single neural generator site and that systematically vary in amplitude across conditions, time, [and] individuals," (Luck, 2014, p. 68). Thus, an ERP component is a portion of the overall waveform—often, a peak deflection in the waveform—that captures the brain response (or set of processes) that are of interest. ERPs are typically described in terms of their amplitude (measured in

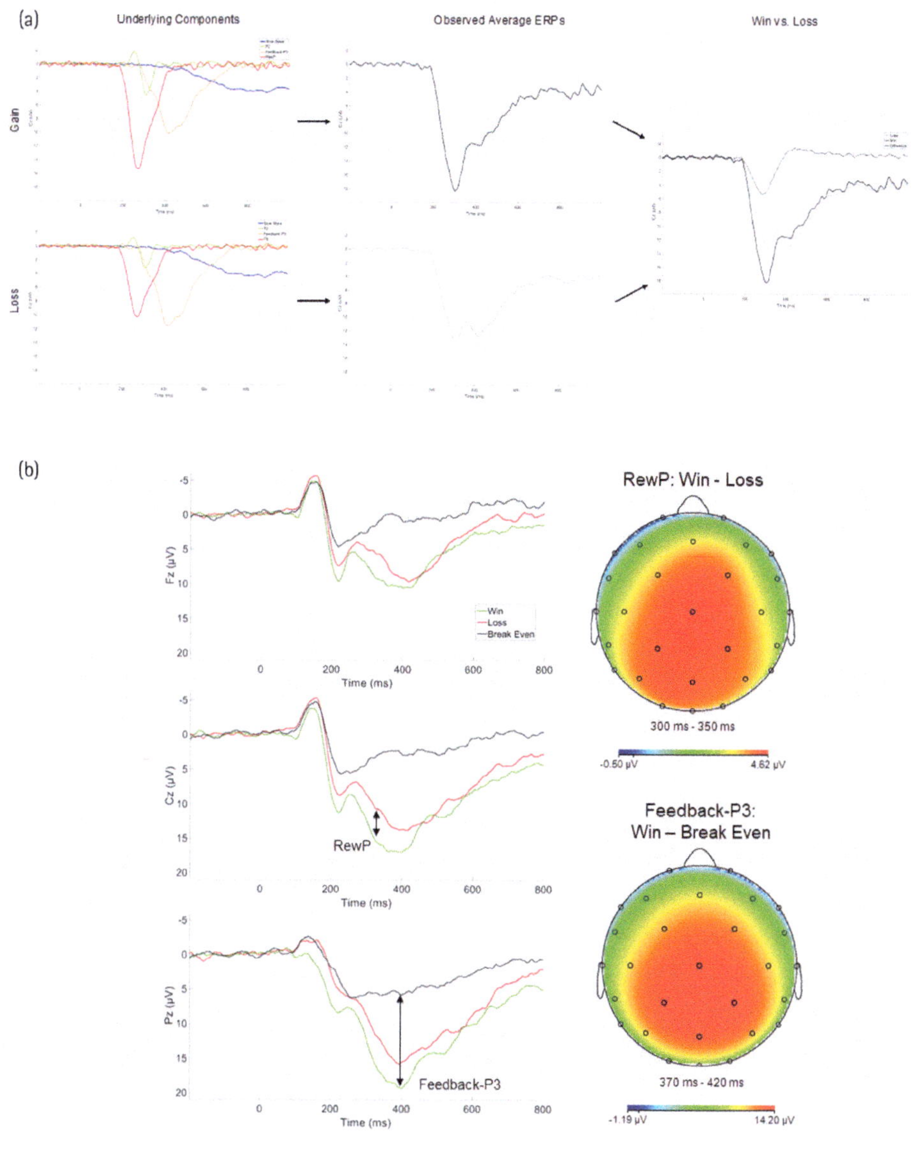

FIGURE 5.1 (A) Example data depicting how multiple underlying ERP components (left) contribute to the observed grand averaged ERP waveforms (middle), which are then compared across conditions (right). (B) ERPs elicited by feedback delivery on the social incentive delay task.

microvolts; µV), polarity (positive or negative), latency (measured in milliseconds; ms), and scalp topography (where on the scalp the component is maximal). Amplitude refers to the difference between activity occurring at some point following an event of interest and an average pre-event baseline period (e.g., 200 ms prior to stimulus onset), and latency refers to the time from stimulus onset to some specific peak activity. Naming conventions then frequently reflect both the polarity and latency of the component. Thus, the P3 is the third major positive-going peak in the ERP waveform after a stimulus is presented (and often peaks in the neighborhood of 300 ms, hence the common alternative name "P300"). At times, ERP components receive functionally descriptive names (e.g., the error-related negativity or ERN).

Because they measure the electrical activity of the brain, the speed of which approaches the speed of light, ERPs capture neural responses in the time-frame in which cognition occurs. This millisecond temporal resolution makes it theoretically possible to isolate dozens of individual neural processes that occur in even very close temporal proximity. ERPs have thus been particularly useful in studies requiring high temporal resolution to identify, for example, the transition from sensory-driven stimulus processing to higher-order cognitive functions (De Cesarei & Codispoti, 2006; Wiens et al., 2011), or specific cognitive or affective impairments in a given diagnostic group (Duncan et al., 2009; Kuperberg et al., 2018).

In comparison with other neuroimaging modalities, EEG techniques are well-suited as assessment tools. EEG/ERP data collection is relatively efficient, economical, noninvasive, (Kappenman & Luck, 2016), and is well-tolerated by most participants across the developmental spectrum (De Haan, 2013). While ERP studies have traditionally been conducted in tightly controlled environments to minimize noise and electrical interference, advances in technology have allowed for progress in ERP data collection *outside* the laboratory and in remote field settings (Tarullo et al., 2017). Additionally, few contraindications exist for EEG research: for example, people with braces, pregnant women, and awake infants—all of whom are frequently excluded from MRI studies—typically *can* participate in EEG studies, making it feasible to collect neuroimaging data from large and relatively diverse groups of participants. Finally, multiple studies over the years have explored the psychometric properties of ERPs and found them to be highly reliable measures of neural activity, comparable to other common assessment methods both in terms of internal consistency and test-retest reliability (Baldwin, Larson, & Clayson, 2015; Ethridge & Weinberg, 2018; Foti et al., 2016; Kujawa et al., 2018; Weinberg & Hajcak, 2011).

Naturally, these strengths are accompanied by a number of limitations. The assumption underlying measurement of ERP components in the time domain is that (a) each component represents distinct sensory and/or cognitive processes (or a small set of related processes), and (b) components reflect the activity of a single brain region (or a small network of closely related brain regions). In line with these assumptions, ERP components are typically scored based on where on the scalp and when in time specific peaks in the waveform occur, taking the average amplitude within a specified time window or the peak deflection (Figure 5.1A). This technique reduces the multidimensional EEG signal down to two dimensions, and has many advantages (Cohen, 2014; Luck, 2014). Yet the reality of neural activity is that multiple sensory, cognitive, affective, and motor processes can and do occur simultaneously. Because the observed trial-averaged waveform represents the sum of all activity measured at a particular site on the head within a particular time-window, ERPs representing these unique processes are summed together in the waveform. Figure 5.1B shows how multiple observed components that overlap both spatially and temporally contribute to the grand averaged waveform. Traditional component-scored methods make it difficult to isolate the contribution each *underlying* component makes to the observed average ERP.

Additionally, the electrical signals captured by ERPs are conducted through the brain, meninges, skull, and scalp, and this signal is subject to spread as it seeks paths of low resistance; precise identification of primary neural contributors to any one component is therefore often difficult, particularly for brain regions that are relatively far from the scalp. Furthermore, neural activity recorded from an electrode exterior to the skull can reflect the simultaneous and summed activation of many, many thousands—even millions—of neurons (Luck, 2014). Combined with the difficulty of effectively isolating different components, this can often make source localization of time-domain ERP components a dicey proposition (Cohen, 2014).

Processing techniques that isolate unique sources of systematic variance within the trial-averaged waveform may allow for both more accurate identification of distinct neuroelectric signals and better description of their anatomical origins. These signals can be differentiated based on their temporal and spatial variance, as is done with typical time-window averages as well as more advanced techniques like principal component analysis (PCA) and independent components analysis (ICA; Dien, Spencer, & Donchin, 2003; Foti, Weinberg, Dien, & Hajcak, 2011; Spencer, Dien, & Donchin, 2001). Critically, distinct neural signals can also be differentiated based on their *spectral* properties. This is because the electrical activity measured by the EEG contains rhythmic oscillations. These oscillations reflect fluctuations in the activity of populations of neurons, and the different properties of these oscillations can be helpful in differentiating cognitive operations.

Oscillations are described by their *frequency*, *power*, and *phase*. *Frequency* describes the speed of the oscillation, or how many times a sine wave repeats, or cycles, in a given period of time, and is measured in hertz (Hz). A wave that repeats four times per second has a frequency of 4 Hz, a wave that repeats 50 times per second has a frequency of 50 Hz, and a wave that repeats once every two seconds has a frequency of 0.5 Hz. *Power* is a measure of how much energy is present in a frequency band and is represented as the amplitude (or height) of the peaks, squared. Finally, *phase* is measured in degrees, or radians, and is a measure of when in time any given part of the sine wave exists. EEG data, as well as the ERPs derived from these EEG data, are composed of oscillations at multiple frequencies, each present with different relative power at different time-points. Time-frequency techniques then attempt to deconvolve, or separate, these signals, to identify, for example, how much power is present at a given frequency at specific points in time.

There are many different signal processing techniques available for decomposing neuroelectric signals to describe the spectral characteristics of ERPs, including moving window Fourier transforms, wavelet transforms, and Cohen's class of time-frequency distributions. These all involve representing a given ERP waveform in terms of a set of sine waves, each of which is characterized by different frequencies, phases, and amplitudes. Each technique uses different mathematical formulae to do so and makes different assumptions about the nature of the oscillatory signal presumed to underlie the ERP waveform (e.g., Bruns, 2004; Cohen, 2014; Luck, 2014). For a full discussion of these—and their relative strengths and weaknesses—see e.g., Aviyente, this volume, Keil & Thigpen, this volume? Voytek, this volume. However, a general caveat is that many of these techniques also involve a loss of temporal precision—one of the chief advantages of the ERP technique (Cohen, 2014; Luck, 2014)—though this matters more for some experiments than others, and there are techniques available to improve temporal precision (Aviyente et al., 2006).

In terms of their application to ERPs, time-frequency techniques are particularly useful when the signals overlap *both* temporally and spatially, therefore making these signals indistinguishable in the time domain (Bernat, Malone, Williams, Patrick, & Iacono, 2007; Bernat, Nelson, Steele, Gehring, & Patrick, 2011; Bernat, Williams, & Gehring, 2005; Cohen, Elger, & Ranganath, 2007; Herrmann, Rach, Vosskuhl, & Strüber, 2014; Kolev, Demiralp, Yordanova, Ademoglu, & Isoglu-Alkaç, 1997). In such instances, assessment of the differing spectral characteristics of overlapping ERP components is useful for distinguishing multiple underlying processes that give rise to the observed waveform. Additionally, insofar as populations of neurons in different regions of the brain may fire at different frequencies, the time-frequency components identified in this way may correspond more closely to distinct sources underlying event-related brain activity (Foti, Weinberg, Bernat, & Proudfit, 2015).

In what follows, we review literature demonstrating the ways in which time-frequency signal processing techniques have been useful in shedding new light on several common ERP components, helping to answer research questions that would be difficult to address solely within the time domain. We focus here on the mismatch negativity (MMN), the P3, the error-related negativity (ERN), and the feedback-related negativity (FN)/reward positivity (RewP). We conclude with a practical example of how to apply time-frequency techniques to ERP data and present new analyses of some of our prior time-domain work with the FN/RewP.

5.2 The Mismatch Negativity (MMN)

The temporal resolution of ERPs makes them well suited for studying the earliest stages of sensory processing. One widely studied sensory ERP component is the MMN, a neural index of change detection that is automatically elicited by a deviant stimulus presented within a repetitive sequence (Näätänen, 1995; Näätänen, Paavilainen, Rinne, & Alho, 2007). While the MMN can be elicited within any sensory modality, it is commonly studied as part of auditory processing. For example, within a sequence of repeating standard tones, the presentation of a tone that differs with regard to pitch, duration, or some other stimulus characteristics will automatically elicit the MMN. The auditory MMN typically occurs between 150 and 250 ms following the deviant stimulus and is maximal at frontocentral electrodes, with a smaller positive-going potential often apparent at temporal/mastoid electrodes. While MMN morphology is dependent upon the stimulus characteristics, studies generally indicate that the MMN emanates from a combination of activity in supratemporal and frontal cortical regions (i.e., "temporal" and "frontal" MMN subcomponents), likely related to pre-perceptual stimulus processing and involuntary attentional switch, respectively (Alho, 1995). In addition to the large basic neuroscience literature applying the MMN to the study of sensory functioning, individual differences in MMN amplitude have been examined with regard to cognitive decline in aging, and cognitive impairment in psychiatric disorders (Näätänen et al., 2011). For example, the MMN is reduced by approximately one standard deviation among individuals with schizophrenia (Umbricht & Krljes, 2005). This impact of schizophrenia on MMN amplitude is equivalent to approximately 30 years of cognitive aging (i.e., the MMN of an individual with schizophrenia at age 20 is comparable to that of a non-psychotic individual at age 50; (Kiang, Braff, Sprock, & Light, 2009). Overall, traditional time-domain analyses of MMN amplitude have been useful for examining the time course of early sensory processing and deficits therein.

Complementing these time-domain studies, time-frequency studies have been useful for isolating the temporal and frontal subcomponents of the MMN, as well as contextualizing individual differences in MMN amplitude. The frontal portion of the MMN has been linked primarily to an increase in theta power (i.e., the amplitude of theta waves across trials), whereas the temporal portion has been linked to theta *phase coherence* (i.e., the alignment of theta waves across trials; also see Chapter 9) but not to theta power (Fuentemilla, Marco-Pallarés, Münte, & Grau, 2008; Ko et al., 2012). Time-frequency approaches have also helped clarify the nature of impaired sensory processing in schizophrenia. While a reduced frontal MMN in schizophrenia is well-documented, there is some evidence from time-domain analyses that the temporal (mastoid) subcomponent may be less affected (Baldeweg, Klugman, Gruzelier, & Hirsch, 2002). Subsequent studies in schizophrenia have used time-frequency approaches to clarify how abnormal MMN amplitude is explained by alterations in theta power and phase coherence. For example, one study found that time-domain MMN amplitude was strongly correlated with frontal theta power among healthy controls but not among individuals with schizophrenia, suggesting a decoupling between these signals (Hong, Moran, Du, O'Donnell, & Summerfelt, 2012). Other work shows that reduced MMN in schizophrenia is characterized by reductions in both theta power and phase coherence, with differential deficits based on the type of deviant stimulus (Lee et al., 2017). Thus, time-frequency decomposition of the MMN has been useful for teasing apart distinct neurological signals involved in sensory processing that would be difficult to capture within the time domain, with relevance to both basic science and clinical applications.

5.3 The P3

The P3 was first reported by Sutton and colleagues (1965) and has since been among the most well-researched components in the ERP literature (Polich, 2012). It is among the canonical ERP components in that it is commonly observed across a wide range of stimuli and laboratory tasks, generally manifesting as the third major positive-going deflection in the waveform and maximal at parietal electrodes. A P3 is typically elicited by stimuli that are relatively salient in the local context due to being infrequent, unexpected, or because they are designated "targets" that are task-relevant. Thus, the P3 is often present within the waveform for most ERP studies, even if it is not the component of primary interest. When the P3 is among the primary outcomes measures, one of the most common laboratory paradigms used is the "oddball" task (Donchin, Ritter, & McCallum, 1978). During an oddball task, participants are asked to respond to or otherwise keep track of a designated target stimulus and otherwise disregard other non-target stimuli (i.e., standard stimuli). In the case of a two-stimulus, auditory oddball task, for example, two different tones may be presented within an ongoing sequence, with differential likelihood (e.g., 80% for standards, 20% for targets). Participants are required to distinguish between these tones by responding to the occurrence of the target (e.g., button press or mentally counting) and not responding to the standard (Polich & Kok, 1995). Discriminating this infrequent target stimulus from the frequently occurring standard elicits a robust P3, which is increased (i.e., more positive) for the target vs. standard stimuli (Polich, 2012). Many studies also distinguish between the P3a and P3b subcomponents, which are clearly differentiated on three-stimuli oddball tasks which include frequent standard, infrequent target, and infrequent novel stimuli (i.e., a third stimulus type that is novel and task-irrelevant). In this case, the classic parietal P3 in response to target stimuli is referred to as the P3b, in order to distinguish it from an overlapping, frontocentral positivity to novel stimuli that is referred to as the P3a, (or Novelty P3; Simons, Graham, Miles, & Chen, 2001; N. Squires, Squires, & Hillyard, 1975). Broadly, the P3a is thought to reflect frontal lobe functioning in response to novelty, whereas the P3/P3b is thought to reflect temporal-parietal brain activity associated with attention and memory processing (Polich, 2007). For the remainder of this chapter, we generally use the term "P3" rather than "P3b."

Some theoretical accounts of the P3 within the oddball task posit that it captures the updating of an individual's mental representation prompted by incoming stimuli (i.e., Context Updating Theory; Donchin, 1981). Following an initial sensory input, it is thought that an attention-driven comparison between the current stimulus and previous mental representation in working memory is made. If no change in stimulus attributes have been distinguished, the prior mental representation or "schema" of the stimulus is maintained, thereby evoking sensory potentials. However, when a new stimulus attribute is perceived (e.g., a target) the mental representation of the stimulus context in working memory is updated to elicit the P3 (Donchin et al., 1986).

P3 amplitude has been shown to be influenced by a number of factors, including cognitive load demands (i.e., reduced P3 amplitude when under high cognitive load) and target stimulus probability (i.e., increased P3 amplitude to relatively rare stimuli). For example, people who performed a primary task with varying cognitive demands while also engaging in a secondary oddball task showed that increasing the primary tasks' difficulty attenuated P3 amplitude to targets on the oddball task (Isreal, Chesney, Wickens, & Donchin, 1980; Wickens, Kramer, Vanasse, & Donchin, 1983). Furthermore, detecting a target from a standard stimulus within the oddball paradigm elicits a robust P3 amplitude that potentiates as the global and local sequence probability for the target decreases (Duncan-Johnson & Donchin, 1977; K. Squires, Wickens, Squires, & Donchin, 1976).

While there is a long history of examining P3 amplitude using traditional time-domain techniques, studies have also applied time-frequency techniques to decompose spectral properties of the P3. This is relevant in part because, in comparison to standard stimuli, target stimuli on an oddball task often elicit broad modulation of the ERP waveform that spans the P2-N2-P3-Slow Wave complex. Thus, the P3 occurs in the context of these other, overlapping neural signals that are also sensitive to stimulus properties, and for some questions it may be helpful to isolate the unique portions of the waveform that reflect context updating vs. other aspects of stimulus processing. Initial work in this domain showed that the oddball response is comprised of a progression from theta- to delta-band activity (Başar-Eroglu, Başar, Demiralp, & Schürmann, 1992; Kolev et al., 1997). Subsequent analyses applied principal components analyses to time-frequency plots in order to isolate a specific spectral subcomponent. The oddball response is characterized by an early, low-frequency component (1 Hz at 150 ms), followed by multiple delta-theta responses corresponding to the rise and peak of the time-domain P3 (1–3 Hz, from 400 to 600 ms), and ultimately resolving with a second low-frequency component (1 Hz from 600 to 800 ms) (Bernat et al., 2007).

Decomposing the oddball response in this manner has also been fruitful for clarifying the nature of individual differences in P3 amplitude in clinical populations. For example, it is well-established that time-domain P3 amplitude is reduced among individuals with externalizing psychopathology, such as alcohol use disorder (Polich, Pollock, & Bloom, 1994). In one study examining a clinically heterogeneous community sample, reduced time-domain P3 was common across conduct disorder, attention-deficit/hyperactivity disorder, oppositional defiant disorder, and substance use disorder (Gilmore, Malone, Bernat, & Iacono, 2010). As expected, the diagnostic groups were also generally associated with reduced power in multiple spectral subcomponents spanning 0.5–3 Hz and 200–900 ms. Critically, reduced power in a specific delta subcomponent (1 Hz, 400–600) differentiated the externalizing groups from controls, even after considering time-domain P3 amplitude. That is, this precise spectral subcomponent exhibited a stronger relationship with the clinical phenotype than the time-domain P3, which by definition is a composite of these multiple spectral subcomponents. Other work has examined reduced P3 amplitude in schizophrenia, which is well-documented (Jeon & Polich, 2003). Incorporating time-frequency analyses, however, showed that *total* delta-band activity to oddball stimuli (including phase-locked and non-phase-locked) is intact among individuals with schizophrenia (Ergen, Marbach, Brand, Başar-Eroğlu, & Demiralp, 2008). This has potentially important implications for the interpretation of reduced time-domain P3: rather than a reduced neural response per se, the time-frequency analyses suggest that schizophrenia may be characterized by greater temporal jitter in the neural response, which manifests as a reduced peak in the trial-averaged waveform. Together, these studies show how a reduced P3 amplitude in clinical populations can be explained by multiple abnormalities in the neural signal, which are not readily apparent in the time domain.

5.4 The Error-Related Negativity (ERN)

Rapidly identifying the mistakes that we make and altering our behavior in response to these errors is critical for adaptive functioning. The error-related negativity (ERN; also referred to as the error negativity or Ne) is an ERP component that has commonly been used to study neural processes associated with identifying and adapting to errors (Falkenstein, Hohnsbein, Hoormann, & Blanke, 1991; Gehring, Goss, Coles, Meyer, & Donchin, 1993). The ERN is a response-locked ERP often elicited in speeded reaction tasks that peaks approximately 100 ms following erroneous responses, is maximal at frontocentral electrode sites, and is thought to be generated in medial frontal cortex regions including the anterior cingulate cortex (ACC; for reviews see Gehring et al., 2011; Holroyd & Coles, 2002; Olvet & Hajcak, 2008). Time-frequency analyses have contributed to our knowledge of the unique structure of the ERN (Bernat et al., 2005; Munneke, Nap, Schippers, & Cohen, 2015; Riesel, Weinberg, Moran, & Hajcak, 2012; Yordanova, Falkenstein, Hohnsbein, & Kolev, 2004), have elucidated possible mechanisms by which it is generated (Luu, Tucker, & Makeig, 2004; Trujillo & Allen, 2007), and have provided empirical support for some existing theories of its functional significance (Cavanagh, Cohen, & Allen, 2009; Cavanagh, Zambrano-Vazquez, & Allen, 2012; Luu & Tucker, 2001; Luu, Tucker, Derryberry, Reed, & Poulsen, 2003).

With regard to the structure of the component, time-frequency analyses have identified how the ERN is both linked to, and separable from, other related ERP components (Cavanagh et al., 2012; Di Gregorio, Maier, & Steinhauser, 2018; Gehring & Willoughby, 2004; Steele et al., 2016; Yordanova et al., 2004). For instance, it has been debated whether neural responses to errors (ERN) and to correct responses (correct-related negativity; CRN or Nc), which overlap in time and scalp topography, reflect the same or distinct processes (Vidal, Hasbroucq, Grapperon, & Bonnet, 2000). While evidence indicates that the ERN and CRN demonstrate similar characteristics in the time-domain, frequency analyses have identified error-specific signals with unique scalp topographies (Yordanova et al., 2004), suggesting that the ERN and CRN are not identical processes and may have distinct neural generators. Specifically, Yordanova and colleagues (2004) identified a delta-frequency component (1.5–3.5 Hz) that was specific to erroneous responses, as well as a theta-frequency component (4–8 Hz) that emerged for both erroneous and correct responses but demonstrated different scalp topographies depending on accuracy and response side (left or right hand). Similarly, time-frequency analyses have been used to demonstrate distinctions between the ERN and the error-related positivity (Pe) by demonstrating that the Pe can emerge in the absence of an ERN (Di Gregorio et al., 2018; Steele et al., 2016), as well as between the ERN and the feedback negativity (FN, see Section 5.5) by identifying distinct scalp topographies of the ERN and FN (Gehring & Willoughby, 2004). Nevertheless, theta frequency activity that is common to several ERP components (ERN, CRN, FN, and N2) has been interpreted to mean that each of these ERPs reflects similar (though not identical) neural processes related to performance monitoring and behavioral control (Cavanagh & Shackman, 2015; Cavanagh et al., 2012).

Time-frequency analyses are also uniquely placed to enhance our understanding of the mechanisms by which the ERN is generated. The classic view of ERPs suggests that they reflect phasic bursts of neural activation, while more recently it has been suggested that ERPs might reflect phase-resetting of ongoing oscillatory activity (see Keil & Thigpen, this volume). Although some argue that time-frequency techniques cannot distinguish between these two hypotheses (Yeung, Bogacz, Holroyd, & Cohen, 2004; Yeung, Bogacz, Holroyd, Nieuwenhuis, & Cohen, 2007), evidence suggests that the ERN is best explained by the combination of an amplitude increase and phase-resetting (Trujillo & Allen, 2007). This is an important contribution, partly because understanding how the ERN is generated provides insight into its functional significance. For example, Cavanagh and colleagues (2009) suggested that theta oscillatory dynamics reflected in the ERN may represent a mechanism by which neural regions responsible for performance monitoring and for cognitive control communicate with each other (see Chapter 3 for a discussion of neural oscillations; Keil & Thigpen).

As with all of the ERPs discussed in this chapter, understanding how the ERN is associated with individual difference variables is an active area of research. Time-window analyses of ERN responses have identified links between the ERN and multiple forms of psychopathology, including generalized anxiety disorder (GAD) (Weinberg, Olvet, & Hajcak, 2010), obsessive compulsive disorder (Endrass et al., 2010; Riesel, Kathmann, & Endrass, 2014), substance use disorder (Euser, Evans, Greaves-Lord, Huizink, & Franken, 2013), and depression (Weinberg, Liu, & Shankman, 2016). Time-frequency analyses have begun to meaningfully advance our understanding of such individual differences. For instance, Cavanagh and colleagues (2017) determined that error-related theta power was a unique predictor of GAD status over and above the time-window scored ERN, and when they included theta network dynamics in their model, they were able to classify clinical participants with an impressive 66% accuracy. These data suggest that time-frequency analyses may have a powerful role to play in using neural data to classify clinical populations. While time-frequency analyses have led to several critical advances in our understanding of the ERN, as noted here, this is likely to remain a fruitful avenue of continued study.

5.5 The Feedback Negativity (FN)/ Reward Positivity (RewP)

In addition to endogenous performance monitoring, optimization of behavior depends on the ability to discriminate positive (e.g., monetary gain, correct performance feedback, positive interpersonal feedback) from negative (e.g., monetary losses, incorrect performance feedback, negative interpersonal feedback) external environmental feedback. A substantial body of research on neural discrimination of favorable vs. unfavorable feedback has used an ERP component called, variously, the feedback negativity (FN), the feedback-related negativity (FRN), or the reward positivity (RewP) (Foti & Weinberg, 2018; Foti et al., 2011; Hajcak, Moser, Holroyd, & Simons, 2006; Holroyd, Nieuwenhuis, Yeung, & Cohen, 2003; Miltner, Braun, & Coles, 1997). This component peaks approximately 250–300 ms at frontocentral recording sites following the presentation of feedback (Miltner et al., 1997), and has often been studied in the context of laboratory gambling tasks, in which individuals make responses in order to win or lose money on each trial. Feedback indicating monetary gains, losses, or non-gains is presented following response selection.

Traditionally, the FN/RewP was viewed as a negative-going component (thus the emphasis on "negativity" in the name) and a member of a broader class of medial-frontal negativities (MFNs) that includes the ERN (Gehring & Willoughby, 2004; Miltner et al., 1997). This negative peak, which like other MFNs has been thought to be generated by the ACC (Gehring & Willoughby, 2002; Potts, Martin, Burton, & Montague, 2006) via phasic input from the midbrain dopamine system (Holroyd & Coles, 2002), was thought to be observed following unfavorable outcomes (e.g., feedback indicating monetary losses) and absent following favorable outcomes (e.g., feedback indicating monetary gain). This perspective suggests a binary differentiation of losses from non-losses, reflecting the activity of a single cognitive process (Hajcak et al., 2006; Kreussel et al., 2011). More recent data suggests, however, that the difference between neural responses to losses and gains may also be driven in part by a substantial positive-going deflection in the waveform elicited by favorable outcomes that is absent following unfavorable outcomes (i.e., the reward positivity; Baker & Holroyd, 2011; Bernat et al., 2011; Bogdan, Santesso, Fagerness, Perlis, & Pizzagalli, 2011; Carlson, Foti, Harmon-Jones, Mujica-Parodi, & Hajcak, 2011; Foti et al., 2011; Harper, Olson, Nelson, & Bernat, 2011; Hewig et al., 2010; Holroyd, Krigolson, & Lee, 2011; Holroyd, Pakzad-Vaezi, & Krigolson, 2008).

A possible explanation for these conflicting views is that the trial-averaged component following feedback reflects the activity of two independent but overlapping processes: a) a negative deflection in the waveform, which is enhanced following loss feedback but not reward feedback, and b) a positive deflection that is enhanced following reward feedback and decreased following losses (e.g., Bernat, Nelson, & Baskin-Sommers, 2015; Bernat et al., 2011; Carlson et al., 2011; Foti et al., 2011). An increased (underlying) negative-going component and a decreased (underlying) positive-going component might then summate to create the *observed* negative-going component following feedback indicating losses that had typically been observed in the time domain (i.e., the classic FN/FRN/MFN).

Studies using time-frequency methods to explore neural activity in this time-window have tended to bear this out, as there is evidence that activity in both the theta- and delta-bands make unique contributions to the amplitude of the FN/RewP (Bernat et al., 2015; Bernat et al., 2011). In particular, theta activity in this time range appears to be primarily sensitive to loss/negative outcomes, with losses eliciting enhanced activity compared to gains in this frequency range (Bernat, Nelson, Holroyd, Gehring, & Patrick, 2008; Bernat et al., 2011; Gheza, De Raedt, Baeken, & Pourtois, 2018; Harper et al., 2011; L. Nelson, Patrick, Collins, Lang, & Bernat, 2011; Olson, Harper, Golosheykin, Bernat, & Anokhin, 2011; Webb et al., 2017). Delta has more often been linked to activity in the time-range of the P3 that follows the FN/RewP at more parietal sites (Gehring & Willoughby, 2002; Holroyd & Coles, 2002; Miltner et al., 1997). However, there is increasing evidence that delta activity also underlies the observed reward positivity occurring in the time-range and spatial location of the FN/RewP (Bernat et al., 2015; Bernat et al., 2008; Foti, Weinberg, Bernat, & Proudfit, 2014; Harper et al., 2011). Indeed, delta activity in this time-range is enhanced for gains compared to losses (Bernat et al., 2015; Bernat et al., 2008; Bernat et al., 2011; Cavanagh, 2015; Cavanagh, Masters, Bath, & Frank, 2014; Foti, Weinberg, et al., 2014; Gheza et al., 2018; Harper et al., 2011; Leicht et al., 2013; L. Nelson et al., 2011; Olson et al., 2011; Pornpattananangkul & Nusslock, 2016; Webb et al., 2017), and appears to drive the reward positivity observed in the trial-averaged waveforms (Bernat et al., 2015; Foti et al., 2011; Harper et al., 2011).

The activity of these two frequency bands also appears to be dissociable, insofar as the differences between loss and gain activity in the theta and delta band tend to be at best weakly correlated (e.g., Bernat et al., 2015; Bernat et al., 2011; Foti, Weinberg, et al., 2014). Moreover, source analysis suggests unique generators, with loss-related theta localizing to the ACC and gain-related delta to a possible source in the basal ganglia (Foti, Weinberg, et al., 2014). Combined, these data suggest that changes in theta and delta are not yoked expressions of the same underlying process, but instead may represent distinct cognitive processes and contributions to the FN/RewP (Bernat et al., 2015; Bernat et al., 2008; Bernat et al., 2011; Gheza et al., 2018; Harper et al., 2011; Olson et al., 2011). Consistent with this, theta response to losses appears to reflect a relatively low-level response to negative outcomes that is frequently insensitive to stimulus parameters and experimental context (Bernat et al., 2015; Bernat et al., 2011; Harper et al., 2011; Watts, Bachman, & Bernat, 2017; Watts & Bernat, 2018). This increased theta (Section 5.4) appears to then act as a signal to recruit attentional and executive resources to respond to mistakes, novelty, and negative feedback (Aviyente, Tootell, & Bernat, 2017; Cavanagh & Frank, 2014; Cavanagh et al., 2012; Van Noordt, Campopiano, & Segalowitz, 2016; van Noordt, Desjardins, Gogo, Tekok-Kilic, & Segalowitz, 2017). In contrast, in monetary reward paradigms, delta appears sensitive not only to loss vs. gain distinctions, but also higher-level secondary stimulus attributes, such as magnitude (Bernat et al., 2015), context (Bernat et al., 2015; Watts & Bernat, 2018), expectancy violations (Cavanagh, 2015;

Gheza et al., 2018; Watts et al., 2017), and reward uncertainty (Gheza et al., 2018) and thus may reflect more elaborative processing of the feedback beyond the most salient dimension of the stimuli (see, however: Leicht et al., 2013).

Identification of dissociable and functionally distinct processes in the time-window of the FN/RewP has also proven to be useful in studies seeking to understand individual differences in neural responses to feedback, including describing developmental influences on more specific neural processes, as well as identifying specific cognitive-affective deficits in clinical samples. For instance, there is evidence from MRI research that brain regions supporting the neural response to rewards undergo considerable developmental changes from childhood through adolescence and into adulthood. In particular, fMRI studies tend to find an adolescent-specific peak in striatal activation to reward feedback (Braams, van Duijvenvoorde, Peper, & Crone, 2015; J. R. Cohen et al., 2010; Ernst et al., 2005; Galvan et al., 2006; Somerville, Hare, & Casey, 2011; Van Leijenhorst et al., 2009). Most studies examining developmental changes in the FN/RewP, however, have failed to find evidence for an adolescent-specific peak, or indeed even significant developmental changes (Kujawa et al., 2018; Larson, South, Krauskopf, Clawson, & Crowley, 2011; Lukie, Montazer-Hojat, & Holroyd, 2014; Santesso, Dzyundzyak, & Segalowitz, 2011); but see also (Arbel, McCarty, Goldman, Donchin, & Brumback, 2018). In a recent study of 8–17-year-old participants, however, Bowers and colleagues (2018) examined both time-domain ERPs and time-frequency-derived theta and delta band activity. They found that, consistent with previous work, the time-domain-scored RewP was not associated with participants' age. Yet, theta power—which was enhanced for losses relative to gains—decreased with age, whereas delta power—which was greater for gains than losses—increased. In a study investigating whether family history of psychopathology might influence these normative developmental effects, we also collected a sample of never-depressed daughters of mothers with or without a history of depression (Ethridge et al., 2021). In this sample, we found that the association between delta power and daughters' developmental stage differed depending on maternal risk status. For daughters of mothers who had never been depressed, increased development was associated with increases in delta power following rewards. For daughters of mothers with a history of depression, increased development was associated with *decreased* delta power following rewards, suggesting high-risk daughters may become increasingly vulnerable across the course of adolescence. This effect was not observed for power in the theta frequency.

Work identifying functional differences between delta and theta activation in response to external feedback has laid the foundation for identifying specific cognitive and affective deficits in various psychopathological groups. For instance, there is extensive data suggesting that individuals high on externalizing-proneness (including alcohol and substance use/abuse, rule-breaking, and personality measures) show broad and nonspecific amplitude reductions in multiple ERPs (Hall, Bernat, & Patrick, 2007; Polich et al., 1994). This work has been further clarified by time-frequency decompositions suggesting these individuals do not in fact show deficits related to theta power elicited by loss feedback; instead, these group differences seem to be driven by reductions in delta activity elicited by rewarding feedback (Bernat et al., 2011).

A great deal of work has also examined the FN/RewP in individuals with depression or at risk for depression (including work from our own groups). Multiple previous studies employing monetary guessing tasks have reported that depressed participants as well as those at risk for depression are characterized by abnormal FN/RewPs (Bress, Smith, Foti, Klein, & Hajcak, 2011; Foti, Carlson, Sauder, & Proudfit, 2014; Foti & Hajcak, 2009; Kujawa, Proudfit, & Klein, 2014; Weinberg, Liu, Hajcak, & Shankman, 2015; Weinberg & Shankman, 2016). However, because many of these studies measured the FN/RewP as a difference score, it is unclear whether these results reflect aberrant neural response to rewards, to losses, or to both. Time-frequency investigations have been helpful in clarifying the locus of dysfunction. For instance, in one study we worked on, greater symptoms of depression were associated specifically with more blunted reward-related delta—no such association was found with loss-related theta (Foti et al., 2015). Similarly, another study indicated that blunted delta activity following reward feedback prospectively predicted the onset of depression in an adolescent sample—independently of other risk factors (B. Nelson et al., 2018). We also found that stress exposure—an important predictor of depression—specifically blunted reward-related delta, and not theta, power; and that decreased delta power prior to an acute stressor predicted heightened physiological responses to that stressor (Ethridge et al., 2020).

Additionally, reduced reward-related delta may be helpful in identifying distinct symptom profiles. A recent study found that, within a depressed group, symptoms of anxiety and depression showed dissociable correlations with punishment-elicited theta power and reward-elicited delta power (Cavanagh, Bismark, Frank, & Allen, 2018), such that symptoms of depression were uniquely associated with decreased reward-elicited delta power. In contrast, one study found depression to be associated with blunted midline theta modulation (Mueller, Panitz, Pizzagalli, Hermann, & Wacker, 2015), while Webb and colleagues (2017) found that depressed adolescents showed *increased* loss-related theta, but no differences in gain-related delta. A theme across these studies is that the time-domain FN/RewP generally represents a composite of frontocentral theta- and delta-band activity. The relative contribution of these frequency bands to the observed ERP score, however, will be different across tasks, sample characteristics, and study contexts, such that a change in FN/RewP amplitude can be explained by modulation of theta activity, delta activity, or a combination of both—a question which time-frequency decompositions can answer directly.

5.6 Example: Social Reward Processing

As a practical example of how to apply time-frequency analyses to averaged ERP data, we revisit recently published findings on social reward processing (Ait Oumeziane,

Schryer-Praga, & Foti, 2017). ERP data were collected from 27 adults during a social incentive delay task, a modified version of the commonly used monetary incentive delay task (Knutson, Westdorp, Kaiser, & Hommer, 2000; B. K. Novak, Novak, Lynam, & Foti, 2016; K. D. Novak & Foti, 2015). The task is designed to tease apart anticipatory and consummatory stages of reward processing. On each trial, participants are first presented with a cue indicating whether it is an incentive or neutral trial. On incentive trials, participants have the opportunity to earn a reward by rapidly responding to a target stimulus (and are punished for slow reaction times), whereas on neutral trials participants break even regardless of their reaction time. Following their behavioral response to the target stimulus, participants are then shown feedback indicating the result of that trial (i.e., win or loss on incentive trials, break-even on neutral trials). On the traditional monetary version of the task the rewards are nominal amounts of money, yet more recently a social reward version has been developed in which the "wins" are instead positive social feedback putatively administered by the experimenter. In a recent study, it was shown that social reward feedback elicits a RewP and feedback-P3 (described as such in order to differentiate it from the oddball P3, described earlier) that is of similar morphology and amplitude to monetary reward feedback (Ait Oumeziane et al., 2017). These original analyses focused exclusively on traditional time-domain approaches to scoring the RewP and feedback-P3. Here, we revisit these data using time-frequency analysis.

The time-domain ERP waveforms and scalp topographies are presented in Figure 5.1A. Consistent with previous research, social reward ("wins") elicited a more positive-going waveform than social punishment ("losses"). The RewP difference score (contrasting the valence of uncertain feedback: win vs. loss) peaked at 325 ms, whereas the feedback-P3 difference score (contrasting the salience of uncertain vs. certain feedback: win vs. break-even, loss vs. break-even) peaked at approximately 400 ms. A challenge in isolating the RewP and feedback-P3 in this case, however, is that the waveform to social rewards is modulated within a relatively broad time window of approximately 200–800 ms at centroparietal electrodes, thereby encompassing the P2-RewP-P3-Slow Wave complex. It is unclear whether the difference between conditions represents multiple, overlapping effects or is instead better interpreted as a single, sustained increase in the ERP waveform. These possibilities are difficult to disentangle using traditional time-window averages but can be addressed using time-frequency analysis. Building upon previous time-frequency decompositions of the RewP (Bernat et al., 2011; Foti et al., 2015), we sought to isolate three distinct neural responses:

1. RewP-delta: an increase in delta power to wins vs. losses at approximately 300 ms and at central electrodes;
2. FN-theta: an increase in theta power to losses vs. wins at approximately 300 ms and at frontocentral electrodes; and
3. Feedback-P3-delta: an increase in delta power to wins vs. break-even outcomes at approximately 400 ms and at parietal electrodes.

For illustrative purposes, we focus here on the P3 to wins; a similar pattern of results was also observed for the P3 to losses.

Standard signal processing procedures were applied to the data, including re-referencing to the mastoid electrodes, filtering from 0.1–30 Hz, ocular correction, and artifact rejection (for details, see Ait Oumeziane et al., 2017). The continuous EEG was segmented relative to the onset of feedback stimuli using a relatively broad time window of −1500 to 1500 ms, allowing for edge artifacts (i.e., distortion of the signal near the edges of the window following time-frequency transform; note this is a broader time-window than is typical for most ERP research). ERP averages were created for each of the three conditions (win, loss, break-even), and then the time-frequency transform was applied. Complex Morlet wavelets were calculated within BrainVision Analyzer (Brain Products), using a frequency range of 0.5–20 Hz and linear steps of 0.25 Hz. A relatively narrow frequency range was chosen here due to the a priori focus on activity in the delta and theta bands, whereas a broader range could be chosen for more exploratory analyses. The Morlet parameter was set at c = 3.5 (i.e., 3.5 cycles in the wavelet), and wavelet functions were normalized using Gabor normalization. Baseline correction was performed for each wavelet layer using a baseline window of −500 to −300 ms; the average amplitude in the baseline window was subtracted from each time point in the layer. These time-frequency transforms were applied to the averaged ERP data for each subject and then averaged across subjects to create the grand averaged spectrogram.

Of interest were two contrasts: wins vs. losses, which ought to elicit reward-related delta (RewP) and loss-related theta (FN) from approximately 200–400 ms; and win vs. break even, which ought to elicit delta-band activity in the time range of the feedback-P3 (300–500 ms). The spectrogram for the win vs. loss contrast is presented in Figure 5.2. As expected, activity in the delta frequency band was increased for wins vs. losses (the red portion of the spectrogram within the delta band), which peaked at approximately 2.5 Hz and 295 ms. Overlapping with this response was an increase in activity within the theta frequency band for losses vs. wins (the blue portion of the spectrogram within the theta band), which peaked at approximately 5.25 Hz and at 235 ms. Both of these responses occurred within the time range of the FN/RewP but at distinct frequency bands, presenting the opportunity to extract separate scores for each. We scored the delta and theta responses by extracting wavelet layers centered at 2.5 Hz and 5.25 Hz, respectively. Wavelet layers yield a waveform representing power at that frequency band over time, which can then be scored by taking a time-window average of peak score akin to time-domain ERPs (Figure 5.2, right). We scored each response as the average power within a 50-ms window surrounding the peak of the win minus loss difference, which was somewhat different for each frequency band: 270–320 ms for delta and 210–260 ms for theta. Notably, these wavelets each exhibited more focal scalp topographies than the time-domain FN/RewP, with maxima at frontocentral electrodes. To maintain consistency with the time-domain FN/RewP, we scored the delta and theta responses at electrode Cz. The effects of condition (win vs. loss) were significant for both delta ($t(26) = 3.04$, $p < .01$) and theta ($t(26) = 3.11$, $p < .01$). Critically, the difference scores

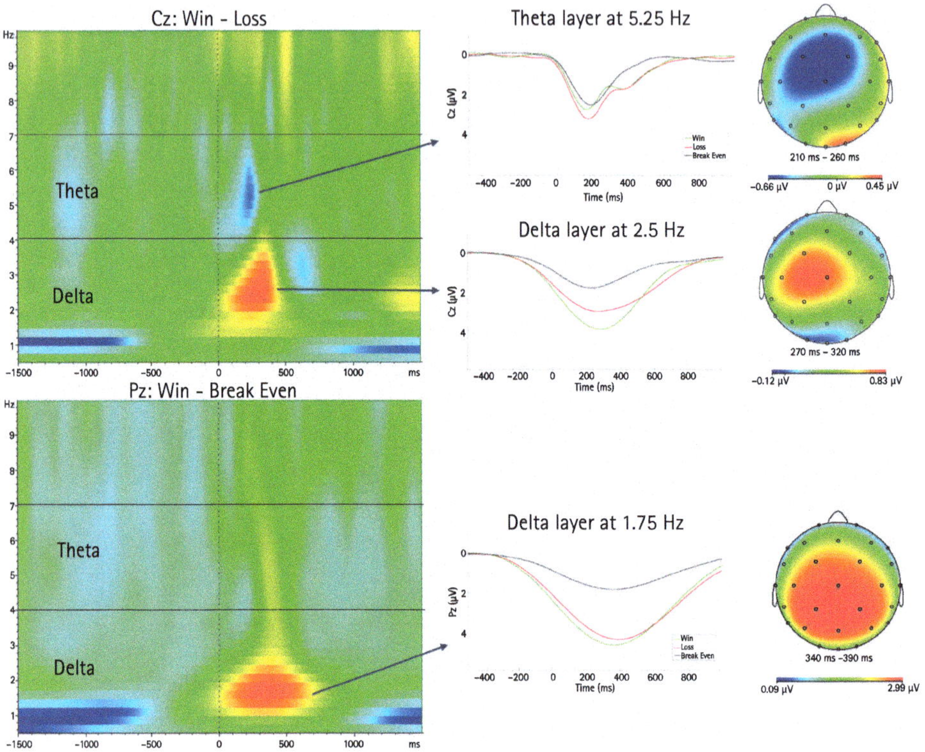

FIGURE 5.2 Time-frequency analysis of the reward positivity elicited by the social incentive delay task. Top: The contrast of wins versus losses, isolating delta- and theta-band activity corresponding to the time-domain FN/RewP. Bottom: The contrast of wins versus break-even outcomes, isolating delta-band activity corresponding to the feedback-P3.

for delta and theta were also both significantly correlated with the time-domain RewP difference score (Table 5.1). As is common in the literature, the delta and theta effects were not significantly correlated with one another, further suggesting that they each capture a distinct portion of the FN/RewP.

The analogous spectrogram for the win vs. break-even contrast is presented in Figure 5.2. In comparison with the FN/RewP analyses, here we expected to find prominent delta-band activity corresponding to the feedback-P3. As seen in the spectrogram, activity in the delta band was indeed increased for wins vs. break-even outcomes, an effect which peaked at 1.75 Hz and at 365 ms. We extracted the wavelet layer at 1.75 Hz and scored this delta response as the average power from 340–390 ms. This wavelet exhibited a scalp topography highly similar to the time-domain feedback-P3, with a peak at centroparietal electrodes. To be consistent with our time-domain analyses, we scored this delta effect at electrode Pz. The increase in delta-band activity for win vs. break-even outcomes was statistically significant ($t(26) = 7.90$, $p < .001$), and the delta-band difference score was significantly correlated with the time-domain feedback-P3 difference score (Table 5.1).

Table 5.1 Correlations between time-domain and time-frequency scores of social reward processing

		Reward Positivity			Feedback-P3	
		Time Domain	Delta (2.5 Hz)	Theta (5.25 Hz)	Time Domain	Delta (1.75 Hz)
FN/RewP	Time Domain	–				
	Delta (2.5 Hz)	**.39**	–			
	Theta (5.25 Hz)	**–.40**	–.04	–		
Feedback-P3	Time Domain	**.47**	.14	–.12	–	
	Delta (1.75 Hz)	.16	.00	–.14	**.71**	–

Note: FN/RewP variables are the contrast of win vs. loss; feedback-P3 variables are the contrast of win vs. break-even. Coefficients are Spearman's rho. Values in bold are significant at $p < .05$.

It is notable that superior separation of the FN/RewP and feedback-P3 was achieved in this case by using time-frequency analyses as compared to traditional time-domain analyses. Specifically, the time-domain FN/RewP and feedback-P3 were significantly correlated with one another despite being scored at non-overlapping time windows (300–350 ms and 370–420 ms) and electrodes (Cz and Pz). This was not the case for the time-frequency measures: the FN-theta and RewP-delta scores were not significantly correlated with the time-domain feedback-P3 score, and P3-delta scores were not significantly correlated with the time-domain FN/RewP scores. We further tested this pattern of specificity using multiple regression, predicting each time-domain ERP from the three time-frequency variables. When entered as simultaneous predictors, time-domain FN/RewP amplitude was significantly predicted by a combination of RewP-delta ($\beta = .46, p < .05$) and FN-theta ($\beta = -.38, p < .05$), but not P3-delta ($\beta = .11, p = .53$). Conversely, time-domain feedback-P3 amplitude was significantly predicted only by P3-delta ($\beta = .66, p < .001$) and not by RewP-delta ($\beta = .15, p= .33$) or FN-theta ($\beta= -.15, p = .28$). Thus, time-frequency analysis facilitated the isolation of three distinct neural signals involved in social reward processing that would have been difficult to achieve solely within the time domain, where signal overlap is more problematic in this experimental context.

The example described shows how time-frequency analyses can aid in the interpretation of traditional ERP effects. In these data, positive social feedback modulated the ERP waveform in a time-range spanning the FN/RewP and P3, which overlapped with each other. The traditional time-domain approach of scoring these ERPs as the average amplitude within separate time windows could not adequately address the problem of component overlap. Time-frequency decomposition, on the other hand, was able to isolate three distinct signals: RewP-delta, FN-theta, and P3-delta. Similar to the time domain, these three signals were all sensitive to outcome type: RewP-delta and FN-theta were modulated by outcome valence, and P3-delta was modulated by outcome certainty. In contrast with the time domain, however, these three signals achieved superior separation: RewP-delta and FN-theta each captured unique portions of the time-domain FN/RewP component, and they were unrelated to the time-domain P3 component. Likewise, P3-delta was strongly related to the time-domain P3 component but was unrelated to the time-domain FN/RewP component. Overall, time-frequency analyses help us rule out the interpretation that social reward broadly increases the ERP waveform in a nonspecific fashion; instead, this broad modulation clearly represents a composite of multiple neural signals. This also lays the groundwork for more precise characterization of individual differences, whereby we would expect specific associations with the time-frequency variables (which are relatively uncorrelated) as compared to the time-domain variables (which have substantial overlap).

Comprehensive descriptions of different methodological approaches are covered in other chapters (e.g., Chapter 4, this volume). However, one important step to consider is whether to conduct these analyses on single-trial or averaged data, as was done in our practical description earlier. As noted, ERPs are typically derived from averages composed of data from multiple trials. This practice reflects the typically low signal-to-noise ratio of ERPs: the amplitude of many ERPs is no more than a few μV, while the amplitude of "background" neuroelectric signals (see chapter 6, this volume), other electrophysiological signals, and electrical interference from non-biological sources can be closer to tens of μV. When averaged together, this presumably randomly distributed noise cancels itself out, whereas the systematic signal, or ERPs, in the data will not, resulting in a legible ERP component. Single-trial analysis of ERP data typically requires relatively large-amplitude components (e.g., the P3). However, time-frequency techniques can in some cases make data more amenable to single-trial analysis, as electrical "noise" often has distinct spectral properties from the signal of interest. For instance, in studies interested in how error or feedback processing might influence trial-level adjustments of behavior, time-frequency techniques could allow the researcher to focus narrowly on theta and delta power on each trial, with higher frequency activity (e.g., alpha, 60-Hz line noise) isolated from these signals. Nonetheless, it is common practice to conduct time-frequency decompositions on averages of many trials. And because averages improve signal-to-noise ratio, they are also a helpful basis for identifying spectral power at different frequency ranges, particularly in exploratory research where the spectral characteristics of ERPs are less well-understood. As is often the case, the optimal method will depend on the research question.

5.7 Conclusions

ERPs are powerful and flexible tools for understanding sensory, cognitive, and affective processes in the brain, but they are not without limitations. As discussed, time-frequency techniques can be helpful in addressing some of these limitations. They can better leverage the multi-dimensional nature of EEG data—and, in principle, better represent the underlying neural signals—by accounting for not only voltage changes over time and site on the scalp, but also frequency, power, and phase, and may reveal multiple dissociable processes folded within the time-window-scored ERP. However, time-frequency techniques are not a magic bullet. As is discussed at length in other chapters, time-frequency decompositions can reduce temporal precision, a chief advantage of the ERP technique (though, as described in Chapter 4, there are techniques to minimize this loss). When applied to trial-averaged ERP data, time-frequency decompositions are best thought of conceptually as isolating potentially relevant ERP subcomponents. That is, it should be possible to "recreate" the pattern of observed ERP findings within the time-frequency domain, perhaps based on a combination of activity across multiple frequency bands. The time-frequency measures may be more precise than their time-domain counterparts to the extent that they isolate the relevant neural signal of interest, but in cases where there is no apparent modulation of the ERP waveform across experimental conditions, it is unlikely that time-frequency approaches will uncover "new" findings. Therefore, we encourage the reader to think of time-frequency decompositions applied to ERP data as the conceptual equivalent of deriving subscales within a self-report questionnaire, which can often enhance the precision of measurement and help clarify the nature of associations with other measures. This is particularly useful where there are already well-established associations between a time-domain ERP and an external measure (e.g., clinical diagnosis, behavior, or personality trait), whereby time-frequency decompositions can help explain the nature of those associations. Thus, time-frequency analyses can be an effective approach for revisiting published data or extending the results of prior studies.

REFERENCES

Ait Oumeziane, B., Schryer-Praga, J., & Foti, D. (2017). "why don't they 'like'me more?": Comparing the time courses of social and monetary reward processing. *Neuropsychologia, 107*, 48–59.

Alho, K. (1995). Cerebral generators of mismatch negativity (MMN) and its magnetic counterpart (MMNm) elicited by sound changes. *Ear and hearing, 16*(1), 38–51.

Arbel, Y., McCarty, K. N., Goldman, M., Donchin, E., & Brumback, T. (2018). Developmental changes in the feedback related negativity from 8 to 14 years. *International Journal of Psychophysiology.*

Aviyente, S., Tootell, A., & Bernat, E. M. (2017). Time-frequency phase-synchrony approaches with ERPs. *International Journal of Psychophysiology, 111*, 88–97.

Baker, T., & Holroyd, C. (2011). Dissociated roles of the anterior cingulate cortex in reward and conflict processing as revealed by the feedback error-related negativity and N200. *Biological Psychology, 87*, 25–34.

Baldeweg, T., Klugman, A., Gruzelier, J. H., & Hirsch, S. R. (2002). Impairment in frontal but not temporal components of mismatch negativity in schizophrenia. *International Journal of Psychophysiology, 43*(2), 111–122.

Baldwin, S. A., Larson, M. J., & Clayson, P. E. (2015). The dependability of electrophysiological measurements of performance monitoring in a clinical sample: A generalizability and decision analysis of the ERN and P e. *Psychophysiology, 52*(6), 790–800.

Başar-Eroglu, C., Başar, E., Demiralp, T., & Schürmann, M. (1992). P300-response: possible psychophysiological correlates in delta and theta frequency channels. A review. *International Journal of Psychophysiology, 13*(2), 161–179.

Bernat, E., Malone, S., Williams, W., Patrick, C., & Iacono, W. (2007). Decomposing delta, theta, and alpha time-frequency ERP activity from a visual oddball task using PCA. *International Journal of Psychophysiology, 64*(1), 62–74.

Bernat, E., Nelson, L., & Baskin-Sommers, A. (2015). Time-Frequency Theta and Delta Measures Index Separable Components of Feedback Processing in a Gambling Task. *Psychophysiology, 52*(5), 626–637.

Bernat, E., Nelson, L., Holroyd, C., Gehring, W., & Patrick, C. (2008). *Separating cognitive processes with principal components analysis of EEG time-frequency distributions.* Paper presented at the Proceedings of SPIE.

Bernat, E., Nelson, L., Steele, V., Gehring, W., & Patrick, C. (2011). Externalizing psychopathology and gain–loss feedback in a simulated gambling task: Dissociable components of brain response revealed by time-frequency analysis. *Journal of abnormal psychology, 120*(2), 352.

Bernat, E., Williams, W., & Gehring, W. (2005). Decomposing ERP time-frequency energy using PCA. *Clinical Neurophysiology, 116*(6), 1314–1334.

Bogdan, R., Santesso, D., Fagerness, J., Perlis, R., & Pizzagalli, D. (2011). Corticotropin-Releasing Hormone Receptor Type 1 (CRHR1) Genetic Variation and Stress Interact to Influence Reward Learning. *The Journal of Neuroscience, 31*(37), 13246–13254.

Bress, J., Smith, E., Foti, D., Klein, D., & Hajcak, G. (2011). Neural response to reward and depressive symptoms in late childhood to early adolescence. *Biological Psychology, 89*(1), 156–162. doi:doi.org/10.1016/j.biopsycho.2011.10.004

Carlson, J., Foti, D., Harmon-Jones, E., Mujica-Parodi, L., & Hajcak, G. (2011). The neural processing of rewards in the human striatum: Correlation of fMRI and event-related potentials. *NeuroImage, 57*, 1608–1616. doi:10.1016/j.neuroimage.2011.05.037

Cavanagh, J. F. (2015). Cortical delta activity reflects reward prediction error and related behavioral adjustments, but at different times. *NeuroImage, 110*, 205–216.

Cavanagh, J. F., Bismark, A. W., Frank, M. J., & Allen, J. J. (2018). Multiple Dissociations between Comorbid Depression and Anxiety on Reward and Punishment Processing: Evidence from Computationally Informed EEG. *Computational Psychiatry*, 1–17.

Cavanagh, J. F., Cohen, M. X., & Allen, J. J. B. (2009). Prelude to and resolution of an error: EEG phase synchrony reveals cognitive control dynamics during action monitoring. *Journal of Neuroscience, 29*(1), 98–105. doi:10.1523/jneurosci.4137-08.2009

Cavanagh, J. F., & Frank, M. J. (2014). Frontal theta as a mechanism for cognitive control. *Trends in cognitive sciences, 18*(8), 414–421.

Cavanagh, J. F., Masters, S. E., Bath, K., & Frank, M. J. (2014). Conflict acts as an implicit cost in reinforcement learning. *Nature communications, 5*, 5394.

Cavanagh, J. F., & Shackman, A. J. (2015). Frontal midline theta reflects anxiety and cognitive control: Meta-analytic evidence. *Journal of Physiology – Paris, 109*, 3–15. doi:10.1016/j.jphysparis.2014.04.003

Cavanagh, J. F., Zambrano-Vazquez, L., & Allen, J. J. (2012). Theta lingua franca: A common mid-frontal substrate for action monitoring processes. *Psychophysiology, 49*(2), 220–238.

Cohen, M. (2014). *Analyzing neural time series data: theory and practice*: MIT press.

Cohen, M., Elger, C., & Ranganath, C. (2007). Reward expectation modulates feedback-related negativity and EEG spectra. *NeuroImage, 35*(2), 968–978.

Di Gregorio, F., Maier, M. E., & Steinhauser, M. (2018). Errors can elicit an error positivity in the absence of an error negativity: Evidence for independent systems of human error monitoring. *NeuroImage, 172*, 427–436. doi:10.1016/j.neuroimage.2018.01.081

Dien, J., Spencer, K. M., & Donchin, E. (2003). Localization of the event-related potential novelty response as defined by principal components analysis. *Cognitive Brain Research, 17*(3), 637–650.

Donchin, E. (1981). Surprise!... surprise? *Psychophysiology, 18*(5), 493–513.

Donchin, E., Ritter, W., & McCallum, W. (1978). Cognitive psychophysiology: The endogenous components of the ERP. *Event-related brain potentials in man*, 349–411.

Duncan-Johnson, C. C., & Donchin, E. (1977). On quantifying surprise: The variation of event-related potentials with subjective probability. *Psychophysiology, 14*(5), 456–467.

Endrass, T., Schuermann, B., Kaufmann, C., Spielberg, R., Kniesche, R., & Kathmann, N. (2010). Performance monitoring and error significance in patients with obsessive-compulsive disorder. *Biological Psychology, 84*, 257–263. doi:10.1016/j.biopsycho.2010.02.002

Ergen, M., Marbach, S., Brand, A., Başar-Eroğlu, C., & Demiralp, T. (2008). P3 and delta band responses in visual oddball paradigm in schizophrenia. *Neuroscience letters, 440*(3), 304–308.

Ethridge, P., & Weinberg, A. (2018). Psychometric Properties of Neural Responses to Monetary and Social Rewards across Development. *International Journal of Psychophysiology, in press.*

Euser, A. S., Evans, B. E., Greaves-Lord, K., Huizink, A. C., & Franken, I. H. A. (2013). Diminished error-related brain activity as a promising endophenotype for substance use disorders: Evidence from high-risk offspring. *Addiction Biology, 18*(970-984). doi:10.1111/adb.12002

Falkenstein, M., Hohnsbein, J., Hoormann, J., & Blanke, L. (1991). Effects of crossmodal divided attention on late ERP components: II. Error processing in choice reaction tasks. *Electroencephalography & Clinical Neurophysiology, 78*(6), 447–455. doi:10.1016/0013-4694(91)90062-9

Foti, D., Carlson, J., Sauder, C., & Proudfit, G. (2014). Reward dysfunction in major depression: Multimodal neuroimaging evidence for refining the melancholic phenotype. *NeuroImage, 101*, 50–58.

Foti, D., & Hajcak, G. (2009). Depression and reduced sensitivity to non-rewards versus rewards. *Biological Psychology, 81*(1), 1–8. doi:10.1016/j.biopsycho.2008.12.004,

Foti, D., Perlman, G., Hajcak, G., Mohanty, A., Jackson, F., & Kotov, R. (2016). Impaired error processing in late-phase psychosis: Four-year stability and relationships with negative symptoms. *Schizophrenia research, 176*(2-3), 520–526.

Foti, D., & Weinberg, A. (2018). Reward and feedback processing: State of the field, best practices, and future directions. In.

Foti, D., Weinberg, A., Bernat, E., & Proudfit, G. (2014). Anterior cingulate activity to monetary loss and basal ganglia activity to monetary gain uniquely contribute to the feedback negativity. *Clinical Neurophysiology.*

Foti, D., Weinberg, A., Bernat, E., & Proudfit, G. (2015). Anterior cingulate activity to monetary loss and basal ganglia activity to monetary gain uniquely contribute to the feedback negativity. *Clin Neurophysiol, 126*(7), 1338–1347. doi:10.1016/j.clinph.2014.08.025

Foti, D., Weinberg, A., Dien, J., & Hajcak, G. (2011). Event-related potential activity in the basal ganglia differentiates rewards from nonrewards: Temporospatial principal components analysis and source localization of the feedback negativity. *Human brain mapping, 32*(12), 2207–2216.

Fuentemilla, L., Marco-Pallarés, J., Münte, T., & Grau, C. (2008). Theta EEG oscillatory activity and auditory change detection. *Brain research, 1220*, 93–101.

Gehring, W., Goss, B., Coles, M. G. H., Meyer, D. E., & Donchin, E. (1993). A neural system for error detection and compensation. *Psychological Science, 4*(6), 385–390. doi:10.1111/j.1467-9280.1993.tb00586.x

Gehring, W., Liu, Y., Orr, J. M., & Carp, J. (Eds.). (2011). *The error-related negativity (ERN/Ne)*. New York, NY: Oxford University Press.

Gehring, W., & Willoughby, A. (2002). The medial frontal cortex and the rapid processing of monetary gains and losses. *Science, 295*(5563), 2279–2282. doi:10.1126/science.1066893

Gehring, W., & Willoughby, A. (2004). Are all medial frontal negativities created equal? Toward a richer empirical basis for theories of action monitoring. *Errors, conflicts, and the brain. Current opinions on performance monitoring, 14*, 20.

Gheza, D., De Raedt, R., Baeken, C., & Pourtois, G. (2018). Integration of reward with cost anticipation during performance monitoring revealed by ERPs and EEG spectral perturbations. *NeuroImage, 173*, 153–164.

Gilmore, C., Malone, S., Bernat, E., & Iacono, W. (2010). Relationship between the P3 event-related potential, its associated time-frequency components, and externalizing psychopathology. *Psychophysiology, 47*(1), 123–132.

Hajcak, G., Moser, J., Holroyd, C., & Simons, R. (2006). The feedback-related negativity reflects the binary evaluation of good versus bad outcomes. *Biological Psychology, 71*(2), 148–154. doi:10.1016/j.biopsycho.2005.04.001

Hall, J. R., Bernat, E. M., & Patrick, C. J. (2007). Externalizing psychopathology and the error-related negativity. *Psychological science, 18*(4), 326–333.

Harper, J., Olson, L., Nelson, L., & Bernat, E. (2011). *Delta-Band Reward Processing Occurring During the Feedback-Related Negativity (FRN)*. Poster presented at annual meeting of the Society for Psychophysiological Research. Boston, MA.

Herrmann, C. S., Rach, S., Vosskuhl, J., & Strüber, D. (2014). Time–frequency analysis of event-related potentials: a brief tutorial. *Brain topography, 27*(4), 438–450.

Hewig, J., Kretschmer, N., Trippe, R., Hecht, H., Coles, M., Holroyd, C., & Miltner, W. (2010). Hypersensitivity to reward in problem gamblers. *Biological psychiatry, 67*(8), 781–783.

Holroyd, C., & Coles, M. (2002). The neural basis of human error processing: Reinforcement learning, dopamine, and the error-related negativity. *Psychological review, 109*(4), 679–709. doi:10.1037//0033-295X.109.4.67

Holroyd, C., Krigolson, O., & Lee, S. (2011). Reward positivity elicited by predictive cues. *Neuroreport, 22*(5), 249.

Holroyd, C., Nieuwenhuis, S., Yeung, N., & Cohen, J. (2003). Errors in reward prediction are reflected in the event-related brain potential. *Neuroreport, 14*(18), 2481–2484. doi:10.10 97/01.wnr.0 0 0 0 0 9 9 6 01.414 03.a

Holroyd, C., Pakzad-Vaezi, K., & Krigolson, O. (2008). The feedback correct-related positivity: sensitivity of the event-related brain potential to unexpected positive feedback. *Psychophysiology, 45*(5), 688–697. doi:10.1111/j.1469-8986.2008.00668.x

Hong, L. E., Moran, L. V., Du, X., O'Donnell, P., & Summerfelt, A. (2012). Mismatch negativity and low frequency oscillations in schizophrenia families. *Clinical Neurophysiology, 123*(10), 1980–1988.

Isreal, J. B., Chesney, G. L., Wickens, C. D., & Donchin, E. (1980). P300 and tracking difficulty: Evidence for multiple resources in dual-task performance. *Psychophysiology, 17*(3), 259–273.

Jeon, Y. W., & Polich, J. (2003). Meta-analysis of P300 and schizophrenia: Patients, paradigms, and practical implications. *Psychophysiology, 40*(5), 684–701.

Kappenman, E. S., & Luck, S. J. (2016). Best practices for event-related potential research in clinical populations. *Biological Psychiatry: Cognitive Neuroscience and Neuroimaging, 1*(2), 110–115.

Kiang, M., Braff, D. L., Sprock, J., & Light, G. A. (2009). The relationship between preattentive sensory processing deficits and age in schizophrenia patients. *Clinical Neurophysiology, 120*(11), 1949–1957.

Knutson, B., Westdorp, A., Kaiser, E., & Hommer, D. (2000). FMRI visualization of brain activity during a monetary incentive delay task. *NeuroImage, 12*(1), 20–27.

Ko, D., Kwon, S., Lee, G.-T., Im, C. H., Kim, K. H., & Jung, K.-Y. (2012). Theta oscillation related to the auditory discrimination process in mismatch negativity: oddball versus control paradigm. *Journal of Clinical Neurology, 8*(1), 35–42.

Kolev, V., Demiralp, T., Yordanova, J., Ademoglu, A., & Isoglu-Alkaç, Ü. (1997). Time–frequency analysis reveals multiple functional components during oddball P300. *Neuroreport, 8*(8), 2061–2065.

Kreussel, L., Hewig, J., Kretschmer, N., Hecht, H., Coles, M., & Miltner, W. (2011). The influence of the magnitude, probability, and valence of potential wins and losses on the amplitude of the feedback negativity. *Psychophysiology, 49*, 207–219. doi:10.1111/j.1469-8986.2011.01291.x

Kujawa, A., Carroll, A., Mumper, E., Mukherjee, D., Kessel, E., Olino, T., ... Klein, D. (2018). A longitudinal examination of event-related potentials sensitive to monetary reward and loss feedback from late childhood to middle adolescence. *International Journal of Psychophysiology, 132*, 323–330.

Kujawa, A., Proudfit, G., & Klein, D. (2014). Neural reactivity to rewards and losses in offspring of mothers and fathers with histories of depressive and anxiety disorders. *Journal of abnormal psychology, 123*(2), 287.

Larson, M. J., South, M., Krauskopf, E., Clawson, A., & Crowley, M. J. (2011). Feedback and reward processing in high-functioning autism. *Psychiatry research, 187*(1-2), 198–203.

Lee, M., Sehatpour, P., Hoptman, M. J., Lakatos, P., Dias, E. C., Kantrowitz, J. T., ... Javitt, D. C. (2017). Neural mechanisms of mismatch negativity dysfunction in schizophrenia. *Molecular psychiatry, 22*(11), 1585.

Leicht, G., Troschütz, S., Andreou, C., Karamatskos, E., Ertl, M., Naber, D., & Mulert, C. (2013). Relationship between oscillatory neuronal activity during reward processing and trait impulsivity and sensation seeking. *PloS one, 8*(12), e83414.

Luck, S. J. (2014). *An introduction to the event-related potential technique*: MIT press.

Lukie, C. N., Montazer-Hojat, S., & Holroyd, C. B. (2014). Developmental changes in the reward positivity: An electrophysiological trajectory of reward processing. *Developmental cognitive neuroscience, 9*, 191–199.

Luu, P., & Tucker, D. M. (2001). Regulating action: Alternating activation of midline frontal and motor cortical networks. *Clinical Neurophysiology, 112*(7), 1295–1306. doi:10.1016/S1388-2457(01)00559-4

Luu, P., Tucker, D. M., Derryberry, D., Reed, M., & Poulsen, C. (2003). Electrophysiological responses to errors and feedback in the process of action regulation. *Psychological Science, 14*(1), 47–53. doi:10.1111/1467-9280.01417

Luu, P., Tucker, D. M., & Makeig, S. (2004). Frontal midline theta and the error-related negativity: Neurophysiological mechanisms of action regulation. *Clinical Neurophysiology, 115*, 1821–1835. doi:10.1016/j.clinph.2004.03.031

Miltner, W., Braun, C., & Coles, M. (1997). Event-related brain potentials following incorrect feedback in a time-estimation task: Evidence for a "generic" neural system for error detection. *Journal of Cognitive Neuroscience, 9*(6), 788–798. doi:10.1162/jocn.1997.9.6.788

Mueller, E. M., Panitz, C., Pizzagalli, D. A., Hermann, C., & Wacker, J. (2015). Midline theta dissociates agentic extraversion and anhedonic depression. *Personality and Individual Differences, 79*, 172–177.

Munneke, G.-J., Nap, T. S., Schippers, E. E., & Cohen, M. X. (2015). A statistical comparison of EEG time- and time-frequency domain representations of error processing. *Brain Research, 1618*(222-230). doi:10.1016/j.brainres.2015.05.030

Näätänen, R. (1995). The mismatch negativity: a powerful tool for cognitive neuroscience. *Ear and hearing, 16*(1), 6–18.

Näätänen, R., Kujala, T., Kreegipuu, K., Carlson, S., Escera, C., Baldeweg, T., & Ponton, C. (2011). The mismatch negativity: an index of cognitive decline in neuropsychiatric and neurological diseases and in ageing. *Brain, 134*(12), 3435–3453.

Näätänen, R., Paavilainen, P., Rinne, T., & Alho, K. (2007). The mismatch negativity (MMN) in basic research of central auditory processing: a review. *Clinical Neurophysiology, 118*(12), 2544–2590.

Nelson, B., Infantolino, Z., Klein, D., Perlman, G., Kotov, R., & Hajcak, G. (2018). Time-frequency reward-related delta prospectively predicts the development of adolescent-onset depression. *Biological Psychiatry: Cognitive Neuroscience and Neuroimaging, 3*(1), 41–49.

Nelson, L., Patrick, C., Collins, P., Lang, A., & Bernat, E. (2011). Alcohol impairs brain reactivity to explicit loss feedback. *Psychopharmacology*, 1–10.

Novak, B. K., Novak, K. D., Lynam, D. R., & Foti, D. (2016). Individual differences in the time course of reward processing: stage-specific links with depression and impulsivity. *Biological Psychology, 119*, 79–90.

Novak, K. D., & Foti, D. (2015). Teasing apart the anticipatory and consummatory processing of monetary incentives: An event-related potential study of reward dynamics. *Psychophysiology, 52*(11), 1470–1482.

Olson, L., Harper, J., Golosheykin, S., Bernat, E., & Anokhin, A. (2011). *Development of the Feedback Negativity in early adolescence: A longitudinal study of time-frequency theta and delta components.* Poster presented at the 2011 meeting of the Society for Psychophysiological Research.

Olvet, D. M., & Hajcak, G. (2008). The error-related negativity (ERN) and psychopathology: Toward an endophenotype. *Clinical Psychology Review, 28*(8), 1343–1354. doi:10.1016/j.cpr.2008.07.003

Polich, J. (2007). Updating P300: an integrative theory of P3a and P3b. *Clinical Neurophysiology, 118*(10), 2128–2148.

Polich, J. (2012). Neuropsychology of P300. *Oxford handbook of event-related potential components*, 159–188.

Polich, J., & Kok, A. (1995). Cognitive and biological determinants of P300: an integrative review. *Biological Psychology, 41*(2), 103–146.

Polich, J., Pollock, V., & Bloom, F. (1994). Meta-analysis of P300 amplitude from males at risk for alcoholism. *Psychological bulletin, 115*(1), 55.

Pornpattananangkul, N., & Nusslock, R. (2016). Willing to wait: Elevated reward-processing EEG activity associated with a greater preference for larger-but-delayed rewards. *Neuropsychologia, 91*, 141–162.

Potts, G., Martin, L., Burton, P., & Montague, P. (2006). When things are better or worse than expected: the medial frontal cortex and the allocation of processing resources. *Journal of Cognitive Neuroscience, 18*(7), 1112–1119.

Riesel, A., Kathmann, N., & Endrass, T. (2014). Overactive performance monitoring in obsessive-compulsive disorder is independent of symptom expression. *European Archives of Psychiatry and Clinical Neuroscience, 264*, 707–717. doi:10.1007/s00406-014-0499-3

Riesel, A., Weinberg, A., Moran, T., & Hajcak, G. (2012). Time course of the error-potentiated startle and its relationship to error-related brain activity. *Journal of Psychophysiology, 27*(2), 51–59. doi:10.1027/0269-8803/a000093

Santesso, D. L., Dzyundzyak, A., & Segalowitz, S. J. (2011). Age, sex and individual differences in punishment sensitivity: Factors influencing the feedback-related negativity. *Psychophysiology, 48*(11), 1481–1489.

Simons, R. F., Graham, F. K., Miles, M. A., & Chen, X. (2001). On the relationship of P3a and the Novelty-P3. *Biological Psychology, 56*(3), 207–218.

Spencer, K. M., Dien, J., & Donchin, E. (2001). Spatiotemporal analysis of the late ERP responses to deviant stimuli. *Psychophysiology, 38*(2), 343–358.

Squires, K., Wickens, C., Squires, N., & Donchin, E. (1976). The effect of stimulus sequence on the waveform of the cortical event-related potential. *Science, 193*(4258), 1142–1146.

Squires, N., Squires, K., & Hillyard, S. (1975). Two varieties of long-latency positive waves evoked by unpredictable auditory stimuli in man. *Electroencephalography and clinical neurophysiology, 38*(4), 387–401.

Steele, V. R., Anderson, N. E., Claus, E. D., Bernat, E. M., Rao, V., Assaf, M., … Kiehl, K. A. (2016). Neuroimaging measures of error-processing: Extracting reliable signals from event-related potentials and functional magnetic resonance imaging. *NeuroImage, 132*, 247–260. doi:10.1016/j.neuroimage.2016.02.046

Tarullo, A. R., Obradović, J., Keehn, B., Rasheed, M. A., Siyal, S., Nelson, C. A., & Yousafzai, A. K. (2017). Gamma power in rural Pakistani children: Links to executive function and verbal ability. *Developmental cognitive neuroscience, 26*, 1–8.

Trujillo, L. T., & Allen, J. J. B. (2007). Theta EEG dynamics of the error-related negativity. *Clinical Neurophysiology, 118*, 645–668. doi:10.1016/j.clinph.2006.11.009

Umbricht, D., & Krljes, S. (2005). Mismatch negativity in schizophrenia: a meta-analysis. *Schizophrenia research, 76*(1), 1–23.

Van Noordt, S. J., Campopiano, A., & Segalowitz, S. J. (2016). A functional classification of medial frontal negativity ERPs: Theta oscillations and single subject effects. *Psychophysiology, 53*(9), 1317–1334.

van Noordt, S. J., Desjardins, J. A., Gogo, C. E., Tekok-Kilic, A., & Segalowitz, S. J. (2017). Cognitive control in the eye of the beholder: electrocortical theta and alpha modulation during response preparation in a cued saccade task. *NeuroImage, 145*, 82–95.

Vidal, F., Hasbroucq, T., Grapperon, J., & Bonnet, M. (2000). Is the 'error negativity' specific to errors? *Biological Psychology, 51*, 109–128. doi:10.1016/S0301-0511(99)00032-0

Watts, A. T., Bachman, M. D., & Bernat, E. M. (2017). Expectancy effects in feedback processing are explained primarily by time-frequency delta not theta. *Biological Psychology, 129*, 242–252.

Watts, A. T., & Bernat, E. M. (2018). Effects of reward context on feedback processing as indexed by time-frequency analysis. *Psychophysiology*, e13195.

Webb, C., Auerbach, R., Bondy, E., Stanton, C., Foti, D., & Pizzagalli, D. (2017). Abnormal neural responses to feedback in depressed adolescents. *Journal of abnormal psychology, 126*(1), 19.

Weinberg, A., & Hajcak, G. (2011). Longer term test–retest reliability of error-related brain activity. *Psychophysiology, 48*(10), 1420–1425.

Weinberg, A., Liu, H., Hajcak, G., & Shankman, S. (2015). Blunted neural response to rewards as a vulnerability factor for depression: Results from a family study. *Journal of abnormal psychology, 124*(4), 878.

Weinberg, A., Liu, H., & Shankman, S. A. (2016). Blunted neural response to errors as a trait marker of melancholic depression. *Biological Psychology, 113*, 100–107. doi:10.1016/j.biopsycho.2015.11.012

Weinberg, A., Olvet, D. M., & Hajcak, G. (2010). Increased error-related brain activity in generalized anxiety disorder. *Biological Psychology, 85*, 472–480. doi:10.1016/j.biopsycho.2010.09.011

Weinberg, A., & Shankman, S. (2016). Blunted reward processing in remitted melancholic depression. *Clinical Psychological Science, in press.*

Wickens, C., Kramer, A., Vanasse, L., & Donchin, E. (1983). Performance of concurrent tasks: a psychophysiological analysis of the reciprocity of information-processing resources. *Science, 221*(4615), 1080–1082.

Yeung, N., Bogacz, R., Holroyd, C. B., & Cohen, J. D. (2004). Detection and synchronized oscillations in the electroencephalogram: An evaluation of methods. *Psychophysiology, 41*, 822–832. doi:10.1111/j.1469-8986.2004.00239.x

Yeung, N., Bogacz, R., Holroyd, C. B., Nieuwenhuis, S., & Cohen, J. D. (2007). Theta phase resetting and the error-related negativity. *Psychophysiology, 44*, 39–49. doi:10.1111/j.1469-8986.2006.00482.x

Yordanova, J., Falkenstein, M., Hohnsbein, J., & Kolev, V. (2004). Parallel systems of error processing in the brain. *NeuroImage, 22*, 590–602. doi:10.1016/j.neuroimage.2004.01.040

CHAPTER 6

THE RELATIONSHIP BETWEEN EVOKED AND INDUCED EEG/MEG CHANGES

Going Beyond Labels

ALI MAZAHERI

6.1 INTRODUCTION

For more than a century now researchers have been examining the electrical potentials and magnetic fields measured at the scalp to understand what is happening inside our brains when we perform various cognitive tasks. Researchers' primary approach is to characterize how the electro/magnetoencephalogram (E/MEG) signal changes in response to a particular "event", whether it be a button press or the onset of an auditory tone. These changes are historically labelled as either *evoked* or *induced*, with each label making assumptions about the origins of the change. The rationale behind evoked activity is that the brain produces a new response as a consequence of processing the event. This response is both time-locked and phase-locked to the experimental event. Induced activity, on the other hand, assumes that the brain has ongoing brain activity (i.e., activity that is always there) independent of any additive activity, and the event modulates this ongoing activity, in a time-locked, but not necessarily phase-locked manner. My hope is that the reader, through what I discuss in this chapter, will understand that these "labels" (while at times useful) can paint an incomplete picture of what is going on in the brain during cognitive processing. According to Andy Warhol, "The moment you label something, you take a step—I mean, you can never go back again to seeing it unlabeled". I further argue that for the field to move forward in gaining a richer understanding of the link between brain and cognition, we need to rethink how we label the different types of EEG responses.

6.2 EVOKED AND INDUCED: THE ASSUMPTION AND LABELS

The evoked potential approach assumes a large component of the electrophysiological signals detected at the scalp is not related to the processing of the phenomena under investigation. Here, we need to average multiple trials (Figure 6.1) of EEG epochs centered around the experimental event to extract the "event"-related EEG signal, called the event-related potential (ERP) for EEG measurements, and event-related fields (ERF) for magnetoencephalography (MEG) measurements. The ERP/F reflects neural activity precisely *time* and *phase-locked* in response to an event. The peaks and troughs in the ERP waveform, which often follow a stereotypical temporal pattern of positive and negative voltage deflections, are classified as "components". Researchers theorize that these components map onto various task-relevant cognitive processes (Kappenman & Luck, 2011).

The evoked potential approach explicitly ignores the ongoing activity present in the EEG, as well as changes that although are time-locked to an experimental event, are not necessarily phase-locked to it. This is because of the rather critical assumption (and one this chapter spends considerable effort arguing against) that non-phase locked activity disappears in the averaging of event locked data epochs due to the deconstructive interference of random phases (Figure 6.1B). Capturing changes to the ongoing activity in EEG, as well as responses that are time-locked but not necessarily phase-locked requires averaging the time-frequency spectra of multiple EEG trials centered on the experimental event (Figure 6.1C). The time-frequency characterization of the ongoing EEG activity works particularly well since the signals contain rhythms, that is, oscillatory activity in characteristic frequency ranges (i.e., bands) including theta (3–7 Hz), alpha (8–13), beta (14–20 Hz), and gamma (30–100 Hz), with each band often exhibiting specific spatial distributions over the scalp (Siegel et al., 2012). The amplitude of an oscillation refers to the size of its (positive or negative) peak relative to some baseline.

The experimentally driven increase in the amplitude of a frequency band is often referred to as an event-related synchronization (ERS). The ERS terminology is based on the fact that when the activity of neurons becomes synchronized, the spatial summation of the post-synaptic potentials results in an amplitude increase (Pfurtscheller & Lopes da Silva, 1999). Conversely, desynchronization of the neuronal population firing results in the cancellation of post-synaptic potentials, and as such, a drop in oscillatory amplitude within a frequency band, sometimes referred to as event-related desynchronization (ERD). Much like the ERP components, the task-related changes

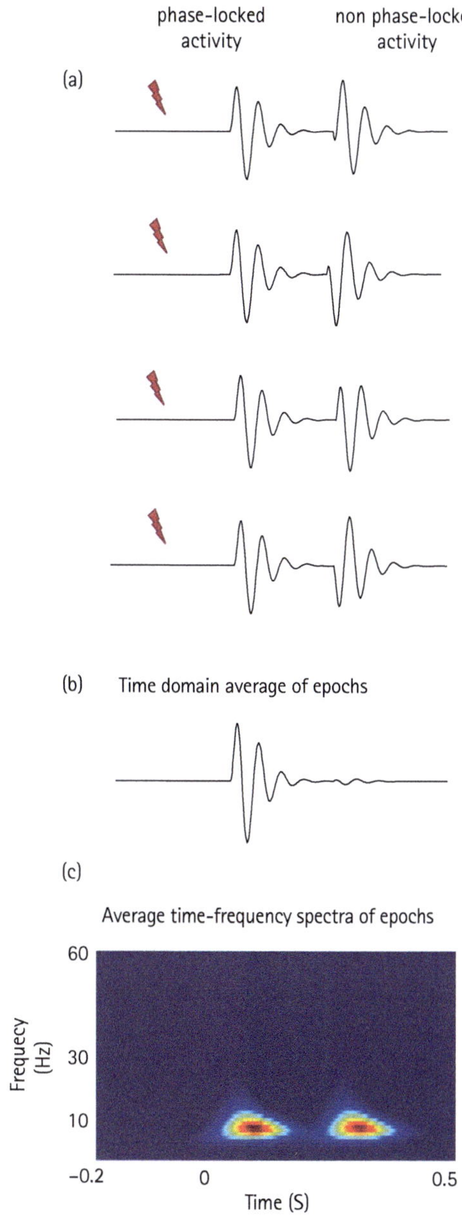

FIGURE 6.1 (A) The onset of an event (e.g., auditory stimulus) can *evoke* activity that is both phase and time-locked to the onset of the event as well as *induce* activity that is time-locked but not phase-locked. (B) Time-domain averaging of multiple data epochs would result in the attenuation of the non-phase locked activity due to destructive interference. The activity remaining after the averaging reflects the brain's transient phase-locked response to an event.

in oscillatory amplitudes map on to different facets of cognition (Siegel et al., 2012; Hanslmayr et al., 2016).

6.3 Going beyond Evoked and induced: the origin of the changes

Besides the methodologically different approaches needed to extract each type of change, evoked and induced activity are also thought to reflect different processes occurring in the brain in response to an outside event. One rather now-classic framework (Pfurtscheller & Lopes da Silva, 1999) proposes that evoked changes (i.e., ERPs) are assumed to occur because of event locked changes of afferent activity into cortical neurons. On the other hand changes in oscillatory power of the ongoing EEG are hypothesized to emerge due to the interaction of neurons and interneurons that control the frequency components of the ongoing activity.

6.4 UNLabeling the labels: the relationship between evoked and ongoing activity

The rather traditional view of evoked and ongoing activity (Figure 6.2) is that they reflect rather separate distinct neural phenomena. According to this view, the evoked activity that always has a consistent phase-locked to the onset of an experimental event rides on top of the ongoing activity. This is also sometimes referred to as the "additive view" of how ERPs are generated. Taken to the extreme, it is possible to view the evoked activity as completely independent of the ongoing activity (figure 6.3A). An alternate theory, referred to as a phase-resetting theory (figure 6.3B), postulates that there is no additive evoked activity elicited by the onset of an event, but that rather, the ongoing activity adjusts its phase to the onset of the experimental event (Makeig et al., 2002). Here, by averaging trials locked to the event, the ongoing activity before the onset of the event which has random phases is averaged out, while the event-related phase perturbed activity emerges as the evoked response.

Given that the predominant ongoing activity in the EEG signal is the alpha rhythm, it is believed that its phase-reset (or adjustment) to the onset of the experimental event plays a particular role in the formation of evoked responses (Makeig et al., 2002; Klimesch et al., 2007; Gruber et al., 2005; Hanslmayr et al., 2006). However, the phase-reset of the ongoing rhythms is not exclusive to the alpha activity, with the theta rhythm

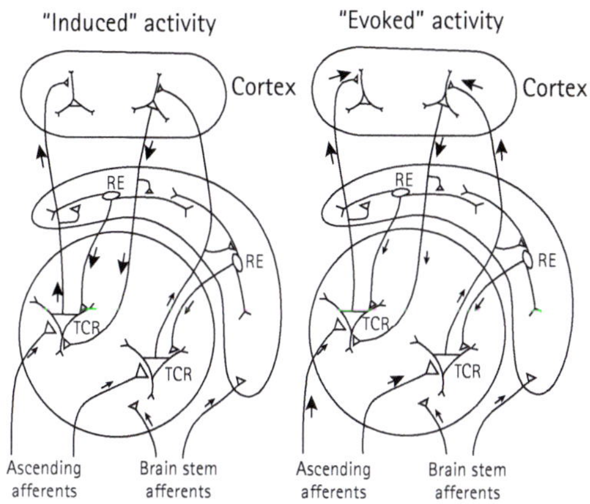

FIGURE 6.2 Schema for the generation of induced (ERD/ERS) and evoked (ERP) activity whereby the former is highly frequency-specific.

Adapted from Pfurtscheller & Lopes da Silva, 1999.

also proposed to be involved in the formation of specific evoked responses such as the error-related negativity (Luu et al., 2004).

There has been a fair amount of controversy over whether phase-resetting can account for the formation of ERPs (Mazaheri & Jensen, 2006). The primary evidence for the occurrence of a phase-reset is that the phase of the ongoing activity at the time of the evoked response would be consistent across trials. However, the addition of a signal with a consistent phase across trials (i.e., a traditional additive evoked response) would also make the phase of the ongoing activity appear consistent across

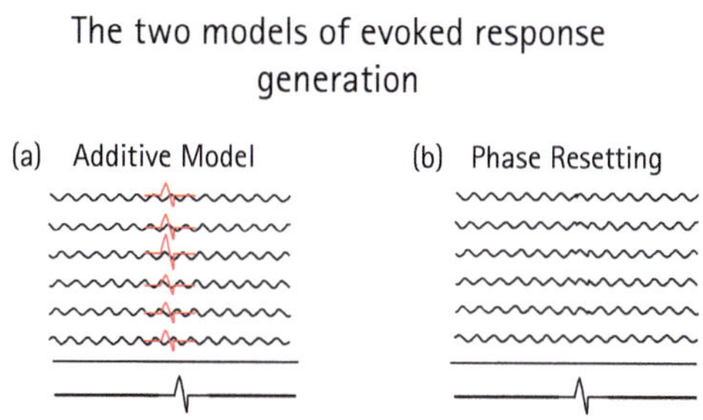

FIGURE 6.3 The additive versus phase-resetting theory of evoked response generation. (A) The additive theory assumes that evoked and ongoing activities are distinct neuronal phenomena. The experimental event "evokes" an additive, phase-locked response in each trial. (B) According to the phase-resetting view, the ongoing and evoked activity are the same neuronal phenomena, with no "new" additive response. Here the phases of the ongoing background oscillations become aligned (phase-reset or partial phase-reset) to an experimental event. The phase-locked (i.e., adjusted) oscillatory activity emerges as an evoked component when averaging the event-locked trials.

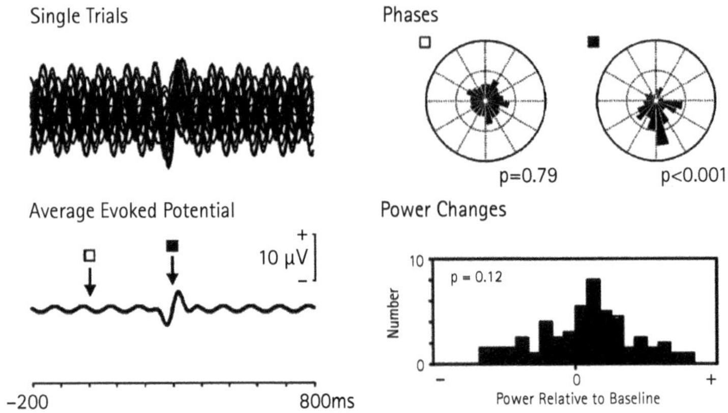

FIGURE 6.4 Model of generation. The upper left of the figure shows 20 superimposed trials of a model evoked potential consisting of a single cycle of activity added to ongoing activity of the same frequency with variable phase and amplitude. Below is given the average evoked potential over 100 trials. At the upper right are the polar plots showing the phase distributions of the frequency of the evoked potential (and background activity) during the baseline and at the middle of the evoked potential. There is significant phase synchronization at the time of the evoked potential. At the bottom is a histogram of the power measurements in the middle of the evoked potential across the 100 trials (because of the Mortlet filtering effect this gives the maximum power). There is no significant change in power.

Reprinted by permission from Mazaheri & Picton, 2005.

trials (Figure 6.4) (Mazaheri & Picton, 2005; van Diepen & Mazaheri, 2018; Yeung et al., 2004).

Moreover, a rather convincing argument has been raised that the additive and phase-resetting model cannot be mathematically distinguished at the scalp level without invasive electrophysiological recordings (Telenczuk et al., 2010).

While the additive and phase-resetting theories offer a contradictory account of evoked and ongoing activity, they do share two common elements. Both theories assume across-trials averaging results in the attenuation of ongoing activity. However, this assumption has been challenged. There is now compelling evidence that the alpha rhythm, the dominant ongoing signal detected at the scalp, is non-sinusoidal, and across-trials averaging never really makes it go away. This observation greatly blurs the line between evoked and ongoing activity. Furthermore, it is important to note that the additive and phase-resetting debate has exclusively focused on the early-stimulus evoked responses (P1, N1, or the ERN) and fails to provide a complete account for the brain responses occurring 200 ms after an experimental event. These sustained responses (Figure 6.5), often lasting 100–200 ms, are believed to reflect neural processing related to high-level cognitive constructs ranging from working memory representation (Ikkai et al., 2010) to language comprehension (Kutas & Federmeier, 2011).

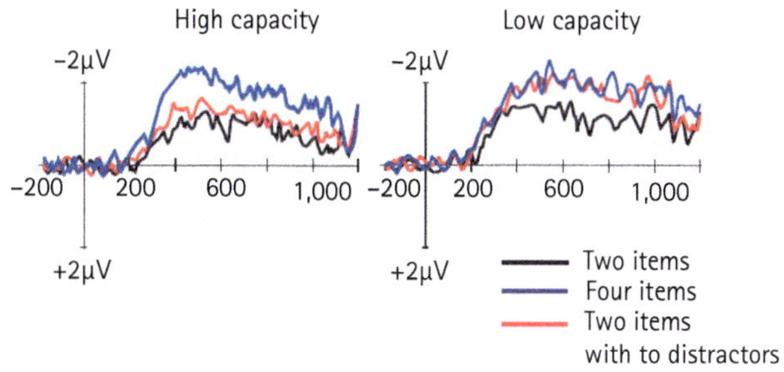

FIGURE 6.5 Grand averaged ERP difference waves (contralateral activity minus ipsilateral activity) timelocked to the memory array averaged across the lateral occipital and posterior parietal electrode sites and divided across the high and low memory capacity groups.

From Vogel et al., 2005.

6.4.1 Amplitude Asymmetry/Baseline Shifts—A Unifying Perspective?

Ongoing activity has been assumed to average out because it has traditionally been viewed to be amplitude symmetric in nature, that is, its peaks and troughs modulate at the same rate (Figure 6.6A).

While we are still seeking a complete understanding of the neural origin of the scalp electrophysiological signals (Cohen, 2017), the general consensus is that they are generated through synchronized post-synaptic current in the dendrites of pyramidal cells (Hämäläinen et al., 1993). Here the EEG reflects the potentials by these currents, while MEG captures the magnetic fields. For an oscillation to have symmetric amplitude fluctuations the intracellular currents propagating forward towards the soma (here let us arbitrarily designate this as the peak of the oscillation) must have the same magnitude as the current coming back from the soma (here assume the trough of the oscillation). However, given the asymmetric placement of channels responsible for the depolarization and repolarization current it is unlikely that the two currents would have the same magnitude when summed up across many synchronized neurons with the same orientation (Mazaheri & Jensen, 2008; Nikulin et al., 2007).

An alternative way to view ongoing activity is that is it amplitude asymmetric, with greater variability in amplitude fluctuations at the peak versus the trough (Figure 6.6B). One critical consequence of amplitude asymmetric ongoing activity is that it will not simply average out to zero when summed across trials. Moreover, any systematic suppression or enhancement of the amplitude of the ongoing activity time-locked to an event would result in the amplitude envelope of the ongoing activity emerging as a slow evoked response when averaging across trials.

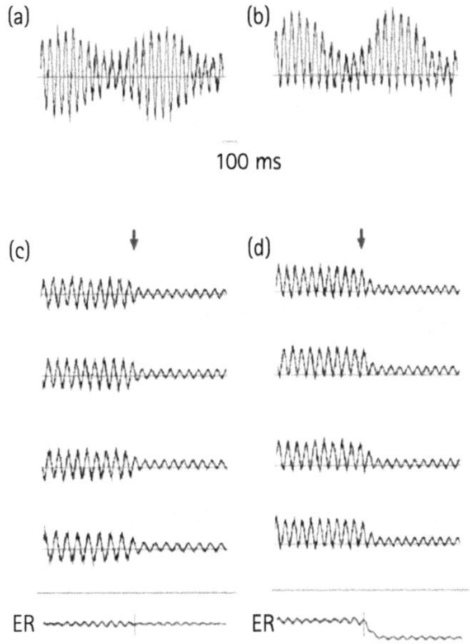

FIGURE 6.6 The amplitude modulation of neuronal oscillatory activity is conventionally viewed as being symmetric at approximately zero. (B) We propose that the amplitude modulations of the oscillatory activity are asymmetric such that the peaks are more strongly modulated than the troughs. For the 10 Hz alpha activity, this could be explained by bouts of activity every ~100 ms. (C) The conventional view ignoring asymmetric modulations of oscillatory activity would mean that averaging across trials (the arrow representing the start of the evoked response) would not result in the generation of slow fields. (D) As a direct consequence of amplitude asymmetry, a depression (or increase) in alpha activity in response to a stimulus will result in the generation of slow fields when multiple trials are averaged. Adapted from Mazaheri and Jensen (2008).

6.4.2 Empirical Evidence Supporting Amplitude Asymmetry

A seminal study by Nikulin and colleagues (2007) provided evidence that the ongoing alpha rhythm is "amplitude asymmetric", specifically referred by them as having a "zero-mean". They went further to propose that a critical consequence of an amplitude asymmetric ongoing rhythm is that any systematic fluctuations in its amplitude would show up as slow responses (they referred to these as baseline shifts) when averaged across trials (Figure 6.7).

Following up Nikulin and colleagues, Mazaheri and Jensen (2008) developed a simple measure to quantify the amplitude of an oscillation by comparing the variance of its peaks with the variance of the troughs (see Figure 6.8).

Moreover, we were able to demonstrate that the degree of amplitude asymmetry of an oscillation is directly related to the amplitude of the evoked response generated by its modulation. Specifically, we presented a simple check-board stimulus across many trials, and then separated the trials into high amplitude of post-stimulus activity, and

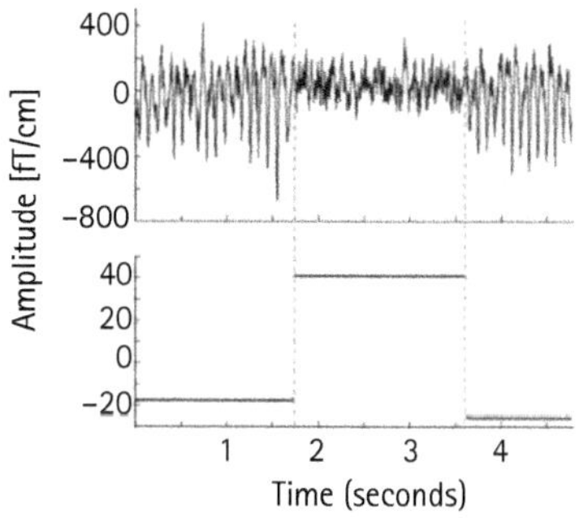

FIGURE 6.7 Baseline shifts in ongoing oscillations. Upper trace: spatially filtered (with independent component analysis) broadband signal from a channel above the right sensorimotor area during rest. Lower trace: the mean values in three time intervals. Clearly, there are baseline shifts in the ongoing activity associated with oscillations changing from large to small and back to large amplitude. If many epochs with similar amplitude dynamics are averaged, oscillatory patterns would disappear whereas the baseline shifts would remain leading to the appearance of an evoked response.

Nikulin et al., 2007.

low amplitude. We found that despite the stimulus being the same, the sorting of the trials based on alpha amplitude resulted in the formation of slow-evoked responses (Figure 6.9). Across participants the amplitude, and polarity of these slow responses was highly correlated with the direction of the amplitude asymmetry of the ongoing alpha activity. Thus we were able to demonstrate (albeit with simple grating stimuli) that it was (in principle) possible to form slow-evoked responses in the trial averaged EEG epochs if there were systematic changes in the amplitude of the ongoing alpha activity.

Mazaheri and Jensen (2010) proposed four prerequisites for linking modulations of oscillatory activity to evoked component generation.

1. The ongoing MEG/EEG oscillations must be modulated in amplitude by the stimuli or event.
2. This amplitude modulation of the ongoing activity must correlate with the time course of the evoked response (over trials or subjects).
3. The ongoing oscillations must have an amplitude asymmetry.
4. The magnitude and/or polarity of the amplitude asymmetry must relate to the amplitude and/or polarity of the evoked responses (over trials or subjects).

6.4.3 Making the Past as Important as the Future

One rather intriguing consequence of having ongoing activity that never averages out is that the amplitude of the pre-event oscillatory activity could modulate the amplitude of the post-event-related potentials, when baseline subtracting the event-related potentials (Figure 6.10).

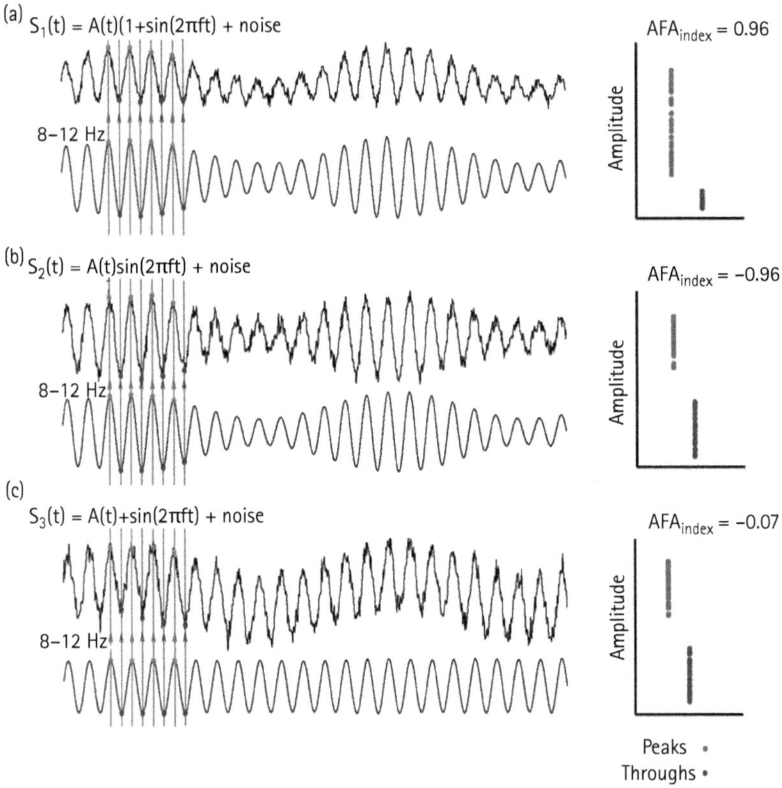

FIGURE 6.8 Various simulations in which surrogate signals were used to test the AFAindex. (A) The signal, s1(t), was designed to have an amplitude asymmetry. The amplitude modulation was determined by a slower signal A(t). Clearly the peaks (red dots) are more modulated than the troughs (blue dots) yielding a strong AFAindex. (B) The signal, s2(t), was designed such that the slow modulations, A(t), affected the alpha rhythm in a multiplicative manner. Thus peaks and troughs are modulated symmetrically over time yielding an AFAindex close to 0. (C) In signal s3(t) slow modulations were added to the alpha oscillations (DC-like offset of the signal). This affected peaks and troughs in the same direction producing an AFAindex close to 0.

Adapted from Mazaheri & Jensen, 2008.

Iemi and colleagues' (2019) comprehensive study systematically examined the relationship of pre-stimulus power of oscillatory, amplitude asymmetry, and the formation of evoked responses. In particular, Iemi and colleagues' (2019) focused on differentiating the impact of pre-stimulus functional inhibition (a sensory state being in a less-responsive state) from amplitude asymmetry on both the early and late sensory evoked responses.

Here, with functional inhibition, the authors were referring to the currently widely held view that an increase in alpha activity in a sensory system reflects its functional inhibition and consequently results in attenuated evoked responses (evidence recently reviewed in Van Diepen et al., 2019). They found that the early evoked (<0.200 s: e.g., the C1/N1 components) were indeed modulated by the amplitude of the pre-stimulus

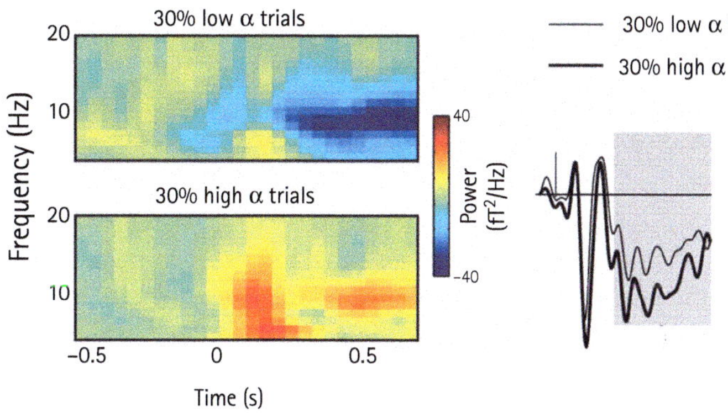

FIGURE 6.9 Time-frequency representations of the trials with the 30% lowest and 30% highest modulations of alpha power (TFRs baseline corrected; $-0.6 < t < -0.1$ s) in a representative subject. The respective ERFs (right) reveal a clear difference in the sustained modulation with respect to low- (thin line) and high-alpha-power changes (thick line).
Adapted from Mazaheri & Jensen, 2008.

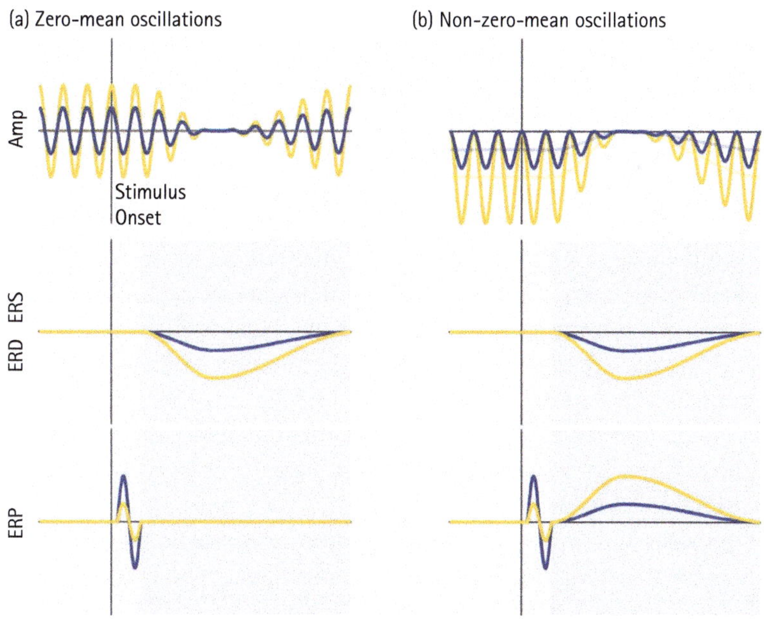

FIGURE 6.10 Ongoing activity (amplitude), event-related oscillations (ERS/ERD) and potentials (ERP) (see text for details).

The vertical line indicates stimulus onset, while the horizontal line indicates zero signal strength. Yellow and blue represent states of strong and weak prestimulus power, respectively. Reprinted from Iemi et al, 2019.

alpha activity, independent of the direction of amplitude asymmetry. However, they found a strong relationship between amplitude asymmetry of the pre-stimulus activity and the late evoked components. These results taken together suggest high pre-stimulus alpha likely causes a suppression of early evoked responses since the neurons producing these responses are in an inhibited stated, while the amplitude of asymmetry property of alpha activity impacts the formation of the later slow evoked responses.

The results discussed so far strongly question the old dogma that ongoing and evoked activity are distinct independent neural phenomena. However, while the studies demonstrate that it is possible to generate slow evoked potentials through modulations of the amplitude of the ongoing activity (without any "new" additive activity), it is still unclear if this mechanism applies to cognitively relevant event-related responses.

6.4.4 *Can* Amplitude Asymmetry Explain the Emergence of the Most Cognitive of ERPs?

As mentioned earlier, the CDA is a slow sustained response proposed to reflect the neural representation of an item in working memory. It is often elicited through a paradigm where participants are presented with a bilateral array of colored squares and instructed to memorize the location of the items in the hemifield indicated by the arrow (i.e., test array). The success of memorizing the items in the test array is then subsequently assessed a second later through the presentation of another array that is either identical to the test array or missing one of the items. The CDA is derived by averaging epochs locked to the onset of the test array and subtracting the contralateral ERPs from the left.

The amplitude of the CDA is modulated by the number of items held in working memory (Vogel et al., 2005) However, the neural origins of the CDA are still rather a mystery. Moreover, the same paradigm has also been found to elicit robust modulations of alpha activity that are also modulated by the number of items held in working memory (Sauseng et al., 2009). In addition, just like the CDA, the degree of lateralized alpha modulation also seems to correlate with the individual differences in working memory. This suggests some overlap between the neural processes underlying the CDA and the alpha modulation.

A study published by Van Dijk and colleagues (2010) explored the link between changes in alpha activity and the CDA, and found them to be quite linked together. Specifically, they observed that both the degree of alpha suppression across individuals, as well as their degree of alpha amplitude asymmetry correlated very strongly with the amplitude of the CDA. Moreover, the alpha modulation and the CDA had a remarkably similar topography over the scalp. These observations taken together could suggest that the CDA and the alpha modulation during the period that the items are held in working memory are one and the same thing.

What are the consequences of re-labelling the CDA as a change in ongoing activity rather than a purely additive response? For one thing, this could have profound implications on how we believe the brain carries out working memory processes. As mentioned, one popular view of the role of alpha modulation in cognition is the suppression of task-irrelevant regions (Van Diepen et al., 2019). Thus, the CDA, rather than being an additive neural process involved in memory maintenance, could instead be reflecting the inhibition of task-irrelevant brain areas. Additionally, unifying ongoing and event-related activity has the potential to mechanistically account for some rather intriguing ERP findings, for which the origins of the responses remain a mystery. For example, a now-classic study (Otten et al., 2006) found that the amplitude of slow event-related potentials locked to the onset of a cue, but peaking before the onset of a word to be remembered, could predict if the word was later remembered. By linking the slow ERPs to the modulation of ongoing alpha activity, one simple interpretation of the observed difference between the remembered versus forgotten words could be that alpha activity is higher (i.e., the brain is in a more inhibited state) prior to the onset of forgotten words. This is indeed in line with several experiments observing pre-stimulus alpha oscillations to modulate perception (van Dijk et al., 2008) as well as reflect slips of sustained attention (Bengson et al., 2012).

6.4.5 *Does* Amplitude Asymmetry Explain the Emergence of Most Cognitive of ERPs?

While I hope that thus far this chapter demonstrates that modulations of ongoing activity that is amplitude asymmetric can produce sustained ERPs, the jury is still out on whether the ongoing and evoked activity, particularly the slow late components, are one and the same. Fukuda and colleagues (2015) challenged this view by observing that, while alpha modulation and CDA are tightly linked, they do appear to uniquely contribute to individual differences between working memory capacity. Specifically, the authors reasoned if the alpha suppression and CDA are two sides of the same neural phenomena they should also show the same relationship to individual differences in working memory performance. However, they found that each signal appeared to uniquely contribute to individual differences in working memory capacity.

More recently, Bae & Luck (2018) went further and used a decoding approach to investigate the specific roles alpha modulation and the slow sustained response could play in attention and working memory. They found modulations in the ongoing alpha activity to be associated with the spatial location of attended stimuli, whereas the amplitude and spatial distribution of the slow-sustained ERPs were sensitive to orientation. Interestingly, they proposed that the ERP and alpha modulation, while serving distinct roles, reflect attentional mechanisms that prevent interference, rather than the actual WM representation.

While these studies certainly do not rule out that the modulation of ongoing activity could be a significant contributor to the formation evoked responses, they do suggest the presence of additive activity involved in WM maintenance. In addition, the mechanism underlying amplitude asymmetry of alpha activity is also applicable to other frequency bands. This means that while the alpha rhythm is the predominant oscillation making up the ongoing activity, there could also be other rhythms present, such as the delta and theta rhythms (Stefanics et al., 2010), whose event-related modulation likely impacts the formation of evoked responses.

6.5 FINAL THOUGHTS

I would certainly not advocate any researchers to dismiss the event-related averaging approach in exchange for looking at changes in the brain's ongoing activity. However, strictly viewing ongoing activity and evoked responses as separate unique entities is implicitly believing the brain was doing nothing before the onset of the experimental event. Such a view is particularly limited when it comes to trying to understand how the brain tries to make sense of the outside world.

As an example of how the brain endeavors to make sense of the world, one rather influential theory proposes that the brain is constantly making predictions about what is going to happen next (reviewed in detail in Friston, 2010). Specifically, this theory, referred to as "predictive coding", postulates that brain sets expectations and predictions about upcoming sensory input and then subsequently updates these expectations after the onset of the sensory input. Here, the discrepancy between the expectation and actual sensory input is referred to as prediction error. While evoked responses can reveal information about the degree of prediction error and the perceived mismatch between expectation and reality, they are not directly informative about the neurophysiology of the predictive processes themselves, since, by definition, the evoked response emerges *after* the sensory input. By removing the separate labels (going back to the Warhol quote) of evoked responses and ongoing activity, it is possible to get a richer, but at the same time more parsimonious, picture of the neural processes underlying cognition.

Finally, I paraphrase Warhol for one last time: I hope some of the mystery behind event-related responses and ongoing activity *is* gone, but the amazement *is* just starting.

REFERENCES

Bae, G.-Y. & Luck, S. J. (2018). Dissociable decoding of spatial attention and working memory from EEG oscillations and sustained potentials. *The Journal of Neuroscience*, *38*(2), 409–422.

Bengson, J. J., Mangun, G. R., & Mazaheri, A. (2012). The neural markers of an imminent failure of response inhibition. *NeuroImage*, *59*(2), 1534–1539.

Cohen, M. X. (2017). Where does EEG come from and what does it mean? *Trends in Neuroscience*, *40*(4), 208–218.

Friston, K. (2010). The free-energy principle: A unified brain theory? *Nature Reviews Neuroscience*, *11*(2), 127–138.

Fukuda, K., Mance, I., & Vogel, E. K. (2015). α-power modulation and event-related slow wave provide dissociable correlates of visual working memory. *The Journal of Neuroscience*, *35*(41), 14009–14016.

Gruber, W. R., Klimesch, W., Sauseng, P., & Doppelmayr, M. (2005). Alpha phase synchronization predicts P1 and N1 latency and amplitude size. *Cerebral Cortex*, *15*(4), 371–377.

Hämäläinen, M., Hari, R., Ilmoniemi, R. J., Knuutila, J., & Lounasmaa, O. V. (1993). Magnetoencephalography: Theory, instrumentation, and applications to noninvasive studies of the working human brain. *Reviews of Modern Physics*, *65*, 413.

Hanslmayr, S., Klimesch, W., Sauseng, P., Gruber, W., Doppelmayr, M., Freunberger, R., Pecherstorfer, T., & Birbaumer, N. (2006). Alpha phase reset contributes to the generation of ERPs. *Cerebral Cortex*, *17*(1), 1–8.

Hanslmayr, S., Staresina, B. P., & Bowman, H. (2016). Oscillations and episodic memory: Addressing the synchronization/desynchronization conundrum. *Trends in Neuroscience*, *39*(1), 16–25.

Iemi, L., Busch, N. A., Laudini, A., Haegens, S., Samaha, J., Villringer, A., & Nikulin, V. V. (2019). Multiple mechanisms link prestimulus neural oscillations to sensory responses. *Elife*, *8*, e43620.

Ikkai, A., McCollough, A. W., & Vogel, E. K. (2010). Contralateral delay activity provides a neural measure of the number of representations in visual working memory. *Journal of Neurophysiology*, 103(4), 1963–1968.

Kappenman, E. S., & Luck, S. J. (Eds.). (2011). *Oxford handbook of event-related potential components*. Oxford University Press.

Klimesch, W., Sauseng, P., Hanslmayr, S., Gruber, W., & Freunberger, R. (2007). Event-related phase reorganization may explain evoked neural dynamics. *Neuroscience & Biobehavioral Reviews*, *31*(7), 1003–1016.

Kutas, M. & Federmeier, K. D. (2011). Thirty years and counting: Finding meaning in the N400 component of the event-related brain potential (ERP). *Annual Review of Psychology*, *62*, 621–647.

Luu, P., Tucker, D. M., & Makeig, S. (2004). Frontal midline theta and the error-related negativity: neurophysiological mechanisms of action regulation. *Clinical Neurophysiology*, *115*(8), 1821–1835.

Makeig, S., Westerfield, M., Jung, T. P., Enghoff, S., Townsend, J., Courchesne, E., & Sejnowski, T. J. (2002). Dynamic brain sources of visual evoked responses. *Science*, *295*(5555), 690–694.

Mazaheri, A. & Jensen, O. (2006). Posterior α activity is not phase-reset by visual stimuli. *Proceedings of the National Academy of Sciences of the United States of America*, *103*(8), 2948–2952.

Mazaheri, A. & Jensen, O. (2008). Asymmetric amplitude modulations of brain oscillations generate slow evoked responses. *The Journal of Neuroscience*, *28*(31), 7781–7787.

Mazaheri, A. & Jensen, O. (2010). Rhythmic pulsing: Linking ongoing brain activity with evoked responses. *Frontiers in Human Neuroscience* [online], *4*(177). doi: 10.3389/fnhum.2010.00177

Mazaheri, A. & Picton, T. W. (2005). EEG spectral dynamics during discrimination of auditory and visual targets. *Brain Research Cognitive Brain Research*, *24*(1), 81–96.

Nikulin, V. V., Linkenkaer-Hansen, K., Nolte, G., Lemm, S., Müller, K. R., Ilmoniemi, R. J., & Curio, G. (2007). A novel mechanism for evoked responses in the human brain. *European Journal of Neuroscience*, *25*(10), 3146–3154.

Otten, L. J., Quayle, A. H., Akram, S., Ditewig, T. A., & Rugg, M. D. (2006). Brain activity before an event predicts later recollection. *Nature Neuroscience*, *9*(4), 489–491.

Pfurtscheller, G. & Lopes da Silva, F. H. (1999). Event-related EEG/MEG synchronization and desynchronization: Basic principles. *Clinical Neurophysiology, 110*(11), 1842–1857.

Sauseng, P., Klimesch, W., Heise, K. F., Gruber, W. R., Holz, E., Karim, A. A., Glennon, M., ... Hummel, F. C. (2009). Brain oscillatory substrates of visual short-term memory capacity. *Current Biology, 19*(21), 1846–1852.

Siegel, M., Donner, T. H. & Engel, A. K. (2012). Spectral fingerprints of large-scale neuronal interactions. *Nature Reviews Neuroscience, 13*(2), 121–134.

Stefanics, G., Hangya, B., Hernádi, I., Winkler, I., Lakatos, P., & Ulbert, I. (2010). Phase entrainment of human delta oscillations can mediate the effects of expectation on reaction speed. *The Journal of Neuroscience, 30*(41), 13578–13585.

Telenczuk, B., Nikulin, V. V. & Curio, G. (2010). Role of neuronal synchrony in the generation of evoked EEG/MEG responses. *Journal of Neurophysiology, 104*(6), 3557–3567.

van Diepen, R. M., Foxe, J. J., & Mazaheri, A. (2019). The functional role of alpha-band activity in attentional processing: The current zeitgeist and future outlook. *Current Opinion in Psychology, 29*, 229–238.

van Diepen, R. M. & Mazaheri, A. (2018). The caveats of observing inter-trial phase-coherence in cognitive neuroscience. *Scientific Reports, 8*(1), 2990.

van Dijk, H., Schoffelen, J. M., Oostenveld, R., & Jensen, O. (2008). Prestimulus oscillatory activity in the alpha band predicts visual discrimination ability. *The Journal of Neuroscience, 28*(8), 1816–1823.

van Dijk, H., van der Werf, J., Mazaheri, A., Medendorp, W. P., & Jensen, O. (2010). Modulations in oscillatory activity with amplitude asymmetry can produce cognitively relevant event-related responses. *Proceedings of the National Academy of Sciences of the United States of America, 107*(2), 900–905.

Vogel, E. K., McCollough, A. W., & Machizawa, M. G. (2005). Neural measures reveal individual differences in controlling access to working memory. *Nature, 438*(7067), 500–503.

Yeung, N., Bogacz, R., Holroyd, C. B., & Cohen, J. D. (2004). Detection of synchronized oscillations in the electroencephalogram: An evaluation of methods. *Psychophysiology, 41*(6), 822–832.

CHAPTER 7

FREQUENCY ANALYSIS OF THE MONKEY NEOCORTICAL LOCAL FIELD POTENTIAL

STEVEN L. BRESSLER

7.1 Introduction

The neocortex of the macaque monkey is very similar to that of the human in its architectonics. The six neocortical laminae show similar variation with region in the monkey and human. In fact, it appears that all mammalian species share a common microstructure, which makes distinguishing neuroanatomical slices from different mammalian species under the microscope nearly impossible. What appears to be more different between the macaque monkey and human neocortices is at the macroscopic level. Overall, the number of neocortical areas is larger in the human brain, and the between-area connectivity is more complex. However, the macroscopic structure of the two species is highly similar in certain systems, for example, the visual system.

The similarity between the human and macaque monkey neocortex was first recognized for the visual system, where neocortical oscillations are a prominent product of the visual architecture of both humans and monkeys. Visual neocortical local field potentials (LFPs) show oscillations in both low-frequency (delta, theta, and alpha) and high-frequency (beta and gamma) bands. Many studies report the importance of the neocortical LFP oscillations in the boundary frequency region between the low beta (13–20 Hz) and alpha (8–12 Hz) frequency bands for top-down neocortical processing in vision (Liang et al., 2002; Bressler et al., 1993; Bressler et al., 2007; Engel & Fries, 2010; Bressler & Richter, 2015; Bastos et al., 2015; Richter et al.,2018), and in the theta (4–7 Hz) and gamma (above >30 Hz) bands for bottom-up visual processing (Markov et al., 2013; Bastos et al., 2015; Michalareas et al., 2016).

A great deal of research has also involved oscillatory activity in the prefrontal cortex of the macaque monkey, which has important analogies to the human prefrontal cortex, which is the most developed of all the mammalian species and is therefore unique. The macaque monkey prefrontal cortex is clearly different from that of the human, but it shares many of the same features. As working memory is an essential component of different cognitive functions requiring the prefrontal cortex in both macaque monkeys and humans (Miller et al., 2018), studies of working memory in the macaque monkey prefrontal cortex are essential for understanding human working memory. Monkey studies first demonstrated a role for prefrontal neocortical oscillatory activity in working memory (Siegel et al.,2009), and these provided the theory that neocortical oscillations are important for working memory processes in the beta-alpha and gamma bands (Pesaran et al., 2002; Salazar et al., 2012; Antzoulatos & Miller, 2016). This speculation has now been verified (discussed later; see also Lundqvist et al.,2016, 2018; Miller et al., 2018).

The study of working memory oscillations in the monkey prefrontal cortex gives impetus to the investigation of prefrontal neocortical oscillations in human working memory (D'Esposito et al., 1995), and recent human studies demonstrate the importance of prefrontal oscillatory activity in human working memory (Jensen et al., 2007; Alekseichuk et al., 2016). Research by Miller and colleagues (2018) involving macaque monkeys verifies that working memory oscillations exist in the theta, alpha-beta, and gamma frequency bands. These oscillations are best studied in monkeys, which demonstrate excellent working memory capability, have associated oscillations that can be studied invasively, and which have a prefrontal neuroanatomy that is similar to humans. Questions about working memory and high-frequency (beta and gamma) oscillations tend to be addressed using monkeys due to the relatively low signal-to-noise ratio at higher frequencies in the human electroencephalogram (EEG) (Crone et al., 2006). The purpose of this report is to present evidence on neocortical oscillations from the macaque monkey engaged in cognitive functions that rely on working memory.

7.2 Dendrites, Oscillations, and Cognition

Transmembrane ionic currents contribute to the extracellular field. Synaptic activity is often the most important physiological source of extracellular current flow. The dendrites and soma of a neuron are treelike and have an electrically conducting interior surrounded by a permeable but insulating membrane. Transmembrane current flows into the dendrites of neocortical neurons at excitatory synapses due to electromotive forces created by neurotransmitter molecules acting on postsynaptic receptor molecules, or out of the synaptic regions at inhibitory synapses. In either case, current flows down the length of the dendritic shaft, across the dendritic membrane, and completes a current loop through the extracellular space. The extracellular current loops give rise to extracellular potential differences that are detected as the LFP (Freeman,

1975; Buzsaki et al., 2012). The MEG is thought to arise from those same currents passing through the dendritic shafts of the same neurons (Murakami & Okada, 2006). Oscillatory activity is a prominent feature of both electric and magnetic signals, higher frequency oscillations likely reflecting interactions of excitatory and inhibitory neurons (Kopell, 2000).

There is growing recognition that the dendrites of neocortical neurons are thus essential for the genesis of neocortical oscillations, and there is speculation that the dynamics of cognitive processing, including working memory (Voytek & Knight, 2015), depends on oscillations. However, the recording and characterization of dendritic activity is problematic. It is usually not possible to record from single dendrites of neurons in the brain due to their thinness, or from the dendritic trees of single neurons due to the lack of extracellular potentials from single-neuron dendrites. Also, neither the single-neuron dendritic branch response nor the single-neuron dendritic tree response can be extracted from the compound dendritic response. For these reasons, much of neurophysiology has focused on the action potential (spike) as the essential neuronal signal. This focus is not because of the neuron doctrine, which, holding only that the neuron is the central signaling cell in the nervous system, is neutral with respect to the parts of the neuron that carry out particular aspects of that signaling.

Dendritic activity, and hence oscillatory activity, is typically recorded at the population level—from groups of neurons rather than from single neurons. In most neurons, the resultant sum of synaptic actions from an entire dendritic tree, contributed to by thousands of synaptic potentials, is delivered to the initial segment of the axon, where it causes the resultant axonal membrane potential to be graded in intensity. The membrane potential of the initial segment has a low threshold for generating a spike because voltage-sensitive Na + membrane channels are concentrated there. The recorded axonal membrane potential may be supra-threshold if the membrane potential is above the threshold, in which case spike trains are generated and travel down the axon, or subthreshold, in which case the axon may be affected but no spike trains are generated. The spike trains may contribute to the neuronal synchronization underlying the LFP (Murthy & Fetz, 1996).

The EEG, electrocorticogram (ECoG), intracranial EEG (iEEG), magnetoencephalogram (MEG), and LFP signals all are generated by the dendritic activity of neuronal populations, and all display oscillations. Except for the MEG, which is magnetic, all these signals are electric and are recorded with respect to a reference potential. The EEG and MEG are usually recorded from outside the cranium, whereas the ECoG, iEEG, and LFP are recorded from inside the cranium. The ECoG is recorded from outside the cortical tissue, whereas the iEEG and LFP are recorded from inside the cortex. The iEEG is recorded from macroscopic electrodes and the LFP is usually recorded with indwelling microelectrodes. The LFP recording may be monopolar, in which case a single microelectrode records the LFP, or it may be bipolar, in which case the LFP is the difference in potential between two nearby microelectrodes. In fact, the LFP is often recorded from the same microelectrodes used to record neuronal spikes, and at the same time (Perelman & Ginosar, 2007; Salazar et al., 2012). However, the LFP typically shows oscillations whereas they are usually not obvious in spike activity.

Since these signal types all reveal oscillatory activity, which is considered by many to be essential for cognition in the brain (Voytek & Knight, 2015), and are typically present when humans perform cognitive tasks, the dendritic activity of neurons in the brain should not be overlooked in the search for neural correlates of cognition, even though it cannot currently be studied in the single neuron. Oscillations in the summed dendritic activity of neuronal populations are strongly related to cognitive function in the neocortex (Donner & Siegel, 2011).

7.3 Phase Coupling and Causality in Sensorimotor Systems

The somatosensory-motor system provides a useful example of neocortical coordination during a cognitive task performed by the macaque monkey. LFP oscillations recorded from somatosensory and motor sites in the macaque monkey neocortex are phase coupled in the beta frequency band during the time in the task that a self-generated hand press cues the monkey by somatosensation that a visual stimulus (0 msec) is soon to appear on a visual display screen. The stimulus is subsequently perceptually discriminated as part of a visual pattern discrimination task requiring hand musculature control (Brovelli et al., 2004). LFPs from specific site pairs in primary and secondary somatosensory cortices, and primary and secondary motor cortex, become beta-phase-coupled (phase synchronized) when the hand press cue is active (Figure 7.1). The presence of beta oscillations in sensorimotor cortex is consistent with the theory that beta oscillations signal maintenance of the current sensorimotor state (Engel & Fries, 2010) since sensorimotor coordination (LFP phase coupling) is likely present throughout this time. The pattern of somatosensory-motor site-pair coupling is consistent with execution of the hand press cue: somatosensory input is fed to the primary somatosensory cortex and motor output is transmitted from the primary motor cortex to the motor spinal cord to execute the hand press. In addition to phase coupling, conditional spectral Wiener–Granger (WG) causality (Ding et al., 2006) was also measured on the same set of sensorimotor LFP recordings. The resulting pattern of sensorimotor neocortical site-pair neural causality also supports a sensorimotor feedback loop for execution of the hand press cue, with influences that are directed to the primary somatosensory cortex from somatosensory inputs, and from there to the primary motor cortex, with feedback loops between primary and secondary somatosensory cortices, and between primary and secondary motor cortices, signaling modulatory influences (Figure 7.1).

The fact that the observed oscillations are in the beta frequency band suggests that sensorimotor neocortical neurons are coordinated in anticipation of making a

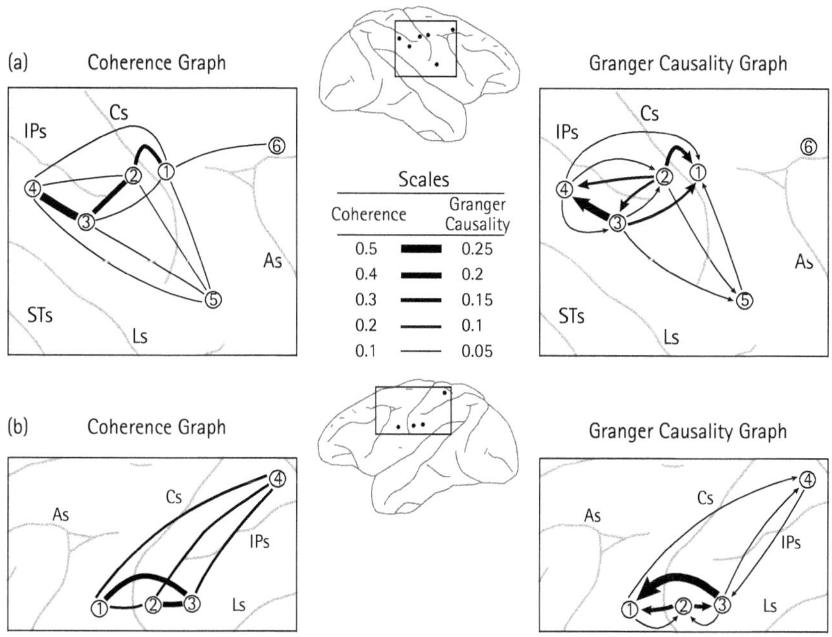

FIGURE 7.1 Prestimulus beta-frequency coherence and (conditional spectral Wiener-) Granger causality graphs derived from the sensorimotor cortices of two monkeys (M1, left hemisphere; M2, right hemisphere). A self-generated hand press cues the monkey that a visual stimulus is soon to appear on a visual display screen. The stimulus is subsequently perceptually discriminated as part of a visual pattern discrimination task (Brovelli et al., 2004). In each case, the pattern of synchronization (coherence) and Granger causality of beta-band oscillations from primary and secondary somatosensory and motor cortices is consistent with execution of the hand press cue: somatosensory input is fed to the primary somatosensory cortex and motor output is transmitted from the primary motor cortex to the motor spinal cord to execute the hand press.

sensorimotor decision in similar manner to the coordination of motor neocortical neurons during steady muscle contractions and steady holding periods following phasic movements as reported by numerous literature studies (e.g., Engel & Fries, 2010). Thus, when the monkey makes a hand press on the lever, sensorimotor neocortical neurons are coordinated in anticipation of the monkey making the subsequent sensorimotor decision. The observation of conditional spectral WG causality from neurons at one sensorimotor neocortical site to those at another site appears to indicate more than the observation of phase coupling alone by suggesting that sensorimotor neocortical neurons causally influence each other as part of that decision. Finally, the reported results are consistent with accumulating evidence that oscillations in the beta frequency band have a distinct physiological role and that they provide an effective means of controlling spike timing, thereby coordinating information transfer across brain regions and supporting spike-timing dependent plasticity (Engel & Fries, 2010).

7.4 Phase Coupling and Causality in Visual Neocortex

LFPs, simultaneously recorded from the striate and extrastriate (V4—visual area 4 and TEO—temporal-occipital area) visual cortices prior to appearance of the visual stimulus in the same visual pattern discrimination task (Bressler & Richter, 2015), also show oscillatory activity in the beta frequency range. The same phase-coupling and neural causality metrics have been computed from the visual LFPs as from somatosensory-motor LFPs. The main finding is that extrastriate–striate site pairs are beta-frequency phase coupled and carried by strong top-down (extrastriate-to-striate) beta influences, in anticipation of visual processing (Figures 7.2 and 7.3). Furthermore, behavioral context is conveyed to primary visual cortex prior to appearance of the visual stimulus (Richter et al., 2018). Thus, information about the visual stimulus appears in the visual cortex

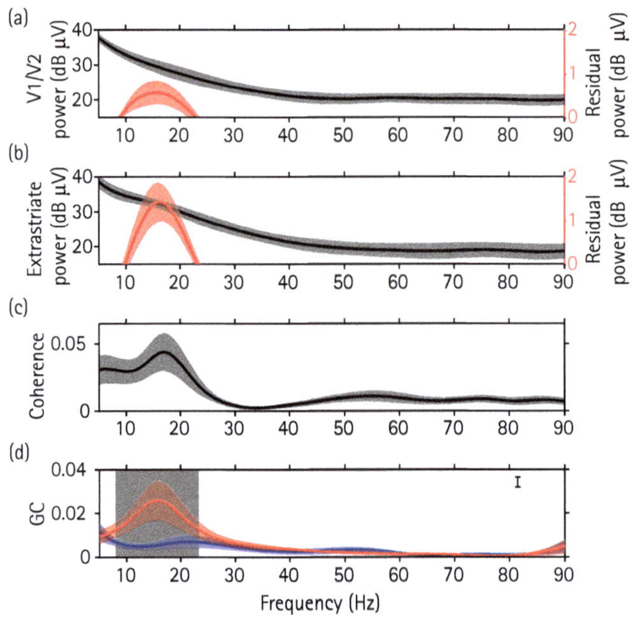

FIGURE 7.2 Prestimulus beta-frequency power, coherence, and (conditional spectral Wiener-Granger causality spectra of V1/V2 and V4/TEO LFPs. (A) Average striate power spectrum over sites (black line ± s.e.m.), and the residual power spectrum after 1/f removal (red line ± s.e.m.) for V1 sites. (B) Average extrastriate power spectrum over sites and monkeys (black line ± s.e.m.), and the residual power spectrum after 1/f removal (red line ± s.e.m.) for the V4/TEO sites. (C) Average coherence spectrum over V1/V2-extrastriate site pairs ± s.e.m. for V1/V2-extrastriate pairs. (D) Average top-down (red line ± s.e.m.), and bottom-up (blue line ± s.e.m.) GC spectra for V1-extrastriate pairs. Shaded grey rectangular region denotes the frequencies (8–23 Hz) where top-down and bottom-up sGC were significantly different (p<0.001).

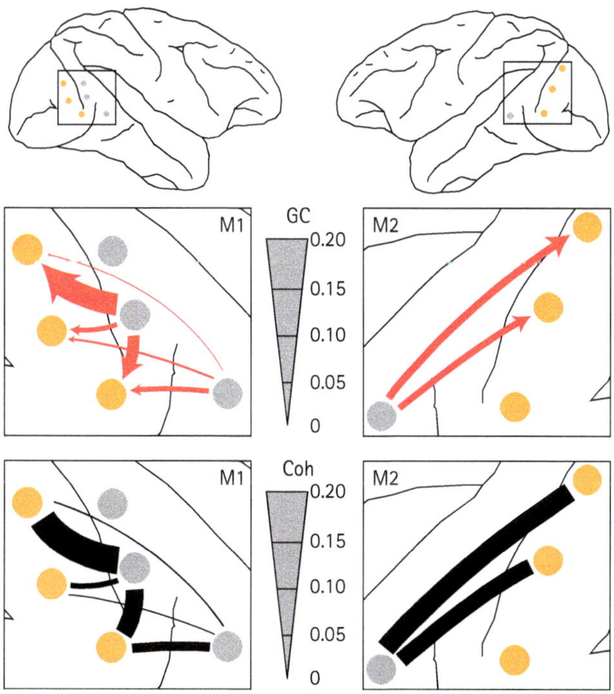

FIGURE 7.3 Prestimulus beta-frequency coherence and top-down (conditional spectral Wiener-)Granger causality maps. Top: Maps of the recording sites for M1 and M2. V1/V2 electrode locations are marked by yellow circles, and extrastriate (V4 and TEO) locations by gray circles. Middle: enlarged maps of visual cortex showing top-down sGC at 16 Hz as arrows for V1/V2-extrastriate pairs. Bottom: corresponding maps of coherence for the same site pairs. Thickness of the top-down sGC arrows and coherence bars is proportional to the magnitude of sGC or coherence at 16 Hz.

before the actual onset of the visual stimulus. This result validates previous proposals that top-down visual processing depends on interareal synchronization in visual neocortex (von Stein et al., 2000), and suggests that it acts to "prime" visual cortex to prepare it for receiving the visual stimulus.

The same basic methodology, namely conditional spectral WG causality analysis, has subsequently been applied to visual neocortical LFPs in macaque monkeys performing a visuospatial attention task (Bastos et al., 2015). In the visual system, neocortical areas are seen to interact in both bottom-up and top-down directions, with bottom-up gamma-band influences conveying sensory signals, and top-down beta-band influences modulating those bottom-up influences according to behavioral context. In the macaque monkey visual neocortex, bottom-up influences are carried by theta-band (~4 Hz) and gamma-band (~60–80 Hz) synchronization, and top-down influences by beta-band (~14–18 Hz) synchronization (Bastos et al. 2015). Furthermore, neocortical hierarchies (Felleman & Van Essen, 1991; Hilgetag et al., 1996) created based on asymmetries in these directional influences are similar to those based on asymmetries in neuroanatomical connections (Markov et al., 2014).

Effective interareal interaction in the visual system depends on local (within-area) synchronization of oscillations that are layer specific. Supragranular layers show local gamma-band synchronization, whereas infragranular layers show local alpha-/beta-band synchronization (Buffalo et al., 2011; Xing et al., 2012; Roberts et al., 2013). Since supragranular layers primarily send bottom-up projections, and infragranular layers primarily send top-down projections, it is proposed that interareal synchronization in the gamma band mediates bottom-up influences and that the beta band mediates top-down influences (Wang, 2010; Bastos et al., 2012; van Kerkoerle et al., 2014).

Metrics of the bottom-up or top-down character of interareal connections have been used to create neocortical hierarchies. A functional metric of directed influence is computed as the *d*irected influence *a*symmetry *i*ndex (DAI), based on the conditional spectral WG causality in bottom-up and top-down directions. A neuroanatomical metric of directional influence is computed as the proportion of supragranular labeled neurons to the sum of supragranular and infragranular labeled neurons (SLN). The DAI and SLN metrics quantify the degree to which an interareal projection is top-down or bottom-up. These hierarchies are generally similar whether created from conditional spectral WG causality or from neuroanatomical metrics. The functional hierarchies further demonstrate that bottom-up influences utilize theta and gamma bands, whereas top-down influences utilize the beta band. If directional influences represent interareal communication, and, as has been speculated (Bastos et al., 2015), increasing oscillation frequency entails increasing communication throughput, then gamma-band communication might be expected for bottom-up influences, since they are expected to require higher-throughput communication.

The finding of top-down beta influences in the visual system (Bastos et al., 2015) is consistent with Richter and colleagues' (2018) results that top-down beta signaling conveys behavioral context to the visual cortex, and with the general conclusion that beta-band synchrony signals the "status-quo" in neocortex (Engel & Fries, 2010). In fact, a distributed network of phase-coupled beta-band oscillations might easily signal predicted sensory and motor events (Bastos et al., 2012; Bressler & Richter, 2015).

7.5 Oscillations in Prefrontal Neocortex

Monkey LFP oscillations have been reported during working memory in the theta, alpha, beta, and gamma bands. The gamma band is associated with sensory information, and gamma-band power correlates with the number of objects held in working memory, whereas the beta band is associated with top-down information, and beta-band synchrony correlates with task rules (Liang et al., 2002; Miller et al., 2018; Richter et al., 2018). Gamma bursting is anti-correlated with alpha-beta bursting (Lundqvist et al., 2016).

Information in prefrontal neuronal spiking has been linked to brief bursts in the gamma band in monkeys performing a working memory task (Lundqvist et al., 2016). Prefrontal laminar LFP data has been obtained from monkeys using prefrontal laminar probes (Bastos et al., 2018). Gamma-band activity is strongest in the superficial layers, and beta and alpha LFP power is strongest in the deep layers. In keeping with previous studies (Lundqvist et al., 2016), gamma-band bursting is most informative about working memory in the superficial layers.

Gamma-band bursts, varying in time and frequency, were reported to accompany both the encoding and re-activation of sensory information (Bastos et al., 2015). The conclusion is that gamma bursts could gate access to, and prevent sensory interference with, working memory, since only the neuronal activity that was associated with working memory encoding and decoding was correlated with gamma-band burst rate changes. Bursts in the beta band were also brief and variable, but they reflected a "default state" that was interrupted by encoding and decoding.

7.6 Phase Coupling and Causality in the Fronto-Parietal Network

The prefrontal cortex is not the only region of neocortex involved in working memory. To test the involvement of posterior parietal cortex in working memory, LFPs were recorded from distributed sites in the prefrontal cortex and in the posterior parietal cortex of macaque monkeys performing a working memory delayed-match-to-sample visual identity task (Salazar et al., 2012). Phase coupling between prefrontal and posterior parietal cortices, examined by computing spectral coherence between prefrontal-posterior parietal LFP pairs, was found in the beta frequency band. Analysis of prefrontal-posterior parietal interactions in working memory by conditional spectral WG causality applied to the delay period of the delayed match-to-sample visual identity task showed that the beta frequency band was dominant. Causal influences in the beta band were roughly balanced between the two directions, that is, beta causal influences in the prefrontal-to-posterior parietal and posterior parietal-to-prefrontal directions were roughly the same (slightly greater from posterior parietal cortex to prefrontal cortex). This finding indicates that prefrontal and posterior parietal cortices are roughly in balance during working memory task performance.

7.7 Conclusions

This chapter focused on neocortical oscillations in the macaque monkey related to cognitive processes in visual pattern discrimination, visuospatial attention, and visual working memory tasks. In visual processing, beta-frequency oscillatory influences predominate in the top-down direction prior to arrival of the visual stimulus, suggesting that beta-frequency oscillations convey task rule-related information from higher to lower areas of visual neocortex for visual processing during vision.

In visual working memory, the prefrontal cortex and posterior parietal cortex in both monkeys and humans are expected to be linked together as parts of the fronto-parietal network (Salazar et al., 2012; Alekseichuk et al., 2016). Beta-frequency oscillations are found to link posterior parietal and prefrontal cortex, with slightly greater causal influence from the posterior parietal cortex to the prefrontal cortex. Oscillations in the gamma and theta frequency bands are also prominent in neocortex in working memory. The gamma band is associated with sensory information in monkey working memory. The theta band is associated with human working memory (Raghavachari et al., 2001). Potential topics for future investigation are the presence of gamma oscillations in the human prefrontal cortex during working memory (Howard et al., 2003), and their relation to theta oscillations (Jokisch & Jensen, 2007).

Oscillations in different frequency bands are exhibited in the monkey neocortex in relation to cognitive function. These oscillations are similar to those in humans, and therefore should be considered when trying to understand the role of oscillations in human cognition. Prominent among the frequency bands at play in cognition involving visual and prefrontal systems are theta, alpha-beta, and gamma. These bands are foremost among those found in visual cortex, prefrontal cortex, and posterior parietal cortex in relation to visual pattern discrimination and visual working memory.

References

Alekseichuk, I., Turi, Z., Amador de Lara, G., Antal, A., & Paulus, W. (2016). Spatial working memory in humans depends on theta and high gamma synchronization in the prefrontal cortex. *Current Biology, 26*, 1513–1521.

Antzoulatos, E. & Miller, E. (2016). Synchronous beta rhythms of frontoparietal networks support only behaviorally relevant representations. *eLife, 5*, e17822.doi: 10.7554/eLife.17822

Bastos, A., Litvak, V., Moran, R., Bosman, C., Fries, P., & Friston, K. (2015). A DCM study of spectral asymmetries in feedforward and feedback connections between visual areas V1 and V4 in the monkey. *Neuroimage, 108*, 460–475.

Bastos, A., Loonis, R., Kornblith, S., Lundqvist, M., & Miller, E. (2018). Laminar recordings in frontal cortex suggest distinct layers for maintenance and control of working memory. *Proceedings of the National Academy of Sciences of the United States of America, 115*, 1117–1122.

Bastos, A., Usrey, W., Adams, R., Mangun, G., Fries, P., & Friston, K. (2012). Canonical microcircuits for predictive coding. *Neuron, 76*, 695–711.

Bressler, S., Coppola, R., & Nakamura, R. (1993). Episodic multiregional cortical coherence at multiple frequencies during visual task performance. *Nature, 366*, 153–156.

Bressler, S., Richter, C., Chen, Y., & Ding, M. (2007). Cortical functional network organization from autoregressive modeling of local field potential oscillations. *Statistics in Medicine, 26*, 3875–3885.

Bressler, S. & Richter, C. (2015). Interareal oscillatory synchronization in top-down neocortical processing. *Current Opinion in Neurobiology, 31C*, 62–66.

Brovelli, A., Ding, M., Ledberg, A., Chen, Y., Nakamura, R., & Bressler, S. (2004). Beta oscillations in a large-scale cortical network: Directional influences revealed by Granger causality. *Proceedings of the National Academy of Sciences of the United States of America, 101*, 9849–9854.

Buffalo, E., Fries, P., Landman, R., Buschman, T., & Desimone, R. (2011). Laminar differences in gamma and alpha coherence in the ventral stream. *Proceedings of the National Academy of Sciences of the United States of America, 108*, 11262–11267.

Buzsaki, G., Anastassiou, C., & Koch, C. (2012). The origin of extracellular fields and currents—EEG, ECoG, LFP and spikes. *Nature Reviews Neuroscience, 13*, 407–420.

Crone, N., Sinai, A., & Korzeniewska, A. (2006). High-frequency gamma oscillations and human brain mapping with electrocorticography. *Progress in Brain Research, 159*, 275–295.

D'Esposito M, Detre J, Alsop D, Shin R, Atlas S, Grossman M (1995) The neural basis of the central executive system in working memory. *Nature, 378*, 279–281.

Ding, M., Chen, Y., & Bressler, S. (2006). Granger causality: Basic theory and application to neuroscience. In B. Schelter, M. Winterhalder, & J. Timmer (Eds.), *Handbook of time series analysis: Recent theoretical developments and applications*, (pp. 437–460). Wiley.

Donner, T. & Siegal, M. (2011). A framework for local cortical oscillation patterns. *Trends in Cognitive Science, 15*, 191–199.

Engel, A. & Fries, P. (2010). Beta-band oscillations – signaling the status quo? *Current Opinion in Neurobiology, 20*, 156–165.

Felleman, D. & Van Essen, D. (1991). Distributed hierarchical processing in the primate cerebral cortex. *Cerebral Cortex, 1*, 1–47.

Freeman, W. (1975). *Mass action in the nervous system*. Academic.

Hilgetag, C., O'Neill, M., & Young, M. (1996). Indeterminate organization of the visual system. *Science, 271*, 776–777.

Howard, M., Rizzuto, D., Caplan, J., Madsen, J., Lisman, J., Aschenbrenner-Scheibe, R., Schulze-Bonhage, A., & Kahana, M. (2003). Gamma oscillations correlate with working memory load in humans. *Cerebral Cortex, 13*, 1369–1374.

Jensen, O., Kaiser, J., & Lachaux, J. (2007). Human gamma-frequency oscillations associated with attention and memory. *Trends in Neuroscience, 30*, 317–324.

Jokisch, D. & Jensen, O. (2007). Modulation of gamma and alpha activity during a working memory task engaging the dorsal or ventral stream. *Journal of Neuroscience, 27*, 3244–3251.

Kopell, N. (2000). We got rhythm: Dynamical systems of the nervous system. *Notices of the American Mathematical Society, 47*, 6–16.

Liang, H., Bressler, S., Ding, M., Truccolo, W., & Nakamura, R. (2002). Synchronized activity in prefrontal cortex during anticipation of visuomotor processing. *Neuroreports, 13*, 2011–2015.

Lundqvist, M., Rose, J., Herman, P., Brincat, S., Buschman, T., & Miller, E. (2016). Gamma and beta bursts underlie working memory. *Neuron, 90*, 152–164.

Lundqvist, M., Herman, P., Warden, M., Brincat, S., & Miller, E. (2018). Gamma and beta bursts during working memory readout suggest roles in its volitional control. *Nature Communications, 9*, 394.

Markov, N., Ercsey-Ravasz, M., Van Essen, D., Knoblauch, K., Toroczkai, Z., & Kennedy, H. (2013). Cortical high-density counterstream architectures. *Science, 342*, 1238406.

Markov, N., Vezoli, J., Chameau, P., Falchier, A., Quilodran, R., Huissod, C., Lamy, C., Misery, P., ... Kennedy, H. (2014). Anatomy of hierarchy: Feedforward and feedback pathways in macaque visual cortex. *Journal of Comparative Neurology, 522*, 225–229.

Michalareas, G., Vezoli, J., Van Pelt, S., Schoffelen, J-M., Kennedy, H., & Fries, P. (2016). Alpha-beta and gamma rhythms subserve feedback and feedforward influences among human visual cortical areas. *Neuron, 46*, 60528.

Miller, E., Lundqvist, M., & Bastos, A. (2018). Working memory 2.0. *Neuron, 100*, 463–475.

Murakami, S. & Okada, Y. (2006). Contributions of principal neocortical neurons to magnetoencephalography and electroencephalography signals. *Journal of Physiology, 575*, 925–936.

Murthy, V. & Fetz, E. (1996). Synchronization of neurons during local field potential oscillations in sensorimotor cortex of awake monkeys. *Journal of Neurophysiology, 76*, 3968–3982.

Perelman, Y. & Ginosar, R. (2007). An integrated system for multichannel recording with spike/LFP separation, integrated A/D conversion, and threshold detection. *IEEE Transactions on Biomedical Engineering, 54*, 130–137.

Pesaran, B., Pezaris, J., Sahani, M., Mitra, P., & Anderson, R. (2002). Temporal structure in neuronal activity during working memory in macaque parietal cortex. *Nature Neuroscience, 5*, 805–81.

Raghavachari, S., Kahana, M., Rizzuto, D., Caplan, J., Kirschen, M., Bourgeois, B., Madsen, J., & Lisman, J. (2001). Gating of human theta oscillations by a working memory task. *Journal of Neuroscience, 21*, 3175–3183.

Richter, C., Copolla, R., & Bressler, S. (2018). Top-down beta oscillatory signaling conveys behavioral context in early visual cortex. *Scientific Reports, 8*, 6991.

Roberts, M., Lowet, E., Brunet, N., Ter Wal, M., Tiesinga, P., Fries, P., & De Weerd, P. (2013). Robust gamma coherence between macaque V1 and V2 by dynamic frequency matching. *Neuron, 78*, 523–536

Salazar, R., Dotson, N., Bressler, S., & Gray, C. (2012). Content-specific fronto-parietal synchronization during visual working memory. *Science, 338*, 1097–1100

Siegel, M., Warden, M., & Miller, E. (2009). Phase-dependent neuronal coding of objects in short-term memory. *Proceedings of the National Academy of Sciences of the United States of America, 106*, 21341–21346

Van Kerkoerle, T., Self, M., Dagnino, B., Gariel-Mathis, M-A., Poort, J., van der Togt, C., & Roelfsema, P. (2014). Alpha and gamma oscillations characterize feedback and feedforward processing in monkey visual cortex. *Proceedings of the National Academy of Sciences of the United States of America, 111*, 14332–14341.

von Stein, A., Chiang, C., & Konig, P. (2000). Top-down processing mediated by interareal synchronization. *Proceedings of the National Academy of Sciences of the United States of America, 97*, 14748–14753.

Voytek, B. & Knight, R. (2015). Dynamic network communication as a unifying neural basis for cognition, development, aging, and disease. *Biol Psych 77*, 1089–1097.

Wang, X-J. (2010). Neurophysiological and computational principles of cortical rhythms in cognition. *Physiological Reviews, 90*, 1195–1268.

Xing, B., Guo, J., Meng, X., Wei, S-G., & Li, S-B. (2012). The dopamine D1 but not D3 receptor plays a fundamental role in spatial working memory and BDNF expression in prefrontal cortex of mice. *Behavioral Brain Research, 235*, 36–41.

PART II

CHAPTER 8

GAMMA ACTIVITY IN SENSORY AND COGNITIVE PROCESSING

DANIEL STRÜBER AND CHRISTOPH S. HERRMANN

8.1 Introduction and History of Research on Gamma Oscillations

When the German psychiatrist Hans Berger reported his discovery of the human EEG in 1929, he described two types of waves with the larger ones oscillating at 10–11 Hz and the smaller ones at 20–30 Hz (Berger, 1929)—the well-known alpha- and beta-waves, respectively. However, in view of the results from the first Fourier analysis carried out on the human EEG by his physicist co-worker Dietsch (1932), Berger had to acknowledge that his beta-waves contained a lot more components than originally thought with frequencies up to 125 Hz (Berger, 1934;1936). By comparing spontaneous EEG recordings with recordings during mental calculation, Berger (1937) observed effects predominantly in the 40–90 Hz range which led him to conclude that beta-waves of this specific (high-)frequency band "are the physical effects that accompany mental processes". One year later, the term "gamma waves" was proposed for higher frequencies at 35–45 Hz, that is, beyond the traditional beta-band (Jasper & Andrews, 1938). However, this first-time labeling went mostly unnoticed at that time and "gamma" was later reintroduced by other authors (e.g., Başar & Özesmi, 1972; Bressler & Freeman, 1980). Instead of "gamma", the term "40-Hz oscillation" is also widely used.

The first experimental evidence for a possible role of gamma oscillations in early sensory processing in the mammalian brain was provided by Adrian (1942; 1950), who recorded oscillatory responses of the olfactory bulb of hedgehogs, cats, and rabbits to odorous substances. He obtained oscillatory responses in the 30–60 Hz frequency range which he termed "induced waves" to differentiate these events from "intrinsic waves", signaling spontaneous activity (Adrian, 1950). Inspired by Adrian's suggestion that neurons responding preferentially to certain odors might be spatially organized, Freeman collected multi-channel EEG recordings from the surface of the olfactory bulb of rabbits trained to respond to different odorants (Freeman & Skarda, 1985; Freeman, 1975). These analyses revealed odor-specific spatial amplitude patterns of transient gamma bursts at 40–80 Hz occurring between inhalation and response, reflecting a spatial code of Adrian's induced waves that seemed to underlie the discrimination of odors.

In addition to the pioneering studies of Freeman on the olfactory bulb of the rabbit, Başar and colleagues observed "gamma resonance" phenomena in the hippocampus and other structures of the cat brain in response to auditory stimulation (Başar et al., 1975; Başar et al., 1976; Başar & Özesmi, 1972), that is, an amplification of the gamma frequency component of the evoked potential in relation to the spontaneous gamma activity just before stimulation. Resonance phenomena have also been studied by recording the evoked potential in response to repetitive stimulation at different frequencies (so-called steady-state evoked potentials, Regan & Spekreijse, 1986). If the amplitude of the steady-state response peaks at a certain frequency, this component is interpreted as the resonance frequency of the underlying oscillator. For gamma oscillations in the human visual cortex, Regan (1968) first demonstrated a peak frequency of 45–55 Hz by using photic stimulation flickering between 5 and 60 Hz. Later on, Herrmann (2001) applied flickering light at frequencies from 1 to 100 Hz and found clear resonance phenomena in the gamma range. For the auditory modality, Galambos and colleagues (1981) reported a comparable phenomenon at 40 Hz in response to tone bursts at different rates, which the authors termed the auditory "40-Hz event-related potential".

Whereas most of the early studies on gamma activity were concerned with its role in sensory processing, the first attempts to relate gamma to more cognitive processes lacking a strict relation to stimulation were in the late 1970's and early 1980's. One prominent example is the work of Sheer (1984), who used 40 Hz EEG as a measure for changing states of focused arousal during cognitive tasks in humans (e.g., Spydell et al., 1979). Similarly in the cat neocortex, fronto-parietal gamma band coherence was shown to increase during states of expectancy and focused attention (Bouyer et al., 1981). These studies pioneered the current notion of a role for gamma oscillations in top-down attentional networks (Engel et al., 2001).

In the late 1980's, Gray & Singer (1989) revolutionized the research on gamma activity by observing synchronous gamma oscillations in spiking activity of the cat visual cortex in response to light stimuli. Gray and collaborators (1989) extended their initial finding of short-range gamma synchronization to spatially more segregated cells. They demonstrated that neurons from distant parts of the visual cortex oscillate synchronously if presented with a coherent stimulus like a single long bar (or two shorter bars moving in the same direction), but not in case of two independent stimuli like two bars moving in opposite directions. This finding was highly influential because it suggested that gamma synchrony could serve as a mechanism for binding together different features to form a coherent object (see, for reviews, Engel et al., 1997; Singer & Gray, 1995).

Such binding-related synchronization of oscillatory gamma responses between neural assemblies has not only been observed in the visual cortex of the anaesthetized cats (Eckhorn et al., 1988; Engel et al., 1991; Gray et al., 1989), but also in awake monkeys (Frien et al., 1994; Kreiter & Singer, 1996). This work stimulated a wealth of human EEG studies on gamma oscillations and their possible role in visual feature-binding and object representation (see, for review, Tallon-Baudry & Bertrand, 1999). However, it soon turned out that gamma oscillations are not related to a single function like visual grouping but that they subserve a multiplicity of cognitive functions, including attention, learning, memory, and language processing, to name just a few (for reviews, see Herrmann et al., 2010; Jensen et al., 2007; Pulvermüller et al., 1997; Tallon-Baudry, 2009).

Given this large variety of cognitive gamma correlates, the question arose whether there might be a common fundamental process to which the many functions can be reduced to. In this context, it has been proposed that basic memory mechanisms could underlie functions like attention, object representation, and language (Herrmann et al., 2004). Going further by including "non-cognitive" functions of gamma oscillations recorded from subcortical areas of the mammalian brain and even from invertebrate ganglia, it has been argued that gamma oscillations represent a universal operator acting in concert with other oscillatory systems to control the integrative brain functions at all sensory and cognitive levels (Başar, 2013; Başar-Eroğlu, Strüber, Schürmann et al. 1996).

However, Whitham and colleagues (2007) challenged research on EEG gamma activity by demonstrating that most of the EEG gamma activity disappeared from the spectrum as a result of total neuromuscular blockade. In the following year, Yuval-Greenberg and colleagues' (2008) work questioned the widely held assumption that scalp-recorded induced gamma activity is of cortical origin. They demonstrated that this type of gamma activity coincides with the occurrence of miniature eye movement. The possibility that large parts of the EEG gamma activity were artefactual rather than neural activity had presented a major challenge to the EEG gamma community, and it is likely that many earlier reports on the role of induced gamma activity are heavily contaminated by muscular activity. Nevertheless, this does not mean that all previous research is invalidated because there are several methodological issues that need to be considered when evaluating how severe specific data sets are affected. There is clear evidence in the literature that neural gamma activity can, and has been, detected by means of EEG (Schwartzman & Kranczioch, 2011). Since then, research aims to develop artefact removal techniques in order to regain confidence into the functional role of EEG gamma activity (see Sections 8.8, 8.9).

8.2 Types of Gamma Activity

Gamma oscillations are part of the total EEG energy to a varying degree at any moment in time. Such spontaneous or ongoing activity occurs independently of any external stimulation and may signal alterations of internal states like arousal or alertness (Strüber et al., 2001). If gamma oscillations are related to the processing of external stimuli, a distinction is made between "evoked" and "induced" activity (Başar-Eroğlu, Strüber, Schürmann, et al., 1996). Evoked gamma responses are strictly time- and phase-locked to stimulus onset, that is, each single trial occurs at the same latency with zero phase lag across trials, therefore summing up after averaging. Thus, evoked activity can be analyzed by transforming the averaged single-trials (i.e., the averaged evoked potential, ERP) into the frequency domain (see Figure 8.1, left panel). This type of gamma activity is typically observed within an early time window of ~50–150 ms following stimulation onset. In contrast, induced activity occurs later, following stimulation by at least 200–300 ms, and it cancels out almost completely if averaged due to its inter-trial variation of phase and latency. Therefore, to detect induced gamma oscillations, each single trial needs to be subjected to a wavelet transform, and then the resulting absolute power

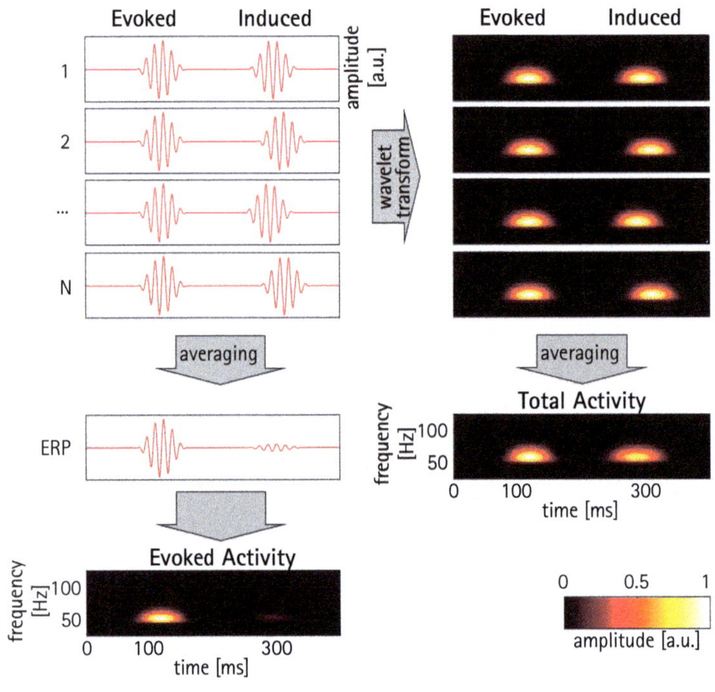

FIGURE 8.1 Analysis of evoked and induced gamma activity. Left: Averaging all single trials containing evoked and induced gamma oscillations yields the ERP. Transforming the ERP into the time-frequency domain (wavelet transform) leaves only the evoked gamma activity, because the induced gamma activity cancels out in the ERP due to phase jitter. Right: Averaging the absolute values of each single trial's wavelet transform yields both evoked and induced activity (i.e., the total activity). Thus, evoked activity is obtained with both types of analyses, whereas induced gamma activity appears only in the average of the single trial wavelet transforms.

Adapted from Herrmann et al., 2014.

NB: The latency jitter of the induced activity induces temporal smearing in the total activity. a.u., arbitrary units.

values are averaged across trials without cancellation (see Figure 8.1, right panel). Note, however, that this average contains not only the induced but also the evoked activity and has, therefore, been termed "total activity" (Herrmann et al., 2014). Within this plot of total activity, all activity that is not also present in the evoked plot, can be referred to as induced activity. While Figure 8.1 depicts the differences between the different types of gamma activity in principle, Figure 8.2 represents real data from human experiments.

In traditional EEG research, the range of gamma frequencies is given as 30–80 Hz with evoked activity often oscillating nearby 40 Hz, whereas induced responses can also reach higher frequencies (Herrmann, Munk, et al., 2004). However, in the context of subdural electrocorticography (ECoG) in epilepsy patients, a broad range of very high gamma frequencies (around 80–200 Hz) was discovered unexpectedly and called "high gamma" to distinguish this broadband response from the traditional "low gamma" oscillations in narrower bands (for a review, see Crone et al., 2011). Although

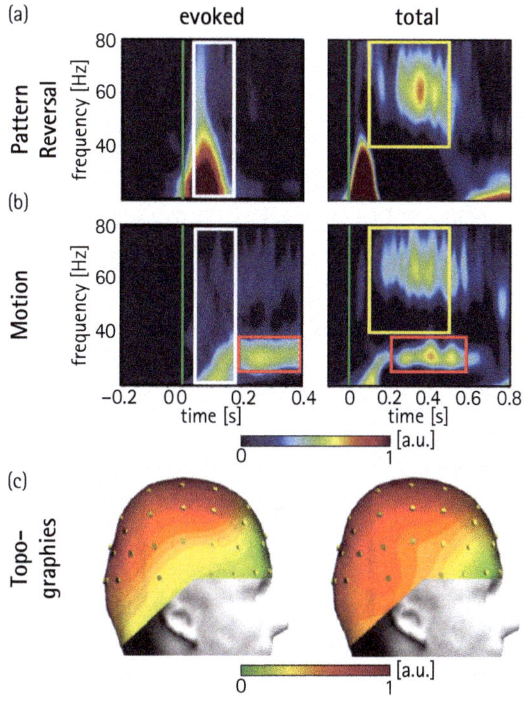

FIGURE 8.2 Time-frequency plots and topographies of evoked and induced gamma activity. White box: Evoked gamma activity for pattern reversal (A) and motion (B). Yellow box: Induced gamma activity for pattern reversal (A) and motion (B) within the plot of total activity. Red box: Steady-state visual evoked potential (SSVEP) in the motion condition as a response to the single images representing the motion of the gratings. The green line represent stimulus onset. C: Topographic maps of evoked (left) and induced (right) gamma activity during an object recognition task. a.u.: arbitrary units.

(A, B) Adapted from Naue et al., 2011; (C) Adapted from Busch, Herrmann, et al., 2006.

this broadband activity often overlaps with the band-limited (low) gamma oscillations, it is possible that such broadband responses reflect mere increases of (non-oscillatory) gamma power and/or spiking activity rather than a true oscillation (Buzsáki & Wang, 2012; Ray & Maunsell, 2011), since the latter would be indicated as a clear peak in the power spectrum. There is a recent debate about how to assess and interpret high gamma activity in the absence of a spectral peak (Brunet et al., 2014; Hermes et al., 2015). Further research is needed to elucidate the distinction between "low and high" or "narrow- and broadband" gamma activity in terms of their neurophysiological underpinnings and functional roles. In this context, note that the observation of high gamma responses is not bound by its superior signal-to-noise ratio to invasive depth recordings and ECoG in patients, as gamma responses in the 90-250 Hz range have also been recorded from conventional scalp EEGs of healthy volunteers (Darvas et al., 2010; Lenz, Jeschke, et al., 2008).

8.3 Evoked gamma activity

In light of the latency difference between evoked (~50–150 ms) and induced gamma activity (~200–300 ms), it has been suggested that evoked activity reflects an early processing stage at the level of primary visual cortices (Zaehle et al., 2009), whereas induced activities occur at later stages (Herrmann et al., 2010). Findings of evoked gamma activity modulated by low-level physical features of visual stimuli support this view (for the sake of brevity, we focus on the visual domain in the remainder of this chapter). However, there is a remarkable overlap between evoked and induced gamma activity in reflecting sensory vs. cognitive effects, as demonstrated in the sequel.

Regarding stimulus-driven effects, early evoked gamma activity has been observed to increase with stimulus size and with central as compared to peripheral stimulation (Busch et al., 2004). Also high visual contrast (Schadow et al., 2007) and low spatial frequencies (Fründ et al., 2007) of simple grating stimuli increased the power of evoked gamma activity. With regard to motion, evoked gamma power did not differ between stationary and moving gratings (Naue et al., 2011; Swettenham et al., 2009). However, inverting the black and white stripes of a stationary grating (pattern reversal) led to a threefold increase of the evoked gamma amplitude compared to stationary and moving gratings, probably due to related contrast effects (Naue et al., 2011, see Figure 8.2).

These exemplary findings not only indicate an early processing stage of evoked gamma responses but also implicate that finding evoked gamma activity requires a corresponding design of physical stimulus parameters. Relatedly, before a cognitive effect on evoked gamma activity can be inferred from two experimental conditions, care must be taken that physically identical stimuli were used. Otherwise, differences of the evoked gamma response cannot unequivocally be attributed to the cognitive process under study.

Cognitive effects on the evoked gamma activity have been observed in visual discrimination tasks where participants had to attend to a target and to ignore distractor stimuli that were physically more or less similar to the target (Herrmann & Mecklinger, 2001; Herrmann et al., 1999). The results not only demonstrated a target-effect (i.e., higher evoked gamma activity to attended stimuli), but also a graded effect of similarity between target and distractor stimuli (i.e., the more features a distractor shared with the target the stronger was the evoked gamma activity). This led to the suggestion that each stimulus is compared to a short-term memory template of the target and that the degree of matching between stimulus and template determines the strength of the evoked gamma activity. Similarly for long-term memory (LTM), a match between line drawings of real-world objects and their well-consolidated memory representation evoked stronger gamma responses than the perception of unfamiliar (non-)objects without an LTM representation (Herrmann et al., 2004).

Modulations of evoked gamma activity have also been reported in the context of visual semantic memory (Oppermann et al., 2012). Simultaneously presented pairs of conceptually coherent scenes (e.g., mouse–cheese) evoked stronger gamma-band responses than semantically unrelated object pairs (e.g., camel–magnet). This effect occurred within a widespread network of bilateral occipital as well as right temporal and frontal regions in a time window between 70–130 ms after stimulus onset, indicating a role of early gamma activity for a rapid memory-based extraction of the gist of a scene.

Together, studies on early evoked gamma oscillations demonstrate not only their sensitivity to low-level stimulus features but also an influence of cognitive effects occurring as early as 50–150 ms after stimulus onset, thereby revealing an early interaction between bottom-up and top-down processes in the gamma band (Busch et al., 2006).

8.4 Induced (total) gamma activity

Especially the induced type of gamma activity has been consistently related to higher cognitive processes (Kaiser & Lutzenberger, 2005), motivated by the initial findings in animals suggesting a possible role of induced synchronous neural discharges in bottom-up feature binding (see Section 8.1; Eckhorn et al., 1988; Gray et al., 1989). By using protocols very similar to these animal studies, their main finding of an increase in the strength of induced gamma synchrony during passive viewing of coherent moving bars could be replicated in the human scalp EEG (Lutzenberger et al., 1995; Müller et al., 1996). Increases of induced gamma activity have also been reported in response to coherent versus incoherent static stimuli during a visual discrimination task. In a classical study, presenting an illusory Kanizsa triangle, a real triangle, and a no-triangle stimulus with the black inducer disks rotated outwards, resulted in specific enhancements of the induced gamma activity for the coherent triangles (illusory and real) as compared to the no-triangle stimuli (Tallon-Baudry et al., 1996).

In addition to these relatively low-level feature binding processes originating from visual cortices, induced gamma activity has been observed also in higher stages of perceptual processing. For instance, at occipito-temporal and frontal areas during rehearsal of an object representation in short-term memory (Tallon-Baudry et al., 1998), and at occipito-parietal areas during the top-down activation of internal object representations (Tallon-Baudry et al., 1997), as well as during selective spatial attention to moving stimuli (Gruber et al., 1999). Together, those findings are in line with the interpretation of induced gamma activity as a signature of both bottom-up and top-down related binding processes involved in object representation (Tallon-Baudry & Bertrand, 1999).

Consistent with this "representational hypothesis" of induced gamma activity are studies demonstrating augmented gamma power and phase coupling during retrieval from LTM for familiar objects compared to unfamiliar ones (e.g., Busch et al., 2006; Gruber et al., 2006). Furthermore, in a recognition memory task, Gruber and colleagues (2004) showed that induced gamma activity during the encoding phase was higher for subsequently recognized words as compared to forgotten words.

Summerfield and Mangels (2006) describe similar enhancements of induced gamma activity during successful encoding, although they compared later memory performance between predictable items that were encoded under top-down attentional control and unpredictable items to which attention could not be oriented in a top-down manner during encoding. Enhanced gamma activity over frontal regions predicted successful memory formation for the predictable items only, indicating a role of frontal gamma activity in attentional top-down mechanisms facilitating memory formation (Summerfield & Mangels, 2006).

Interestingly, a frontal gamma-band enhancement was also reported during multistable visual perception (Başar-Eroğlu, Strüber, Kruse, et al., 1996), that is, spontaneously switching between different interpretations of an invariant stimulus pattern like the famous Necker cube. This frontal gamma activity was stronger for observers experiencing more figure reversals compared to those with a lower reversal rate (Strüber et al., 2000; Strüber et al., 2001), and when observers were required to control their reversal rate intentionally (Mathes et al., 2006). Together, these findings might indicate that frontal gamma band activity represents top-down influences of attentional selection on feature binding of relevant object representations during perceptual reversals (Engel et al., 2001).

Overall, these example studies on induced gamma activity demonstrate a variety of gamma-related functions ranging from basic perceptual to high-level cognitive processes like learning, memory, and attention. On the one hand, such a diversity of functional correlates speaks against a role of induced gamma activity for a single specific cognitive function. On the other hand, the processes of perception, recognition, and attention are closely intertwined and, therefore, might rely on the same neurophysiological mechanisms inherent in gamma oscillations that subserve these functions in a dynamic task-dependent manner (Fries, 2009; Tallon-Baudry, 2009; Varela et al., 2001).

8.5 Neurophysiological mechanisms

How do gamma oscillations exert their functions in basic sensory and higher cognitive functions and what are the neurophysiological mechanisms underlying oscillatory gamma activity?

The central concept which is thought to underlie the functional relevance of gamma activity is referred to as "oscillatory synchrony", that is, the periodic co-occurrence of electrical impulses from a group of neurons on a fine temporal scale. In the context of EEG scalp recordings, the obtained electrophysiological data inevitably reflect synchronized activity from large neuronal populations, because otherwise the weak synaptic currents would not be measurable at the scalp level. Local oscillatory synchrony resulting from within-area interactions would then appear as increased power at individual electrodes, whereas oscillatory synchrony related to large-scale integration is best characterized by phase coherence between two distant sources (Siegel et al., 2012; Varela et al., 2001).

Oscillatory synchrony is thought to temporally structure the occurrence of spike trains without changing the mean firing rates of neurons and, thereby, to have a functional role in the processing of incoming inputs and the emergence of functional networks by gating the information flow (Salinas & Sejnowski, 2001). Converging evidence from experimental work in animals (in vivo and in vitro) and modeling approaches suggests that cortical gamma oscillations and their local synchronization result from interactions of reciprocally connected excitatory pyramidal cells and inhibitory gamma-aminobutyric acid (GABA)-ergic interneurons (Bartos et al., 2007; Buzsáki & Wang, 2012; Wang, 2010; Whittington et al., 2011; Whittington et al., 2000).

In such a network, the pyramidal cells activate the interneurons, which self-generate synchronized gamma oscillations. This gamma-synchronized activity of the interneurons is then imposed onto the pyramidal cells, resulting in rhythmic inhibition of the pyramidal cells, which in turn leads to a rhythmic synchronization of their discharges. Thus, during each cycle of these excitatory-inhibitory feedback loops, the time window during which the pyramidal cells are able to discharge is restricted by the decay time of their inhibitory input from the fast-spiking interneurons, introducing a phase delay of a few milliseconds between pyramidal and interneuron discharges (Fries et al, 2007). From this basic mechanism of gamma synchronization several mechanistic consequences arise that are thought to be instrumental for the formation of neuronal cell assemblies, cortical signal transmission, and, thereby, the implementation of cognitive functions (Bosman et al., 2014; Cannon et al., 2014; Vinck & Bosman, 2016).

One consequence of gamma synchronization that relates to assembly formation is its involvement in the regulation of synaptic plasticity. The precise timing of pre- and post-synaptic processes emerging from gamma synchronization acts on a time scale of a few tens of milliseconds, which is relevant for spike-timing-dependent plasticity and, thus, for the induction of long-term potentiation or depression (Fell & Axmacher,

2011; Paulsen & Sejnowski, 2000; Sejnowski & Paulsen, 2006) and associative learning (Miltner et al., 1999). It has been proposed that the period of gamma oscillations (~25 msec) is "designed" to match the time course of calcium fluctuations in dendrites and, therefore, to facilitate learning (Bibbig et al., 2001).

With regard to local within-area interactions, synchronization in the gamma-band with its high temporal precision in the millisecond range is hypothesized to increase the impact of presynaptic neurons on their target cells because the gamma-synchronized spike packages arrive close together, thus summing up more effectively to initiate postsynaptic discharges. This effect of precise spike timing is referred to as feedforward coincidence detection (Fries, 2009; Salinas & Sejnowski, 2001), which has been linked to the integration (binding) of different stimulus features during object recognition (Bosman et al., 2014).

This concept has also been transferred to inter-areal communication between multiple groups of neurons, each oscillating in the gamma frequency range. In this context, the "communication through coherence" (CTC) hypothesis (Fries, 2005; 2015) claims that the communication between two groups of neurons can be facilitated by gamma synchronization in the two groups, if the spikes from the presynaptic group (sender) arrive at the postsynaptic group (receiver) at the appropriate phase, that is, during a minimal amount of GABAergic inhibition. This phase-coupling of oscillations constitutes a window of "opportunity" in which neural networks jointly involved in signal processing can communicate. This mechanism allows for gain modulation of pre-synaptic inputs that compete for activating higher-level post-synaptic targets as is the case with selective attention. In this scenario, visual attention might selectively increase the effective strength of those synaptic inputs from lower level neurons that process attended stimuli at the expense of inputs from the non-attended stimuli (Bosman et al., 2014; Fries, 2015). In this way, gamma synchronization might dynamically route the information flow between higher and lower level areas within the visual hierarchy as has been observed locally between multiple visual areas (see, for review, Bosman et al., 2014).

However, it is less clear whether gamma synchronization also serves as a mechanism for large-scale integration across distant brain regions that also include, for example, frontal or parietal areas. It has been suggested that the spatial distance between interacting brain areas and, hence, the conduction delay may define the communication frequency, with gamma oscillations acting more locally and lower frequencies more globally (Kopell et al., 2000; von Stein & Sarnthein, 2000). A reason for this may be that, in contrast to the millisecond precision of gamma oscillations, lower-frequency bands are more robust to spike timing delays (Buschman & Miller, 2007). Indeed, several findings in monkeys have shown that long-distance top-down processes are carried by inter-areal synchrony in the alpha and low-beta frequency range, whereas gamma oscillations index a local encoding of information and the bottom-up transfer of low-level sensory information to higher-level areas (for reviews, see, Bressler & Richter, 2015; Gregoriou et al., 2015; Siegel et al., 2012; Wang, 2010).

On the other hand, there is also increasing evidence for a role of gamma synchrony in large-scale interaction during various perceptual and attentional processes (see for a review, Gregoriou et al., 2015), indicating that physical distance might not be the only factor that determines the frequency band used for inter-areal synchronization. Moreover, gamma-synchronized long-range signal transmission seems to work both upstream (i.e., bottom-up) and downstream (i.e., top-down) within the cortical hierarchy. For example, long-range gamma coupling between prefrontal and visual areas during directed attention was found to be initiated frontally, thereby signaling top-down attentional influences on the visual cortex (Gregoriou et al., 2009). Another study reported long-range bottom-up directed gamma synchrony between posterior parietal and prefrontal regions during automatic attention driven by salient stimuli (Buschman & Miller, 2007). Intriguingly, the same areas synchronized in top-down direction if attention was focused volitionally, but in this case in the beta-range (Buschman & Miller, 2007), indicating that different frequencies are used for feedforward and feedback signaling.

Such frequency-specific differences in the direction of inter-areal interaction (i.e., bottom-up vs. top-down) have been related to the different directions of information flow in the cortical layers with feedforward and feedback connections originating primarily in superficial and deep layers, respectively (Felleman & Van Essen, 1991; Markov et al., 2014). Accumulating evidence suggests that alpha/beta oscillations originate in deeper layers of the visual cortex and support feedback signaling, whereas gamma synchrony emerges from superficial layers and signals feedforward information flow (Buffalo et al., 2011; Michalareas et al., 2016; van Kerkoerle et al., 2014; von Stein et al., 2000). Together, these findings clearly suggest a role for gamma synchrony in local bottom-up interaction between visual cortices (Siegel et al., 2012). However, it remains an open question how this function can be reconciled with the documented long-range gamma synchrony during top-down attention (Gregoriou et al., 2015).

Overall, the recurrent excitatory-inhibitory network interactions underlying local gamma synchronization are thought to establish low-level circuit functions (e.g., synaptic plasticity, coincidence detection, gain modulation, phase coding, dynamic routing) that may act as elementary building blocks of cognitive functions (e.g., visual feature integration, selective attention, learning, and memory) in a task-specific combination (Bosman et al., 2014; Siegel et al., 2012). Nevertheless, there are multiple methodological concerns that need to be considered to reliably assess EEG gamma activity as outlined in the next section.

8.6 Methodological aspects

Scalp recorded gamma activity is subject to a variety of high-frequency artefacts which might seriously affect recording, analysis, and interpretation of the data. Therefore, these artefacts need to be either prevented by appropriate recording conditions or to be removed by post hoc analysis techniques. There are two main sources of artefacts. First, technical artefacts like power line noise, the screen refresh rate, or the rate at which single images are presented in order to show a movie (cf. Figure 8.2 right panel, SSVEP). Second, physiological artefacts like eye movements and electromyogenic (EMG) activity generated by muscles from the scalp, face, and neck (see, for review, Nottage & Horder, 2016).

8.7 Technical artefacts

Power line noise occurs at a frequency of 50 Hz in Europe and 60 Hz in the US; both frequencies are in the gamma range. Therefore, any power supply like electric cables, wall sockets, or electrically operated equipment inside the recording cabin (lamps) results in a 50-or 60-Hz peak in the EEG power spectrum. To effectively avoid frequency interference, experiments should be conducted in an electrically shielded room with all devices inside the cabin operated on batteries. The stimulation monitor can be placed outside the cabin behind an electrically shielded window and a fiberoptic cable can be used to transfer the EEG data to a computer outside the recording cabin. If it is not possible to prevent power line noise from being recorded, a low-pass or "notch" filter at 50 or 60 Hz might be applied to the data. However, the use of such filters is problematic because they might induce "ringing" (i.e., the induction of spurious oscillations by transients in the EEG) and distort the phase of neural oscillations. Alternative strategies have been developed on the basis of noise cancellation (see Nottage, 2010 for details).

In addition to power line noise, every screen refresh generates an electromagnetic signal which might be seen as a narrow band in the EEG gamma range, typically at anterior electrodes. The exact frequency depends on the primary refresh rate and its harmonics. Nottage & Horder (2016) explore possible ways to remove this type of artefact.

8.8 Physiological artefacts

The difficulty of properly removing EMG artifacts is particularly important for studies on gamma activity, given the fact that the broad frequency spectrum of EMG activity substantially overlaps with the gamma frequency range. An amplitude maximum of EMG contamination in the gamma range (40–80 Hz) was found at temporal sites during phasic contraction of facial muscles (Goncharova et al., 2003). Because of their high amplitude, occasionally occurring phasic muscle contractions (e.g., chewing or jaw clenching) can be detected relatively easily by visual inspection or mathematical algorithms and then be omitted from further analysis (e.g., Fitzgibbon et al., 2015). However, in addition to large phasic contractions, the head and neck muscles are constantly active to maintain posture, which might result in tonic EMG activity throughout an EEG session.

Critically, low-amplitude tonic muscle activity occurs even in "relaxed" states and is, thus, difficult to detect in scalp recordings but still contributes significantly to EMG contamination in the gamma range, as has been shown in pharmacologically paralyzed human volunteers (Whitham et al., 2007). Paralysis compared to pre-paralyzed states of the same participants was found to result in a marked reduction of spectral power in the gamma range, with 84% of the power derived from EMG, mostly over peripheral scalp regions. Moreover, the execution of different cognitive tasks induced a broadband increase in the gamma frequency range in unparalyzed but not paralyzed participants, indicating that EMG activity was responsible for the task-related spectral power increases (Whitham et al., 2008). However, when employing a standard oddball task that allows for stimulus-locked data analysis, significant gamma activity was identified in both the pre-paralyzed and the paralyzed condition, with stronger activity for rare target than for more frequent standard stimuli (Pope et al., 2009).

While these studies clearly demonstrate a large contribution of EMG activity to the overall EEG spectral power, they also show that (at least under experimental conditions where time- and stimulus-locked analysis is possible, and in contrast to, for example, analyzing states of varying cognitive load during mental arithmetic) neuronal gamma activity can still be measured at the scalp EEG. Thus, it seems possible to extract neural gamma activity from EMG noise if it is analyzed in response to discrete stimuli (e.g., coherent stimuli vs. incoherent visual objects). For this, however, methods for the effective removal of EMG artefacts are required.

There is a multitude of methods for dealing with scalp EMG (see, for review, Nottage & Horder, 2016). One novel approach that specifically addresses the tonic nature of scalp and neck muscle artefacts uses mathematical modeling to fit individual muscle spikes and subtract these from the signal, resulting in an effective correction of the gamma activity associated with a self-paced motor task (Nottage et al., 2013). Recently, Janani and colleagues (2018) used datasets that are free of muscle activity due to paralysis (taken from Whitham et al., 2007, 2008) to identify limitations of traditional approaches and, then, to evaluate the improvements of a newly developed algorithm for tonic muscle artefact removal. With this method, high-frequency EMG artefacts were reduced considerably, although a residual artefact still remains (compared to paralysis). Given the variety of techniques that are based on different mathematical concepts, the authors suggested using a combination of complementary algorithms for a further improvement of EMG artefact removal (Janani et al., 2018).

Ocular activity also generates muscle-related artefacts. The main sources of eye movement artefacts are blinks and saccades. It is standard practice to measure the electro-oculogram (EOG) from two channels of the left and right eye. For rejecting eye blinks and large saccades, an amplitude threshold is usually set (e.g., 50 μV), which is then used by an algorithm to exclude contaminated EEG trials in any channel from further analysis. This automatic amplitude threshold procedure is then complemented by visual inspection of all epochs. For algorithms based on eye tracking data that detect and correct eye blinks and other ocular artifacts in a fully automated fashion, see Plöchl et al., 2012.

Blink artefacts elicit a large potential with a predominantly frontal topography and are easy to identify in the vertical electro-oculogram (VEOG). Saccades are generated by a pair of extra-ocular muscles which contract to move the eyeball, thereby inducing a saccadic spike potential in the EEG at the onset of each saccade (Thickbroom & Mastaglia, 1985). Saccadic spike potentials are too small to be detected by standard amplitude thresholds for automatic EOG-artefact removal, especially in case of microsaccades which are produced during attempted fixation (Martinez-Conde et al., 2013). In their seminal paper, Yuval-Greenberg and colleagues (2008) suspected that high-frequency EEG activity that had been regarded as induced gamma activity in fact reflects microsaccade-induced muscle artefacts. Naturally, this report triggered an intense debate within the field of induced gamma activity research (Schwartzman & Kranczioch, 2011). Section 8.9 summarizes the main points.

8.9 Microsaccades and Induced Gamma Activity

Microsaccades are small (up to 1 degree), jerk-like saccades with a duration of about 25 ms and an average rate of 1–2 per second (Martinez-Conde et al., 2013). There are at least three important similarities between microsaccades and induced gamma activity that need to be considered. First, in response to a sudden change in stimulus input, the microsaccade rate shows a characteristic biphasic time course, with an early inhibition phase peaking at 100–150 ms after stimulus onset, followed by a rebound phase peaking between 200 and 400 ms after stimulus onset (Engbert, 2006). Thus, the rebound phase of microsaccades corresponds to the time window where induced gamma activity typically appears.

Second, the rate of microsaccades is modulated by perceptual and cognitive factors that have also been linked to induced gamma activity, including attention and memory-related processes (Engbert & Kliegl, 2003; Valsecchi et al., 2007; Valsecchi et al., 2009), physical stimulus features like color and luminance contrast (Rolfs et al., 2008), and the coherence of objects (Yuval-Greenberg et al., 2008). Thus, microsaccade rate represents a true confound in experiments on gamma activity since modulations of microsaccade rate and gamma activity might generate an identical pattern of results across experimental conditions. For example, both microsaccade rate and induced gamma power show increased responses to a coherent object in comparison to an incoherent object.

Third, depending on the location of the reference electrode, the frontally occurring microsaccade-induced saccadic spike potential translates into a broadband (~20–90 Hz) gamma power increase with a maximum at centro-parietal and occipital electrodes, that is, regions where it strongly coincides with induced gamma activity (Reva & Aftanas, 2004; Yuval-Greenberg et al., 2008). Thus, saccadic spike potentials mimic not only the frequency content of induced gamma activity but also its typical topography.

Notably, this can be avoided by using MEG, since it does not require a reference (Gruber et al., 2008).

These similarities between microsaccades and induced gamma activity regarding their sensitivity to cognitive effects as well as their temporal, topographical and spectral properties have been convincingly demonstrated and have led to the suggestion that—at least in some cases—induced gamma activity reflect saccadic spike potentials rather than a neural response (Yuval-Greenberg et al., 2008). This also concerns earlier EEG findings that might have been erroneously interpreted as neuronal gamma activity, when in fact they reflected microsaccade-induced spike potential artefacts (Keren et al., 2010). How strong previous EEG findings might be affected by microsaccades depends on several aspects, for example, the exact time window of the effects (in relation to the saccadic rebound), the broadness of the gamma response (in relation to the 20–90 Hz spike potential), and the choice of the reference electrode, given its influence on the topography, to name but a few (for further details, see Schwartzman & Kranczioch, 2011).

Notably, evoked gamma activity is not affected by microsaccade-induced spike potential artefacts. One reason is that evoked gamma activity occurs much earlier than the microsaccade rebound after sensory stimulation. Another reason is that spike potentials, in contrast to the evoked gamma activity, are not tightly time-locked to the stimulus and, thus, cancel out during averaging of the evoked potential (Yuval-Greenberg & Deouell, 2009).

Although Yuval-Greenberg and colleagues (2008) did not deny the existence of induced gamma activity and its role in perception and cognition in general, their findings posed a challenge to develop methods for separating the effects of microsaccades and induced gamma responses in EEG scalp recordings. One immediate consequence of this study was the necessity to use eye tracking with sufficient spatial and temporal resolution for detecting microsaccades as small as 0.15° visual angle, given that even precise fixation of a continuously present fixation point does not preclude microsaccade-related brain activity (Dimigen et al., 2009). The rationale behind a combined recording of EEG and eye tracking data is to demonstrate that differences of induced gamma activity and microsaccade rate do not covary across conditions and, thus, effects of induced gamma activity cannot easily be explained by microsaccadic muscle potentials. Indeed, such differential modulations of gamma activity and microsaccades time courses have been reported, for example, during object motion (Naue et al., 2011) and memory-based object recognition (Hassler et al., 2013).

However, high-resolution eye trackers are expensive and not always available. Alternatively, there are methods available allowing for offline artefact correction that identify saccadic spike potentials via EOG sensors, that is, without the need of an eye tracker (Hassler et al., 2011; Keren et al., 2010; Nottage, 2010). However, Keren and colleagues (2010) used eye tracking in addition to EOG-based detection of microsaccades to evaluate the hit and false alarm rates of several spatiotemporal filters that were applied to the EOG data to identify the sharp amplitude increase at the onset of saccadic spike potentials. With this method, detection rates of 80% on average could be achieved for microsaccades of at least 0.2°, which were then efficiently attenuated by means of mathematical algorithms. For artifact correction methods based on combined

EEG and eye tracking recordings, see also Plöchl et al. (2012). Nottage (2010) and Hassler and colleagues (2011) used different mathematical approaches but both have been demonstrated to effectively remove microsaccadic muscle artefacts, resulting in a more sustained and narrow-band gamma signal of the residual activity compared to the frequency plot of the saccadic spike potential (see, for review, Nottage & Horder, 2016).

Although the available methods are relatively easy to apply and effective in reducing saccadic spike potentials, they do not remove the artefact completely and they do not attenuate saccade-related visual brain activity (Dimigen et al., 2009; Plöchl et al., 2012). Also, the fundamental problem of the confounding co-modulation of induced gamma activity and microsaccade rate cannot be solved by artefact correction techniques. Therefore, it has been suggested recently to reduce the occurrence of microsaccades and related confounds in the EEG at the level of experimental design instead of applying off-line correction methods (Tal & Yuval-Greenberg, 2018). Considerably reducing the incidence of microsaccades during experimentation would allow rejecting microsaccades rather than just correcting artefactual trials and still leaving enough (then artefact-free) trials for analysis. As a first step, Tal and Yuval-Greenberg (2018) were able to reduce the average number of saccades (including microsaccades) by 10–25% through manipulation of different task characteristics (e.g., by adding a foveal task, whereas the stimulus of interest was parafoveal). However, it remains to be seen how this approach can be applied to the diverse experimental setups covering the full range of induced gamma activity related cognitive processes.

In summary, together with other muscle artefacts, microsaccades impose a serious difficulty on the analysis and interpretation of induced gamma activity in human EEG. To separate activity related to saccadic spike potentials from induced gamma activity, using high-resolution eye tracking is recommended as the most reliable way to identify microsaccades. However, there are also techniques available to use EOG data for saccadic spike potential detection. In a second step, off-line artefact removal methods should be used to correct artefactual trials. Rejecting all artefactual trials is currently not possible due to the high prevalence of microsaccades in typical visual experiments, but there are suggestions how to reduce the number of microsaccades through experimental design. Current evidence from artefact-corrected data confirms the influence of microsaccades on induced gamma activity, but also indicates the existence of induced gamma activity that survived artifact suppression, thereby replicating earlier findings on the functional roles of induced gamma activity in, for instance, object representation (Hassler et al., 2013; 2011). Finally, depending on the research question at hand, a further option to prevent microsaccadic contamination of gamma activity might be to restrict the analyses to the evoked gamma activity, which is not affected (e.g., Lally et al., 2014).

8.10 Clinical relevance

Given its role in multiple cognitive processes and neural information integration, disturbed gamma activity might reflect an important pathophysiological mechanism underlying the mental deficits of patients suffering from diverse neuropsychiatric disorders (Başar, 2013; Herrmann & Demiralp, 2005; Uhlhaas & Singer, 2006). One of the most extensively studied disorders in relation to gamma activity is schizophrenia (Uhlhaas & Singer, 2010) (see also Chapter 18), which is characterized by positive symptoms like hallucinations, delusions, and thought disorders, as well as negative symptoms like reduced affect, motivation, and behavior. Positive symptoms have been generally linked to increased gamma activity, whereas negative symptoms correlate with a decrease of gamma activity compared to healthy controls (Herrmann & Demiralp, 2005). Moreover, as a group, schizophrenic patients have deficits in both basic sensory and cognitive functions, which have been related to evoked and induced gamma activity.

For instance, early evoked gamma activity in response to auditory stimulation was found to be reduced in schizophrenic patients receiving medication (Leicht et al., 2010; Lenz et al., 2011), but not in unmedicated patients (Gallinat et al., 2004). Başar-Eroğlu and colleagues (2011) reported that auditorily evoked gamma responses did not differ between schizophrenic patients and healthy controls, but on the single-trial level, gamma responses were higher in patients compared to controls. Modality-specific effects were found for chronic medicated patients, demonstrating reduced early evoked gamma activity for visual but not auditory stimulation (Spencer et al., 2008). Seemingly, more research is needed to get a clearer picture regarding the role of early evoked gamma activity in sensory processing of schizophrenic patients.

With regard to induced gamma activity in schizophrenic patients, most of the evidence points to impaired gamma activity related to higher perceptual and cognitive functions (see, for review, Uhlhaas & Singer, 2010; 2012). This includes, for example, cognitive control in first-episode schizophrenia patients with and without medication (Minzenberg et al., 2010), and working memory (Haenschel et al., 2009), although one study found working memory-related gamma amplitudes to be preserved in patients but lacking any relation to task difficulty, as present in normal controls (Basar-Eroğlu et al., 2007). Reduced amplitudes or reduced phase locking of gamma oscillations during working memory and other cognitive tasks have been found predominantly in frontal areas which corresponds to the frequently reported hypo-frontality in schizophrenia (see, for review, Senkowski & Gallinat, 2015).

Interestingly, a recent review on gamma activity in first episode psychosis patients and people at high risk for psychosis found a similar decrease of evoked and induced gamma activity as has been reported for chronic schizophrenia (Reilly et al., 2018), indicating that reduced gamma activity in the early phase of psychosis might serve as a biomarker or endophenotype allowing for an early intervention before the disorder has fully developed. This, however, would require a more standardized procedure regarding the employed experimental paradigms and measures of gamma activity.

Alternatively, spontaneous or resting-state gamma activity might be used as a biomarker of schizophrenia, which would bear the advantage that it can be recorded without engagement in a task (i.e., spontaneous activity) or as pre-stimulus baseline activity (i.e., resting-state). Indeed, there is evidence for altered spontaneous/resting gamma activity in schizophrenia. But in contrast to the reported reduction of evoked and induced gamma activity, this type of gamma activity has been found to be predominantly increased in chronic schizophrenic patients (Spencer, 2012; see, for review, White & Siegel, 2016) as well as in first episode patients and high-risk populations (Reilly et al., 2018). While it has been proposed that increased spontaneous/resting gamma activity contributes to positive symptoms like hallucinations (White & Siegel, 2016), its relation to negative symptoms and reduced gamma activity during stimulus processing is less clear. Thus, different computational models have been discussed regarding the neurophysiological mechanisms underlying gamma activity increases during resting state and gamma decreases during stimulus processing in schizophrenia (Jadi et al., 2016). In general, abnormal gamma activity in schizophrenia has been related to dysfunctional circuit mechanisms responsible for generating gamma oscillations, that is, a reduced input from inhibitory GABAergic interneurons to pyramidal cells and/or abnormal glutamatergic input from the pyramidal cells to the interneurons (see, for review, McNally & McCarley, 2016), resulting in a disturbed balance between excitatory and inhibitory mechanisms (E/I balance).

Also in autism spectrum disorders (ASD), abnormal gamma oscillations have been suggested as a potential biomarker for a dysfunctional E/I balance in children with ASD (Stroganova et al., 2015; Uhlhaas & Singer, 2012). Similar to patients with schizophrenia, children with ASD have been consistently characterized by reduced gamma activity during sensory stimulation with overlapping perceptual deficits (see, for review, Uhlhaas & Singer, 2012). Interestingly, children with ASD also show an increase in resting-state gamma activity (van Diessen et al., 2015; Wang et al., 2013) but, unlike schizophrenic patients, do not experience hallucinations. However, despite the substantial overlap between schizophrenia and ASD regarding the dysfunctional E/I balance and the resulting impairments in cognitive integration, there are important differences with regard to the developmental timing of the changing E/I balance, which might explain some of the unique features of each disorder (Uhlhaas & Singer, 2012; White & Siegel, 2016).

Another neurodevelopmental disorder that is predominantly diagnosed in childhood and adolescence is attention-deficit/hyperactivity disorder (ADHD). According to its core symptoms (inattention, hyperactivity, impulsivity), one might expect a reduction of gamma activity due to the attention deficit and/or a gamma increase due to the hyperactivity component. In one of the first studies on evoked gamma activity in ADHD children, Yordanova and colleagues (2001) reported larger and more strongly phase-locked early evoked gamma activity in ADHD children during an auditory detection task involving a motor response. Gamma augmentations did not differentiate between attended and ignored stimuli. This finding was interpreted as reflecting impaired early auditory processing due to deficient motor inhibition in ADHD children (Yordanova et al., 2001).

Enhanced evoked gamma activity in children with ADHD was also found during stimulus encoding of a visual memory task (Lenz, Krauel, et al., 2008). However, in contrast to healthy children, the encoding-related gamma activity was enhanced unspecifically, that is, it was unrelated to subsequent recognition performance. Similarly, early evoked gamma activity did not reflect the difference in long-term memory representation between familiar and unfamiliar objects in a simple forced-choice reaction task (Lenz et al., 2010).

In a study by Prehn-Kristensen and colleagues (2015), the distractibility of ADHD patients was assessed in a delayed-match-to-sample paradigm by presenting a distractor stimulus during the delay period, which should be ignored. Children with ADHD but not healthy controls exhibited an increase of occipital early evoked gamma activity in response to the distractor, indicating a higher distractibility resulting in disturbed working memory maintenance already on a very early level of interference processing (Prehn-Kristensen et al., 2015).

Together, these findings might indicate that attentional problems in ADHD are related to early sensory processing deficits as reflected by unspecific evoked gamma increases that might be interpreted as an increase of noise. This may be partially caused by the fact that the same genetic variations within the dopaminergic pathway that have been associated with ADHD, also contribute to enhancements of gamma activity in healthy volunteers (Demiralp et al., 2007; Herrmann & Demiralp, 2005). However, this hyperactivity-based interpretation might be restricted to gamma activity during stimulus or cognitive processing since spontaneous gamma activity was shown to be diminished in children (Barry et al., 2010) and adults (Tombor et al., 2019) with ADHD.

Gamma oscillations have also been associated with neurodegenerative diseases like Parkinson's disease (PD) and Alzheimer's dementia (AD) (Nimmrich et al., 2015). For PD, however, oscillations in the beta range are more directly related to the clinical symptoms than gamma activity and might serve as a biomarker, whereas studies on gamma oscillations in AD yielded divergent findings of reduced as well as enhanced amplitudes, indicating the need for further research before clear statements about the usefulness of gamma oscillations as a biomarker for AD can be made (see, for review, Nimmrich et al., 2015). In two recent studies on AD, Başar and colleagues employed a target detection task and simple visual stimulation to analyze gamma activity within three sub-bands over four time windows and found a decreased response to visual stimuli and a delayed gamma response to target stimuli compared to healthy controls (Başar et al., 2016), as well as an abnormal increase of gamma coherence for AD patients in both conditions (Başar et al., 2017).

Overall, AD and many other neuropsychiatric disorders are characterized by aberrant gamma activity pointing to dysfunctional oscillatory network activities that are regulated by inhibitory interneurons (see, for reviews, Palop & Mucke, 2016; Ruden et al., 2021). While such an interneuron impairment might indeed represent a common mechanism of cognitive dysfunction in diverse neuropsychiatric disorders, the use of gamma oscillations as a single oscillatory biomarker is limited due to its lack of specificity (see, however, Lenz et al., 2011 for a study on the specificity of gamma activity in neuropsychiatric disorders). Therefore, it might be important to look for interactions with other frequency bands to create oscillatory biomarkers that are more specific for each disorder (Başar et al., 2013; Başar, Schmiedt-Fehr, et al., 2016; Uhlhaas & Singer, 2012).

Notably, however, there is recent evidence from a mouse model of AD that impaired gamma oscillations are specifically related to a hallmark of AD, that is, the abnormal aggregation of plaque-forming proteins (amyloid-beta protein) in the brain (Iaccarino et al., 2016). The authors found that inducing gamma oscillations by a 40-Hz flickering light led to a major reduction of plaques in the visual cortex of the exposed mice and to an activation of microglia (i.e., immune cells in the brain) that degrade the plaques (Iaccarino et al., 2016). Crucially, later studies using different forms of sensory gamma stimulation in multiple AD mouse models could not only demonstrate a clear link between gamma activity and cellular metabolism but also improvements of cognitive functions like learning and memory, suggesting a therapeutic role of gamma entrainment in AD (see, for review, Adaikkan & Tsai, 2020). There is emerging evidence that similar cognitive effects can be achieved in humans by employing non-invasive brain stimulation techniques that could, in contrast to rhythmic sensory stimulation, target higher-order cognitive brain areas linked more directly to core symptoms of the disorder than sensory cortices (Benussi et al., 2021; see, for reviews, Bréchet et al., 2021, and Strüber & Herrmann, 2020).

8.11 Challenges and future directions

There are a number of challenges in relation to the ubiquitous role of gamma oscillations for cognitive functions. For instance, it has been criticized that the power of gamma oscillations is low and inconsistent and that its frequency and power depend on low-level stimulus features like size or contrast, which seems to be incompatible with the proposed role of gamma synchronization in cortical processing (Ray & Maunsell, 2015). However, there is also recent reconciling evidence in support of a functional role for gamma synchronization (Singer, 2018).

One related issue regarding the functions of gamma activity in sensory and cognitive processing as well as in clinical contexts is that most of the evidence is correlative in nature. Therefore, it remains unclear whether gamma oscillations and their synchronization are truly relevant for information processing in the brain or whether they merely reflect a by-product of brain organization. It has been argued that even gamma oscillations as such may be a functional epiphenomenon arising from network activities supporting the excitatory-inhibitory balance necessary for normal brain functioning (Merker, 2013). Although this argumentation might reflect a "category mistake" since gamma oscillations cannot be functionally separated from the circuit mechanisms that generate them (Bosman et al., 2014), the necessity remains to go beyond correlations and to demonstrate that gamma activity is causally involved in cortical information processing and cognitive functions.

Causal evidence for a role of gamma oscillation in cognitive processes can be provided by demonstrating that modulating gamma power or coherence improves cognitive outcomes. There are several established methods to modulate brain oscillations, including rhythmic sensory (steady-state) stimulation, EEG neurofeedback, and non-invasive brain stimulation like repetitive transcranial magnetic stimulation (rTMS; see, for a methods overview, Herrmann et al., 2016). Whereas rTMS is not considered safe for stimulation in the gamma frequency range due to its risk of inducing epileptic seizure (Rossi et al., 2009), steady-state visual stimulation has been successful in entraining gamma oscillations (Herrmann, 2001; Regan, 1968) and improving binding-related perceptual performance (Elliott et al., 2000). However, whereas most human EEG studies using rhythmic visual stimulation obtained effects restricted to early visual cortex, EEG-based neurofeedback allows for direct neurocognitive modulation of those brain regions that are crucially involved in specific tasks (see, for review, Enriquez-Geppert et al., 2013).

Neurofeedback training to increase EEG gamma oscillations has been shown to improve perceptual processing (Salari et al., 2014), cognitive control and intelligence measures (Keizer et al., 2010; Keizer, Verschoor, et al., 2010), indicating a functional role of gamma activity for these cognitive processes. However, gamma neurofeedback is technically challenging due to its susceptibility to muscle and ocular artefacts (see Section 8.8). Indeed, there are reports that participants of a neurofeedback training were able to control the feedback signal by activating their head and neck muscles, which might serve as a caveat when designing neurofeedback studies, especially if EEG recordings are used (Merkel et al., 2018). Furthermore, neurofeedback is an endogenous technique that needs to be learned and not all participants might be able to do this, especially patients.

A recent method of non-invasive brain stimulation, called transcranial alternating current stimulation (tACS) avoids these problems by applying sinusoidal currents to stimulate the brain exogenously (see Chapter 22). Its effectiveness in modulating gamma oscillations during motor, sensory and a wide range of high-level cognitive processes has been demonstrated recently (see, for a review, Strüber & Herrmann, 2020). Thus, we propose that tACS might be the appropriate method not only to provide causal evidence for a role of gamma oscillations in brain processing but also to improve gamma activity-related sensory and cognitive functioning in healthy volunteers and clinical populations (Herrmann & Strüber, 2017; Strüber & Herrmann, 2020).

REFERENCES

Adaikkan, C. & Tsai, L.-H. (2020). Gamma entrainment: Impact on neurocircuits, glia, and therapeutic opportunities. *Trends in Neurosciences, 43*(1), 24–41. https://doi.org/10.1016/j.tins.2019.11.001

Adrian, E. D. (1942). Olfactory reactions in the brain of the hedgehog. *Journal of Physiology, 100*(4), 459–473.

Adrian, E. D. (1950). The electrical activity of the mammalian olfactory bulb. *Electroencephalography and Clinical Neurophysiology*, 2, 377–388.

Barry, R. J., Clarke, A. R., Hajos, M., McCarthy, R., Selikowitz, M., & Dupuy, F. E. (2010). Resting-state EEG gamma activity in children with attention-deficit/hyperactivity disorder. *Clinical Neurophysiology, 121*(11), 1871–1877. http://doi.org/10.1016/j.clinph.2010.04.022

Bartos, M., Vida, I., & Jonas, P. (2007). Synaptic mechanisms of synchronized gamma oscillations in inhibitory interneuron networks. *Nature Reviews Neuroscience*, *8*(1), 45–56. http://doi.org/10.1038/nrn2044

Başar, E. (2013). A review of gamma oscillations in healthy subjects and in cognitive impairment. *International Journal of Psychophysiology*, *90*(2), 99–117. http://doi.org/10.1016/j.ijpsycho.2013.07.005

Başar, E., Başar-Eroğlu, C., Güntekin, B., & Yener, G. G. (2013). Brain's alpha, beta, gamma, delta, and theta oscillations in neuropsychiatric diseases: Proposal for biomarker strategies. *Supplements to Clinical Neurophysiology*, *62*, 19–54. http://doi.org/10.1016/B978-0-7020-5307-8.00002-8

Başar, E., Emek-Savaş, D. D., Güntekin, B., & Yener, G. G. (2016). Delay of cognitive gamma responses in Alzheimer's disease. *NeuroImage: Clinical*, *11*, 106–115. http://doi.org/10.1016/j.nicl.2016.01.015

Başar, E., Femir, B., Emek-Savaş, D. D., Güntekin, B., & Yener, G. G. (2017). Increased long distance event-related gamma band connectivity in Alzheimer's disease. *NeuroImage: Clinical*, *14*, 580–590. http://doi.org/10.1016/j.nicl.2017.02.021

Başar, E., Gönder, A., Özesmi, C., & Ungan, P. (1975). Dynamics of brain rhythmic and evoked potentials. II. Studies in the auditory pathway, reticular formation and hippocampus during the waking stage. *Biological Cybernetics*, *20*(3-4), 145–160.

Başar, E., Gönder, A., & Ungan, P. (1976). Important relation between EEG and brain evoked potentials. I. Resonance phenomena in subdural structures of the cat. *Biological Cybernetics*, *25*(1), 27–40.

Başar, E. & Özesmi, C. (1972). The hippocampal EEG-activity and a systems analytical interpretation of averaged evoked potentials of the brain. *Kybernetik*, *12*, 45–54.

Başar, E., Schmiedt-Fehr, C., Mathes, B., Femir, B., Emek-Savaş, D. D., Tülay, E., Tan, D., Düzgün, A., ... Başar-Eroğlu, C. (2016). What does the broken brain say to the neuroscientist? Oscillations and connectivity in schizophrenia, Alzheimer's disease, and bipolar disorder. *International Journal of Psychophysiology*, *103*, 135–148. http://doi.org/10.1016/j.ijpsycho.2015.02.004

Başar-Eroğlu, C., Brand, A., Hildebrandt, H., Karolina Kedzior, K., Mathes, B., & Schmiedt, C. (2007). Working memory related gamma oscillations in schizophrenia patients. *International Journal of Psychophysiology*, *64*(1), 39–45. http://doi.org/10.1016/j.ijpsycho.2006.07.007

Başar-Eroğlu, C., Mathes, B., Brand, A., & Schmiedt-Fehr, C. (2011). Occipital γ response to auditory stimulation in patients with schizophrenia. *International Journal of Psychophysiology*, *79*, 3–8.

Başar-Eroğlu, C., Strüber, D., Kruse, P., Başar, E., & Stadler, M. (1996). Frontal gamma-band enhancement during multistable visual perception. *International Journal of Psychophysiology*, *24*(1-2), 113–125. http://doi.org/10.1016/S0167-8760(96)00055-4

Başar-Eroğlu, C., Strüber, D., Schürmann, M., Stadler, M., & Başar, E. (1996). Gamma-band responses in the brain: A short review of psychophysiological correlates and functional significance. *International Journal of Psychophysiology*, *24*(1-2), 101–112. http://doi.org/10.1016/S0167-8760(96)00051-7

Benussi, A., Cantoni, V, Cotelli, M. S., Cotelli, M., Brattini, C., Datta, A., Thomas, C., Santarnecchi, E., Pascual-Leone, A., & Borroni, B. (2021). Exposure to gamma tACS in Alzheimer's disease: A randomized, double-blind, sham-controlled, crossover, pilot study. *Brain Stimulation*, *14*, 531–540. https://doi.org/10.1016/j.brs.2021.03.007

Berger, H. (1929). Über das Elektrenkephalogramm des Menschen, I. Mitteilung. *Archiv Für Psychiatrie Und Nervenkrankheiten*, *87*(4), 527–570.

Berger, H. (1934). Über das Elektrenkephalogramm des Menschen, IX. Mitteilung. *Archiv Für Psychiatrie Und Nervenkrankheiten*, *102*, 538–557. http://doi.org/10.1007/BF01797193

Berger, H. (1936). Über das Elektrenkephalogramm des Menschen, XI. Mitteilung. *Archiv Für Psychiatrie Und Nervenkrankheiten*, *104*, 678–689. http://doi.org/10.1007/BF01797193

Berger, H. (1937). Über das Elektrenkephalogramm des Menschen, XII. Mitteilung. *Archiv Für Psychiatrie Und Nervenkrankheiten*, *106*, 165–187. http://doi.org/10.1007/BF01797193

Bibbig, A., Faulkner, H. J., Whittington, M. A., & Traub, R. D. (2001). Self-organized synaptic plasticity contributes to the shaping of gamma and beta oscillations in vitro. *Journal of Neuroscience*, *21*(22), 9053–9067. http://doi.org/21/22/9053 [pii]

Bosman, C. A., Lansink, C. S., & Pennartz, C. M. A. (2014). Functions of gamma-band synchronization in cognition: From single circuits to functional diversity across cortical and subcortical systems. *European Journal of Neuroscience*, *39*(11), 1982–1999. http://doi.org/10.1111/ejn.12606

Bouyer, J. J., Montaron, M. F., & Rougeul, A. (1981). Fast fronto-parietal rhythms during combined focused attentive behaviour and immobility in cat: Cortical and thalamic localizations. *Electroencephalography and Clinical Neurophysiology*, *51*(3), 244–252. http://doi.org/10.1016/0013-4694(81)90138-3

Bréchet, L., Michel, C. M., Schacter, D. L., & Pascual-Leone, A. (2021). Improving autobiographical memory in Alzheimer's disease by transcranial alternating current stimulation. *Current Opinion in Behavioral Sciences*, 40, 64–71. https://doi.org/10.1016/j.cobeha.2021.01.003

Bressler, S. L. & Freeman, W. J. (1980). Frequency analysis of olfactory system EEG in cat, rabbit, and rat. *Electroencephalography and Clinical Neurophysiology*, *50*(1-2), 19–24. http://doi.org/10.1016/0013-4694(80)90319-3

Bressler, S. L. & Richter, C. G. (2015). Interareal oscillatory synchronization in top-down neocortical processing. *Current Opinion in Neurobiology*, *31*, 62–66. http://doi.org/10.1016/j.conb.2014.08.010

Brunet, N., Vinck, M., Bosman, C. A., Singer, W., & Fries, P. (2014). Gamma or no gamma, that is the question. *Trends in Cognitive Sciences*, *18*(10), 507–509. http://doi.org/10.1016/j.tics.2014.08.006

Buffalo, E. A., Fries, P., Landman, R., Buschman, T. J., & Desimone, R. (2011). Laminar differences in gamma and alpha coherence in the ventral stream. *Proceedings of the National Academy of Sciences of the United States of America*, *108*(27), 11262–11267.

Busch, N. A., Debener, S., Kranczioch, C., Engel, A. K., & Herrmann, C. S. (2004). Size matters: Effects of stimulus size, duration and eccentricity on the visual gamma-band response. *Clinical Neurophysiology*, *115*(8), 1810–1820. http://doi.org/10.1016/j.clinph.2004.03.015

Busch, N. A., Herrmann, C. S., Müller, M. M., Lenz, D., & Gruber, T. (2006). A cross-laboratory study of event-related gamma activity in a standard object recognition paradigm. *NeuroImage*, *33*(4), 1169–1177. http://doi.org/10.1016/j.neuroimage.2006.07.034

Busch, N. A., Schadow, J., Fründ, I., & Herrmann, C. S. (2006). Time-frequency analysis of target detection reveals an early interface between bottom-up and top-down processes in the gamma-band. *NeuroImage*, *29*(4), 1106–1116. http://doi.org/10.1016/j.neuroimage.2005.09.009

Buschman, T. J. & Miller, E. K. (2007). Top-down versus bottom-up control of attention in the prefrontal and posterior parietal cortices. *Science, 315*(5820), 1860–1862. http://doi.org/10.1126/science.1138071

Buzsáki, G. & Wang, X.-J. (2012). Mechanisms of gamma oscillations. *Annual Review of Neuroscience, 35*(1), 203–225. http://doi.org/10.1146/annurev-neuro-062111-150444

Cannon, J., McCarthy, M. M., Lee, S., Lee, J., Börgers, C., Whittington, M. A., & Kopell, N. (2014). Neurosystems: Brain rhythms and cognitive processing. *European Journal of Neuroscience, 39*(5), 705–719. http://doi.org/10.1111/ejn.12453

Crone, N. E., Korzeniewska, A., & Franaszczuk, P. J. (2011). Cortical gamma responses: Searching high and low. *International Journal of Psychophysiology, 79*(1), 9–15. http://doi.org/10.1016/j.ijpsycho.2010.10.013

Darvas, F., Scherer, R., Ojemann, J. G., Rao, R. P., Miller, K. J., & Sorensen, L. B. (2010). High gamma mapping using EEG. *NeuroImage, 49*(1), 930–938. http://doi.org/10.1016/j.neuroimage.2009.08.041

Demiralp, T., Herrmann, C. S., Erdal, M. E., Ergenoglu, T., Keskin, Y. H., Ergen, M., & Beydagi, H. (2007). DRD4 and DAT1 polymorphisms modulate human gamma band responses. *Cerebral Cortex, 17*(5), 1007–1019. http://doi.org/10.1093/cercor/bhl011

Dietsch, G. (1932). Fourier-Analyse von Elektrencephalogrammen des Menschen. *Pflügers Archiv für die gesamte Physiologie des Menschen und der Tiere, 230*, 106–112. http://doi.org/10.1007/BF01751972

Dimigen, O., Valsecchi, M., Sommer, W., & Kliegl, R. (2009). Human microsaccade-related visual brain responses. *Journal of Neuroscience, 29*(39), 12321–12331. http://doi.org/10.1523/JNEUROSCI.0911-09.2009

Eckhorn, R., Bauer, R., Jordan, W., Brosch, M., Kruse, W., Munk, M., & Reitboeck, H. J. (1988). Coherent oscillations: A mechanism of feature linking in the visual cortex? *Biological Cybernetics, 60*, 121–130.

Elliott, M. A., Herrmann, C. S., Mecklinger, A., & Müller, H. J. (2000). The loci of oscillatory visual-object priming: A combined electroencephalographic and reaction-time study. *International Journal of Psychophysiology, 38*(3), 225–241. http://doi.org/10.1016/S0167-8760(00)00167-7

Engbert, R. (2006). Microsaccades: A microcosm for research on oculomotor control, attention, and visual perception. *Progress in Brain Research, 154*, 177–192. http://doi.org/10.1016/S0079-6123(06)54009-9

Engbert, R. & Kliegl, R. (2003). Microsaccades uncover the orientation of covert attention. *Vision Research, 43*(9), 1035–1045. http://doi.org/10.1016/S0042-6989(03)00084-1

Engel, A. K., Fries, P., & Singer, W. (2001). Dynamic predictions: oscillations and synchrony in top-down processing. *Nature Reviews Neuroscience, 2*(10), 704–716. http://doi.org/10.1038/35094565

Engel, A. K., Roelfsema, P. R., Fries, P., Brecht, M., & Singer, W. (1997). Role of the temporal domain for response selection and perceptual binding. *Cerebral Cortex, 7*(6), 571–582. http://doi.org/10.1093/cercor/7.6.571

Engel, A., König, P., Kreiter, A. K., & Singer, W. (1991). Interhemispheric synchronization of oscillatory neuronal responses in cat visual cortex. *Science, 252*(5009), 1177–1179. http://doi.org/10.1126/science.252.5009.1177

Enriquez-Geppert, S., Huster, R. J., & Herrmann, C. S. (2013). Boosting brain functions: Improving executive functions with behavioral training, neurostimulation, and neurofeedback. *International Journal of Psychophysiology*, *88*(1), 1–16. http://doi.org/10.1016/j.ijpsycho.2013.02.001

Fell, J. & Axmacher, N. (2011). The role of phase synchronization in memory processes. *Nature Reviews Neuroscience*, *12*(2), 105–118. http://doi.org/10.1038/nrn2979

Felleman, D. J. & Van Essen, D. C. (1991). Distributed hierarchical processing in the primate cerebral cortex. *Cerebral Cortex*, *1*, 1–47.

Fitzgibbon, S. P., DeLosAngeles, D., Lewis, T. W., Powers, D. M., Whitham, E. M., Willoughby, J. O., & Pope, K. J. (2015). Surface Laplacian of scalp electrical signals and independent component analysis resolve EMG contamination of electroencephalogram. *International Journal of Psychophysiology*, *97*(3), 277–284.

Freeman, W. J. (1975). *Mass action in the nervous system*. Academic Press.

Freeman, W. J. & Skarda, C. A. (1985). Spatial EEG patterns, non-linear dynamics and perception: the neo-Sherringtonian view. *Brain Research Reviews*, *10*, 147–175.

Frien, A., Eckhorn, R., Bauer, R., Woelbern, T., & Kehr, H. (1994). Stimulus-specific fast oscillations at zero phase between visual areas V1 and V2 of awake monkey. *Neuroreport*, *5*(17), 2273–2277.

Fries, P. (2005). A mechanism for cognitive dynamics: Neuronal communication through neuronal coherence. *Trends in Cognitive Sciences*, *9*(10), 474–480. http://doi.org/10.1016/j.tics.2005.08.011

Fries, P. (2009). Neuronal gamma-band synchronization as a fundamental process in cortical computation. *Annual Review of Neuroscience*, *32*(1), 209–224. http://doi.org/10.1146/annurev.neuro.051508.135603

Fries, P. (2015). Rhythms for cognition: Communication through coherence. *Neuron*, *88*(1), 220–235. http://doi.org/10.1016/j.neuron.2015.09.034

Fries, P., Nikolić, D., & Singer, W. (2007). The gamma cycle. *Trends in Neurosciences*, *30*(7), 309–316. http://doi.org/10.1016/j.tins.2007.05.005

Fründ, I., Busch, N. A., Körner, U., Schadow, J., & Herrmann, C. S. (2007). EEG oscillations in the gamma and alpha range respond differently to spatial frequency. *Vision Research*, *47*(15), 2086–2098. http://doi.org/10.1016/j.visres.2007.03.022

Galambos, R., Makeig, S., & Talmachoff, P. J. (1981). A 40-Hz auditory potential recorded from the human scalp. *Proceedings of the National Academy of Sciences of the United States of America*, *78*(4), 2643–2647. http://doi.org/10.1073/pnas.78.4.2643

Gallinat, J., Winterer, G., Herrmann, C. S., & Senkowski, D. (2004). Reduced oscillatory gamma-band responses in unmedicated schizophrenic patients indicate impaired frontal network processing. *Clinical Neurophysiology*, *115*(8), 1863–1874. http://doi.org/10.1016/j.clinph.2004.03.013

Goncharova, I. I., McFarland, D. J., Vaughan, T. M., & Wolpaw, J. R. (2003). EMG contamination of EEG: Spectral and topographical characteristics. *Clinical Neurophysiology*, *114*(9), 1580–1593.

Gray, C. M., König, P., Engel, A. K., & Singer, W. (1989). Oscillatory responses in cat visual cortex exhibit inter-columnar synchronization which reflects global stimulus properties. *Nature*, *338*, 334–337.

Gray, C. M. & Singer, W. (1989). Stimulus-specific neuronal oscillations in orientation columns of cat visual cortex. *Proceedings of the National Academy of Sciences of the United States of America*, *86*(5), 1698–1702. http://doi.org/10.1073/pnas.86.5.1698

Gregoriou, G. G., Gotts, S. J., Zhou, H., & Desimone, R. (2009). High-frequency, long-range coupling between prefrontal and visual cortex during attention. *Science*, *324*(5931), 1207–1210. http://doi.org/10.1126/science.1171402

Gregoriou, G. G., Paneri, S., & Sapountzis, P. (2015). Oscillatory synchrony as a mechanism of attentional processing. *Brain Research, 1626*, 165–182. http://doi.org/10.1016/j.brainres.2015.02.004

Gruber, T., Maess, B., Trujillo-Barreto, N. J., & Müller, M. M. (2008). Sources of synchronized induced gamma-band responses during a simple object recognition task: A replication study in human MEG. *Brain Research, 1196*, 74–84. http://doi.org/10.1016/j.brainres.2007.12.037

Gruber, T., Müller, M. M., Keil, A., & Elbert, T. (1999). Selective visual-spatial attention induced gamma band responses in the human EEG. *Clinical Neurophysiology, 110*(1999), 2074–2085.

Gruber, T., Trujillo-Barreto, N. J., Giabbiconi, C. M., Valdés-Sosa, P. A., & Müller, M. M. (2006). Brain electrical tomography (BET) analysis of induced gamma band responses during a simple object recognition task. *NeuroImage, 29*(3), 888–900. http://doi.org/10.1016/j.neuroimage.2005.09.004

Gruber, T., Tsivilis, D., Montaldi, D., & Müller, M. M. (2004). Induced gamma band responses: An early marker of memory encoding and retrieval. *NeuroReport, 15*(11), 1837–1841. http://doi.org/10.1097/01.wnr.0000137077.26010.12

Haenschel, C., Bittner, R. A., Waltz, J., Haertling, F., Wibral, M., Singer, W., ... Rodriguez, E. (2009). Cortical oscillatory activity is critical for working memory as revealed by deficits in early-onset schizophrenia. *Journal of Neuroscience, 29*(30), 9481–9489. http://doi.org/10.1523/JNEUROSCI.1428-09.2009

Hassler, U., Friese, U., Martens, U., Trujillo-Barreto, N., & Gruber, T. (2013). Repetition priming effects dissociate between miniature eye movements and induced gamma-band responses in the human electroencephalogram. *European Journal of Neuroscience, 38*(3), 2425–2433. http://doi.org/10.1111/ejn.12244

Hassler, U., Trujillo Barreto, N., & Gruber, T. (2011). Induced gamma band responses in human EEG after the control of miniature saccadic artifacts. *NeuroImage, 57*(4), 1411–1421. http://doi.org/10.1016/j.neuroimage.2011.05.062

Hermes, D., Miller, K. J., Wandell, B. A., & Winawer, J. (2015). Gamma oscillations in visual cortex: The stimulus matters. *Trends in Cognitive Sciences, 19*(2), 57–58. http://doi.org/10.1016/j.tics.2014.12.009

Herrmann, C. S. (2001). Human EEG responses to 1-100 Hz flicker: Resonance phenomena in visual cortex and their potential correlation to cognitive phenomena. *Experimental Brain Research, 137*(3-4), 346–353. http://doi.org/10.1007/s002210100682

Herrmann, C. S. & Demiralp, T. (2005). Human EEG gamma oscillations in neuropsychiatric disorders. *Clinical Neurophysiology, 116*(12), 2719–2733. http://doi.org/10.1016/j.clinph.2005.07.007

Herrmann, C. S., Fründ, I., & Lenz, D. (2010). Human gamma-band activity: A review on cognitive and behavioral correlates and network models. *Neuroscience and Biobehavioral Reviews, 34*(7), 981–992. http://doi.org/10.1016/j.neubiorev.2009.09.001

Herrmann, C. S., Lenz, D., Junge, S., Busch, N. A., & Maess, B. (2004). Memory-matches evoke human gamma-responses. *BMC Neuroscience, 5*, 13.

Herrmann, C. S. & Mecklinger, A. (2001). Gamma activity in human EEG is related to highspeed memory comparisons during object selective attention. *Visual Cognition, 8*(3-5), 593–608. http://doi.org/10.1080/13506280143000142

Herrmann, C. S., Mecklinger, A., & Pfeifer, E. (1999). Gamma responses and ERPs in a visual classification task. *Clinical Neurophysiology, 110*(4), 636–642. http://doi.org/10.1016/S1388-2457(99)00002-4

Herrmann, C. S., Munk, M. H. J., & Engel, A. K. (2004). Cognitive functions of gamma-band activity: Memory match and utilization. *Trends in Cognitive Sciences*, *8*(8), 347–355. http://doi.org/10.1016/j.tics.2004.06.006

Herrmann, C. S., Rach, S., Vosskuhl, J., & Strüber, D. (2014). Time-frequency analysis of event-related potentials: A brief tutorial. *Brain Topography*, *27*(4), 438–450. http://doi.org/10.1007/s10548-013-0327-5

Herrmann, C. S. & Strüber, D. (2017). What can transcranial alternating current stimulation tell us about brain oscillations? *Current Behavioral Neuroscience Reports*, *4*(2), 128–137. http://doi.org/10.1007/s40473-017-0114-9

Herrmann, C. S., Strüber, D., Helfrich, R. F., & Engel, A. K. (2016). EEG oscillations: From correlation to causality. *International Journal of Psychophysiology*, *103*, 12–21. http://doi.org/10.1016/j.ijpsycho.2015.02.003

Iaccarino, H. F., Singer, A. C., Martorell, A. J., Rudenko, A., Gao, F., Gillingham, T. Z., ... Tsai, L. H. (2016). Gamma frequency entrainment attenuates amyloid load and modifies microglia. *Nature*, *540*(7632), 230–235. http://doi.org/10.1038/nature20587

Jadi, M. P., Behrens, M. M., & Sejnowski, T. J. (2016). Abnormal gamma oscillations in N-Methyl-D-Aspartate receptor hypofunction models of schizophrenia. *Biological Psychiatry*, *79*(9), 716–726. http://doi.org/10.1016/j.biopsych.2015.07.005

Janani, A. S., Grummett, T. S., Lewis, T. W., Fitzgibbon, S. P., Whitham, E. M., DelosAngeles, D., ... Pope, K. J. (2018). Improved artefact removal from EEG using canonical correlation analysis and spectral slope. *Journal of Neuroscience Methods*, *298*, 1–15. http://doi.org/10.1016/j.jneumeth.2018.01.004

Jasper, H. H. & Andrews, H. L. (1938). Electro-encephalography III. Normal differentiation of occipital and precentral regions in man. *Archives of Neurology & Psychiatry*, *39*(1), 96–115.

Jensen, O., Kaiser, J., & Lachaux, J. P. (2007). Human gamma-frequency oscillations associated with attention and memory. *Trends in Neurosciences*, *30*(7), 317–324. http://doi.org/10.1016/j.tins.2007.05.001

Kaiser, J. & Lutzenberger, W. (2005). Human gamma-band activity: A window to cognitive processing. *NeuroReport*, *16*(3), 207–211. http://doi.org/10.1097/00001756-200502280-00001

Keizer, A. W., Verment, R. S., & Hommel, B. (2010). Enhancing cognitive control through neurofeedback: A role of gamma-band activity in managing episodic retrieval. *NeuroImage*, *49*(4), 3404–3413. http://doi.org/10.1016/j.neuroimage.2009.11.023

Keizer, A. W., Verschoor, M., Verment, R. S., & Hommel, B. (2010). The effect of gamma enhancing neurofeedback on the control of feature bindings and intelligence measures. *International Journal of Psychophysiology*, *75*(1), 25–32. http://doi.org/10.1016/j.ijpsycho.2009.10.011

Keren, A. S., Yuval-Greenberg, S., & Deouell, L. Y. (2010). Saccadic spike potentials in gamma-band EEG: Characterization, detection and suppression. *NeuroImage*, *49*(3), 2248–2263. http://doi.org/10.1016/j.neuroimage.2009.10.057

Kopell, N., Ermentrout, G. B., Whittington, M. A., & Traub, R. D. (2000). Gamma rhythms and beta rhythms have different synchronization properties. *Proceedings of the National Academy of Sciences of the United States of America*, *97*(4), 1867–1872. http://doi.org/10.1073/pnas.97.4.1867

Kreiter, A. K. & Singer, W. (1996). Stimulus-dependent synchronization of neuronal responses in the visual cortex of the awake macaque monkey. *Journal of Neuroscience*, *16*(7), 2381–2396.

Lally, N., Mullins, P. G., Roberts, M. V., Price, D., Gruber, T., & Haenschel, C. (2014). Glutamatergic correlates of gamma-band oscillatory activity during cognition: A concurrent ER-MRS and EEG study. *NeuroImage*, *85*, 823–833. http://doi.org/10.1016/j.neuroimage.2013.07.049

Leicht, G., Kirsch, V., Giegling, I., Karch, S., Hantschk, I., Möller, H. J., ... Mulert, C. (2010). Reduced early auditory evoked gamma-band response in patients with schizophrenia. *Biological Psychiatry*, *67*(3), 224–231. http://doi.org/10.1016/j.biopsych.2009.07.033

Lenz, D., Fischer, S., Schadow, J., Bogerts, B., & Herrmann, C. S. (2011). Altered evoked gamma-band responses as a neurophysiological marker of schizophrenia? *International Journal of Psychophysiology*, *79*(1), 25–31. http://doi.org/10.1016/j.ijpsycho.2010.08.002

Lenz, D., Jeschke, M., Schadow, J., Naue, N., Ohl, F. W., & Herrmann, C. S. (2008). Human EEG very high frequency oscillations reflect the number of matches with a template in auditory short-term memory. *Brain Research*, *1220*, 81–92. http://doi.org/10.1016/j.brainres.2007.10.053

Lenz, D., Krauel, K., Flechtner, H. H., Schadow, J., Hinrichs, H., & Herrmann, C. S. (2010). Altered evoked gamma-band responses reveal impaired early visual processing in ADHD children. *Neuropsychologia*, *48*(7), 1985–1993. http://doi.org/10.1016/j.neuropsychologia.2010.03.019

Lenz, D., Krauel, K., Schadow, J., Baving, L., Duzel, E., & Herrmann, C. S. (2008). Enhanced gamma-band activity in ADHD patients lacks correlation with memory performance found in healthy children. *Brain Research*, *1235*, 117–132. http://doi.org/10.1016/j.brainres.2008.06.023

Lutzenberger, W., Pulvermüller, F., Elbert, T., & Birbaumer, N. (1995). Visual stimulation alters local 40-Hz responses in humans: an EEG-study. *Neuroscience Letters*, *183*(1–2), 39–42. http://doi.org/10.1016/0304-3940(94)11109-V

Markov, N. T., Vezoli, J., Chameau, P., Falchier, A., Quilodran, R., Huissoud, C., ... Kennedy, H. (2014). Anatomy of hierarchy: Feedforward and feedback pathways in macaque visual cortex. *Journal of Comparative Neurology*, *522*(1), 225–259. http://doi.org/10.1002/cne.23458

Martinez-Conde, S., Otero-Millan, J., & Macknik, S. L. (2013). The impact of microsaccades on vision: Towards a unified theory of saccadic function. *Nature Reviews Neuroscience*, *14*(2), 83–96. http://doi.org/10.1038/nrn3405

Mathes, B., Strüber, D., Stadler, M. A., & Başar-Eroğlu, C. (2006). Voluntary control of Necker cube reversals modulates the EEG delta- and gamma-band response. *Neuroscience Letters*, *402*(1-2), 145–149. http://doi.org/10.1016/j.neulet.2006.03.063

McNally, J. M. & McCarley, R. W. (2016). Gamma band oscillations: A key to understanding schizophrenia symptoms and neural circuit abnormalities. *Current Opinion in Psychiatry*, *29*(3), 202–210. http://doi.org/10.1097/YCO.0000000000000244.Gamma

Merkel, N., Wibral, M., Bland, G., & Singer, W. (2018). Endogenously generated gamma-band oscillations in early visual cortex: A neurofeedback study. *Human Brain Mapping*, *39*(9), 3487–3502. http://doi.org/10.1002/hbm.24189

Merker, B. (2013). Cortical gamma oscillations: The functional key is activation, not cognition. *Neuroscience and Biobehavioral Reviews*, *37*(3), 401–417. http://doi.org/10.1016/j.neubiorev.2013.01.013

Michalareas, G., Vezoli, J., van Pelt, S., Schoffelen, J. M., Kennedy, H., & Fries, P. (2016). Alpha-beta and gamma rhythms subserve feedback and feedforward influences among human visual cortical areas. *Neuron*, *89*(2), 384–397. http://doi.org/10.1016/j.neuron.2015.12.018

Miltner, W. H. R., Braun, C., Arnold, M., Witte, H., & Taub, E. (1999). Coherence of gamma-band EEG activity as a basis for associative learning. *Nature*, *397*(6718), 434–436. http://doi.org/10.1038/17126

Minzenberg, M. J., Firl, A. J., Yoon, J. H., Gomes, G. C., Reinking, C., & Carter, C. S. (2010). Gamma oscillatory power is impaired during cognitive control independent of medication status in first-episode schizophrenia. *Neuropsychopharmacology*, *35*(13), 2590–2599. http://doi.org/10.1038/npp.2010.150

Müller, M. M., Bosch, J., Elbert, T., Kreiter, A., Sosa, M., Sosa, P., & Rockstroh, B. (1996). Visually induced gamma-band responses in human electroencephalographic activity: A link to animal studies. *Experimental Brain Research*, *112*(1), 96–102. http://doi.org/10.1007/BF00227182

Naue, N., Strüber, D., Fründ, I., Schadow, J., Lenz, D., Rach, S., Körner, U., & Herrmann, C. S. (2011). Gamma in motion: Pattern reversal elicits stronger gamma-band responses than motion. *NeuroImage*, *55*(2), 808–817. http://doi.org/10.1016/j.neuroimage.2010.11.053

Nimmrich, V., Draguhn, A., & Axmacher, N. (2015). Neuronal network oscillations in neurodegenerative diseases. *Neuromolecular Medicine*, *17*(3), 270–284. http://doi.org/10.1007/s12017-015-8355-9

Nottage, J. F. (2010). Uncovering gamma in visual tasks. *Brain Topography*, *23*(1), 58–71. http://doi.org/10.1007/s10548-009-0129-y

Nottage, J. F. & Horder, J. (2016). State-of-the-art analysis of high-frequency (gamma range) electroencephalography in humans. *Neuropsychobiology*, *72*(3-4), 219–228. http://doi.org/10.1159/000382023

Nottage, J. F., Morrison, P. D., Williams, S. C. R., & Ffytche, D. H. (2013). A novel method for reducing the effect of tonic muscle activity on the gamma band of the scalp EEG. *Brain Topography*, *26*(1), 50–61. http://doi.org/10.1007/s10548-012-0255-9

Oppermann, F., Hassler, U., Jescheniak, J. D., & Gruber, T. (2012). The rapid extraction of gist—early neural correlates of high-level visual processing. *Journal of Cognitive Neuroscience*, *24*(2), 521–529.

Palop, J. J. & Mucke, L. (2016). Network abnormalities and interneuron dysfunction in Alzheimer disease. *Nature Reviews Neuroscience*, *17*(12), 777–792. http://doi.org/10.1038/nrn.2016.141

Paulsen, O. & Sejnowski, T. J. (2000). Natural patterns of activity and long-term synaptic plasticity. *Current Opinion in Neurobiology*, *10*(2), 172–179. http://doi.org/10.1016/S0959-4388(00)00076-3

Plöchl, M., Ossandón, J. P., & König, P. (2012). Combining EEG and eye tracking: Identification, characterization, and correction of eye movement artifacts in electroencephalographic data. *Frontiers in Human Neuroscience*, *6*, 278. http://doi.org/10.3389/fnhum.2012.00278

Pope, K. J., Fitzgibbon, S. P., Lewis, T. W., Whitham, E. M., & Willoughby, J. O. (2009). Relation of gamma oscillations in scalp recordings to muscular activity. *Brain Topography*, *22*(1), 13–17. http://doi.org/10.1007/s10548-009-0081-x

Prehn-Kristensen, A., Wiesner, C. D., & Baving, L. (2015). Early gamma-band activity during interference predicts working memory distractibility in ADHD. *Journal of Attention Disorders*, *19*(11), 971–976. http://doi.org/10.1177/1087054712459887

Pulvermüller, F., Birbaumer, N., Lutzenberger, W., & Mohr, B. (1997). High-frequency brain activity: Its possible role in attention, perception and language processing. *Progress in Neurobiology*, *52*(5), 427–445.

Ray, S. & Maunsell, J. H. R. (2011). Different origins of gamma rhythm and high-gamma activity in macaque visual cortex. *PLoS Biology*, *9*(4), e1000610. http://doi.org/10.1371/journal.pbio.1000610

Ray, S. & Maunsell, J. H. R. (2015). Do gamma oscillations play a role in cerebral cortex? *Trends in Cognitive Sciences*, *19*(2), 78–85. http://doi.org/10.1016/j.tics.2014.12.002

Regan, D. (1968). A high frequency mechanism which underlies visual evoked potentials. *Electroencephalography and Clinical Neurophysiology*, 25, 231–237.

Regan, D. & Spekreijse, H. (1986). Evoked potentials in vision research 1961–86. *Vision Research*, *26*(9), 1461–1480. http://doi.org/10.1016/B978-0-12-385157-4.00529-7

Reilly, T. J., Nottage, J. F., Studerus, E., Rutigliano, G., De Micheli, A. I., Fusar-Poli, P., & McGuire, P. (2018). Gamma band oscillations in the early phase of psychosis: A systematic review. *Neuroscience & Biobehavioral Reviews*, *90*, 381–399. http://doi.org/10.1016/j.neubiorev.2018.04.006

Reva, N. V. & Aftanas, L. I. (2004). The coincidence between late non-phase-locked gamma synchronization response and saccadic eye movements. *International Journal of Psychophysiology*, *51*(3), 215–222. http://doi.org/10.1016/j.ijpsycho.2003.09.005

Rolfs, M., Kliegl, R., & Engbert, R. (2008). Toward a model of microsaccade generation: The case of microsaccadic inhibition. *Journal of Vision*, *8*(5), 1–23.

Rossi, S., Hallett, M., Rossini, P. M., Pascual-Leone, A., Avanzini, G., Bestmann, S., … Ziemann, U. (2009). Safety, ethical considerations, and application guidelines for the use of transcranial magnetic stimulation in clinical practice and research. *Clinical Neurophysiology*, *120*(12), 2008–2039. http://doi.org/10.1016/j.clinph.2009.08.016

Ruden, J. B., Dugan, L. L., Konradi, C. (2021). Parvalbumin interneuron vulnerability and brain disorders. *Neuropsychopharmacology*, 46, 279–287. https://doi.org/10.1038/s41386-020-0778-9

Salari, N., Büchel, C., & Rose, M. (2014). Neurofeedback training of gamma band oscillations improves perceptual processing. *Experimental Brain Research*, *232*(10), 3353–3361. http://doi.org/10.1007/s00221-014-4023-9

Salinas, E. & Sejnowski, T. J. (2001). Correlated neuronal activity and the flow of neural information. *Nature Reviews Neuroscience*, 2, 539–550. http://doi.org/10.1038/35086012

Schadow, J., Lenz, D., Thaerig, S., Busch, N. A., Fründ, I., Rieger, J. W., & Herrmann, C. S. (2007). Stimulus intensity affects early sensory processing: Visual contrast modulates evoked gamma-band activity in human EEG. *International Journal of Psychophysiology*, *66*(1), 28–36. http://doi.org/10.1016/j.ijpsycho.2007.05.010

Schwartzman, D. J. & Kranczioch, C. (2011). In the blink of an eye: The contribution of microsaccadic activity to the induced gamma band response. *International Journal of Psychophysiology*, *79*(1), 73–82. http://doi.org/10.1016/j.ijpsycho.2010.10.006

Sejnowski, T. J. & Paulsen, O. (2006). Network oscillations: Emerging computational principles. *Journal of Neuroscience*, *26*(6), 1673–1676. http://doi.org/10.1523/JNEUROSCI.3737-05d.2006

Senkowski, D. & Gallinat, J. (2015). Dysfunctional prefrontal gamma-band oscillations reflect working memory and other cognitive deficits in schizophrenia. *Biological Psychiatry*, *77*(12), 1010–1019.

Sheer, D. E. (1984). Focused arousal, 40-Hz EEG, and dysfunction. In T. Elbert, B. Rockstroh, W. Lutzenberger, & N. Birbaumer (Eds.), *Self-regulation of the brain and behavior* (pp. 64–84). Springer.

Siegel, M., Donner, T. H., & Engel, A. K. (2012). Spectral fingerprints of large-scale neuronal interactions. *Nature Reviews Neuroscience*, *13*(2), 121–134. http://doi.org/10.1038/nrn3137

Singer, W. (2018). Neuronal oscillations: Unavoidable and useful? *European Journal of Neuroscience*, *48*(7), 2389–2398. http://doi.org/10.1111/ejn.13796

Singer, W. & Gray, C. M. (1995). Visual feature integration and the temporal correlation hypothesis. *Annual Review of Neuroscience, 18*, 555–586. http://doi.org/10.1146/annurev.ne.18.030195.003011

Spencer, K. M. (2012). Baseline gamma power during auditory steady-state stimulation in schizophrenia. *Frontiers in Human Neuroscience*, 5, 190. http://doi.org/10.3389/fnhum.2011.00190

Spencer, K. M., Niznikiewicz, M. A., Shenton, M. E., & McCarley, R. W. (2008). Sensory-evoked gamma oscillations in chronic schizophrenia. *Biological Psychiatry, 63*(8), 744–747.

Spydell, J. D., Ford, J. D., & Sheer, D. E. (1979). Task dependent cerebral lateralization of the 40 Hertz EEG rhythm. *Psychophysiology, 16*(4), 347–350.

Stroganova, T. A., Butorina, A. V., Sysoeva, O. V., Prokofyev, A. O., Nikolaeva, A. Y., Tsetlin, M. M., & Orekhova, E. V. (2015). Altered modulation of gamma oscillation frequency by speed of visual motion in children with autism spectrum disorders. *Journal of Neurodevelopmental Disorders, 7*(1), 1–17. http://doi.org/10.1186/s11689-015-9121-x

Strüber, D., Başar-Eroğlu, C., Hoff, E., & Stadler, M. A. (2000). Reversal-rate dependent differences in the EEG gamma-band during multistable visual perception. *International Journal of Psychophysiology, 38*(3), 243–252.

Strüber, D., Başar-Eroğlu, C., Miener, M., & Stadler, M. A. (2001). EEG gamma-band response during the perception of Necker cube reversals. *Visual Cognition, 8*(3), 609–621. http://doi.org/10.1080/13506280143000151

Strüber, D. & Herrmann, C. S. (2020). Modulation of gamma oscillations as a possible therapeutic tool for neuropsychiatric diseases: A review and perspective. *International Journal of Psychophysiology, 152*, 15–25. https://doi.org/10.1016/j.ijpsycho.2020.03.003

Summerfield, C. & Mangels, J. A. (2006). Dissociable neural mechanisms for encoding predictable and unpredictable events. *Journal of Cognitive Neuroscience, 18*(7), 1120–1132.

Swettenham, J. B., Muthukumaraswamy, S. D., & Singh, K. D. (2009). Spectral properties of induced and evoked gamma oscillations in human early visual cortex to moving and stationary stimuli. *Journal of Neurophysiology, 102*(2), 1241–1253. http://doi.org/10.1152/jn.91044.2008

Tal, N. & Yuval-Greenberg, S. (2018). Reducing saccadic artifacts and confounds in brain imaging studies through experimental design. *Psychophysiology, 55*(11), 1–20. http://doi.org/10.1111/psyp.13215

Tallon-Baudry, C. (2009). The roles of gamma-band oscillatory synchrony in human visual cognition. *Frontiers in Bioscience, 14*, 321–332.

Tallon-Baudry, C. & Bertrand, O. (1999). Oscillatory gamma activity in humans and its role in object representation. *Trends in Cognitive Sciences, 3*(4), 151–162.

Tallon-Baudry, C., Bertrand, O., Delpuech, C., & Pernier, J. (1997). Oscillatory γ-band (30–70 Hz) activity induced by a visual search task in humans. *The Journal of Neuroscience, 17*(2), 722–734. https://doi.org/10.1523/JNEUROSCI.17-02-00722.1997

Tallon-Baudry, C., Bertrand, O., Delpuech, C., & Pernier, J. (1996). Stimulus specificity of phase-locked and non-phase-locked 40 Hz visual responses in human. *The Journal of Neuroscience, 16*(13), 4240–4249. http://doi.org/10.1016/j.neuropsychologia.2011.02.038

Tallon-Baudry, C., Bertrand, O., Peronnet, F., & Pernier, J. (1998). Induced gamma-band activity during the delay of a visual short-term memory task in humans. *The Journal of Neuroscience, 18*(11), 4244–4254. http://doi.org/20026318

Thickbroom, G. W. & Mastaglia, L. (1985). Presaccadic "spike" potential: Investigation of topography and source. *Brain Research, 339*, 271–280.

Tombor, L., Kakuszi, B., Papp, S., Réthelyi, J., Bitter, I., & Czobor, P. (2019). Decreased resting gamma activity in adult attention deficit/hyperactivity disorder. *The World Journal of Biological Psychiatry, 20*(9), *691-702*. http://doi.org/10.1080/15622975.2018.1441547

Uhlhaas, P. J. & Singer, W. (2006). Neural synchrony in brain disorders: Relevance for cognitive dysfunctions and pathophysiology. *Neuron, 52*(1), 155–168. http://doi.org/10.1016/j.neuron.2006.09.020

Uhlhaas, P. J. & Singer, W. (2010). Abnormal neural oscillations and synchrony in schizophrenia. *Nature Reviews Neuroscience, 11*(2), 100–113. http://doi.org/10.1038/nrn2774

Uhlhaas, P. J. & Singer, W. (2012). Neuronal dynamics and neuropsychiatric disorders: Toward a translational paradigm for dysfunctional large-scale networks. *Neuron, 75*(6), 963–980. http://doi.org/10.1016/j.neuron.2012.09.004

Valsecchi, M., Betta, E., & Turatto, M. (2007). Visual oddballs induce prolonged microsaccadic inhibition. *Experimental Brain Research, 177*(2), 196–208. http://doi.org/10.1007/s00221-006-0665-6

Valsecchi, M., Dimigen, O., Kliegl, R., Sommer, W., & Turatto, M. (2009). Microsaccadic inhibition and P300 enhancement in a visual oddball task. *Psychophysiology, 46*(3), 635–644. http://doi.org/10.1111/j.1469-8986.2009.00791.x

Van Diessen, E., Senders, J., Jansen, F. E., Boersma, M., & Bruining, H. (2015). Increased power of resting-state gamma oscillations in autism spectrum disorder detected by routine electroencephalography. *European Archives of Psychiatry and Clinical Neuroscience, 265*(6), 537–540. http://doi.org/10.1007/s00406-014-0527-3

Van Kerkoerle, T., Self, M. W., Dagnino, B., Gariel-Mathis, M.-A., Poort, J., van der Togt, C., & Roelfsema, P. R. (2014). Alpha and gamma oscillations characterize feedback and feedforward processing in monkey visual cortex. *Proceedings of the National Academy of Sciences of the United States of America, 111*(40), 14332–14341.

Varela, F., Lachaux, J. P., Rodriguez, E., & Martinerie, J. (2001). The brainweb: Phase synchronization and large-scale integration. *Nature Reviews Neuroscience, 2*(4), 229–239. http://doi.org/10.1038/35067550

Vinck, M. & Bosman, C. A. (2016). More gamma more predictions: Gamma-synchronization as a key mechanism for efficient integration of classical receptive field inputs with surround predictions. *Frontiers in Systems Neuroscience*, 10, 35. http://doi.org/10.3389/fnsys.2016.00035

Von Stein, A., Chiang, C., & König, P. (2000). Top-down processing mediated by interareal synchronization. *Proceedings of the National Academy of Sciences of the United States of America, 97*(26), 14748–14753. http://doi.org/10.1073/pnas.97.26.14748

Von Stein, A. & Sarnthein, J. (2000). Different frequencies for different scales of cortical integration: From local gamma to long range alpha/theta synchronization. *International Journal of Psychophysiology, 38*(3), 301–313. http://doi.org/10.1016/S0167-8760(00)00172-0

Wang, J., Barstein, J., Ethridge, L. E., Mosconi, M. W., Takarae, Y., & Sweeney, J. A. (2013). Resting state EEG abnormalities in autism spectrum disorders. *Journal of Neurodevelopmental Disorders, 5*(1), 24. http://doi.org/10.1186/1866-1955-5-24

Wang, X.-J. (2010). Neurophysiological and computational principles of cortical rhythms in cognition. *Physiological Reviews, 90*(3), 1195–1268. http://doi.org/10.1152/physrev.00035.2008

White, R. S. & Siegel, S. J. (2016). Cellular and circuit models of increased resting-state network gamma activity in schizophrenia. *Neuroscience, 321*, 66–76.

Whitham, E. M., Lewis, T., Pope, K. J., Fitzgibbon, S. P., Clark, C. R., Loveless, S., . . . Willoughby, J. O. (2008). Thinking activates EMG in scalp electrical recordings. *Clinical Neurophysiology, 119*(5), 1166–1175. http://doi.org/10.1016/j.clinph.2008.01.024

Whitham, E. M., Pope, K. J., Fitzgibbon, S. P., Lewis, T., Clark, C. R., Loveless, S., . . . Willoughby, J. O. (2007). Scalp electrical recording during paralysis: Quantitative evidence that EEG frequencies above 20 Hz are contaminated by EMG. *Clinical Neurophysiology*, *118*(8), 1877–1888. http://doi.org/10.1016/j.clinph.2007.04.027

Whittington, M. A., Cunningham, M. O., LeBeau, F. E. N., Racca, C., & Traub, R. D. (2011). Multiple origins of the cortical gamma rhythm. *Developmental Neurobiology*, *71*(1), 92–106. http://doi.org/10.1002/dneu.20814

Whittington, M. A., Traub, R. D., Kopell, N., Ermentrout, B., & Buhl, E. H. (2000). Inhibition-based rhythms: Experimental and mathematical observations on network dynamics. *International Journal of Psychophysiology*, *38*(3), 315–336.

Yordanova, J., Banaschewski, T., Kolev, V., Woerner, W., & Rothenberger, A. (2001). Abnormal early stages of task stimulus processing in children with attention-deficit hyperactivity disorder: Evidence from event-related gamma oscillations. *Clinical Neurophysiology*, *112*(6), 1096–1108. http://doi.org/10.1016/S1388-2457(01)00524-7

Yuval-Greenberg, S. & Deouell, L. Y. (2009). The broadband-transient induced gamma-band response in scalp EEG reflects the execution of saccades. *Brain Topography*, *22*(1), 3–6. http://doi.org/10.1007/s10548-009-0077-6

Yuval-Greenberg, S., Tomer, O., Keren, A. S., Nelken, I., & Deouell, L. Y. (2008). Transient induced gamma-band response in EEG as a manifestation of miniature saccades. *Neuron*, *58*(3), 429–441. http://doi.org/10.1016/j.neuron.2008.03.027

Zaehle, T., Fründ, I., Schadow, J., Thärig, S., Schoenfeld, M. A., & Herrmann, C. S. (2009). Inter- and intra-individual covariations of hemodynamic and oscillatory gamma responses in the human cortex. *Frontiers in Human Neuroscience*, *3*, 8. http://doi.org/10.3389/neuro.09.008.2009

CHAPTER 9

FRONTAL MIDLINE THETA AS A MODEL SPECIMEN OF CORTICAL THETA

JAMES F. CAVANAGH AND MICHAEL X COHEN

9.1 History

MULTIPLE distinct rhythmic processes in the approximate 4–8 Hz theta-band range have been historically observed in the continuous EEG. Widely distributed theta-band rhythmicity is commonly observed during stage 1 sleep. Abnormal background presence or hemispheric asymmetries in this rhythm are used to infer neural dysfunction in clinical practice. Yet nearly 70 years of observation have also detailed a task-related increase in frontal midline theta (FMT), usually observed over electrode Fz or its nearest 10/20 system neighbors during effortful mental processes (Arellano & Schwab, 1950; Brazier & Casby, 1952; Inouye, Shinosaki, Iyama, Matsumoto, Toi, et al., 1994; Sato, 1952). Since this FMT feature is cognitively induced, it has been considered distinct from arousal- or clinically related theta rhythms. These spatially specified and elicitation-dependent distinctions are critical to make when describing any frequency-based EEG activity: manifest characteristics like a common frequency range can be shared between different latent processes.

Towards the end of the twentieth century, a growing quantitative approach to EEG was facilitated by advancements in signal processing methods, including event-related potentials (ERPs) and Fourier-based analyses, improvements in computational power to implement those analyses, and increased data quality through better scalp EEG equipment and direct intracranial human recordings during cognitive performance. Following these technical revolutions, an emerging consensus deduced that cortically generated theta-band activities are reliably associated with general cognitive performance (Başar, 1998b), including specific processes like working memory (Asada et al., 1999; Gevins et al., 1997; Ishii et al., 1999), spatial navigation (Kahana et al., 1999), or episodic memory encoding and retrieval (Jacobs et al., 2006; Kahana et al., 1999; Klimesch, 1999; Klimesch et al., 2000; Nyhus & Curran, 2010; Rizzuto et al., 2006; Sauseng et al., 2004). These findings indicated that theta is generated across human neocortical areas (Caplan et al., 2003; Jacobs et al., 2006; Raghavachari et al., 2006) and that it reflects a multitude of active cortical processes. Here we reiterate our caveat for determining the relationship between a brain rhythm and a specific cognitive process: rhythmic processes reflect common neural operations that have distinct representational content depending on the generative neural system. Theta-band dynamics reflect a non-specific marker of active cortical operation. The frontal midline variant of theta is particularly prevalent in the human EEG, making it a good example for understanding broader cortical theta activities.

Experimental findings have accumulated an increasingly well-defined set of processes associated with FMT. Talairach and colleagues (1973) described how electrical stimulation of the human anterior cingulate cortex (ACC) elicited motor actions that were integrated with environmental context, sometimes accompanied by FMT oscillatory activities. Throughout the 1990s, the qualitative appearance of scalp-recorded FMT during cognitive effort could be reliably evoked (Asada et al., 1999; Inouye et al., 1994; Inouye, Shinosaki, Iyama, Matsumoto, Toi, et al., 1994) and a corresponding relationship with anxiety was often observed (Mizuki et al., 1997; Mizuki et al., 1992; Mizuki et al., 1996). These two *effective* and *affective* facets of vigilance have been increasingly associated with FMT throughout the past few decades. Paralleling the evolving capabilities of experimental neurophysiology, a growing area of research has moved beyond qualitative assessment towards a detailed quantification of the generators, elicitors, and moderators involved in the genesis of the FMT rhythm.

9.2 Characteristics

Mental effort is a reliable elicitor of FMT (Smit et al., 2004; Smit et al., 2005). FMT increases during perseverance and decreases during fatigue (Wascher et al., 2014). Working memory load scales with FMT power (Gevins & Smith, 2000; Ishii et al., 1999; Itthipuripat et al., 2013; Onton et al., 2005; Sauseng et al., 2010), although it should be examined if this relationship is simply due to increased effort or if it represents specific information content (see Hsieh et al., 2011; Roberts et al., 2013). Memory encoding and retrieval are associated with broad cortical theta, including FMT (Hsieh & Ranganath, 2014; Jacobs et al., 2006; Kahana et al., 1999; Klimesch, 1999; Klimesch et al., 2000; Nyhus & Curran, 2010; Rizzuto et al., 2003, 2006; Sauseng et al., 2004). It remains unknown if the role of FMT is specific to control processes in memory rather than encoding per se (see Hanslmayr et al., 2010; Staudigl et al., 2010), particularly since other cortical theta is specifically associated with encoding (see Rizzuto et al., 2006; Wang et al., 2018).

FMT has been localized to broad medial frontal cortical areas, including the ACC and the midcingulate cortex (MCC) using MEG (Beaton et al., 2018; Ishii et al., 1999; Jensen

& Tesche, 2002) and EEG (Cohen & Ridderinkhof, 2013; Gevins & Smith, 2000; Gevins et al., 1999; Gevins et al., 1997; Onton et al., 2005). The MCC generates oscillations in the theta band in human intracranial recording (Cohen et al., 2008; Wang et al., 2005) as well as in non-human primates (Tsujimoto et al., 2010; Tsujimoto et al., 2006; Womelsdorf, Johnston, et al., 2010; Womelsdorf, Vinck, et al., 2010). Recent reviews have summarized distinct functional aspects of FMT, including its modulators (Mitchell, McNaughton, Flanagan, & Kirk, 2008), its role in memory (Hsieh & Ranganath, 2014), and its broader role in cognitive control (Cavanagh & Frank, 2014). We turn now to this defining area of cognitive control and the history of linking FMT processes to frontal midline ERP components intricately related to the need for control.

9.2.1 ERP Findings Linked to the FMT Response

Figure 9.1 shows frontal midline ERPs associated with a common theta-band substrate. The discovery of the response-locked frontal midline ERP feature known as the error-related negativity (ERN) indirectly motivated a renewed interest in FMT dynamics (Falkenstein et al., 1991; Gehring et al., 1993). Not long after its discovery, the ERN was characterized as being particularly sensitive to anxiety (Gehring et al., 2000) with a presumed generative source in ACC (Dehaene et al., 1994; Ishii et al., 1999; Van Veen & Carter, 2002). Importantly, the ERN was shown to have a spectral response in the theta band (Luu & Tucker, 2001; Luu et al., 2004), although concurrent delta-band activities have also been noted (Yordanova et al., 2004) and recently highlighted as functionally distinct processes (Cohen & Donner, 2013).

A smaller voltage negativity was also observed on non-error trials (Coles et al., 2001; Vidal et al., 2003; Vidal et al., 2000), which is sometimes called the correct-related negativity (CRN). While this EEG feature was sometimes interpreted as an artifact (Coles et al., 2001), it is now known to be reliably observed during action commission, even in the absence of task demands (Cavanagh et al., 2012). CRN amplitudes mirror variations in performance monitoring during manual responses, as they are larger under conditions of increased task difficulty (Hajcak et al., 2005) and uncertainty (Pailing & Segalowitz, 2004) and they are diminished on correct trials immediately preceding errors (Allain et al., 2004; Cavanagh et al., 2009).

A stimulus-locked error signal following punishing or error feedback was characterized soon after the discovery of the ERN (Miltner et al., 1997), with varied theoretical accounts arguing for a common process with the response-locked ERN (Holroyd & Coles, 2002; Holroyd et al., 2002) or a similar but distinct comparator process (Gehring & Willoughby, 2002). This component developed many different names, including the feedback-related negativity (FRN), the feedback error-related negativity (fERN), or the mediofrontal negativity (MFN). However, the process reflected by these terms all share common features with the well-known frontal midline N2 component (Folstein & Van Petten, 2008; Holroyd, 2002), arguing against any specificity of information content in these processes. Indeed, the N2 component was already known to

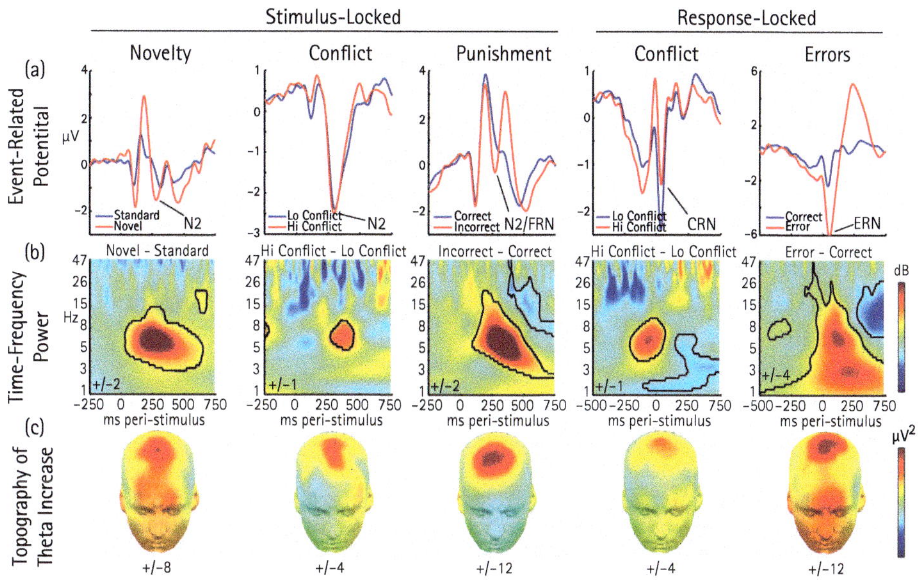

FIGURE 9.1 The need for cognitive control is associated with a similar frontal midline theta signature across a variety of eliciting events. (A) Phase-locked EEG activities (ERPs). While these ERP components (i.e., peaks and troughs in the signal locked to particular external events and averaged across trials) are related to learning and adaptive control, they represent a small fraction of ongoing neural dynamics. (B) Time-frequency plots show richer spectral dynamics of event-related neuro-electrical activity by averaging activities regardless of phase-locking. Here, significant increases in power to novelty, conflict, punishment, and error are outlined in black, revealing a common frontal midline theta band feature during events that signal a need for control. (C) Scalp topography of event-related frontal midline theta activity. The distribution of theta power bursts is consistently maximal over the frontal midline.
N2, a component elicited by novelty or stimulus/response conflict; Feedback related negativity (FRN), A similar N2-like component elicited by external feedback signaling that one's actions were incorrect or yielded a loss; Correct-related negativity (CRN), a small, obligatory component evoked by motor responses even when these are correct according to the task and enhanced by response conflict; Error related negativity (ERN), A component evoked by motor commission errors.

have a spectral representation in the theta band (Başar-Eroğlu et al., 1992; Başar, 1998a; Yordanova et al., 2002). The ERN and the FRN/N2 are estimated to have MCC sources via EEG source estimation (Gehring et al., 2012; Gruendler et al., 2011; van Noordt & Segalowitz, 2012; Walsh & Anderson, 2012) and EEG-informed fMRI (Becker et al., 2014; Debener et al., 2005; Edwards et al., 2012; Hauser et al., 2014; Huster et al., 2011), suggesting at least some common processes linking the two.

A parsimonious summary could propose that both the stimulus- and response-related fronto-central negativities reflect common features of the processing demands of the MCC. These features are varied across systems related to cognitive and motor control, attention, and reinforcement learning, but are especially sensitive to mismatch signals of conflict, punishment, and error in the service of behavioral adaptation. The commonality of these theta-band processes led to an integrative theory of a common language, a *theta lingua franca*, for the realization of the need for control (Cavanagh et al., 2009; 2012). Stimulus- and response-locked obligatory theta-band phase dynamics were proposed to represent a biophysical mechanism for the common temporal organization of neural processes during stimulus or response processes. Variation on this theme, such as power enhancement, reflects the realization of these reactive responses (Figure 9.1). These computations appear to be used to merge attentive, affective, and cognitive functions with motor selection in order to utilize environmental context during action monitoring. FMT therefore appears to reflect general operations of the MCC during action monitoring, particularly as an initial orienting response to a novel or surprising event (Wessel, 2018).

9.2.2 Do ERP Theta and FMT Reflect the Same Process?

Recent findings have begun to parse some of the overlapping constructs in the broad *theta lingua franca* perspective. FMT dynamics occur in the "background" during cognitive performance, emerging in a scale-free manner over varied time scales and thus not only to punctate evoked orienting responses (Cohen, 2016). Conflict appears to be specifically represented in the theta band, whereas error-specific features are associated with an additional delta-band response (Cohen & Donner, 2013; Cohen & van Gaal, 2014; Yordanova et al., 2004). Conflict can have many different definitions, with formal mathematical models equating to a type of surprise (Berlyne, 1957; Botvinick et al., 2001; Wiecki & Frank, 2013) and informal definitions comprising a homology of difficulty or effort.

In line with Shackman and colleagues' (2011) adaptive control hypothesis, FMT closely aligns with the proposed domain-general (effective and affective) role of the MCC for selecting actions under uncertainty (Cavanagh & Shackman, 2014). Meta-analytic evidence supports the hypothesis that cognitive effort and anxiety are both related to FMT and associated ERP features (Cavanagh & Shackman, 2014; Moser et al., 2013). Highly anxious individuals appear to utilize punishment and error information more effectively than a control sample. Enhanced FMT mediates the relationship between anxiety and risky decision making (Schmidt et al., 2018), and it correlates with an enhanced ability to learn from punishment (Cavanagh et al., 2018) and adjust behavior following errors (Cavanagh, Meyer, Hajcak, et al., 2017).

In sum, definitions of conflict, difficulty, and effort can all be equivalently applied to situations that elicit FMT as well as ERP components with theta-dominant substrates. In line with the adaptive control hypothesis, FMT appears to mediate the increased affective and effective vigilance leading to more avoidant behaviors, particularly in highly anxious individuals. However, our overarching caveat should be noted again: there are a multitude of theta band responses, even over the frontal midline, and each can reflect a variety of processes (Töllner et al., 2017). For example, Cohen and Donner (2013) found that the response-conflict-related N2 ERP component was uncorrelated with conflict-related FMT, and that removing the phase-locked (ERP) component of the signal did not affect the conflict-related FMT. This is consistent with other reports that have reported a qualitative absence of phase-locking associated with response-conflict FMT (Nigbur et al., 2012; Pastötter et al., 2010). These findings can be contrasted with similar analyses of the FMT response to erroneous button presses, which contains theta-band phase-locking (Trujillo & Allen, 2007) that is significantly affected by removing the ERP (Munneke et al., 2015).

9.3 Translational Underpinnings

In some fields of neuroscience research, the term "theta" implicitly implies rodent hippocampal activity (~4–12 Hz) that has been associated with learning, memory, and spatial navigation (Buzsáki, 2006). Rodent hippocampal theta is not a unitary construct (Colgin, 2013; Pignatelli et al., 2012), with separate theta rhythms occurring due to septal drive as well as an intrinsic generative process that seems to be common to many types of cortical and sub-cortical excitatory-inhibitory networks (Womelsdorf et al., 2014).

Mediofrontal spikes are phase-locked to both mediofrontal theta as well as hippocampal theta (Benchenane et al., 2010; Hyman et al., 2011; Jones & Wilson, 2005b; 2005a; Paz et al., 2008; Pignatelli et al., 2012; Siapas et al., 2005). However, this evidence of hippocampal interaction with frontal theta processes could reflect a generic phenomenon whereby cortical theta synchronizes disparate neural areas in a global workspace, possibly via travelling waves (Lubenov & Siapas, 2009; Zhang et al., 2018). Many cortical areas have shown phase-synchronous relationships with frontal theta, including visual cortex (Lee et al., 2005; Liebe et al., 2012; Phillips et al., 2013), amygdala (Taub et al., 2018), and ventral tegmental areas (Fujisawa & Buzsáki, 2011). The ubiquity of theta-band findings across species has led to the suggestion that FMT reflects a non-specific mechanism for organizing neural processes around "decision points", such as action selection (Womelsdorf, Vinck, et al., 2010).

The translational potential for using theta to infer similar cognitive processes between species is promising but needs additional clarification. Some studies show that non-human primates have similar error, conflict, and feedback ERPs at the skull (Phillips & Everling, 2014), the scalp (Godlove et al., 2011), the dura (Vezoli & Procyk, 2009), and within the cingulate cortex (Emeric et al., 2010), although the spectral representation of these signals has not been defined. Rats have a FMT-dominant control network that is transiently instantiated following an imperative tone, affording a chance to causally manipulate this network and draw parallel conclusions to humans (Narayanan et al., 2013) although this is non-specifically spectrally localized to the theta band compared to typical human EEG findings. This common FMT electrophysiological response is diminished in Parkinson's patients as well as in a dopamine depletion rodent model (Parker et al., 2015), suggesting a novel model of cognitive dysfunction in Parkinsonism. Ample evidence suggests that FMT is sensitive to dopamine in humans, but it appears to also be sensitive to other monoamines like norepinephrine and acetylcholine (see review by Jocham & Ullsperger, 2009) so the specificity of neuromodulator influence is likely to be low.

Given this pervasive dominance of theta-band activities across mammalian cognitive processes, one may wonder if there is something special about this frequency. Indeed, many mammalian motor activities occur within a broadly-defined theta frequency (Cohen, 2014; Colgin, 2013). Reflexive movements like sniffing (Macrides et al., 1982), whisking (Berg & Kleinfeld, 2002), licking (Amarante et al., 2017), giggling (Luschei, 2006), and shivering (Petajan & Williams, 1972), as well as controlled processes like saccade initiation (Jutras et al., 2013), typing (Yamaguchi et al., 2013), and speaking (Pellegrino et al., 2011), all occur with theta rhythmicity. This theta dominance of the speech rate may have emerged as a consequence of motor properties of mouth movements, which themselves emerged from reflexive sucking, chewing, and licking patterns (MacNeilage & Davis, 2001). Yet there are theta rhythms in attentive processes as well. Saccades reset hippocampal theta rhythms in monkeys (Jutras et al., 2013), which appear to facilitate the creation of grid-based representations of space (Killian et al., 2012). Sustained attention has rhythmic fluctuations at an approximate theta frequency (Helfrich et al., 2018; Huang et al., 2015), although this may be a general low-frequency phenomenon and may not be specific to theta (VanRullen, 2016).

In sum, cognitive processes and controlled motor activities appear to emerge from existing phylogenetic scaffolds that use a basic pattern generator for action initiation. For whatever reason, theta rhythmicity may have first been leveraged to optimize skeleto-motor action integration. The common occurrence of theta across arousal, motor, cognitive, and attentive processes suggests a *degeneracy* of function: many different processes evoke a theta band correlate. Still, there might be a common computational advantage of particular temporal pattern.

9.3.1 Theta Phase Dynamics and Decision Integration

Theta oscillatory dynamics are a possible computational mechanism by which expectations and outcomes can be compared. Oscillations alter the membrane potential of neurons "tuned" to the oscillation frequency, forcing windows of time where any given neuron is either more (trough) or less (peak) likely to be excited. Neurons participating in this given frequency perturbation are more likely to interact, exchange information, and modulate synaptic plasticity together (Fries, 2015). Since oscillations reflect the action of a fundamental, energy-efficient physical principle, functional dynamics of information processing may be inferred from the qualities inherent to oscillations: timing, prediction, storage, and communication.

The 1/f characteristic of brain oscillations suggests that small-amplitude fast-oscillating networks are a characteristic of more local operations, whereas slow, large rhythms link areas over a greater spatial distance. Sinusoidal (harmonic) oscillators, such as the theta rhythm, are good time keepers due to the ability to predict future states based on knowledge of the oscillation period and the phase at any given moment

(Buzsáki, 2006). Pulsatile (non-harmonic, or relaxation) oscillators, such as the neuronal membrane potential, may be synchronized easily by harmonic oscillators which can act to separate information transfer and information intake (Buzsáki & Draguhn, 2004). These characteristics of timing and selective entrainment make oscillations a substrate for ongoing processing (storage) and phase reset (input enhancement), two necessary components of expectation and prediction (Buzsáki, 2006).

Across networks, oscillations within one network may entrain other neurons by locking them into the same oscillatory phase (Fries, 2005; Womelsdorf et al., 2007), to a harmonic oscillatory phase (Gruber et al., 2005), or by coupling a higher frequency power increase to lower frequency phase (Canolty et al., 2006; Jensen & Lisman, 2005). Coherent oscillations between distal areas are thought to reflect entrained inter-regional activity, which serves to increase the coordination of spike timing across spatially separate neural networks (Fries, 2005). Thus, slower oscillations like theta can act as a reader of fast activities (Buzsáki, 2010; Jensen & Colgin, 2007; Womelsdorf, Vinck, et al., 2010), for example integrating phonemes into words during speech processing (Giraud et al., 2007; Giraud & Poeppel, 2012). Theta phase thus appears to be a general neural mechanism for coupling disparate local gamma band sequences (Canolty & Knight, 2010; Fukai, 1999; Jensen & Lisman, 2005; Solomon et al., 2017; Tang et al., 2016; White et al., 2000). This pattern of theta-gamma coupling even forms the basis of some current network-level models of cognitive control (Gratton, 2018; Verguts, 2017).

9.3.2 Theta phase dynamics and cognitive control

If FMT power bursts signal a generic need for control (Figure 9.1), theta-band phase synchrony between frontal midline and distal sites may communicate how to implement that control (Figure 9.2a). In the more than ten years since the initial publication of this simultaneous discovery by two separate groups (Cavanagh et al., 2009; Hanslmayr et al., 2008), there have been over 20 replications of this effect of transient similarity in theta-specific phase angle following a variety of events indicating a need for control. This finding further validates a theta-based similarity between conflict, punishment, error, and working-memory evoked FMT as a common *lingua franca* for implementing control.

Other types of statistical interaction have also been shown in theta-band networks, primarily using Granger prediction where midline areas lead distal areas (Cohen & van Gaal, 2013; Popov et al., 2018; Rajan et al., 2018; Zavala et al., 2016; Zavala et al., 2014). Intracranial recordings in humans have validated the hypothesis that cingulate theta couples with gamma, and that this process leads dlPFC, MFC, and OFC during the need for cognitive control (Bartoli et al., 2017; Oehrn et al., 2014; Rothé et al., 2011; Smith et al., 2015; Tang et al., 2016). Similar medio-occipital phase synchrony has been observed in monkeys (Phillips et al., 2013) and medio-motor synchrony has been observed in rats (Narayanan et al., 2013). These theta-phase networks can be boosted by transcranial stimulation (Reinhart & Woodman, 2014; Reinhart, 2017), and diminished

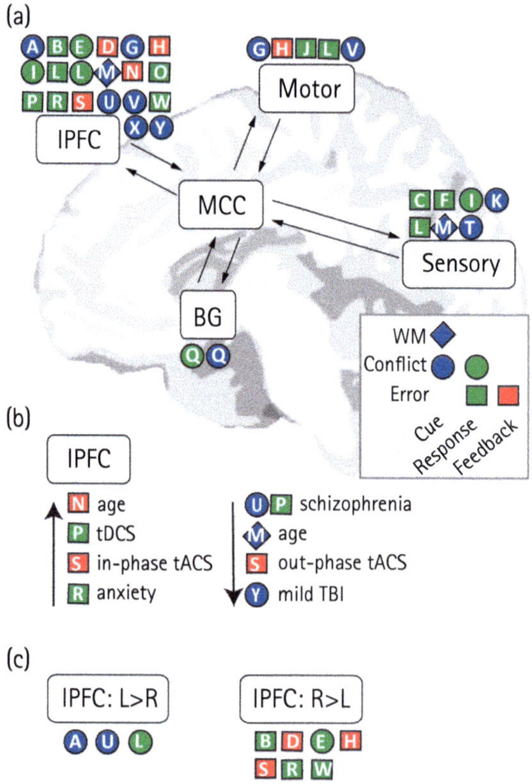

FIGURE 9.2 Theta band phase consistency between mid-frontal and distal sites is transiently increased following events that indicate a need for control. (A) Twenty five separate studies (A through Y) have replicated the finding of theta-band phase synchrony between the frontal midline (presumably MCC) and varied cortical areas, including lateral prefrontal cortex (lPFC), motor cortex, sensory cortices, and basal ganglia (BG). WM = working memory. (B) A variety of moderators affect medio-lateral phase synchrony, with increases due to age, anxiety and transcranial direct or alternating current stimulation (tD/ACS), and decreases due to schizophrenia, age, anti-phase tACS, and mild traumatic brain injury (TBI). (C) Studies oftentimes report a nominal pattern of hemisphericity in medio-lateral phase synchrony. Errors and punishments consistently evoked a right>left pattern, whereas conflict tended towards a left>right pattern. However, the studies contributing to the left>right pattern were more complex and involved proactive and reactive processes compared to error realization, which is rather straightforward and reactive. Future studies should formally investigate the moderators of hemisphericity in medio-lateral phase synchrony.

Citations for figure 9.2:

A: Hanslmayr et al., 2008; B: Cavanagh et al., 2009; C: Cohen et al., 2009; D: Cavanagh et al., 2010; E: Cohen & Cavanagh, 2011; F: Cohen & van Gaal, 2013; G: Nigbur et al., 2012; H: van de Vijver et al., 2011; I: van Driel et al., 2012; J: Narayanan et al., 2013; K: Anguera et al., 2013; L: Cohen & van Gaal, 2014; M: Tóth et al., 2014; N: Van de Vijver et al., 2014; O: Moran et al., 2014; P: Reinhart et al., 2015; Q: Zavala et al., 2013; R: Cavanagh et al., 2017; S: Reinhart, 2017; T: Vissers et al., 2018: U: Ryman et al., 2018; V: Swart et al., 2018; W: Buzzell et al., 2018; X: Oehrn et al., 2014; Y: Cavanagh et al., 2020.

by out-of-phase alternating current (Reinhart, 2017). These networks are also altered in psychiatric distress, being increased in anxiety (Cavanagh et al., 2017) and diminished in schizophrenia (Ryman et al., 2018), see Figure 9.2b. Many questions remain to be addressed about the function of this theta-band phase synchrony, but there is correlative (Anguera et al., 2013; Cavanagh et al., 2009; Cavanagh et al., 2017; Swart et al., 2018) and causal evidence (Narayanan et al., 2013; Reinhart, 2017; Reinhart et al., 2015) that this network directly affects behaviors related to the ability to learn from and adapt to the need for control.

Many studies have observed a right-sided dominance of medio-lateral phase synchrony, but this issue of hemisphericity requires further rigorous testing. Figure 9.2c sorts studies that reported even a nominal pattern of hemisphericity, and it can be seen that errors and punishments consistently evoked a right>left bias, whereas conflict tended towards a left>right bias. Although this distinction appears straightforward, it is likely overly simplistic to suggest a simple conflict vs. error dissociation on hemisphericity. The studies contributing to the left>right pattern were more complex and involved both proactive and reactive processes whereas error realization is rather straightforward and reactive. Future studies should be formally test for hemispheric bias in medio-lateral phase synchrony to better address questions about the functional role of this signal.

9.4 Clinical applications

FMT and related ERP features have compelling characteristics for clinical advancement. The majority of the units of analysis in the National Institute of Mental Health (NIMH) Research Domain Criteria (Insel et al., 2010) are EEG-based, and many of these are FMT-family responses. Lower FMT appears to be a reliable endophenotype for substance abuse and externalizing disorders (Gilmore et al., 2010; Kamarajan et al., 2015; Kang et al., 2012; Rangaswamy et al., 2007; Zlojutro et al., 2011). Higher FMT is reliably associated with anxious temperament (Cavanagh & Shackman, 2014; Moser et al., 2013; Riesel et al., 2017), and ERN amplitude can even predict treatment response in anxiety disorder patients (Gorka et al., 2018) . Future studies should derive the sensitivity and specificity of FMT to determine the biomarker potential in select clinical applications.

FMT can be elicited in simple tasks that are viable within a clinical environment. Aberrant auditory orienting responses have already been advanced as candidate biomarkers, like diminished mismatch negativity (MMN) in schizophrenia (Javitt et al., 2018; Light et al., 2015) or diminished novelty habituation in Parkinson's disease (Cavanagh, Kumar, et al., 2018). The MMN is a theta-dominant response with separable frontal and temporal processes (Fuentemilla et al., 2008; Ko et al., 2012) that may interact via theta-band phase synchrony (Choi et al., 2013). The neural systems underlying auditory novelty detection are well detailed in rodent models (Escera &

Malmierca, 2014; Featherstone et al., 2018; Lee et al., 2018), facilitating cross-species translation. Auditory-evoked responses are already routinely used in brainstem auditory testing for hearing acuity in newborns, demonstrating that clinical infrastructure and expertise exists for applying relevant tasks to patient groups when using EEG as a diagnostic tool.

9.5 Broader Impact and Future Directions

FMT is a well-characterized candidate mechanism underlying the ability to realize and communicate the need for cognitive control. Further tests of this theory will need to integrate findings from preclinical animal models, computational accounts of information representation, broader human imaging literature, and the sensitivity and specificity diagnostics for human patient groups. Fortunately, electrophysiology is routinely utilized in vitro, in vivo, and in outpatient neurological clinics, making it uniquely positioned at the crossroads of many sub-fields of neuroscience.

References

Allain, S., Carbonnell, L., Falkenstein, M., Burle, B., & Vidal, F. (2004). The modulation of the Ne-like wave on correct responses foreshadows errors. *Neuroscience Letters*, *372*(1–2), 161–166. https://doi.org/S0304-3940(04)01171-1 [pii]10.1016/j.neulet.2004.09.036

Amarante, L. M., Caetano, M. S., & Laubach, M. (2017). Medial frontal theta is entrained to rewarded actions. *The Journal of Neuroscience*, *37*(44), 1965–1917. https://doi.org/10.1523/JNEUROSCI.1965-17.2017

Anguera, J. A., Boccanfuso, J., Rintoul, J. L., Al-Hashimi, O., Faraji, F., Janowich, J., . . . Gazzaley, A. (2013). Video game training enhances cognitive control in older adults. *Nature*, *501*(7465), 97–101. https://doi.org/10.1038/nature12486

Arellano, A. P. & Schwab, R. S. (1950). Scalp and basal recording during mental activity. Proceedings of the 1st International Congress of Psychiatry, September 18–27, Paris.

Asada, H., Fukuda, Y., Tsunoda, S., Yamaguchi, M., & Tonoike, M. (1999). Frontal midline theta rhythms reflect alternative activation of prefrontal cortex and anterior cingulate cortex in humans. *Neuroscience Letters*, *274*(1), 29–32. https://doi.org/S0304-3940(99)00679-5

Bartoli, E., Conner, C. R., Kadipasaoglu, C. M., Yellapantula, S., Rollo, M. J., Carter, C. S., & Tandon, N. (2017). Temporal dynamics of human frontal and cingulate neural activity during conflict and cognitive control. *Cerebral Cortex*, *28*(11), 3842–3856. https://doi.org/10.1093/cercor/bhx245

Başar-Eroğlu, C., Başar, E., Demiralp, T., & Schurmann, M. (1992). P300-response: Possible psychophysiological correlates in delta and theta frequency channels. A review. *International Journal of Psychophysiology*, *13*, 161–179.

Başar, E. (1998a). Brain function and oscillations: Brain oscillations–principles and approaches. Springer-Verlag. https://doi.org/10.1007/978-3-642-72192-2

Başar, E. (1998b). Brain function and oscillations: Integrative brain function–neurophysiology and cognitive processes. Springer-Verlag. https://doi.org/10.1007/978-3-642-59893-7

Beaton, L. E., Azma, S., & Marinkovic, K. (2018). When the brain changes its mind: Oscillatory dynamics of conflict processing and response switching in a flanker task during alcohol challenge. *PLoS One*, *13*(1), 1–24. https://doi.org/10.1371/journal.pone.0191200

Becker, M. P. I., Nitsch, A. M., Miltner, W. H. R., & Straube, T. (2014). A single-trial estimation of the feedback-related negativity and its relation to BOLD responses in a time-estimation task. *The Journal of Neuroscience*, *34*(8), 3005–3012. https://doi.org/10.1523/JNEUROSCI.3684-13.2014

Benchenane, K., Peyrache, A., Khamassi, M., Tierney, P. L., Gioanni, Y., Battaglia, F. P., & Wiener, S. I. (2010). Coherent theta oscillations and reorganization of spike timing in the hippocampal-prefrontal network upon learning. *Neuron*, *66*(6), 921–936. https://doi.org/10.1016/j.neuron.2010.05.013

Berg, R. W. & Kleinfeld, D. (2002). Rhythmic whisking by rat: Retraction as well as protraction of the vibrissae is under active muscular control. *Journal of Neurophysiology*, *89*(1), 104–117. https://doi.org/10.1152/jn.00600.2002

Berlyne, D. E. (1957). Uncertainty and conflict: A point of contact between information-theory and behavior-theory concepts. *Psychological Review*, *64*(6), 329–339.

Botvinick, M. M., Braver, T. S., Barch, D. M., Carter, C. S., & Cohen, J. D. (2001). Conflict monitoring and cognitive control. *Psychological Review*, *108*(3), 624–652. https://pubmed.ncbi.nlm.nih.gov/11488380/

Brazier, M. A. B. & Casby, J. U. (1952). Cross-correlation and autocorrelation studies of electroencephalographic potentials. *Electroencephalography and Clinical Neurophysiology*, *4*, 201–211.

Buzsáki, G. (2006). Rhythms of the brain. Oxford University Press.

Buzsáki, G. (2010). Neural syntax: Cell assemblies, synapsembles, and readers. *Neuron*, *68*(3), 362–385. https://doi.org/10.1016/j.neuron.2010.09.023

Buzsáki, G. & Draguhn, A. (2004). Neuronal oscillations in cortical networks. *Science*, *304*(5679), 1926–1929. https://doi.org/10.1126/science.1099745 304/5679/1926

Buzzell, G. A., Barker, T. V, Troller-Renfree, S. V, Bernat, E. M., Bowers, M. E., Morales, S., ... Fox, N. A. (2018). Adolescent cognitive control, theta oscillations, and social motivation. *BioRxiv* [online], 366831. https://doi.org/10.1101/366831

Canolty, R. T., Edwards, E., Dalal, S. S., Soltani, M., Nagarajan, S. S., Kirsch, H. E., ... Knight, R. T. (2006). High gamma power is phase-locked to theta oscillations in human neocortex. *Science*, *313*(5793), 1626–1628. https://doi.org/10.1126/science.1128115

Canolty, R. T. & Knight, R. T. (2010). The functional role of cross-frequency coupling. *Trends in Cognitive Sciences*, *14*(11), 506–515. https://doi.org/10.1016/j.tics.2010.09.001

Caplan, J. B., Madsen, J. R., Schulze-Bonhage, A., Aschenbrenner-Scheibe, R., Newman, E. L., & Kahana, M. J. (2003). Human theta oscillations related to sensorimotor integration and spatial learning. *Journal of Neuroscience*, *23*(11), 4726–4736. https://doi.org/23/11/4726

Cavanagh, J. F., Bismark, A. W., Frank, M. J., & Allen, J. J. B. (2018). Multiple dissociations between comorbid depression and anxiety on reward and punishment processing: Evidence from computationally informed EEG. *Computational Psychiatry*, *3*, 1–17. doi: 10.1162/cpsy_a_00024.

Cavanagh, J. F., Cohen, M. X., & Allen, J. J. B. (2009). Prelude to and resolution of an error: EEG phase synchrony reveals cognitive control dynamics during action monitoring. *The Journal of Neuroscience*, *29*(1), 98–105. https://doi.org/10.1523/JNEUROSCI.4137-08.2009

Cavanagh, J. F. & Frank, M. J. (2014). Frontal theta as a mechanism for cognitive control. *Trends in Cognitive Sciences*, *18*(8), 1–8. https://doi.org/10.1016/j.tics.2014.04.012

Cavanagh, J. F., Frank, M. J., Klein, T. J., & Allen, J. J. B. (2010). Frontal theta links prediction errors to behavioral adaptation in reinforcement learning. *NeuroImage*, *49*(4), 3198–3209. https://doi.org/10.1016/j.neuroimage.2009.11.080

Cavanagh, J. F., Kumar, P., Mueller, A. A., Pirio Richardson, S., Mueen, A., & Richardson, S. P. (2018). Diminished EEG habituation to novel events effectively classifies Parkinson's patients. *Clinical Neurophysiology*, *129*(2), 409–418. https://doi.org/10.1016/j.clinph.2017.11.023

Cavanagh, J. F., Meyer, A., & Hajcak, G. (2017). Error-specific cognitive control alterations in generalized anxiety disorder. *Biological Psychiatry: Cognitive Neuroscience and Neuroimaging*, *2*(5), 413–420. https://doi.org/10.1016/j.bpsc.2017.01.004

Cavanagh, J. F., Rieger, R. E., Wilson, J. K., Gill, D., Fullerton, L., Brandt, E., & Mayer, A. R. (2020). Joint analysis of frontal theta synchrony and white matter following mild traumatic brain injury. *Brain Imaging and Behavior*, *14*(6), 2210–2223.

Cavanagh, J. F., & Shackman, A. J. (2014). Frontal midline theta reflects anxiety and cognitive control: Meta-analytic evidence. *Journal of Physiology*, *109*(1–3), 3–15. https://doi.org/10.1016/j.jphysparis.2014.04.003

Cavanagh, J. F., Zambrano-Vazquez, L., & Allen, J. J. B. (2012). Theta lingua franca: A common mid-frontal substrate for action monitoring processes. *Psychophysiology*, *49*(2), 220–238. https://doi.org/10.1111/j.1469-8986.2011.01293.x

Choi, J. W., Lee, J. K., Ko, D., Lee, G. T., Jung, K. Y., & Kim, K. H. (2013). Fronto-temporal interactions in the theta-band during auditory deviant processing. *Neuroscience Letters*, *548*, 120–125. https://doi.org/10.1016/j.neulet.2013.05.079

Cohen, M. X. (2014). A neural microcircuit for cognitive conflict detection and signaling. *Trends in Neurosciences*, *37*(9), 480–490. https://doi.org/10.1016/j.tins.2014.06.004

Cohen, M. X. (2016). Midfrontal theta tracks action monitoring over multiple interactive time scales. *NeuroImage*, *141*, 262–272. https://doi.org/10.1016/J.NEUROIMAGE.2016.07.054

Cohen, M. X. & Cavanagh, J. F. (2011). Single-trial regression elucidates the role of prefrontal theta oscillations in response conflict. *Frontiers in Psychology* [online], 2, 30. https://doi.org/10.3389/fpsyg.2011.00030

Cohen, M. X. & Donner, T. H. (2013). Midfrontal conflict-related theta-band power reflects neural oscillations that predict behavior. *Journal of Neurophysiology*, *110*(12), 2752–2763. https://doi.org/10.1152/jn.00479.2013

Cohen, M. X. & Ridderinkhof, K. R. (2013). EEG source reconstruction reveals frontal-parietal dynamics of spatial conflict processing. *PLoS One*, *8*(2), e57293. https://doi.org/10.1371/journal.pone.0057293

Cohen, M. X., Ridderinkhof, K. R., Haupt, S., Elger, C. E., & Fell, J. (2008). Medial frontal cortex and response conflict: Evidence from human intracranial EEG and medial frontal cortex lesion. *Brain Research*, *1238*, 127–142. https://doi.org/S0006-8993(08)01872-6 [pii]10.1016/j.brainres.2008.07.114

Cohen, M. X, & van Gaal, S. (2013). Dynamic interactions between large-scale brain networks predict behavioral adaptation after perceptual errors. *Cerebral Cortex (New York, N.Y.: 1991)*, *23*(5), 1061–1072. https://doi.org/10.1093/cercor/bhs069

Cohen, M. X, & van Gaal, S. (2014). Subthreshold muscle twitches dissociate oscillatory neural signatures of conflicts from errors. *NeuroImage*, *86*, 503–513. https://doi.org/10.1016/J.NEUROIMAGE.2013.10.033

Cohen, M. X, van Gaal, S., Ridderinkhof, K. R., & Lamme, V. A. F. (2009). Unconscious errors enhance prefrontal-occipital oscillatory synchrony. *Frontiers in Human Neuroscience*, *3*, 54. https://doi.org/10.3389/neuro.09.054.2009

Coles, M. G., Scheffers, M. K., & Holroyd, C. B. (2001). Why is there an ERN/Ne on correct trials? Response representations, stimulus-related components, and the theory of error-processing. *Biological Psychology*, *56*(3), 173–189. https://doi.org/10.1016/S0301-0511(01)00076-X

Colgin, L. L. (2013). Mechanisms and functions of theta rhythms. *Annual Review of Neuroscience*, *36*(1), 295–312. https://doi.org/10.1146/annurev-neuro-062012-170330

Debener, S., Ullsperger, M., Siegel, M., Fiehler, K., von Cramon, D. Y., & Engel, A. K. (2005). Trial-by-trial coupling of concurrent electroencephalogram and functional magnetic resonance imaging identifies the dynamics of performance monitoring. *Journal of Neuroscience*, 25(50), 11730–11737. https://doi.org/25/50/11730 [pii]10.1523/JNEUROSCI.3286-05.2005

Dehaene, S., Posner, M. I., & Tucker, D. M. (1994). Localization of a neural system for error detection and compensation. *Psychological Science*, *5*(5),303–305.

Edwards, B. G., Calhoun, V. D., & Kiehl, K. A. (2012). Joint ICA of ERP and fMRI during error-monitoring. *NeuroImage*, *59*(2), 1896–1903. https://doi.org/10.1016/j.neuroimage.2011.08.088

Emeric, E. E., Leslie, M. W., Pouget, P., & Schall, J. D. (2010). Performance monitoring local field potentials in the medial frontal cortex of primates: Supplementary eye field. *Journal of Neurophysiology*, *104*, 1523–1537. https://doi.org/jn.01001.2009

Escera, C. & Malmierca, M. S. (2014). The auditory novelty system: An attempt to integrate human and animal research. *Psychophysiology*, *51*(2), 111–123. https://doi.org/10.1111/psyp.12156

Falkenstein, M., Hohnsbein, J., Hoormann, J., & Blanke, L. (1991). Effects of crossmodal divided attention on late ERP components. II. Error processing in choice reaction tasks. *Electroencephalography and Clinical Neurophysiology*, *78*(6), 447–455. https://pubmed.ncbi.nlm.nih.gov/1712280/

Featherstone, R. E., Melnychenko, O., & Siegel, S. J. (2018). Mismatch negativity in preclinical models of schizophrenia. *Schizophrenia Research*, *191*, 35–42. https://doi.org/10.1016/j.schres.2017.07.039

Folstein, J. R. & Van Petten, C. (2008). Influence of cognitive control and mismatch on the N2 component of the ERP: a review. *Psychophysiology*, *45*(1), 152–170. https://doi.org/PSYP602 [pii]10.1111/j.1469-8986.2007.00602.x

Fries, P. (2005). A mechanism for cognitive dynamics: neuronal communication through neuronal coherence. *Trends in Cognitive Science*, *9*(10), 474–480. https://doi.org/S1364-6613(05)00242-1 [pii]10.1016/j.tics.2005.08.011

Fries, P. (2015). Rhythms for cognition: Communication through coherence. *Neuron*, *88*(1), 220–235. https://doi.org/10.1016/j.neuron.2015.09.034

Fuentemilla, L., Marco-Pallarés, J., Münte, T. F., & Grau, C. (2008). Theta EEG oscillatory activity and auditory change detection. *Brain Research*, *1220*, 93–101. https://doi.org/10.1016/j.brainres.2007.07.079

Fujisawa, S. & Buzsáki, G. (2011). A 4 Hz oscillation adaptively synchronizes prefrontal, VTA, and hippocampal activities. *Neuron*, *72*(1), 153–165. https://doi.org/10.1016/j.neuron.2011.08.018

Fukai, T. (1999). Sequence generation in arbitrary temporal patterns from theta-nested gamma oscillations: A model of the basal ganglia-thalamo-cortical loops. *Neural Networks*, *12*(7–8), 975–987. https://pubmed.ncbi.nlm.nih.gov/12662640/

Gehring, W. J., Goss, B., Coles, M. G. H., Meyer, D. E., & Donchin, E. (1993). A neural system for error-detection and compensation. *Psychological Science*, *4*(6), 385–390.

Gehring, W. J., Himle, J., & Nisenson, L. G. (2000). Action-monitoring dysfunction in obsessive-compulsive disorder. *Psychological Science*, *11*(1), 1–6. https://pubmed.ncbi.nlm.nih.gov/11228836/

Gehring, W. J., Liu, Y., Orr, J. M., & Carp, J. (2012). The error-related negativity (ERN/Ne). In S. J. Luck & E. Kappenman (Eds.), *Oxford handbook of event-related potential components* (pp. 231–291). Oxford University Press.

Gehring, W. J. & Willoughby, A. R. (2002). The medial frontal cortex and the rapid processing of monetary gains and losses. *Science*, *295*(5563), 2279–2282. https://doi.org/10.1126/science.1066893295/5563/2279

Gemba, H., Sasaki, K., & Brooks, V. B. (1986). "Error" potentials in limbic cortex (anterior cingulate area 24) of monkeys during motor learning. *Neuroscience Letters*, *70*(2), 223–227. https://doi.org/0304-3940(86)90467-2

Gevins, A. & Smith, M. E. (2000). Neurophysiological measures of working memory and individual differences in cognitive ability and cognitive style. *Cerebral Cortex*, *10*(9), 829–839. https://pubmed.ncbi.nlm.nih.gov/10982744/

Gevins, A., Smith, M. E., McEvoy, L. K., Leong, H., & Le, J. (1999). Electroencephalographic imaging of higher brain function. *Philosophical Transactions of the Royal Society of London Series B: Biological Sciences*, *354*(1387), 1125–1133. https://doi.org/10.1098/rstb.1999.0468

Gevins, A., Smith, M. E., McEvoy, L., & Yu, D. (1997). High-resolution EEG mapping of cortical activation related to working memory: Effects of task difficulty, type of processing, and practice. *Cerebral Cortex*, *7*(4), 374–385. https://pubmed.ncbi.nlm.nih.gov/9177767/

Gilmore, C. S., Malone, S. M., & Iacono, W. G. (2010). Brain electrophysiological endophenotypes for externalizing psychopathology: A multivariate approach. *Behavior Genetics*, *40*(2), 186–200. https://doi.org/10.1007/s10519-010-9343-3

Giraud, A.-L. L., Kleinschmidt, A., Poeppel, D., Lund, T. E., Frackowiak, R. S. J. J., & Laufs, H. (2007). Endogenous cortical rhythms determine cerebral specialization for speech perception and production. *Neuron*, *56*(6), 1127–1134. https://doi.org/10.1016/j.neuron.2007.09.038

Giraud, A.-L. L., & Poeppel, D. (2012). Cortical oscillations and speech processing: emerging computational principles and operations. *Nature Neuroscience*, *15*(4), 511–517. https://doi.org/10.1038/nn.3063

Godlove, D. C., Emeric, E. E., Segovis, C. M., Young, M. S., Schall, J. D., & Woodman, G. F. (2011). Event-related potentials elicited by errors during the stop-signal task. I. Macaque monkeys. *The Journal of Neuroscience*, *31*(44), 15640–15649. https://doi.org/10.1523/JNEUROSCI.3349-11.2011

Gorka, S. M., Burkhouse, K. L., Klumpp, H., Kennedy, A. E., Afshar, K., Francis, J., ... Phan, K. L. (2018). Error-related brain activity as a treatment moderator and index of symptom change during cognitive-behavioral therapy or selective serotonin reuptake inhibitors. *Neuropsychopharmacology*, *43*(6), 1355–1363. https://doi.org/10.1038/npp.2017.289

Gratton, G. (2018). Brain reflections: A circuit-based framework for understanding information processing and cognitive control. *Psychophysiology*, *55*(3), 1–26. https://doi.org/10.1111/psyp.13038

Gruber, W. R., Klimesch, W., Sauseng, P., & Doppelmayr, M. (2005). Alpha phase synchronization predicts P1 and N1 latency and amplitude size. *Cerebral Cortex*, *15*(4), 371–377. https://doi.org/10.1093/cercor/bhh139

Gruendler, T. O. J., Ullsperger, M., & Huster, R. J. (2011). Event-related potential correlates of performance-monitoring in a lateralized time-estimation task. *PloS One*, *6*(10), e25591. https://doi.org/10.1371/journal.pone.0025591

Hajcak, G., Moser, J. S., Yeung, N., & Simons, R. F. (2005). On the ERN and the significance of errors. *Psychophysiology*, *42*(2), 151–160.

Hanslmayr, S., Pastotter, B., Bauml, K. H., Gruber, S., Wimber, M., & Klimesch, W. (2008). The electrophysiological dynamics of interference during the Stroop task. *Journal of Cognitive Neuroscience*, *20*(2), 215–225. https://doi.org/10.1162/jocn.2008.20020

Hanslmayr, S., Staudigl, T., Aslan, A., & Bauml, K. H. (2010). Theta oscillations predict the detrimental effects of memory retrieval. *Cognitive, Affective, and Behavioral Neuroscience*, *10*(3), 329–338. doi: 10.3758/cabn.10.3.329

Hauser, T. U., Iannaccone, R., Stämpfli, P., Drechsler, R., Brandeis, D., Walitza, S., & Brem, S. (2014). The feedback-related negativity (FRN) revisited: New insights into the localization, meaning and network organization. *NeuroImage*, *84*, 159–168. https://doi.org/10.1016/j.neuroimage.2013.08.028

Helfrich, R. F., Fiebelkorn, I. C., Szczepanski, S. M., Lin, J. J., Parvizi, J., Knight, R. T., & Kastner, S. (2018). Neural mechanisms of sustained attention are rhythmic. *Neuron*, *99*(4), 854–865.e5. https://doi.org/10.1016/j.neuron.2018.07.032

Holroyd, C. B. (2002). A note on the oddball N200 and the feedback ERN. In M. Ullsperger & M. Falkenstein (Eds.), *Errors, conflicts and the brain. Current opinions on performance monitoring* (pp. 211–218). Max Planck Institute for Human Cognitive and Brain Sciences.

Holroyd, C. B. & Coles, M. G. (2002). The neural basis of human error processing: reinforcement learning, dopamine, and the error-related negativity. *Psychological Review*, *109*(4), 679–709. https://pubmed.ncbi.nlm.nih.gov/12374324/

Holroyd, C. B., Coles, M. G. H., Nieuwenhuis, S., Gehring, W. J., & Willoughby, A. R. (2002). Medial prefrontal cortex and error potentials. *Science*, *296*(5573), 1610–1611.

Hsieh, L.-T., Ekstrom, A. D., & Ranganath, C. (2011). Neural oscillations associated with item and temporal order maintenance in working memory. *The Journal of Neuroscience*, *31*(30), 10803–10810. https://doi.org/10.1523/JNEUROSCI.0828-11.2011

Hsieh, L. T. & Ranganath, C. (2014). Frontal midline theta oscillations during working memory maintenance and episodic encoding and retrieval. *NeuroImage*, *85*, 721–729. https://doi.org/10.1016/j.neuroimage.2013.08.003

Huang, Y., Chen, L., & Luo, H. (2015). Behavioral oscillation in priming: Competing perceptual predictions conveyed in alternating theta-band rhythms. *Journal of Neuroscience*, *35*(6), 2830–2837. https://doi.org/10.1523/JNEUROSCI.4294-14.2015

Huster, R. J., Eichele, T., Enriquez-Geppert, S., Wollbrink, A, Kugel, H., Konrad, C., & Pantev, C. (2011). Multimodal imaging of functional networks and event-related potentials in performance monitoring. *NeuroImage*, *56*(3), 1588–1597. https://doi.org/10.1016/j.neuroimage.2011.03.039

Hyman, J. M., Hasselmo, M. E., & Seamans, J. K. (2011). What is the functional relevance of prefrontal cortex entrainment to hippocampal theta rhythms? *Frontiers in Neuroscience*, *5*, 1–13. https://doi.org/10.3389/fnins.2011.00024

Inouye, T., Shinosaki, K., Iyama, A., Matsumoto, Y., & Toi, S. (1994). Moving potential field of frontal midline theta activity during a mental task. *Cognitive Brain Research*, *2*, 87–92.

Inouye, T., Shinosaki, K., Iyama, A., Matsumoto, Y., Toi, S., & Ishihar, T. (1994). Potential flow of frontal midline theta activity during a mental task in the human electroencephalogram. *Neuroscience Letters*, *169*, 145–148.

Insel, T., Cuthbert, B., Garvey, M., Heinssen, R., Pine, D. S., Quinn, K., ... Wang, P. (2010). Research domain criteria (RDoC): Toward a new classification framework for research on mental disorders. *American Journal of Psychiatry*, *167*(7), 748–751. https://doi.org/167/7/748 [pii]10.1176/appi.ajp.2010.09091379

Ishii, R., Shinosaki, K., Ukai, S., Inouye, T., Ishihara, T., Yoshimine, T., ... Takeda, M. (1999). Medial prefrontal cortex generates frontal midline theta rhythm. *Neuroreport*, *10*(4), 675–679. https://pubmed.ncbi.nlm.nih.gov/10208529/

Itthipuripat, S., Wessel, J. R., & Aron, A. R. (2013). Frontal theta is a signature of successful working memory manipulation. *Experimental Brain Research*, *224*(2), 255–262. https://doi.org/10.1007/s00221-012-3305-3

Jacobs, J., Hwang, G., Curran, T., & Kahana, M. J. (2006). EEG oscillations and recognition memory: Theta correlates of memory retrieval and decision making. *NeuroImage*, *32*(2), 978–987. https://pubmed.ncbi.nlm.nih.gov/16843012/

Javitt, D. C., Lee, M., Kantrowitz, J. T., & Martinez, A. (2018). Mismatch negativity as a biomarker of theta band oscillatory dysfunction in schizophrenia. *Schizophrenia Research*, *191*, 51–60. https://doi.org/10.1016/j.schres.2017.06.023

Jensen, O. & Tesche, C. D. (2002). Frontal theta activity in humans increases with memory load in a working memory task. *European Journal of Neuroscience*, *15*(8), 1395–1399. https://pubmed.ncbi.nlm.nih.gov/11994134/

Jensen, O. & Colgin, L. L. (2007). Cross-frequency coupling between neuronal oscillations. *Trends in Cognitive Sciences*, *11*(7), 267–269. https://doi.org/10.1016/j.tics.2007.05.003

Jensen, O. & Lisman, J. E. (2005). Hippocampal sequence-encoding driven by a cortical multi-item working memory buffer. *Trends in Neurosciences*, *28*(2), 67–72. https://doi.org/10.1016/j.tins.2004.12.001

Jocham, G. & Ullsperger, M. (2009). Neuropharmacology of performance monitoring. *Neuroscience and Biobehavioral Reviews*, *33*(1), 48–60. https://pubmed.ncbi.nlm.nih.gov/18789964/

Jones, M. W. & Wilson, M. A. (2005a). Phase precession of medial prefrontal cortical activity relative to the hippocampal theta rhythm. *Hippocampus*, *15*(7), 867–873. https://doi.org/10.1002/hipo.20119

Jones, M. W. & Wilson, M. A. (2005b). Theta rhythms coordinate hippocampal-prefrontal interactions in a spatial memory task. *PLoS Biology*, *3*(12), e402. https://doi.org/10.1371/journal.pbio.0030402

Jutras, M. J., Fries, P., & Buffalo, E. A. (2013). Oscillatory activity in the monkey hippocampus during visual exploration and memory formation. *Proceedings of the National Academy of Sciences of the United States of America*, *110*(32), 13144–13149.

Kahana, M. J., Sekuler, R., Caplan, J. B., Kirschen, M., & Madsen, J. R. (1999). Human theta oscillations exhibit task dependence during virtual maze navigation. *Nature*, *399*(6738), 781–784. https://doi.org/10.1038/21645

Kamarajan, C., Pandey, A. K., Chorlian, D. B., Manz, N., Stimus, A. T., Anokhin, A. P., ... Porjesz, B. (2015). Deficient event-related theta oscillations in individuals at risk for alcoholism: A study of reward processing and impulsivity features. *PLoS One*, *10*(11), 1–32. https://doi.org/10.1371/journal.pone.0142659

Kang, S. J., Rangaswamy, M., Manz, N., Wang, J. C., Wetherill, L., Hinrichs, T., ... Porjesz, B. (2012). Family-based genome-wide association study of frontal theta oscillations identifies potassium channel gene KCNJ6. *Genes, Brain and Behavior*, *11*(6), 712–719. https://doi.org/10.1111/j.1601-183X.2012.00803.x

Killian, N. J., Jutras, M. J., & Buffalo, E. A. (2012). A map of visual space in the primate entorhinal cortex. *Nature, 491*(7426), 761–764. https://doi.org/10.1038/nature11587

Klimesch, W. (1999). EEG alpha and theta oscillations reflect cognitive and memory performance: a review and analysis. *Brain Research: Brain Research Reviews, 29*(2–3), 169–195. https://pubmed.ncbi.nlm.nih.gov/10209231/

Klimesch, W., Doppelmayr, M., Schwaiger, J., Winkler, T., & Gruber, W. (2000). Theta oscillations and the ERP old/new effect: Independent phenomena? *Clinical Neurophysiology, 111*(5), 781–793.

Ko, D., Kwon, S., Lee, G. T., Im, C. H., Kim, K. H., & Jung, K. Y. (2012). Theta oscillation related to the auditory discrimination process in mismatch negativity: Oddball versus control paradigm. *Journal of Clinical Neurology (Korea), 8*(1), 35–42. https://doi.org/10.3988/jcn.2012.8.1.35

Lee, H., Simpson, G. V., Logothetis, N. K., & Rainer, G. (2005). Phase locking of single neuron activity to theta oscillations during working memory in monkey extrastriate visual cortex. *Neuron, 45*(1), 147–156. https://doi.org/10.1016/j.neuron.2004.12.025

Lee, M., Balla, A., Sershen, H., Sehatpour, P., Lakatos, P., & Javitt, D. C. (2018). Rodent mismatch negativity/theta neuro-oscillatory response as a translational neurophysiological biomarker for N-methyl-D-aspartate receptor-based new treatment development in schizophrenia. *Neuropsychopharmacology, 43*(3), 571–582. https://doi.org/10.1038/npp.2017.176

Liebe, S., Hoerzer, G. M., Logothetis, N. K., & Rainer, G. (2012). Theta coupling between V4 and prefrontal cortex predicts visual short-term memory performance. *Nature Neuroscience, 15*(3), 456–462, S1-2. https://doi.org/10.1038/nn.3038

Light, G. A., Swerdlow, N. R., Thomas, M. L., Calkins, M. E., Green, M. F., Greenwood, T. A., … Turetsky, B. I. (2015). Validation of mismatch negativity and P3a for use in multisite studies of schizophrenia: Characterization of demographic, clinical, cognitive, and functional correlates in COGS-2. *Schizophrenia Research, 163*(1–3), 63–72. https://doi.org/10.1016/j.schres.2014.09.042

Lubenov, E. V. & Siapas, A. G. (2009). Hippocampal theta oscillations are travelling waves. *Nature, 459*(7246), 534–539. https://doi.org/10.1038/nature08010

Luschei, E. S. (2006). Patterns of laryngeal electromyography and the activity of the respiratory system during spontaneous laughter. *Journal of Neurophysiology, 96*(1), 442–450. https://doi.org/10.1152/jn.00102.2006

Luu, P. & Tucker, D. M. (2001). Regulating action: Alternating activation of midline frontal and motor cortical networks. *Clinical Neurophysiology, 112*(7), 1295–1306. https://doi.org/S1388-2457(01)00559-4 [pii]

Luu, P., Tucker, D. M., & Makeig, S. (2004). Frontal midline theta and the error-related negativity: Neurophysiological mechanisms of action regulation. *Clinical Neurophysiology, 115*(8), 1821–1835. https://doi.org/10.1016/j.clinph.2004.03.031S1388245704001294 [pii]

MacNeilage, P. F. & Davis, B. L. (2001). Motor mechanisms in speech ontogeny: Phylogenetic, neurobiological and linguistic implications. *Current Opinion in Neurobiology, 11*(6), 696–700. https://doi.org/10.1016/S0959-4388(01)00271-9

Macrides, F., Eichenbaum, H., & Forbes, W. B. (1982). Temporal relationship between sniffing and the limbic theta rhythm during odor discrimination reversal learning. *Journal of Neuroscience, 2*(12), 1705–1717.

Miltner, W. H. R., Braun, C. H., & Coles, M. G. H. (1997). Event-related brain potentials following incorrect feedback in a time-estimation task: Evidence for a "generic" neural system for error detection. *Journal of Cognitive Neuroscience, 9*(6), 788–798.

Mitchell, D. J., McNaughton, N., Flanagan, D., & Kirk, I. J. (2008). Frontal-midline theta from the perspective of hippocampal "theta". *Progress in Neurobiology*, *86*(3), 156–185. https://doi.org/10.1016/j.pneurobio.2008.09.005

Mizuki, Y., Kajimura, N., Kai, S., Suetsugi, M., Ushijima, I., & Yamada, M. (1992). Differential responses to mental stress in high and low anxious normal humans assessed by frontal midline theta activity. *International Journal of Psychophysiology*, *12*, 169–178.

Mizuki, Y., Suetsugi, M., Ushijima, I., & Yamada, M. (1996). Differential effects of noradrenergic drugs on anxiety and arousal in healthy volunteers with high and low anxiety. *Progress in Neuro-Psychopharmacology & Biological Psychiatry*, *20*(8), 1353–1367.

Mizuki, Y, Suetsugi, M., Ushijima, I., & Yamada, M. (1997). Differential effects of dopaminergic drugs on anxiety and arousal in healthy volunteers with high and low anxiety. *Progress in Neuro-Psychopharmacology & Biological Psychiatry*, *21*(4), 573–590. http://www.ncbi.nlm.nih.gov/pubmed/9194141

Moran, T. P., Bernat, E. M., Aviyente, S., Schroder, H. S., & Moser, J. S. (2014). Sending mixed signals: Worry is associated with enhanced initial error processing but reduced call for subsequent cognitive control. *Social Cognitive and Affective Neuroscience*, *10*(11), 1548–1556. https://doi.org/10.1093/scan/nsv046

Moser, J. S., Moran, T. P., Schroder, H. S., Donnellan, M. B., & Yeung, N. (2013). On the relationship between anxiety and error monitoring: A meta-analysis and conceptual framework. *Frontiers in Human Neuroscience*, *7*, 466. https://doi.org/10.3389/fnhum.2013.00466

Munneke, G.-J., Nap, T. S., Schippers, E. E., & Cohen, M. X. (2015). A statistical comparison of EEG time- and time–frequency domain representations of error processing. *Brain Research*, *1618*, 222–230. https://doi.org/10.1016/j.brainres.2015.05.030

Narayanan, N. S., Cavanagh, J. F., Frank, M. J., & Laubach, M. (2013). Common medial frontal mechanisms of adaptive control in humans and rodents. *Nature Neuroscience*, *16*, 1–10. https://doi.org/10.1038/nn.3549

Nigbur, R., Cohen, M. X., Ridderinkhof, K. R., & Stürmer, B. (2012). Theta dynamics reveal domain-specific control over stimulus and response conflict. *Journal of Cognitive Neuroscience*, *24*(5), 1264–1274.

Nyhus, E., & Curran, T. (2010). Functional role of gamma and theta oscillations in episodic memory. *Neuroscience and Biobehavioral Reviews*, *34*(7), 1023–1035. https://doi.org/10.1016/j.neubiorev.2009.12.014

Oehrn, C. R., Hanslmayr, S., Fell, J., Deuker, L., Kremers, N. A., Do Lam, A. T., … Axmacher, N. (2014). Neural communication patterns underlying conflict detection, resolution, and adaptation. *Journal of Neuroscience*, *34*(31), 10438–10452. https://doi.org/10.1523/JNEUROSCI.3099-13.2014

Onton, J., Delorme, A., & Makeig, S. (2005). Frontal midline EEG dynamics during working memory. *NeuroImage*, *27*(2), 341–356. https://pubmed.ncbi.nlm.nih.gov/15927487/

Pailing, P. E. & Segalowitz, S. J. (2004). The effects of uncertainty in error monitoring on associated ERPs. *Brain and Cognition*, *56*(2), 215–233.

Parker, K. L., Chen, K.-H., Kingyon, J. R., Cavanagh, J. F., & Narayanan, N. S. (2015). Medial frontal ~4-Hz activity in humans and rodents is attenuated in PD patients and in rodents with cortical dopamine depletion. *Journal of Neurophysiology*, *114*(2), 1310–1320. https://doi.org/10.1152/jn.00412.2015

Pastötter, B., Hanslmayr, S., & Bauml, K.-H. T. (2010). Conflict processing in the anterior cingulate cortex constrains response priming. *NeuroImage*, *50*(4), 1599–1605. https://doi.org/10.1016/j.neuroimage.2010.01.095

Paz, R., Bauer, E. P., & Pare, D. (2008). Theta synchronizes the activity of medial prefrontal neurons during learning. *Learning & Memory*, *15*(7), 524–531. https://doi.org/10.1101/lm.932408

Pellegrino, F., Coupé, C., & Marsico, E. (2011). Across-language perspective on speech information rate. *Language*, *87*(3), 539–558. https://doi.org/10.1353/lan.2011.0057

Petajan, J. H. & Williams, D. D. (1972). Behavior of single motor units during pre-shivering tone and shivering tremor. *American Journal of Physical Medicine*, *51*(1), 16–22.

Phillips, J. M. & Everling, S. (2014). Event-related potentials associated with performance monitoring in non-human primates. *NeuroImage*, *97*, 308–320. https://doi.org/10.1016/j.neuroimage.2014.04.028

Phillips, J. M., Vinck, M., Everling, S., & Womelsdorf, T. (2013). A long-range fronto-parietal 5- to 10-Hz network predicts "top-down" controlled guidance in a task-switch paradigm. Cerebral Cortex, *24*(8), 1996–2008. https://doi.org/10.1093/cercor/bht050

Pignatelli, M., Beyeler, A., & Leinekugel, X. (2012). Neural circuits underlying the generation of theta oscillations. *Journal of Physiology*, *106*(3–4), 81–92. https://doi.org/10.1016/j.jphysparis.2011.09.007

Popov, T., Westner, B. U., Silton, R. L., Sass, S. M., Spielberg, J. M., Rockstroh, B., ... Miller, G. A. (2018). Time course of brain network reconfiguration supporting inhibitory control. *The Journal of Neuroscience*, *38*(18), 2639–17. https://doi.org/10.1523/JNEUROSCI.2639-17.2018

Raghavachari, S., Lisman, J. E., Tully, M., Madsen, J. R., Bromfield, E. B., & Kahana, M. J. (2006). Theta oscillations in human cortex during a working-memory task: Evidence for local generators. *Journal of Neurophysiology*, 95(3), 1630–1638. doi: 10.1152/jn.00409.2005

Rajan, A., Siegel, S. N., Liu, Y., Bengson, J., Mangun, G. R., & Ding, M. (2018). Theta oscillations index frontal decision-making and mediate reciprocal frontal–parietal interactions in willed attention. *Cerebral Cortex*, *29*(7), 2832–2843 . https://doi.org/10.1093/cercor/bhy149

Rangaswamy, M., Jones, K. A., Porjesz, B., Chorlian, D. B., Padmanabhapillai, A., Kamarajan, C., ... Begleiter, H. (2007). Delta and theta oscillations as risk markers in adolescent offspring of alcoholics. *International Journal of Psychophysiology*, *63*(1), 3–15. https://doi.org/10.1016/j.ijpsycho.2006.10.003

Reinhart, R. M. G. & Woodman, G. F. (2014). Causal control of medial-frontal cortex governs electrophysiological and behavioral indices of performance monitoring and learning. *Journal of Neuroscience*, *34*(12), 4214–4227. https://doi.org/10.1523/JNEUROSCI.5421-13.2014

Reinhart, R. M. G. (2017). Disruption and rescue of interareal theta phase coupling and adaptive behavior. *Proceedings of the National Academy of Sciences of the United States of America*, *114*(43), 11542–11547. https://doi.org/10.1073/pnas.1710257114

Reinhart, R. M. G., Zhu, J., Park, S., & Woodman, G. F. (2015). Synchronizing theta oscillations with direct-current stimulation strengthens adaptive control in the human brain. *Proceedings of the National Academy of Sciences of the United States of America*, *112*(30), 9448–9453. https://doi.org/10.1073/pnas.1504196112

Riesel, A., Goldhahn, S., & Kathmann, N. (2017). Hyperactive performance monitoring as a transdiagnostic marker: Results from health anxiety in comparison to obsessive–compulsive disorder. *Neuropsychologia*, *96*, 1–8. https://doi.org/10.1016/j.neuropsychologia.2016.12.029

Rizzuto, D. S., Madsen, J. R., Bromfield, E. B., Schulze-Bonhage, A., & Kahana, M. J. (2006). Human neocortical oscillations exhibit theta phase differences between encoding and retrieval. *NeuroImage*, *31*(3), 1352–1358. doi: 10.1016/j.neuroimage.2006.01.009

Rizzuto, D. S., Madsen, J. R., Bromfield, E. B., Schulze-Bonhage, A., Seelig, D., Aschenbrenner-Scheibe, R., & Kahana, M. J. (2003). Reset of human neocortical oscillations during a working memory task. *Proceedings of the National Academy of Sciences of the United States of America*, *100*(13), 7931–7936. https://doi.org/10.1073/pnas.0732061100732061100 [pii]

Roberts, B. M., Hsieh, L. T., & Ranganath, C. (2013). Oscillatory activity during maintenance of spatial and temporal information in working memory. *Neuropsychologia*, *51*(2), 349–357. https://doi.org/10.1016/j.neuropsychologia.2012.10.009

Rothé, M., Quilodran, R., Sallet, J., & Procyk, E. (2011). Coordination of high gamma activity in anterior cingulate and lateral prefrontal cortical areas during adaptation. *The Journal of Neuroscience*, *31*(31), 11110–11117. https://doi.org/10.1523/JNEUROSCI.1016-11.2011

Ryman, S. G., Cavanagh, J. F., Wertz, C. J., Shaff, N. A., Dodd, A. B., Stevens, B., . . . Mayer, A. R. (2018). Impaired midline theta power and connectivity during proactive cognitive control in schizophrenia. *Biological Psychiatry*, *84*(9), 675–683. https://doi.org/10.1016/j.biopsych.2018.04.021

Sato, K. (1952). On the theta rhythms in healthy human EEG. *Folia Psychiatrica et Neurologica*, *5*(4), 283–297.

Sauseng, P., Griesmayr, B., Freunberger, R., & Klimesch, W. (2010). Control mechanisms in working memory: A possible function of EEG theta oscillations. *Neuroscience and Biobehavioral Reviews*, *34*(7), 1015–1022. doi: 10.1016/j.neubiorev.2009.12.006

Sauseng, P., Klimesch, W., Doppelmayr, M., Hanslmayr, S., Schabus, M., & Gruber, W. R. (2004). Theta coupling in the human electroencephalogram during a working memory task. *Neuroscience Letters*, *354*(2), 123–126. doi: 10.1016/j.neulet.2003.10.002

Schmidt, B., Kanis, H., Holroyd, C. B., Miltner, W. H. R., & Hewig, J. (2018). Anxious gambling: Anxiety is associated with higher frontal midline theta predicting less risky decisions. *Psychophysiology*, *55*(10), 1–9. https://doi.org/10.1111/psyp.13210

Shackman, A. J., Salomons, T. V, Slagter, H. A, Fox, A. S., Winter, J. J., & Davidson, R. J. (2011). The integration of negative affect, pain and cognitive control in the cingulate cortex. *Nature Reviews. Neuroscience*, *12*(3), 154–167. https://doi.org/10.1038/nrn2994

Siapas, A. G., Lubenov, E. V, & Wilson, M. A. (2005). Prefrontal phase locking to hippocampal theta oscillations. *Neuron*, *46*(1), 141–151. https://doi.org/10.1016/j.neuron.2005.02.028

Smit, A. S., Eling, P. A T. M., & Coenen, A. M. L. (2004). Mental effort affects vigilance enduringly: after-effects in EEG and behavior. *International Journal of Psychophysiology*, *53*(3), 239–243. https://doi.org/10.1016/j.ijpsycho.2004.04.005

Smit, A. S., Eling, P. A T. M., Hopman, M. T., & Coenen, A. M. L. (2005). Mental and physical effort affect vigilance differently. *International Journal of Psychophysiology*, *57*(3), 211–217. https://doi.org/10.1016/j.ijpsycho.2005.02.001

Smith, E. H., Banks, G. P., Mikell, C. B., Cash, S. S., Patel, S. R., Eskandar, E. N., & Sheth, S. A. (2015). Frequency-dependent representation of reinforcement-related information in the human medial and lateral prefrontal cortex. *Journal of Neuroscience*, *35*(48), 15827–15836. https://doi.org/10.1523/JNEUROSCI.1864-15.2015

Solomon, E. A., Kragel, J. E., Sperling, M. R., Sharan, A., Worrell, G., Kucewicz, M., . . . Kahana, M. J. (2017). Widespread theta synchrony and high-frequency desynchronization underlies enhanced cognition. *Nature Communications*, *8*, 1704. https://doi.org/10.1038/s41467-017-01763-2

Staudigl, T., Hanslmayr, S., & Bauml, K. H. (2010). Theta oscillations reflect the dynamics of interference in episodic memory retrieval. *Journal of Neuroscience*, *30*(34), 11356–11362. doi:10.1523/JNEUROSCI.0637-10.2010

Swart, J. C., Frank, M. J., Määttä, J. I., Jensen, O., Cools, R., & den Ouden, H. E. M. (2018). Frontal network dynamics reflect neurocomputational mechanisms for reducing maladaptive biases in motivated action. *PLoS Biology* [online], 1–25. https://doi.org/10.1371/journal.pbio.2005979

Talairach, J., Bancaud, J., Geier, S., Bordas-Ferrer, M., Bonis, A., Szikla, G., & Rusu, M. (1973). The cingulate gyrus and human behaviour. *Electroencephalography and Clinical Neurophysiology*, *34*(1), 45–52. doi: 10.1016/0013-4694(73)90149-1

Tang, H., Yu, H. Y., Chou, C. C., Crone, N. E., Madsen, J. R., Anderson, W. S., & Kreiman, G. (2016). Cascade of neural processing orchestrates cognitive control in human frontal cortex. *ELife*, *5*, e12352. https://doi.org/10.7554/eLife.12352

Taub, A. H., Perets, R., Kahana, E., & Paz, R. (2018). Oscillations synchronize amygdala-to-prefrontal primate circuits during aversive learning. *Neuron*, *97*(2), 291–298.e3. https://doi.org/10.1016/j.neuron.2017.11.042

Töllner, T., Wang, Y., Makeig, S., Müller, H. J., Jung, T.-P., & Gramann, K. (2017). Two independent frontal midline theta oscillations during conflict detection and adaptation in a Simon-type manual reaching task. *The Journal of Neuroscience*, *37*(9), 2504–2515. https://doi.org/10.1523/JNEUROSCI.1752-16.2017

Tóth, B., Kardos, Z., File, B., Boha, R., Stam, C. J., & Molnár, M. (2014). Frontal midline theta connectivity is related to efficiency of WM maintenance and is affected by aging. *Neurobiology of Learning and Memory*, *114*, 58–69. https://doi.org/10.1016/j.nlm.2014.04.009

Trujillo, L. T. & Allen, J. J. (2007). Theta EEG dynamics of the error-related negativity. *Clinical Neurophysiology*, *118*(3), 645–668. doi: 10.1016/j.clinph.2006.11.009

Tsujimoto, S., Genovesio, A., & Wise, S. P. (2010). Evaluating self-generated decisions in frontal pole cortex of monkeys. *Nature Neuroscience*, *13*(1), 120–126. https://doi.org/10.1038/nn.2453

Tsujimoto, T., Shimazu, H., & Isomura, Y. (2006). Direct recording of theta oscillations in primate prefrontal and anterior cingulate cortices. *Journal of Neurophysiology*, *95*(5), 2987–3000. doi:10.1152/jn.00730.2005

van de Vijver, I., Cohen, M. X., & Ridderinkhof, K. R. (2014). Aging affects medial but not anterior frontal learning-related theta oscillations. *Neurobiology of Aging*, *35*(3), 692–704. https://doi.org/10.1016/j.neurobiolaging.2013.09.006

van de Vijver, I., Ridderinkhof, K. R., & Cohen, M. X. (2011). Frontal oscillatory dynamics predict feedback learning and action adjustment. *Journal of Cognitive Neuroscience*, *23*(12), 4106–4121. https://doi.org/10.1162/jocn_a_00110

van Driel, J., Ridderinkhof, K. R., & Cohen, M. X. (2012). Not all errors are alike: Theta and alpha EEG dynamics relate to differences in error-processing dynamics. *The Journal of Neuroscience*, *32*(47), 16795–16806. https://doi.org/10.1523/JNEUROSCI.0802-12.2012

van Noordt, S. J. R. & Segalowitz, S. J. (2012). Performance monitoring and the medial prefrontal cortex: A review of individual differences and context effects as a window on self-regulation. *Frontiers in Human Neuroscience*, *6*, 197. https://doi.org/10.3389/fnhum.2012.00197

Van Veen, V. & Carter, C. S. (2002). The timing of action-monitoring processes in the anterior cingulate cortex. *Journal of Cognitive Neuroscience*, *14*(4), 593–602. https://doi.org/10.1162/08989290260045837

VanRullen, R. (2016). Perceptual cycles. *Trends in Cognitive Sciences*, *20*(10), 723–735. https://doi.org/10.1016/j.tics.2016.07.006

Verguts, T. (2017). Binding by random bursts: A computational model of cognitive control. *Journal of Cognitive Neuroscience*, *29*(6), 1103–1118. https://doi.org/10.1162/jocn_a_01117

Vezoli, J. & Procyk, E. (2009). Frontal feedback-related potentials in nonhuman primates: modulation during learning and under haloperidol. *Journal of Neuroscience*, *29*(50), 15675–15683. doi:10.1523/JNEUROSCI.4943-09.2009

Vidal, F., Burle, B., Bonnet, M., Grapperon, J., & Hasbroucq, T. (2003). Error negativity on correct trials: a reexamination of available data. *Biological Psychology*, *64*(3), 265–282. https://doi.org/S0301051103000978

Vidal, F., Hasbroucq, T., Grapperon, J., & Bonnet, M. (2000). Is the "error negativity" specific to errors? *Biological Psychology*, *51*(2–3), 109–128. https://doi.org/10.1016/S0301-0511(99)00032-0

Vissers, M. E., Ridderinkhof, K. R., Cohen, M. X., & Slagter, H. A. (2018). Oscillatory mechanisms of response conflict elicited by color and motion direction: An individual differences approach. *Journal of Cognitive Neuroscience*, *30*(4), 468–481. https://doi.org/10.1162/jocn_a_01222

Walsh, M. M. & Anderson, J. R. (2012). Learning from experience: Event-related potential correlates of reward processing, neural adaptation, and behavioral choice. *Neuroscience and Biobehavioral Reviews*, *36*(8), 1870–1884. https://doi.org/10.1016/j.neubiorev.2012.05.008

Wang, C., Ulbert, I., Schomer, D. L., Marinkovic, K., & Halgren, E. (2005). Responses of human anterior cingulate cortex microdomains to error detection, conflict monitoring, stimulus-response mapping, familiarity, and orienting. *Journal of Neuroscience*, *25*(3), 604–613. doi:10.1523/JNEUROSCI.4151-04.2005

Wang, D., Clouter, A., Chen, Q., Shapiro, K. L., & Hanslmayr, S. (2018). Single-trial phase entrainment of theta oscillations in sensory regions predicts human associative memory performance. *The Journal of Neuroscience*, *38*(28), 0349–18. https://doi.org/10.1523/JNEUROSCI.0349-18.2018

Wascher, E., Rasch, B., Sänger, J., Hoffmann, S., Schneider, D., Rinkenauer, G., … Gutberlet, I. (2014). Frontal theta activity reflects distinct aspects of mental fatigue. *Biological Psychology*, *96*, 57–65. https://doi.org/10.1016/j.biopsycho.2013.11.010

Wessel, J. R. (2018). An adaptive orienting theory of error processing. *Psychophysiology*, *55*(3), 1–21. https://doi.org/10.1111/psyp.13041

White, J. A., Banks, M. I., Pearce, R. A., & Kopell, N. J. (2000). Networks of interneurons with fast and slow gamma-aminobutyric acid type A (GABAA) kinetics provide substrate for mixed gamma–theta rhythm. *Proceedings of the National Academy of Sciences of the United States of America*, *97*(14), 8128–8133.

Wiecki, T. V. & Frank, M. J. (2013). A computational model of inhibitory control in frontal cortex and basal ganglia. *Psychological Review*, *120*(2), 329–355.

Womelsdorf, T., Johnston, K., Vinck, M., & Everling, S. (2010). Theta-activity in anterior cingulate cortex predicts task rules and their adjustments following errors. *Proceedings of the National Academy of Sciences of the United States of America*, *107*(11), 5248–5253. doi: 10.1073/pnas.0906194107

Womelsdorf, T., Schoffelen, J. M., Oostenveld, R., Singer, W., Desimone, R., Engel, A. K., & Fries, P. (2007). Modulation of neuronal interactions through neuronal synchronization. *Science*, *316*(5831), 1609–1612. doi:10.1126/science.1139597

Womelsdorf, T., Valiante, T. A., Sahin, N. T., Miller, K. J., & Tiesinga, P. (2014). Dynamic circuit motifs underlying rhythmic gain control, gating and integration. *Nature Neuroscience*, *17*(8), 1031–1039. https://doi.org/10.1038/nn.3764

Womelsdorf, T., Vinck, M., Leung, L. S., & Everling, S. (2010). Selective theta-synchronization of choice-relevant information subserves goal-directed behavior. *Frontiers in Human Neuroscience*, *4*, 210. https://doi.org/10.3389/fnhum.2010.00210

Yamaguchi, M., Crump, M. J. C., & Logan, G. D. (2013). Speed-accuracy trade-off in skilled typewriting: Decomposing the contributions of hierarchical control loops. *Journal of Experimental Psychology: Human Perception and Performance*, *39*(3), 678–699.

Yordanova, J, Falkenstein, M., Hohnsbein, J., & Kolev, V. (2004). Parallel systems of error processing in the brain. *NeuroImage*, *22*(2), 590–602. https://doi.org/10.1016/j.neuroimage.2004.01.040S1053811904000850 [pii]

Yordanova, Juliana, Kolev, V., Rosso, O. A., Schürmann, M., Sakowitz, O. W., Özgören, M., & Basar, E. (2002). Wavelet entropy analysis of event-related potentials indicates modality-independent theta dominance. *Journal of Neuroscience Methods*, *117*(1), 99–109. https://doi.org/10.1016/S0165-0270(02)00095-X

Zavala, B. A., Tan, H., Little, S., Ashkan, K., Hariz, M., Foltynie, T., . . . Brown, P. (2014). Midline frontal cortex low-frequency activity drives subthalamic nucleus oscillations during conflict. *Journal of Neuroscience*, *34*(21), 7322–7333. https://doi.org/10.1523/JNEUROSCI.1169-14.2014

Zavala, B., Brittain, J.-S., Jenkinson, N., Ashkan, K., Foltynie, T., Limousin, P., . . . Brown, P. (2013). Subthalamic nucleus local field potential activity during the Eriksen Flanker task reveals a novel role for theta phase during conflict monitoring. *The Journal of Neuroscience*, *33*(37), 14758–14766. https://doi.org/10.1523/JNEUROSCI.1036-13.2013

Zavala, B., Tan, H., Little, S., Ashkan, K., Green, A. L., Aziz, T., . . . Brown, P. (2016). Decisions made with less evidence involve higher levels of corticosubthalamic nucleus theta band synchrony. *Journal of Cognitive Neuroscience*, *28*(6), 811–825. https://doi.org/10.1162/jocn_a_00934

Zhang, H., Watrous, A. J., Patel, A., & Jacobs, J. (2018). Theta and alpha oscillations are traveling waves in the human neocortex. *Neuron*, *98*(6), 1269–1281.e4. https://doi.org/10.1016/j.neuron.2018.05.019

Zlojutro, M., Manz, N., Rangaswamy, M., Xuei, X., Flury-Wetherill, L., Koller, D., . . . Almasy, L. (2011). Genome-wide association study of theta band event-related oscillations identifies serotonin receptor gene HTR7 influencing risk of alcohol dependence. *American Journal of Medical Genetics, Part B: Neuropsychiatric Genetics*, *156*(1), 44–58. https://doi.org/10.1002/ajmg.b.31136

CHAPTER 10

THE ROLE OF ALPHA AND BETA OSCILLATIONS IN THE HUMAN EEG DURING PERCEPTION AND MEMORY PROCESSES

SEBASTIAN MICHELMANN, BENJAMIN GRIFFITHS, AND SIMON HANSLMAYR

10.1 Alpha oscillations and their relation to cognition

In 1929 a German physician named Hans Berger recorded the first human EEG. The first thing he noticed was a regular oscillation with a frequency of 10 Hz (Berger, 1929), which he termed alpha oscillations. Many decades after Berger discovered these alpha oscillations researchers use them as a first quality check of their EEG signal. After attaching electrodes to the subject's head, the researcher typically asks the subject to close their eyes and relax; the researcher then sees beautiful alpha waves with a frequency of around 10 Hz being maximal over posterior channels. When the subject then opened their eyes, alpha oscillations largely reduce. This phenomenon is extremely reliable, such that not seeing the reduction in alpha amplitude when subjects open their eyes would typically indicate that something went wrong. Reductions in alpha amplitude occur in all ranges of cognitive tasks such as visual processing (Adrian, 1944), auditory processing (Krause et al., 1994; Obleser & Weisz, 2012), somatosensory processing (Crone et al., 1998), memory encoding (Hanslmayr et al., 2009; Klimesch et al., 1996), memory retrieval (Burgess & Gruzelier, 2000; Waldhauser et al., 2016), working memory maintenance (Sauseng et al., 2009), decision making (Pornpattananangkul et al., 2019), and motor preparation and execution (Pfurtscheller et al., 1997). The exact frequency at which power reductions are maximal varies between tasks and often involves the faster beta oscillation around 15 Hz, which can be considered the "fast" sibling of alpha. In this chapter we therefore do not distinguish between alpha and beta oscillations and refer to these oscillations as alpha/beta. Suppression of alpha/beta oscillations is not only observed in humans, but also in a wide range of animals, for example, non-human primates (Haegens et al., 2011), dogs (Lopes da Silva et al., 1980), cats (von Stein et al., 2000), rodents (Wiest & Nicolelis, 2003), and even insects (Popov & Szyszka, 2019). This ubiquity of alpha/beta power suppression across cognitive domains and across the animal kingdom indicates that alpha/beta power reductions are a signature of an extremely general mechanism, which is called upon in almost any cognitive task and has been retained over millions of years of evolution. What could this mechanism be? This chapter attempts to answer this question.

Before delving into the different physiological interpretations of alpha oscillations it is worth pointing out a theoretical caveat that has become evident in cognitive neuroscience and the way we can avoid these problems. Historically the job description for a cognitive neuroscientist was to pick a cognitive phenomenon (i.e., attention, memory) and then find the "neural correlate" of that phenomenon. To demonstrate this approach let us consider a short thought experiment involving two hypothetical cognitive neuroscientists: AC and BD. Researcher AC is interested in attention. She runs several meticulously controlled experiments that all manipulate certain aspects of attention while she records EEG. Across this series of experiments, she finds that alpha oscillations are very consistently modulated by attention. She goes to a conference, presents her results, and concludes that alpha oscillations are *the* neural correlate of attention. At that same conference, researcher BD, who is interested in memory, also presents his results. In a series of carefully controlled experiments he shows that alpha oscillations are reliably modulated by various memory processes. He concludes his talk with saying that alpha oscillations are *the* neural correlate of memory.

So, who is right? What cognitive function do alpha oscillations represent: memory or attention? The answer is both are wrong. The error that both scientists make is to assume that there is a one-to-one mapping between neural phenomena and cognitive functions. This error has become known under the term "reverse inference error", that is, observing neural signature X has no predictive value for cognitive process Y to occur (Poldrack, 2011). With respect to alpha oscillations, given their ubiquity in terms of cognitive tasks (from attention to decision making) and species (from human to honeybees) it is evident that the task of trying to attach one particular label from cognitive psychology (attention vs. memory) to them is bound to fail. An alternative approach is needed if we want to truly understand the function of alpha oscillations. This alternative approach needs to avoid limiting itself by definitions of cognitive processes. Instead, this approach needs to embrace the fact that a given neural phenomenon can be of service to many different cognitive processes in many different species. One such alternative is to assume that different oscillations implement canonical computations (Siegel et al., 2012; Womelsdorf et al., 2014), which are basic neural operations that are called upon by different cognitive processes. Regulating the balance between excitation and inhibition, for instance, would be one such basic operation. Another basic operation is to enable the neural representation of information rich content. These operations arguably are required by almost any cognitive task and species. The difference from this approach to the traditional cognitive neuroscience approach is not in using different labels, but to try to understand the computational utility of a neural operation for a given cognitive process (see Buzsaki, 2019 for a similar line of argument). Instead of asking what the neural correlate of attention is, we ask how a decrease in alpha oscillations can be of service for attention, memory, decision making, etc. Section 10.2 provides a brief overview of the behaviour of alpha oscillations in terms of frequency, power, and phase. It is critical to understand these terms first before we can consider the physiological interpretations of alpha oscillations.

10.2 Signal properties of alpha: Frequency, Power, and Phase

10.2.1 Frequency

Analyzing alpha oscillations typically involves transforming an EEG signal into time-frequency space (see Cohen, 2014 for an excellent overview of the different methods and hands-on tutorials). Figure 10.1A shows a typical example of a raw EEG trace in a healthy human subject. It is easy to spot the decrease in alpha amplitudes at time 0 (when a stimulus was presented). Brain oscillations are defined by three physical properties: (i) frequency (Figure 10.1B), (ii) amplitude (Figure 10.1C), and (iii) phase (Figure 10.1D). Different brain networks are hypothesized to oscillate at different frequencies (Keitel & Gross, 2016), with small networks oscillating at fast frequencies (>40 Hz) and large networks oscillating at slower frequencies (<20 Hz) (von Stein & Sarnthein, 2000). Small and large here refers to the number of neurons involved in generating the signal. This anatomical property is reflected in the 1/F power ratio of EEG signals, which refers to the drop in signal power with increasing frequency. Alpha oscillations are remarkable because they stand out from the 1/F pattern (Figure 10.1B), which shows that they are a particularly strong oscillation recruiting large pools of neurons. Frequencies are typically used to distinguish between different types of oscillations, that is, theta ~ 4 Hz, beta ~15 Hz, gamma ~ 40 Hz, etc. The frequency within an oscillation has been shown to be linked to interindividual variability in memory processes (Cohen, 2011) or intelligence (Anokhin & Vogel, 1996). Recent work demonstrates that the frequency of alpha can also change within a subject, from trial-to-trial (Haegens et al., 2014) and may reflect cortical excitability (Cohen, 2014). Therefore, it is important to consider that the frequency of alpha may vary from individual to individual and even from trial to trial within a participant. Sometimes it can even be challenging to distinguish a "fast theta" from a "slow alpha".

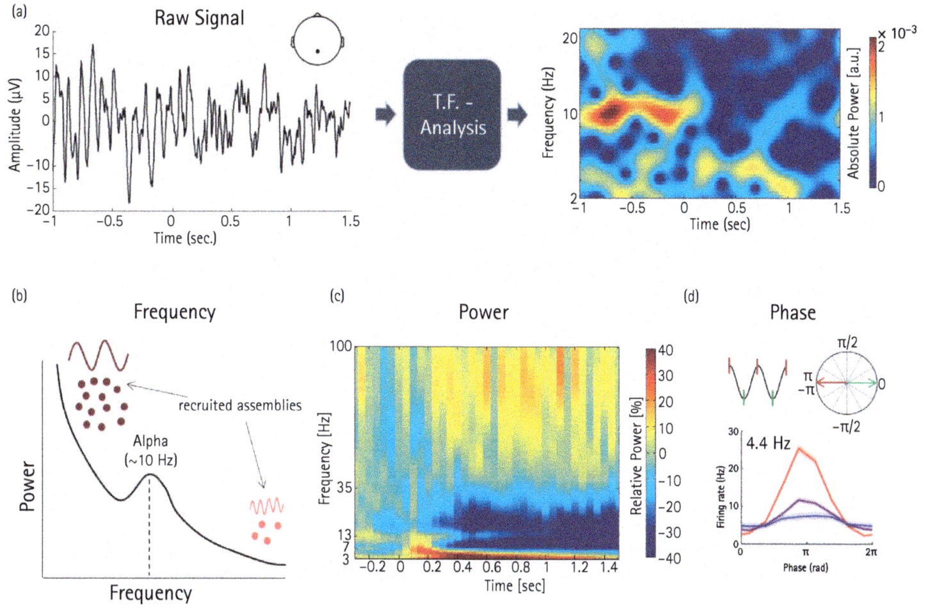

FIGURE 10.1 Alpha oscillations and their parameters. (A) An example of a raw signal as recorded with a parietal EEG electrode is shown on the left. A stimulus was presented at time 0. The plot on the right shows the results of a time-frequency analysis in which power is depicted for each time-point (x-axis) and frequency band (y-axis; a.u. = arbitrary units). (B) A schematic of a typical EEG power spectrum is shown, with frequency on the x-axis and power on the y-axis. The inverse relationship between the size of neural assemblies and power is depicted. Note the peak at the alpha frequency which violates the 1/F relation between power and frequency. (C) A typical time-frequency plot showing event related power increases (hot colors) and decreases (cold colors) during processing of verbal information. Note the power increases in theta (3–7 Hz) and gamma (35–100 Hz) and the power decreases in alpha (8–12 Hz) and beta (13–35 Hz). (D) The relationship between EEG phase (top) and firing rates (bottom) is shown. The differently colored lines show phase modulations in trials with high (red), medium (purple) or low (blue) power.

(A) Reprinted with permission from Hanslmayr et al., 2011. (C) is modified and reprinted with permission from Hanslmayr et al., 2012. (D) modified and reprinted with permission from Jacobs et al., 2007.

10.2.2 Power

The power of an oscillation refers to its signal strength. It is usually calculated by taking the square of the signal. In human EEG, a signal is generated by the summation of several millions of postsynaptic potentials (inhibitory and excitatory) over an area of some cm^2 (Pfurtscheller et al., 1996). Importantly, the current changes induced by postsynaptic potentials are tiny and we record these potentials with an electrode that is attached to the scalp, separated by various layers of bone, cerebrospinal fluid, skin, etc., from the brain. Therefore, in order to detect a signal that is large enough to be visible in the EEG, tens of thousands of post-synaptic potentials must come together in time. To give an analogy, recording EEG is a bit like recording the noise in a football stadium with one big microphone hanging from the ceiling. Using this coarse signal we cannot stand the chance to tune in onto an individual conversation between two fans, but we can use the signal to tell whether a goal was scored (because thousands of fans start shouting at the same time). Accordingly, the strength of an oscillation is assumed to reflect the degree of synchrony between inhibitory or excitatory postsynaptic potentials to an underlying neural assembly. Power increases indicate increased local synchrony whereas power decreases indicate de-synchronized local activity. This idea is reflected in the classic work of Pfurtscheller & Aranibar (1977), who coined the terms event-related synchronization and de-synchronization (ERS/ERD), which denotes power increases and decreases in response to an event or stimulus, respectively. In EEG experiments absolute power is usually transformed into power changes in response to a baseline (e.g., prestimulus interval). Figure 10.1C shows a typical example of such data with stimulus driven power increases in the lower (1–8 Hz: delta/theta) and higher (40–100 Hz; gamma) frequency ranges, and power decreases in the middle frequency ranges (8–35 Hz; alpha/beta).

10.2.3 Phase

The phase of an oscillation specifies the current position in a given cycle (Figure 10.1D), that is whether the oscillation exhibits a peak, trough, or zero crossing. Because the EEG reflects the sum of postsynaptic input to a given neuron, we can assume that it impinges on the neuron's probability to fire an action potential. Figure 10.1D shows a single neuron recorded in the human brain, and that the neuron is more likely to fire in the trough of the oscillation, which indeed resembles the time point of maximally coinciding excitatory input (note that the LFP in this case is measured in the extracellular space; therefore, negativity indexes excitation). This same figure also shows that the modulation of firing rate is stronger for trials of high power (red) and lower for trials of low power (blue). We can therefore assume that the phase of alpha oscillations (or any other oscillation for that matter) represent discrete time windows for neural firing, and that this synchronizing effect scales with power. This is a very useful computational property as it gives alpha oscillations the power to control the timing of neural firing in large groups of neurons. Like the conductor of an orchestra, who tells individual instruments when to play a particular note (and when to not play a note), alpha oscillations can synchronize large groups of neurons to temporally structure neural processing. This aspect becomes particularly important in Section 10.5, which covers the role of alpha phase for sampling information and replaying this information from memory.

10.3 Alpha Oscillations: Passive Idling versus Active Inhibition

Alpha oscillations decrease in amplitude when a subject is engaged in a task as opposed to when a subject is resting (especially with eyes closed). This goes against intuition as one would usually expect to see an increase in brain signal strength when a subject performs a challenging task, not a decrease. How can we functionally interpret this negative relationship between alpha power and cognitive processing? Up until the early 2000s the prevailing view was that alpha oscillations reflect a state of "idling" or rest. In their article, Pfurtscheller and colleagues (1996) give the example of the motor cortex, in particular the visual cortex and the hand area during a reading task or a motor task requiring finger movements, respectively. During reading, the visual cortex displays profound alpha-power decreases, whereas the hand motor area shows an increase in alpha power. This picture switches when the subject engages in a motor task requiring finger movement. Within the idling hypothesis, one would interpret the increased alpha power over areas that are not required by the task as "nil work", that is, a passive resonance phenomenon of a part of cortex that has "nothing to do" (Adrian & Matthews, 1934). Since almost any cognitive task always involves specific activation of some regions and de-activation of other regions (Fox et al., 2005) the EEG would always reflect some areas that show alpha power decreases (or desynchronization) and some areas that show alpha-power increases (or synchronization). The important emphasis of the idling hypothesis is on the passive aspect of alpha synchronization which has no functional role per se.

Klimesch and colleagues (2007) and Jensen and Mazaheri (2010) presented a contrasting view to the idling hypothesis and suggested that alpha oscillations play a critical role in cognitive processing. The seminal findings that led to this interpretation were studies showing that alpha power *increased* with increasing cognitive load in a working memory task (Jensen et al., 2002; Klimesch et al., 1999). This alpha- power increase with cognitive load is difficult to reconcile with the idling hypothesis, which led to the active inhibition hypothesis (Jensen & Mazaheri, 2010; Klimesch et al., 2007). Within the active inhibition hypothesis an increase in alpha power reflects an active inhibition process that serves to silence a particular region that is task irrelevant. This silencing of task-irrelevant areas ensures that information is processed selectively in task-relevant areas and protects the processing of this information from interference or noise. A critical prediction that the inhibition account made was that an increase in alpha oscillations narrows the time windows for neurons to fire (see red vs. blue lines in Figure 10.1D) and therefore reduces neural firing. Thus, periods of high alpha power should coincide with low neural firing, whereas periods of low alpha power should coincide with high neural firing. This prediction was confirmed in a non-human primate study (Haegens et al., 2011), which suggests that an increase in alpha power in a given area acts as a "silencing mechanism" which muffles neural assemblies that otherwise might interfere.

Coming back to our example, alpha synchronization of the hand area while reading this chapter ensures that you can focus on the text instead of moving around (or thinking about movements). The more challenging a task, the more we need to tune out task-irrelevant activity. The inhibition account has gained considerable support over the last decades because it can accommodate the findings in the literature better than the idling hypothesis. To give two examples, if subjects maintain visual content in working memory that was presented in the left hemifield, therefore being processed in the right occipital cortex, alpha power increases over the left occipital cortex (Sauseng et al., 2009). Externally enhancing alpha power with repetitive transcranial magnetic stimulation over the irrelevant hemisphere then increases working memory performance. Similar evidence comes from Bonnefond and Jensen (2012), who demonstrated that subjects actively increase their alpha power in anticipation of a task-irrelevant distractor presented during working memory maintenance. The more the subjects upregulated alpha power the better the performance on the working memory task. These results rule out a passive perspective of alpha oscillations and instead suggest that alpha oscillations are very much an active process that regulates neural activity to ensure selective information processing. Within this perspective, alpha oscillations serve the function of a filter that tunes out task-irrelevant information to render the task-relevant signal more salient.

10.4 Alpha Power Decreases and the Representation of Information

The active inhibition account has been extremely useful in interpreting the role of alpha power increases, or alpha synchronization during cognitive processing. This is because any cognitive process requires selective information processing; alpha oscillations, by inhibiting task irrelevant neural assemblies, ensure such selective information processing. This functional interpretation of alpha oscillations is broad enough to accommodate the fact that modulations of alpha oscillations are observed in a variety of cognitive tasks and species. Returning to the thought experiment from earlier, the attention scientist AC and memory scientist BD would interpret their findings to show that both memory and attention crucially rely on an active filtering mechanism. Indeed, alpha power increases over areas that hold the representation of *task-irrelevant* information, regardless of whether this information is currently perceived (Thut et al., 2006; Worden et al., 2000), held in working memory (Sauseng et al., 2009), or stored in long-term memory (Waldhauser et al., 2012). The emphasis of the active inhibition hypothesis is on alpha synchronization and its computational utility in terms of providing a filter mechanism for selective information processing. What is less clear from this perspective, however, is what the computational utility of alpha power decreases are for information processing (other than allowing for increased neural spike rates through less inhibition). We therefore proposed an additional theory, which is complementary to the inhibition account. Within this framework we emphasized the role of alpha power decreases in allowing for high-fidelity information to be represented in neural assemblies. A key assumption of this account rests on the fact the alpha/beta power decreases represent periods of de-correlated neural firing (see Murthy & Fetz, 1996 for such a demonstration for beta oscillations in the motor cortex).

One way of interpreting the functional utility of de-correlated firing in alpha for cognitive processing applies tenets of information theory (Shannon & Weaver, 1949) to neural oscillations. This framework is known as the "information-via-desynchronization hypothesis" (Hanslmayr et al., 2012), which proposes that synchronized alpha/beta states are inherently bad for information representation as neuronal activity is highly redundant. Take the simplified instance of two presynaptic neurons that act upon the same postsynaptic neuron: if one presynaptic neuron fires in perfect synchrony with another, what can this neuron add to the neuronal code that its synchronous partner does not already (Schneidman et al., 2011)? If we expand this principle to networks of neurons, we can postulate that highly synchronous networks are detrimental to information processing, because they only represent redundancies. To overcome this limitation therefore, the networks must desynchronize. Through desynchronization, the underlying neural code can be more complex and hence convey a more detailed representation of information. The event-related desynchronization so commonly seen in alpha oscillation may be a prime example of this phenomenon.

Numerous studies support the idea that alpha/beta power decreases reflect the representation of information within the cortex. One of the most direct lines of support comes from a recent simultaneous EEG-fMRI experiment by Griffiths and colleagues (2019), who asked participants to associate video clips with words, and to later recall the clips using the words as a cue (Figure 10.2). For each trial, the researchers quantified the amount of visual information present in the cortex by conducting representational similarity analysis (RSA) on the fMRI data (Kriegeskorte et al., 2008). RSA is based on correlations of neural patterns and reasons that representations of the same content should elicit neural patterns that are more alike than the patterns elicited by representations of different content (Kriegeskorte et al., 2008). The researchers then asked whether the power of the alpha and beta frequencies (8–30Hz) correlated with the quantity of information (calculated via RSA) represented on a given trial. Indeed, they found evidence to suggest a parametric link between alpha/beta power and information: as power decreased, information increased (Figure 10.2b). Similarly, Hanslmayr and colleagues (2009) presented participants with words and asked them to engage in semantic processing (i.e., does this word represent a living entity or a non-living entity?) or shallow processing (i.e., does the first letter of the word precede the last in the alphabet?). As semantic processing involves much greater levels of information processing (you must not only process the letters, but also what those letters mean), the researchers hypothesized that alpha/beta power decreases would be greater during this type of processing. Their results revealed just that—suggesting that alpha/beta power decreases scale with the depth of information processing (see Fellner et al., 2018 for similar results contrasting familiar and unfamiliar stimuli). In conjunction with a number of other studies, these results strongly implicate alpha/beta power reductions in information processing (Hanslmayr et al., 2012).

While the alpha idling theory assumes that synchronous alpha oscillations mark a default state in which the cortex does nothing (Pfurtscheller et al., 1996), the alpha inhibition theory highlights the active role of alpha synchronization in the suppression of

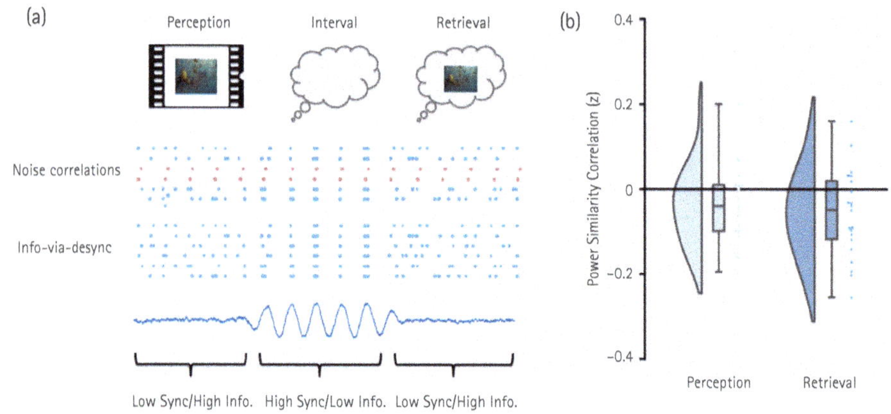

FIGURE 10.2 Alpha power and information processing. (A) Infographic depicting theories behind alpha and information processing. Perception and memory retrieval involve the processing of large quantities of information. The noise correlation account proposes that when task irrelevant neurons (in blue) synchronize (e.g., during the interval), they mask the signal generated by task-relevant neurons (in red). When the task-irrelevant neurons desynchronize however (i.e., during perception/retrieval), the signal can be detected above the background noise. The information-via-desynchronization account proposes that oscillatory desynchronization allows a more complex neural code to be generated. Such a complex code is necessary to process the highly complex information encountered in daily life. In both instances, oscillatory desynchronization benefits information processing (B) Reproduction of the results by Griffiths and colleagues. Power decreases during both perception and retrieval negatively correlate with the amount of stimulus specific information present on that trial.

task irrelevant information (Jensen & Mazaheri, 2010; Klimesch et al., 2007). It therefore attributes an operative function to alpha synchrony. The information via desynchronization hypothesis goes beyond the inhibition theory, in that it highlights the active role of power decreases for the processing of information. Therein, power decreases are not a mere absence of inhibition, but rather functionally involved in neural computations. The crucial insight is that in order to process complex information or represent information rich content, synchronous neural activity does not provide enough coding space. Desynchronous neural activity on the other hand, which is marked by power decreases in the alpha/beta band, provides the required coding space through locally decoupled neural assemblies. The information via desynchronization theory therefore stresses that, in order to perform cognitive operations that work on complex and information-rich content, alpha/beta power must decrease.

Another way to interpret the inverse relation between alpha power and information representation is offered by studies investigating "noise-correlations", which refers to correlated firing of task-irrelevant neurons—a process that can be detrimental to information representation (Mitchell et al., 2009). If two task-irrelevant neurons fire together, their noise is amplified. Expand this principle to hundreds or thousands of neurons and their noise becomes deafening. In such quantities, these noise correlations mask the signal generated by neurons critical to the task at hand (Averbeck et al., 2006; see also Figure 10.2a), leading to an impaired ability in processing and representing information. To rectify this situation, the magnitude of noise correlations needs to be attenuated. How, though, do noise correlations relate to alpha oscillations? Given that the summed electric potential of the correlated neurons creates a spike in the amplitude of local field potential (Averbeck et al., 2006), repeated and rhythmic patterns of noise correlations would create repeated and rhythmic increases in the amplitude of local field potential (LFP). As alpha oscillations dominate the neocortex, one may speculate that the rhythmic noise correlations may resonate within this frequency band. Under this assumption, periods of high-amplitude alpha oscillations would reflect periods of numerous noise correlations where information processing is inhibited. This interpretation conforms with the active inhibition account because it would allow alpha power increases to suppress task-irrelevant information. Periods of alpha desynchrony, in contrast, would reflect periods of limited noise correlations, where there is a greatest potential for information representation.

In summary, there are two theoretical arguments that implicate alpha power decreases in the representation of stimulus-specific information. The information-via-desynchronization account suggests that alpha power reductions facilitate the evolution of more complex neural codes, which then allows for the representation of high-fidelity information in the cortex. The noise correlation account suggests that alpha power decreases reduce the background noise in the cortex, allowing for key signals to be more clearly communicated. Currently, empirical evidence supports both ideas by demonstrating that alpha power decreases scale with the quantity of information present in the cortex.

10.5 Alpha Phase: Information Sampling and Replay

Recent evidence suggests that the continuous input that our brain receives from the outside world is not sampled continuously, but in discrete rhythmic steps around the alpha frequency (VanRullen, 2016). These studies show that the probability of detecting a briefly presented visual stimulus fluctuates rhythmically (VanRullen et al., 2007). This attentional sampling process has been suggested to operate at a frequency of roughly 8 Hz (Landau & Fries, 2012), even when sustained attention to the same object is maintained (Fiebelkorn & Kastner, 2019; Fiebelkorn et al., 2013). Overall, there appears to be a close correlation between the phase of an oscillation in the lower alpha band (around 8 Hz) and the perception of an incoming stimulus. In the perception of continuous and dynamic stimuli, the stimulus identity can be reliably decoded from the ongoing phase of neural oscillations. Therein, an individual continuous stimulus is associated with a unique time course of neural activity. Specifically, the stimulus entrains the cortical rhythm in a content-specific way that organizes neural firing (Ng et al., 2013). Interestingly, the phase of the ongoing oscillation contains more information about stimulus identity than its power (Schyns et al., 2011).

Considering this information sampling role of alpha phase in perception, an imminent question is whether or not the reinstatement of information from memory relies on the replay of that information sampled in the alpha frequency. To this end, alpha phase could index similar mechanisms during attention and memory, just like alpha power decreases. But how can we measure the information sampling and the replay thereof in the phase of a recorded EEG?

Similar to assessing the similarity of neural patterns in space, as is usually done in fMRI, one can assess the similarity of neural patterns in phase (over time) via the use of RSA methods (discussed earlier). Established measures of connectivity can help quantify the similarity of neural patterns in the phase of the alpha-frequency band (Greenblatt et al., 2012). Connectivity measures usually assess the similarity between different channels in their time course of activity. This typically measures shared information between regions. Yet, measures of connectivity will lend themselves perfectly to quantify representational similarity between encoding and retrieval (Figure 10.3a).

In a first study, we applied such similarity measures of phase in a human EEG experiment where subjects were instructed to encode and replay dynamic visual (short movie clips) or auditory stimuli (short melodies played by different instruments). Indeed, content-specific patterns of oscillatory phase in the lower alpha band during perception represent the identity of the visual or auditory stimuli in the visual and auditory cortex, respectively. Strikingly, these phase patterns reappear when representations of short video clips and short sound clips are replayed from memory (Michelmann et al., 2016). Importantly, this replay takes place in the absence of the dynamic stimulus itself and can be localized in sensory specific areas (Figure 10.3c). A recent study replicated this effect in the visual domain and demonstrated that the reinstatement of temporal patterns is only observed when content is successfully recalled, that is, temporal pattern reinstatement is implicated in successful memory (Michelmann et al., 2019). Another study replicated this effect during sleep, indicating that replay of stimulus-specific phase patterns supports memory consolidation (Schreiner et al., 2018). Further evidence documents that such content-specific phase patterns are also replayed when an association with a previously shown dynamic stimulus is formed (Michelmann et al., 2018).

Interestingly, the reinstatement of temporal patterns is not always beneficial for memory but can in some cases interfere with memory as demonstrated by studies which manipulated the contextual overlap between encoding and retrieval. For instance, (Staudigl et al., 2015) manipulated the context in which a word is learned and remembered by playing the same video clip in the background behind the word at encoding and at retrieval (context match); or playing a different video clip in the background behind the word at encoding and at retrieval (context mismatch). The importance of reinstatement of contextual information was observed via temporal and spatial pattern similarity in the beta frequency band. Specifically, higher pattern similarity was associated with better memory performance when contexts where matching. On the

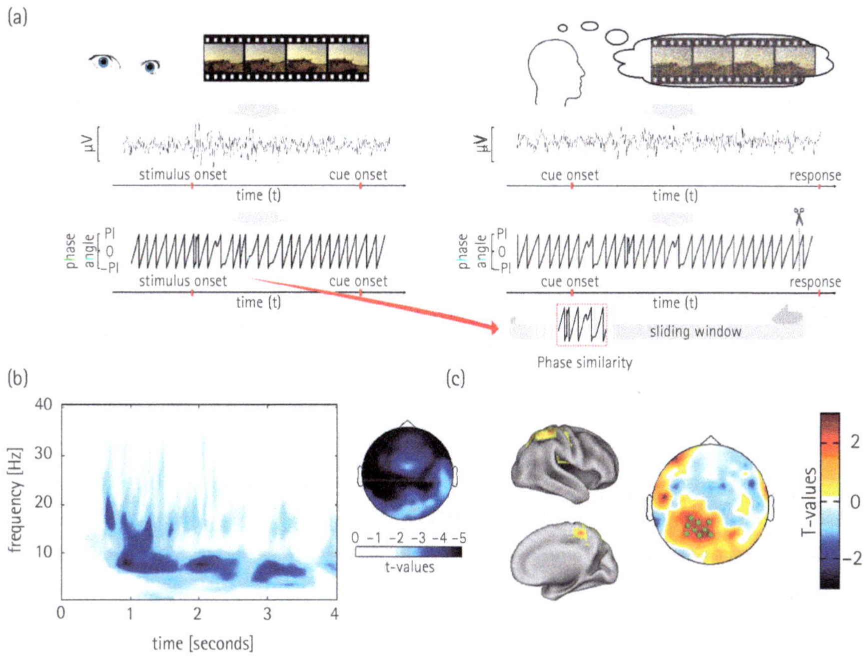

FIGURE 10.3 Alpha power decreases during memory retrieval code stimulus-specific information. (A) Subjects first encoded a video (left). They later retrieved a vivid representation of that video from memory (right). The phase time course was extracted from the EEG during encoding and retrieval in order to calculate a similarity measure between encoding and retrieval. (B) During retrieval, strong and sustained alpha power decreases were observed. (C) Reactivation of stimulus-specific information, as measured with phase similarity, could be detected in the alpha frequency band with a maximum in parietal regions. This reactivation was localized in parietal cortex (C, left).

Reproduced from Michelmann et al., 2016.

other hand, non-overlapping contexts were characterized by more dissimilar patterns for remembered vs. forgotten words. These findings were recently replicated in a study using sensory modality (i.e., presenting a word visually or aurally) as a context match/mismatch manipulation (Staudigl & Hanslmayr, 2019). The results of these studies are in line with the notion that context reinstatement is only helpful when the context has not changed and substantiate the importance of temporal patterns for content representation.

Most of the studies referenced earlier found that stimulus-specific information was coded in the phase of the lower alpha band, around 8 Hz (Michelmann et al., 2016, 2018; Michelmann et al., 2019; Staudigl & Hanslmayr, 2019). Corroborating evidence for a special role of 8 Hz in the sampling of memories comes from Kerren and colleagues (2018), who showed decoded content-specific representations in memory from spatial patterns of activity, and that these patterns fluctuate at a frequency of 8 Hz. The sampling of information at an 8 Hz rhythm therefore seems to underlie both, rhythmic sampling of information during perception (VanRullen et al., 2007) and the replay of that information during memory reinstatement.

10.6 Linking alpha phase and power

Michelmann and colleagues (2016) showed that the frequency band that contains content-specific temporal patterns is also the one that displays the most prominent power decreases during retrieval (Figure 10.3b), and furthermore, they observed an interaction such that sensory areas that were involved in the reinstatement of auditory and visual temporal patterns also expressed stronger power decreases in the respective condition. This suggests that power decreases and the representation of information in oscillatory phase are not two separate processes but rather are intertwined. This raises the question as to how the two signal properties of alpha—power and phase—interact in the service of information representation.

We suggest that a decrease of power in an ongoing oscillation renders a signal less stationary (and therefore also less predictable) and thereby allows for a flexible adjustment of phase (Hanslmayr et al., 2016). These phase adjustments, or deviations from stationarity, make it possible that time courses in phase can represent stimulus specific patterns, and to replay these patterns from memory (Figure 10.4C). From an information theoretic view, a stationary oscillation without phase adjustments wouldn't

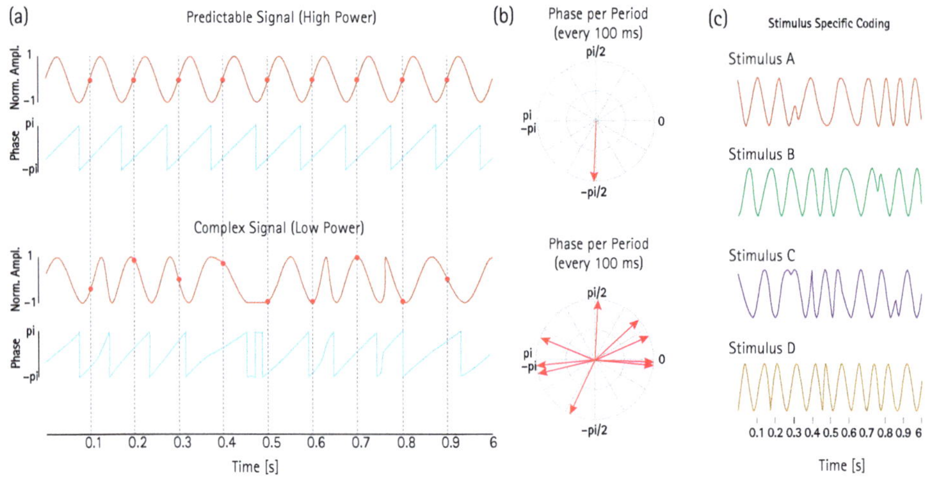

FIGURE 10.4 Information coding properties for stationary and non-stationary signals is illustrated. (A) A stationary (high power) signal (orange) is shown together with its phase (blue) in the upper row. The lower row shows a less stationary signal. (B) Phases for each signal are shown every 100 ms (indicated by ticks in A). (C) Non-stationary signals allow for stimulus specific coding by assigning a different time course to each stimulus.

be able to represent much information in its phase because the time course of a stationary signal is perfectly predictable once the phase at one time point is known. For example, the signal in the top row of Figure 10.4A visits the same phase every 100 ms (top row in Figure 10.4B). Since information theory quantifies information as the inverse of the predictability (i.e., negative logarithm in the case of Shannon's Entropy, see Shannon & Weaver, 1949) we can infer that this signal has little potential to carry information. Indeed, if we were to code the identity of a stimulus in such a perfectly predictable signal, we would not be able to distinguish between different stimuli. In contrast, a non-stationary signal (lower row of Figure 10.4A), which has phase modulations, can carry much more information. In this case, the phase cannot be predicted from previous time points (lower panel in Figure 10.4B). This allows us to code different stimuli by assigning a phase time course to each stimulus (Figure 10.4C). This simple relationship between the power of a signal and predictability of phase time courses elegantly unifies the findings described in this chapter. To this end, a signal with high alpha power leads to a more stationary time course and thus inhibits information coding. In contrast, a signal with lower alpha power allows for a less stationary time course and therefore the coding of information. Importantly, this idea is in line with the general notion of an inhibitory role of alpha power increases but goes beyond the previous work in ascribing specific computational roles to power and phase in the service of representing information.

10.7 Concluding Remarks

This chapter showed that alpha oscillations are ubiquitous as they are modulated by almost any task and can be observed in almost any animal. It therefore follows that alpha oscillations must perform a basic neural operation, which is of service for many cognitive operations. After an overview of the idling vs. inhibition hypothesis we then discussed the computational utility of alpha power reductions for information representation. We reviewed studies that demonstrated that alpha power reductions are intimately linked to information coding—specifically, these studies showed that stimulus-specific information is coded in the phase of alpha. Importantly, several studies demonstrated that this phase information is replayed when a reminder to that stimulus is presented. Finally, we illustrated how power decreases allow for less stationary phase time courses and consequently for information representation. Together, we conclude that alpha power decreases do play an important role for information representation—a neural operation needed in almost any task and in almost any animal.

An important crucial future question refers to the nature of the relationship between alpha power decreases and its role for coding information. If indeed, as we here suggest, alpha power decreases are mechanisms for representing information we should be able to manipulate perception, maintenance, and retrieval of information by directly manipulating alpha oscillations via brain stimulation techniques (Hanslmayr et al.,

2019). Such a demonstration of a causal relationship between alpha power decreases and the representation of information is crucial in order to show that alpha oscillations are indeed a mechanism for information representation, instead of a mere epiphenomenon.

References

Adrian, E. D. (1944). Brain rhythms. *Nature, 153*, 360–362.

Adrian, E. D. & Matthews, B. H. (1934). The interpretation of potential waves in the cortex. *Journal of Physiology, 81*(4), 440–471.

Anokhin, A. & Vogel, F. (1996). EEG alpha rhythm frequency and intelligence in normal adults. *Intelligence, 23*(1), 1–14. doi:10.1016/S0160-2896(96)80002-X

Averbeck, B. B., Latham, P. E., & Pouget, A. (2006). Neural correlations, population coding and computation. *Nature Reviews Neuroscience, 7*(5), 358–366. doi:10.1038/nrn1888

Berger, H. (1929). Über das Elektrenkephalogramm des Menschen. *Archiv für Psychiatrie und Nervenkrankheiten, 87*, 527–570.

Bonnefond, M. & Jensen, O. (2012). Alpha oscillations serve to protect working memory maintenance against anticipated distracters. *Current Biology, 22*(20), 1969–1974. doi:10.1016/j.cub.2012.08.029

Burgess, A. P. & Gruzelier, J. H. (2000). Short duration power changes in the EEG during recognition memory for words and faces. *Psychophysiology, 37*(5), 596–606.

Buzsaki, G. (2019). *The Brain from inside out*. Oxford University Press.

Cohen, M. X. (2011). Hippocampal-prefrontal connectivity predicts midfrontal oscillations and long-term memory performance. *Current Biology, 21*(22), 1900–1905. doi:10.1016/j.cub.2011.09.036

Cohen, M. X. (2014). *Analyzing neural time series data: Theory and practice*. MIT Press.

Cohen, M. X. (2014). Fluctuations in oscillation frequency control spike timing and coordinate neural networks. *Journal of Neuroscience, 34*(27), 8988–8998. doi:10.1523/JNEUROSCI.0261-14.2014

Crone, N. E., Miglioretti, D. L., Gordon, B., Sieracki, J. M., Wilson, M. T., Uematsu, S., & Lesser, R. P. (1998). Functional mapping of human sensorimotor cortex with electrocorticographic spectral analysis. I. Alpha and beta event-related desynchronization. *Brain, 121 (Pt 12)*, 2271–2299.

Fellner, M. C., Gollwitzer, S., Rampp, S., Kreiselmeyr, G., Bush, D., Diehl, B., . . . Hanslmayr, S. (2018). Spectral fingerprints or spectral tilt? Evidence for distinct oscillatory signatures of memory formation. https://journals.plos.org/plosbiology/article?id=10.1371/journal.pbio.3000403

Fiebelkorn, I. C. & Kastner, S. (2019). A rhythmic theory of attention. *Trends in Cognitive Sciences, 23*(2), 87–101. doi:10.1016/j.tics.2018.11.009

Fiebelkorn, I. C., Saalmann, Y. B., & Kastner, S. (2013). Rhythmic sampling within and between objects despite sustained attention at a cued location. *Current Biology, 23*(24), 2553–2558. doi:10.1016/j.cub.2013.10.063

Fox, M. D., Snyder, A. Z., Vincent, J. L., Corbetta, M., Van Essen, D. C., & Raichle, M. E. (2005). The human brain is intrinsically organized into dynamic, anticorrelated functional networks. *Proceedings of the National Academy of Sciences of the United States of America, 102*(27), 9673–9678. doi:10.1073/pnas.0504136102

Greenblatt, R. E., Pflieger, M. E., & Ossadtchi, A. E. (2012). Connectivity measures applied to human brain electrophysiological data. *Journal of Neuroscience Methods, 207*(1), 1–16. doi:10.1016/j.jneumeth.2012.02.025

Griffiths, B. J., Mayhew, S. D., Mullinger, K. J., Jorge, J., Charest, I., Wimber, M., & Hanslmayr, S. (2019). Alpha/beta power decreases track the fidelity of stimulus-specific information. https://elifesciences.org/articles/49562

Haegens, S., Cousijn, H., Wallis, G., Harrison, P. J., & Nobre, A. C. (2014). Inter- and intra-individual variability in alpha peak frequency. *NeuroImage*, *92*, 46–55. doi:10.1016/j.neuroimage.2014.01.049

Haegens, S., Nacher, V., Luna, R., Romo, R., & Jensen, O. (2011). alpha-Oscillations in the monkey sensorimotor network influence discrimination performance by rhythmical inhibition of neuronal spiking. *Proceedings of the National Academy of Sciences of the United States of America*, *108*(48), 19377–19382. doi:10.1073/pnas.1117190108

Hanslmayr, S., Axmacher, N., & Inman, C. S. (2019). Modulating human memory via entrainment of brain oscillations. *Trends in Neuroscience*, *42*(7), 485–499.

Hanslmayr, S., Gross, J., Klimesch, W., & Shapiro, K. L. (2011). The role of α oscillations in temporal attention. *Brain Research Reviews*, *67*(1–2), 331–343.

Hanslmayr, S., Spitzer, B., & Bauml, K. H. (2009). Brain oscillations dissociate between semantic and nonsemantic encoding of episodic memories. *Cerebral Cortex*, *19*(7), 1631–1640. doi:10.1093/cercor/bhn197

Hanslmayr, S., Staresina, B. P., & Bowman, H. (2016). Oscillations and episodic memory: Addressing the synchronization/desynchronization conundrum. *Trends in Neuroscience*, *39*(1), 16–25. doi:10.1016/j.tins.2015.11.004

Hanslmayr, S., Staudigl, T., & Fellner, M. C. (2012). Oscillatory power decreases and long-term memory: The information via desynchronization hypothesis. *Frontiers in Human Neuroscience*, *6*, 74. doi:10.3389/fnhum.2012.00074

Jacobs, J., Kahana, M. J., Ekstrom, A. D., & Fried, I. (2007). Brain oscillations control timing of single-neuron activity in humans. *Journal of Neuroscience*, *27*(14), 3839–3844.

Jensen, O., Gelfand, J., Kounios, J., & Lisman, J. E. (2002). Oscillations in the alpha band (9-12 Hz) increase with memory load during retention in a short-term memory task. *Cerebral Cortex*, *12*(8), 877–882.

Jensen, O. & Mazaheri, A. (2010). Shaping functional architecture by oscillatory alpha activity: gating by inhibition. *Frontiers in Human Neuroscience*, *4*, 186. doi:10.3389/fnhum.2010.00186

Keitel, A. & Gross, J. (2016). Individual human brain areas can be identified from their characteristic spectral activation fingerprints. *PLoS Biology*, *14*(6), e1002498. doi:10.1371/journal.pbio.1002498

Kerren, C., Linde-Domingo, J., Hanslmayr, S., & Wimber, M. (2018). An optimal oscillatory phase for pattern reactivation during memory retrieval. *Current Biology*, *28*(21), 3383–3392 e3386. doi:10.1016/j.cub.2018.08.065

Klimesch, W., Doppelmayr, M., Schwaiger, J., Auinger, P., & Winkler, T. (1999). 'Paradoxical' alpha synchronization in a memory task. *Brain Research: Cognitive Brain Research*, *7*(4), 493–501.

Klimesch, W., Sauseng, P., & Hanslmayr, S. (2007). EEG alpha oscillations: the inhibition-timing hypothesis. *Brain Research Reviews*, *53*(1), 63–88. doi:10.1016/j.brainresrev.2006.06.003

Klimesch, W., Schimke, H., Doppelmayr, M., Ripper, B., Schwaiger, J., & Pfurtscheller, G. (1996). Event-related desynchronization (ERD) and the Dm effect: Does alpha desynchronization during encoding predict later recall performance? *International Journal of Psychophysiology*, *24*(1-2), 47–60.

Krause, C. M., Lang, H. A., Laine, M., Helle, S. I., Kuusisto, M. J., & Porn, B. (1994). Event-related desynchronization evoked by auditory stimuli. *Brain Topography*, *7*(2), 107–112.

Kriegeskorte, N., Mur, M., & Bandettini, P. (2008). Representational similarity analysis - connecting the branches of systems neuroscience. *Frontiers in Systems Neuroscience*, 2, 4. doi:10.3389/neuro.06.004.2008

Landau, A. N. & Fries, P. (2012). Attention samples stimuli rhythmically. *Current Biology*, *22*(11), 1000–1004. doi:10.1016/j.cub.2012.03.054

Lopes da Silva, F. H., Vos, J. E., Mooibroek, J., & Van Rotterdam, A. (1980). Relative contributions of intracortical and thalamo-cortical processes in the generation of alpha rhythms, revealed by partial coherence analysis. *Electroencephalography and Clinical Neurophysiology*, *50*(5-6), 449–456.

Michelmann, S., Bowman, H., & Hanslmayr, S. (2016). The temporal signature of memories: Identification of a general mechanism for dynamic memory replay in humans. *PLoS Biology*, *14*(8), e1002528. doi:10.1371/journal.pbio.1002528

Michelmann, S., Bowman, H., & Hanslmayr, S. (2018). Replay of stimulus-specific temporal patterns during associative memory formation. *Journal of Cognitive Neuroscience*, *30*(11), 1577–1589. doi:10.1162/jocn_a_01304

Michelmann, S., Staresina, B. P., Bowman, H., & Hanslmayr, S. (2019). Speed of time-compressed forward replay flexibly changes in human episodic memory. *Nature Human Behaviour*, *3*(2), 143–154. doi:10.1038/s41562-018-0491-4

Mitchell, J. F., Sundberg, K. A., & Reynolds, J. H. (2009). Spatial attention decorrelates intrinsic activity fluctuations in macaque area V4. *Neuron*, *63*(6), 879–888. doi:10.1016/j.neuron.2009.09.013

Murthy, V. N. & Fetz, E. E. (1996). Synchronization of neurons during local field potential oscillations in sensorimotor cortex of awake monkeys. *Journal of Neurophysiology*, *76*(6), 3968–3982. doi:10.1152/jn.1996.76.6.3968

Ng, B. S., Logothetis, N. K., & Kayser, C. (2013). EEG phase patterns reflect the selectivity of neural firing. *Cerebral Cortex*, *23*(2), 389–398. doi:10.1093/cercor/bhs031

Obleser, J. & Weisz, N. (2012). Suppressed alpha oscillations predict intelligibility of speech and its acoustic details. *Cerebral Cortex*, *22*(11), 2466–2477. doi:10.1093/cercor/bhr325

Pfurtscheller, G. & Aranibar, A. (1977). Event-related cortical desynchronization detected by power measurements of scalp EEG. *Electroencephalography and Clinical Neurophysiology*, *42*(6), 817–826.

Pfurtscheller, G., Neuper, C., Andrew, C., & Edlinger, G. (1997). Foot and hand area mu rhythms. *International Journal of Psychophysiology*, *26*(1–3), 121–135.

Pfurtscheller, G., Stancak, A., Jr., & Neuper, C. (1996). Event-related synchronization (ERS) in the alpha band--an electrophysiological correlate of cortical idling: a review. *International Journal of Psychophysiology*, *24*(1-2), 39–46.

Poldrack, R. A. (2011). Inferring mental states from neuroimaging data: from reverse inference to large-scale decoding. *Neuron*, *72*(5), 692–697. doi:10.1016/j.neuron.2011.11.001

Popov, T., & Szyszka, P. (2019). Alpha oscillations govern interhemispheric spike timing coordination in the honeybee brain. https://royalsocietypublishing.org/doi/10.1098/rspb.2020.0115#:~:text=What%20is%20the%20function%20of,information%20flow%20during%20spontaneous%20activity.

Pornpattananangkul, N., Grogans, S., Yu, R., & Nusslock, R. (2019). Single-trial EEG dissociates motivation and conflict processes during decision-making under risk. *NeuroImage*, *188*, 483–501. doi:10.1016/j.neuroimage.2018.12.029

Sauseng, P., Klimesch, W., Heise, K. F., Gruber, W. R., Holz, E., Karim, A. A., ... Hummel, F. C. (2009). Brain oscillatory substrates of visual short-term memory capacity. *Current Biology*, *19*(21), 1846–1852. doi:10.1016/j.cub.2009.08.062

Schneidman, E., Puchalla, J. L., Segev, R., Harris, R. A., Bialek, W., & Berry, M. J., 2nd. (2011). Synergy from silence in a combinatorial neural code. *Journal of Neuroscience*, *31*(44), 15732–15741. doi:10.1523/JNEUROSCI.0301-09.2011

Schreiner, T., Doeller, C. F., Jensen, O., Rasch, B., & Staudigl, T. (2018). Theta phase-coordinated memory reactivation reoccurs in a slow-oscillatory rhythm during NREM sleep. *Cell Reports*, *25*(2), 296–301. doi:10.1016/j.celrep.2018.09.037

Schyns, P. G., Thut, G., & Gross, J. (2011). Cracking the code of oscillatory activity. *PLoS Biology*, *9*(5), e1001064. doi:10.1371/journal.pbio.1001064

Shannon, C. E. & Weaver, W. (1949). *A mathematical theory of communication*. University of Illinois Press.

Siegel, M., Donner, T. H., & Engel, A. K. (2012). Spectral fingerprints of large-scale neuronal interactions. *Nature Reviews Neuroscience*, *13*(2), 121–134. doi:10.1038/nrn3137

Staudigl, T. & Hanslmayr, S. (2019). Reactivation of neural patterns during memory reinstatement supports encoding specificity. *Cognitive Neuroscience*, *10*(4), 175–185.

Staudigl, T., Vollmar, C., Noachtar, S., & Hanslmayr, S. (2015). Temporal-pattern similarity analysis reveals the beneficial and detrimental effects of context reinstatement on human memory. *Journal of Neuroscience*, *35*(13), 5373–5384. doi:10.1523/JNEUROSCI.4198-14.2015

Thut, G., Nietzel, A., Brandt, S. A., & Pascual-Leone, A. (2006). Alpha-band electroencephalographic activity over occipital cortex indexes visuospatial attention bias and predicts visual target detection. *Journal of Neuroscience*, *26*(37), 9494–9502. doi:10.1523/JNEUROSCI.0875-06.2006

VanRullen, R. (2016). Perceptual cycles. *Trends in Cognitive Sciences*, *20*(10), 723–735. doi:10.1016/j.tics.2016.07.006

VanRullen, R., Carlson, T., & Cavanagh, P. (2007). The blinking spotlight of attention. *Proceedings of the National Academy of Sciences of the United States of America*, *104*(49), 19204–19209. doi:10.1073/pnas.0707316104

von Stein, A., Chiang, C., & Konig, P. (2000). Top-down processing mediated by interareal synchronization. *Proceedings of the National Academy of Sciences of the United States of America*, *97*(26), 14748–14753. doi:10.1073/pnas.97.26.14748

von Stein, A. & Sarnthein, J. (2000). Different frequencies for different scales of cortical integration: from local gamma to long range alpha/theta synchronization. *International Journal of Psychophysiology*, *38*(3), 301–313.

Waldhauser, G. T., Braun, V., & Hanslmayr, S. (2016). Episodic memory retrieval functionally relies on very rapid reactivation of sensory information. *Journal of Neuroscience*, *36*(1), 251–260. doi:10.1523/JNEUROSCI.2101-15.2016

Waldhauser, G. T., Johansson, M., & Hanslmayr, S. (2012). alpha/beta oscillations indicate inhibition of interfering visual memories. *Journal of Neuroscience*, *32*(6), 1953–1961. doi:10.1523/JNEUROSCI.4201-11.2012

Wiest, M. C. & Nicolelis, M. A. (2003). Behavioral detection of tactile stimuli during 7-12 Hz cortical oscillations in awake rats. *Nature Neuroscience*, *6*(9), 913–914. doi:10.1038/nn1107

Womelsdorf, T., Valiante, T. A., Sahin, N. T., Miller, K. J., & Tiesinga, P. (2014). Dynamic circuit motifs underlying rhythmic gain control, gating and integration. *Nature Neuroscience*, *17*(8), 1031–1039. doi:10.1038/nn.3764

Worden, M. S., Foxe, J. J., Wang, N., & Simpson, G. V. (2000). Anticipatory biasing of visuospatial attention indexed by retinotopically specific alpha-band electroencephalography increases over occipital cortex. *Journal of Neuroscience*, *20*(6), RC63.

CHAPTER 11

THEORY AND RESEARCH ON ASYMMETRIC FRONTAL CORTICAL ACTIVITY AS ASSESSED BY EEG FREQUENCY ANALYSES

EDDIE HARMON-JONES, TAYLOR POPP, AND PHILIP A. GABLE

FREQUENCY analyses of electroencephalographic (EEG) activity have played a major role in research and theory concerning the role of left versus right frontal cortical activity in emotion- and motivation-related phenomena. In this work, researchers have focused primarily on alpha power (8–13 Hz) activity in the left versus right frontal regions. This focus on alpha power was based on other research suggesting that alpha power is inversely related to cortical activity (Cook et al., 1998; Davidson et al., 1990).

Much of this research has used the difference between the left and right frontal region as the variable of interest. The use of difference scores is based on past research using a variety of methods that suggest with regard to emotion- and motivation-related variables, one frontal hemispheric region may be inhibiting the opposite frontal hemispheric region (Schutter, 2009; Schutter et al., 2001). The corpus callosum, the largest white matter bundle connecting the left and right hemispheres, likely plays a critical role in this asymmetric functioning of the frontal cortices (Schutter & Harmon-Jones, 2013). Given this, researchers typically subtract left frontal alpha power from right frontal alpha power (after the values are log transformed to normalize the distributions, e.g., log F4 minus log F3). The resulting metric is then referred to as *relative left or relative right frontal activity*, depending on whether the difference score is a more positive (relative left) or a more negative (relative right) numeric value.

11.1 Definitions of Psychological Concepts: Affective Valence and Motivational Direction

We begin by providing our definitions of the psychological concepts of affective valence and motivational direction. Affective valence denotes whether an affective trait or state is positive or negative. Although affective valence can be defined in a number of ways (Harmon-Jones & Gable, 2018; Lazarus, 1991), we define it as the subjective feel of the affect (emotion or mood; affect encompasses both of these latter concepts). Affective experiences that organisms like feeling are positive, and affective experiences that organisms dislike feeling are negative (Harmon-Jones et al., 2011).

Motivational direction refers to the urge to move toward (approach) or away (withdrawal) from something (Harmon-Jones et al., 2013; Gable & Dreisbach, 2021). Many theories suggest that approach motivation is triggered by rewards, desired outcomes, or positive goals, and that withdrawal (avoidance) motivation is triggered by punishments, undesired outcomes, or negative goals. The anticipations of these end-products may inspire much approach and withdrawal motivation, but these anticipations are not the only causes of approach and withdrawal motivation and they should therefore not be part of the definition. That is, approach and withdrawal motivation can occur without being prompted by these anticipations (for an extended discussion, see Harmon-Jones et al., 2013).

11.2 Affective Valence and Asymmetric Frontal Cortical Activity

Beginning in the 1930s, research using a variety of methods suggested that the left and right frontal cortices are involved in different emotional (affective) responses. For example, lesion studies as well as experiments that injected a barbiturate derivative (amytal) into one of the internal carotid arteries (to suppress the activity of one hemisphere) showed that the loss of activity in the left frontal cortex was associated with depressed affect, whereas loss of activity in the right frontal cortex was associated with manic affect and euphoria (Alema et al., 1961; Black, 1975; Gainotti, 1972; Gasparrini et al., 1978; Goldstein, 1939; Perria et al., 1961; Robinson & Price, 1982; Rossi & Rosadini, 1967; Sackeim et al., 1982; Terzian & Cecotto, 1959). These outcomes are likely due to the release of one hemisphere from contralateral inhibitory forces (Schutter & Harmon-Jones, 2013). For example, when the right hemisphere was deactivated by damage or amytal, the left hemisphere became more uninhibited and more active, which caused manic affect.

These results can be interpreted in one of two ways. According to the affective valence model of frontal asymmetry, the left frontal cortical region is involved in the experience and expression of positive affect, whereas the right frontal cortical region is involved in the experience and expression of negative affect. According to the motivational direction model of frontal asymmetry, the left frontal cortical region is involved in the experience and expression of approach motivation, whereas the right frontal cortical region is involved in the experience and expression of withdrawal motivation. Until the 1990s, the two conceptual models were considered to yield the same predictions because approach motivation was conceived of as being associated with positive affect and withdrawal motivation was conceived of as being associated with negative affect. In the late 1990s, researchers began using anger to tease predictions from these two conceptual models apart, because anger is an approach-motivated but negative affect (Harmon-Jones, 2003a). The following sections briefly review the research on the two models along with research on anger and frontal asymmetry.

11.3 Resting Asymmetric Frontal Activity and Trait Affective Valence

These early results inspired EEG researchers to examine EEG alpha power over the frontal cortex and test whether it was associated with affective variables. Most of the first EEG studies on asymmetric frontal asymmetry tested individuals in a resting, baseline state (i.e., sitting quietly in the lab for four to eight minutes). Researchers assumed that this resting, baseline EEG would tap into a personality trait, and they then related EEG frontal asymmetry with other personality or individual difference measures.

Several studies found that depression was correlated with less relative left frontal cortical activity (e.g., Allen et al., 1993; Henriques & Davidson, 1990; Jacobs & Snyder, 1996; Schaffer et al., 1983; see meta-analysis by Thibodeau et al., 2006). Other studies suggested that trait negative affect was correlated with greater relative right frontal activity, whereas trait positive affect was correlated with greater relative left frontal activity (Tomarken et al., 1992).

11.4 Manipulations of Asymmetric Frontal Cortical Activity and Affective Reactions

The studies assessing resting baseline asymmetric frontal cortical activity were correlational; therefore, it is impossible to infer that the differences in asymmetric frontal cortical activity were causally involved in the observed affective processes. In an effort to provide evidence that would allow causal inferences, some researchers manipulated asymmetric frontal cortical activity and tested whether manipulation influenced other affective variables.

11.4.1 Neurofeedback

Studies have used neurofeedback training to manipulate asymmetric frontal cortical activity (e.g., Allen et al., 2001; Harmon-Jones et al., 2008). In this research, operant conditioning is used to produce certain patterns of brainwaves. Reward feedback (e.g., a simple tone) corresponding to "desired" patterns of brainwave activity is presented to participants. Over numerous presentations of this reward feedback, the brain learns to produce the desired brainwave activity, such as greater relative left frontal cortical activity. These neurofeedback-induced changes in brainwaves occur via nonconscious learning processes (Kamiya, 1979; Siniatchkin et al., 2000).

In one example experiment, relative right versus relative left frontal activity was manipulated using neurofeedback (Allen et al., 2001). In the experiment, participants were instructed to attempt to make a particular tone play as much as possible. Asymmetric frontal cortical activity (i.e., alpha power at F4 minus alpha power at F3) was calculated during the first second of each two-second time period. This calculated value was then compared against a criterion value that had been set for that session (based on the individual's previous asymmetry index). If the calculated value was larger than the criterion value in the desired direction, a "reward" tone was played; if it was not, a "non-reward" tone was played. After several days of neurofeedback training, participants then viewed emotionally evocative film clips as zygomatic and corrugator muscle region activity was recorded. Results revealed that the neurofeedback training influenced asymmetric frontal cortical activity in the predicted direction (for a replication, see Quaedflieg et al., 2016). Moreover, the manipulated increase in relative right frontal cortical activity, as compared to relative left frontal cortical activity, caused less zygomatic and more corrugator muscle region activity in response to all film clips. These results suggest that asymmetric frontal cortical activity is *causally* involved in emotional responses.

11.4.2 Unilateral hand contractions

Asymmetric frontal cortical activity has also been manipulated using unilateral hand contractions. Because unilateral hand contractions cause increased activation of the contralateral motor region (Hellige, 1993), whose activation may spread to frontal regions, unilateral hand contractions may influence emotional responses (Schiff & Lamon, 1989; 1994). Indeed, research revealed that contractions of the left hand cause increased feelings of sadness and more negative perceptions and judgments, whereas contractions of the right hand cause increased feelings of positive affect and more positive perceptions and judgments (Schiff & Lamon, 1989, 1994).

Based on this research, experiments have tested whether unilateral hand contractions would influence EEG and other emotional responses. In one experiment, participants were instructed to squeeze a ball in their left or right hand for several minutes (Harmon-Jones, 2006). Then, their self-reported affective responses were measured in response to a radio editorial designed to evoke a moderate amount of positive affect. Results revealed that the contraction of one hand increased cortical activity over the central and frontal regions of the contralateral hemisphere (as measured by EEG alpha suppression; see Gable et al., 2013 for replication). Moreover, this manipulation influenced positive affect, such that right-hand contractions caused more positive affect than the left-hand contractions.

11.4.3 Situational Manipulations of Positive and Negative Affect and Asymmetric Frontal Cortical Responses

Additional research has measured asymmetric frontal cortical activity in response to manipulations of positive and negative affect. For example, Davidson and Fox (1982) found that 10-month-old infants evidenced greater relative left frontal cortical activity to film clips of an actress displaying happy facial expressions as compared to sad facial expressions. Ekman and Davidson (1993) found that adults evidenced greater relative left frontal activity when they were induced to make facial expressions of genuine smiles of joy as compared to facial expressions of non-genuine smiles of joy. Coan, Allen, and Harmon-Jones (2001) found that adults evidenced less relative left frontal activity when they were induced to make facial expressions of fear as compared to several other types of emotional facial expressions.

11.5 Asymmetric Frontal Cortical Activity and Motivational Direction

Several studies suggest that the left frontal cortical region is involved in the experience and expression of positive affect, whereas the right frontal cortical region is involved in the experience and expression of negative affect. This research suggested that asymmetric frontal cortical activity reflected positive or negative affect, or affective valence. Research testing this affective valence model happened concurrently with other research testing a motivational direction model, which posits that relative left frontal cortical activity is associated with approach motivation and that relative right frontal cortical activity is associated with withdrawal motivation. For several years, the affective valence and motivational direction models were viewed as compatible models that made identical predictions. This compatibility resulted from viewing positive affect as being always associated with approach motivation and negative affect as being always associated with withdrawal motivation.

11.5.1 Resting Asymmetric Frontal Activity and Trait Motivational Direction

Many studies measured trait approach and withdrawal motivation with Carver and White's (1994) behavioral inhibition/behavioral activation system (BIS/BAS) questionnaires. The approach motivation (BAS) questionnaire has items such as "I go out of my way to get things I want" and "I crave excitement and new sensations". The withdrawal motivation (BIS) questionnaire has items such as "I worry about making mistakes" and "I have very few fears compared to my friends (reverse scored)."

The first two studies found that individual differences in self-reported trait approach motivation were correlated with greater relative left frontal cortical activity during resting baseline sessions (Harmon-Jones & Allen, 1997; Sutton & Davidson, 1997). One of these studies also found that trait "withdrawal" motivation correlated with greater relative right frontal activity during resting baseline (Sutton & Davidson, 1997), whereas the other study found no significant correlation between trait withdrawal motivation and relative right frontal activity (Harmon-Jones & Allen, 1997).

Additional studies replicated this latter pattern of results (Amodio et al., 2008; Amodio et al., 2004; Coan & Allen, 2003), and an EEG source localization study suggested that this trait approach/relative left frontal cortical activity correlation was due to activation in the middle frontal gyrus (BA11; De Pascalis et al., 2013).

11.5.2 Situational Manipulations of Motivation and Asymmetric Frontal Cortical Responses

11.5.2.1 *Pictures of Motivationally Significant Stimuli*

To test whether approach motivational intensity would predict relative left frontal cortical activity to appetitive stimuli, studies have been conducted that first measured individuals' self-reported liking for dessert and the time since they had last eaten. These measures were included to assess individual differences in approach motivation relevant to the stimuli to be presented. Next, participants viewed pictures of desserts and neutral objects while EEG was recorded. Results revealed that individuals with more intense approach motivation (i.e., longer time since eating, more liking for dessert) toward pictures of desserts evidenced more relative left frontal activity to dessert stimuli but not to neutral stimuli (Gable & Harmon-Jones, 2008; Harmon-Jones & Gable, 2009). In these studies, the appetitive pictures alone did not cause a significant increase in relative left frontal cortical activity. This lack of a main effect of picture type on asymmetric frontal activity is consistent with some other research (Elgavish et al., 2003; Hagemann et al., 1998; see also reviews by Murphy et al., 2003 and Pizzagalli et al., 2003). The lack of a main effect may have occurred because pictures evoke different amounts of approach (withdrawal) motivation across participants. Some individuals may respond with no motivation, and others may respond with much motivation (Harmon-Jones, 2007).

Other research has suggested that some pictorial stimuli may evoke intense approach (or withdrawal) motivation in all participants. For example, Schöne and colleagues (2016) found that erotic stimuli caused greater relative left frontal activity than comparison stimuli of extreme sports, dressed women, and daily activities. The observed difference in relative left frontal activity between erotic and extreme sports stimuli is particularly interesting because both sets of stimuli evoked high and equal levels of self-reported positive affect and arousal.

11.5.2.2 *Positive Affects Varying in Approach and Asymmetric Frontal Cortical Activity*

This research suggests that relative left frontal cortical activity reflects the intensity of approach motivation. However, most of this research could be interpreted by the valence model as well, which would simply posit that the results are due to positive affect instead of approach motivation. One way to tease these two conceptual models apart is to manipulate the intensity of approach motivation while holding positive affect constant. The motivational direction model would predict that positive affects high in approach motivation should evoke greater relative left frontal cortical activity than positive affects low in approach motivation. The affective valence model would instead predict that both types of positive affect should evoke equal levels of relative left frontal cortical activity.

One experiment tested these competing predictions by having individuals write about: 1) the steps needed to obtain a desired goal (positive action-oriented); 2) a normal day (neutral); or 3) a past event that made them feel good without personal action (positive inaction-oriented; Harmon-Jones et al., 2008, Experiment 2). The two positive conditions evoked greater levels of self-reported general positive affect than the neutral condition. However, the positive action-oriented condition caused greater relative left frontal cortical activity than the other two conditions did. Thus, these results suggest that high approach-motivated positive affect caused greater relative left frontal cortical activity than low approach-motivated positive affect.

11.5.2.3 *Facial Expressions of Approach-Motivated Positive Affect*

Based on research that had revealed that facial expressions of determination were identified as being high in positive approach motivation (Harmon-Jones et al., 2011), subsequent research instructed individuals to make a facial expression that expressed "determination" or a low approach-motivated positive expression of "satisfaction" or a neutral expression (Price et al., 2013). Then, individuals performed a task that assessed behavioral persistence. As predicted by the motivational model (but not the affective valence model), individuals who made the determination facial expression (high approach-motivated positive affect) had greater relative left frontal activity than individuals in the other two conditions. In addition, within the determination condition, greater relative left frontal cortical activity was associated with greater behavioral persistence.

11.5.2.4 *Using Whole Body Posture to Manipulate Approach Motivation*

Based on previous research that had shown that whole body posture can manipulate levels of approach motivation (e.g., Harmon-Jones & Peterson, 2009; Price & Harmon-Jones, 2011), research was conducted to test whether this manipulation would influence relative left frontal cortical activity. In this study, seated individuals were instructed to either lean forward as one might do while reaching for a desired object (high approach) or to recline fully in a reclining chair (low approach; Harmon-Jones et al., 2011). Then, while in one of these two positions, the individuals viewed approach-oriented (dessert) and neutral (rock) pictures using virtual reality glasses, so that the viewing distance was held constant. As predicted, individuals in the high approach leaning forward posture evidenced greater relative left frontal activity to the approach-oriented stimuli as compared to the neutral stimuli. Individuals in the low approach reclining posture did not evidence a difference in relative left frontal activity to the two picture types.

11.5.2.5 *Final Thoughts*

The reviewed evidence suggests that relative left frontal cortical activity is associated with approach motivation. This association occurs even when approach motivation is not confounded with positive affect. That is, in approach motivation, some positive affects are low and some are high, but the research has revealed that it is primarily the high approach-motivated positive affects that are associated with greater relative left frontal activity.

11.6 Anger and Asymmetric Frontal Cortical Activity

To better understand the emotive functions of asymmetric frontal cortical activity, anger has been examined, as anger provides a way of removing the "natural" confound between affective valence and motivational direction. That is, anger is often posited to be a negative affect (Harmon-Jones, 2004; Harmon-Jones et al., 2011) associated with approach motivation, as recognized in classic work (e.g., Darwin, 1872; Blanchard & Blanchard, 1984; Ekman & Friesen, 1975; Lagerspetz, 1969; Plutchik, 1980; Young, 1943) as well as more contemporary work (e.g., Berkowitz & Harmon-Jones, 2004;

Carver & Harmon-Jones, 2009; Harmon-Jones et al., 2009; Harmon-Jones et al., 2013). To give just a few examples, research has revealed that individual differences in BAS (Carver & White, 1994) are positively correlated with individual differences in anger (Harmon-Jones, 2003b; Smits & Kuppens, 2005), greater anger responses to situational anger manipulations (Carver, 2004), greater aggressive inclinations especially when approach motivation is activated (Harmon-Jones & Peterson, 2008), and more attentional engagement to angry faces, as in approach-based dominance confrontations (Putman et al., 2004).

If asymmetric frontal cortical activity reflects affective valence, then anger should be associated with relative right frontal activity because anger is a negative affect. In contrast, if asymmetric frontal cortical activity reflects motivational direction, then anger should be associated with relative left frontal activity because anger is approach motivated. Research has tested these competing predictions.

11.6.1 Resting Asymmetric Frontal Cortical Activity and Trait Anger

Studies with adolescents (Harmon-Jones & Allen, 1998) and young adults (Harmon-Jones, 2004) have found that individual differences in anger relate positively with relative left frontal cortical activity, assessed during resting baseline. This latter study also measured attitudes toward anger and revealed that anger was regarded as a negative affect and that these negative attitudes toward anger did not correlate with asymmetric frontal activity. Other studies have revealed that trait anger correlates positively with relative left frontal activity in incarcerated violent offenders (Keune et al., 2012), and that trait aggression correlates positively with relative left frontal activity in adults with ADHD (Keune et al., 2011). Following from research showing that anger and jealousy are positively correlated in some circumstances, research has revealed that one-year-old infants with greater relative left frontal activity during a resting baseline display more jealous responses when their mothers attend to a social rival (Mize & Jones, 2012).

11.6.2 Manipulation of Asymmetric Frontal Cortical Activity and Anger Reactions

Other research has tested whether the manipulation of asymmetric frontal cortical activity via unilateral hand contractions will influence anger-related responses. In one experiment, participants contracted their right or left hand using the methods described earlier (Harmon-Jones, 2006). Then, they received insulting interpersonal feedback and then played a reaction time game against the person who had ostensibly insulted them. The reaction time game provided an assessment of behavioral aggression, as participants could select the volume and duration of noise blast to give to the other participants on trials they won. Replicating previous research, the contraction of the right hand led to greater relative left central and frontal cortical activity than the contraction of left hand. More importantly, as compared to the left hand contractions, the right hand contractions led to more aggression behavior (Peterson et al., 2008).

Additional research revealed that compared to left-hand contractions, right-hand contractions led to greater self-reported anger to being socially ostracized in a Cyberball game (Peterson et al., 2011). These results suggest that manipulated increases in relative left frontal cortical activity lead to increased anger-related responses.

11.6.3 State Anger and Relative Left Frontal Cortical Activity

Other research has tested whether situationally manipulated anger influences relative left frontal cortical activity. In one experiment, participants who were insulted (i.e., received insulting feedback on an essay they had written) responded with greater relative left frontal activity than individuals who were not insulted (i.e., received mildly positive feedback; Harmon-Jones & Sigelman, 2001). Moreover, within the insult condition, self-reported anger to the feedback and behavioral aggression toward the person who provided the feedback were both positively correlated with relative left frontal activity. Other experiments have conceptually replicated these results (e.g., Harmon-Jones et al., 2004; Jensen-Campbell et al., 2007; Verona et al., 2009). Research has extended these early results by showing that lab-induced social rejections can induce feelings of jealousy and anger as well as increased relative left frontal cortical activity (Harmon-Jones et al., 2009).

Individual differences have also been found to moderate these anger-related relative left frontal cortical activity responses. For instance, one study found that, as compared to individuals with no affective disorders, individuals with borderline personality disorder had greater relative left frontal activity to social rejection, whereas individuals with major depression had greater relative right frontal activity to social rejection (Beeney et al., 2014). In addition, individual differences in BAS positively correlated with greater relative left frontal cortical activity to anger induced via pictures offensive to the sample of individuals (Americans viewing anti-American images such as flag burning; Gable & Poole, 2014).

11.6.4 Manipulating Approach Motivation Independently of Anger

Anger is presumably related to relative left frontal cortical activity because anger is associated with approach motivation. Even more compelling evidence for the association between approach motivation and relative left frontal cortical activity could be provided by manipulating approach motivation independent of anger. That is, even though anger is often approach motivated, anger is not perfectly associated with approach motivation and thus it should be possible to manipulate the two constructs separately.

For example, one variable that should influence (approach) motivational intensity is coping potential or the difficulty of engaging in behavior. That is, motivational intensity increases with perceived task difficulty up to the point where the task is perceived as impossible, and then motivational intensity drops. So, when a task is perceived as impossible, or coping potential is very low, motivational intensity should be low (Brehm, 1999; Brehm & Self, 1989).

In one experiment testing these hypotheses, individuals were induced to believe that they would or would not be able to engage in action that might resolve the anger-evoking event (i.e., sign petitions to halt a university tuition increase; Harmon-Jones et al., 2003). Results revealed that individuals who were angered and believed they could engage in action had greater relative left frontal activity than individuals who were angered and believed they could not engage in action (the tuition increase had already been approved). Moreover, in the action-possible condition, greater relative left frontal activity in response to the angering event correlated with more approach-related behavior (i.e., signing the petition and taking more petitions to have others sign). Other studies have conceptually replicated this effect of anger and approach action possibility on relative left frontal activity using pictorial stimuli to evoke anger (Harmon-Jones et al., 2006).

These results could be interpreted to indicate that greater relative left frontal cortical activity only occurs to anger evocations when individuals are given explicit approach motivational opportunities. Other research suggests, however, that these explicit approach motivational opportunities increase relative left frontal activity but are not necessary for it to occur. That is, individuals who were exposed to pictures that evoked anger and were given no explicit approach opportunities had increased relative left frontal cortical activity if they were high in trait anger (Harmon-Jones, 2007). Thus, explicit opportunities for approach-motivated behavior are not necessary to cause relative left frontal activity during anger. Anger can evoke approach motivation without explicit approach motivational opportunities being immediately present.

Other experiments have manipulated approach motivation independently of anger and found that the approach motivation drives the increase in relative left frontal activity. For example, one experiment used a manipulation of whole-body posture to manipulate approach motivation. When individuals are in a supine body position (lying flat on their backs), they are likely to be less approach motivated. In this anger experiment, individuals were interpersonally insulted while sitting upright or while in a supine posture. Results revealed that individuals who were insulted while in the supine posture had lower relative left frontal cortical activity than individuals who were insulted while sitting upright (and the latter condition had greater relative left frontal activity than an upright-neutral-no-insult comparison condition; Harmon-Jones & Peterson, 2009). Interestingly, the results also revealed that both the supine and upright insult conditions reported feeling angrier than the neutral comparison condition, and these two insult conditions did not differ from each other. Thus, this research suggests that anger relates to relative left frontal cortical activity because of approach motivation; when approach motivation was decreased with a supine body posture, individuals who were angered did not have the typical increase in relative left frontal activity. However, they did have the same level of angry feelings as those who were in the standard upright insult condition, suggesting that angry feelings are not inevitably associated with relative left frontal activity or approach motivation.

11.6.5 Anger and Avoidance Motivation

A correlational study (Hewig et al., 2004) measured trait anger-out, trait anger-in, and trait anger-control using the State-Trait Anger Expression Questionnaire (Spielberger, 1988). Resting baseline asymmetric frontal cortical activity was also assessed. Trait anger-out likely measures approach-motivated anger (e.g., "When angry or furious, I lose my temper"; Spielberger et al., 1995, p. 57). Trait anger-in measures the extent to which individuals hold anger in (e.g., "When angry or furious, I keep things in"). Trait anger-control measures the extent to which individuals control their anger (e.g., "When angry or furious, I control my angry feelings."). Trait anger-out correlated with greater relative left frontal activity. Trait anger-control correlated with greater relative right frontal activity. Trait anger-in was not significantly correlated with asymmetric frontal cortical activity. The researchers suggested that anger-control was associated with relative right frontal activity because anger-control was associated with anger withdrawal.

Anger may evoke withdrawal when anger is mixed with concerns about being punished. For example, some individuals may become angry over social policies that pressure one to behave less racially prejudiced. At the same time, however, they may experience anxiety about expressing anger in these situations because they fear social censure. A study designed such a situation in the lab and found that anger in this situation was associated with greater relative right frontal cortical activity (Zinner et al., 2008). In this study, anger in response to the situation correlated with more spontaneous eye blinking, suggesting that anger was also associated with suppressing emotions (Gross & Levenson, 1993). Moreover, self-reported anger was associated with self-reported anxiety. These latter results support the idea that the social situation had also evoked concerns about being socially censured for feeling angry. Although these results are correlational, the authors suggest that anger may be associated with relative right frontal cortical activity when the motivation to withdraw is also high.

11.7 Effortful Control and Relative Right Frontal Activity

As noted, models of frontal asymmetry link approach motivation with greater relative left frontal activity. Both the affective valence model and the motivational direction model of frontal asymmetry associate withdrawal motivation with greater relative right frontal activity. Despite evidence supporting this model, many studies have failed to replicate the relationship between right frontal activity and withdrawal motivated traits and states (Amodio et al., 2008; Berkman & Lieberman, 2010; Coan & Allen, 2003; Coan et al., 2001; De Pascalis et al., 2013; Henriques & Davidson, 2000; Hewig et al., 2004, 2006; Jackson et al., 2003; Keune et al., 2012; Kline et al., 2000; Pizzagalli et al., 2005; Quirin et al., 2013; Wacker et al., 2008; Wacker et al., 2010). This has led researchers to question what could be causing the inconsistencies in the literature. Some suggest that withdrawal motivation is a complex system that is difficult to accurately measure independently of confounding variables (Amodio et al., 2008; Coan & Allen, 2004). Others suggest that another system entirely accounts for greater right frontal asymmetry.

A model presented by Gable and colleagues (2018) suggested that right frontal activity is more related to regulatory control than to withdrawal motivation. Regulatory control is thought to be carried out by the revised BIS (r-BIS; Gray & McNaughton, 2000), which may act as a governing body over conflicts between the approach and withdrawal systems. Activation of r-BIS leads to enhanced attention to, memory for, and detection of negatively valenced information, allowing it to manage conflicts by enhancing aversion to one motivated behavior over the other (Heym et al., 2008). This can occur as either the suppression of a behavioral response or an override of motivational impulses (Aron et al., 2004, 2014; Carver & Connor-Smith, 2010; Hester & Garavan, 2004, 2009). As such, r-BIS is related to effortful control, constraint, self-control, inhibition, conflict monitoring, and error detection (Carver & Connor-Smith, 2010; Carver et al., 2008; Derryberry & Rothbart, 1997; Gray & McNaughton, 2000; Kochanska & Knaack, 2003; Nigg, 2006; Rothbart & Rueda, 2005). Low functioning r-BIS is thought to be related to impulsive behavior, deficits in inhibitory control, and externalizing disorders such as substance abuse (Enticott et al., 2006; Logan et al., 1997). Unusually high functioning of r-BIS, on the other hand, may be related to passive avoidance, anxious inaction, and internalizing disorders such as generalized anxiety disorder (Carver et al., 2008; DeYoung, 2015; Eisenberg et al., 2004; Rothbart et al., 2004; Strack & Deutsch, 2004; Valiente et al., 2003).

While much research has examined the relationship between asymmetric frontal cortical activity and motivational systems, few studies have investigated the connections between regulatory control (r-BIS) and asymmetric frontal activity (Gable et al., 2015; Grimshaw & Carmel, 2014; Neal & Gable, 2016; Wacker et al., 2003). R-BIS acts as a controlling agent over approach and withdrawal behaviors. The present model suggests that greater relative right frontal activity is indicative of greater r-BIS functioning whereas reduced relative right frontal activity suggests reduced r-BIS functioning.

11.7.1 Trait r-BIS and Frontal EEG Activity

R-BIS functioning is a stable individual difference measured via personality questionnaires targeting traits such as impulsivity, sensation seeking, and inhibition. When r-BIS is hyperactive, traits such as neuroticism and anxiety caused by passive avoidance become prevalent, often resulting from approach-avoidant conflicts (DeYoung, 2015). When r-BIS is hypoactive, on the other hand, approach and withdrawal may be unregulated and individuals may be less able to inhibit motivational tendencies. Impulsivity is thought to index inverse functioning of r-BIS because r-BIS is strongly related to inhibition, effortful control, and overall executive functioning (Bari & Robbins, 2013; Bickel et al., 2012; Eisenberg et al., 2004).

Impulsivity can be measured with a number of personality traits. For instance, positive urgency measures impulsivity through the failure of r-BIS to inhibit approach tendencies, which ultimately leads to rash action during intense positive states (Cyders et al., 2010; Zapolski et al., 2009). To examine relationships between impulsive personality traits and asymmetrical frontal cortical activity, Gable and colleagues (2015) conducted a resting frontal EEG study investigating the association between positive urgency and frontal activity. Results showed a relationship between positive urgency and greater relative left frontal activity. This suggests that reduced relative right frontal activity is associated with reduced r-BIS function. Using standardized low-resolution brain electromagnetic tomography (SLORETA; Pascual-Marqui, 2002), source localization results indicated that lower relative right frontal activity was due to reduced activity in the right inferior frontal gyrus (rIFG). This further suggests that it is reduced right frontal activity, rather than increased left frontal activity, that was driving the observed asymmetric activity.

Neal and Gable (2016) investigated the relationship between frontal activity and tenets of impulsivity that are not related to positive affect or approach motivation. Participants completed the UPPS-P scale, which assesses negative urgency, lack of premeditation, and lack of perseverance, along with positive urgency (Cyders & Smith, 2007; Whiteside et al., 2005). Participants' resting EEG activity was then assessed. Negative urgency (i.e., rash behavior in negative emotional contexts) was related to heightened left frontal activity at rest, suggesting that the relationship between impulsivity and relative left frontal activity was not driven by positive emotionality. Non-emotional impulsive traits (i.e., lack of premeditation and lack of perseverance) were also associated with reduced right frontal activity. This relationship between impulsivity and right frontal activity remained unchanged when controlling for trait approach motivation. Source localization for the relationship between these tenets of impulsivity showed reduced activity in the right cingulate gyrus and the right medial frontal gyrus. Taken together, these results suggest that impulsivity is associated with reduced right frontal activity, independent of affective valence.

This study did not find a relationship between trait sensation seeking and frontal activity. However, the type of sensation seeking scale used (UPPS-P measure of trait sensation seeking) does not correlate well with other subscales of the UPPS-P, and some researchers have suggested that it may reflect a construct other than impulsivity (Simons et al., 2010). Santesso et al. (2008), however, used the Zuckerman Sensation Seeking Scale to measure trait sensation seeking and found that it was related to greater left frontal (reduced right frontal) activity.

Overall, these findings suggest that deficits in persistence and inhibiting behavior are associated with reduced right frontal activity. Source localization of the relationship between r-BIS and frontal asymmetry suggests that the asymmetry is driven by reduced activity in the right medial and lateral frontal areas, including areas of the right prefrontal cortex.

11.7.2 Evidence of r-BIS Functioning in Frontal EEG Activity

Source localization studies have also investigated the source of activity relating to behavioral measures of control. Gianotti and colleagues (2009) collected resting EEG data before having participants engage in a behavioral risk task. When participants engaged in greater risk-taking behavior, activity was localized to diminished baseline activity in the right lateral prefrontal cortex. Those whose resting activation of the right prefrontal cortex was less stable demonstrated less supervisory control of risky behavior. In a later study, resting EEG activity was compared to participants' subsequent acceptance of unfair offers in the ultimatum game (Knoch et al., 2010). When presented with an unfair offer, participants could either accept the offer or punish the opponent for giving them the offer. Acceptance of unfair offers was assumed to reflect an ability to exercise control over the initial emotional response in order to maximize economic benefits. Results showed a positive correlation between baseline right frontal activity and increased acceptance of most unfair offers. This activity was also localized to the right lateral prefrontal cortex.

Studies investigating regulation of anger have demonstrated that effortful control of anger relates to greater right frontal activity. Hewig and colleagues (2004) found that the extent to which individuals control their anger (State-Trait Anger Expression Questionnaire; Spielberger, 1988) related to greater relative right frontal asymmetry during resting baseline. The researchers suggested that anger-control was associated with relative right frontal activity because anger-control was associated with anger withdrawal, but based on the evidence, it seems likely that greater right frontal activity is linked with greater anger-control because it is associated with greater effortful control. Other work has demonstrated that suppressing anger, when it is socially inappropriate to express it, causes greater right frontal activity (Zinner et al., 2008). Although greater right frontal activity may be associated with the motivation to withdraw in some anger states, these results also support that greater right frontal activity is associated with effortful control stemming from emotional suppression of anger.

Functioning of r-BIS can also be connected to drug and alcohol reactivity. Greater relative left frontal (reduced right frontal) activity has been connected to drug-cue reactivity such as alcohol exposure (Myrick et al., 2004) and cocaine cravings (van de Laar et al., 2004). These increases in left frontal activity in response to substance cues are thought to be indicative of appetitive responses evoked from substance-related stimuli (Carter & Tiffany, 1999). Hypoactivation of rBIS, however, can also cause increased responsiveness toward alcohol. Mechin and colleagues (2016) investigated whether it is trait impulsivity or trait approach motivation that drives the increase in relative left frontal activity in response to alcohol-related cues. To determine this, they had participants complete the UPPS-P Behavioral Impulsivity Scale (Cyders & Smith, 2007; Whiteside et al., 2005), the BIS/BAS scales (Carver & White, 1994), and questions about their drinking habits. They then collected EEG data while participants viewed alcohol-related and neutral picture cues. Results suggested that the reduction in right frontal activity toward alcohol cues was due to trait impulsivity rather than trait approach motivation. The relationship between impulsivity and alcohol picture presentation remained constant when controlling for drinking behaviors and asymmetric frontal activity in response to neutral pictures. These results suggest that r-BIS moderates asymmetric frontal activity to alcohol cues while BAS does not.

Impulsivity is presumably related to relative right frontal cortical activity because impulsivity is associated with reduced effortful control. Evidence demonstrating increases in relative right frontal activity when individuals demonstrate greater effortful control and increases in relative left frontal activity when individuals demonstrate greater impulsivity could provide more compelling evidence for this model. Neal and Gable (2019) used a Balloon Analogue Risk Task (BART) to test whether impulsive or controlled behaviors influence asymmetric frontal activity. In this study, EEG recordings were collected as participants performed the BART while simultaneously viewing alcohol stimuli designed to enhance impulsive tendencies. Participants could win money by successfully blowing up a virtual balloon. With each pump of the balloon, more money could be won, but the likelihood of the balloon popping increased. Each trial ended with either the participant cashing out on the balloon or the balloon popping.

EEG data collected during this task showed that asymmetry scores shifted throughout alcohol trials. Frontal activity would shift to greater relative left frontal activity in the last half of trials where the balloon popped (impulsive behavior) while activity would shift to the right on the last half of trials where the participant cashed out (successful inhibition). These shifts were localized to reduced activity in the rIFG and lIFG, respectively. This suggests that increased right frontal activity is indicative of impulse control and diminished right frontal activity leads to more impulsive behavior.

The right frontal cortex has also been found to play an important role in the r-BIS functions of error detection, emotion regulation, and self-control. When an incorrect or inappropriate behavior is enacted, the error-related negativity (ERN) is evoked in response. Individuals with greater levels of behavioral inhibition, anxiety, and emotion regulation tend to have increased ERN amplitudes, suggesting that those with higher functioning r-BIS show greater neural responses in response to conflict monitoring (Amodio et al., 2008; Proudfit et al., 2013; Teper & Inzlicht, 2013). Greater relative right frontal activity at baseline has been linked to increased ERN amplitudes, while greater relative left frontal activity has been linked to reduced ERNs (Nash et al., 2012). Taken together, these results suggest that greater relative right frontal activity is related to greater r-BIS functioning in terms of conflict detection in response to errors.

Recently, Lacey and colleagues (2020) further connected emotion regulation to right frontal activity. In one experiment, participants listened to anxiety-inducing and neutral sound clips and were told to either listen naturally or suppress their emotional reactions to the clips. Participants recorded their level of effort when suppressing their reactions and noted their affective experience on each trial. Results showed that, when participants recorded higher levels of effortful control, right frontal activity was increased. However, experience of negative emotion was not associated with this increase in right frontal activity. In a second experiment, participants were shown negative and neutral pictures and told that looking at the images for a long time would earn them money, while choosing to escape from looking at the pictures would earn them no reward. In this study, right frontal activity was associated with looking at negative pictures for a longer time during reward trials, but not during non-reward trials. Both studies indicate that it was not the negative affect associated with the aversive stimuli, but rather the effort to engage the negative stimuli that was related to an increase in right frontal activity.

Schmeichel and colleagues (2016) evaluated the relationship between self-control and asymmetrical frontal activity. In this study, participants either had their self-control depleted or not, and then underwent EEG recording while viewing positive pictures. Among individuals who were relatively higher in trait BAS than BIS, those with depleted (vs. non-depleted) self-control showed increased left frontal activity in response to positive pictures. Among those with no relative difference in BIS and BAS scores, those with depleted self-control showed decreased left frontal activity in response to positive pictures. This suggests that when r-BIS is depleted, those who generally exhibit more approach motivation show greater left frontal activity in response to rewarding stimuli. An enhanced r-BIS, on the other hand, may be able to lower approach motivation in those with generally hyperactive BAS and hypoactive BIS.

Overall, these studies suggest that greater right frontal activity may be associated with processes involving emotion regulation while lower right frontal activity may be associated with reduced self-control and hindered error monitoring. Additionally, both situational contexts where control must be utilized and control-related personality traits have been related to right frontal activity as measured with EEG.

11.8 Additional Models of Frontal Asymmetry

Researchers have presented additional models to explain frontal asymmetry and its relationship to emotional and motivational variables. The Bilateral BAS Model (Hewig et al, 2004) and the Activation vs. Inhibition Model (Wacker et al., 2003) each propose variants of how approach, withdrawal, and inhibition influences frontal asymmetry.

11.8.1 Bilateral BAS Model

Hewig's bilateral BAS model suggests that both the approach and withdrawal systems encompass the basic mechanisms of active behavior and are therefore both subsystems of the BAS. When active behavior is initiated, then, bilateral activation of the frontal cortices is expected. Hewig and colleagues (2004) first addressed the model in their study analyzing resting frontal alpha band asymmetry and its relationships with affective valence, motivational direction, and behavioral activation. Resting state data obtained from each participant was related to measures of trait anger and different anger styles (STAXI, Spielberger, 1988), affective valence (PANAS), aggression (Aggression Questionnaire), and behavioral activation (BIS/BAS). Data analyses indicated that while valence was not significantly related to frontal asymmetry, relative left frontal cortical activity was positively correlated with outwardly expressed anger ("anger-out," Spielberger, 1988), trait anger, and aggression, while relative right frontal cortical activity was positively correlated with lowering arousal in anger-inducing situations ("anger-control," Spielberger, 1988). Additionally, greater bilateral frontal cortical activity was associated specifically with higher BAS scores. These results together suggest that greater levels of frontal asymmetry are driven by motivational tendencies and that bilateral frontal activation is driven by the combination of the approach and withdrawal subsystems of the BAS. A later study (Hewig et al., 2006) using similar methods found comparable results. Those participants who showed greater bilateral activation while in a resting state had higher BAS scores. The authors suggest that this result indicates that BAS scores may encompass both approach- and withdrawal-motivated behaviors.

In 2006, Hewig and colleagues conducted a go/no-go experiment to discern the relationship between positive and negative reinforcement and BAS scores. In this study, participants were first shown a cue indicating whether success on the given trial could add to the participant's money (positive reinforcement) or stop money from being subtracted (negative reinforcement). According to Gray's model of BIS/BAS (e.g., 1982), both types of reinforcement should activate the BAS, as both are a form of reward or non-punishment. Participants then completed a go/no-go task in which no-go trials were considered to be a form of passive avoidance, therefore activating the BIS. After the task, participants were shown monetary feedback based on their performance. Results showed that all participants showed greater bilateral frontal activation in response to reinforcement trials—regardless of whether it was positive or negative reinforcement—and that this effect was strongest in those with higher trait BAS scores. When these results are considered with prior research suggesting that approach motivation aligns with greater relative left frontal activity and withdrawal motivation aligns with greater relative right frontal activity, the suggestion is that the two are subsystems of the overarching BAS, once again supporting Hewig and colleagues' bilateral BAS model.

Rodrigues and colleagues (2018) found further support for the bilateral BAS model using a virtual reality paradigm. In this study, participants navigated a virtual T maze in which they were to respond to various events. These events could be positive, negative, or neutral in valence. Positive trials encouraged participants to move toward the event while negative trials encouraged withdrawal from the event. Approach-avoidance conflict trials involved having a positive stimulus guarded by a negative stimulus and approach-approach conflict trials involved choosing between two positive stimuli. There were also two control trial types. In each trial, participants chose in which direction they would like to move while EEG was recorded. Left frontal activity increased when participants chose an approach direction while right frontal activity increased when participants chose a withdrawal direction. Further, bilateral frontal activity increased during any choice in behavior relative to the choice to not move in any direction. The authors argue that these results support the role of frontal asymmetry in behavioral approach or avoidance motivation, as well as the role of bilateral frontal activation in active behavior.

11.8.2 Activation vs. Inhibition Model

Wacker and colleagues (Wacker et al., 2003; Wacker et al., 2008, 2010) suggest that the driving force behind frontal asymmetry is whether any motivational system is being activated or inhibited. Consistent with Gray's revised behavioral inhibition system/behavioral activation system (BIS/BAS) model (Gray & McNaughton, 2000), BAS is activated in response to approach motivation and rewarding stimuli, whereas the fight-flight-freeze system (FFFS) responds to withdrawal motivation and aversive stimuli. BIS responds to conflict between these two systems, acting as conflict monitor, shifting attention to allow for more efficient goal direction in conflicting situations (i.e., situations with more than one appetitive goal or a goal that requires combating aversive obstacles). Wacker and colleagues argue that all forms of behavioral activation are controlled by BAS and FFFS and are associated with greater relative left frontal activation, whereas behavioral inhibition and conflict monitoring are controlled by BIS and are associated with greater relative right frontal activation.

In an early study, Wacker and colleagues (2003) compared the viability of the BIS/BAS, motivational direction, and valence models of frontal asymmetry. Participants completed a mental image script task in which emotion (fear vs. anger) and motivation (approach vs. withdrawal) were manipulated and the participants reported the degree to which they agreed that the outcome of each task was the best option (degree of conflict). Results did not fall in line with the predictions of either the motivational direction or valence models of anterior asymmetry, but the degree of conflict experienced by participants was positively correlated with relative right frontal activation. Assuming that agreement ratings were a valid measure of BIS activation, these results support the model of frontal asymmetry suggested by Wacker and colleagues.

Similar results were found in a later study by Wacker and colleagues (2008), which directly compared the BIS/BAS model of anterior activation (BBMAA) to the motivational direction model. During a mental image script task similar to that used by Wacker and colleagues (2003), participants showed greater relative right frontal activation in response to scripts designed to target BIS than to those targeting FFFS. Self-reported measures of FFFS activation, conversely, were associated with greater relative left frontal activation.

The relationship between trait BIS and frontal asymmetry was investigated by Wacker and colleagues (2010) via a go/no-go task. Such a paradigm was used because no-go tasks have been suggested to be a viable measure of the conflict and behavioral inhibition functions of Wacker's revised BIS. Consistent with this assumption, those participants who had greater trait BIS showed greater relative right frontal activation in response to no-go trials than to go trials.

11.9 Psychopathology and Frontal Asymmetry

Because frontal alpha asymmetry plays a role in the experience of motivation and emotion, it follows that it would also factor into the experience of mental illness, particularly mood disorders. The psychological conditions most frequently investigated in relation to frontal asymmetry are depression, anxiety, and bipolar disorder.

11.9.1 Depression

Depression is one of the most commonly studied conditions in relation to frontal asymmetry. Depressive symptomology includes experiences such as decreased response to reward, lack of positive affect, and greater tendencies toward withdrawal from triggering activities. Davidson and colleagues (2002) pointed out that these symptoms can often be described as deficiencies in approach motivation and hyperactive withdrawal motivation, and these trends are associated with decreased relative left frontal cortical activity (greater relative right frontal activity). A number of studies have found a relationship between depressive symptoms and reduced left-frontal activity during a resting state. For instance, Schaffer and colleagues (1983) found a negative correlation between scores on the Beck Depression Inventory (BDI) and relative left frontal activity. Subsequent studies have found similar relationships between relative left frontal activity at rest and self-reported (Diego et al., 2001) or clinically diagnosed depression (Henriques & Davidson, 1990; Smith et al., 2018; Stewart et al., 2010). Henriques and Davidson (1990) even found that participants who had previously been depressed showed lower relative left frontal activity at rest compared to those who had never been depressed, and that these patterns of asymmetry were comparable to those in participants experiencing acute depressive symptoms. These results suggest that frontal asymmetry may be a state-independent marker of depression.

It is of note that diminished relative left frontal activity in depressed individuals is particularly robust in women compared to men (Stewart et al., 2010). Stewart and Allen (2018) found evidence supporting this notion in a sample with no history of major depressive disorder. Participants engaged in resting state EEG recordings before returning one year later to report depressive symptomology during the worst two-week period experienced throughout the one-year interim. It was found that women—but not men—with lower relative left frontal activity at baseline reported greater degrees of depressive symptomology during the interim period.

While resting data provides a means of studying the relationship between frontal asymmetry and depression, some studies have found a null relationship between depression and frontal asymmetry (Harmon-Jones et al., 2002; Metzger et al., 2004; Tomarken & Davidson, 1994; McFarland et al., 2006). These null results have led some researchers to argue that state-measures of frontal asymmetry may be more reliably associated with depression. For example, the capability model of frontal asymmetry suggests that cortical activity measured during emotional challenges is more indicative of predispositions toward psychopathology than cortical activity measured at rest (Stewart et al., 2014). Lower relative left frontal activity during a facial emotion task was found to be indicative of a history of major depressive disorder (Stewart et al., 2011). Participants prone to depressive symptomology have also exhibited lower relative left frontal activity during tasks that evoke anger (Harmon-Jones et al., 2002) and sadness (Nitschke et al., 2004). Those with early-onset depression also exhibited lower relative left frontal activity during an approach-related reward paradigm (Shankman et al., 2007).

Frontal asymmetry may act as more of a risk factor than an indicator of current depression. In one study, Possel and colleagues (2008) measured resting frontal asymmetry in adolescent boys and related it to depressive symptoms experienced throughout the year following the EEG recording. Results suggested that lower baseline left frontal activity was predictive of melancholic depressive symptoms and increased right frontal activity was predictive of non-melancholic depressive symptoms one year later, but depressive state was not predictive of baseline asymmetry. This suggests that frontal asymmetry may be indicative of a predisposition to depression. Mitchell and Possel (2012) conducted a similar study comparing depression and resting asymmetry in a nonclinical population and found that individuals with lower baseline left frontal activity were more likely to develop depressive symptoms over the next year. Nusslock and colleagues' (2011) results echoed these findings, suggesting that cognitive vulnerability to depression was both associated with lower relative left frontal activity at baseline and predicted onset of depressive symptoms one year later.

In addition, frontal asymmetry has been found to be a predictor of treatment response in depressed individuals. Greater relative left frontal activity prior to treatment predicted more successful response to fluoxetine (Bruder et al., 2001), as well as to escitalopram and sertraline in women (Arns et al., 2015).

11.9.2 Anxiety

Anxiety symptoms also appear to be related to greater right frontal activity. In a sample of participants consisting of healthy controls, those with major depression in remission, those with acute depression without comorbid anxiety disorder, and those with acute depression with a comorbid anxiety disorder, the only group difference observed was greater relative right frontal activity in the group with both major depression and a comorbid anxiety disorder compared to healthy controls (Feldmann et al., 2018). Other findings suggested that those with both anxiety and depression show frontal asymmetry patterns similar to those with depression alone (Mathersul et al., 2008). Nusslock and colleagues (2018) found asymmetry patterns to be most similar between healthy controls and those with comorbid depression and anxiety. In this study, women with a history of childhood onset depression without anxiety diagnoses showed reduced left frontal activity, which is consistent with past research on depression. However, women with a history of childhood onset depression and with pathological levels of anxious apprehension (in the form of generalized anxiety disorder, obsessive compulsive disorder, or separation anxiety disorder) showed resting asymmetry patterns statistically indistinguishable from healthy controls. These results highlight the role of comorbid depression and anxiety in its complex relationship with frontal asymmetry.

Similar results have been found in individuals with clinical diagnoses of anxiety disorders. Adults with panic disorder, for example, showed increased right frontal activity in both resting and anxiety-provoking contexts relative to controls (Wiedemann et al., 1999). Individuals with social phobia also showed increased right frontal activity compared to controls in response to anticipation of giving a speech (Davidson et al., 2000). Additionally, participants with post-traumatic stress disorder showed greater state-dependent right frontal activity when they were presented with trauma-relevant stimuli (Meyer et al., 2015). When participants engaged in an emotional Stroop task consisting of human faces depicting various emotions, those participants with greater levels of trait anxiety showed increased right frontal activity in response to fearful faces than did those with lower levels of trait anxiety (Avram et al., 2010). Similarly, participants who scored higher in trait anxiety showed greater right frontal activity during anxiety-provoking situations (Balconi & Pagani, 2014; Cole et al., 2012; Crost et al., 2008).

Inconsistent patterns of frontal asymmetry have been found in children with anxiety disorders. In a study comparing resting data of boys and girls aged 8 years and 11 years, anxious girls aged 8 and 11 showed decreased relative left frontal activity at rest, while their healthy control counterparts showed no significant frontal asymmetry at 8 years and greater relative left frontal activity at 11 years. While these results are consistent with research on adults with anxiety disorders, the young boys in the study showed inconsistent patterns. Anxious 8-year-old boys showed no significant frontal asymmetry and anxious 11-year-old boys showed greater relative left frontal activity while healthy boys showed greater relative right frontal activity (Baving et al.,, 2002). These results suggest that patterns of frontal asymmetry in the context of anxiety are not consistent across gender and age; further research is needed to better understand this relationship.

Inconsistent patterns have also been found when comparing various subtypes of anxiety. For instance, at baseline, participants with generalized anxiety disorder and increased worry show greater relative left frontal activity while those with high trait anxiety and low worry show lower relative left frontal activity (Smith et al., 2016). Crost and colleagues (2008) also found that those with more anxiety show greater relative right frontal activity in response to social threat while those higher in defensiveness show greater relative left frontal activity in response to social threat. It has further been argued that symptoms of anxious arousal (e.g., panic) are correlated with lower relative left frontal activity and symptoms of anxious apprehension (e.g., worry) are correlated with greater relative left frontal activity. As these results suggest, symptoms of anxiety have complex relationships with patterns of frontal asymmetry.

11.9.3 Bipolar Disorder

Individuals with bipolar disorder experience heightened approach motivation and hypersensitivity to goal- and reward-relevant cues during episodes of mania/hypomania (Alloy & Abramson, 2010; Johnson, 2005; Urosevic et al., 2010). This trend is indicative of increased BAS sensitivity in individuals with bipolar disorder. Indeed, self-reported BAS sensitivity scores are higher in those with bipolar I disorder (Meyer et al., 2001; Salavert et al., 2007), bipolar II disorder, and cyclothymia (Alloy et al., 2008), as well as those prone to hypomanic symptomology (Meyer et al., 1999). Since increased BAS sensitivity and approach motivation are positively correlated with relative left frontal cortical activity, it is expected that this trend should be found in individuals with bipolar disorder. Kano and colleagues (1992) collected resting EEG data from participants with bipolar disorder and found greater left frontal cortical activation, suggesting that those experiencing manic symptoms (as opposed to depressive symptoms) have greater relative left frontal activity at baseline.

Nusslock and colleagues (2012) also measured resting EEG frontal asymmetry in individuals with cyclothymia and followed up approximately 5 years later to observe changes in bipolar course. Those with greater levels of left frontal activity at baseline had a greater likelihood of converting to more severe diagnoses over the interim period (i.e., conversion from cyclothymia or bipolar II to bipolar I). Greater relative left frontal activity at baseline was also predictive of earlier age-of-onset of a first bipolar spectrum episode, an indicator of severity of bipolar disorder. This relationship was observed even when controlling for mood state and medication status at the time of EEG recording.

An early study by Harmon-Jones and colleagues (2002) also found associations between bipolar disorder and frontal asymmetry in a task-based paradigm. In this study, participants completed the General Behavior Inventory (GBI; Depue et al., 1989) to assess potential risk for developing bipolar or depressive disorders. Then, EEG data was collected while participants engaged in an anger-evoking task. The authors hypothesized that individuals with hypomania/mania symptoms would have greater relative left frontal activity when angered, based on prior research (Depue & Iacono, 1989). They also hypothesized that the opposite pattern (a decrease in left frontal activity) would be seen in participants with depressive symptoms. As predicted, those with hypomanic/manic symptoms showed increased left frontal activity during the anger-inducing situation while those with depressive symptoms showed decreased left frontal activity. These results suggest that the greater approach tendencies experienced by those with hypomanic/manic symptoms and the diminished approach tendencies experienced by those with depressive symptoms manifest through variations in frontal asymmetry.

Harmon-Jones and colleagues (2008) tested the BAS dysregulation theory of bipolar disorder by measuring frontal asymmetry in response to tasks of varying difficulty. Participants had either a bipolar spectrum diagnosis or no major psychopathology. During EEG data collection, participants engaged in an anagram task during which they were given cues indicating the difficulty (i.e., easy, medium, or hard) of the upcoming trial. They were also told whether they could receive money (win) or avoid losing money (loss) by successfully completing the upcoming trial. Results indicated that those with bipolar disorder had greater relative left frontal activity in preparation for hard/win trials, while control participants showed a decrease in relative left frontal activity in anticipation of the same trial type. Greater relative left frontal activity was also correlated with self-reported hypomanic/manic experience during the task in individuals with bipolar disorder. These results therefore provide evidence supporting BAS dysregulation theory. They also suggest that increased relative left frontal activity, which may be related to manic symptoms, can be triggered by more difficult and rewarding stimuli.

11.10 Discussion of Reviewed Research

11.10.1 Non-Significant Associations of Frontal Asymmetry and Affect/Motivation

Some studies have reported non-significant correlations of resting baseline frontal asymmetry with trait affective valence/motivational direction measures (Reid et al.,

1998). A meta-analysis of the association between affect-related personality traits and frontal asymmetry found that some associations were non-existent, and some were statistically significant in the predicted direction, but with small effect sizes (Kuper et al., 2019). We believe it is important to consider the complexity of human traits and neurophysiology before drawing conclusions about the interpretation of null-effects and small effect sizes of psychological phenomena. Next, we discuss the concerns of interpreting effect sizes and reasons some frontal asymmetry effects may be "small".

As it stands, effect sizes are generally misrepresented and misunderstood by researchers and readers alike. Funder and Ozer (2019) note that, even when effect sizes are reported and, even more rarely, are interpreted, those interpretations are based on fundamentally flawed standards. The two most common means of interpreting effect sizes are via Cohen's standards and squaring the correlation coefficient r. Funder and Ozer (2019) argue that the former of these techniques is nonsensical and is used out of context of any sort of comparison (Cohen agrees that these standards are not ideal; Cohen, 1977, 1988), while the latter is misleading. Squaring the correlation coefficient changes the scale upon which the effect size is measured, making the effect appear less impactful than it may be in reality.

More foundational to the issue of effect sizes is the field's tendency to dismiss small effect sizes without proper consideration. As Funder and Ozer (2019) point out, the reality is that small effect sizes are the most believable. The human experience is complex, and it is ultimately unrealistic to expect that any one phenomenon being studied (neural or behavioral) will explain the bulk of any human behavior. In addition, small effect sizes are not necessarily associated with small consequences. Immediate but frequent phenomena with seemingly small effect sizes can cumulate to have greater implications in the long run of the individual or the population (Funder & Ozer, 2019). "Small" effect sizes should not be dismissed; rather, they should be reported explicitly in the context of their overall consequence. The results of Kuper and colleagues' (2019) meta-analysis also revealed much heterogeneity in results from individual studies, suggesting that situational variables may influence the relationship between resting, baseline frontal asymmetry and self-reported personality traits.

The inconsistency observed in the literature concerning frontal asymmetry could be a result of several things that do not necessarily indicate a lack of relationship with motivation. Motivational tendencies can be sensitive to individual situations, and their relationship to frontal asymmetry could be masked by situational variables. For instance, some research suggests that in baseline EEG recording sessions, half of the variation in the data is due to trait influences and half is due to state influences (Hagemann et al., 2002; Hagemann et al., 2005).

The relationship between trait BAS and resting relative left frontal activity may be influenced by the specific circumstances of an experiment's procedure. Wacker and colleagues (2013) demonstrated this by analyzing the effect of attractiveness of opposite-sex experimenter on the correlation between trait BAS and frontal asymmetry. Results suggested that the correlation between BAS and greater relative left frontal activity was present primarily in cases when male participants interacted with attractive female experimenters. The attractiveness of the experimenter seems to have encouraged approach motivation in the participants, thereby strengthening the relationship between trait BAS and resting relative left frontal activity.

A similar pattern can be seen when reward and effort are manipulated. Hughes and colleagues (2015) conducted an experiment in which low-effort trials elicited low rewards while high-effort trials elicited high rewards. Participants showed greater left frontal activity during high-effort high-reward trials than during low-effort low-rewards trials. These results suggest that the motivation and incentive anticipation involved in a situation can impact measures of frontal asymmetry.

Situational variables outside of the lab space can also impact frontal asymmetry measures. Peterson and Harmon-Jones (2009) found that participants had greater right frontal activity when they participated in the study during fall mornings. Because depression may be experienced more during the fall months (King et al., 2000) and cortisol levels are high in the mornings (King et al., 2000), the variation in frontal asymmetry may have been due to reductions in approach motivation associated with depression and increases in inhibition or withdrawal motivation associated with higher cortisol.

Despite some inconsistencies in the literature, there is sufficient evidence to suggest a relationship between trait motivational tendencies and frontal asymmetry. There are at least three possible explanations for this trend in resting frontal asymmetry data. First, resting frontal asymmetry may reflect neural tendencies of the individual; when they are not engaged in a specific task, their frontal asymmetry scores reflect their trait-based tendencies toward approach and effortful control. Second, resting frontal asymmetry could reflect the individual's response to being in a novel environment. If participants feel curiosity (approach) when introduced to the lab space, they exhibit greater relative left frontal activity; if participants feel discomfort while continuing to engage in the new setting (and engage effortful control), they exhibit greater relative right frontal activity. Third, participants' thought processes during resting data collection could be driving frontal asymmetry variability. When not given any instruction, if participants tend to think about their goals and aspirations, they may exhibit greater relative left frontal activity; if participants tend to think about anxiety evoking situations, they may exhibit greater relative right frontal activity. The underlying mechanisms driving baseline frontal asymmetry are not yet understood and require further research.

11.11 CONCLUSION

The motivational model of frontal asymmetry suggests that relative left frontal activity is associated with approach motivation while right frontal activity is associated with withdrawal motivation. The former assumption of this model is supported by the literature, both through possible motivation confounds associated with the affective valence model and through more direct manipulations of motivation. Anger—an emotion that is both negative and approach motivated—is associated with greater left frontal activity at the state and trait level. Further support stems from research connecting higher BAS scores with greater baseline left frontal activity. The connection between BIS scores and baseline right frontal activity, however, has been widely contested.

This inconsistency in the literature concerning withdrawal motivation and right frontal has led to the development of the effortful control model of right frontal asymmetry. The effortful control model suggests that right frontal activity is associated with control and regulatory processing. Research supporting the effortful control model has linked right frontal activity with increased anxious inaction, error detection, and emotional control-related behaviors; right frontal activity is also negatively correlated with state and trait impulsivity. Past work linking right frontal activity and withdrawal motivation may have activated effortful control, because withdrawal manipulations may have co-activated effortful control to stay engaged with the aversive stimuli, but also because activation of r-BIS increases the cognitive load devoted to negative stimuli.

Psychopathologies influence patterns of frontal asymmetry. For instance, individuals with depression show reduced levels of left frontal activity at rest and during emotionally salient tasks. Reduced left frontal activity may be a risk factor for depression in nonclinical populations. Those with anxiety generally show increased levels of right frontal activity at rest and during tasks that elicit fear and/or anxiety; however, this relationship is more complicated in children and when comparing subtypes of anxiety. Those with bipolar disorder who experience manic symptoms exhibit greater left frontal activity at rest and during anger-evoking or difficult rewarding tasks. These patterns are consistent with the motivational direction and effortful control models of frontal asymmetry.

While there have been failures to replicate some past frontal asymmetry findings, these failures to replicate may be due to the complex nature of the systems underlying frontal asymmetry. Phenomena such as affect and motivation change by the moment and can be swayed by situational effects. Despite this, frontal asymmetry research has a long and well-established history in EEG frequency research and continues to spark new predictions, models, and discoveries.

REFERENCES

Alema, G., Rosadini, G., & Rossi, G. F. (1961). Preliminary experiments on the effects of the intracarotid introduction of sodium amytal in parkinsonian syndromes. *Bollettino della Società Italiana di Biologia Sperimentale*, *37*, 1036–1037.

Allen, J. J. B., Harmon-Jones, E., & Cavender, J. H. (2001). Manipulation of frontal EEG asymmetry through biofeedback alters self-reported emotional responses and facial EMG. *Psychophysiology*, *38*, 685–693.

Allen, J. J. B., Iacono, W. G., Depue, R. A., & Arbisi, P. (1993). Regional electroencephalographic asymmetries in bipolar seasonal affective disorder before and after exposure to bright light. *Biological Psychiatry*, *33*, 642–646.

Alloy, L. B., Abramson, L. Y., Walshaw, P. D., Cogswell, A., Grandin, L. D., Hughes, M. E., ... & Hogan, M. E. (2008). Behavioral approach system and behavioral inhibition system sensitivities and bipolar spectrum disorders: Prospective prediction of bipolar mood episodes. *Bipolar Disorders, 10*(2), 310–322.

Alloy, L. B., & Abramson, L. Y. (2010). The role of the behavioral approach system (BAS) in bipolar spectrum disorders. *Current Directions in Psychological Science, 19*(3), 189–194.

Amodio, D. M., Master, S. L. Yee, C. M., & Taylor, S. E. (2008). Neurocognitive components of behavioral inhibition and activation systems: Implications for theories of self-regulation. *Psychophysiology, 45*, 11–19.

Amodio, D. M., Shah, J. Y., Sigelman, J., Brazy, P. C., & Harmon-Jones, E. (2004). Implicit regulatory focus associated with resting frontal cortical asymmetry. *Journal of Experimental Social Psychology, 40*, 225–232.

Arns, M., Bruder, G., Hegerl, U., Spooner, C., Palmer, D. M., Etkin, A., ... & Gordon, E. (2016). EEG alpha asymmetry as a gender-specific predictor of outcome to acute treatment with different antidepressant medications in the randomized iSPOT-D study. *Clinical Neurophysiology, 127*(1), 509–519.

Aron, A. R., Robbins, T. W., & Poldrack, R. A. (2004). Inhibition and the right inferior frontal cortex. *Trends in Cognitive Sciences, 8*(4), 170–177. doi:10.1016/j.tics.2004.02.010

Aron, A. R., Robbins, T. W., & Poldrack, R. A. (2014). Inhibition and the right inferior frontal cortex: One decade on. *Trends in Cognitive Sciences, 18*(4), 177–185. doi:10.1016/j.tics.2013.12.003

Avram, J., Balteş, F. R., Miclea, M., & Miu, A. C. (2010). Frontal EEG activation asymmetry reflects cognitive biases in anxiety: Evidence from an emotional face Stroop task. *Applied Psychophysiology and Biofeedback, 35*(4), 285–292. doi:10.1007/ s10484-010-9138-6

Balconi, M. & Pagani, S. (2014). Personality correlates (BAS-BIS), self-perception of social ranking, and cortical (alpha frequency band) modulation in peer-group comparison. *Physiology & Behavior, 133*, 207–215. doi:10.1016/j.physbeh.2014.05.043

Bari, A. & Robbins, T. W. (2013). Inhibition and impulsivity: Behavioral and neural basis of response control. *Progress in Neurobiology, 108*, 44–79. doi:10.1016/j.pneurobio.2013.06.005

Baving, L., Laucht, M., & Schmidt, M. H. (2002). Frontal brain activation in anxious school children. *Journal of Child Psychology and Psychiatry, and Allied Disciplines, 43*(2), 265–274. https://doi.org/10.1111/1469-7610.00019

Beeney, J. E., Levy, K. N., Gatzke-Kopp, L. M., & Hallquist, M. N. (2014). EEG asymmetry in borderline personality disorder and depression following rejection. *Personality Disorders: Theory, Research, and Treatment, 5*, 178–185.

Berkman, E. T. & Lieberman, M. D. (2010). Approaching the bad and avoiding the good: Lateral prefrontal cortical asymmetry distinguishes between action and valence. *Journal of Cognitive Neuroscience, 22*(9), 1970–1979. doi:10.1162/jocn.2009.21317

Berkowitz, L. & Harmon-Jones, E. (2004). Toward an understanding of the determinants of anger. *Emotion, 4*, 107–130.

Bickel, W. K., Jarmolowicz, D. P., Mueller, E. T., Gatchalian, K. M., & McClure, S. M. (2012). Are executive function and impulsivity antipodes? A conceptual reconstruction with special reference to addiction. *Psychopharmacology, 221*(3), 361–387. doi:10.1007/ s00213-012-2689-x

Black, E W. (1975). Unilateral brain lesions and MMPI performance: A preliminary study. *Perceptual and Motor Skills, 40*, 87–93.

Blanchard R. J. & Blanchard D. C. (1984). Affect and aggression: An animal model applied to human behavior. In R. J. Blanchard, & D. C. Blanchard (Eds.), *Advances in the study of aggression* (pp. 1–62). Academic Press.

Brehm, J. W. (1999). The intensity of emotion. *Personality and Social Psychology Review, 3*, 2–22.

Brehm, J. W., & Self, E. A. (1989). The intensity of motivation. *Annual Review of Psychology, 40*, 109–131.

Bruder, G. E., Stewart, J. W., Tenke, C. E., McGrath, P. J., Leite, P., Bhattacharya, N., & Quitkin, F. M. (2001). Electroencephalographic and perceptual asymmetry differences between responders and nonresponders to an SSRI antidepressant. *Biological psychiatry, 49*(5), 416–425.

Carter, B. L. & Tiffany, S. T. (1999). Meta-analysis of cue-reactivity in addiction research. *Addiction, 94*(3), 327–340. doi:10.1046/ j.1360-0443.1999.9433273.x

Carver, C. S. (2004). Negative affects deriving from the behavioral approach system. *Emotion, 4*, 3–22.

Carver, C. S. & Connor-Smith, J. (2010). Personality and coping. *Annual Review of Psychology, 61*, 679–704. doi:10.1146/annurev.psych.093008.100352

Carver, C. S. & Harmon-Jones, E. (2009). Anger is an approach-related affect: Evidence and implications. *Psychological Bulletin, 135*, 183–204.

Carver, C. S. & White, T. L. (1994). Behavioral inhibition, behavioral activation, and affective responses to impending reward and punishment: The BIS/BAS scales. *Journal of Personality and Social Psychology, 67*, 319–333.

Carver, C. S., Johnson, S. L., & Joormann, J. (2008). Serotonergic function, two-mode models of self-regulation, and vulnerability to depression: What depression has in common with impulsive aggression. *Psychological bulletin, 134*(6), 912. doi:10.1037/ a0013740

Coan, J. A. & Allen, J. J. (2003). Frontal EEG asymmetry and the behavioral activation and inhibition systems. *Psychophysiology, 40*, 106–114.

Coan, J. A. & Allen, J. J. (2004). Frontal EEG asymmetry as a moderator and mediator of emotion. *Biological Psychology, 67*, 7–50.

Coan, J. A., Allen, J. J. B., & Harmon-Jones, E. (2001). Voluntary facial expression and hemispheric asymmetry over the frontal cortex. *Psychophysiology, 38*, 912–925.

Cohen, J. (1977). *Statistical power analysis for the behavioral sciences* (rev. ed.). Academic Press.

Cohen, J. (1988). *Statistical power analysis for the behavioral sciences* (2nd ed.). Erlbaum.

Cole, C., Zapp, D. J., Nelson, S. K., & Perez-Edgar, K. (2012). Speech presentation cues moderate frontal EEG asymmetry in socially withdrawn young adults. *Brain and Cognition, 78*(2), 156–162. doi:10.1016/j.bandc.2011.10.013

Cook, I. A., O'Hara, R., Uijtdehaage, S. H. J., Mandelkern, M., & Leuchter, A. F. (1998). Assessing the accuracy of topographic EEG mapping for determining local brain function. *Electroencephalography and Clinical Neurophysiology, 107*, 408–414.

Crost, N. W., Pauls, C. A., & Wacker, J. (2008). Defensiveness and anxiety predict frontal EEG asymmetry only in specific situational contexts. *Biological Psychology, 78*(1), 43–52. doi:10.1016/j. biopsycho.2007.12.008

Cyders, M. A., & Smith, G. T. (2007). Mood-based rash action and its components: Positive and negative urgency. *Personality and Individual Differences, 43*(4), 839–850. doi:10.1016/ j.paid.2007. 02.008

Cyders, M. A., Zapolski, T. C., Combs, J. L., Settles, R. F., Fillmore, M. T., & Smith, G. T. (2010). Experimental effect of positive urgency on negative outcomes from risk taking and on increased alcohol consumption. *Psychology of Addictive Behaviors, 24*(3), 367–375.

Darwin, C. (1872/1965). *The expressions of the emotions in man and animals.* Oxford University Press.

Davidson, R. J. & Fox, N. (1982). Asymmetric brain activity discriminates between positive and negative affective stimuli in 10 month old infants. *Science, 218*, 1235–1237.

Davidson, R. J., Chapman, J. P., Chapman, L. J., & Henriques, J. B. (1990). Asymmetric brain electrical-activity discriminates between psychometrically-matched verbal and spatial cognitive tasks. *Psychophysiology*, *27*, 528–543.

Davidson, R. J., Marshall, J. R., Tomarken, A. J., & Henriques, J. B. (2000). While a phobic waits: Regional brain electrical and autonomic activity in social phobics during anticipation of public speaking. *Biological Psychiatry*, *47*(2), 85–95. https://doi.org/10.1016/s0006-3223(99)00222-x

Davidson, R. J., Pizzagalli, D., Nitschke, J. B., & Putnam, K. (2002). Depression: Perspectives from affective neuroscience. *Annual Review of Psychology*, *53*, 545–574. doi:10.1146/annurev.psych.53.100901.135148

De Pascalis, V., Cozzuto, G., Caprara, G. V., & Alessandri, G. (2013). Relations among EEG-alpha asymmetry, BIS/BAS, and dispositional optimism. *Biological Psychology*, *94*, 198–209.

Depue, R. A., Krauss, S., Spoont, M. R., & Arbisi, P. (1989). General Behavior Inventory identification of unipolar and bipolar affective conditions in a nonclinical university population. *Journal of Abnormal Psychology*, *98*(2), 117–126.

Depue, R.A. & Iacono, W.G. (1989). Neurobehavioral aspects of affective disorders. *Annual Review of Psychology*, *40*, 457–492.

Derryberry, D. & Rothbart, M. K. (1997). Reactive and effortful processes in the organization of temperament. *Development and Psychopathology*, *9*(4), 633–652. doi:10.1017/S0954579497001375

DeYoung, C. G. (2015). Cybernetic big five theory. *Journal of Research in Personality*, *56*, 33–58. doi:10.1016/j.jrp.2014.07.004

Diego, M. A., Field, T., & Hernandez-Reif, M. (2001). CES-D depression scores are correlated with frontal EEG alpha asymmetry. *Depression and Anxiety*, *13*(1), 32–37.

Eisenberg, N., Spinrad, T. L., Fabes, R. A., Reiser, M., Cumberland, A., Shepard, S. A., ... Thompson, M. (2004). The relations of effortful control and impulsivity to children's resiliency and adjustment. *Child Development*, *75*(1), 25–46. doi:10.1111/j.1467- 8624.2004.00652.x

Ekman, P. & Davidson, R. J. (1993). Voluntary smiling changes regional brain activity. *Psychological Science*, *4*, 342–345.

Ekman, P. & Friesen, W. V. (1975). *Unmasking the face: A guide to recognizing emotions from facial clues*. Prentice-Hall.

Elgavish, E., Halpern, D., Dikman, Z., & Allen, J. J. B. (2003). Does frontal EEG asymmetry moderate or mediate responses to the international affective picture system (IAPS)? *Psychophysiology*, *40*, S38.

Enticott, P. G., Ogloff, J. R., & Bradshaw, J. L. (2006). Associations between laboratory measures of executive inhibitory control and self-reported impulsivity. *Personality and Individual Differences*, *41*(2), 285–294. doi:10.1016/j.paid.2006.01.011

Feldmann, L., Piechaczek, C. E., Grünewald, B. D., Pehl, V., Bartling, J., Frey, M., ... & Greimel, E. (2018). Resting frontal EEG asymmetry in adolescents with major depression: impact of disease state and comorbid anxiety disorder. *Clinical Neurophysiology*, *129*(12), 2577–2585.

Funder, D. C. & Ozer, D. J. (2019). Evaluating effect size in psychological research: Sense and nonsense. *Advances in Methods and Practices in Psychological Science*, *2*(2), 156–168.

Gable, P. A. & Dreisbach, G. (2021). Approach motivation and positive affect. *Current Opinion in Behavioral Sciences*, *39*, 203–208.

Gable, P. A. & Harmon-Jones, E. (2008). Relative left frontal activation to appetitive stimuli: Considering the role of individual differences. *Psychophysiology*, *45*, 275–278.

Gable, P. A. & Poole, B. D. (2014). Influence of trait behavioral inhibition and behavioral approach motivation systems on the LPP and frontal asymmetry to anger pictures. *Social Cognitive and Affective Neuroscience*, *9*, 182–190.

Gable, P. A., Mechin, N. C., Hicks, J. A., & Adams, D. L. (2015). Supervisory control system and frontal asymmetry: Neurophysiological traits of emotion-based impulsivity. *Social, Cognitive, and Affective Neuroscience*, *10*(10), 1310–315.

Gable, P. A., Mechin, N., Hicks, J.A., & Adams, D.A. (2015). Positive urgency and left-frontal asymmetry: Neurophysiological traits of emotion-based impulsivity. *Social Cognitive and Affective Neuroscience*, *10*, 1310–1315

Gable, P. A., Neal, L. B., & Threadgill, A. H. (2018). Regulatory behavior and frontal activity: Considering the role of revised-BIS in relative right frontal asymmetry. *Psychophysiology*, *55*(1), e12910.

Gable, P. A., Poole, B. D., & Cook, M. S. (2013). Asymmetrical hemisphere activation enhances global-local processing. *Brain and Cognition*, *83*, 337–341.

Gainotti, G. (1972). Emotional behavior and hemispheric side of the lesion. *Cortex*, *8*, 41–55.

Gasparrini, W. G., Satz, P., Heilman, K., & Coolidge, F. L. (1978). Hemispheric asymmetries of affective processing as determined by the Minnesota Multiphasic Personality Inventory. *Journal of Neurology, Neurosurgery and Psychiatry*, *41*, 470–473.

Gianotti, L. R., Knoch, D., Faber, P. L., Lehmann, D., Pascual-Marqui, R. D., Diezi, C., ... & Fehr, E. (2009). Tonic activity level in the right prefrontal cortex predicts individuals' risk taking. *Psychological Science*, *20*, 33–38.

Goldstein, K. (1939). *The organism: An holistic approach to biology, derived from pathological data in man*. American Books.

Gray, J. (1982). Précis of *The neuropsychology of anxiety: An enquiry into the functions of the septo-hippocampal system. Behavioral and Brain Sciences*, 5(3), 469–484. doi:10.1017/S0140525X00013066

Gray, J. A. & McNaughton, N. (2000). *The neuropsychology of anxiety: An enquiry into the functions of the septo-hippocampal system* (2nd ed.). Oxford University Press.

Grimshaw, G. M. & Carmel, D. (2014). An asymmetric inhibition model of hemispheric differences in emotional processing. *Frontiers in Psychology*, 5, 489. doi:10.3389/fpsyg.2014.00489

Gross, J. J. & Levenson, R. W. (1993). Emotional suppression: Physiology, self-report, and expressive behavior. *Journal of Personality and Social Psychology*, *64*, 970–986. doi:10.1037/0022-3514.64.6.970

Hagemann, D., Hewig, J., Seifert, J., Naumann, E., & Bartussek, D. (2005). The latent state-trait structure of resting EEG asymmetry: Replication and extension. *Psychophysiology*, *42*, 740–752.

Hagemann, D., Naumann, E., Becker, G., Maier, S., & Bartussek, D. (1998). Frontal brain asymmetry and affective style: A conceptual replication. *Psychophysiology*, *35*, 372–388.

Hagemann, D., Naumann, E., Thayer, J. F., & Bartussek, D. (2002). Does resting EEG asymmetry reflect a trait? An application of latent state-trait theory. *Journal of Personality and Social Psychology*, *82*, 619–641.

Harmon-Jones, C., Schmeichel, B. J., Mennitt, E., & Harmon-Jones, E. (2011). The expression of determination: Similarities between anger and approach-related positive affect. *Journal of Personality and Social Psychology*, *100*, 172–181. doi:10.1037/a0020966

Harmon-Jones, E. (2003a). Clarifying the emotive functions of asymmetrical frontal cortical activity. *Psychophysiology*, *40*, 838–848.

Harmon-Jones, E. (2003b). Anger and the behavioural approach system. *Personality and Individual Differences*, *35*, 995–1005.

Harmon-Jones, E. (2004). On the relationship of anterior brain activity and anger: Examining the role of attitude toward anger. *Cognition and Emotion*, *18*, 337–361.

Harmon-Jones, E. (2006). Unilateral right-hand contractions cause contralateral alpha power suppression and approach motivational affective experience. *Psychophysiology*, *43*, 598–603.

Harmon-Jones, E. (2007). Trait anger predicts relative left frontal cortical activation to anger-inducing stimuli. *International Journal of Psychophysiology*, *66*, 154–160.

Harmon-Jones, E., Abramson, L. Y., Nusslock, R., Sigelman, J. D., Urosevic, S., Turonie, L. D., Alloy, L. B., & Fearn, M. (2008). Effect of bipolar disorder on left frontal cortical responses to goals differing in valence and task difficulty. *Biological Psychiatry*, *63*, 693–698.

Harmon-Jones, E., Abramson, L. Y., Sigelman, J., Bohlig, A., Hogan, M. E., & Harmon-Jones, C. (2002). Proneness to hypomania/mania or depression and asymmetric frontal cortical responses to an anger-evoking event. *Journal of Personality and Social Psychology*, *82*, 610–618.

Harmon-Jones, E. & Allen, J. J. B. (1997). Behavioral activation sensitivity and resting frontal EEG asymmetry: Covariation of putative indicators related to risk for mood disorders. *Journal of Abnormal Psychology*, *106*, 159–163.

Harmon-Jones, E. & Allen, J. J. B. (1998). Anger and prefrontal brain activity: EEG asymmetry consistent with approach motivation despite negative affective valence. *Journal of Personality and Social Psychology*, *74*, 1310–1316.

Harmon-Jones, E. & Gable, P. A. (2009). Neural activity underlying the effect of approach-motivated positive affect on narrowed attention. *Psychological Science*, *20*, 406–409.

Harmon-Jones, E. & Gable, P. A. (2018). On the role of asymmetric frontal cortical activity in approach and withdrawal motivation: An updated review of the evidence. *Psychophysiology*, *55*(1), e12879.

Harmon-Jones, E., Gable, P. A., & Price, T. F. (2011). Leaning embodies desire: Evidence that leaning forward increases relative left frontal activation to appetitive stimuli. *Biological Psychology*, *87*, 311–313.

Harmon-Jones, E., Harmon-Jones, C., Abramson, L. Y., & Peterson, C. K. (2009). PANAS positive activation is associated with anger. *Emotion*, *9*, 183–196.

Harmon-Jones, E., Harmon-Jones, C., Amodio, D. M., & Gable, P. A. (2011). Attitudes toward emotions. *Journal of Personality and Social Psychology*, *101*, 1332–1350. doi:10.1037/a0024951

Harmon-Jones, E., Harmon-Jones, C., Fearn, M., Sigelman, J. D., & Johnson, P. (2008). Left frontal cortical activation and spreading of alternatives: Tests of the action-based model of dissonance. *Journal of Personality and Social Psychology*, *94*, 1–15. doi:10.1037/0022-3514.94.1.1

Harmon-Jones, E., Harmon-Jones, C., & Price, T. F. (2013). What is approach motivation? *Emotion Review*, *5*, 291–295. doi:10.1177/1754073913477509

Harmon-Jones, E., Lueck, L., Fearn, M., & Harmon-Jones, C. (2006). The effect of personal relevance and approach-related action expectation on relative left frontal cortical activity. *Psychological Science*, *17*, 434–440.

Harmon-Jones, E. & Peterson, C. K. (2008). Effect of trait and state approach motivation on aggressive inclinations. *Journal of Research in Personality*, *42*, 1381–1385.

Harmon-Jones, E. & Peterson, C. K. (2009). Supine body position reduces neural response to anger evocation. *Psychological Science*, *20*, 1209–1210. doi:10.1111/j.1467-9280.2009.02416.x

Harmon-Jones, E., Peterson, C. K., & Harris, C. R. (2009). Jealousy: Novel methods and neural correlates. *Emotion*, *9*, 113–117.

Harmon-Jones, E. & Sigelman, J. D. (2001). State anger and prefrontal brain activity: Evidence that insult-related relative left-prefrontal activation is associated with experienced anger and aggression. *Journal of Personality and Social Psychology*, *80*, 797–803.

Harmon-Jones, E., Sigelman, J. D., Bohlig, A., & Harmon-Jones, C. (2003). Anger, coping, and frontal cortical activity: The effect of coping potential on anger-induced left frontal activity. *Cognition and Emotion*, *17*, 1–24.

Harmon-Jones, E., Vaughn-Scott, K., Mohr, S., Sigelman, J., & Harmon-Jones, C. (2004). The effect of manipulated sympathy and anger on left and right frontal cortical activity. *Emotion*, *4*, 95–101.

Hellige, J. B. (1993). *Hemispheric asymmetry: What's right and what's left*. Harvard University Press.

Henriques, J. B. & Davidson, R. J. (1990). Regional brain electrical asymmetries discriminate between previously depressed and healthy control subjects. *Journal of Abnormal Psychology*, *99*, 22–31.

Henriques, J. B. & Davidson, R. J. (2000). Decreased responsiveness to reward in depression. *Cognition & Emotion*, *14*(5), 711–724. doi:10.1080/026999300050117684

Hester, R. & Garavan, H. (2004). Executive dysfunction in cocaine addiction: Evidence for discordant frontal, cingulate, and cerebellar activity. *The Journal of Neuroscience*, *24*(49), 11017–11022. doi:10.1523/JNEUROSCI.3321-04.2004

Hester, R. & Garavan, H. (2009). Neural mechanisms underlying drug-related cue distraction in active cocaine users. *Pharmacology Biochemistry and Behavior*, *93*(3), 270–277. doi:10.1016/j. pbb.2008.12.009

Hewig, J., Hagemann, D., Seifert, J., Naumann, E., & Bartussek, D. (2004). On the selective relation of frontal cortical activity and anger-out versus anger-control. *Journal of Personality and Social Psychology*, *87*, 926–939.

Hewig, J., Hagemann, D., Seifert, J., Naumann, E., & Bartussek, D. (2006). The relation of cortical activity and BIS/BAS on the trait level. *Biological Psychology*, *71*(1), 42–53. doi:10.1016/j. biopsycho.2005.01.006

Heym, N., Ferguson, E., & Lawrence, C. (2008). An evaluation of the relationship between Gray's revised RST and Eysenck's PEN: Distinguishing BIS and FFFS in Carver and White's BIS/BAS scales. *Personality and Individual Differences*, *45*(8), 709–715. doi:10.1016/ j.paid.2008.07.013

Hughes, D. M., Yates, M. J., Morton, E. E., & Smillie, L. D. (2015). Asymmetric frontal cortical activity predicts effort expenditure for reward. *Social Cognitive and Affective Neuroscience*, *10*, 1015–1019.

Jackson, D. C., Mueller, C. J., Dolski, I., Dalton, K. M., Nitschke, J. B., Urry, H. L., ... & Davidson, R. J. (2003). Now you feel it, now you don't: Frontal brain electrical asymmetry and individual differences in emotion regulation. *Psychological Science*, *14*(6), 612–617. doi:10.1046/ j.0956-7976.2003.psci_1473.x

Jacobs, G. D. & Snyder, D. (1996). Frontal brain asymmetry predicts affective style in men. *Behavioral Neuroscience*, *110*, 3–6.

Jensen-Campbell, L. A., Knack, J. M., Waldrip, A. M., & Campbell, S. D. (2007). Do big five personality traits associated with self-control influence the regulation of anger and aggression? *Journal of Research in Personality*, *41*, 403–424.

Johnson, S. L. (2005). Mania and dysregulation in goal pursuit: A review. *Clinical Psychology Review*, 25, 241–262.

Kamiya, J. (1979). Autoregulation of the EEG alpha rhythm: A program for the study of consciousness. In S. A. E. Peper & M. Quinn (Eds.), *Mind/body integration: Essential readings in biofeedback* (pp. 289–297). Plenum Press.

Kano, K., Nakamura, M., Matsuoka, T., Iida, H., & Nakajima, T. (1992). The topographical features of EEGs in patients with affective disorders. *Electroencephalography and Clinical Neurophysiology*, 83, 124–129.

Keune, P. M., Bostanov, V., Kotchoubey, B., & Hautzinger, M. (2012). Mindfulness versus rumination and behavioral inhibition: A perspective from research on frontal brain asymmetry. *Personality and Individual Differences*, 53(3), 323–328. doi:10.1016/j. paid.2012.03.034

Keune, P. M., Schönenberg, M., Wyckoff, S., Mayer, K., Riemann, S., Hautzinger, M., & Strehl, U. (2011). Frontal alpha-asymmetry in adults with attention deficit hyperactivity disorder: Replication and specification. *Biological Psychology*, 87, 306–310.

Keune, P. M., van der Heiden, L., Várkuti, B., Konicar, L., Veit, R., & Birbaumer, N. (2012). Prefrontal brain asymmetry and aggression in imprisoned violent offenders. *Neuroscience Letters*, 515, 191–195.

King, J. A., Rosal, M. C., Ma, Y., Reed, G., Kelly, T., Stanek, III, E.J., & Ockene, I. S. (2000). Sequence and seasonal effects of salivary cortisol. *Behavioral Medicine*, 26, 67–73.

Kline, J. P., Blackhart, G. C., Woodward, K. M., Williams, S. R., & Schwartz, G. E. (2000). Anterior electroencephalographic asymmetry changes in elderly women in response to a pleasant and an unpleasant odor. *Biological Psychology*, 52(3), 241–250. doi:10.1016/S0301-0511(99)00046-0

Knoch, D., Gianotti, L. R., Baumgartner, T., & Fehr, E. (2010). A neural marker of costly punishment behavior. *Psychological Science*, 21(3), 337–342. doi:10.1177/0956797609360750

Kochanska, G. & Knaack, A. (2003). Effortful control as a personality characteristic of young children: Antecedents, correlates, and consequences. *Journal of Personality*, 71(6), 1087–1112. doi:10.1111/1467-6494.7106008

Kuper, N., Käckenmester, W., & Wacker, J. (2019). Resting frontal EEG asymmetry and personality traits: A meta-analysis. *European Journal of Personality*, 33(2), 154–175.

Lacey, M. F., Neal, L. B., & Gable, P. A. (2020). Effortful control of motivation, not withdrawal motivation, relates to greater right frontal asymmetry. *International Journal of Psychophysiology*, 147, 18–25.

Lagerspetz, K. M. J. (1969). Aggression and aggressiveness in laboratory mice. In S. Garattini, & E. B. Sigg (Eds.), *Aggressive Behavior* (pp. 77–85). Wiley.

Lazarus, R. S. (1991). *Emotion and adaptation*. Oxford University Press.

Logan, G. D., Schachar, R. J., & Tannock, R. (1997). Impulsivity and inhibitory control. *Psychological Science*, 8(1), 60–64. doi:10.1111/j.1467-9280.1997.tb00545.x

Mathersul, D., Williams, L. M., Hopkinson, P. J., & Kemp, A. H. (2008). Investigating models of affect: Relationships among EEG alpha asymmetry, depression, and anxiety. *Emotion*, 8(4), 560–572

McFarland, B. R., Shankman, S. A., Tenke, C. E., Bruder, G. E., & Klein, D. N. (2006). Behavioral activation system deficits predict the six-month course of depression. *Journal of Affective Disorders*, 91(2–3), 229–234.

Mechin, N., Gable, P. A., & Hicks, J. A. (2016). Frontal asymmetry and alcohol cue reactivity: Influence of core personality systems. *Psychophysiology*, 53(8), 1224–1231. doi:10.1111/psyp.12659

Metzger, L. J., Paige, S. R., Carson, M. A., Lasko, N. B., Paulus, L. A., Pitman, R. K., & Orr, S. P. (2004). PTSD arousal and depression symptoms associated with increased right-sided parietal EEG asymmetry. *Journal of Abnormal Psychology, 113*(2), 324–329.

Meyer, B., Johnson, S. L., & Carver, C. S. (1999). Exploring behavioral activation and inhibition sensitivities among college students at risk for bipolar spectrum symptomatology. *Journal of Psychopathology and Behavioral Assessment, 21*, 275–292.

Meyer, B., Johnson, S.L., & Winters, R. (2001). Responsiveness to threat and incentive in bipolar disorder: Relations of the BIS/BAS scales with symptoms. *Journal of Psychopathology and Behavioral Assessment, 23*, 133–143.

Meyer, T., Smeets, T., Giesbrecht, T., Quaedflieg, C. W., Smulders, F. T., Meijer, E. H., & Merckelbach, H. L. (2015). The role of frontal EEG asymmetry in post-traumatic stress disorder. *Biological Psychology, 108*, 62–77. doi:10.1016/j.biopsycho.2015.03.018

Mitchell, A. M. & Pössel, P. (2012). Frontal brain activity pattern predicts depression in adolescent boys. *Biological Psychology, 89*(2), 525–527.

Mize, K. D. & Jones, N. A. (2012). Infant physiological and behavioral responses to loss of maternal attention to a social-rival. *International Journal of Psychophysiology, 83*(1), 16–23. doi:10.1016/j.ijpsycho.2011.09.018

Murphy, F. C., Nimmo-Smith, I., & Lawrence, A. D. (2003). Functional neuroanatomy of emotion: A meta-analysis. *Cognitive, Affective, & Behavioral Neuroscience, 3*, 207–233.

Myrick, H., Anton, R. F., Li, X., Henderson, S., Drobes, D., Voronin, K., & George, M. S. (2004). Differential brain activity in alcoholics and social drinkers to alcohol cues: Relationship to craving. *Neuropsychopharmacology, 29*(2), 393–402. doi:10.1038/sj.npp.1300295

Nash, K., Inzlicht, M., & McGregor, I. (2012). Approach-related left prefrontal EEG asymmetry predicts muted error-related negativity. *Biological Psychology, 91*, 96–102.

Neal, L. B. & Gable, P. A. (2016). Neurophysiological markers of multiple facets of impulsivity. *Biological Psychology, 115*, 64–68. doi:10.1016/j.biopsycho.2012.05.005

Neal, L. B. & Gable, P. A. (2019). Shifts in frontal asymmetry underlying impulsive and controlled decision-making. *Biological Psychology, 140*, 28–34.

Nigg, J. T. (2006). Temperament and developmental psychopathology. *Journal of Child Psychology and Psychiatry, 47*(3-4), 395–422. doi:10.1111/j.1469-7610.2006.01612.x

Nitschke, J. B., Heller, W., Etienne, M. A., & Miller, G. A. (2004). Prefrontal cortex activity differentiates processes affecting memory in depression. *Biological Psychology, 67*(1–2), 125–143.

Nusslock, R., Harmon-Jones, E., Alloy, L. B., Urosevic, S., Goldstein, K., & Abramson, L. Y. (2012). Elevated left mid-frontal cortical activity prospectively predicts conversion to bipolar *I disorder. Journal of Abnormal Psychology, 121*(3), 592–601. doi:10.1037/a0028973

Nusslock, R., Shackman, A. J., Harmon-Jones, E., Alloy, L. B., Coan, J. A., & Abramson, L. Y. (2011). Cognitive vulnerability and frontal brain asymmetry: Common predictors of first prospective depressive episode. *Journal of Abnormal Psychology, 120*, 497–503.

Nusslock, R., Shackman, A. J., McMenamin, B. W., Greischar, L. L., Davidson, R. J., & Kovacs, M. (2018). Comorbid anxiety moderates the relationship between depression history and prefrontal EEG asymmetry. *Psychophysiology, 55*(1), e12953.

Pascual-Marqui, R. D. (2002). Standardized low-resolution brain electromagnetic tomography (sLORETA): Technical details. *Methods and Findings in Experimental and Clinical Pharmacology, 24* (Suppl D), 5–12.

Perria, P., Rosadini, G., & Rossi, G. F. (1961). Determination of side of cerebral dominance with Amobarbital. *Archives of Neurology, 4*, 175–181.

Peterson, C. K. & Harmon-Jones, E. (2009). Circadian and seasonal variability of resting frontal EEG asymmetry. *Biological Psychology, 80*, 315–320.

Peterson, C. K., Gravens, L., & Harmon-Jones, E. (2011). Asymmetric frontal cortical activity and negative affective responses to ostracism. *Social Cognitive Affective Neuroscience, 6*, 277–285. doi:10.1093/scan/nsq027

Peterson, C. K., Shackman, A. J., & Harmon-Jones, E. (2008). The role of asymmetric frontal cortical activity in aggression. *Psychophysiology, 45*, 86–92.

Pizzagalli, D. A., Sherwood, R. J., Henriques, J. B., & Davidson, R. J. (2005). Frontal brain asymmetry and reward responsiveness: A source-localization study. *Psychological Science, 16*(10), 805–813. doi:10.1111/j.1467-9280.2005.01618.x

Pizzagalli, D., Shackman, A. J., & Davidson, R. J. (2003). The functional neuroimaging of human emotion: Asymmetric contributions of cortical and subcortical circuitry. In K. Hugdahl & R. J. Davidson (Eds.), *The asymmetric brain* (pp. 511–532). MIT Press.

Plutchik, R. (1980). *Emotion: A psychoevolutionary synthesis.* HarperCollins College Division.

Pössel, P., Lo, H., Fritz, A., & Seemann, S. (2008). A longitudinal study of cortical EEG activity in adolescents. *Biological Psychology, 78*(2), 173–178. https://doi.org/https://doi.org/10.1016/j.biopsycho.2008.02.004

Price, T. F. & Harmon-Jones, E. (2011). Approach motivational body postures lean toward left frontal brain activity. *Psychophysiology, 48*, 718–722.

Price, T. F., Hortensius, R., & Harmon-Jones, E. (2013). Neural and behavioral associations of manipulated determination facial expressions. *Biological Psychology, 94*, 221–227. http://dx.doi.org/10.1016/j.biopsycho.2013.06.001

Proudfit, G. H., Inzlicht, M., & Mennin, D. S. (2013). Anxiety and error monitoring: The importance of motivation and emotion. *Frontiers in Human Neuroscience, 7*, 636. doi:10.3389/fnhum.2013.00636

Putman, P., Hermans, E., & van Honk, J. (2004). Emotional Stroop performance for masked angry faces: It's BAS, not BIS. *Emotion, 4*, 305–311.

Quaedflieg, C. W. E. M., Smulders, F. T. Y., Meyer, T., Peeters, F. P. M. L., Merckelbach, H. L. G. J., & Smeets, T. (2016). The validity of individual frontal alpha asymmetry EEG neurofeedback. *Social Cognitive and Affective Neuroscience, 11*, 33–43. Doi: 10.1093/scan/nsv090

Quirin, M., Gruber, T., Kuhl, J., & Düsing, R. (2013). Is love right? Prefrontal resting brain asymmetry is related to the affiliation motive. *Frontiers in Human Neuroscience, 7*, 902. doi:10.3389/ fnhum.2013.00902

Reid, S. A., Duke, L. M., & Allen, J. J. B. (1998). Resting frontal electroencephalographic asymmetry in depression: Inconsistencies suggest the need to identify mediating factors. *Psychophysiology, 35*, 389–404.

Robinson, R. G. & Price, T. R. (1982). Post-stroke depressive disorders: A follow-up study of 103 patients. *Stroke, 13*, 635–641.

Rodrigues, J., Müller, M., Mühlberger, A., & Hewig, J. (2018). Mind the movement: Frontal asymmetry stands for behavioral motivation, bilateral frontal activation for behavior. *Psychophysiology, 55*(1), e12908. https://doi.org/10.1111/psyp.12908

Rossi, G. F. & Rosadini, G. R. (1967). Experimental analyses of cerebral dominance in man. In D. H. Millikan & F. L. Darley (Eds.), *Brain mechanisms underlying speech and language* (pp. 167–184). Grune & Stratton.

Rothbart, M. K. & Rueda, M. R. (2005). The development of effortful control. In U. Mayr, E. Awh, & S. W. Keele (Eds.), *Developing individuality in the human brain: A tribute to Michael I. Posner* (pp. 167–188). American Psychological Association.

Rothbart, M. K., Ellis, L. K., & Posner, M. I. (2004). Temperament and self-regulation. In K. D. Vohs & R. F. Baumeister (Eds.), *Handbook of self-regulation: Research, theory, and applications* (pp. 284–299). Guilford Press.

Sackeim, H., Greenberg, M. S., Weimen, A. L., Gur, R. C., Hungerbuhler, J. P., & Geschwind, N. (1982). Hemispheric asymmetry in the expression of positive and negative emotions: Neurologic evidence. *Archives of Neurology*, *39*, 210–218.

Salavert, J., Caseras, X., Torrubia, R., Furest, S., Arranz, B., Dueñas, R., & San, L. (2007). The functioning of the Behavioral Activation and Inhibition Systems in bipolar I euthymic patients and its influence in subsequent episodes over an eighteen-month period. *Personality and Individual Differences*, *42*(7), 1323–1331.

Santesso, D. L., Segalowitz, S. J., Ashbaugh, A. R., Antony, M. M., McCabe, R. E., & Schmidt, L. A. (2008). Frontal EEG asymmetry and sensation seeking in young adults. *Biological Psychology*, *78*, 164–172.

Schaffer, C. E., Davidson, R. J., & Saron, C. (1983). Frontal and parietal electroencephalogram asymmetry in depressed and nondepressed subjects. *Biological Psychiatry*, *18*, 753–762.

Schiff, B. B. & Lamon, M. (1989). Inducing emotion by unilateral contraction of facial muscles: A new look at hemispheric specialization and the experience of emotion. *Neuropsychologia*, *27*, 923–935.

Schiff, B. B. & Lamon, M. (1994). Inducing emotion by unilateral contraction of hand muscles. *Cortex*, *30*, 247–254.

Schmeichel, B. J., Crowell, A., & Harmon-Jones, E. (2016). Exercising self-control increases relative left frontal cortical activation. *Social Cognitive and Affective Neuroscience*, *11*, 282–288.

Schöne, B., Schomberg, J., Gruber, T., & Quirin, M. (2016). Event-related frontal alpha asymmetries: Electrophysiological correlates of approach motivation. *Experimental Brain Research*, *234*(2), 559–567. doi:10.1007/s00221-015-4483-6

Schutter, D. J. L. G. (2009). Transcranial magnetic stimulation. In E. Harmon-Jones & J. S. Beer (Eds.), Methods in social neuroscience (pp. 233–258). Guilford Press.

Schutter, D. J. L. G. & Harmon-Jones, E. (2013). The corpus callosum: A commissural road to anger and aggression. *Neuroscience & Biobehavioral Reviews*, *37*, 2481–2488. http://dx.doi.org/10.1016/j.neubiorev.2013.07.013

Schutter, D. J. L. G., van Honk, J., d'Alfonso, A. A. L., Postma, A., & de Haan, E.H.F. (2001). Effects of slow rTMS at the right dorsolateral prefrontal cortex on EEG asymmetry and mood. *Neuroreport*, *12*, 445–447.

Shankman, S. A., Klein, D. N., Tenke, C. E., & Bruder, G. E. (2007). Reward sensitivity in depression: A biobehavioral study. *Journal of Abnormal Psychology*, *116*(1), 95–104

Simons, J. S., Dvorak, R. D., Batien, B. D., & Wray, T. B. (2010). Event-level associations between affect, alcohol intoxication, and acute dependence symptoms: Effects of urgency, self-control, and drinking experience. *Addictive Behaviors*, *35*(12), 1045–1053. doi:10.1016/j.addbeh.2010.07.001

Siniatchkin, M., Kropp, P., & Gerber, W-D. (2000). Neurofeedback—The significance of reinforcement and the search for an appropriate strategy for the success of self-regulation. *Applied Psychophysiology and Biofeedback*, *25*, 167–175.

Smith E. E., Cavanagh, J. F., & Allen, J. J. B. (2018). Intracranial source activity (eLORETA) related to scalp-level asymmetry scores and depression status. *Psychophysiology*, *55*(1), 119–134. https://doi.org/10.1111/psyp.13019

Smith, E. E., Zambrano-Vazquez, L., & Allen, J. J. B. (2016). Patterns of alpha asymmetry in those with elevated worry, trait anxiety, and obsessive-compulsive symptoms: A test of the worry and avoidance models of alpha asymmetry. *Neuropsychologia*, *85*, 118–126. https://doi.org/10.1016/j.neuropsychologia.2016.03.010

Smits, D. J. M. & Kuppens, P. (2005). The relations between anger, coping with anger, and aggression, and the BIS/BAS system. *Personality and Individual Differences*, *39*, 783–793.

Spielberger, C. D. (1988). *State-trait anger expression inventory*. Psychological Assessment Resources.

Spielberger, C. D., Reheiser, E. C., & Sydeman, S. J. (1995). Measuring the experience, expression, and control of anger. In H. Kassinove (Ed.), *Anger disorders: Definition, diagnosis, and treatment* (pp. 49–67). Taylor & Francis.

Stewart, J. L. & Allen, J. J. (2018). Resting frontal brain asymmetry is linked to future depressive symptoms in women. *Biological Psychology*, *136*, 161–167.

Stewart, J. L., Bismark, A. W., Towers, D. N., Coan, J. A., & Allen, J. J. B. (2010). Resting frontal EEG asymmetry as an endophenotype for depression risk: Sex-specific patterns of frontal brain asymmetry. *Journal of Abnormal Psychology*, *119*(3), 502–512.

Stewart, J. L., Coan, J. A., Towers, D. N., & Allen, J. J. B. (2011). Frontal EEG asymmetry during emotional challenge differentiates individuals with and without lifetime major depressive disorder. *Journal of Affective Disorders*, *129*(1), 167–174. https://doi.org/https://doi.org/10.1016/j.jad.2010.08.029

Stewart, J. L., Coan, J. A., Towers, D. N., & Allen, J. J. B. (2014). Resting and task-elicited prefrontal EEG alpha asymmetry in depression: Support for the capability model. *Psychophysiology*, *51*(5), 446–455. https://doi.org/10.1111/psyp.12191

Strack, F. & Deutsch, R. (2004). Reflective and impulsive determinants of social behavior. *Personality and Social Psychology Review*, *8*(3), 220–247. doi:10.1207/s15327957pspr0803_1

Sutton, S. K. & Davidson, R. J. (1997). Prefrontal brain asymmetry: A biological substrate of the behavioral approach and inhibition systems. *Psychological Science*, *8*, 204–210.

Teper, R. & Inzlicht, M. (2013). Meditation, mindfulness and executive control: The importance of emotional acceptance and brain-based performance monitoring. *Social, Cognitive, and Affective Neuroscience*, *8*(1), 85–92. doi:10.1093/scan/nss045

Terzian, H. & Cecotto, C. (1959). Determination and study of hemisphere dominance by means of intracarotid sodium amytal injection in man: II. Electroencephalographic effects. *Bollettino della Societa Italiana Sperimentale*, *35*, 1626–1630.

Thibodeau, R., Jorgensen, R. S., & Kim, S. (2006). Depression, anxiety, and resting frontal EEG asymmetry: A meta-analytic review. *Journal of Abnormal Psychology*, *115*(4), 715–729. doi:10.1037/ 0021-843X.115.4.715.

Tomarken, A. J. & Davidson, R. J. (1994). Frontal brain activation in repressors and nonrepressors. *Journal of Abnormal Psychology*, *103*(2), 339–349. doi:10.1037/0021-843X.103.2.339

Tomarken, A. J., Davidson, R. J., Wheeler, R. E., & Doss, R. (1992). Individual differences in anterior brain asymmetry and fundamental dimensions of emotion. *Journal of Personality and Social Psychology*, *62*, 676–687.

Urosevic, S., Abramson, L. Y., Alloy, L. B., Nusslock, R., Harmon-Jones, E., Bender, R., & Hogan, M. E. (2010). Increased rates of events that activate or deactivate the Behavioral Approach System, but not events related to goal attainment, in bipolar spectrum disorders. *Journal of Abnormal Psychology*, *119*, 610–615.

Valiente, C., Eisenberg, N., Smith, C. L., Reiser, M., Fabes, R. A., Losoya, S., ... Murphy, B. C. (2003). The relations of effortful control and reactive control to children's externalizing problems: A longitudinal assessment. *Journal of Personality*, *71*(6), 1171–1196. doi:10.1111/1467-6494.7106011

van de Laar, M. C., Licht, R., Franken, I. H. A., & Hendriks, V. M. (2004). Event-related potentials indicate motivational relevance of cocaine cues in abstinent cocaine addicts. *Psychopharmacology*, *177*, 121–129.

Verona, E., Sadeh, N., & Curtin, J. J. (2009). Stress-induced asymmetric frontal brain activity and aggression risk. *Journal of Abnormal Psychology*, *118*, 131–145.

Wacker, J., Chavanon, M. L., Leue, A., & Stemmler, G. (2008). Is running away right? The behavioral activation-behavioral inhibition model of anterior asymmetry. *Emotion*, *8*(2), 232–249. doi:10.1037/1528-3542.8.2.232

Wacker, J., Chavanon, M. L., & Stemmler, G. (2010). Resting EEG signatures of agentic extraversion: New results and meta-analytic integration. *Journal of Research in Personality*, *44*(2), 167–179. doi:10.1016/j.jrp.2009.12.004

Wacker, J., Heldmann, M., & Stemmler, G. (2003). Separating emotion and motivational direction in fear and anger: Effects on frontal asymmetry. *Emotion*, *3*, 167–193.

Wacker, J., Mueller, E. M., Pizzagalli, D. A., Hennig, J., & Stemmler, G. (2013). Dopamine-D2-receptor blockade reverses the association between trait approach motivation and frontal asymmetry in an approach-motivation context. *Psychological Science*, *24*, 489–497.

Whiteside, S. P., Lynam, D. R., Miller, J. D., & Reynolds, S. K. (2005). Validation of the UPPS impulsive behaviour scale: A four-factor model of impulsivity. *European Journal of Personality*, *19*(7), 559–574. doi:10.1002/per.556

Wiedemann, G., Pauli, P., Dengler, W., Lutzenberger, W., Birbaumer, N., & Buchkremer, G. (1999). Frontal brain asymmetry as a biological substrate of emotions in patients with panic disorders. *Archives of GENERAL PSYCHIATRY*, *56*(1), 78–84.

Young, P. T. (1943). *Emotion in man and animal: Its nature and relation to attitude and motive.* John Wiley and Sons.

Zapolski, T. C., Cyders, M. A., & Smith, G. T. (2009). Positive urgency predicts illegal drug use and risky sexual behavior. *Psychology of Addictive Behaviors*, *23*(2), 348. doi:10.1037/a0014684

Zinner, L. R., Brodish, A. B., Devine, P. G., & Harmon-Jones, E. (2008). Anger and asymmetric frontal cortical activity: Evidence for an anger-withdrawal relationship. *Cognition & Emotion*, *22*, 1081–1093.

CHAPTER 12

OSCILLATORY ACTIVITY IN SENSORIMOTOR FUNCTION

BERNADETTE C. M. VAN WIJK

12.1 Understanding the Sensorimotor System

Movement is quintessential for most living organisms including humans. It is through movement that we feed ourselves, reproduce, communicate, perceive, and affect the external world. Most of our daily movements are performed without giving them much thought. We lift our arm to reach for a cup of coffee, grasp our fingers around it with enough force to hold the cup, and bring it to our mouth in order to drink without spilling—all seemingly without any effort. Even this arguably simple action requires transformation of an action goal into a movement trajectory, activation and deactivation of several muscle groups at appropriate moments in both space and time, and relies on accurate integration of visual information on the shape and position of the cup, proprioceptive input from muscles and joints on position and velocity of the limbs, and haptic feedback from which we can infer the cup's weight, to plan and adjust muscle activations. Fortunately, we can acquire complex motor skills through practice. We can learn how to coordinate our finger movements to efficiently type a text on a computer keyboard, or even to play the most virtuoso piano pieces.

The sensorimotor system is the network of motor and sensory regions in the brain and spinal cord thought to be involved in the planning, initiation, and execution of movements. In humans, it comprises primary somatosensory cortex (S1), primary motor cortex (M1), dorsal and ventral premotor cortex (PMd, PMv), (pre-)supplementary motor area (SMA), cingulate motor areas (CMA), cerebellum, the basal ganglia, several nuclei in brainstem and thalamus, motoneurons and interneurons in the spinal cord, and muscle afferents (Figure 12.1). Roughly speaking, one can distinguish a medial pathway of fiber tracts descending from cortex and brainstem nuclei down the spinal

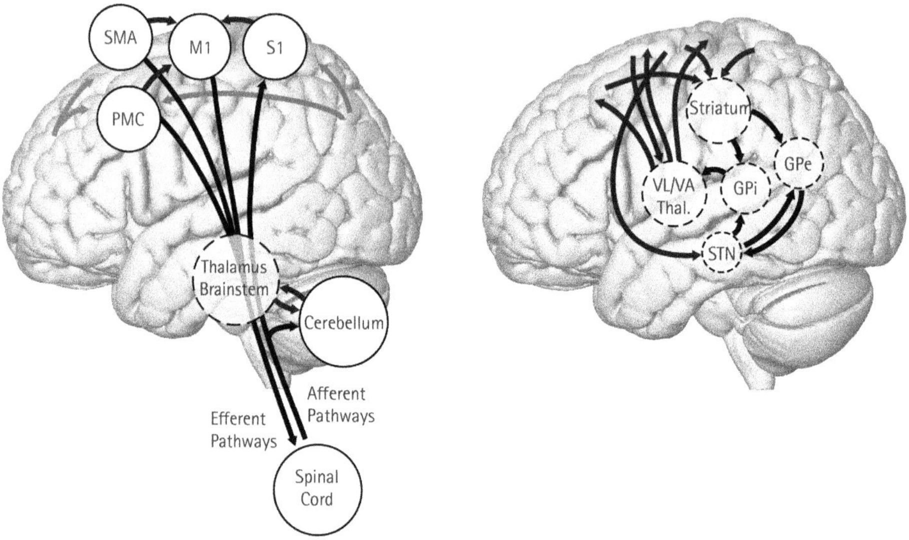

FIGURE 12.1 Schematic overview of major cortical and subcortical pathways that comprise the sensorimotor system. Left: primary and secondary motor cortex project directly to the spinal cord and via brainstem nuclei. Afferent proprioceptive information travels to primary somatosensory cortex via thalamus and enters the cerebellum directly. Right: The basal ganglia consist of the striatum, internal and external pallidum (GPi, GPe), and subthalamic nucleus (STN). Together with the thalamus they form circuits with various cortical regions, of which the ventral lateral and ventral anterior thalamus (VL, VA) project to primary and secondary motor cortex.

cord that innervate axial muscles for the regulation of posture and balance, and a lateral tract innervating distal muscles. The sensorimotor system has a bilateral organization, with most corticospinal tracts crossing at the level of the medulla. This suggests that movements of the right arm and leg are primarily controlled by the left hemisphere, and vice versa. Movements and actions are intricately linked with perception and cognition; therefore, the sensorimotor system must uphold strong ties with other circuits. As such, the distinction between systems is not always clear and some sensorimotor regions have been ascribed more cognitive roles.

As much as one would like to understand how a gymnast performs a twisting somersault, the study of (supraspinal) neural control of movement in laboratory settings is typically restricted to what is often referred to as "simple" uni- or bimanual movements like finger tapping or isometric force generation. Electroencephalography (EEG), magnetoencephalography (MEG), and electrocorticography (ECoG) are predominantly sensitive to synchronous synaptic transmission in cortical regions, while electromyography (EMG) may provide insight into the activation of motoneurons in the spinal cord via the spread of action potential currents that can be picked up non-invasively with electrodes placed on the skin over the muscle belly. Increasing use of deep brain stimulation (DBS) for the treatment of movement disorders allows for invasively recording local field potentials (LFPs) from basal ganglia and thalamic structures in humans. It is even possible to combine recording techniques in a single experiment to investigate distant interactions between different parts of the sensorimotor system. For example, simultaneous use of EEG/MEG and EMG allows for the detection of functional interactions between cortex and spinal cord, and invasive LFP recordings from DBS electrodes can be combined with non-invasive EEG or MEG to detect functional interactions between cortex and basal ganglia structures like the subthalamic nucleus (STN) and internal pallidum (GPi). Performed movements can be co-registered via button-presses, force sensors, trackpads, joy sticks, or motion-capturing systems. By investigating how recorded neural signals change during the experimental task or due to pathology one could relate oscillatory activity to motor function.

Movement-related neural population activity appears in several frequency bands (Figure 12.2). Many studies contributed to understanding the functional role of these frequency bands in individual sensorimotor regions. Via measures of coupled oscillatory activity (e.g., *coherence*) one could get a sense of functional connectivity within the system. This chapter highlights several research findings and (ongoing) debates on which we build our current knowledge. While it covers many topics, it is meant as introductory text rather than an exhaustive literature review. Although we have studied oscillations in sensorimotor function for many decades, it is exciting to see that the field is still evolving with new technologies, signal processing techniques, and novel experimental paradigms.

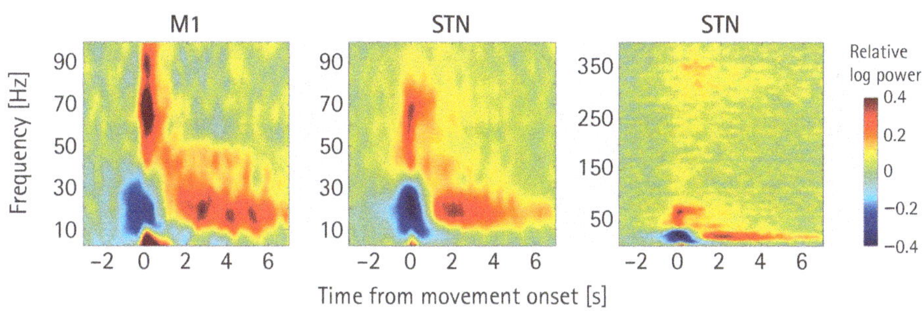

FIGURE 12.2 Movement-related time-frequency spectra for primary motor cortex (M1) and subthalamic nucleus (STN). Panels show the relative change in spectral power with respect to a pre-movement baseline window. Participants performed a simple self-paced button press movement with three fingers simultaneously at time point 0. Beta power starts to decrease before movement onset (ERD) and shows a clear post-movement rebound (ERS). Gamma power increases around movement onset. In STN, another movement-related increase with smaller amplitude can be observed in the range for high-frequency oscillations (HFOs). Note how time-frequency modulations look very similar for M1 and STN, suggesting generic principles of information processing.

Spectra are based on MEG and DBS-LFP recordings from eleven Parkinson's disease patients on dopaminergic medication taken from Litvak et al., 2012.

12.2 A TRADITIONAL VIEW ON FREQUENCY BANDS

Ever since the first recordings, researchers have divided EEG/MEG time series into frequency bands that are presumed to have their own functional roles. This section focuses on the properties of frequency bands that are thought to be central to sensorimotor function.

12.2.1 The Difference Between Alpha, Mu, and Beta

The alpha rhythm was the first rhythm observed by Berger (1929). Soon after, it was discovered that the alpha rhythm is suppressed when closing the eyes, has its origin in the occipital lobes (Adrian & Matthews, 1934), and that another alpha rhythm can be observed close to the central sulcus that, together with a beta rhythm, is suppressed upon tactile stimulation (Jasper & Andrews, 1938). Later studies showed that the central rhythms are also modulated by passive, voluntary, and imagined movements (Chatrian et al., 1959; Gastaut, 1952; Jasper & Penfield, 1949). The alpha rhythm close to the central sulcus (or "Rolandic fissure") was named *mu*—after the arch-like shape of the oscillation. However, in the current literature mu and alpha are used interchangeably. Mu/alpha oscillations are usually defined to have a peak frequency between 8–13 Hz. Beta oscillations are considered to occur within the approximate range of 13–30 Hz. Both alpha/mu and beta oscillations show a distinctive decrease in amplitude prior to and during movement execution (event-related desynchronization, ERD), followed by a rebound that exceeds baseline after the movement has been terminated (event-related synchronization, ERS) (Pfurtscheller & Lopes Da Silva, 1999). Still, dissimilarities in time course, spatial origin, and coupling to other regions indicate that the two bands are functionally distinct.

Beta ERD might start as early as 2 s before movement initiation and is often thought to reflect a general state of motor preparation (Neuper & Pfurtscheller, 2001). It starts over sensorimotor cortex contralateral to the moving hand and becomes bilateral during movement. Alpha/mu ERD starts a bit later in time and is spatially more diffuse and somatotopically less specific (Crone et al., 1998b). Beta ERS is often more pronounced and occurs at a shorter latency compared alpha ERS (Alegre et al., 2003; Erbil & Ungan, 2007; Pfurtscheller et al., 1996; Salmelin & Hari, 1994). The two rhythms are likely to originate from distinct sources as different source localization methods ascribed beta modulations to precentral (motor) cortex and mu/alpha modulations to postcentral (somatosensory) cortex (Cheyne et al., 2003; Ritter et al., 2009; Salmelin et al., 1995; Salmelin & Hari, 1994). As explained in the subsequent sections, the beta band often shows the most specific modulations with experimental conditions and pathology and dominates the functional connectivity profile for sensorimotor cortex. For these reasons, beta oscillations are considered most associated with motor control.

12.2.2 A Prominent Role for Beta Oscillations

The primary functional role of beta oscillations seems to be the facilitation and inhibition of movements. Beta ERD already starts during the movement preparation phase and appears stronger when the action that needs to be prepared is known in advance. The ERD is lateralized with stronger suppression over contralateral M1 when the response hand is known, also resulting in faster reaction times (Doyle et al., 2005b; van Wijk et al., 2009). On the other hand, ERS only appears after the movement has been terminated, and therefore has been proposed to reflect a period of inactivation to recover from previous activation (Pfurtscheller et al., 1996), or processing of afferent somatosensory information (Cassim et al., 2001). Chen and colleagues (1998) tracked the time course of corticospinal excitability around self-paced and instructed movements using transcranial magnetic stimulation. Pulses were delivered at different time points around movement onset to see how the amplitude of motor evoked potentials varied during the time windows of ERD and ERS. An increase in corticospinal excitability was found in the final 100 ms of the pre-movement preparation phase and around 100 ms after movement onset, which are time periods during which ERD is often strongest. By contrast, a decrease in corticospinal excitability was found between 500 and 1000 ms after movement, which is around the time of ERS.

Beta oscillations are not simply an idling state, but also have a functional impact on the initiation and stabilization of motor output. Movements are performed more slowly when initiated during periods of high-amplitude beta oscillations in the ongoing EEG (Gilbertson et al., 2005). Entrainment of cortical beta oscillations by transcranial alternating-current stimulation at 20 Hz has a similar effect (Pogosyan et al., 2009). Cortico-spinal beta phase synchronization increases when the current motor output needs to be maintained (van Wijk et al., 2009), for example, in anticipation of upcoming perturbations to finger position (Androulidakis et al., 2007a). This implies that both a down- and up-regulation of beta oscillations is employed by the nervous system to tune motor output.

What happens at a neurobiological level that determines the amplitude of beta oscillations? Motor cortical architecture consists of intricately and reciprocally connected excitatory pyramidal cells and inhibitory interneurons within and across layers (Keller, 1993). In order to determine the contribution of individual receptor types to network activity they can be activated or blocked with pharmacological agents in in vitro slice preparations of animal cortex. This revealed $GABA_A$ receptors to be critical for the generation of beta oscillations together with an influence of gap junctions, but not AMPA receptors (Yamawaki et al., 2008). There are strong indications that beta oscillations originate in layer V while also clearly present in layer II. IPSPs on layer 5 pyramidal cells are phase locked with the LFP beta oscillation, as is the spiking of pyramidal cells but at much sparser rates (Lacey et al., 2014). The dependence of beta oscillation amplitude on $GABA_A$ receptors has also been established through in vivo pharmacological studies in humans. After administration of benzodiazepines, a class of GABAergic drugs often prescribed as anticonvulsants, sedatives, or muscle relaxants, spectral beta power in EEG/MEG appears enhanced (Baker & Baker, 2003; Hall et al., 2010; Jensen et al., 2005). Along these lines, computational modelling work suggests that stronger synaptic inputs between pyramidal cells in different layers might underlie the beta suppression that is observed during movement (Bhatt et al., 2016).

Despite the prominence of beta band oscillations in sensorimotor function, movement-related modulations only seem to encode general motor aspects. More forceful movements induce stronger alpha/mu and beta ERD (Mima et al., 1999; Stančák et al., 1997; Stančák & Pfurtscheller, 1996) as do movements that are performed with a higher frequency (Toma et al., 2002) or involving more complex sequences (Hummel et al., 2003; Manganotti et al., 1998). Movements with more muscle mass involved do not influence ERD but rather lead to stronger beta ERS (Pfurtscheller et al., 1998; Stančák et al., 2000). Movement duration has little effect on either ERD or ERS (Cassim et al., 2000; Stančák & Pfurtscheller, 1996). It is however difficult to pinpoint ERD and ERS patterns to specific motor parameters.

12.2.3 Prokinetic Gamma

To find a mechanistic link between beta oscillations and the control of movement trajectory and muscle force, it can be useful to consider the behavior of individual neurons. Pyramidal tract neurons (PTNs) in infragranular layers project to motoneurons of individual muscles and groups of muscles in the spinal cord, and appear to be functionally organized in small clusters (Asanuma et al., 1979). They only form a minority of cells in motor cortex and form intricate connections with other neurons within and between cortical modules that together encode movement patterns (Keller, 1993). Microelectrode recordings in sensorimotor cortices of the macaque monkey revealed spikes of individual neurons to be phase-locked to the LFP beta oscillation during time periods when the oscillation is well pronounced (Baker et al., 1997; Denker et al., 2007; Murthy & Fetz, 1996a). Although the phase locking of individual PTNs to the beta oscillation may only be weak, summation over a population of neurons can give a clearer picture (Baker et al., 2003). We also know that the encoding of parameters such as hand position, direction of motion, velocity, and force emerges from the joint firing rates of a group of neurons that are individually tuned (Ashe & Georgopoulos, 1994; Fu et al., 1995; Georgopoulos et al., 1986; Moran & Schwartz, 2017; Paninski et al., 2004). This is termed *population coding*.

When movements are executed, phase locking of spikes with the beta oscillation drops and spike rates strongly increase (Baker et al., 2001; Spinks et al., 2008). There is a crude inverse correlation between spike rates and LFP beta power. Spike rates reach frequencies above 30 Hz (Baker et al., 2001; Grammont and Riehle, 2003) and may well correspond to the brief increases in gamma power (~30–100 Hz) that can be observed in EEG and MEG recordings in a time window around movement onset (Cheyne et al., 2008; Muthukumaraswamy, 2010; Ohara et al., 2001; Pfurtscheller et al., 1993; Pfurtscheller & Neuper, 1992). Unlike the alpha/mu and beta ERD, the increase in gamma only occurs in the hemisphere contralateral to the moving body part and is somatotopically more focused (Crone et al., 1998a; Miller et al., 2007; Szurhaj et al., 2005). Some studies report the gamma amplitude increase to vary with movement direction in invasive recordings (Leuthardt et al., 2004; Rickert et al., 2005). However, it remains to be seen whether gamma power in EEG and MEG recordings has the same specificity. Taken together, one may hypothesize that the beta oscillation inhibits, or constrains, neuronal spiking. That is, a suppression of the beta rhythm can lead to an increase in excitability of individual neurons that together encode movement parameters.

12.2.4 High-Frequency Oscillations

The latest addition to the repertoire of movement-related oscillations is a spectral component with a clear peak in the 150–400 Hz frequency range, aptly coined *high-frequency oscillations* (HFO). HFO have been observed in LFP recordings from DBS electrodes in the STN and internal pallidum. They are of interest to sensorimotor function as they show a characteristic increase in amplitude during movement (Foffani et al., 2003; Litvak et al., 2012; López-Azcárate et al., 2010; Tan et al., 2013; Tsiokos et al., 2013), like the gamma band. Their peak frequency is modulated by dopaminergic medication (López-Azcárate et al., 2010; Özkurt et al., 2011; van Wijk et al., 2016) and indicative of tremor symptoms in patients with Parkinson's disease (Hirschmann et al., 2016). Yet, a similar movement-related HFO component in the cortex has not been reported although this may simply be due to the poor signal-to-noise ratio of commonly used EEG/MEG compared to more focal, invasive techniques.

12.3 Coherence as a Measure of Functional Connectivity within the Sensorimotor System

Any understanding of sensorimotor function will not be complete if the information exchange between regions that comprise the sensorimotor system is ignored. Rhythmic synchronization in the form of oscillations has been recognized as a means by which neural populations selectively gate their sensitivity to input from other populations (Fries, 2005). The degree of synchronization between neural populations can be estimated by computing *coherence* between time series as a measure of functional connectivity. This measure can be understood as spectral counterpart of conventional correlation.

12.3.1 Cortico-Spinal Coherence

Corticospinal coherence, also called corticomuscular coherence, refers to the synchronization between oscillations in cortex and motoneuron activity in the spinal cord. Firing of motoneurons yields motor unit action potentials that induce contraction of muscle fibers. Repetitive firing is necessary in order to build up force. The rate at which this occurs ranges from 6 Hz in rest to 35 Hz during forceful isometric contractions and bursts of 80–120 Hz in case of rapid, ballistic movements (Freund, 1983). As such, corticospinal coherence may reflect whether synchronized activity in cortex also reaches the muscles.

Several studies report weak but significant beta-band coherence between the EMG of hand or foot muscles and EEG over contralateral sensorimotor regions during sustained isometric muscle contractions (Conway et al., 1995; Gross et al., 2000; Halliday et al., 1998; Salenius et al., 1997; van Wijk et al., 2012). Coherence levels increase for muscle contractions with higher force levels (Chakarov et al., 2009; Witte et al., 2007) and peak frequencies shift into the gamma range during maximal force production (Brown et al., 1998; Mima et al., 1999). Task-dependent coherence has also been found for the EEG with different muscle groups, thereby forming functional synergies (Zandvoort et al., 2019), or between EMG recordings of different muscles themselves (Boonstra et al., 2016). While beta-band corticospinal coherence diminishes during movement (Baker et al., 1997; Kilner et al., 2000), alpha-band coherence between active muscle groups increases (Boonstra et al., 2009). Typically, no significant coherence can be observed in rest unless a strong tremor is present (Hellwig et al., 2001).

Corticospinal beta coherence can be modulated by cognitive factors. The division of attention during dual task performance decreases coherence levels (Johnson et al., 2011; Kristeva-Feige et al., 2002; Safri et al., 2007), whereas task instructions to maintain force output at a target level with high precision lead to increased coherence (Kristeva et al., 2007). Using a pre-cued choice reaction time task, Van Wijk and colleagues (2009) demonstrated that corticospinal coherence can be up-regulated in anticipation of an upcoming movement decision. Key to their experimental paradigm was the pre-activation of muscles in each trial via a precision grip before stimuli were displayed. This led to a build-up of corticospinal coherence that allows for modulations to be observable. After presentation of a warning cue of the likely upcoming response hand, corticospinal synchronization increased for the non-selected hand, while cortical beta power decreased for the selected hand. These and other findings jointly suggest that the role of beta oscillations to facilitate or inhibit movements extends to the level of the spinal cord.

Does corticospinal coherence reflect mere entrainment of motoneurons by cortical output? Feedback signals from muscle afferents ascend the spinal cord to somatosensory cortex, which through strong connections with primary motor cortex, closes the corticospinal loop. Riddle and Baker (2005) sought to tackle this question experimentally by cooling subjects' arms. Cooling caused an additional time delay (inferred from phase-frequency regression) that was about twice the conduction time in one direction.

The authors therefore concluded that both ascending and descending pathways contribute to the occurrence of corticospinal coherence. One may determine the directionality of coupling by using measures such as Granger causality. This similarly revealed significant contributions of both ascending and descending pathways (Witham et al., 2011). It is worth noting that corticospinal coherence does not strictly follow modulations of cortical power. Baker and Baker (2003) demonstrated this by recording subjects before and after intake of diazepam, a benzodiazepine that enhances inhibitory post-synaptic potentials via $GABA_A$ receptors. EEG beta power doubled in amplitude, but corticospinal coherence was little altered. Interactions between cortex and spinal cord might therefore be more complex than a simple one-way drive.

12.3.2 Cortico-Subcortical Coherence

With EEG and MEG being most sensitive to cortical sources, it is easy to forget the contribution of more deeply located brain structures. Many of these are highly relevant for sensorimotor and cognitive functions. DBS treatment allows for invasively recording electrophysiological activity from subcortical structures in humans. The subthalamic nucleus (STN) and internal pallidum (GPi) are primary targets for the treatment of Parkinson's disease and dystonia, the pedunculopontine nucleus (PPN) for postural instability and gait freezing, and the ventrolateral thalamus (including ventral intermediate nucleus) for tremor. It is possible, albeit challenging, to combine LFP recordings from DBS electrodes with simultaneous EEG or MEG to study cortico-subcortical interactions.

Early simultaneous LFP-MEG studies sought to map functional connectivity with cortical regions across different frequency ranges during rest. Litvak and colleagues (2011) and Hirschmann and colleagues (2011) showed converging evidence for the presence of two spatially distinct and frequency-specific networks for the STN and ipsilateral cortical areas: alpha-band coherence with temporoparietal cortex and beta-band coherence with pre-motor cortex. Coherence values for these sources were little altered with dopaminergic medication. Using directionality analysis, cortical activity was found to drive STN activity for both frequency bands (Litvak et al., 2011). Both research labs followed up on these studies by investigating movement-related coherence but found more conflicting results. Litvak and colleagues (2012) reported an increase in gamma band coherence during movement execution that was further increased by dopaminergic mediation. Beta coherence increased during the post-movement beta rebound period but was unaltered by medication. In a separate study, they describe a reduction of alpha-band coherence during movement that was more pronounced with medication (Oswal et al., 2013). By contrast, Hirschmann and colleagues (2013) found that medication reduced beta-band coherence during movement but not alpha.

Simultaneous LFP-MEG recordings identified functional networks also for other subcortical structures. Activity in PPN appeared coherent with that of the brainstem

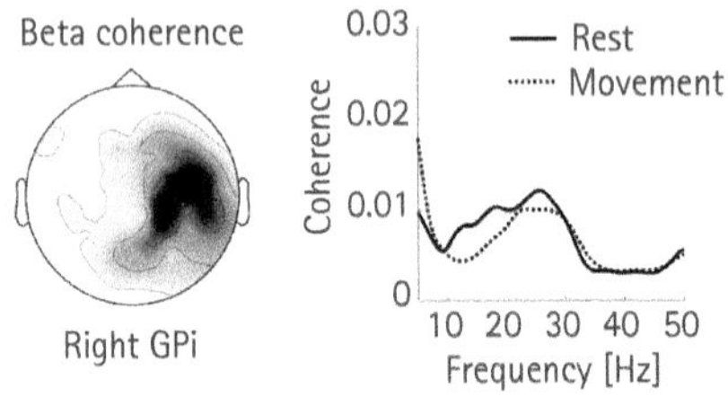

FIGURE 12.3 Beta band coherence between the right internal pallidum (GPi) and ipsilateral sensorimotor cortex. Left: Topography with darker colors indicating stronger coherence values. Right: Coherence is reduced during movement compared to rest.

Both panels are based on MEG and DBS-LFP recordings from eight dystonia patients taken from Van Wijk et al., 2017.

and cingulate cortex in the alpha band, and with medial cortical motor areas in the beta band (Jha et al., 2017). For GPi, coherence was found with temporal cortex in the theta band, cerebellum in the alpha band, and sensorimotor cortex in the beta band (Neumann et al., 2015). This study was conducted in dystonia patients, for whom disease severity significantly correlated only with theta band coherence. Beta-band coherence was unrelated to symptoms but showed a movement-related reduction (van Wijk et al., 2017; see Figure 12.3). The lack of clear alterations of cortico-subcortical coherence by medication or correlation with clinical symptoms suggests that this form of functional coupling might be physiological rather than disease-related. Experimental paradigms with more refined task designs will be needed to unravel their functional roles.

12.3.3 Cortico-Cortical Coherence

Many motor tasks call upon the coordinated activation of multiple cortical regions. Researchers have used coherence analysis with the aim to find traces of functional coupling between EEG/MEG time series recorded at different locations. Some of these findings confirmed inter-regional coupling patterns as one would expect from postulated sensorimotor functions of individual brain regions. For example, the observation of higher coherence values between sensorimotor cortex and mesial premotor areas for internally compared to externally paced movement is in line with the presumed involvement of SMA in self-initiated movements (Gerloff et al., 1998; Serrien, 2008). Other findings are arguably more open for interpretation, like the increased level of interhemispheric coherence between sensorimotor regions during more complex unimanual and bimanual tasks (Gerloff et al., 1998; Gross et al., 2005; Manganotti et al., 1998; Mima et al., 2000), or the observed gamma band coherence between cerebellar hemispheres during bimanual finger tapping (Pollok et al., 2007). Notably, cortico-cortical coherence can be used to reveal the integration of information from sensory modalities during motor tasks. Classen and colleagues (1998) demonstrated that beta-band coherence between motor and visual cortex is higher when participants make use of visual information to perform a visuomotor tracking task compared to when the visual stimulus is a mere distractor. Similarly, significant alpha-band coherence between auditory cortices and the motor network can be found when finger tapping is paced by a metronome (Pollok et al., 2005).

Coherence analysis between EEG/MEG time series is easily prone to volume conduction: multiple electrodes or sensors pick up the same neural activity leading to an overestimation of functional connectivity. This prompted the development of alternative connectivity measures like the *imaginary part of coherency* (Nolte et al., 2004), where the real part of the cross-spectrum is ignored as it may contain volume conduction artifacts. The authors applied this measure to EEG data of a unimanual finger tapping task and demonstrated a weak but significant 20 Hz coupling from contralateral to ipsilateral sensorimotor cortex in the time period before movement onset and in the reverse direction after movement. Another measure is the *phase lag index* (Stam et al., 2007), which eliminates instantaneous coupling by looking at the asymmetry of the relative phase distribution. Hillebrand and colleagues (2012) showed that this measure removes volume conduction effects that are still present after projecting MEG data to source space and identified a strongly connected sensorimotor network in the beta band. The disadvantage of these measures is that any true coupling with zero time-lag is also ignored, therefore their estimates are on the conservative side.

Invasive recordings reduce the problem of volume conduction if electrodes can be placed sufficiently far apart. Murthy and Fetz (1996b) studied the occurrence of synchronized oscillations in LFP signals from bilateral sensorimotor areas in the macaque monkey. Oscillations occurred spontaneously at rest but only infrequently. They appeared more often during exploratory arm movements with fine use of the fingers, during which they were frequently synchronized between M1 and S1, and between bilateral M1s. Both occurred at near zero time-lags. Oscillations, however, did not seem to occur at consistent points in time related to the movements. Other studies did observe a systematic increase in synchronous oscillations between M1, S1, and premotor areas of the same hemisphere or between bilateral M1s around or before movement onset in more standardized tasks (Cardoso de Oliveira et al., 2001; Ohara et al., 2001; Sanes & Donoghue, 1993), therefore providing support for the involvement of oscillations in functional coupling between cortical areas.

12.4 Long-term changes in oscillatory activity

Measures of sensorimotor beta oscillations appear to be highly consistent within individuals across repeated recording sessions (Espenhahn et al., 2017). However, their characteristics might change over time due to several factors.

12.4.1 Altered Oscillations Due to Development and Aging

Physiological changes occur throughout our body as we grow older. Alterations in grey matter volume, myelination, structural connectivity, neurotransmitters, receptors, and skull conductivity (when recording EEG) potentially affect amplitudes and peak frequencies of neural oscillations. In a cross-sectional study, Heinrichs-Graham and colleagues (2018) investigated the age-dependency of movement-related oscillations by measuring MEG of 57 healthy participants between the ages of 9 and 75 years. Absolute (baseline) beta power showed a quadratic relation with age, with smallest values for young adults, somewhat larger values for children, and clearly largest values for older adults. The magnitude of beta ERD linearly increased across all ages, whereas post-movement ERS was most pronounced for young adults. Similar patterns were found by other studies (Gaetz et al., 2010; Schmiedt-Fehr et al., 2016) that also reported a reduction of peak frequency for ERD (Rossiter et al., 2014) or ERS (Espenhahn et al., 2019) in older adults. Gamma ERS around movement onset has been reported to decrease in amplitude from childhood to adolescence (Trevarrow et al., 2019). While gamma ERS typically only occurs in contralateral sensorimotor cortex in adults, it may also be observed over ipsilateral cortex in younger children. Huo and colleagues (2011) reported ipsilateral gamma ERS for 12 out of 20 children in the 6–9-year-old age group and significantly fewer instances for older age groups. These findings are likely related to maturation of the corpus collosum and development of transcallosal inhibition that still continues at that age (Müller et al., 1997). Changes in oscillatory activity are hence evident throughout the lifespan; this should be considered when the experimental design includes between-group comparisons.

12.4.2 Altered Oscillations Due to Learning

On short time scales, oscillatory activity may depend on experience in the context of sensorimotor learning. Synaptic plasticity can lead to a strengthening or weakening of synaptic transmissions that may affect the amplitude of oscillations at a neural population level. Such reorganization has been observed during the course of learning new visuomotor tracking tasks or bimanual coordination patterns.

Tan and colleagues (2014) showed that the magnitude of post-movement beta ERS depends on kinematic error in a reaching task with rotational perturbations. In these experiments, single-trial beta ERS was negatively correlated with the angular discrepancy between the target movement trajectory and the actual performed trajectory. This effect was strongest when the perturbation angle was constant over trials compared to random. Hence, post-movement beta ERS does not merely reflect afferent feedback of the performed movement; rather, it represents a combination of prediction error and uncertainty of expected sensory consequences (Tan et al., 2016). Alayrangues and colleagues (2019) suggested that reorganization of the internal representation itself is not reflected by the post-movement ERS but by lower beta amplitudes during the preparatory period of movements in the *next* trial.

Motor skill learning is also associated with stronger beta ERD (Andres et al., 1999; Boonstra et al., 2007) and ERS (Moisello et al., 2015) in sensorimotor cortex, and more pronounced alpha band modulations in the cerebellum (Houweling et al., 2008) during task performance. More widespread changes in spectral power may also occur in these frequency bands that are stronger in young compared to older adults (Rueda-Delgado et al., 2019). Learning bimanual patterns impacts on functional connectivity within the sensorimotor network. In a study by Andres et al. (1999), healthy adult participants performed a 30-minute training session during which they learned to fuse two overlearned unimanual finger-tapping sequences into a new bimanual sequence. Interhemispheric coherence between sensorimotor regions in the alpha and beta band was increased during early stages of the training phase and decreased again after the bimanual sequence was learned. Interhemispheric interactions might be initially recruited to learn a new bimanual coordination pattern but are seemingly less needed after the pattern has been acquired. By contrast, Houweling and colleagues (2010) showed that the time course of bimanual learning is reflected in the build-up of corticospinal phase synchronization. There, it appears as if the synchronization is needed to perform the task adequately.

Changes in oscillatory activity can also be observed directly after a training session in the amplitude of ongoing beta oscillations. Espenhahn and colleagues (2019) recorded EEG before and after a 30-minute training session of a unimanual visuomotor tracking task. Absolute beta power during rest was significantly higher after training and returned to pre-training levels the following day. A similar effect was found by Moisello and colleagues (2015) and was suggested to reflect a reduction in cortical excitability after extended use. It would be of interest to conduct more longitudinal studies to see whether learning induces long-lasting changes to oscillatory activity or whether the system returns to baseline levels when the newly learned skills are fully consolidated.

12.4.3 Altered Oscillations in Movement Disorders

Given the contributions of beta oscillations in facilitating and inhibiting movements, it comes as no surprise that this rhythm may deviate in patients with movement disorders. This is most evident in Parkinson's disease, a neurodegenerative disorder where a loss of dopaminergic cells in substantia nigra may be accompanied by symptoms of rigidity (muscle stiffness), bradykinesia (slowness of movement), tremor, posture and balance problems, slurred speech, and several non-motor symptoms. The disease is initially treated with dopaminergic medication, but in a more advanced stage patients might be referred to DBS. Electrodes with multiple contacts are typically implanted in either STN or GPi in both left and right hemispheres. Both targets are known for the effect of 130-Hz stimulation on improvement of motor function (Deuschl et al., 2006;

Weaver et al., 2009). Besides stimulation, the electrodes can be used to record LFPs, though recordings are typically limited to a time window of about a week after electrode implantation when wires are still externalized. Such recordings have revealed elevated levels of beta oscillations in patients off medication during rest (Figure 12.4; e.g., Brown et al., 2001; Levy et al., 2002; Priori et al., 2004), and some studies report less-pronounced beta ERD during self-paced movement (Doyle et al., 2005a).

One of the challenges with invasive recordings in patients is to determine the degree to which observed activities are physiological or disease-related. Acquiring the same recordings from healthy subjects for comparison is not an option as they would need to undergo the same surgical procedures. Still there are strong reasons to believe that excessive beta oscillations in the STN are a marker of Parkinsonism:

1. The amplitude of beta oscillations correlates with severity of bradykinesia and rigidity as measured with the Unified Parkinson's Disease Rating Scale (Neumann et al., 2016; van Wijk et al., 2016).
2. The amplitude of beta oscillations decreases after dopaminergic medication or DBS in parallel with clinical improvement (Kühn et al., 2009, 2008, 2006b; Ray et al., 2008).
3. The amplitude of beta oscillations in STN and GPi is higher in patients with Parkinson's disease than in patients with dystonia (Piña-Fuentes et al., 2019).
4. Stimulation at 20 Hz instead of the clinically effective 130 Hz impedes motor performance (Chen et al., 2007), therefore underscoring a causal role of excessive beta oscillations in motor impairment.

While bradykinesia and rigidity are associated with beta oscillations, tremor typically manifests at frequencies below 12 Hz. Several forms of physiological and pathological tremors with central or peripheral origins can be distinguished, all characterized by involuntary rhythmic movements in one or more body parts (Deuschl et al., 2001; McAuley & Marsden, 2000). Essential tremor is the most common movement disorder,

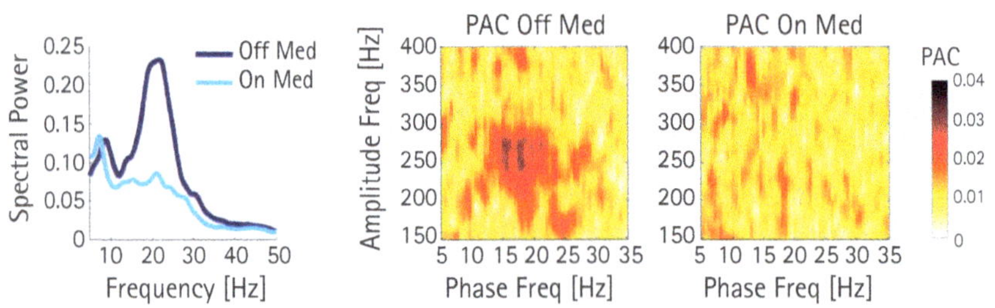

FIGURE 12.4 Abnormal subthalamic nucleus oscillatory activity in Parkinson's disease. Left: Spectral power in the beta frequency range is elevated when the patient is withdrawn from dopaminergic medication. Right: Phase-amplitude coupling (PAC) between beta and high-frequency oscillations is another marker of motor impairment.

Example of single subject DBS-LFP recordings taken from Van Wijk et al., 2016.

an action tremor that occurs with voluntary muscle contraction. Neural entrainment between inferior olive, cerebellum, thalamus, and motor cortex is believed to underlie its emergence (Raethjen and Deuschl, 2012). Activity at the tremor frequency within this network has been found to be coherent with EMG (Hellwig et al., 2001; Schnitzler et al., 2009). By contrast, Parkinsonian tremor presents during rest and may have its origin in the basal ganglia (Deuschl et al., 2001). Both Parkinsonian and essential tremor can be suppressed by 130-Hz DBS in the ventrolateral thalamus (Benabid et al., 1996). Stimulation at the tremor frequency could either enhance or reduce tremor amplitude, depending on the phase at which stimulation pulses are delivered (Cagnan et al., 2013).

Dystonia is another hyperkinetic movement disorder that can be associated with increased power for frequencies below the beta band in GPi (Chen et al., 2006; Liu et al., 2008; Sharott et al., 2008; Silberstein et al., 2003). This arguably reflects diminished and more irregular neuronal firing that is considered distinctive of the disorder (Hendrix & Vitek, 2012). Clinical symptoms include twisting movements and abnormal posture resulting from involuntary sustained and sometimes repetitive muscle contractions (Fahn, 1988). Also in dystonia symptoms might be suppressed via DBS but improvement may not become evident before several weeks or months of continuous stimulation (Vidailhet et al., 2005). Barow et al. (2014) showed that 4–12 Hz pallidal activity is reduced upon DBS with more immediate effects for phasic compared to tonic dystonia subtypes.

Elevated levels of oscillatory activity seem to be a common feature of a number of movement disorders. Effective treatments are frequently associated with a reduction of this activity although causal relations are often difficult to establish.

12.5 A DYNAMIC VIEW ON FREQUENCY BANDS

Until now, averaging of spectral power within frequency bands or across long time windows has been common practice in the field. Recently, dynamical properties of oscillations have received more attention. For example, beta oscillations are not present with constant amplitude but appear in short bursts of varying amplitude and duration (Feingold et al., 2015; Murthy and Fetz, 1996b). In Parkinson's disease, bursts with long duration occur more frequently in the STN of patients with larger clinical impairment and also have a higher amplitude than short bursts (Tinkhauser et al., 2017a). Dopaminergic medication significantly reduces the number of long bursts (Tinkhauser et al., 2017b). These new insights help to further constrain computational models, like that proposed by Sherman and colleagues (2016) for S1, which explore the synaptic mechanisms by which beta bursts emerge.

Another example is the notion that functions may arise from the interaction between frequency bands as opposed to individual frequency bands forming parallel communication channels. Using recordings from rat motor cortex, Igarashi et al. (2013)

demonstrated gamma oscillations comprised of a slow and fast component that were both coupled with an ongoing theta rhythm. Slow gamma oscillations were pronounced when rats were holding on to a lever and were phase-locked to peaks of the theta oscillation, whereas fast gamma oscillations emerged around the time of pulling the lever and were phase-locked to a trough of the theta oscillation. These findings are reminiscent of *phase precession* in hippocampal cortex where firing of place cells in the gamma range is nested in the ongoing theta rhythm. With individual gamma cycles representing a particular location in space, these progressively shift forward to earlier phases of the theta cycle as the rat navigates towards that location (O'Keefe & Recce, 1993). The nesting of oscillations allows for the encoding of near and far locations, i.e. information that is not contained in the individual frequencies. On the other hand, cross-frequency coupling might also arise due to pathology. We have found the strength of phase-amplitude coupling between beta and HFOs in the STN to correlate with the severity of bradykinesia and rigidity symptoms in Parkinson's disease (Figure 12.4; van Wijk et al., 2016).

Cross-frequency coupling could also be indicative of non-sinusoidal waveforms as these would appear at higher frequencies in the spectrum (van Wijk, 2017). Cole et al. (2017) demonstrated that the waveform shape of beta oscillations could indeed explain beta-gamma phase-amplitude coupling patterns observed in M1 of Parkinson's patients (de Hemptinne et al., 2013). It is interesting that DBS reduces the asymmetry of the beta oscillation waveform (de Hemptinne et al., 2015) as it hints at less-synchronous synaptic input from the basal ganglia. A focus on the detailed time dynamics, cross-frequency coupling, and waveform shape of oscillations holds great promise to further unravel their role in sensorimotor function.

12.6 Cognitive Aspects of Sensorimotor Function

We move to interact with the external world in a meaningful way. Our actions depend on goals and intentions, which in turn depend on the context we are in. Cognition and motor control are highly intertwined. Several lines of research indeed suggest that sensorimotor regions contribute beyond the classical view on motor control.

First of all, (pre-)motor cortex is already activated during the process of action selection instead of merely being informed on the final outcome that needs to be executed (Cisek and Kalaska, 2005). Donner and colleagues (2009) demonstrated that lateralized beta suppression and gamma increase in motor cortex could predict on a single trial level which left/right choice participants were going to make during a perceptual detection task. Instructions to emphasize decision speed versus accuracy decrease prestimulus M1 beta levels, which might explain the upsurge of errors under speed stress (Pastötter et al., 2012; Steinemann et al., 2018). Pre-decision activation of the motor system is therefore likely to speed-up or even influence behavioral performance.

The ability to *inhibit* actions is seen as an important marker of cognitive control. Pro-active inhibition and reactive stopping of actions are often studied with go/no-go and stop-signal paradigms, respectively. Both are thought to rely on the "stopping network" of inferior frontal gyrus (IFG), pre-SMA, and STN (Aron, 2011). In IFG, trials with a stop signal induce an increase in beta power that is larger for successful vs. unsuccessful stop trials (Swann et al., 2009). In STN, task-related beta suppression is weaker for successful compared to unsuccessful stops (Kühn et al., 2004; Wessel et al., 2016). In turn, successful stops are also characterized by a weaker beta suppression in M1 compared to unsuccessful stops (Swann et al., 2009). More generally, fronto-basal ganglia circuits are considered to act as a brake to stop actions, resolve conflict, or process surprising events (Aron et al., 2016).

Execution of movements is not strictly necessary for the sensorimotor system to become active. Beta ERD can be observed in contralateral M1 and the STN when participants are merely imagining that they are making a movement albeit with lower magnitude compared to real movements (Kühn et al., 2006a; Pfurtscheller and Neuper, 1997; Schnitzler et al., 1997). Researchers have gratefully taken advantage of this phenomenon for the development of *brain computer interfaces* (Wolpaw et al., 2002). Even patients in an advanced stage of ALS have been able to learn how to control vertical cursor movement on a computer screen by up- or down-regulating the amplitude of sensorimotor EEG rhythms via motor imagery (Kübler et al., 2005). With the use of machine learning it is possible to reduce the training time needed to successfully operate such a device to just 20 minutes in healthy individuals (Blankertz et al., 2007). Controlling a robot arm or computer cursor by imagining movement of your own limb is appealing as it relies to a large extent on how the brain controls real movements.

Observing movements performed by others can also induce alpha/mu and beta suppression in human sensorimotor cortex (Babiloni et al., 2002; Cochin et al., 1998; Hari et al., 1998; Kilner et al., 2009). The suppression is slightly larger when the action is target-directed (Avanzini et al., 2012; Muthukumaraswamy et al., 2004) and when the observer is more experienced with the action him/herself (Cannon et al., 2014). A credible hypothesis for the activation of motor regions during perception is that the observed action needs to be internally simulated in order to understand the other person's actions and intentions (Jeannerod, 2001). As such, the discovery of *mirror neurons* in (pre-) motor regions (Cattaneo & Rizzolatti, 2009) means that considerable ground is shared with the field of social neuroscience.

12.7 Conclusions

Oscillations are omnipresent in the sensorimotor system and show distinct modulations with experimental tasks and pathology. In particular beta oscillations seem to have a prominent role by means of their association with movement facilitation and inhibition. Movement-related beta ERD and ERS are among the most robust time-frequency patterns in EEG/MEG studies. Intriguingly, the cortex is not alone in displaying these characteristic modulations. Very similar movement-related beta ERD/ERS and gamma ERS can be observed in LFP recordings from the STN (Alegre et al., 2005; Androulidakis et al., 2007b; Cassidy et al., 2002; Kühn et al., 2004; Litvak et al., 2012), GPi (Brücke et al., 2008; Talakoub et al., 2016; Tsang et al., 2012), and the ventral lateral thalamus (Brücke et al., 2013; Klostermann et al., 2007; Paradiso et al., 2004). This implies that these modulations are generic principles of information processing that might have different functional meaning in different regions of the sensorimotor network.

Over the years, the main focus of the field at large has been on oscillations in contralateral M1. While simple flexion/extension movements of the fingers indeed seem to be fairly restricted to activation in the primary sensorimotor cortices, more complex rhythmic or bimanual movement patterns recruit additional regions like ipsilateral M1, SMA, premotor cortex, cerebellum, and also primary and secondary sensory and association cortices if the task involves strong visual or auditory components (Heinrichs-Graham & Wilson, 2015; Houweling et al., 2008; Hummel et al., 2003; Pollok et al., 2005; Rueda-Delgado et al., 2014). The contribution of subcortical structures is more difficult to study in humans but should not be overlooked. Intriguingly, lesions to motor cortex do not affect execution of a task-specific motor sequence in rats once the sequence has been learned, suggesting reliance on subcortical controllers with projections to the spinal cord (Kawai et al., 2015). Simultaneous MEG and DBS-LFP recordings have revealed a number of spatially and spectrally distinct cortico-subcortical networks that appear to be disease-unrelated. The functional relevance of these networks deserves further exploration in future studies.

Although we have learned a great deal about the neural control of movement by studying oscillations, several aspects remain unexplained. Important questions that are still open are mostly of mechanistic nature. For example, how does beta ERD lead to muscle activations? What causes beta ERD to start and initiate movement in the first place? Why is desynchronization so widespread even for simple finger movement? Why are movement-related modulations so similar across various parts of the motor system? How are movement plans translated into motor commands? How are they encoded? Oscillations reflect the summed activity of numerous neurons, from which it might be difficult to infer details of individual muscle control especially with non-invasive techniques such as EEG and MEG. Instead, they are more likely to reflect general states of activation or deactivation. I believe there is still much more to gain from studying oscillations through the combination of recordings techniques, the use of advanced signal processing algorithms and the development of computational models. Fortunately, grasping a cup of coffee is much easier than understanding how we do it.

REFERENCES

Adrian, E. D. & Matthews, B. H. C. (1934). The Berger rhythm: Potential changes from the occipital lobes in man. *Brain*, 57, 355–385. https://doi.org/10.1093/brain/57.4.355

Alayrangues, J., Torrecillos, F., Jahani, A., & Malfait, N. (2019). Error-related modulations of the sensorimotor post-movement and foreperiod beta-band activities arise from distinct neural substrates and do not reflect efferent signal processing. *NeuroImage 184*, 10–24. https://doi.org/10.1016/j.neuroimage.2018.09.013

Alegre, M., Alonso-Frech, F., Rodríguez-Oroz, M. C., Guridi, J., Zamarbide, I., Valencia, M., ... Artieda, J. (2005). Movement-related changes in oscillatory activity in the human subthalamic nucleus: Ipsilateral vs. contralateral movements. *European Journal of Neuroscience, 22*, 2315–2324. https://doi.org/10.1111/j.1460-9568.2005.04409.x

Alegre, M., Labarga, A., Gurtubay, I., Iriarte, J., Malanda, A., & Artieda, J. (2003). Movement-related changes in cortical oscillatory activity in ballistic, sustained and negative movements. *Experimental Brain Research, 148*, 17–25. https://doi.org/10.1007/s00221-002-1255-x

Andres, F. G., Mima, T., Schulman, A. E., Dichgans, J., Hallett, M., & Gerloff, C. (1999). Functional coupling of human cortical sensorimotor areas during bimanual skill acquisition. *Brain, 122*, 855–870. https://doi.org/10.1093/brain/122.5.855

Androulidakis, A. G., Doyle, L. M. F., Yarrow, K., Litvak, V., Gilbertson, T. P., & Brown, P. (2007a). Anticipatory changes in beta synchrony in the human corticospinal system and associated improvements in task performance. *European Journal of Neuroscience, 25*, 3758–3765. https://doi.org/10.1111/j.1460-9568.2007.05620.x

Androulidakis, A. G., Kühn, A. A., Chen, C. C., Blomstedt, P., Kempf, F., Kupsch, A., ... Brown, P. (2007b). Dopaminergic therapy promotes lateralized motor activity in the subthalamic area in Parkinson's disease. *Brain, 130*, 457–68. https://doi.org/10.1093/brain/awl358

Aron, A. R. (2011). From reactive to proactive and selective control: Developing a richer model for stopping inappropriate responses. *Biological Psychiatry, 69*, e55–e68. https://doi.org/10.1016/j.biopsych.2010.07.024

Aron, A. R., Herz, D. M., Brown, P., Forstmann, B. U., & Zaghloul, K. (2016). Frontosubthalamic circuits for control of action and cognition. *Journal of Neuroscience, 36*, 11489–11495. https://doi.org/10.1523/JNEUROSCI.2348-16.2016

Asanuma, H., Zarzecki, P., Jankowska, E., Hongo, T., & Marcus, S. (1979). Projection of individual pyramidal tract neurons to lumbar motor nuclei of the monkey. *Experimental Brain Research, 34*, 73–89. https://doi.org/103742

Ashe, J. & Georgopoulos, A. P. (1994). Movement parameters and neural activity in motor cortex and area 5. *Cerebral Cortex, 4*, 590–600. https://doi.org/10.1093/cercor/4.6.590

Avanzini, P., Fabbri-Destro, M., Dalla Volta, R., Daprati, E., Rizzolatti, G., & Cantalupo, G. (2012). The dynamics of sensorimotor cortical oscillations during the observation of hand movements: An EEG study. *PLoS One, 7*, e37534. https://doi.org/10.1371/journal.pone.0037534

Babiloni, C., Babiloni, F., Carducci, F., Cincotti, F., Cocozza, G., Del Percio, C., Moretti, D. V., & Rossini, P. M. (2002). Human cortical electroencephalography (EEG) rhythms during the observation of simple aimless movements: A high-resolution EEG study. *NeuroImage, 17*, 559–572. https://doi.org/10.1006/nimg.2002.1192

Baker, M. R. & Baker, S. N. (2003). The effect of diazepam on motor cortical oscillations and corticomuscular coherence studied in man. *Journal of Physiology, 546*, 931–942. https://doi.org/10.1113/jphysiol.2002.029553

Baker, S. N., Olivier, E., & Lemon, R. N. (1997). Coherent oscillations in monkey motor cortex and hand muscle EMG show task-dependent modulation. *Journal of Physiology, 501*, 225–241. https://doi.org/10.1111/j.1469-7793.1997.225bo.x

Baker, S. N., Pinches, E. M., & Lemon, R. N. (2003). Synchronization in monkey motor cortex during a precision grip task. II. Effect of oscillatory activity on corticospinal output. *Journal of Neurophysiology*, *89*, 1941–1953. https://doi.org/10.1152/jn.00832.2002

Baker, S. N., Spinks, R., Jackson, A., & Lemon, R. N. (2001). Synchronization in monkey motor cortex during a precision grip task. I. Task-dependent modulation in single-unit synchrony. *Journal of Neurophysiology*, *85*, 869–885. https://doi.org/10.1152/jn.00832.2002

Barow, E., Neumann, W. J., Brücke, C., Huebl, J., Horn, A., Brown, P., ... Kühn, A.A., 2014. Deep brain stimulation suppresses pallidal low frequency activity in patients with phasic dystonic movements. *Brain*, *137*, 3012–3024. https://doi.org/10.1093/brain/awu258

Benabid, A. L., Pollak, P., Gao, D., Hoffmann, D., Limousin, P., Gay, E., Payen, I., & Benazzouz, A. (1996). Chronic electrical stimulation of the ventralis intermedius nucleus of the thalamus as a treatment of movement disorders. *Journal of Neurosurgery*, *84*, 203–214. https://doi.org/10.3171/jns.1996.84.2.0203

Berger, H. (1929). Über das Elektroenkephalogramm des Menschen. *Archiv für Psychiatrie und Nervenkrankheiten*, *87*, 527–570.

Bhatt, M. B., Bowen, S., Rossiter, H. E., Dupont-Hadwen, J., Moran, R.J., Friston, K.J., Ward, N.S., 2016. Computational modelling of movement-related beta-oscillatory dynamics in human motor cortex. *Neuroimage* *133*, 224–232. https://doi.org/10.1016/j.neuroimage.2016.02.078

Blankertz, B., Dornhege, G., Krauledat, M., Müller, K.-R., & Curio, G. (2007). The non-invasive Berlin Brain–Computer Interface: Fast acquisition of effective performance in untrained subjects. *NeuroImage*, *37*, 539–550. https://doi.org/10.1016/j.neuroimage.2007.01.051

Boonstra, T. W., Daffertshofer, A., Breakspear, M., & Beek, P. J. (2007). Multivariate time-frequency analysis of electromagnetic brain activity during bimanual motor learning. *NeuroImage*, *36*, 370–377. https://doi.org/10.1016/j.neuroimage.2007.03.012

Boonstra, T. W., Danna-Dos-Santos, A., Xie, H.-B., Roerdink, M., Stins, J. F., & Breakspear, M. (2016). Muscle networks: Connectivity analysis of EMG activity during postural control. *Scientific Reports*, *5*, 17830. https://doi.org/10.1038/srep17830

Boonstra, T. W., van Wijk, B. C. M., Praamstra, P., & Daffertshofer, A. (2009). Corticomuscular and bilateral EMG coherence reflect distinct aspects of neural synchronization. *Neuroscience Letters*, *463*, 17–21. https://doi.org/10.1016/j.neulet.2009.07.043

Brown, P., Oliviero, A., Mazzone, P., Insola, A., Tonali, P., & Di Lazzaro, V. (2001). Dopamine dependency of oscillations between subthalamic nucleus and pallidum in Parkinson's disease. *Journal of Neuroscience*, *21*, 1033–1038. https://doi.org/10.1523/JNEUROSCI.21-03-01033.2001

Brown, P., Salenius, S., Rothwell, J. C., & Hari, R. (1998). Cortical correlate of the Piper rhythm in humans. *Journal of Neurophysiology*, *80*, 2911–2917. https://doi.org/10.1152/jn.1998.80.6.2911

Brücke, C., Bock, A., Huebl, J., Krauss, J. K., Schönecker, T., Schneider, G. H., Brown, P., & Kühn, A. A. (2013). Thalamic gamma oscillations correlate with reaction time in a go/no-go task in patients with essential tremor. *NeuroImage*, *75*, 36–45. https://doi.org/10.1016/j.neuroimage.2013.02.038

Brücke, C., Kempf, F., Kupsch, A., Schneider, G. H., Krauss, J.K., Aziz, T., ... Kühn, A. A. (2008). Movement-related synchronization of gamma activity is lateralized in patients with dystonia. *European Journal of Neuroscience*, *27*, 2322–2329. https://doi.org/10.1111/j.1460-9568.2008.06203.x

Cagnan, H., Brittain, J.-S., Little, S., Foltynie, T., Limousin, P., Zrinzo, L., ... Brown, P. (2013). Phase dependent modulation of tremor amplitude in essential tremor through thalamic stimulation. *Brain, 136*, 3062–3075. https://doi.org/10.1093/brain/awt239

Cannon, E. N., Yoo, K. H., Vanderwert, R. E., Ferrari, P. F., Woodward, A.L., & Fox, N. A. (2014). Action experience, more than observation, influences mu rhythm desynchronization. *PLoS One, 9*, e92002. https://doi.org/10.1371/journal.pone.0092002

Cardoso de Oliveira, S., Gribova, A., Donchin, O., Bergman, H., & Vaadia, E. (2001). Neural interactions between motor cortical hemispheres during bimanual and unimanual arm movements. *European Journal of Neuroscience, 14*, 1881–1896. https://doi.org/10.1046/j.0953-816X.2001.01801.x

Cassidy, M., Mazzone, P., Oliviero, A., Insola, A., Tonali, P., Di Lazzaro, V., & Brown, P. (2002). Movement-related changes in synchronization in the human basal ganglia. *Brain, 125*, 1235–1246. https://doi.org/10.1093/brain/awf135

Cassim, F., Monaca, C., Szurhaj, W., Bourriez, J.-L., Defebvre, L., Derambure, P., & Guieu, J.-D. (2001). Does post-movement beta synchronization reflect an idling motor cortex? *Neuroreport, 12*, 3859–3863. https://doi.org/10.1097/00001756-200112040-00051

Cassim, F., Szurhaj, W., Sediri, H., Devos, D., Bourriez, J.-L., Poirot, I., ... Guieu, J.-D. (2000). Brief and sustained movements: Differences in event-related (de)synchronization (ERD/ERS) patterns. *Clinical Neurophysiology, 111*, 2032–2039. https://doi.org/10.1016/S1388-2457(00)00455-7

Cattaneo, L. & Rizzolatti, G. (2009). The mirror neuron system. *Archives of Neurology, 66*, 557–560. https://doi.org/10.1001/archneurol.2009.41

Chakarov, V., Naranjo, J.R., Schulte-Monting, J., Omlor, W., Huethe, F., & Kristeva, R. (2009). Beta-range EEG-EMG coherence with isometric compensation for increasing modulated low-level forces. *Journal of Neurophysiology, 102*, 1115–1120. https://doi.org/10.1152/jn.91095.2008

Chatrian, G. E., Petersen, M. C., & Lazarte, J. A. (1959). The blocking of the Rolandic wicket rhythm and some central changes related to movement. *Electroencephalography and Clinical Neurophysiology, 11*, 497–510. https://doi.org/10.1016/0013-4694(59)90048-3

Chen, C.C., Kühn, A.A., Hoffmann, K.T., Kupsch, A., Schneider, G.H., Trottenberg, T., ... Brown, P. (2006). Oscillatory pallidal local field potential activity correlates with involuntary EMG in dystonia. *Neurology, 66*, 418–420. https://doi.org/10.1212/01.wnl.0000196470.00165.7d

Chen, C. C., Litvak, V., Gilbertson, T., Kühn, A., Lu, C., Lee, S., ... Hariz, M. (2007). Excessive synchronization of basal ganglia neurons at 20 Hz slows movement in Parkinson's disease. *Experimental Neurology, 205*, 214–221. https://doi.org/10.1016/j.expneurol.2007.01.027

Chen, R., Yaseen, Z., Cohen, L. G., & Hallett, M. (1998). Time course of corticospinal excitability in reaction time and self-paced movements. *Annals of Neurology, 44*, 317–325. https://doi.org/10.1002/ana.410440306

Cheyne, D., Bells, S., Ferrari, P., Gaetz, W., & Bostan, A. C. (2008). Self-paced movements induce high-frequency gamma oscillations in primary motor cortex. *NeuroImage, 42*, 332–342. https://doi.org/10.1016/j.neuroimage.2008.04.178

Cheyne, D., Gaetz, W., Garnero, L., Lachaux, J.-P., Ducorps, A., Schwartz, D., & Varela, F. J. (2003). Neuromagnetic imaging of cortical oscillations accompanying tactile stimulation. *Cognitive Brain Research, 17*, 599–611. https://doi.org/10.1016/S0926-6410(03)00173-3

Cisek, P. & Kalaska, J. F. (2005). Neural correlates of reaching decisions in dorsal premotor cortex: specification of multiple direction choices and final selection of action. *Neuron, 45*, 801–814. https://doi.org/10.1016/j.neuron.2005.01.027

Classen, J., Gerloff, C., Honda, M., & Hallett, M. (1998) Integrative visuomotor behavior is associated with interregionally coherent oscillations in the human brain. *Journal of Neurophysiology*, *79*, 1567–1573. https://doi.org/10.1152/jn.1998.79.3.1567
Cochin, S., Barthelemy, C., Lejeune, B., Roux, S., & Martineau, J. (1998). Perception of motion and qEEG activity in human adults. *Electroencephalography and Clinical Neurophysiology*, *107*, 287–295. https://doi.org/10.1016/S0013-4694(98)00071-6
Cole, S. R., van der Meij, R., Peterson, E. J., de Hemptinne, C., Starr, P. A., & Voytek, B. (2017). Nonsinusoidal beta oscillations reflect cortical pathophysiology in Parkinson's disease. *Journal of Neuroscience*, *37*, 2208–2216. https://doi.org/10.1523/JNEUROSCI.2208-16.2017
Conway, B. A., Halliday, D. M., Farmer, S. F., Shahani, U., Maas, P., Weir, A. I., & Rosenberg, J. R. (1995). Synchronization between motor cortex and spinal motoneuronal pool during the performance of a maintained motor task in man. *Journal of Physiology*, *489*, 917–924. https://doi.org/10.1113/jphysiol.1995.sp021104
Crone, N. E., Miglioretti, D. L., Gordon, B., & Lesser, R. P. (1998a). Functional mapping of human sensorimotor cortex with electrocorticographic spectral analysis. II. Event-related synchronization in the gamma band. *Brain*, *121*, 2301–2315. https://doi.org/10.1093/brain/121.12.2301
Crone, N. E., Miglioretti, D. L., Gordon, B., Sieracki, J. M., Wilson, M. T., Uematsu, S., & Lesser, R. P. (1998b). Functional mapping of human sensorimotor cortex with electrocorticographic spectral analysis. I. Alpha and beta event-related desynchronization. *Brain*, *121*, 2271–2299. https://doi.org/10.1093/brain/121.12.2271
de Hemptinne, C., Ryapolova-Webb, E. S., Air, E. L., Garcia, P. A., Miller, K. J., Ojemann, J. G.,, ... Starr, P. A. (2013). Exaggerated phase-amplitude coupling in the primary motor cortex in Parkinson disease. *Proceedings of the National Academy of Sciences of the United States of America*, *110*, 4780–4785. https://doi.org/10.1073/pnas.1214546110
de Hemptinne, C., Swann, N. C., Ostrem, J. L., Ryapolova-Webb, E. S., San Luciano, M., Galifianakis, N. B., & Starr, P. A. (2015). Therapeutic deep brain stimulation reduces cortical phase-amplitude coupling in Parkinson's disease. *Nature Neuroscience*, *18*, 779–786. https://doi.org/10.1038/nn.3997
Denker, M., Roux, S., Timme, M., Riehle, A., & Grün, S. (2007). Phase synchronization between LFP and spiking activity in motor cortex during movement preparation. *Neurocomputing*, *70*, 2096–2101. https://doi.org/10.1016/j.neucom.2006.10.088
Deuschl, G., Raethjen, J., Lindemann, M., & Krack, P. (2001). The pathophysiology of tremor. *Muscle Nerve*, *24*, 716–735. https://doi.org/10.1002/mus.1063
Deuschl, G., Schade-Brittinger, C., Krack, P., Volkmann, J., Schäfer, H., Bötzel, K., ... Voges, J. (2006). A randomized trial of deep-brain stimulation for Parkinson's disease. *New England Journal of Medicine*, *355*, 896–908. https://doi.org/10.1056/nejmoa060281
Donner, T. H., Siegel, M., Fries, P., & Engel, A. K. (2009). Buildup of choice-predictive activity in human motor cortex during perceptual decision making. *Current Biology*, *19*, 1581–1585. https://doi.org/10.1016/j.cub.2009.07.066
Doyle, L. M. F., Kühn, A. A., Hariz, M., Kupsch, A., Schneider, G. H., & Brown, P. (2005a). Levodopa-induced modulation of subthalamic beta oscillations during self-paced movements in patients with Parkinson's disease. *European Journal of Neuroscience*,. *21*, 1403–1412. https://doi.org/10.1111/j.1460-9568.2005.03969.x
Doyle, L. M. F., Yarrow, K., & Brown, P. (2005b). Lateralization of event-related beta desynchronization in the EEG during pre-cued reaction time tasks. *Clinical Neurophysiology*, *116*, 1879–1888. https://doi.org/10.1016/j.clinph.2005.03.017

Erbil, N. & Ungan, P. (2007). Changes in the alpha and beta amplitudes of the central EEG during the onset, continuation, and offset of long-duration repetitive hand movements. *Brain Research, 1169*, 44–56. https://doi.org/10.1016/j.brainres.2007.07.014

Espenhahn, S., de Berker, A. O., van Wijk, B. C. M., Rossiter, H. E., & Ward, N. S. (2017). Movement-related beta oscillations show high intra-individual reliability. *NeuroImage, 147*, 175–185. https://doi.org/10.1016/j.neuroimage.2016.12.025

Espenhahn, S., van Wijk, B. C. M., Rossiter, H. E., de Berker, A. O., Redman, N. D., Rondina, J., Diedrichsen, J., & Ward, N. S. (2019). Cortical beta oscillations are associated with motor performance following visuomotor learning. *NeuroImage, 195*, 340–353. https://doi.org/10.1016/j.neuroimage.2019.03.079

Fahn, S. (1988). Concept and classification of dystonia. *Advances in Neurology, 50*, 1–8.

Feingold, J., Gibson, D. J., DePasquale, B., & Graybiel, A. M. (2015). Bursts of beta oscillation differentiate postperformance activity in the striatum and motor cortex of monkeys performing movement tasks. *Proceedings of the National Academy of Sciences of the United States of America, 112*, 13687–92. https://doi.org/10.1073/pnas.1517629112

Foffani, G., Priori, A., Egidi, M., Rampini, P., Tamma, F., Caputo, E., . . . Barbieri, S. (2003). 300-Hz subthalamic oscillations in Parkinson's disease. *Brain, 126*, 2153–2163. https://doi.org/10.1093/brain/awg229

Freund, H. J. (1983). Motor unit and muscle activity in voluntary motor control. *Physiological Reviews, 63*, 387–436. https://doi.org/10.1152/physrev.1983.63.2.387

Fries, P. (2005). A mechanism for cognitive dynamics: Neuronal communication through neuronal coherence. *Trends in Cognitive Science, 9*, 474–480. https://doi.org/10.1016/j.tics.2005.08.011

Fu, Q. G., Flament, D., Coltz, J. D., & Ebner, T. J. (1995). Temporal encoding of movement kinematics in the discharge of primate primary motor and premotor neurons. *Journal of Neurophysiology, 73*, 836–854. https://doi.org/10.1152/jn.1995.73.2.836

Gaetz, W., MacDonald, M., Cheyne, D., & Snead, O. C. (2010). Neuromagnetic imaging of movement-related cortical oscillations in children and adults: Age predicts post-movement beta rebound. *NeuroImage, 51*, 792–807. https://doi.org/10.1016/j.neuroimage.2010.01.077

Gastaut, H. (1952). Etude électrocorticographique de la réactivité des rythmes rolandiques. *Revue neurologique, 87*, 176–82.

Georgopoulos, A., Schwartz, A., & Kettner, R. (1986). Neuronal population coding of movement direction. *Science, 233*, 1416–1419. https://doi.org/10.1126/science.3749885

Gerloff, C., Richard, J., Hadley, J., Schulman, A.E., Honda, M., Hallett, M., 1998. Functional coupling and regional activation of human cortical motor areas during simple, internally paced and externally paced finger movements. *Brain 121*, 1513–1531. https://doi.org/10.1093/brain/121.8.1513

Gilbertson, T., Lalo, E., Doyle, L., Lazzaro, V. Di, Cioni, B., & Brown, P. (2005). Existing motor state is favored at the expense of new movement during 13–35 Hz oscillatory synchrony in the human corticospinal system. *Journal of Neuroscience, 25*, 7771–7779. https://doi.org/10.1523/JNEUROSCI.1762-05.2005

Grammont, F. & Riehle, A. (2003). Spike synchronization and firing rate in a population of motor cortical neurons in relation to movement direction and reaction time. *Biological Cybernetics, 88*, 360–373. https://doi.org/10.1007/s00422-002-0385-3

Gross, J., Pollok, B., Dirks, M., Timmermann, L., Butz, M., & Schnitzler, A. (2005). Task-dependent oscillations during unimanual and bimanual movements in the human primary motor cortex and SMA studied with magnetoencephalography. *NeuroImage*, *26*, 91–98. https://doi.org/10.1016/j.neuroimage.2005.01.025

Gross, J., Tass, P. A., Salenius, S., Hari, R., Freund, H. J., & Schnitzler, A. (2000). Cortico-muscular synchronization during isometric muscle contraction in humans as revealed by magnetoencephalography. *Journal of Physiology*, *527*, 623–631. https://doi.org/10.1111/j.1469-7793.2000.00623.x

Hall, S. D., Barnes, G. R., Furlong, P. L., Seri, S., & Hillebrand, A. (2010). Neuronal network pharmacodynamics of GABAergic modulation in the human cortex determined using pharmaco-magnetoencephalography. *Human Brain Mapping*, *31*, 581–594. https://doi.org/10.1002/hbm.20889

Halliday, D. M., Conway, B. A., Farmer, S. F., and Rosenberg, J. R. (1998). Using electroencephalography to study functional coupling between cortical activity and electromyograms during voluntary contractions in humans. *Neuroscience Letters*, *241*, 5–8. https://doi.org/10.1016/S0304-3940(97)00964-6

Hari, R., Forss, N., Avikainen, S., Kirveskari, E., Salenius, S., & Rizzolatti, G. (1998). Activation of human primary motor cortex during action observation: A neuromagnetic study. *Proceedings of the National Academy of Sciences of the United States of America*, *95*, 15061–15065. https://doi.org/10.1073/pnas.95.25.15061

Heinrichs-Graham, E., McDermott, T. J., Mills, M. S., Wiesman, A. I., Wang, Y.-P., Stephen, J. M., Calhoun, V. D., & Wilson, T. W. (2018). The lifespan trajectory of neural oscillatory activity in the motor system. *Developmental Cognitive Neuroscience*, *30*, 159–168. https://doi.org/10.1016/j.dcn.2018.02.013

Heinrichs-Graham, E. & Wilson, T. W. (2015). Coding complexity in the human motor circuit. *Human Brain Mapping*, *36*, 5155–5167. https://doi.org/10.1002/hbm.23000

Hellwig, B., Häußler, S., Schelter, B., Lauk, M., Guschlbauer, B., Timmer, J., & Lücking, C. H. (2001). Tremor-correlated cortical activity in essential tremor. *The Lancet*, *357*, 519–523. https://doi.org/10.1016/S0140-6736(00)04044-7

Hendrix, C. M. & Vitek, J. L. (2012). Toward a network model of dystonia. *Annals of the New York Academy of Sciences*, *1265*, 46–55. https://doi.org/10.1111/j.1749-6632.2012.06692.x

Hillebrand, A., Barnes, G. R., Bosboom, J. L., Berendse, H. W., & Stam, C. J. (2012). Frequency-dependent functional connectivity within resting-state networks: An atlas-based MEG beamformer solution. *NeuroImage*, *59*, 3909–3921. https://doi.org/10.1016/j.neuroimage.2011.11.005

Hirschmann, J., Butz, M., Hartmann, C. J., Hoogenboom, N., Özkurt, T. E., Vesper, J., Wojtecki, L., & Schnitzler, A. (2016). Parkinsonian rest tremor is associated with modulations of subthalamic high-frequency oscillations. *Movement Disorders*, *31*, 1551–1559. https://doi.org/10.1002/mds.26663

Hirschmann, J., Özkurt, T. E., Butz, M., Homburger, M., Elben, S., Hartmann, C. J., . . . Schnitzler, A. (2011). Distinct oscillatory STN-cortical loops revealed by simultaneous MEG and local field potential recordings in patients with Parkinson's disease. *NeuroImage*, *55*, 1159–1168. https://doi.org/10.1016/j.neuroimage.2010.11.063

Hirschmann, J., Özkurt, T.E., Butz, M., Homburger, M., Elben, S., Hartmann, C.J., . . . Schnitzler, A. (2013). Differential modulation of STN-cortical and cortico-muscular coherence by movement and levodopa in Parkinson's disease. *NeuroImage*, *68*, 203–213. https://doi.org/10.1016/j.neuroimage.2012.11.036

Houweling, S., Daffertshofer, A., van Dijk, B. W., & Beek, P. J. (2008). Neural changes induced by learning a challenging perceptual-motor task. *NeuroImage*, *41*, 1395–1407. https://doi.org/10.1016/j.neuroimage.2008.03.023

Houweling, S., van Dijk, B. W., Beek, P. J., & Daffertshofer, A. (2010). Cortico-spinal synchronization reflects changes in performance when learning a complex bimanual task. *NeuroImage*, *49*, 3269–3275. https://doi.org/10.1016/j.neuroimage.2009.11.017

Hummel, F., Kirsammer, R., & Gerloff, C. (2003). Ipsilateral cortical activation during finger sequences of increasing complexity: representation of movement difficulty or memory load? *Clinical Neurophysiology*, *114*, 605–613. https://doi.org/10.1016/S1388-2457(02)00417-0

Huo, X., Wang, Y., Kotecha, R., Kirtman, E.G., Fujiwara, H., Hemasilpin, N., ... Xiang, J. (2011). High gamma oscillations of sensorimotor cortex during unilateral movement in the developing brain: A MEG study. *Brain Topography*, *23*, 375–384. https://doi.org/10.1007/s10548-010-0151-0

Igarashi, J., Isomura, Y., Arai, K., Harukuni, R., & Fukai, T. (2013). A θ-γ oscillation code for neuronal coordination during motor behavior. *Journal of Neuroscience*, *33*, 18515–18530. https://doi.org/10.1523/JNEUROSCI.2126-13.2013

Jasper, H. & Penfield, W. (1949). Electrocorticograms in man: Effect of voluntary movement upon the electrical activity of the precentral gyrus. *Archiv für Psychiatrie und Nervenkrankheiten*, *183*, 163–174. https://doi.org/10.1007/BF01062488

Jasper, H. H. & Andrews, H. L. (1938). Electro-encephalography. III. Normal differentiation of occipital and precentral regions in man. *Archiv für Psychiatrie und Nervenkrankheiten*, *39*, 96–115.

Jeannerod, M. (2001). Neural simulation of action: A unifying mechanism for motor cognition. *NeuroImage*, S103–S109. https://doi.org/10.1006/nimg.2001.0832

Jensen, O., Goel, P., Kopell, N., Pohja, M., Hari, R., & Ermentrout, B. (2005). On the human sensorimotor-cortex beta rhythm: Sources and modeling. *NeuroImage*, *26*, 347–355. https://doi.org/10.1016/j.neuroimage.2005.02.008

Jha, A., Litvak, V., Taulu, S., Thevathasan, W., Hyam, J.A., Foltynie, T., ... Brown, P. (2017). Functional connectivity of the pedunculopontine nucleus and surrounding region in Parkinson's disease. *Cerebral Cortex*, *27*, 54–67. https://doi.org/10.1093/cercor/bhw340

Johnson, A. N., Wheaton, L. A., & Shinohara, M. (2011). Attenuation of corticomuscular coherence with additional motor or non-motor task. *Clinical Neurophysiology*, *122*, 356–363. https://doi.org/10.1016/j.clinph.2010.06.021

Kawai, R., Markman, T., Poddar, R., Ko, R., Fantana, A. L., Dhawale, A. K., Kampff, A.R., & Ölveczky, B. P. (2015). Motor cortex is required for learning but not for executing a motor skill. *Neuron*, *86*, 800–812. https://doi.org/10.1016/j.neuron.2015.03.024

Keller, A. (1993). Intrinsic synaptic organization of the motor cortex. *Cerebral Cortex*, *3*, 430–441. https://doi.org/10.1093/cercor/3.5.430

Kilner, J. M., Baker, S. N., Salenius, S., Hari, R., & Lemon, R. N. (2000). Human cortical muscle coherence is directly related to specific motor parameters. *Journal of Neuroscience*, *20*, 8838–8845. https://doi.org/10.1523/JNEUROSCI.20-23-08838.2000

Kilner, J. M., Marchant, J. L., & Frith, C. D. (2009). Relationship between activity in human primary motor cortex during action observation and the mirror neuron system. *PLoS One*, *4*, e4925. https://doi.org/10.1371/journal.pone.0004925

Klostermann, F., Nikulin, V. V., Kühn, A. A., Marzinzik, F., Wahl, M., Pogosyan, A., ... Curio, G. (2007). Task-related differential dynamics of EEG alpha- and beta-band synchronization in cortico-basal motor structures. *European Journal of Neuroscience*, *25*, 1604–1615. https://doi.org/10.1111/j.1460-9568.2007.05417.x

Kristeva-Feige, R., Fritsch, C., Timmer, J., & Lücking, C.-H. (2002). Effects of attention and precision of exerted force on beta range EEG-EMG synchronization during a maintained motor contraction task. *Clinical Neurophysiology*, *113*, 124–31. https://doi.org/10.1016/S1388-2457(01)00722-2

Kristeva, R., Patino, L., & Omlor, W. (2007). Beta-range cortical motor spectral power and corticomuscular coherence as a mechanism for effective corticospinal interaction during steady-state motor output. *NeuroImage*, *36*, 785–792. https://doi.org/10.1016/j.neuroimage.2007.03.025

Kübler, A., Nijboer, F., Mellinger, J., Vaughan, T. M., Pawelzik, H., Schalk, G., ... Wolpaw, J. R. (2005). Patients with ALS can use sensorimotor rhythms to operate a brain-computer interface. *Neurology*, *64*, 1775–7. https://doi.org/10.1212/01.WNL.0000158616.43002.6D

Kühn, A. A., Doyle, L., Pogosyan, A., Yarrow, K., Kupsch, A., Schneider, G.-H. H., ... Brown, P. (2006a). Modulation of beta oscillations in the subthalamic area during motor imagery in Parkinson's disease. *Brain*, *129*, 695–706. https://doi.org/10.1093/brain/awh715

Kühn, A. A., Kempf, F., Brucke, C., Gaynor Doyle, L., Martinez-Torres, I., Pogosyan, A., ... Brown, P. (2008). High-frequency stimulation of the subthalamic nucleus suppresses oscillatory activity in patients with Parkinson's disease in parallel with improvement in motor performance. *Journal of Neuroscience*, *28*, 6165–6173. https://doi.org/10.1523/JNEUROSCI.0282-08.2008

Kühn, A. A., Kupsch, A., Schneider, G.-H., & Brown, P. (2006b). Reduction in subthalamic 8-35 Hz oscillatory activity correlates with clinical improvement in Parkinson's disease. *European Journal of Neuroscience*, *23*, 1956–60. https://doi.org/10.1111/j.1460-9568.2006.04717.x

Kühn, A. A., Tsui, A., Aziz, T., Ray, N., Brücke, C., Kupsch, A., Schneider, G.-H., & Brown, P. (2009). Pathological synchronisation in the subthalamic nucleus of patients with Parkinson's disease relates to both bradykinesia and rigidity. *Experimental Neurology*, *215*, 380–7. https://doi.org/10.1016/j.expneurol.2008.11.008

Kühn, A. A., Williams, D., Kupsch, A., Limousin, P., Hariz, M., Schneider, G. H., Yarrow, K., & Brown, P. (2004). Event-related beta desynchronization in human subthalamic nucleus correlates with motor performance. *Brain*, *127*, 735–746. https://doi.org/10.1093/brain/awh106

Lacey, M. G., Gooding-Williams, G., Prokic, E. J., Yamawaki, N., Hall, S. D., Stanford, I. M., & Woodhall, G. L. (2014). Spike firing and IPSPs in layer V pyramidal neurons during beta oscillations in rat primary motor cortex (M1) in vitro. *PLoS One*, *9*, e85109. https://doi.org/10.1371/journal.pone.0085109

Leuthardt, E. C., Schalk, G., Wolpaw, J. R., Ojemann, J. G., & Moran, D. W. (2004). A brain–computer interface using electrocorticographic signals in humans. *Journal of Neural Engineering*, *1*, 63–71. https://doi.org/10.1088/1741-2560/1/2/001

Levy, R., Ashby, P., Hutchison, W. D., Lang, A. E., Lozano, A. M., & Dostrovsky, J. O. (2002). Dependence of subthalamic nucleus oscillations on movement and dopamine in Parkinson's disease. *Brain*, *125*, 1196–1209. https://doi.org/10.1093/brain/awf128

Litvak, V., Eusebio, A., Jha, A., Oostenveld, R., Barnes, G., Foltynie, T., ... Brown, P. (2012). Movement-related changes in local and long-range synchronization in Parkinson's disease revealed by simultaneous magnetoencephalography and intracranial recordings. *Journal of Neuroscience*, *32*, 10541–10553. https://doi.org/10.1523/JNEUROSCI.0767-12.2012

Litvak, V., Jha, A., Eusebio, A., Oostenveld, R., Foltynie, T., Limousin, P., ... Brown, P. (2011). Resting oscillatory cortico-subthalamic connectivity in patients with Parkinson's disease. *Brain, 134*, 359–374. https://doi.org/10.1093/brain/awq332

Liu, X., Wang, S., Yianni, J., Nandi, D., Bain, P. G., Gregory, R., Stein, J. F., & Aziz, T. Z. (2008). The sensory and motor representation of synchronized oscillations in the globus pallidus in patients with primary dystonia. *Brain, 131*, 1562–1573. https://doi.org/10.1093/brain/awn083

López-Azcárate, J., Tainta, M., Rodríguez-Oroz, M. C., Valencia, M., González, R., Guridi, J., ... Alegre, M. (2010). Coupling between beta and high-frequency activity in the human subthalamic nucleus may be a pathophysiological mechanism in Parkinson's disease. *Journal of Neuroscience, 30*, 6667–6677. https://doi.org/10.1523/JNEUROSCI.5459-09.2010

Manganotti, P., Gerloff, C., Toro, C., Katsuta, H., Sadato, N., Zhuang, P., Leocani, L., & Hallett, M. (1998). Task-related coherence and task-related spectral power changes during sequential finger movements. *Electroencephalography and Clinical Neurophysiology, 109*, 50–62. https://doi.org/10.1016/S0924-980X(97)00074-X

McAuley, J. H. & Marsden, C. D. (2000). Physiological and pathological tremors and rhythmic central motor control. *Brain, 123*, 1545–1567. https://doi.org/10.1093/brain/123.8.1545

Miller, K. J., Leuthardt, E. C., Schalk, G., Rao, R. P. N., Anderson, N. R., Moran, D. W., Miller, J. W., & Ojemann, J. G. (2007). Spectral changes in cortical surface potentials during motor movement. *Journal of Neuroscience, 27*, 2424–2432. https://doi.org/10.1523/JNEUROSCI.3886-06.2007

Mima, T., Matsuoka, T., & Hallett, M. (2000). Functional coupling of human right and left cortical motor areas demonstrated with partial coherence analysis. *Neuroscience Letters, 287*, 93–96. https://doi.org/10.1016/S0304-3940(00)01165-4

Mima, T., Simpkins, N., Oluwatimilehin, T., & Hallett, M. (1999). Force level modulates human cortical oscillatory activities. *Neuroscience Letters, 275*, 77–80. https://doi.org/10.1016/S0304-3940(99)00734-X

Moisello, C., Blanco, D., Lin, J., Panday, P., Kelly, S.P., Quartarone, A., ... Ghilardi, M. F. (2015). Practice changes beta power at rest and its modulation during movement in healthy subjects but not in patients with Parkinson's disease. *Brain and Behavior, 5*, e00374. https://doi.org/10.1002/brb3.374

Moran, D. W. & Schwartz, A. B. (2017). Motor cortical representation of speed and direction during reaching. *Journal of Neurophysiology, 82*, 2676–2692. https://doi.org/10.1152/jn.1999.82.5.2676

Müller, K., Kass-Iliyya, F., & Reitz, M. (1997). Ontogeny of ipsilateral corticospinal projections: A developmental study with transcranial magnetic stimulation. *Annals of Neurology, 42*, 705–711. https://doi.org/10.1002/ana.410420506

Murthy, V. N. & Fetz, E. E. (1996a). Synchronization of neurons during local field potential oscillations in sensorimotor cortex of awake monkeys. *Journal of Neurophysiology, 76*, 3968–3982. https://doi.org/10.1152/jn.1996.76.6.3968

Murthy, V. N. & Fetz, E. E. (1996b). Oscillatory activity in sensorimotor cortex of awake monkeys: synchronization of local field potentials and relation to behavior. *Journal of Neurophysiology, 76*, 3949–3967. https://doi.org/10.1152/jn.1996.76.6.3949

Muthukumaraswamy, S. D. (2010). Functional properties of human primary motor cortex gamma oscillations. *Journal of Neurophysiology, 104*, 2873–2885. https://doi.org/10.1152/jn.00607.2010

Muthukumaraswamy, S. D., Johnson, B. W., & McNair, N. A. (2004). Mu rhythm modulation during observation of an object-directed grasp. *Cognitive Brain Research, 19*, 195–201. https://doi.org/10.1016/j.cogbrainres.2003.12.001

Neumann, W.-J., Degen, K., Schneider, G.-H., Brücke, C., Huebl, J., Brown, P., & Kühn, A. A. (2016). Subthalamic synchronized oscillatory activity correlates with motor impairment in patients with Parkinson's disease. *Movement Disorders*, *31*, 1748–1751. https://doi.org/10.1002/mds.26759

Neumann, W.-J., Jha, A., Bock, A., Huebl, J., Horn, A., Schneider, G., ... Kühn, A. A. (2015). Cortico-pallidal oscillatory connectivity in patients with dystonia. *Brain*, *138*, 1894–1906. https://doi.org/10.1093/brain/awv109

Neuper, C. & Pfurtscheller, G. (2001). Event-related dynamics of cortical rhythms: Frequency-specific features and functional correlates. *International Journal of Psychophysiology*, *43*, 41–58. https://doi.org/10.1016/S0167-8760(01)00178-7

Nolte, G., Bai, O., Wheaton, L., Mari, Z., Vorbach, S., & Hallett, M. (2004). Identifying true brain interaction from EEG data using the imaginary part of coherency. *Clinical Neurophysiology*, *115*, 2292–307. https://doi.org/10.1016/j.clinph.2004.04.029

O'Keefe, J. & Recce, M. L. (1993). Phase relationship between hippocampal place units and the EEG theta rhythm. *Hippocampus* 3, 317–330. https://doi.org/10.1002/hipo.450030307

Ohara, S., Mima, T., Baba, K., Ikeda, A., Kunieda, T., Matsumoto, R., ... Shibasaki, H. (2001). Increased synchronization of cortical oscillatory activities between human supplementary motor and primary sensorimotor areas during voluntary movements. *Journal of Neuroscience*, *21*, 9377–9386. https://doi.org/10.1523/JNEUROSCI.21-23-09377.2001

Oswal, A., Brown, P., & Litvak, V. (2013). Movement related dynamics of subthalmo-cortical alpha connectivity in Parkinson's disease. *NeuroImage*, *70*, 132–142. https://doi.org/10.1016/j.neuroimage.2012.12.041

Özkurt, T. E., Butz, M., Homburger, M., Elben, S., Vesper, J., Wojtecki, L., & Schnitzler, A. (2011). High-frequency oscillations in the subthalamic nucleus: A neurophysiological marker of the motor state in Parkinson's disease. *Experimental Neurology*, *229*, 324–31. https://doi.org/10.1016/j.expneurol.2011.02.015

Paninski, L., Fellows, M. R., Hatsopoulos, N. G., & Donoghue, J. P. (2004). Spatiotemporal tuning of motor cortical neurons for hand position and velocity. *Journal of Neurophysiology*, *91*, 515–532. https://doi.org/10.1152/jn.00587.2002

Paradiso, G., Cunic, D., Saint-Cyr, J. A., Hoque, T., Lozano, A. M., Lang, A. E., & Chen, R. (2004). Involvement of human thalamus in the preparation of self-paced movement. *Brain*, *127*, 2717–2731. https://doi.org/10.1093/brain/awh288

Pastötter, B., Berchtold, F., & Bäuml, K.-H. T. (2012). Oscillatory correlates of controlled speed-accuracy tradeoff in a response-conflict task. *Human Brain Mapping*, *33*, 1834–49. https://doi.org/10.1002/hbm.21322

Pfurtscheller, G. & Lopes Da Silva, F. H. (1999). Event-related EEG/MEG synchronization and desynchronization: Basic principles. *Clinical Neurophysiology*, *110*, 1842–1857. https://doi.org/10.1016/S1388-2457(99)00141-8

Pfurtscheller, G. & Neuper, C. (1992). Simultaneous EEG 10 Hz desynchronization and 40 Hz synchronization during finger movements. *Neuroreport*, *3*(12), 1057–1060. https://doi.org/10.1097/00001756-199212000-00006

Pfurtscheller, G. & Neuper, C. (1997). Motor imagery activates primary sensorimotor area in humans. *Neuroscience Letters*, *239*, 65–8. https://doi.org/10.1016/s0304-3940(97)00889-6

Pfurtscheller, G., Neuper, C., & Kalcher, J. (1993). 40-Hz oscillations during motor behavior in man. *Neuroscience Letters*, *164*, 179–182. https://doi.org/10.1016/0304-3940(93)90886-P

Pfurtscheller, G., Stančák, A., & Neuper, C. (1996). Post-movement beta synchronization. A correlate of an idling motor area? *Electroencephalography and Clinical Neurophysiology*, *98*, 281–293. https://doi.org/10.1016/0013-4694(95)00258-8

Pfurtscheller, G., Zalaudek, K., & Neuper, C. (1998). Event-related beta synchronization after wrist, finger and thumb movement. *Electroencephalography and Clinical Neurophysiology, 109*, 154–160. https://doi.org/10.1016/S0924-980X(97)00070-2

Piña-Fuentes, D., van Dijk, J. M. C., Drost, G., van Zijl, J. C., van Laar, T., Tijssen, M. A. J., & Beudel, M. (2019). Direct comparison of oscillatory activity in the motor system of Parkinson's disease and dystonia: A review of the literature and meta-analysis. *Clinical Neurophysiology, 130*, 917–924. https://doi.org/10.1016/j.clinph.2019.02.015

Pogosyan, A., Gaynor, L. D., Eusebio, A., & Brown, P. (2009). Boosting cortical activity at beta-band frequencies slows movement in humans. *Current Biology, 19*, 1637–41. https://doi.org/10.1016/j.cub.2009.07.074

Pollok, B., Butz, M., Gross, J., & Schnitzler, A. (2007). Intercerebellar coupling contributes to bimanual coordination. *Journal of Cognitive Neuroscience, 19*, 704–719. https://doi.org/10.1162/jocn.2007.19.4.704

Pollok, B., Gross, J., Müller, K., Aschersleben, G., & Schnitzler, A. (2005). The cerebral oscillatory network associated with auditorily paced finger movements. *NeuroImage, 24*, 646–655. https://doi.org/10.1016/j.neuroimage.2004.10.009

Priori, A., Foffani, G., Pesenti, A., Tamma, F., Bianchi, A., Pellegrini, M., ... Villani, R. (2004). Rhythm-specific pharmacological modulation of subthalamic activity in Parkinson's disease. *Experimental Neurology, 189*, 369–379. https://doi.org/10.1016/j.expneurol.2004.06.001

Raethjen, J. & Deuschl, G. (2012). The oscillating central network of Essential tremor. *Clinical Neurophysiology, 123*, 61–64. https://doi.org/10.1016/j.clinph.2011.09.024

Ray, N. J., Jenkinson, N., Wang, S., Holland, P., Brittain, J. S., Joint, C., Stein, J. F., & Aziz, T. (2008). Local field potential beta activity in the subthalamic nucleus of patients with Parkinson's disease is associated with improvements in bradykinesia after dopamine and deep brain stimulation. *Experimental Neurology, 213*, 108–113. https://doi.org/10.1016/j.expneurol.2008.05.008

Rickert, J., de Oliveira, S. C., Vaadia, E., Aertsen, A., Rotter, S., & Mehring, C. (2005). Encoding of movement direction in different frequency ranges of motor cortical local field potentials. *Journal of Neuroscience, 25*, 8815–24. https://doi.org/10.1523/JNEUROSCI.0816-05.2005

Riddle, C. N. & Baker, S. N. (2005). Manipulation of peripheral neural feedback loops alters human corticomuscular coherence. *Journal of Physiology, 566*, 625–639. https://doi.org/10.1113/jphysiol.2005.089607

Ritter, P., Moosmann, M., & Villringer, A. (2009). Rolandic alpha and beta EEG rhythms' strengths are inversely related to fMRI-BOLD signal in primary somatosensory and motor cortex. *Human Brain Mapping, 30*, 1168–1187. https://doi.org/10.1002/hbm.20585

Rossiter, H. E., Davis, E. M., Clark, E. V, Boudrias, M.-H., & Ward, N. S. (2014). Beta oscillations reflect changes in motor cortex inhibition in healthy ageing. *NeuroImage, 91*, 360–5. https://doi.org/10.1016/j.neuroimage.2014.01.012

Rueda-Delgado, L. M., Heise, K. F., Daffertshofer, A., Mantini, D., & Swinnen, S. P. (2019). Age-related differences in neural spectral power during motor learning. *Neurobiology of Aging, 77*, 44–57. https://doi.org/10.1016/j.neurobiolaging.2018.12.013

Rueda-Delgado, L. M., Solesio-Jofre, E., Serrien, D. J., Mantini, D., Daffertshofer, A., & Swinnen, S. P. (2014). Understanding bimanual coordination across small time scales from an electrophysiological perspective. *Neuroscience & Biobehavioral* Reviews, *47*, 614–635. https://doi.org/10.1016/j.neubiorev.2014.10.003

Safri, N. M., Murayama, N., Hayashida, Y., & Igasaki, T. (2007). Effects of concurrent visual tasks on cortico-muscular synchronization in humans. *Brain Research, 1155*, 81–92. https://doi.org/10.1016/j.brainres.2007.04.052

Salenius, S., Portin, K., Kajola, M., Salmelin, R., & Hari, R. (1997). Cortical control of human motoneuron firing during isometric contraction. *Journal of Neurophysiology, 77*, 3401–3405. https://doi.org/10.1016/j.conb.2003.10.008

Salmelin, R., Hämäläinen, M., Kajola, M., & Hari, R. (1995). Functional segregation of movement-related rhythmic activity in the human brain. *NeuroImage, 2*, 237–243. https://doi.org/10.1006/nimg.1995.1031

Salmelin, R. & Hari, R. (1994). Spatiotemporal characteristics of sensorimotor neuromagnetic rhythms related to thumb movement. *Neuroscience, 60*, 537–550. https://doi.org/10.1016/0306-4522(94)90263-1

Sanes, J. N. & Donoghue, J. P. (1993). Oscillations in local field potentials of the primate motor cortex during voluntary movement. *Proceedings of the National Academy of Sciences of the United States of America, 90*, 4470–4474. https://doi.org/10.1073/pnas.90.10.4470

Schmiedt-Fehr, C., Mathes, B., Kedilaya, S., Krauss, J., & Başar-Eroğlu, C. (2016). Aging differentially affects alpha and beta sensorimotor rhythms in a go/no-go task. *Clinical Neurophysiology, 127*, 3234–42. https://doi.org/10.1016/j.clinph.2016.07.008

Schnitzler, A., Münks, C., Butz, M., Timmermann, L., & Gross, J. (2009). Synchronized brain network associated with essential tremor as revealed by magnetoencephalography. *Movement Disorders, 24*, 1629–1635. https://doi.org/10.1002/mds.22633

Schnitzler, A., Salenius, S., Salmelin, R., Jousmäki, V., & Hari, R. (1997). Involvement of primary motor cortex in motor imagery: A neuromagnetic study. *NeuroImage, 6*, 201–8. https://doi.org/10.1006/nimg.1997.0286

Serrien, D. J. (2008). The neural dynamics of timed motor tasks: Evidence from a synchronization–continuation paradigm. *European Journal of Neuroscience, 27*, 1553–1560. https://doi.org/10.1111/j.1460-9568.2008.06110.x

Sharott, A., Grosse, P., Kühn, A.A., Salih, F., Engel, A.K., Kupsch, A., . . . Brown, P. (2008). Is the synchronization between pallidal and muscle activity in primary dystonia due to peripheral afferance or a motor drive? *Brain, 131*, 473–484. https://doi.org/10.1093/brain/awm324

Sherman, M. A., Lee, S., Law, R., Haegens, S., Thorn, C. A., Hämäläinen, M. S., Moore, C. I., & Jones, S. R. (2016). Neural mechanisms of transient neocortical beta rhythms: Converging evidence from humans, computational modeling, monkeys, and mice. *Proceedings of the National Academy of Sciences of the United States of America, 113*, E4885–E4894. https://doi.org/10.1073/pnas.1604135113

Silberstein, P., Kühn, A. A., Kupsch, A., Trottenberg, T., Krauss, J. K., Wöhrle, J.C., . . . Brown, P. (2003). Patterning of globus pallidus local field potentials differs between Parkinson's disease and dystonia. *Brain, 126*, 2597–608. https://doi.org/10.1093/brain/awg267

Spinks, R. L., Kraskov, A., Brochier, T., Umilta, M. A., & Lemon, R. N. (2008). Selectivity for grasp in local field potential and single neuron activity recorded simultaneously from M1 and F5 in the awake macaque monkey. *Journal of Neuroscience, 28*, 10961–71. https://doi.org/10.1523/JNEUROSCI.1956-08.2008

Stam, C. J., Nolte, G., & Daffertshofer, A. (2007). Phase lag index: Assessment of functional connectivity from multi-channel EEG and MEG with diminished bias from common sources. *Human Brain Mapping, 28*, 1178–1193. https://doi.org/10.1002/hbm.20346

Stančák, A., Feige, B., Lücking, C. H., & Kristeva-Feige, R. (2000). Oscillatory cortical activity and movement-related potentials in proximal and distal movements. *Clinical Neurophysiology, 111*, 636–650. https://doi.org/10.1016/S1388-2457(99)00310-7

Stančák, A. & Pfurtscheller, G. (1996). Mu-rhythm changes in brisk and slow self-paced finger movements. *Neuroreport, 7*, 1161–1164. https://doi.org/10.1097/00001756-199604260-00013

Stančák, A., Riml, A., & Pfurtscheller, G. (1997). The effects of external load on movement-related changes of the sensorimotor EEG rhythms. *Electroencephalography and Clinical Neurophysiology, 102*, 495–504. https://doi.org/10.1016/S0013-4694(96)96623-0

Steinemann, N. A., O'Connell, R. G., & Kelly, S. P. (2018). Decisions are expedited through multiple neural adjustments spanning the sensorimotor hierarchy. *Nature Communications, 9*, 3627. https://doi.org/10.1038/s41467-018-06117-0

Swann, N., Tandon, N., Canolty, R., Ellmore, T. M., McEvoy, L. K., Dreyer, S., DiSano, M., & Aron, A. R. (2009). Intracranial EEG reveals a time- and frequency-specific role for the right inferior frontal gyrus and primary motor cortex in stopping initiated responses. *Journal of Neuroscience, 29*, 12675–12685. https://doi.org/10.1523/JNEUROSCI.3359-09.2009

Szurhaj, W., Bourriez, J. L., Kahane, P., Chauvel, P., Mauguière, F., & Derambure, P. (2005). Intracerebral study of gamma rhythm reactivity in the sensorimotor cortex. *European Journal of Neuroscience, 21*, 1223–1235. https://doi.org/10.1111/j.1460-9568.2005.03966.x

Talakoub, O., Neagu, B., Udupa, K., Tsang, E., Chen, R., Popovic, M. R., & Wong, W. (2016). Time-course of coherence in the human basal ganglia during voluntary movements. *Scientific Reports, 6*, 34930. https://doi.org/10.1038/srep34930

Tan, H., Jenkinson, N., & Brown, P. (2014). Dynamic neural correlates of motor error monitoring and adaptation during trial-to-trial learning. *Journal of Neuroscience, 34*, 5678–5688. https://doi.org/10.1523/JNEUROSCI.4739-13.2014

Tan, H., Pogosyan, A., Anzak, A., Ashkan, K., Bogdanovic, M., Green, A.L., … Brown, P. (2013). Complementary roles of different oscillatory activities in the subthalamic nucleus in coding motor effort in Parkinsonism. *Experimental Neurology, 248*, 187–195. https://doi.org/10.1016/j.expneurol.2013.06.010

Tan, H., Wade, C., & Brown, P. (2016). Post-movement beta activity in sensorimotor cortex indexes confidence in the estimations from internal models. *Journal of Neuroscience, 36*, 1516–1528. https://doi.org/10.1523/JNEUROSCI.3204-15.2016

Tinkhauser, G., Pogosyan, A., Little, S., Beudel, M., Herz, D. M., Tan, H., & Brown, P. (2017a). The modulatory effect of adaptive deep brain stimulation on beta bursts in Parkinson's disease. *Brain, 140*, 1053–1067. https://doi.org/10.1093/brain/awx010

Tinkhauser, G., Pogosyan, A., Tan, H., Herz, D. M., Kühn, A. A., & Brown, P. (2017b). Beta burst dynamics in Parkinson's disease OFF and ON dopaminergic medication. *Brain, 140*, 2968–2981. https://doi.org/10.1093/brain/awx252

Toma, K., Mima, T., Matsuoka, T., Gerloff, C., Ohnishi, T., Koshy, B., Andres, F., & Hallett, M. (2002). Movement rate effect on activation and functional coupling of motor cortical areas. *Journal of Neurophysiology, 88*, 3377–3385. https://doi.org/10.1152/jn.00281.2002

Trevarrow, M. P., Kurz, M. J., McDermott, T. J., Wiesman, A. I., Mills, M. S., Wang, Y.-P., … Wilson, T. W. (2019). The developmental trajectory of sensorimotor cortical oscillations. *NeuroImage, 184*, 455–461. https://doi.org/10.1016/j.neuroimage.2018.09.018

Tsang, E.W., Hamani, C., Moro, E., Mazzella, F., Lozano, A.M., Hodaie, M., Yeh, I.-J., & Chen, R. (2012). Movement related potentials and oscillatory activities in the human internal globus pallidus during voluntary movements. *Journal of Neurology, Neurosurgery, and Psychiatry, 83*, 91–7. https://doi.org/10.1136/jnnp.2011.243857

Tsiokos, C., Hu, X., & Pouratian, N. (2013). 200–300 Hz movement modulated oscillations in the internal globus pallidus of patients with Parkinson's Disease. *Neurobiological Disorders, 54*, 464–474. https://doi.org/10.1016/j.nbd.2013.01.020

van Wijk, B. C. M. (2017). Is broadband gamma activity pathologically synchronized to the beta rhythm in Parkinson's disease? *Journal of Neuroscience, 37*, 9347–9349. https://doi.org/10.1523/JNEUROSCI.2023-17.2017

van Wijk, B. C. M., Beek, P. J., & Daffertshofer, A. (2012). Differential modulations of ipsilateral and contralateral beta (de)synchronization during unimanual force production. *European Journal of Neuroscience, 36*, 2088–2097. https://doi.org/10.1111/j.1460-9568.2012.08122.x

van Wijk, B .C. M., Beudel, M., Jha, A., Oswal, A., Foltynie, T., Hariz, M. I., … Litvak, V. (2016). Subthalamic nucleus phase-amplitude coupling correlates with motor impairment in Parkinson's disease. *Clinical Neurophysiology, 127*, 2010–2019. https://doi.org/10.1016/j.clinph.2016.01.015

van Wijk, B. C. M., Daffertshofer, A., Roach, N., & Praamstra, P. (2009) A role of beta oscillatory synchrony in biasing response competition? *Cerebral Cortex, 19*, 1294–1302. https://doi.org/10.1093/cercor/bhn174

van Wijk, B. C. M., Neumann, W.-J., Schneider, G.-H., Sander, T. H., Litvak, V., & Kühn, A. A. (2017). Low-beta cortico-pallidal coherence decreases during movement and correlates with overall reaction time. *NeuroImage, 159*, 1–8. https://doi.org/10.1016/j.neuroimage.2017.07.024

Vidailhet, M., Vercueil, L., Houeto, J.-L., Krystkowiak, P., Benabid, A.-L., Cornu, P., … Pollak, P. (2005). Bilateral deep-brain stimulation of the globus pallidus in primary generalized dystonia. *New England Journal of Medicine, 352*, 459–467. https://doi.org/10.1056/nejmoa042187

Weaver, F. M., Follett, K., Stern, M., Hur, K., Harris, C., Marks, … Huang, G.D.; CSP 468 Study Group. (2009). Bilateral deep brain stimulation vs best medical therapy for patients with advanced Parkinson disease: A randomized controlled trial. *Journal of the American Medical Association, 301*, 63–73. https://doi.org/10.1001/jama.2008.929

Wessel, J. R., Ghahremani, A., Udupa, K., Saha, U., Kalia, S. K., Hodaie, M., … Chen, R. (2016). Stop-related subthalamic beta activity indexes global motor suppression in Parkinson's disease. *Movement Disorders, 12*, 1846–1853. https://doi.org/10.1002/mds.26732

Witham, C. L., Riddle, C. N., Baker, M. R., & Baker, S. N. (2011). Contributions of descending and ascending pathways to corticomuscular coherence in humans. *Journal of Physiology, 589*, 3789–3800. https://doi.org/10.1113/jphysiol.2011.211045

Witte, M., Patino, L., Andrykiewicz, A., Hepp-Reymond, M.-C., & Kristeva, R. (2007). Modulation of human corticomuscular beta-range coherence with low-level static forces. *European Journal of Neuroscience, 26*, 3564–3570. https://doi.org/10.1111/j.1460-9568.2007.05942.x

Wolpaw, J. R., Birbaumer, N., McFarland, D. J., Pfurtscheller, G., & Vaughan, T. M. (2002). Brain–computer interfaces for communication and control. *Clinical Neurophysiology, 113*, 767–791. https://doi.org/10.1016/S1388-2457(02)00057-3

Yamawaki, N., Stanford, I. M., Hall, S. D., & Woodhall, G. L. (2008). Pharmacologically induced and stimulus evoked rhythmic neuronal oscillatory activity in the primary motor cortex in vitro. *Neuroscience, 151*, 386–395. https://doi.org/10.1016/j.neuroscience.2007.10.021

Zandvoort, C. S., van Dieën, J. H., Dominici, N., & Daffertshofer, A. (2019). The human sensorimotor cortex fosters muscle synergies through cortico-synergy coherence. *NeuroImage, 199*, 30–37. https://doi.org/10.1016/j.neuroimage.2019.05.041

PART III

CHAPTER 13

EEG FREQUENCY DEVELOPMENT ACROSS INFANCY AND CHILDHOOD

KIMBERLY CUEVAS AND MARTHA ANN BELL

13.1 Introduction

HANS Berger (1929) published the first report of the electroencephalogram (EEG) recorded on an adult scalp. After success with adult recordings, Berger began to record electrical activity from the scalps of a wide age range of individuals and it was Berger's fifth report three years later (1932) that created the field of developmental neurophysiology. Berger said in the earlier part of his fifth report that he had recorded EEG from the scalps of a few children but had not yet recorded EEG from anyone under 5 years of age. His rationale for avoiding younger children was that he was using needle electrodes and did not want to use local anesthesia with such young individuals. Thus, for his initial recordings with infants, Berger used silver foil electrodes on the forehead and occiput, with each the size of the palm of his hand. On each section of silver foil was a piece of flannel cloth soaked in saline. Another piece of silver foil was placed on top so that the flannel would not dry out. He held this on the scalp with a rubber bandage (Berger, 1932).

Berger reported that he attempted to record EEG from six infants between 8 and 13 days old, but there were no characteristic alpha waves that he had come to expect after doing so many adult and older child EEG recordings. He concluded that there was no EEG in the first weeks after birth because the cerebral cortex had not yet assumed its function. A 35-day-old infant (a "healthy and very strong boy") was the youngest in whom Berger could record EEG. The infant lay completely still and began to fall asleep and there was some evidence of the posterior alpha activity that Berger had seen for the past three years in older children and adults. Berger (1932) concluded in his fifth report that one could see EEG in all healthy and strong children who are older than 2 months of age and that by the age of 5 years, children show an EEG tracing comparable to that of an adult.

The EEG recordings Berger accomplished from infants were highly intriguing. Infant EEG had greater amplitude and cycled at a lower frequency than adult EEG. From Berger's time, researchers have assumed that EEG differences among infants, children, and adults reflect differences in brain maturation. Longitudinal research studies, including our own contributions (Bell & Fox, 1992; Cuevas & Bell, 2011), have verified that position. Researchers continue to examine infant and young child EEG as a relatively inexpensive, non-invasive way to study brain development.

This chapter provides an overview of the ontogeny of EEG frequency bands during infancy and early childhood. We include infant correlates of adult frequency bands labeled as delta, theta, alpha family, beta, and gamma. We focus most intensely on the 6–9 Hz "infant alpha" frequency band, as it is the band that has received the most attention in the developmental neurophysiology research literature and the frequency band that we have examined so intensely in our own programs of research and our collaborative efforts. We organize the discussion of infant and child alpha around studies focused on action-perception processes, executive processes, and affective processes. The chapter ends with a brief section on broader impacts of infant and child EEG and future directions.

13.2 Ontogenesis of EEG Bands

This section focuses on the oscillatory rhythms that compose the waking EEG during infancy and childhood. The primary frequency bands of normative EEG oscillatory activity based on the adult literature are delta (1–4 Hz), theta (4–8 Hz), alpha (8–13 Hz), beta (13–30 Hz), and gamma (36–44 Hz). These oscillations often occur at slower rates during early development. However, throughout ontogeny, EEG amplitude is typically inversely related to frequency, with greater amplitude exhibited for slower oscillations. The functional properties of oscillatory rhythms are defined here in terms of event-related changes in EEG power values (*EEG reactivity*) in response to external stimuli and/or psychological (e.g., cognitive, affective) processing. Specifically, the term *synchronization* refers to an increase in EEG power values relative to resting baseline, whereas *desynchronization* refers to a decrease in EEG power values.

We provide an overview of what is known about each of these EEG frequency bands during early development, including functional properties as well as age-related changes in frequency and amplitude. In general, there is a decrease in the amount of low frequency activity (1–5 Hz; i.e., delta, lower-theta) and an increase in intermediate frequency oscillations (6–12 Hz; i.e., upper-theta; alpha) from infancy through childhood (Gibbs & Knott, 1949; Hagne, 1968; Marshall et al., 2002). As there is no standardization of EEG rhythms in the developmental neurophysiology literature as found in adolescent and adult EEG work, EEG researchers have used multiple techniques to determine appropriate frequency band definitions for infants and young children.

13.2.1 Delta Band

Large, slow wave oscillations in the delta band are typically investigated in the context of sleep. Although oscillatory activity within the delta frequency band is present in the waking EEG during early development, descriptions of this band have primarily focused on decreases in delta activity and corresponding increases in higher frequency oscillations as a function of age (Hagne, 1968; Ogawa et al., 1984). Although Lindsley (1938) recorded from only occipital sites, he noted the presence of "type V waves" in subjects from 3 months of age through adulthood, with the average frequency increasing gradually from 0.68 to 1.08 Hz over this developmental period. The delta rhythm was unaffected by stimulation and reached the 1 Hz oscillation rate by 3 years of age.

13.2.2 Theta Band

Oscillatory activity within the theta range is faster than the adjacent delta band. Theta rhythms are prominent in the waking EEG during early development, but the theta frequency range is slightly lower than that of the adult 4–8 Hz theta band. The adult literature has noted that emotional and cognitive processing is associated with increases in theta power (e.g., Walter & Walter, 1949; see also Chapter 15). Theta oscillations have been hypothesized to be linked to reward processing and active learning during early development (Begus & Bonawitz, 2020). Visual perceptual-stimulation techniques have also been used to "entrain" (elicit) neural oscillatory activity at a particular frequency, with specific focus on the functional properties of the theta rhythm (Köster et al., 2019). To our knowledge, however, there have been no systematic examinations of theta's functional properties throughout development. Here, we highlight some of the characteristics of theta reactivity during infancy and childhood.

In the context of examining 7- to 12-month-olds' alpha rhythms, Stroganova and colleagues (Stroganova et al., 1999) identified 3.6–6 Hz activity with unique functional properties. This band exhibited increases in theta power during "internally controlled (anticipatory) attention" as infants were engaged in a peek-a-boo game (Orekhova et al., 1999; Stroganova et al., 1998). Recent analysis of 6- to 12-month-olds' sustained attention, using heart-rate-defined phases of attention, has confirmed and extended these findings. By 10–12 months of age, sustained attention during a video clip was associated with theta (2–6 Hz) synchronization at multiple scalp sites, including frontal pole, central, and parietal (Xie et al., 2018). Event-related theta synchronization during attentional processes may also be related to infants' subsequent memory and cognitive abilities. For instance, increases in 11-month-olds' frontal theta (3–5 Hz) power during object exploration was associated with variability in infants' future object recognition

(Begus et al., 2015). Likewise, measures of frontal theta (3–6 Hz) enhancement during a dynamic non-social video at 6 months of age is associated with non-verbal cognitive abilities at 9 months (Braithwaite et al., 2020). Together, these findings indicate that theta synchronization is intricately linked to aspects of infant attention and memory; however, the precise developmental onset and trajectory of event-related theta reactivity are important open areas of inquiry.

Increases in theta power are also exhibited in response to positive and negative affective processing during infancy (Futagi et al., 1998; Nikitina et al., 1985) as well as nutritive sucking (Lehtonen et al., 2016; Paul et al., 1996). However, little is known about age-related changes in theta reactivity beyond infancy. To this end, Orekhova and colleagues (Orekhova et al., 2006) directly compared infants' and young children's theta reactivity to "theta-provoking" stimuli—toy exploration and attention to social stimulation (i.e., adult speech). Theta scalp topography differed as a function of age and stimulus type; though, both 8- to 12-month-olds and 3- to 6-year-olds exhibited increases in theta (3.6–5.6 Hz and 4–8 Hz, respectively) power in response to these stimuli. Thus, in some affective contexts, the adult theta range is developmentally appropriate by early childhood. Recent evidence also indicates changes in the topography of infants' theta (3–6 Hz) synchronization in response to social stimuli between 6 and 12 months of age (Jones et al., 2015). In sum, theta synchronization is displayed in a variety of cognitive and emotional settings early in development, but the precise nature of the scalp topography of theta responsivity varies as a function of multiple factors, including age and reactive context.

13.2.3 Alpha Band ("Alpha Family")

The majority of developmental research on waking EEG has examined oscillatory activity that falls within the alpha frequency range. The alpha band (or "*alpha family*") includes two rhythms that overlap in frequency, but differ in their functional properties, scalp topographies, and developmental trajectories. Berger (1929) initially identified the *posterior alpha rhythm* (or *occipital alpha rhythm*) in the adult waking EEG. This oscillatory activity is prominent over posterior scalp locations (parietal, occipital) when there is a homogenous visual field (e.g., dark room; eye closure) and it is suppressed by visual stimulation (e.g., flashes of light; eye opening). The *central alpha rhythm*, also known as the *Rolandic mu rhythm* or *pre-central alpha rhythm*, is prominent over the sensorimotor cortex during periods of stillness (i.e., no motor movement). Foundational work with adults has revealed that mu oscillatory activity is suppressed in response to motor movement (Gastaut et al., 1954) with more recent investigations extending these findings to motor imagery and the perception of others' actions (Muthukumaraswamy & Johnson, 2004; Pfurtscheller et al., 2006). Thus, although the mu and posterior alpha rhythms are both within the alpha frequency range, their unique topographic and functional properties indicate that they are independent oscillations. The second half of this chapter reviews developmental work focusing on frontal EEG activity within the alpha band (6–9 Hz) and its associations with affective and cognitive processes. Frontal alpha activity is proposed to have unique neural generators by some, and to be associated with central and/or posterior alpha rhythms by others (see Stroganova & Orekhova, 2007 for discussion).

13.2.3.1 *Posterior Alpha*

Early developmental EEG investigations using visual quantification noted posterior oscillatory activity that had similar rhythmic qualities to the adult posterior alpha rhythm, though the rate of oscillation was slower. Around 3–4 months of age, an occipital rhythm with a frequency of 3–5 Hz has been identified in the waking EEG (Lindsley, 1938, 1939; Smith, 1938, 1941). It increases in both amplitude and frequency during the first postnatal year (6–7 Hz), with the posterior alpha oscillation rate continuing to gradually increase throughout childhood. The occipital alpha amplitude remains fairly stable until 3–4 years when there is a sharp drop followed by a gradual decline to adult levels by 15–16 years (Lindsley, 1939). The adult alpha frequency range is reached during middle childhood, with mean occipital frequencies of 9 and 10 Hz at 8 and 11–16 years, respectively (Lindsley, 1938, 1939; Smith, 1938, 1941). Subsequent investigations using frequency analyzers and spectral analysis have confirmed these early findings (Hagne, 1968; Hagne et al., 1973; Miskovic et al., 2015).

The functional properties of the posterior alpha rhythm were first noted during early development by Lindsley (1938). Visual stimuli (i.e., bright lights in a dark room) blocked occipital oscillatory activity in young participants, including infants who were a "few months" of age. However, scalp recordings were limited to occipital sites and detailed developmental analyses were not reported. Using spectral analysis, Stroganova and colleagues (1999) confirmed and extended these early findings by comparing 7- to 12-month-olds' posterior oscillatory activity during conditions of darkness (i.e., lights off in the testing room) and quite visual attention (i.e., watching soap bubbles). These conditions were designed to be developmentally appropriate analogues of the "eyes closed versus eyes open" procedures used to identify posterior alpha reactivity with older children and adults. Their work indicated that the 5.2–9.6 Hz band exhibited prominent posterior oscillatory activity in the response to the "lights off" condition, and this activity was attenuated during visual attention. Importantly, these properties were not displayed at anterior/central scalp locations, indicating similar functional properties and scalp topography to the adult posterior alpha rhythm.

13.2.3.2 *Central Alpha (Mu)*

Initial visual examinations of EEG in the neonate indicated rhythmic activity at areas over the sensorimotor cortex during sleep. However, this activity was not seen at occipital sites or when neonates were awake (Smith, 1941). By 3–4 months, 7-Hz rhythmic activity has been found at central sites in awake infants (Smith, 1939, 1941). Using spectral analysis, Hagne and colleagues (1973) noted the "hint of a frequency peak" at central sites at 3 weeks of age around 4–5 Hz, which reached 7–8 Hz by 12 months. Similarly, more recent longitudinal work extending from infancy through early childhood, has indicated that the central peak frequency rises from 6–7 Hz at 5 months to 9 Hz by 4 years of age (Marshall et al., 2002). These findings are consistent with Smith's (1941) original report of central alpha activity increasing to 8 Hz at 1.5 years, to 9 Hz at 3.5 years, to 10 Hz at 10–14 years. Prior to reaching adult frequency ranges, the oscillatory rate of the mu rhythm is faster than the posterior alpha rhythm.

Although the mu rhythm is attenuated during various conditions related to action perception and production, the functional properties of this rhythm are most robust and reliably identified via conditions of discrete motor movement. Currently, the most systematic developmental examination of the functional properties of the central alpha rhythm comes from a cross-sectional magnetoencephalography (MEG) study. Age-related changes in the peak frequency of mu rhythm attenuation during a discrete hand movement (i.e., squeezing a pipette) were found from infancy through childhood. There was a linear increase in peak mu frequency from 2.75 to 8.25 Hz between 11 and 47 weeks of age (Berchicci et al., 2011). This was followed by a more gradual increase in peak mu frequency during early childhood (2–5 years; average 8.5 Hz), which was below the adult 10.2-Hz peak. These findings have been confirmed and extended by a recent cross-sectional EEG study of mu rhythm attenuation during reaching-grasping actions. Mu peaks were found at 7.39, 8.81, and 10.51 Hz at 12 months, 4 years, and 18–21 years, respectively (Thorpe et al., 2016). A bourgeoning recent literature examining the infant mu rhythm has found 6–9 Hz attenuation during action perception and/or execution (Cuevas et al., 2014; Marshall & Meltzoff, 2011); a detailed description of the EEG correlates of action-perception processes is provided later in the chapter.

13.2.4 Beta Band

Berger (1929) identified beta waves in conjunction with alpha waves in the adult waking EEG, but as higher frequency oscillations. Similar to the central alpha rhythm, the "precentral" beta rhythm is associated with motor processes (Jasper & Andrews, 1938; Jasper & Penfield, 1949). The adult literature has revealed that the beta rhythm is attenuated during action execution, motor planning, and action perception (Jarvelainen et al., 2004; Pfurtscheller et al., 1997). The beta frequency is often a harmonic of the mu rhythm frequency. Although these rhythms have overlapping functional properties, there are a variety of unique characteristics of each band (see Crone et al., 1998; Hari & Salmelin, 1997).

Initial description of developmental changes in beta rhythm comes from Lindsley (1938), who identified "type IV waves" from the waking EEG activity at occipital sites (the only recording sites); these waves had mean frequencies approximately double the alpha wave frequency throughout development. This oscillatory activity exhibited substantial increases in frequency throughout infancy and childhood: 7 Hz at 3 months, 11.6 Hz at 12 months, and 16 Hz at 4 years. The adult 20-Hz frequency was attained around 11–12 years. The literature is mixed, however, on whether there are age-related increases or decreases in beta band power (Gasser et al., 1988; Matsuura et al., 1985; Ogawa et al., 1984).

Contemporary developmental investigations of the beta rhythm have primarily focused on its involvement in motor processes. Similar to the mu rhythm, discrete motor movement has been used to characterize the beta rhythm's functional properties. This literature is rather limited in comparison to corresponding work on the central alpha rhythm. Further, there is mixed evidence regarding beta rhythm attenuation during infancy and early childhood. Meyer and colleagues (2016) identified age-appropriate beta rhythm frequencies for 8-month-olds (12–15 Hz), 14-month-olds (14–18 Hz), and adults (16–25 Hz) during reaching/grasping movements. Attenuation of beta activity in response to action production was evident only in 14-month-olds and adults, despite the presence of mu rhythm attenuation in all age groups. Likewise, a recent cross-sectional analysis revealed that although 12-month-olds, 4-year-olds, and adults exhibited mu rhythm attenuation during action execution, beta rhythm attenuation was present only for adults (Thorpe et al., 2016).

Other MEG and EEG work with 3- to 6-year-olds has found evidence of decreases in beta oscillatory activity in response to discrete motor movements (EEG: Bryant & Cuevas, 2019; Liao et al., 2015; MEG: Cheyne et al., 2014; Gaetz et al., 2010). However, the magnitude of young children's beta attenuation was smaller than beta changes exhibited by older children (11–13 years) and adults (Gaetz et al., 2010). Taken together, these findings suggest that although there is evidence of beta oscillatory activity early in ontogeny, its functional properties may not be as readily detected during early development as compared to lower frequency oscillations, such as the mu rhythm. Given their overlapping functional properties, concurrent developmental analysis of beta and central alpha rhythms is essential to a more integrative understanding of the network dynamics underlying motor and action-perception processes (see also Section 13.3.1).

13.2.5 Gamma Band

Because of its low amplitude, high-frequency oscillations, systematic analysis of gamma band activity via visual examination was not feasible (Lindsley, 1938). A variety of broadband and sub-band definitions have been used to define the gamma oscillatory activity, but the vast majority of the developmental and adult literature has focused on oscillations near 40 Hz. Work with adults has revealed that increases in gamma band power are associated with both motor and cognitive processes, including perceptual binding and memory (see Başar, 2013, for review).

Early examination of the gamma (35–45 Hz) band with a large sample of 3- to 12-year-olds (N = 707), noted increased power between 3 and 4 years at all recorded sites (frontal, central, occipital) with a distinct peak at frontal sites between 4 and 5 years of age (Takano & Ogawa, 1998). Although occipital gamma power was stable after 4 years of age, there were some age-related decreases in frontal and central gamma power between 4 and 11 years. A recent analysis of gamma (31–50 Hz) oscillations with a comparatively smaller sample of 3- to 38-year-olds (N = 156) also identified age-related decreases in resting gamma power; however, these changes were seen throughout the scalp

(Tierney et al., 2013). Tierney and colleagues proposed that decreases in gray matter—specifically, synaptic density via synaptic pruning (Whitford et al., 2007)—may underlie the widespread reductions in resting gamma power as a function of age. Additional work is needed to characterize the development of gamma power during the first three postnatal years. A recent large-scale longitudinal investigation between 2 to 6 postnatal months (N = 518) found gradual increases in gamma (30–50 Hz) power across all scalp sites (Pivik et al., 2019). Pivik and colleagues hypothesize that ontogenetic changes in GABAergic (γ-aminobutyric acid) interneurons contribute to emerging gamma activity (Paredes et al., 2016; Xu et al., 2011).

During infancy, resting gamma power varies as a function of internal and external factors, such as socioeconomic status, sex, and diet (Pivik et al., 2019; Tomalski et al., 2013, but see Brito et al., 2016). Further, higher levels of resting gamma oscillatory activity are positively associated with both concurrent and future measures of early cognitive development, including language, memory, and executive function (Benasich et al., 2008; Brito et al., 2013; Tarullo et al., 2017). Developmental studies have also revealed changes in gamma band activity associated with ongoing cognitive processing. An initial investigation of gamma's involvement in perceptual binding revealed bursts of gamma oscillatory activity at frontal areas during the perception of illusory objects in 8-month-olds, but not 6-month-olds (Csibra et al., 2000). These findings confirmed behavioral evidence that younger infants were unsuccessful at perceiving illusory objects (Ghim, 1990). Subsequent work also shows increases in 6-month-olds' gamma oscillatory activity over temporal areas in the context of object occlusion events; presumably contributing infants' ability to maintain object representations (Kaufman et al., 2003, 2006). Gamma oscillations are also associated with aspects of language learning during infancy, including recognition of familiar objects with verbal labels (Gliga et al., 2010) and phonemic perceptual narrowing (Ortiz-Mantilla et al., 2016).

13.3 6–9 Hz Alpha Band: Action-Perception, Executive, and Affective Processes

Our primary focus in this section is on the infant and child frontal alpha and central mu rhythms, as the 6–9 Hz alpha band has received the most attention in the developmental literature. We noted at the beginning of this chapter that the developmental neurophysiology literature has no standardization of EEG rhythms as found in adult EEG work. The research we have reviewed thus far shows how individual research teams since Berger's time have worked toward defining specific frequency bands of interest, with clear indication that the traditional frequency band definitions used with adults (e.g., 8–13 Hz for alpha) do not apply to studies of infant and young child EEG.

After six decades of developmental EEG research, the Society for Psychophysiological Research published a set of recommendations for recording and analyzing EEG in research contexts (Pivik et al., 1993). Those guidelines included two suggestions for EEG researchers who focus on infants and young children. The first was wide band analysis, potentially including all frequencies with evidence of power after doing spectral plots. Although some researchers working with infant samples have used this approach in studies of infant emotion (e.g., 3–13 Hz: Diego et al., 2006), the wide band approach was never commonly adopted for use in infant and early childhood research. This may be because of the assumption that there was the potential for the developmental EEG field to eventually standardize frequency bands based on temporal, spatial, amplitude, and functional characteristics much as the adult EEG field has done.

The second suggestion was that spectral plots be examined and frequency bands determined that center around the peaks in the spectrum. This is the approach we used with our first infant EEG data set (Bell & Fox, 1992). We recorded resting baseline EEG from 13 infants from 7 to 12 months of age because we had hypotheses about relations between the development of frontal EEG and performance on a classic infant task (i.e., Piaget's A-not-B task). Spectral plots of the frontal EEG leads (F3, F4) were created for each infant's monthly EEG recording for a total of 91 frontal spectral plots. At each age the plots revealed a dominant frequency around 6–9 Hz. More importantly, the plots revealed month-to-month changes in the peak frequency for some infants. The spectral plots from one infant in our Bell and Fox (1992) data set are shown in Figure 13.1. The four spectral plots represent this infant's EEG data at F3 and F4 from 8 until 11 months of age. One can observe a peak frequency at 7 Hz in the EEG recording made at 8 months of age and the sharing of peak frequency between 7 and 8 Hz at 9 months of age. In the 10-month recording, peak frequency appears to be shifting toward 8 Hz, whereas at 11 months of age there is only one peak at 8 Hz. These spectral plots demonstrate a definite shift in peak frequency on a month-to-month basis during infancy. They also show a shift in this overall dominant frequency band from 5–9 Hz at 8 months to 6–10 Hz at 11 months for this one infant. After examining all 91 spectral plots, we determined that 6–9 Hz frequency band captured the dominant frequency in this sample of infants.

Our selection of the 6–9 Hz frequency band as dominant in our infant longitudinal study was replicated and extended to early childhood by Marshall and colleagues (2002) with a longitudinal data set that included recordings of baseline EEG from a group of 29 children at 5, 10, 14, 24, and 51 months of age. They reported a peak frequency in the 6–9 Hz band that emerged across multiple scalp locations. Although not evident during the earliest recordings at 5 months of age, the 6–9 Hz band was reliable across all scalp locations by 10 months of age and continued to be the dominant frequency band through the 51-month EEG recordings. Thus, with each of these two small longitudinal datasets, the 6–9 Hz band contained infant and child peak frequency, much as adult alpha band of 8–13 Hz contains the typical mature peak frequency at 10 Hz, although there are individual differences in specific peak during adulthood (Klimesch, 1999).

This section highlights our 6–9 Hz alpha band research and selectively reviews other work examining the alpha band's relation to critical action-perception, cognitive,

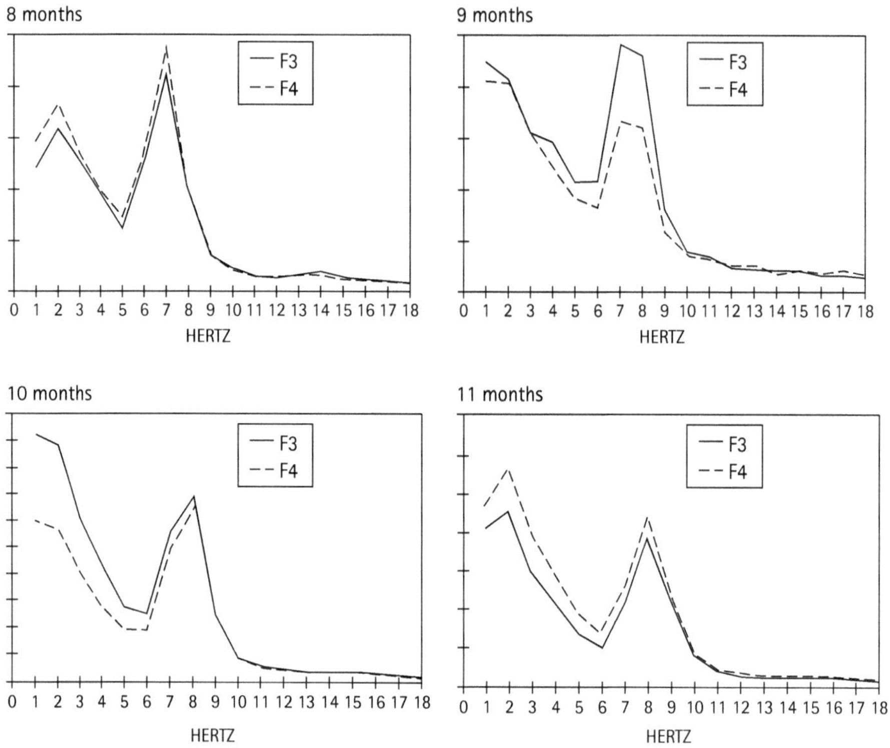

FIGURE 13.1 Month-to-month spectral plots from one infant. Note the change in peak frequency from 8 to 11 months. The *x*-axis is frequency in single hertz (Hz) bins and the *y*-axis is EEG power (mean square microvolts).

From the Bell & Fox, 1992 dataset.

and affective processes during early development. Although 6–9 Hz oscillatory activity has been linked to multiple aspects of early cognition, including future thinking (Blankenship et al., 2018) and memory (Cuevas et al., 2012c), the most extensive area of research has investigated emerging executive functions. Here, we provide a developmental analysis of the characteristics and individual variations in alpha band activity in association with action-perception, executive, and affective processes.

13.3.1 Action-Perception Processes

The functional properties of the mu and beta rhythms during motor movement were established by early investigations with adults. A relatively recent area of inquiry in the developmental literature has focused on these sensorimotor rhythms and their involvement in action-perception processes. Specifically, there is evidence that the mu and beta rhythms exhibit "neural mirroring" properties: oscillatory activity is attenuated during both action execution and observation (Lepage & Théoret, 2006; Liao et al., 2015).

Characterizing the conditions under which the sensorimotor mu and beta rhythms are reactive has been an area of substantial interest for developmental research as neural mirroring systems have been proposed to be involved in action understanding as well as broader socio-cognitive development (e.g., imitation; Marshall & Meltzoff, 2011).

Research with non-human primates has found evidence of neural mirroring during the observation and execution of facial gestures in 1- to 7-day-old rhesus macaque monkeys (Ferrari et al., 2012). Subsequent work has revealed that early experience with the mother, as compared to nursery-rearing, is associated with (1) greater mu rhythm attenuation when observing facial gestures and (2) more frequent imitation of these gestures (Vanderwert et al., 2015). These findings have been recently extended to human infants, with 9-month-olds exhibiting attenuation of the mu rhythm during facial expression perception and production (Rayson et al., 2017). Further, variability in 9-month mu rhythm reactivity was related to their mothers' tendency to mirror their facial gestures at 2 months of age, with more frequent maternal mirroring at 2 months of age being associated with greater action perception mu attenuation at age 9 months. Together, these findings highlight the biopsychosocial underpinnings of emerging action-perception processes.

Another line of research has examined the impact of children's action experience on the neural correlates of action perception. Recent evidence indicates that the effects of early motor experience are evident shortly after birth. Newborn monkeys with more frequent reaching attempts display greater beta rhythm reactivity when observing goal-directed reaching movements (Festante et al., 2018). Likewise, 9- and 12-month-old human infants with more advanced manual motor skills (e.g., reaching, grasping) show enhanced mu rhythm attenuation during manual action perception (Cannon et al., 2016; Upshaw et al., 2016; Yoo et al., 2016). Training paradigms have extended these findings by controlling children's short-term motor or perceptual experience with novel tools/actions. In an action-sound perception paradigm, 10-month-olds exhibited greater mu rhythm attenuation when hearing sounds associated with actions that they had received motor (active training), as compared to observational (passive training), experience (Gerson et al., 2015). A recent extension of this work to 3- to 6-year-olds has revealed greater EEG indices of visual processing (occipital alpha attenuation) as children observed tool-use actions that they had received active training (Bryant & Cuevas, 2019). However, there were no experience-related effects on sensorimotor rhythms, indicating that their role in action understanding might vary depending on age and context.

Other work has taken a broader perspective to consider whether individual differences in motor system activation are related to aspects of socio-cognitive development. For instance, in the context of a goal-directed reaching task, 7-month-olds with greater mu rhythm attenuation were more likely to express the same goal as an adult; thus, selecting the goal toy more often than the non-goal toy (Filippi et al., 2016). These early socially-contingent behavioral and neural responses may provide the foundation for more advanced socio-cognitive skills that emerge during early childhood. Recent evidence indicates that 3- to 5-year-olds' motor skills, action-representation abilities, and theory of mind (i.e., understanding others' mental states and their effects on behavior) are intricately linked for children with greater mu rhythm reactivity when performing goal-directed actions (Bowman et al., 2017). Likewise, 4-year-olds with enhanced beta rhythm reactivity during action perception display more successful cooperation with peers (Endedijk et al., 2017). These findings highlight the complexity of interrelations among action-perception and socio-cognitive processes. Although sensorimotor rhythms were linked with socio-cognitive development, the precise nature of these associations (e.g., beta vs. mu rhythm; action perception EEG vs. action execution EEG) varied. Additional work, including systematic longitudinal and cross-sectional investigations, are essential next steps toward a comprehensive analysis of action-perception processes and their involvement in socio-cognitive development.

13.3.2 Executive Processes

Executive function (EF) is an umbrella term that captures a wide array of higher-order cognitive processes—working memory, inhibitory control, cognitive flexibility—that organize and coordinate behavior to support complex goal-directed actions. This construct has been a major focus of developmental research as proficiency in executive processing is associated with optimal behavioral and academic outcomes (see Diamond, 2013 for review). EF emerges during the first postnatal year, with substantial advances throughout childhood and beyond (see Bell & Cuevas, 2016; Cuevas et al., 2018, for reviews). Individual differences in the developmental progression of EF are linked to biopsychosocial factors, including frontal lobe development, temperament, and co-occurring socialization experiences. Here, we discuss our cross-sectional and longitudinal work examining associations between EF and both resting-state and task-related EEG measures throughout early development.

13.3.2.1 *EF and Resting-State EEG*

One approach to examining associations between EF and neural activity has been to use measures of *resting-state* (or baseline) EEG when participants are awake, calm, still, and not engaged in cognitive tasks. Work with adults and older children accomplish this by measuring EEG during conditions of eyes open and closed; developmentally appropriate modifications to extend these procedures to infancy and early childhood typically include periods of "quiet visual attention" in which participants observe a visual stimulus (e.g., bubbles or images on a screen; see Anderson & Perone, 2018 for review and discussion). Resting-stating measures are often associated with brain maturation and provide information about neural oscillations when individuals are not involved in active cognitive processing. Our EF work with infants and young children has focused primarily on the 6–9 Hz "alpha" band, unless otherwise stated (Table 13.1). Below, we review findings using resting-stating measures of EEG power (oscillatory amplitude) and frontal intra-hemisphere EEG coherence (functional connectivity between two cortical

Table 13.1 Overview of executive function and 6–9 Hz EEG findings

EEG Power & Coherence Findings	Infants	Toddlers & Preschoolers
Resting-state (baseline) EEG is associated with EF (*task performance*)	Bell & Fox, 1992, 1997; MacNeill et al. 2018	Perone et al. 2018;* Wolfe & Bell, 2004
Changes in EEG during executive processing (*baseline-to-task related changes*)	Bell, 2001, 2002; Bell & Wolfe, 2007; Cuevas & Bell, 2011; Cuevas et al., 2012a, 2012d	Bell & Wolfe, 2007; Swingler et al., 2011
Changes in EEG during increased inhibitory demands	Cuevas et al., 2012e	Broomell & Bell, 2017; Cuevas et al., 2016
Executive processing-related EEG is associated with EF	Bell, 2001, 2002; 2012; Cuevas et al., 2012a, 2012d	Cuevas et al., 2016; Morasch & Bell, 2011; Swingler et al., 2011; Watson & Bell, 2013; Wolfe & Bell, 2004, 2007a, 2007b
Resting-state EEG is associated with future EF	Broomell et al. 2019; Kraybill & Bell, 2013	Cuevas et al., 2012b

Note. * = theta/beta ratio.

regions; squared cross-correlation of EEG power at two scalp electrodes; Thatcher et al., 1986).

To examine executive processing during infancy, we have used the A-not-B task, a variant of the delayed response task. Infants observe as a desirable object is hidden; their gaze to the hiding location is broken during a brief delay interval; and then they search for the toy (Bell & Adams, 1999; Cuevas & Bell, 2010). This task requires both working memory (to maintain the item location in active memory) and inhibitory control (to withhold responding to a previously rewarded location). Longitudinal and cross-sectional work has revealed that resting-state frontal-posterior power and coherence are associated with infants' task performance, including how long they can maintain the object location in working memory (Bell & Fox, 1992, 1997; MacNeill et al., 2018). Convergent evidence from diffusion tensor imaging (DTI) indicates that white matter tract microstructure connecting frontal, parietal, and temporal regions is associated with infants' delayed response task performance (Short et al., 2013). Differences in resting-state oscillatory activity and neural networks that are evident during infancy may have cascading effects on the developmental trajectory of EF. For instance, EEG measures of infant resting-state frontal power and frontal-temporal coherence account for variability in toddler and early childhood EF (Broomell et al., 2019; Kraybill & Bell, 2013).

During early childhood, EF tasks require children to withhold prepotent responses and/or keep up with changing task rules (requiring cognitive flexibility and working memory). In Stroop-like tasks, children must do something that conflicts with their natural tendencies, such as saying "day" when shown a picture of a moon (Day-Night task) or making a fist when shown a flat hand (Luria's Hand game). In sorting tasks (Dimensional Change Card Sort), children first place items in groups based on one dimension (e.g., color), and then sort items based on another feature (e.g., shape). Consistent with the infant findings, concurrent and predictive links between EF and resting-state frontal power are also evident during early childhood (Cuevas et al., 2012b; Wolfe & Bell, 2004). A recent cross-sectional investigation with 3- to 9-year-olds examined resting-state oscillatory activity across multiple frequency bands, revealing that variations in resting-state frontal theta/beta ratio are inversely related to EF performance (Perone et al., 2018). Thus, in line with anticipated maturational changes in "slow-" versus "fast-wave" rhythms (see Section 13.2), children with more beta oscillatory resting-state activity in comparison to theta activity, displayed enhanced executive processing. Taken together, these findings suggest that measures of resting-state activity and brain maturation in regions that support EF in older children and adults (e.g., Klingberg, 2006) are associated with individual differences in infant and early childhood EF.

13.3.2.2 *Executive Processing-Related EEG*

Developmental investigations have also examined task-related changes in EEG measures during executive processing in comparison to a baseline/resting-state, thus providing information about changes in brain activation and organization as a function of mental activity. Early childhood EF tasks are similar to the ones described earlier, and looking versions of the infant A-not-B task are used to minimize motor artifacts (as compared to reaching versions) (see Bell & Adams, 1999; Cuevas & Bell, 2010; Morasch & Bell, 2011).

Longitudinal and cross-sectional work indicates that infants and children exhibit task-related changes in EEG power and frontal intra-hemisphere EEG coherence (e.g., Bell & Wolfe, 2007; Cuevas & Bell, 2011; Cuevas et al., 2012a). These findings have been extended to characterize the impact of task difficulty on executive processing-related EEG. For example, 10-month-olds exhibit more frontal EEG coherence on reversal trials of the A-not-B task that require both working memory and inhibitory control as compared to nonreversal, working memory-only trials (Cuevas et al., 2012e). Similarly, 4-year-olds display increases in frontal, temporal, and parietal EEG power in response to added EF demands (i.e., Stroop vs. non-Stroop Day-Night task; Broomell & Bell, 2017); although the scalp distribution of these changes was more widespread for boys than girls despite equivalent task performance (Cuevas et al., 2016). These findings highlight the importance of considering both neural and behavioral measures when examining a particular construct, including the value of testing the role of potential moderators (e.g., sex, socio-economic status, race/ethnicity; see Gatzke-Kopp, 2016 for discussion).

As with resting-state measures, executive processing-related EEG is associated with variations in behavioral indicators of EF. Within-subjects analyses have revealed enhanced EEG power (6–9 Hz) and frontal intra-hemisphere EEG coherence (10–13 Hz) during trials in which infants make correct, as opposed to incorrect, responses (Bell, 2002; Cuevas et al., 2012d). Likewise, between-subjects comparisons indicate: (a) different patterns of EEG power and coherence as a function of EF task performance (Bell, 2001, 2012; Wolfe & Bell, 2004); and (b) that EEG power and coherence measures are unique predictors of EF task performance (Bell, 2012; Cuevas et al., 2012a, 2016; Swingler et al., 2011; Watson & Bell, 2013). In sum, executive-processing related measures of EEG are informative of emerging executive processes, likely reflecting the contributions from frontal-parietal networks.

13.3.3 Affective Processes

Developmental evidence suggests that resting state 6-9 Hz frontal EEG activation patterns are associated with individual differences in general affective style from at least 4 or 5 months of age (e.g., Fox et al., 1992). Resting state frontal EEG patterns also differentiate between infants born to women who experienced prenatal depression versus nondepressed women as early one week after birth at the broader 3–9 Hz band (e.g., Diego et al., 2010). These types of EEG findings are based on the measurement of frontal alpha-band EEG asymmetry, which highlights relative activation patterns between homologous left and right frontal scalp electrodes. Specifically, right frontal asymmetry reflects greater relative right frontal activation (i.e., lower EEG power values at right hemisphere) and left frontal asymmetry reflects the opposite pattern.

Frontal EEG asymmetry is typically conceptualized as reflecting approach and avoidance motivational tendencies across the lifespan (Coan & Allen, 2004; Harmon-Jones & Allen, 1998). During infancy and early childhood, frontal EEG asymmetry is also associated with temperament-based differences in emotion reactivity and emotion regulation (Fox, 1994). Thus, when examining resting state (or baseline) EEG, left frontal asymmetry is linked to approach motivation, positive emotions (plus the approach emotion of anger; Harmon-Jones & Allen, 1998), and better emotion regulation abilities. Right frontal asymmetry is associated with withdrawal motivation, negative emotions, and less skill at emotion regulation. Baseline frontal EEG asymmetry patterns are conceptualized as trait or biomarker indicators of motivational and emotion tendencies, whereas task-related frontal EEG asymmetry patterns are considered to be state indicators of reactivity and regulation (Allen & Reznik, 2015; Coan & Allen, 2004; Diaz & Bell, 2012).

Developmental research has examined links between frontal alpha (6–9 Hz) EEG asymmetry and temperament, which is defined as biologically based individual differences in emotion reactivity and emotion regulation (Rothbart & Bates, 2006). Other work has used resting frontal EEG asymmetry in early development to predict later socioemotional outcomes. A third line of frontal EEG asymmetry research includes parenting behaviors in examining both simple and complex links between child trait asymmetry and later outcomes. We briefly discuss each of these three areas by highlighting our research and selectively reviewing the work of others.

13.3.3.1 *Temperament and Frontal EEG Asymmetry*

Early research on frontal asymmetry and infant temperament focused on resting state 6-9 Hz EEG recordings, with reports that 10-month-old infants with right frontal asymmetry were more likely to later cry during maternal separation than infants with left frontal asymmetry (Davidson & Fox, 1989). In a longitudinal study from 7 to 12 months of age, infants with a longer latency to cry at maternal separation (i.e., did not cry until 25 seconds after mother left the room) had a pattern of left frontal asymmetry at each age. In contrast, infants with a much shorter latency to cry from 7 to 12 months had a pattern of symmetry or right frontal asymmetry across age (Fox et al, 1992).

State-related frontal EEG asymmetry research captures frontal EEG asymmetry during an emotion-eliciting situation that captures infant emotion reactivity. These studies report right frontal asymmetry during (a) the approach of a stranger at age 6 months (Buss et al., 2003), (b) presentation of scary masks at age 10 months (Diaz & Bell, 2012), and (c) maternal restraint of infants' arms at age 10 months (Gartstein et al., 2014). Ten-month-old infants displaying felt smiles during the approach of their mothers exhibited left frontal asymmetry; when those same infants displayed unfelt smiles (i.e., smiles did not involve the action of orbicularis oculi) during the approach of a stranger, they exhibited right frontal EEG asymmetry (Fox & Davidson, 1988).

As noted, resting frontal alpha EEG asymmetry is conceptualized as a trait—a stable marker of approach-avoidance motivation. Stability can be seen across four measures of frontal EEG asymmetry during a two-year time period in a small group of first and second graders (Poole et al., 2018). Work with a large community sample across a period of four years, however, did not show a high level of stability for frontal asymmetry at the group level, but did show a pattern of associations between frontal EEG asymmetry and parent-report of child temperament (Howarth et al., 2016). Specifically, in a sample of children who came to the research lab at 10, 24, 36, and 48 months of age, parent report of temperament-based fear and activity level were highly correlated across infancy, toddlerhood, and early childhood. Frontal EEG asymmetry scores were uncorrelated across age; however, frontal EEG asymmetry at 10 months positively predicted child activity level at age 24 months, suggesting a link between early left frontal asymmetry and later approach temperament. Conversely, child fear at age 36 months negatively predicted frontal EEG asymmetry at age 48 months, suggesting a link between early avoidance temperament and later right frontal asymmetry. With this same sample and considering developmental change at individual frontal EEG leads separately (i.e., not asymmetry scores) from 5 to 72 months, different temperament-frontal EEG patterns emerged for girls and boys (Gartstein et al, 2019). Parent-report of temperament reactivity at 5 months of age and changes in frontal EEG power across age occurred in later development for boys (after 24 months of age) and in earlier development for girls. Furthermore, early left frontal EEG power at 6–9 Hz predicted later change in right hemisphere frontal EEG power for girls, whereas early right hemisphere EEG power predicted subsequent later change in left hemisphere frontal EEG power for boys. These findings might account for reports of sex differences in temperament and related behavior problems (e.g., Else-Quest et al., 2006).

13.3.3.2 *Frontal EEG Asymmetry and Socioemotional Outcomes*

Although frontal EEG asymmetry might not be highly stable at the group level across the early years, it is the case that some children do exhibit stable asymmetry patterns. Infants with stable left or right frontal 6–9 Hz EEG asymmetry at age 10 and 24 months were high in parent-rated externalizing or internalizing behaviors at age 30 months, respectively (Smith & Bell 2010). Infants who changed from left to right, or vice versa, were rated between the two stable groups on both internalizing and externalizing behaviors. Likewise, work with older children, 6-year-olds who had four EEG recordings across two years, identified stable left and stable right frontal alpha EEG asymmetry groups (10–13 Hz; Poole et al., 2018). Parent-report, child-report, and child self-presentation behaviors were assessed for approach-avoidance tendencies. As predicted, the stable left group had higher approach-related behavioral tendencies, whereas the stable right group had higher avoidance-related tendencies. Both studies highlight that stability in frontal EEG asymmetry may be implicated in later behavior problems.

Importantly, one-time frontal EEG asymmetry measures are informative of future outcomes. Temperamental negative affectivity during infancy is related to regulatory problems during toddlerhood only for children with right frontal EEG asymmetry during infancy (Smith et al., 2016). Infant negativity is not associated with later regulatory problems for children with left frontal asymmetry. A similar pattern is seen with older children. Children with right frontal EEG asymmetry at 4 years of age show greater physiological arousal during a self-presentation task at 9 years of age and have greater problems with emotion regulation according to parent report (Hannesdottir et al., 2010). In sum, although not stable for all children, early frontal EEG asymmetry can predict later child outcomes in the socio-emotional domain.

13.3.3.3 *Parenting and Child Frontal EEG Asymmetry*

Previously we noted EEG research that examines frontal asymmetry in context. There is no more important context than the parenting environment. Women with depression exhibit right frontal alpha EEG asymmetry during baseline recordings (e.g., Henriques & Davidson, 1990) and so do their infants (e.g., Dawson et al., 1997). If the women are depressed while pregnant, then their infants exhibit right frontal asymmetry by 1 week of age (Diego et al., 2010). For older infants, prenatal depression and infant frontal EEG asymmetry are linked only for women with postpartum depression as well, with asymmetry stable from 3 to 6 months of age (Lusby et al., 2014).

It is not only maternal depression that is correlated with child frontal EEG asymmetry scores; it is also maternal behavior. We hyperscanned mothers and their 3-year-old children while they played together with puzzles. Mothers who were high in behaviorally coded negativity while playing with their children had children with right frontal asymmetry after controlling for child's own negativity. Conversely, children who exhibited challenging and negative behaviors while playing with mothers had mothers with right frontal EEG asymmetry after controlling for mother's own negativity (Atzaba-Poria et al., 2017). Thus, mother and child frontal EEG asymmetry patterns during dyadic recordings are influenced by partner's level of negativity.

Although there exist associations between frontal 6-9 Hz EEG asymmetry and some aspect of emotion during infancy and early childhood (and beyond), the link between frontal asymmetry and emotion is likely more than a simple association, as we have described them here. It may be that frontal EEG asymmetry is either a mediator or a moderator of emotion (Coan & Allen, 2004; Reznick & Allen, 2018). Two examples from our own work highlight frontal asymmetry as a moderator. Resting frontal EEG asymmetry was recorded in a large group of 5-month-old infants. Behavioral coding included mothers' responsive behaviors during free play with her infant, as well as infants' regulatory behaviors and level of negativity during the arm restraint procedure where mother faces her infant and prevents the infant from moving his/her arms. Maternal responsive behaviors during free play with infants prior to the arm restraint procedure predicted infant regulatory behaviors for infants with left frontal asymmetry but predicted infant negativity for infants with right frontal asymmetry (Swingler et al., 2014). Thus, the same maternal behavior is associated with different infant regulatory behaviors depending on infant frontal EEG asymmetry pattern, suggesting an advantage for infants with both left frontal EEG asymmetry and an environment with responsive parenting.

In this same large group of infants, we examined age 5-month frontal EEG asymmetry as a moderator of the link between maternal behavior and child negative affect at age 24 months. Maternal sensitive and responsive behaviors at age 5 months were associated with less toddler negative affect only for children who also had left frontal EEG asymmetry at age 5 months. The same pattern held when examining maternal behaviors at 24 months of age (Diaz et al., 2019). We interpreted this to mean that the level of toddler negative affect is influenced by the child's own neurophysiology as well as the parenting context.

In sum, frontal alpha EEG asymmetry, conceptualized as approach-avoidance motivation, has been examined in studies of temperament, socio-emotional outcomes, and parenting. The underlying neurological basis for frontal EEG asymmetry is typically focused on reciprocal metabolic connections between prefrontal cortex and the amygdala. Activation of the right amygdala is associated with increased dispositional (i.e., temperament-based) negative affect (Abercrombie et al., 1998); conversely, individuals with greater left frontal metabolic rate exhibit decreased amygdala activation and thus decreased negative affect (Davidson, 2001). Although frontal EEG asymmetry does not measure amygdala activity, it may be measuring the frontal cortical activation linked with amygdala connectivity.

13.3.4 Differential Functions of 6–9 Hz Activity during Cognitive and Affective Processing

There appear to be two patterns for 6–9 Hz activity during infancy and early childhood. In studies of affective processes, decreases in EEG power values in one hemisphere relative to the other are associated with emotion reactivity or emotion regulation. In studies of cognition, increases in EEG power values are associated with higher levels of performance on cognitive tasks. In the adult EEG literature, researchers have long focused on the 8–13 Hz peak in the adult spectrum. It is commonly reported that alpha activity (8–13 Hz) exhibits *desynchronization* (decreased power values) during increased cortical processing (cognitive, affective), although there are some reports of alpha *synchronization* (increased power values) during long-term memory tasks (Klimesch et al., 1999). There are also suggestions that alpha power is related to an inhibitory filter that is reflected in alpha synchronization (Klimesch, 2012).

Researchers have noted that adult theta activity (4–7 Hz) exhibits synchronization during memory and attention tasks (e.g., Burgess & Gruzelier, 2000; Klimesch, 1999). Thus, for the mature EEG signal, unique patterns of fluctuations in power levels at the defined frequency bands are associated with different types of cognitive processing (Cavanaugh, 2019). This type of information is lacking with respect to the EEG signal recorded from infants and young children. Currently, it appears that infant and child 6–9 Hz alpha behaves like adult theta or adult alpha depending on the type of cognitive processing, and definitely behaves like adult alpha during affective processing (Anderson & Perone, 2018; Bell & Cuevas, 2012; Saby & Marshall, 2012). Systematic developmental research is needed to understand the neurophysiological and functional significance of infant and child alpha at 6–9 Hz.

13.4 Broader Impact and Future Directions

The brain's electrical activity consists of multiple, co-occurring oscillatory rhythms that underlie neural computation, communication, and transmission of information between brain networks—functions critical to simple and complex brain processes (Lopes da Silva, 2013). Transient interactions among neural rhythms are critical for the coordination of neural, cognitive, affective, and behavioral processes (Jensen & Colgin, 2007); yet, holistic ontogenetic analysis is largely missing from the field. As shown throughout this chapter, developmental analyses of EEG rhythms typically focus on a single oscillatory rhythm either cross-sectionally or at a particular age. These approaches preclude integrative analysis of co-occurring oscillatory rhythms, including associated ontogenetic changes in neural activity.

The current review focused on individual frequency bands during early development, with emphasis on the infant and child alpha 6–9 Hz band. We were among the first to emphasize this particular frequency band, based on spectral plots of our initial infant longitudinal study (Figure 13.1; Bell & Fox, 1992). Stroganova and colleagues (1999) were among the first to label 6–9 Hz as "alpha" based on their spectral plots of the spatial and amplitude properties of this frequency band when infants were in a room with lights turned off compared with quiet alert baseline. More recently, developmental investigations of the central 6–9 Hz band have identified functional properties of the mu rhythm during action perception and execution (see Cuevas et al., 2014, for review). Furthermore, resting-state 6–9 Hz alpha rhythm measures are associated with concurrent and future affective, cognitive, and motor processes during infancy and early childhood (see Anderson & Perone, 2018, for review). Despite the potential functional significance of alpha (and gamma) resting-state activity, relatively little is known about other co-occurring resting-state oscillatory rhythms, such as theta and beta frequency bands. In sum, what is missing is systematic examination of the alpha family frequency in conjunction with other co-occurring oscillatory rhythms to aid our understanding of their frequency signatures, spatial topography, and functional significance, all within a developmental time course.

Klimesch (2012) proposed that the alpha rhythm plays a central role in inhibition and timing of brain activity, and thus may be importantly involved in activity at other frequency bands. Accordingly, he proposed that for adult EEG, other frequency bands are best defined by a frequency architecture focused on the alpha band and its activity. Because alpha is a central process in modulating resting state and task behaviors in the awake brain, the best way to understand frequency bands is by examining bands that have a harmonic mean with alpha. The neurophysiology literature is missing this type of functional thinking about alpha from a developmental point of view.

At the same time, there is increasing interest in how multiple oscillatory rhythms interact to produce complex brain functions. Cross-frequency coupling and band power ratios are example measures that permit more integrative analysis of resting-state EEG as well as during cognitive, perceptual, affective, and motor processes. Their application in the developmental literature has been limited to date (e.g., Brooker et al., 2016; Perone et al., 2018; Stamoulis et al., 2015), but they provide a promising avenue for understanding ontogenetic changes in neural oscillations and corresponding brain-behavior associations (e.g., Brooker et al., 2021).

Finally, there have been recent initiatives aimed at enhancing the interpretability and scientific rigor of pediatric EEG. One line of work has focused on factors related to attrition and EEG data loss for developmental samples with recommendation for the field (e.g., van der Velde & Junge, 2020). Concurrently, efforts to establish standard automated, freely-available, data preprocessing pipelines designed specifically for pediatric EEG (e.g., MADE and HAPPE) are promising in terms of increasing the amount of useable data as well as improving reliability and cross-lab comparisons (Debnath et al., 2020; Gabard-Durnam et al., 2018). Furthermore, recent evidence suggests that analysis of infant and child oscillatory rhythms will be enhanced via techniques that extract the periodic and aperiodic (1/*f*) components from the neural power spectra (Cellier et al., 2021; Schaworonkow & Voytek, 2021). Even with the aforementioned strategies, analysis of the psychometric properties of developmental EEG measures (e.g., longitudinal stability, reliability, validity) highlights important considerations for pediatric data (Anaya et al., 2021; Vincent et al., 2021). At a time of pre-registration and open-access initiatives, we are particularly excited about the widespread effects these methodological, theoretical, and analytical approaches will have on the field.

Related Topics

Alpha Activity for Inhibiting Distractors to Facilitate Attention and Memory

Asymmetric Activation of the Frontal Cortices

Cross-Frequency/Phase-Amplitude Coupling

Development Across Adolescence and early Adulthood

Frequency Activity over the Motor Cortex: Alpha, Beta, and Mu Suppression

Theta Activity in Working Memory and Performance Monitoring

Acknowledgements: Preparation of this manuscript was supported by Research Excellence Program, University of Connecticut (Cuevas) and College of Science Faculty Fellowship, Virginia Tech (Bell). Much of our research highlighted in this chapter was supported by grants R01 HD049878 and R03 HD043057 (PI: M. A. Bell) from the *Eunice Kennedy Shriver* National Institute of Child Health and Human Development (NICHD). The content of this manuscript is solely the responsibility of the authors and does not necessarily represent the official views of the NICHD or the National Institutes of Health.

References

Abercrombie, H. C., Schaefer, S. M., Larson, K. L., Oakes, T. R., Lindgre, K. A., ... Davidson, R. J. (1998). Metabolic rate in the right amygdala predicts negative affect in depressed patients. *NeuroReport*, 9, 3301–3307.

Allen, J. J. B. & Reznik, S. J. (2015). Frontal EEG asymmetry as a promising marker of depression vulnerability: Summary and methodological considerations. *Current Opinion in Psychology*, 4, 93–97. https://doi.org/10.1016/j.copsyc.2014.12.017

Anaya, B., Ostlund, B., LoBue, V., Buss, K., & Pérez-Edgar, K. (2021). Psychometric properties of infant electroencephalography: Developmental stability, reliability, and construct validity of frontal alpha asymmetry and delta-beta coupling. *Developmental Psychobiology*, 63, e22178. https://doi.org/10.1002/dev.22178

Anderson, A. J. & Perone, S. (2018). Developmental change in the resting state electroencephalogram: Insights into cognition and the brain. *Brain and Cognition*, 126, 40–52. https://doi.org/10.1016/j.bandc.2018.08.001

Atzaba-Poria, N., Deater-Deckard, K., & Bell, M. A. (2017). Mother-child interaction: Links between mother and child frontal electroencephalograph asymmetry and negative behavior. *Child Development*, 88, 544–554. doi:10.111/cdev.12583

Başar, E. (2013). A review of gamma oscillations in healthy subjects and in cognitive impairment. *International Journal of Psychophysiology*, 90, 99–117. http://dx.doi.org/10.1016/j.ijpsycho.2013.07.005

Begus, K. & Bonawitz, E. (2020). The rhythm of learning: Theta oscillations as an index of active learning in infancy. *Developmental Cognitive Neuroscience, 45*, 100810. https://doi.org/10.1016/j.dcn.2020.100810

Begus, K., Southgate, V., & Gliga, T. (2015). Neural mechanisms of infant learning: differences in frontal theta activity during object exploration modulate subsequent object recognition. *Biology Letters, 11*, 20150041. http://dx.doi.org/10.1098/rsbl.2015.0041

Bell, M. A. (2001). Brain electrical activity associated with cognitive processing during a looking version of the A-not-B task. *Infancy, 2*, 311–330.

Bell, M. A. (2002). Power changes in infant EEG frequency bands during a spatial working memory task. *Psychophysiology, 39*, 450–458.

Bell, M. A. (2012). A psychobiological perspective on working memory performance at 8 months of age. *Child Development, 83*, 251–265. http://dx.doi.org/10.1111/j.1467-8624.2011.01684.x

Bell, M. A. & Adams, S. E. (1999). Comparable performance on looking and reaching versions of the A-not-B task at 8 months of age. *Infant Behavior and Development, 22*, 221–235.

Bell, M. A. & Cuevas, K. (2012). The use of EEG to study cognitive development: Issues and practices. *Journal of Cognition and Development, 13*, 281–294. http://dx.doi.org/10.1080/15248372.2012.691143

Bell, M. A. & Cuevas, K. (2016). Psychobiology of executive function in early development. In J. A. Griffin, P. McCardle, & L. S. Freund (Eds.), *Executive function in preschool-age children: Integrating measurement, neurodevelopment, and translational research* (pp. 157–179). APA Press. http://dx.doi.org/10.1037/14797-008

Bell, M. A. & Fox, N. A. (1992). The relations between frontal brain electrical activity and cognitive development during infancy. *Child Development, 63*, 1142–1163.

Bell, M. A. & Fox, N. A. (1997). Individual differences in object permanence performance at 8 months: Locomotor experience and brain electrical activity. *Developmental Psychobiology, 31*, 287–297.

Bell, M. A. & Wolfe, C. D. (2007). Changes in brain functioning from infancy to early childhood: Evidence from EEG power and coherence during working memory tasks. *Developmental Neuropsychology, 31*, 21–38.

Benasich, A. A., Gou, Z., Choudhury, N., & Harris, K. D. (2008). Early cognitive and language skills are linked to resting frontal gamma power across the first 3 years. *Behavioural Brain Research, 195*, 215–222. doi:10.1016/j.bbr.2008.08.049

Berchicci, M., Zhang, T., Romero, L., Peters, A., Annett, R., Teuscher, U., ... Comani, S. (2011). Development of mu rhythm in infants and preschool children. *Developmental Neuroscience, 33*, 130–143. doi:10.1159/000329095

Berger, H. (1929). Über das Elektrenkephalogramm des Menschen. *Archiv für Psychiatrie und Nervenkrankheiten, 87*, 527–570.

Berger, H. (1932). Über das Elektroenzephalogramm des Menschen. Fünfter Bericht. *Archiv für Psychiatrie und Nervenkrankheiten, 98*, 231–254.

Blankenship, T. L., Broomell, A. P. R., & Bell, M. A.(2018). Semantic future thinking and executive functions at age 4: The moderating role of frontal brain electrical activity. *Developmental Psychobiology, 60*, 608–614. https://doi.org/10.1002/dev.21629

Bowman, L. C., Thorpe, S. G., Cannon, E. N., & Fox, N. A. (2017). Action mechanisms for social cognition: Behavioral and neural correlates of developing Theory of Mind. *Developmental Science, 20*, e12447. https://doi.org/10.1111/desc.12447

Braithwaite, E. K., Jones, E. J. H., Johnson, M. H., & Holmboe, K. (2020). Dynamic modulation of frontal theta power predicts cognitive ability in infancy. *Developmental Cognitive Neuroscience, 45*, 100818. https://doi.org/10.1016/j.dcn.2020.100818

Brito, N. H., Fifer, W. P., Myers, M. M., Elliott, A. J., & Noble, K. G. (2016). Associations among family socioeconomic status, EEG power at birth, and cognitive skills during infancy. *Developmental Cognitive Neuroscience, 19*, 144–151. http://dx.doi.org/10.1016/j.dcn.2016.03.004

Brooker, R. J., Mistry-Patel, S., Kling, J. L., & Howe, H. A. (2021). Deriving within-person estimates of delta-beta coupling: A novel measure for identifying individual differences in emotion and neural function in childhood. *Developmental Psychobiology, 63*, e22172. https://doi.org/10.1002/dev.22172

Brooker, R. J., Phelps, R. A., Davidson, R. J., & Goldsmith, H. H. (2016). Contextual differences in delta beta coupling are associated with neuroendocrine reactivity in infants. *Developmental Psychobiology, 58*, 406–418. doi:10.1002/dev.21381

Broomell, A. P. R. & Bell, M. A. (2017). Inclusion of a mixed condition makes the day/night task more analogous to the adult Stroop. *Developmental Neuropsychology, 42*, 241–252. doi:10.1080/87565641.2017.1309655

Broomell, A. P. R., Salva, J., & Bell, M. A. (2019). Infant electroencephalogram coherence and toddler inhibition are associated with social responsiveness at age 4. *Infancy, 24*, 43–56. doi:10.1111/infa.12273

Bryant, L. J. & Cuevas, K. (2019). Effects of active and observational experience on EEG activity during early childhood. *Psychophysiology, 56*, e13360. https://doi.org/10.1111/psyp.13360

Burgess, A. P. & Gruzelier, J. H. (2000). Short duration power changes in the EEG during recognition memory for words and faces. *Psychophysiology, 37*, 596–606. https://doi.org/10.1111/1469-8986.3750596

Buss, K. A., Schumacher, J. R. M., Dolski, I., Kalin, N. H., Goldsmith, H. H., & Davidson R. J. (2003). Right frontal brain activity, cortisol, and withdrawal behavior in 6-month-old infants. *Behavioral Neuroscience, 117*, 11–20. doi:10.1037/0735-7044.117.1.11

Cannon, E. N., Simpson, E. A., Fox, N. A., Vanderwert, R. E., Woodward, A. L., & Ferrari, P. F. (2016). Relations between infants' emerging reach-grasp competence and event-related desynchronization in EEG. *Developmental Science, 19*, 50–62. https://doi.org/10.1111/desc.12295

Cavanaugh, J. F. (2019). Electrophysiology as a theoretical and methodological hub for the neural sciences. *Psychophysiology, 56*, e13314. https://doi.org/10.1111/psyp.13314

Cellier, D., Riddle, J., Petersen, I., & Hwang, K. (2021). The development of theta and alpha neural oscillations from ages 3 to 34 years. *Developmental Cognitive Neuroscience, 50*, 100969. https://doi.org/10.1016/j.dcn.2021.100969

Cheyne, D., Jobst, C., Tesan, G., Crain, S., & Johnson, B. (2014). Movement-related neuromagnetic fields in preschool age children. *Human Brain Mapping, 35*, 4858–4875. doi:10.1002/hbm.22518

Coan, J. A. & Allen, J. J. B. (2004). Frontal EEG asymmetry as a moderator and mediator of emotion. *Biological Psychology, 67*, 7–50. https://doi.org/10.1016/j.biopsycho.2004.03.002

Crone, N. E., Miglioretti, D. L., Gordon, B., Sieracki, J. M., Wilson, M. T., Uematsu, S., & Lesser, R. P. (1998). Functional mapping of human sensorimotor cortex with electrocorticographic spectral analysis. I. Alpha and beta event-related desynchronization. *Brain, 121*, 2271–2299.

Csibra, G., Davis, G., Spratling, M. W., & Johnson, M. H. (2000). Gamma oscillations and object processing in the infant brain. *Science, 290*, 1582–1585. doi:10.1126/science.290.5496.1582

Cuevas, K. & Bell, M. A. (2010). Developmental progression of looking and reaching performance of the A-not-B task. *Developmental Psychology, 46*, 1363–1371. http://dx.doi.org/10.1037/a0020185

Cuevas, K. & Bell, M. A. (2011). EEG and ECG from 5 to 10 months of age: Developmental changes in baseline activation and cognitive processing during a working memory task. *International Journal of Psychophysiology*, *80*, 119–128. http://dx.doi.org/10.1016/j.ijpsycho.2011.02.009

Cuevas, K., Bell, M. A., Marcovitch, S., & Calkins, S. D. (2012a). EEG and heart rate measures of working memory at 5 and 10 months of age. *Developmental Psychology*, *48*, 907–917. http://dx.doi.org/10.1037/a0026448

Cuevas, K., Calkins, S. D., & Bell, M. A. (2016). To Stroop or not to Stroop: Sex-related differences in brain-behavior associations during early childhood. *Psychophysiology*, *53*, 30–40. http://dx.doi.org/10.1111/psyp.12464

Cuevas, K., Cannon, E. N., Yoo, K., & Fox, N. A. (2014). The infant EEG mu rhythm: Methodological considerations and best practices. *Developmental Review*, *34*, 26–43. http://dx.doi.org/10.1016/j.dr.2013.12.001

Cuevas, K., Hubble, M., & Bell, M. A. (2012b). Early childhood predictors of post-kindergarten executive function: Behavior, parent-report, and psychophysiology. *Early Education and Development*, *23*, 59–73. http://dx.doi.org/10.1080/10409289.2011.611441

Cuevas, K., Raj, V., & Bell, M. A. (2012c). A frequency band analysis of two-year-olds' memory processes. *International Journal of Psychophysiology*, *83*, 315–322. http://dx.doi.org/10.1016/j.ijpsycho.2011.11.009

Cuevas, K., Raj, V., & Bell, M. A. (2012d). Functional connectivity and infant spatial working memory: A frequency band analysis. *Psychophysiology*, *49*, 271–280. http://dx.doi.org/10.1111/j.1469-8986.2011.01304.x

Cuevas, K., Rajan, V., & Bryant, L. J. (2018). Emergence of executive function in infancy. In S. A. Wiebe & J. Karbach (Eds.), *Executive function: Development across the life span* (pp. 11–28). Routledge.

Cuevas, K., Swingler, M. M., Bell, M. A., Marcovitch, S., & Calkins, S. D. (2012e). Measures of frontal functioning and the emergence of inhibitory control processes at 10 months of age. *Developmental Cognitive Neuroscience*, *2*, 235–243. http://dx.doi.org/10.1016/j.dcn.2012.01.002

Davidson, R. J. (2001). Toward a biology of personality and emotion. *Annals of the New York Academy of Sciences*, *935*, 191–207.

Davidson, R. J. & Fox, N. A. (1989). Frontal brain asymmetry predicts infants' response to maternal separation. *Journal of Abnormal Psychology*, *98*, 127–131. https://doi.org/10.1037/0021-843X.98.2.127

Dawson, G., Frey, K., Panagiotides, H., Osterling, J., & Hessl, D. (1997). Infants of depressed mothers exhibit atypical frontal brain activity: A replication and extension of previous findings. *Journal of Child Psychology and Psychiatry*, *38*, 179–186. https://doi.org/10.1111/j.1469-7610.1997.tb01852.x

Debnath R., Buzzell G. A., Morales S., Bowers M. E., Leach S. C., Fox N. A. (2020). The Maryland analysis of developmental EEG (MADE) pipeline. *Psychophysiology*, *57*, e13580. https://doi.org/10.1111/psyp.13580

Diamond, A. (2013). Executive functions. *Annual Review of Psychology*, *64*, 135–168. http://dx.doi.org/10.1146/annurev-psych-113011-143750

Diaz, A. & Bell, M. A. (2012). Frontal EEG asymmetry and fear reactivity in different contexts at 10 months. *Developmental Psychobiology*, *54*, 536–545. doi:10.1002/dev.20612

Diaz, A., Swingler, M. M., Tan, L., Smith, C. L., Calkins, S. D., & Bell, M. A. (2019). Infant frontal EEG asymmetry moderates the association between maternal behavior and toddler negative affectivity. *Infant Behavior and Development*, 55, 88–99. https://doi.org/10.1016/j.infbeh.2019.03.002

Diego, M. A., Field, T., Jones, N. A., & Hernandez-Reif, M. (2006). Withdrawn and intrusive maternal interaction style and frontal EEG asymmetry shifts in infants of depressed and non-depressed mothers. *Infant Behavior and Development*, 29, 220–229. https://doi.org/10.1016/j.infbeh.2005.12.002

Diego, M. A., Jones, N. A., & Field, T. (2010). EEG in 1-week, 1-month, and 3-month-old infants of depressed and non-depressed mothers. *Biological Psychology*, 83, 7–14. https://doi.org/10.1016/j.biopsycho.2009.09.007

Else-Quest, N. M., Hude, J. S., Goldsmith, H. H., & Van Hulle, C. A. (2006). Gender differences in temperament: A meta-analysis. *Psychological Bulletin*, 132, 33–72. https://doi.org/10.1037/0033-2909.132.1.33

Endedijk, H. M., Meyer, M., Bekkering, H., Cillessen A. H. N., & Hunnius, S. (2017). Neural mirroring and social interaction: Motor system involvement during action observation relates to early peer cooperation. *Developmental Cognitive Neuroscience*, 24, 33–41. https://doi.org/10.1016/j.dcn.2017.01.001

Ferrari, P. F., Vanderwert, R. E., Paukner, A., Bower, S., Suomi, S. J., & Fox, N. A. (2012). Distinct EEG amplitude suppression to facial gestures as evidence for a mirror mechanism in newborn monkeys. *Journal of Cognitive Neuroscience*, 24, 1165–1172.

Festante, F., Vanderwert, R. E., Sclafani, V., Paukner, A., Simpson, E. A., Suomi, S. J., . . . Ferrari, P. F. (2018). EEG beta desynchronization during hand goal-directed action observation in newborn monkeys and its relation to the emergence of hand motor skills. *Developmental Cognitive Neuroscience*, 30, 142–149. https://doi.org/10.1016/j.dcn.2018.02.010

Filippi, C. A., Cannon, E. N., Fox, N. A., Thorpe, S. G., Ferrari, P. F., & Woodward, A. L. (2016). Motor system activation predicts goal imitation in 7-month-old infants. *Psychological Science*, 27, 675–684. doi:10.1177/0956797616632231

Fox, N. A. (1994). Dynamic cerebral processes underlying emotion regulation. *Monographs of the Society for Research in Child Development*, 59(2–3), 152–166. https://psycnet.apa.org/doi/10.2307/1166143

Fox, N. A., Bell, M. A., & Jones, N. A. (1992). Individual differences in response to stress and cerebral asymmetry. *Developmental Neuropsychology*, 8, 161–184. https://doi.org/10.1080/87565649209540523

Fox, N. A. & Davidson, R. J. (1988). Patterns of brain electrical activity during facial signs of emotion in 10-month-old infants. *Developmental Psychology*, 24, 230–236. https://doi.org/10.1037/0012–1649.24.2.230

Futagi, Y., Ishihara, T., Tsuda, K., Suzuki, Y., & Goto, M. (1998). Theta rhythms associated with sucking, crying, gazing and handling in infants. *Electroencephalography and Clinical Neurophysiology*, 106, 392–399.

Gabard-Durnam, L. J., Mendez Leal, A. S., Wilkinson, C. L., & Levin, A. R. (2018). The Harvard automated processing pipeline for electroencephalography (HAPPE): Standardized processing software for developmental and high-artifact data. *Frontiers in Neuroscience*, 12(97), 1–24. https://doi.org/10.3389/fnins.2018.00097

Gaetz, W., MacDonald, M., Cheyne, D., & Snead, O. C. (2010). Neuromagnetic imaging of movement-related cortical oscillations in children and adults: Age predicts post-movement beta rebound. *NeuroImage*, 51, 792–807. doi:10.1016/j.neuroimage.2010.01.077

Gartstein, M. A., Bell, M. A., & Calkins, S. D. (2014). EEG asymmetry at 10 months of age: Are temperament trait predictors different for boys and girls? *Developmental Psychobiology*, 56, 1327–1340. https://doi.org/10.1002/dev.21212

Gartstein, M. A., Hancock, G. R., Potapova, N. V., Calkins, S. D., & Bell, M. A. (2019). Modeling development of frontal electroencephalogram (EEG) asymmetry: Sex differences and links with temperament. *Developmental Science, 23*, e12891. https://doi.org/10.1111/desc.12891

Gasser, T., Verleger, R., Bacher, P., & Sroka, L. (1988). Development of the EEG of school-age children and adolescents. I. Analysis of band power. *Electroencephalography and Clinical Neurophysiology, 69*, 91–99.

Gastaut, H., Dongier, M., & Courtois, G. (1954). On the significance of "wicket rhythms" ("rythmes en arceau") in psychosomatic medicine. *Electroencephalography and Clinical Neurophysiology, 6*, 687.

Gatzke-Kopp, L. M. (2016). Diversity and representation: Key issues for psychophysiological science. *Psychophysiology, 53*, 3–13. https://doi.org/10.1111/psyp.12566

Gerson, S. A., Bekkering, H., & Hunnius, S. (2015). Short-term motor training, but not observational training, alters neurocognitive mechanisms of action processing in infancy. *Journal of Cognitive Neuroscience, 27*, 1207–1214. https://doi.org/10.1162/jocn_a_00774

Ghim, H.-R. (1990). Evidence for perceptual organization in infants: Perception of subjective contours by young infants. *Infant Behavior and Development, 13*, 221–248. https://doi.org/10.1016/0163-6383(90)90032-4

Gibbs, F. A. & Knott, J. R. (1949). Growth of the electrical activity of the cortex. *Electroencephalography and Clinical Neurophysiology, 1*, 223–229.

Gliga, T., Volein, A., & Csibra, G. (2010). Verbal labels modulate perceptual object processing in 1-year-old children. *Journal of Cognitive Neuroscience, 22*, 2781–2789.

Hagne, I. (1968). Development of the waking EEG in normal infants during the first year of life. A longitudinal study including automatic frequency analysis. In P. Kellaway and I. Petersen (Eds.), *Clinical electroencephalography of children* (pp. 97–118). Grune & Stratton.

Hagne, I., Persson, J., Magnusson, R., & Petersen, I. (1973). Spectral analysis via fast Fourier transform of waking EEG in normal infants. In P. Kellaway and I. Petersen (Eds.), *Automation of clinical electroencephalography* (pp. 103–143). Raven Press.

Hannesdottir, D. K., Doxie, J., Bell, M. A., Ollendick, T. H., & Wolfe, C. D. (2010). A longitudinal study of emotion regulation and anxiety in middle childhood: Associations with frontal EEG asymmetry in early childhood. *Developmental Psychobiology, 52*, 197–204. doi:10.1002/dev.20425

Hari, R., & Salmelin, R. (1997). Human cortical oscillations: a neuromagnetic view through the skull. *Trends in Neurosciences, 20*, 44–49.

Harmon-Jones, E. & Allen, J. J. B. (1998). Anger and frontal brain activity: EEG asymmetry consistent with approach motivation despite negative affective valence. *Journal of Personality and Social Psychology, 74*, 1310–1316. https://doi.org/10.1037/0022-3514.74.5.1310

Henriques, J. B. & Davidson, R. J. (1990). Regional brain electrical asymmetries discriminate between previously depressed and healthy control subjects. *Journal of Abnormal Psychology, 99*, 22–31. https://doi.org/10.1037/0021-843X.99.1.22

Howarth, G. Z., Fettig, N. B., Curby, T. W., & Bell, M. A. (2016). Frontal electroencephalogram asymmetry and temperament across infancy and early childhood: An exploration of stability and bidirectional relations. *Child Development, 87*, 465–476. doi:10.1111/cdev.12466

Jarvelainen, J., Schurmann, M., & Hari, R. (2004). Activation of the human primary motor cortex during observation of tool use. *NeuroImage, 23*, 187–192. doi:10.1016/j.neuroimage.2004.06.010

Jasper, H. H. & Andrews, H. L. (1938). Electro-encephalography. III. Normal differentiation of occipital and precentral regions in man. *Archives of Neurology and Psychiatry, 39*, 96–115.

Jasper, H. & Penfield, W. (1949). Electrocorticograms in man: Effect of voluntary movement upon the electrical activity of the precentral gyrus. *Archiv für Psychiatrie und Zeitschrift Neurologie, 183*, 163–174.

Jensen, O. & Colgin, L. L. (2007). Cross-frequency coupling between neuronal oscillations. *Trends in Cognitive Sciences, 11*, 267–269.

Jones, E. J. H., Venema, K., Lowy, R., Earl, R. K., & Webb, S. J. (2015). Developmental changes in infant brain activity during naturalistic social experiences. *Developmental Psychobiology, 57*, 842–853. doi:10.1002/dev.21336

Kaufman, J., Csibra, G., & Johnson, M. H. (2003). Representing occluded objects in the human infant brain. *Proceedings of the Royal Society of London. Series B: Biological Sciences (Suppl.), 270*, S140–S143. doi:10.1098/rsbl.2003.0067

Kaufman, J., Csibra, G., & Johnson, M. H. (2006). Oscillatory activity in the infant brain reflects object maintenance. *Proceedings of the National Academy of Sciences of the United States of America, 102*, 15271–15274. https://doi.org/10.1073/pnas.0507626102

Klimesch, W. (1999). EEG alpha and theta oscillations reflect cognitive and memory performance: A review and analysis. *Brain Research Reviews, 29*, 169–195. doi:10.1016/S0165-0173(98)00056-3

Klimesch, W. (2012). Alpha-band oscillations, attention, and controlled access to store information. *Trends in Cognitive Sciences, 16*, 606–617. doi:10.1016/j.tics.2012.10.007

Klimesch, W., Doppelmayr, M., Schwaiger, J., Auinger, P., & Winkler, T. (1999)."Paradoxical" alpha synchronization in a memory task. *Cognitive Brain Research, 7*, 493–501. https://doi.org/10.1016/S0926-6410(98)00056-1

Klingberg, T. (2006). Development of a superior frontal-intraparietal network for visuo-spatial working memory. *Neuropsychologia, 44*, 2171–2177.

Köster, M., Langeloh, M., & Hoehl, S. (2019). Visually entrained theta oscillations increase for unexpected events in the infant brain. *Psychological Science, 30*, 1656–1663. https://doi.org/10.1177/0956797619876260

Kraybill, J. H. & Bell, M. A. (2013). Infancy predictors of preschool and post-kindergarten executive function. *Developmental Psychobiology, 55*, 530–538. http://dx.doi.org/10.1002/dev.21057

Lehtonen, J., Valkonen-Korhonen, M., Georgiadis, S., Tarvainen, M. P., Lappi, H., Niskanen, J.-P., ... Karhalainen, P. A. (2016). Nutritive sucking induces age-specific EEG-changes in 0–24 week-old infants. *Infant Behavior and Development, 45*, 98–108. http://dx.doi.org/10.1016/j.infbeh.2016.10.005

Lepage, J. & Théoret, H. (2006). EEG evidence for the presence of an action observation-execution matching system in children. *European Journal of Neuroscience, 23*, 2505–2510. https://doi.org/10.1111/j.1460-9568.2006.04769.x

Liao, Y., Acar, Z. A., Makeig, S., & Deak, G. (2015). EEG imaging of toddlers during dyadic turn-taking: Mu-rhythm modulation while producing or observing social actions. *NeuroImage, 112*, 52–60. http://dx.doi.org/10.1016/j.neuroimage.2015.02.055

Lindsley, D. B. (1938). Electrical potentials of the brain in children and adults. *The Journal of General Psychology, 19*, 285–306.

Lindsley, D. B. (1939). A longitudinal study of the occipital alpha rhythm in normal children: Frequency and amplitude standards. *The Journal of Genetic Psychology, 55*, 197–213.

Lopes da Silva, F. (2013). EEG and MEG: Relevance to neuroscience. *Neuron, 80*, 1112–1128. doi:10.1016/j.neuron.2013.10.017

Lusby, C. M., Goodman, S. H., Bell, M. A., & Newport, D. J. (2014). Electroencephalogram patterns in infants of depressed mothers. *Developmental Psychobiology, 56*, 459–473. doi:10.1002/dev.21112

MacNeill, L. A., Ram, N., Bell, M. A., Fox, N. A., & Perez-Edgar, K. (2018). Trajectories of infants' biobehavioral development: Timing and rate of A-not-B performance gains and EEG maturation. *Child Development, 89*, 711–724. doi:10.1111/cdev.13022

Marshall, P. J., Bar-Haim, Y., & Fox, N. A. (2002). Development of the EEG from 5 months to 4 years of age. *Clinical Neurophysiology, 113*, 1199–1208.

Marshall, P. J. & Meltzoff, A. N. (2011). Neural mirroring systems: Exploring the EEG mu rhythm in human infancy. *Developmental Cognitive Neuroscience, 1*, 110–123. https://doi.org/10.1016/j.dcn.2010.09.001

Matsuura, M., Yamamoto, K., Fukuzawa, H., Okubo, Y., Uesugi, H., Moriiwa, M., ... Shimazono, Y. (1985). Age development and sex differences of various EEG elements in healthy children and adults—Quantification by a computerized wave form recognition method. *Electroencephalography and Clinical Neurophysiology, 60*, 394–406.

Meyer, M., Braukmann, R., Stapel, J. C., Bekkering, H., & Hunnius, S. (2016). Monitoring others' errors: The role of the motor system in early childhood and adulthood. *British Journal of Developmental Psychology, 34*, 66–85. doi:10.1111/bjdp.12101

Miskovic, V., Ma, X., Chou, C.-A., Fan, M., Owens, M., Sayama, H., & Gibb, B. E. (2015). Developmental changes in spontaneous electrocortical activity and network organization from early to late childhood. *NeuroImage, 118*, 237–247. http://dx.doi.org/10.1016/j.neuroimage.2015.06.013

Morasch, K. C. & Bell, M. A. (2011). The role of inhibitory control in behavioral and physiological expressions of toddler executive function. *Journal of Experimental Child Psychology, 108*, 593–606. https://doi.org/10.1016/j.jecp.2010.07.003

Muthukumaraswamy, S. D. & Johnson, B. W. (2004). Changes in Rolandic mu rhythm during observation of a precision grip. *Psychophysiology, 41*, 152–156. doi:10.1046/j.1469-8986.2003.00129.x

Nikitina, G. M., Posikera, I. N., & Stroganova, T. A. (1985). Central organization of emotional reaction of infants during first year of life. *Ontogenesis of the Brain, 4*, 223–227.

Ogawa, T., Sugiyama, A., Ishiwa, S., Suzuki, M., Ishihara, T., & Sato, K. (1984). Ontogenic development of autoregressive component waves of waking EEG in normal infants and children. *Brain & Development, 6*, 289–303.

Orekhova, E. V., Stroganova, T. A., & Posikera, I. N. (1999). Theta synchronization during sustained anticipatory attention in infants over the second half of the first year of life. *International Journal of Psychophysiology, 32*, 151–172.

Orekhova, E. V., Stroganova, T. A., Posikera, I. N., & Elam, M. (2006). EEG theta rhythm in infants and preschool children. *Clinical Neurophysiology, 117*, 1047–1062. doi:10.1016/j.clinph.2005.12.027

Ortiz-Mantilla, S., Hamalainen, J. A., Realpe-Bonilla, T., & Benasich, A. A. (2016). Oscillatory dynamics underlying perceptual narrowing of native phoneme mapping from 6 to 12 months of age. *The Journal of Neuroscience, 36*, 12095–12105. doi:10.1523/JNEUROSCI.1162-16.2016

Paredes, M. F., James, D., Gil-Perotin, S., Kim, H., Cotter, J. A., Ng, C., ... Alvarez-Buylla, A. (2016). Extensive migration of young neurons into the infant human frontal lobe. *Science, 354*, aaf7073. doi:10.1126/science.aaf7073

Paul, K., Dittrichova, J., & Papousek, H. (1996). Infant feeding behavior: Development in patterns and motivation. *Developmental Psychobiology, 29*, 563–576.

Perone, S., Palanisamy, J., & Carlson, S. M. (2018). Age-related change in brain rhythms from early to middle childhood: Links to executive function. *Developmental Science, 21*, e12691. https://doi.org/10.1111/desc.12691

Pfurtscheller, G., Brunner, C., Schlogel, A., & Lopes da Silva, F. H. (2006). Mu rhythm (de) synchronization and EEG single-trial classification of different motor imagery tasks. *NeuroImage, 31*, 153–159. doi:10.1016/j.neuroimage.2005.12.003

Pfurtscheller, G., Stančák, A., & Edlinger, G. (1997). On the existence of different types of central beta rhythms below 30 Hz. *Electroencephalography and Clinical Neurophysiology, 102*, 316–325.

Pivik R. T., Andres, A., Tennal, K. B., Gu, Y., Downs, H., Bellando, B. J. . . . Badger, T. M. (2019). Resting gamma power during the postnatal critical period for GABAergic system development is modulated by infant diet and sex. *International Journal of Psychophysiology, 135*, 73–94. https://doi.org/10.1016/j.ijpsycho.2018.11.004

Pivik, R. T., Broughton, R. J., Coppola, R., Davidson, R. J., Fox, N. A., & Nuwer, M. R. (1993). Guidelines for the recording and quantitative analysis of electroencephalographic activity in research contexts. *Psychophysiology, 30*, 547–558. https://doi.org/10.1111/j.1469-8986.1993.tb02081.x

Poole, K. L., Santesso, D. L., Van Lieshout, R. J., & Schmidt, L. A. (2018). Trajectories of frontal brain activity and socio-emotional development in children. *Developmental Psychobiology, 60*, 353–363. doi:10.1002/dev.21620

Rayson, H., Bonaiuto, J. J., Ferrari, P. F., & Murray, L. (2017). Early maternal mirroring predicts infant motor system activation during facial expression observation. *Scientific Reports, 7*, 11738. doi:10.1038/s41598-017-12097-w

Reznik, S. J. & Allen, J. J. B. (2018). Frontal asymmetry as a mediator and moderator of emotion: An updated review. *Psychophysiology, 55*, e12965. https://doi.org/10.1111/psyp.12965

Rothbart, M. K., & Bates, J. E. (2006). Temperament. In W. Damon & R.M. Lerner (Series Eds.) & N. Eisenberg (Vol. Ed.), *Handbook of child psychology: Vol. 3. Social, emotional, and personality development*, 6th ed. (pp. 99–166). Wiley.

Saby, J. N. & Marshall, P. J. (2012). The utility of EEG band power analysis in the study of infancy and early childhood. *Developmental Neuropsychology, 37*, 253–273. https://doi.org/10.1080/87565641.2011.614663

Schaworonkow, N. & Voytek, B. (2021). Longitudinal changes in aperiodic and periodic activity in electrophysiological recordings in the first seven months of life. *Developmental Cognitive Neuroscience, 47*, 100895. https://doi.org/10.1016/j.dcn.2020.100895

Short, S. J., Elison, J. T., Goldman, B. D., Styner, M., Gu, H., Connelly, M., ... Gilmore, J. H. (2013). Associations between white matter microstructure and infants' working memory. *NeuroImage, 64*, 156–166. https://doi.org/10.1016/j.neuroimage.2012.09.021

Smith, C. L. & Bell, M. A. (2010). Stability in infant frontal asymmetry as a predictor of toddlerhood internalizing and externalizing behaviors. *Developmental Psychobiology, 52*, 158–167. doi:10.1002/dev.20427

Smith, C. L., Diaz, A., Day, K. L., & Bell, M. A. (2016). Infant frontal electroencephalogram asymmetry and negative emotional reactivity as predictors of toddlerhood effortful control. *Journal of Experimental Child Psychology, 142*, 262–273. doi:10.1016/j.jecp.2015.09.031

Smith, J. R. (1938). The electroencephalogram during normal infancy and childhood: II. The nature of the growth of the alpha waves. *The Journal of Genetic Psychology, 53*, 455–469.

Smith, J. R. (1939). The "occipital" and "pre-central" alpha rhythms during the first two years. *The Journal of Psychology, 7*, 223–226.

Smith, J. R. (1941). The frequency growth of the human alpha rhythms during normal infancy and childhood. *The Journal of Psychology, 11*, 177–198.

Stamoulis, C., Vanderwert, R. E., Zeanah, C. H., Fox, N. A., & Nelson, C. A. (2015). Early psychosocial neglect adversely impacts developmental trajectories of brain oscillations and their interactions. *Journal of Cognitive Neuroscience, 27*, 2512–2528. doi:10.1162/jocn_a_00877

Stroganova, T. A. & Orekhova, E. V. (2007). EEG and infant states. In M. de Haan (Ed.), *Infant EEG and event-related potentials* (pp. 251–287). Psychology Press.

Stroganova, T. A., Orekhova, E. V., & Posikera, I. N. (1998). Externally and internally controlled attention in infants: An EEG study. *International Journal of Psychophysiology, 30*, 339–351.

Stroganova, T. A., Orekhova, E. V., & Posikera, I. N. (1999). EEG alpha rhythm in infants. *Clinical Neurophysiology, 110*, 997–1012.

Swingler, M. M., Perry, N. B., Calkins, S. D., & Bell, M. A. (2014). Maternal sensitivity and infant response to frustration: The moderating role of EEG asymmetry. *Infant Behavior & Development, 37*, 523–535. doi:10.1016.j.infbeh.2014.06.010

Swingler, M. M., Willoughby, M. T., & Calkins, S. D. (2011). EEG power and coherence during preschoolers' performance of an executive function battery. *Developmental Psychobiology, 53*, 771–784. doi:10.1002/dev.20588

Takano, T. & Ogawa, T. (1998). Characterization of developmental changes in EEG-gamma band activity during childhood using the autoregressive model. *Acta Paediatrica Japonica, 40*, 446–452.

Tarullo, A. R., Obradovic, J., Keehn, B., Rasheed, M. A., Siyal, S., Nelson, C. A., & Yousafzai, A. K. (2017). Gamma power in rural Pakistani children: Links to executive function and verbal ability. *Developmental Cognitive Neuroscience, 26*, 1–8. http://dx.doi.org/10.1016/j.dcn.2017.03.007

Thatcher, R. W., Krause, P. J., & Hrybyk, M. (1986). Cortico-cortical associations and EEG coherence: A two-compartmental model. *Electroencephalography and Clinical Neurophysiology, 64*, 123–143. https://doi.org/10.1016/0013-4694(86)90107-0

Thorpe, S. G., Cannon, E. N., & Fox, N. A. (2016). Spectral and source structural development of mu and alpha rhythms from infancy through adulthood. *Clinical Neurophysiology, 127*, 254–269. http://dx.doi.org/10.1016/j.clinph.2015.03.004

Tierney, A., Strait, D. L., O'Connell, S., & Kraus, N. (2013). Developmental changes in resting gamma power from age three to adulthood. *Clinical Neurophysiology, 124*, 1040–1042. http://dx.doi.org/10.1016/j.clinph.2012.09.023

Tomalski, P., Moore, D. G., Ribeiro, H., Axelsson, E. L., Murphy, E., Karmiloff-Smith, A., ... Kushnerenko, E. (2013). Socioeconomic status and functional brain development – associations in early infancy. *Developmental Science, 16*, 676–687. doi:10.1111/desc.12079

Upshaw, M. B., Bernier, R. A., & Sommerville, J. A. (2016). Infants' grip strength predicts mu rhythm attenuation during observation of lifting actions with weighted blocks. *Developmental Science, 19*, 195–207. doi:10.1111/desc.12308

van der Velde, B., & Junge, C. (2020). Limiting data loss in infant EEG: putting hunches to the test. *Developmental Cognitive Neuroscience, 45*, 100809. https://doi.org/10.1016/j.dcn.2020.100809

Vanderwert, R. E., Simpson, E. A., Paukner, A., Suomi, S. J., Fox, N. A., & Ferrari, P. F. (2015). Early social experience affects neural activity to affiliative facial gestures in newborn nonhuman primates. *Developmental Neuroscience, 37*, 243–252. doi:10.1159/000381538

Vincent, K. M., Xie, W., & Nelson, C. A. (2021). Using different methods for calculating frontal alpha asymmetry to study its development from infancy to 3 years of age in a large longitudinal sample. *Developmental Psychobiology, 63*, e22163. https://doi.org/10.1002/dev.22163

Walter, V. J. & Walter, W. G. (1949). The central effects of rhythmic sensory stimulation. *Electroencephalography and Clinical Neurophysiology, 1*, 57–86.

Watson, A. J. & Bell, M. A. (2013). Individual differences in inhibitory control skills at three years of age. *Developmental Neuropsychology, 38*, 1–21. https://doi.org/10.1080/87565641.2012.718818

Whitford, T., Rennie, C. J., Grieve, S. M., Clark, C. R., Gordon, E., & Williams, L. M. (2007). Brain maturation in adolescence: Concurrent changes in neuroanatomy and neurophysiology. *Human Brain Mapping, 28*, 228–237. doi:10.1002/hbm.20273

Wolfe, C. D. & Bell, M. A. (2004). Working memory and inhibitory control in early childhood: Contributions from physiology, temperament, and language. *Developmental Psychobiology, 44*, 68–83. doi:10.1002/dev.10152

Wolfe, C.D., & Bell, M.A. (2007a). Sources of variability in working memory in early childhood: A consideration of age, temperament, language, and brain electrical activity. *Cognitive Development, 22*, 431–455.

Wolfe, C.D., & Bell, M.A. (2007b). The integration of cognition and emotion during infancy and early childhood: Regulatory processes associated with the development of working memory. *Brain and Cognition, 65*, 3–13.

Xie, W., Mallin, B. M., & Richards, J. E. (2018). Development of infant sustained attention and its relation to EEG oscillations: an EEG and cortical source analysis study. *Developmental Science, 21*, e12562. https://doi.org/10.1111/desc.12562

Xu, G., Broadbelt, K. G., Haynes, R. L., Folkerth, R. D., Borenstein, N. S., Belliveau, R. A. ... Kinney, H. C. (2011). Late development of the GABAergic system in the human cerebral cortex and white matter. *Journal of Neuropathology & Experimental Neurology, 70*, 841–858. https://doi.org/10.1097/NEN.0b013e31822f471c

Yoo, K. H., Cannon, E. N., Thorpe, S. G., & Fox, N. A. (2016). Desynchronization in EEG during perception of means-end actions and relations with infants' grasping skill. *British Journal of Developmental Psychology, 34*, 24–37. https://doi.org/10.1111/bjdp.12115

CHAPTER 14

DEVELOPMENTAL RESEARCH ON TIME-FREQUENCY ACTIVITY IN ADOLESCENCE AND EARLY ADULTHOOD

STEPHEN M. MALONE, JEREMY HARPER, AND WILLIAM G. IACONO

THE notion that any complex system depends critically on its history (D'Souza & Karmiloff-Smith, 2016) is central to a developmental perspective, which holds that the state of an organism at any moment in its lifetime reflects its history, or trajectory in developmental time. A full understanding of a human characteristic is only possible through a systematic examination of how it came to be. The time-frequency features of an adult study participant's response to target stimuli in a visual oddball task, for instance, also reflect the current state of the dynamic interplay among genetic, biological, psychological, and external forces over time that have given rise to the psychological and neural processes, motivational state, and personality characteristics influencing activity recorded in the laboratory. Understanding the nature of this interplay is critical to understanding the response, whether frontal theta or parietal delta activity.

An approach informed by this type of perspective examines phenotypic variation over time to ask questions about mechanisms underlying a psychological, behavioral, or psychopathological process (Rutter, 1988). A developmental perspective on cognitive control, for example, can help elucidate the mechanisms underlying it by examining how time-frequency features related to cognitive control develop over the course of the lifespan. Discontinuities or inflection points in this process might reflect important transitions, which can be exploited to test hypotheses about causal mechanisms. A developmental approach thus has the potential to inform us about *how* time-frequency features come about, as well as *what* those features are.

14.1 Advantages Inherent in the Developmental Study of Time-Frequency Activity

A developmental approach confers several advantages.

14.1.1 Fine-Grained Analysis

Although a developmental perspective offers a unique view of time-frequency features, developmental research on time-frequency features can also offer purely pragmatic advantages. Time-frequency decompositions of electroencephalographic (EEG) signals into moment-by-moment activity at different frequencies lend themselves to a fine-grained analysis of brain activity, with accumbent advantages for developmental research.

14.1.2 Tracking Developmental Change

Time-frequency components correlate more strongly with age than time-domain components (Bowers et al., 2018). At the same time, change in time-frequency features correlates more strongly with change in time-domain characteristics than participants' chronological age (Yordanova & Kolev, 1996). In both instances, then, time-frequency activity appears to represent something particularly salient developmentally.

14.1.3 Stability of Time-Frequency Measures

Time-frequency features may be more stable over time than ERP amplitudes (DuPuis et al., 2015), which makes them particularly attractive for detecting change in an outcome between two time points. This follows from the fact that statistical power in a paired t-test is a function in part of the standard error of the difference in means, which is in turn a function of the correlation between measurements of the variable at the two time points, $\sigma_{Y1-Y2} = (\sigma_{Y1}{}^2 + \sigma_{Y2}{}^2 - 2\rho_{12}{}^2)^{1/2}$, where σ_{Y1-Y2} is the standard error of the difference in means, $\sigma_{Y1}{}^2$ is the variance at Time 1, $\sigma_{Y1}{}^2$ is the variance at Time 2, and ρ_{12} is the Time 1–Time 2 correlation. Hence, holding sample and effect sizes constant, the greater the Time 1–Time 2 correlation (stability), the greater the power to detect change.

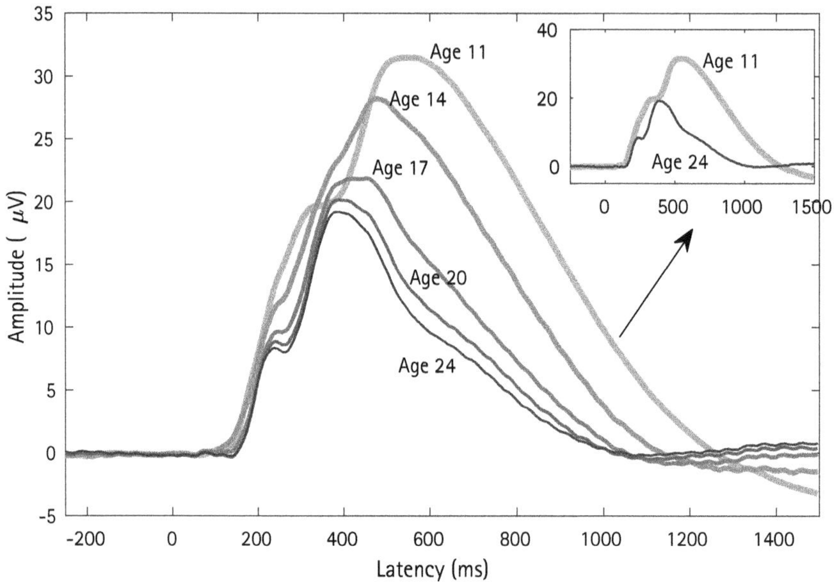

FIGURE 14.1 Grand mean ERP waveforms across five assessment waves spanning preadolescence to early adulthood. ERPs represent the average response to target stimuli at Pz among subjects assessed longitudinally at target ages of 11, 14, 17, 20, and 24. ERPs for successive assessment waves are progressively smaller in amplitude, with the ERP at age 11 being the largest and the ERP at age 24 the smallest. The inset depicts the same ERPs for the age-11 and age-24 assessments only.

14.1.4 Insight into ERP Dynamics

Characterizing stability and change in time-frequency features can help us to understand the developmental course of event-related potential (ERP) responses to experimental events (Malone et al., 2021; Mathes et al., 2016; Yordanova & Kolev, 2008). In Malone and colleagues (2021) we reported results of a longitudinal investigation into time-frequency activity in participants first assessed at the age of 11, with follow-up assessments at 3- to 4-year-intervals to a target age of 24. Participants were administered the oddball task at each assessment wave. Figure 14.1 depicts the grand-average ERP at Pz, the midline parietal electrode location, for the different assessment waves. Amplitude of the ERP decreases monotonically across waves. The morphology of the grand mean ERPs is broadly similar across waves, with an immediately recognizable P3 component. Nevertheless, subtle change in ERP morphology across the different assessments is evident as well. We examined the time-frequency representation of trial-level EEG activity at each assessment wave, described in greater detail later. Figure 14.2 consists of false color maps (heatmaps) of the magnitude of time-frequency activity at each discrete time-frequency bin for the features that best accounted for the EEG data at each wave. The overall similarity in components is striking. Yet differences in structure

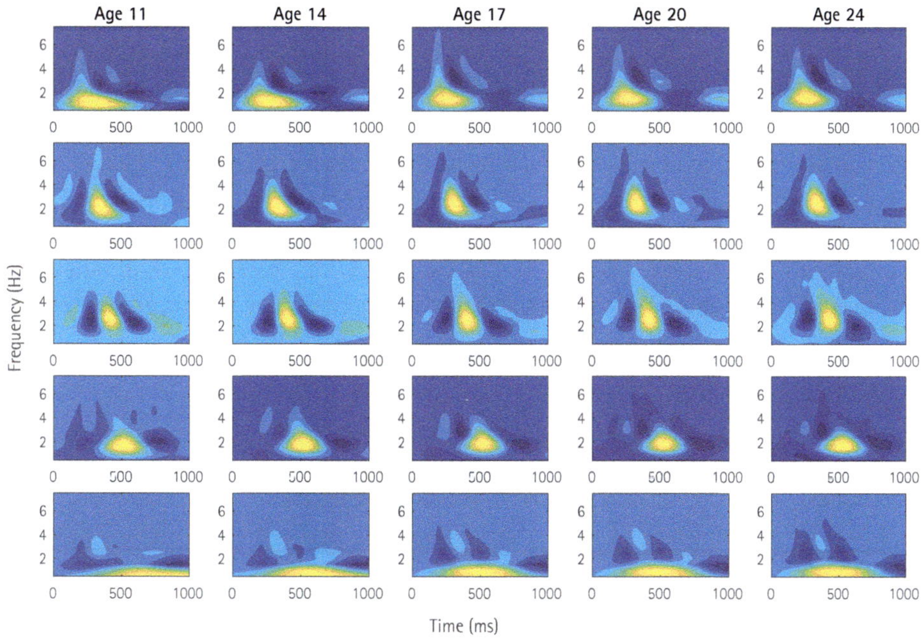

FIGURE 14.2 Component weights (loadings) across assessment waves. The figure consists of false color maps (heatmaps) of weights at each time-frequency bin from PCA of time-frequency energy for all target trials. PCA was conducted separately for each assessment wave. Components were matched by congruence coefficients computed between loadings at successive assessment waves.

across assessments are evident as well, including shifts in the latency of the maximum amount of energy in Component 5 and a trend toward increasing compactness in time-frequency space of Component 1.

Thus, a common set of time-frequency features characterizes the EEG data across development from early adolescence to early adulthood. Yet at the same time the subtle developmental change in ERP morphology across waves evident in Figure 14.1 occurred through small changes in this common set of features. Investigating differences between those components that shifted in time and/or frequency with development and those that did not will contribute to a full understanding of the decision-making processes elicited by a visual oddball task.

14.2 Overview of the Chapter

In the remainder of this chapter we summarize the main findings of research on developmental aspects of time-frequency activity, which highlights the importance of methodological choices for interpreting developmental trends. We then discuss advantages and disadvantages of cross-sectional and longitudinal designs, with a treatment of missing data and planned missingness. We briefly summarize statistical approaches to characterizing developmental time-frequency activity and how the different approaches implicitly or explicitly treat (developmental) time. We discuss methodological issues related to developmental time-frequency research, namely, age differences in task difficulty or degree of contamination by artifacts and variation in pubertal status among study participants. We then present empirical findings from two longitudinal investigations to illustrate what can be gleaned from a longitudinal design. Finally, we conclude with thoughts about how a developmental perspective might profitably influence future work.

14.3 Time-Frequency Activity and Development

This section reviews findings from selected studies that highlight the range of developmental insights contributed by time-frequency research.

14.3.1 Time-Frequency Activity in the Auditory Oddball Paradigm

The vast majority of the early work on age differences and developmental trends in time- and frequency-specific activity consisted of the pioneering work of Yordanova, Kolev, and colleagues (much of which is summarized in Yordanova & Kolev, 2008). Using an auditory oddball paradigm, Yordanova and Kolev conducted a number of studies of age-related change in key time-frequency features: power, or magnitude, modulation of power by stimulus onset, and phase-locking of the EEG response over trials. Several key broad findings emerged from this work. Amplitude in different frequency bands was commonly observed to decrease across age groups of increasing age. This was true of pre-stimulus as well as post-stimulus amplitude. The latency of the maximal theta response also decreased across groups of increasing age. In contrast, the single-sweep wave index (SSWI), a measure of the consistency of responses across trials (Kolev & Yordanova, 1997; Yordanova & Kolev, 1998), increased with age (except in the gamma band). In a finding that anticipates our subsequent treatment of the topic, Yordanova and Kolev found that the *enhancement factor* (Basar, 1980) of event-related oscillatory activity, which reflects change in the amplitude of the signal following stimulus onset relative to pre-stimulus amplitude, increased with age as well. Thus, there was substantial developmental change in time-frequency features, which varied somewhat by frequency band and the particular measure examined. The authors conclude that developmental changes in theta activity in particular may be related to cognitive development.

14.3.2 Time-Frequency Activity Related to Specific Psychological and Cognitive Processes

Building on this earlier work, several recent studies have examined time-frequency activity in a visual oddball task or variant thereof (Chorlian et al., 2015; Chorlian et al., 2017; Malone et al., 2021; Mathes et al., 2016; Wienke et al., 2018). Other work has examined age differences in focused on time-frequency activity related to the development of error or feedback processing (Bowers et al., 2018; Crowley et al., 2014; DuPuis et al., 2015), response preparation (Bender et al., 2005) and inhibition (Hwang et al., 2016; Liu et al., 2014; Papenberg et al., 2013), perception (Ehlers et al., 2016; Uhlhaas et al., 2009), and working memory (Sander et al., 2012). Although several studies have examined synchrony in activity recorded at different pairs of electrodes or between different frequency bands, the measures of interest in this research are largely the same as in the work of Yordanova and Kolev: the magnitude of time-frequency power and intertrial phase consistency (ITPC) (Tallon-Baudry et al., 1996),[1] a measure of the degree to which the timing of time-frequency activity is consistent across trials which is similar to the SSWI.

14.3.3 General Developmental Trends

A review of this work reveals consistent trends. ITPC is positively associated with age or age groups of increasing age. Time-frequency power is also associated with age in a consistent direction, where the direction of association depends on the nature of baseline correction. Studies using raw baseline correction, most often consisting of subtracting the mean level of power in a pre-stimulus period from all post-stimulus power values, typically report a negative association between total power and age. By contrast, studies using baseline normalization, such as a dB transformation of post-stimulus power relative to pre-stimulus power, defined as $10log_{10}A_{posttf} / \overline{A}_{pref}$, where $A_{post_{tf}}$ is power at a particular time-frequency bin in the post-stimulus period and $\overline{A}_{pref}$ is the mean pre-stimulus power at the same frequency, typically report a *positive* association between age and time-frequency power. Simple baseline subtraction is comparable to removing a DC offset in the time series recorded from a single electrode and thus simply centers power values at all frequencies around the same baseline level, whereas baseline normalization expresses post-stimulus activity in terms of proportional change from baseline. If y represents post-stimulus activity and $\overline{x}$, mean pre-stimulus activity, then it is clear

[1] ITPC is also known as intertrial phase coherence or clustering and phase-locking factor. Because time-frequency activity is often expressed as a complex number, it has both magnitude and phase. ITPC is a measure of the degree to which the phase of the complex-valued signal at a given time-frequency bin is consistent across trials.

why this would be true because $\log(y)-\log(\bar{x})=\log\left(\frac{y}{\bar{x}}\right)$. (See chapter 18 in Cohen, 2014, for a treatment of the issue of baseline correction in time-frequency analysis).

Several of the studies using baseline normalization also examined time-frequency power in the pre-stimulus period (without baseline correction), which serves to help reconcile the divergent findings. All report an inverse association between participant age and raw theta and delta power, which contrasts with the positive association with post-stimulus power modulation but mirrors the inverse association between age and power commonly reported in studies using raw baseline correction (e.g., Mathes et al., 2016).

In summary, the phasic brain response to an experimental event (stimulus onset or incorrect or inappropriate response) becomes increasingly aligned temporally with the event, which is reflected in the positive association between age and ITPC commonly observed. Although there are exceptions (Bowers et al., 2018; Crowley et al., 2014), baseline subtraction most often results in finding overall reductions in power with age, which is sometimes interpreted to reflect increased processing efficiency. However, baseline normalization most often results in overall *increases* with age, which is not as easily explained in terms of efficiency. In addition, time-frequency power in the pre-stimulus baseline period also decreases. This suggests that pre- and post-stimulus power both decrease with development, but at different rates. It may be that with development, fewer neurons are recruited in response to an experimental event, leading to reduced power overall, but children and adolescents become better able to maintain an appropriate task set, resulting in greater phase consistency in the EEG response and increased post-stimulus power. Both interpretations are speculative but point to the fact that the different methods of accounting for baseline activity can be complementary.

14.3.4 Specific Age-Related Associations

Age differences in specific aspects of time-frequency activity have been reported in several studies, in addition to the general overall developmental trend. Investigators using a gambling task with participants ranging in age from 8 to 17 found a negative association between theta power and age only on trials in which participants won points, not when they lost them (Bowers et al., 2018). Studies of cognitive control report age differences in specific time-frequency features. Papenberg and colleagues (2013) found that theta ITPC was relatively equal among four age groups (children, adolescents, young adults, and older adults) during the go condition of a go/no-go task, whereas it varied significantly by groups during the no-go condition. Thus, the degree to which ITPC was differentially related to cognitive control and the behavioral demands of making a response versus withholding one depended on age. Liu and colleagues, also using a go/no-go task with children and adolescents, found that ITPC was greatest on go and no-go trials requiring extra effort to withhold a response in order to regain points previously lost only among the oldest adolescents. In a study using MEG (magnetoencephalography), Hwang and colleagues (2016) observed differences between adolescents and young adults in the magnitude of preparatory activity in the frontal eye fields (FEF) bilaterally before antisaccades versus prosaccades. Power between 10 and 18 Hz was lesser in magnitude throughout the preparatory period among adolescents. Preparatory activity in the ventrolateral and dorsolateral prefrontal cortex of the right hemisphere (VLPFC and DLPFC) did not differ between groups, suggesting equivalent levels of motor inhibition and task-rule representations, but coupling between beta activity in DLPFC and alpha activity in the frontal eye fields did. Thus, these three studies taken together indicate that time-frequency activity implicated in different aspects of cognitive control differs across developmental stages.

Finally, researchers have looked for age differences in the scalp distribution of frequency-specific responses, rather than the magnitude of responses. Children show more broadly distributed theta-band activity and smaller magnitude beta- and theta-frequency responses to grammatical errors than adults (Schneider et al., 2018; Schneider et al., 2016), which the authors interpret to indicate that children use semantic information to process grammatical errors, whereas adults use syntactic information. Age differences in the topography or frequency-specific pattern of responses to Mooney faces have been reported as well, taken to reflect a reorganization of the brain response and increasing specialization for face processing with development (Mišić et al., 2014; Uhlhaas et al., 2009).

Thus, there is evidence for both general and specific age-related differences. From a developmental perspective, observing age differences is only the first step along the way to understanding a phenomenon (Rutter, 1988). Age differences in time-frequency features might reflect hormonal differences, differences in cognitive development, or differences in the salience of a monetary reward or other task demands, not simply differences in chronological age. Unpacking findings of mean differences between age groups is likely to be informative about the mechanisms underlying them, which in turn may have implications for better understanding normative cognitive development as well as psychopathological processes.

14.4 Research Design Issues

Several considerations emerge from reviewing the literature on age-related differences in time-frequency features. The first involves use of a cross-sectional or longitudinal design, although these clearly can complement one another. The vast majority of studies we reviewed used a cross-sectional design, which obviously requires less time to conduct than longitudinal studies, a significant practical advantage. However, they can present interpretive difficulties. The primary threat to validity derives from the possibility of cohort or group differences in characteristics that might affect time-frequency responses. Cohort differences are more likely to be problematic in studies of aging than in studies of participants spanning a narrow range of ages. However, cross-sectional designs can nevertheless be susceptible to age-group or age-related differences due to sampling variation, especially if sample sizes are small. As a result, cross-sectional studies can lead researchers to under- or overestimate the true magnitude of change, and we cannot know to what extent, if any, this is true.

Longitudinal designs are clearly superior for studying *change* rather than age-related differences in mean levels, permitting a characterization of normative developmental trajectories and individual differences therein. Change can be characterized by a small number of parameters that represent the nature of developmental trajectories in a meaningful way. There are many possible parametric models that might be appropriate for the phenomenon under study. Furthermore, longitudinal designs allow the researcher to determine whether developmental trajectories are modulated by characteristics of the individual, such as sex, genotype, or risk status for psychopathology (e.g., Carlson & Iacono, 2008; Chorlian et al., 2015; Chorlian et al., 2017; Hill & Shen, 2002). Developmental trajectories may also constitute useful endophenotypes (Iacono & Malone, 2011; Iacono et al., 2017). An important paper by Curran and Bauer (2011) advocates disaggregating between- and within-individual change in longitudinal models when assessing correlates of change.

Yet longitudinal designs suffer from challenges of their own. The most obvious is practical: longitudinal studies take longer to conduct than cross-sectional ones. We return to this issue later. Longitudinal designs are also vulnerable to participant attrition and missed assessments. However, recent developments in statistical methods allow researchers to use all available data, rather than having to resort to crude methods, such as listwise deletion of cases missing any data. Provided the mechanism that generates missing data is such that there is not a relationship between the likelihood of a value being missing and the value itself, missingness can be considered "ignorable" (Little & Rubin, 2002).

Longitudinal studies are vulnerable to possible changes in the orientation or motivation of participants to engage fully as well as the possibility of familiarity or exposure effects. Although perhaps more obvious in regard to neuropsychological tasks than the tasks used to elicit and study time-frequency features, we must not overlook the possibility of familiarity effects in regard to a gambling or go/no-go task. In fact, we have observed likely familiarity effects in an oddball task despite time intervals between assessments of three to four years (Malone et al., 2021). Fortunately, when more than two measurements are available per participant, separating practice or familiarity effects from true developmental change becomes possible (Ferrer et al., 2004). Comparing results from a longitudinal design with those from a cross-sectional study of participants spanning the same age range also permits assessing the degree to which familiarity effects may have influenced the longitudinal results.

14.4.1 Missingness by Design and Cohort-Sequential Designs

One way to increase the efficiency of longitudinal designs that may not be widely recognized is through incorporating planned missingness in the study design. In such designs, each individual participant is observed at a subset of the possible time points or

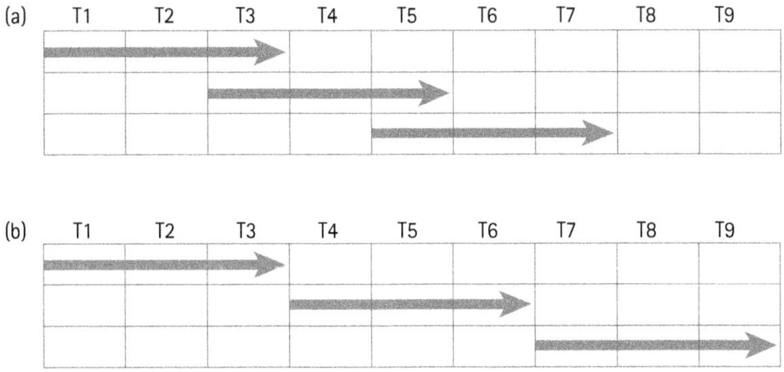

FIGURE 14.3 Schematic of two variants of planned missingness designs. The hypothetical sample in this scenario consists of three age groups of adolescents assessed three times over a two-year period. Panel A depicts a cohort-sequential design, in which each age group shares an assessment age with at least one other, which allows an assessment of cohort differences in a dependent measure. In panel B, the age range spanned by the sample is maximized at the expense of the ability to assess cohort differences. Both are examples of missingness by design.

ages rather than at all of them. This design is illustrated in Figure 14.3, which depicts two possible data collection plans for a hypothetical study of adolescents between the ages of 14 and 20 or older. (What we present here is purely hypothetical; see Brandmaier et al., 2020 for guidelines for planned missingness in latent growth curve designs.) There are three age cohorts in this example, with all participants assessed annually for a period of two years. Subjects in the three respective cohorts were 14, 16, or 18 years old at the initial assessment for the first design plan (Plan A) and 14, 17, and 20 years old at the initial assessment for the second design plan. Plan A depicts a cohort-sequential design, in which participants' mean age at the last assessment for each of the two youngest age groups is the same as the mean age of participants in the next older age group at their initial visit. Overlap in ages allows the researcher to compare cohorts. In Plan B, participants in the three age cohorts are assessed at nonoverlapping ages, effectively increasing the age range spanned by the sample. In both cases, the missing data are missing by virtue of the experimental design, which makes missingness ignorable and yields unbiased parameter estimation. Thus, they allow the researcher to characterize developmental change across the entire age range of the sample despite the fact that each subject is only observed three times.

14.5 METHODOLOGICAL ISSUES

14.5.1 Task Difficulty Versus Task Demands

An experimental task may be more difficult for younger participants than older ones, confounding task demands with task difficulty and necessitating modifications to the experimental procedure to equate task difficulty across participants of different ages. Several of the studies we reviewed adjusted stimulus delivery or response windows to accomplish this. Two used tasks requiring cognitive control, which develops gradually throughout adolescence (Casey et al., 2008; Luciana & Collins, 2012; Somerville & Casey, 2010). One was a "simple" oddball task (Wienke et al., 2018). In all three, some of the participants were as young as eight or nine years old. An additional consideration is the need to equate the number of trials that go into estimating power or ITPC, which affects the reliability of the estimate, if there is age-related variation in the number of trials lost due to artifacts.

14.5.2 Pubertal Stage

The ages of participants spanned middle childhood to late adolescence or adulthood in several studies, which creates a potential confound between age and pubertal status, and thus neuroendocrinological development. A chronological age of 13, say, may not be equivalent for boys and girls with respect to the phenomenon under study. Sample sizes may not be large enough for a careful analysis of effects of puberty, but it should not be overlooked. Indeed, we found that incorporating a measure of pubertal status in growth curve models eliminated apparent sex differences in rates of change in time-frequency power in adolescence. In addition, accounting for individual differences in pubertal status in this manner resulted in superior correspondence between the observed data and predicted trajectories of change in the amplitude of one particular time-frequency component (Malone et al., 2021).

14.6 Statistical analysis

14.6.1 Age as discrete or continuous

Cross-sectional studies with groups of participants at distinctly different ages typically make use of t-tests or analysis of variance (ANOVA) for the purpose of statistical analysis. The focus of analysis is necessarily on differences in mean levels among groups. Repeated measures ANOVAs with longitudinal data in distinct age groups also model mean levels within and between subjects. By contrast, the explicit goal of regression models is to describe the form of change, such as whether it is linear with age (such that change is additive) or nonlinear, as well as the rate of change. Modern linear mixed effects (LME) regression models allow for participants to be measured at different ages or times or even to have been assessed a different number of times, and LMEs also allow a great deal of flexibility with respect to characterizing the covariance structure of repeated measures (within subjects) in addition to characterizing trends in mean levels of the outcome measure. Thus, these models are well suited for longitudinal analysis.[2] An additional advantage of an LME framework over traditional ANOVA models is that it uses all available data as long as missingness is ignorable.

Of course, one is not limited to linear models: a nonlinear or nonparametric model may be appropriate for capturing growth that follows a well-specified form, such as exponential and logistic growth models. Models with a latent age or time basis can be an attractive way to model nonlinearity in mean levels of an outcome; we illustrate the use of such a model in the next section. Kernel regression is a form of locally weighted regression that has been used to model change in time-frequency features (Chorlian et al., 2015). Nonparametric regression can be useful when a simple parametric model is not appropriate or as a preliminary step in choosing an appropriate model (Helwig, 2019). Thus, a wide variety of models are available in regression analysis, whether the data are from a longitudinal design or a cross-sectional one, which should prompt the researcher to think about the most appropriate model of change for the data at hand and the question asked.

14.6.2 The Form of Developmental Change in Time-Frequency Characteristics

Although the form of developmental change is rarely explicitly addressed, there is agreement among the studies examining it that the association between age and time-frequency features in an oddball task is non-additive (nonlinear) in adolescence and early adulthood. Wienke and colleagues (2018) found that a quadratic polynomial function of age best explained the association between age and theta power amplitude modulation and ITPC. Chorlian and colleagues (2015), in a longitudinal study of 2,170 subjects assessed every two years and spanning the ages of 12 to 25, observed striking sex differences in the pattern of change in theta power elicited by a novelty oddball task. A steeper decline in power was observed among males than females, with a subsequent leveling off between approximately 21 and 25. Power declined more gradually among females but continued to do so even when power had leveled off among males. We also observed nonlinear change in a large sample studied longitudinally and spanning essentially the same age rage, described in the next section (Malone et al., 2021).

14.7 An Empirical Exploration

The preceding sections discussed several issues involved in developmental research of time-frequency activity: the utility of longitudinal designs; planned and accidental

[2] ANOVA models can now incorporate a variety of residual covariance structures, which overcomes some of their limitations (Liu et al., 2012).

missingness; influences of puberty; and the appeal of regression approaches. This section illustrates the relevance of these issues to time-frequency research. Our presentation is largely descriptive and intended to be illustrative, with a focus on longitudinal methods, to serve as something of a counterpoint to the existing literature, which is predominantly cross-sectional. First, we briefly summarize results of our recent paper (Malone et al., 2021), some of which we discussed earlier. We then present preliminary results of an investigation extending these results in a second independent sample with high-density EEG recordings. Each is based on a population-based cohort from the Minnesota Center for Twin and Family Research (MCTFR) (Wilson et al., 2019) consisting of twin pairs (and their parents). Approximately 60% are monozygotic, or identical, twins. Twins are same-sex pairs who were raised together who have been followed since a target age of 11 at approximately three- to five-year intervals. Participants in both cohorts were administered the same task—Begleiter's rotated heads oddball task (Begleiter, Porjesz, Bihari, & Kissin, 1984).

In Malone and colleagues (2021) we report findings from a longitudinal investigation into time-frequency activity in 1,692 twin subjects (917 of them female). The majority (N = 1,512; 760 females) comprised the younger cohort of the MCTFR (Iacono et al., 1999). Subjects were first assessed at the age of 11, with four follow-up assessments at three- to four-year-intervals to a target age of 24. An additional 205 participants (172 females) from the separate ES cohort (Keyes et al., 2009) who completed an identical intake assessment were included. EEG activity was collected by means of a Grass Neurodata 12 system at three or four scalp electrodes, with linked ears serving as reference. A pair of electrodes arranged in a transverse montage above and to the side of one eye allowed for identification of eye blinks and other EOG activity.

Time-frequency energy was derived from individual EEG signals by means of the reduced interference distribution (RID; Williams, 2001; Chapter 4.) Meaningful features in time-frequency energy on target trials were identified by means of principal component analysis (PCA) (Bernat et al., 2007; Bernat et al., 2005) via singular value decomposition on centered columns of the "matricized" or unfolded matrix of baseline-corrected time-frequency energy values (see Figure 14.2). PCA accounted well for the data and yielded components that were highly similar from one time point to the next. That is, after ordering components based on their timing, we computed Tucker congruence coefficients (Tucker, 1951) between all pairs of components for a given assessment and the successive assessment. These were uniformly large and approached 1 for matched pairs of components, indicating nearly perfect congruence, whereas the off-diagonal elements in each matrix of congruence coefficients were small and approached 0, indicating lack of congruence between the unmatched pairs.

This degree of component congruence allowed the use of scores on matched components to characterize longitudinal change. Examining plots of mean scores across waves as well as individual trajectories of randomly selected individuals suggested that piecewise linear regression models might be appropriate. LME models, treating assessment wave as nested within individuals who were in turn nested within twin pair, with component scores as the dependent measure accounted well for the observed trends in the data. The change point, or knot, in each regression was identified empirically and ranged from 16 to 19.5. Developmental trajectories in time-frequency power were characterized by an initial decrease in (baseline-corrected) raw power, followed by a flatter trajectory in later adolescence and early adulthood. Although the rate of change was smaller in magnitude for the second phase of these models, it was still significant (and negative). Data for one representative component are plotted in Figure 14.4. Panel (a) depicts the component structure across waves. This recapitulates the first row of Figure 14.2 and indicates that the component structure became less dispersed in time-frequency space with development. Panel (b) consists of violin plots representing the distribution of scores across waves, separately for male and female participants. This illustrates the overall trend for this component, its similarity for participants of both

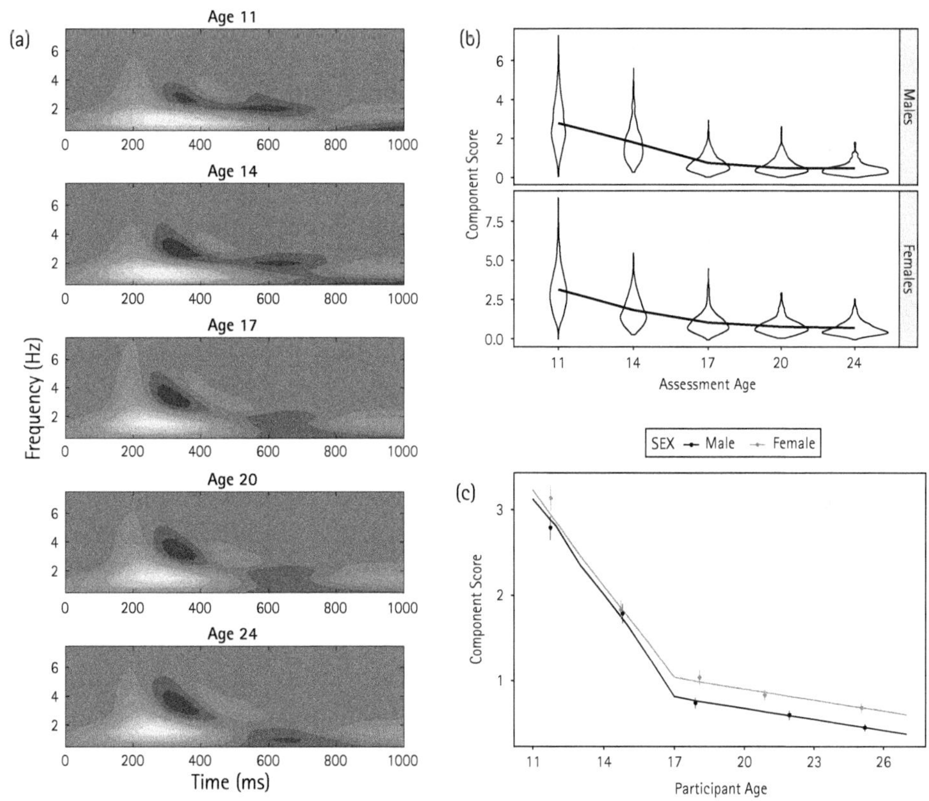

FIGURE 14.4 Component scores from a PCA analysis of time-frequency energy plotted against assessment age. Subjects are the same as in Figure 14.1. Data are for Component 1 in Figure 14.2, chosen to illustrate key results. Panel A depicts the component weights. This is identical to the first row of time-frequency weights in Figure 14.2. Panel B consists of violin plots of the distribution of Component 1 scores across assessment waves, separately for the two sexes. A nonparametric smoother (loess) represents the average developmental trend. Panel C depicts trajectories implied by a piecewise linear model fit to these scores together with the observed mean scores at each wave and 95% confidence intervals around them.

sexes and a striking decrease in variance with development. Finally, Panel (c) depicts the mean trajectory implied by our piecewise linear regression model along with observed means and confidence intervals. The (nonlinear) model-implied trajectory accounts for the observed data well.

That the magnitude of time-frequency activity could be modeled by piecewise linear trajectories suggests that there were distinct developmental phases in this sample with respect to the brain response to target stimuli in the oddball paradigm used. These results also illustrate one of the advantages of LME models: only 485 participants had data from all five assessments, yet all data were used in fitting models, including age-11 data from the approximately 200 ES twins, who were assessed at subsequent visits using a different EEG recording system (described next) and as a result contributed only a single observation in this investigation.

14.7.1 Extending These Findings

We extended these findings using high-density EEG data from the ES cohort of the MCTFR (Keyes et al., 2009), which comprises 998 twins, first assessed at the target age of 11 and subsequently followed up at target ages of 14 and 17. Participants completed the same rotated heads task as participants in the MTFS (programmed in E-Prime software) and EEG were recorded from 61 scalp electrodes using a Biosemi ActiveTwo system. An additional four electrodes were placed just above and beside each eye for the purpose of identifying vertical and horizontal EOG activity. Data were screened for eye blinks, horizontal eye movement and other artifacts using an in-house pipeline (see Burwell et al., 2014; Harper et al., 2019).[3] A total of 942 subjects had usable data: 601 at age 11 (270 females; ages 10.9 to 12.9, with a mean of 11.8); 717 at the first follow-up (347 females; ages 13.6 to 16.7, with a mean of 14.9); 763 at the second follow-up (375 females; ages 16.8 to 19.2, with a mean of 17.8). As described above, the initial assessment of 204 twins followed the MTFS protocol, whereas they completed this version of the task at the two follow-ups. Approximately 80% had data from at least two of the three assessments.

A time-frequency representation of EEG responses to target stimuli was obtained by means of a complex Morlet wavelet convolution. Frequencies ranged from 1 to 40 Hz, with the number of cycles ranging from 3 to 8 (with 25 logarithmic steps for both). The resulting signal was downsampled in time to 64 Hz and an estimate of power at each time point obtained as the modulus of the complex signal at that time point. Total power was averaged over 25 subsets of trials sampled randomly without replacement from each experimental condition (target or nontarget) so as to equate the number of trials across conditions. Power for each discrete frequency bin was represented by means of a dB transformation relative to the average activity in that frequency between 450 and 250 ms prior to stimulus onset, which complements the approach in Malone and colleagues (2021). In addition, we computed ITPC, as in Tallon-Baudry and colleagues (1996).

[3] https://www.github.com/sjburwell/eeg_commander

We again identified meaningful time-frequency feature by means of PCA on centered columns of the matrix of time-frequency power values for all subject-electrodes. Scree plots suggested that eight components were appropriate for characterizing the data at each assessment wave, which accounted for 75–80% of the total variance at each wave. As in Malone and colleagues (2021), components were varimax-rotated in order to obtain a more meaningful solution, and we computed Tucker's congruence coefficient for pairs of components that were visually matched between successive assessment waves. Coefficients were uniformly large in magnitude, ranging from 0.96 to 0.99 for all pairs of matched components, whether for age-11 and age-14 solutions or age-14 and age-17 solutions. Components were highly similar across assessment waves, as in our previous investigation.

14.7.2 Characterizing the Results

We selected delta and theta components for presentation. False color maps of the components are plotted in the left-hand column of Figure 14.5, which illustrate how similar the components are across assessment waves. Nevertheless, the delta component shifts in time, appearing earlier in the response window at the later follow-up assessments, while both components become more compact in time. Both trends echo findings from the previous investigation. The topographic distribution of theta power (see the head plots inset in each heatmap) is also more diffuse at age 11.

Figure 14.5 also presents mean levels of component scores at the different assessment waves, averaged over electrodes. ITPC scores were obtained by multiplying ITPC values for each subject-electrode by the weights (loadings) for delta and theta components, respectively, and summing weighted values. Trajectories are plotted in the right-hand portion of panel (a) and (b) in Figure 14.5. These are somewhat nonlinear, distinctly so in the case of theta ITPC. To keep the numbers of parameters to a minimum, we used a latent age basis model. This is a form of structural equation model (SEM) similar to latent growth curve (LGC) models (themselves similar to regression-based approaches). In both, change in a dependent measure is modeled in terms of a latent intercept and slope (or slopes, if a higher-order polynomial is used to model time). Time (or age) enters the model via the slope loadings, rather than being an observed variable as in regression models. However, whereas in the LGC model the slope parameters are fixed, in the latent basis model one or more are freely estimated, which allows one to capture nonlinearity in developmental trajectory.[4] Rate of change is proportional over time, as determined by the age basis coefficients (Grimm et al., 2011).

Models were estimated using Mplus version 7.2 (Muthén & Muthén, 1998–2012) via MplusAutomation (Hallquist & Wiley, 2018), an R package that serves as a flexible wrapper to Mplus. Mplus permits using a subject's actual age at each assessment

[4] If there are T time points, T − 2 parameters can be estimated; two parameters must be fixed to identify these models. With only three time points, there is a single free parameter, but the number of parameters estimated can increase substantially with more time points.

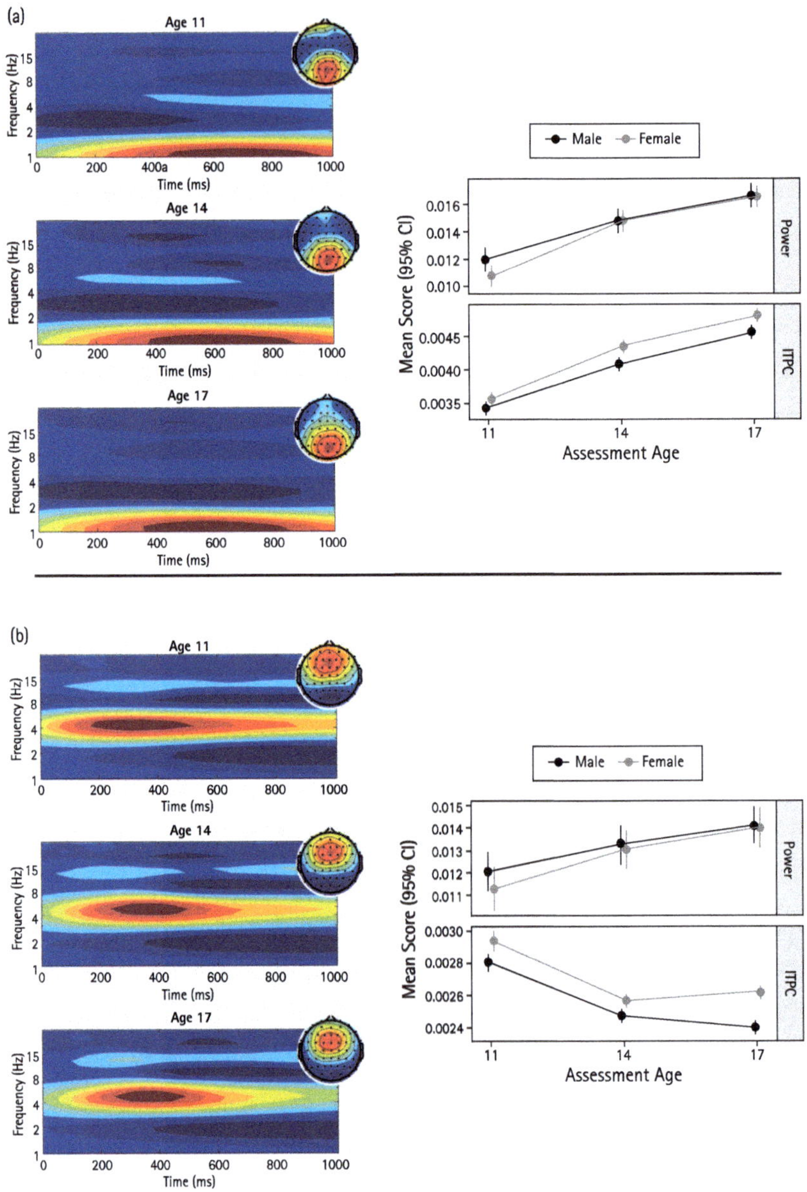

FIGURE 14.5 Developmental course and topographic distribution of time-frequency components. Loadings of two components from a PCA of dB-transformed time-frequency power at each of three assessment waves in a large sample of adolescents in a longitudinal investigation. Component loadings at each wave are depicted in the form of a heat map representing the magnitude (and sign) of the component weights on each time-frequency bin. Insets in the upper right-hand corner of each heat map depict the mean score on the two components respectively at each electrode on a schematic representation of a head. The color map indicates the magnitude of mean score. Red indicates the largest mean values, whereas blue indicates the smallest. The right-hand plot in each panel depicts the mean component score, averaged over all electrodes, at each assessment wave, separately for males and females.

wave, rather than assuming they are exactly the same, as SEM software often requires. Because our interest is in individual trajectories we used a cluster-robust sandwich estimator via the COMPLEX samples option to derive appropriate standard errors given that this sample comprises twins. In light of evidence of sex differences in level or shape of trajectories of time-frequency activity (Chorlian et al., 2015; Malone et al., 2021), we allowed for (estimated) sex differences in intercept and slope. Only the sex effect on the intercept of theta ITPC was significant, $p = .005$, which is evident in the difference in overall level between males and females in panel (b) of Figure 14.5. The rate of change in all four measures could be considered roughly equivalent for males and females (p-values $\geq .110$). Loadings for the age-11 and age-17 data on the slope latent variable were fixed to the average age at the age-11 and age-17 assessments, respectively, in order to identify the model, whereas the slope loading on age-14 scores (or, strictly speaking, the offset in years from the average age at this assessment) was estimated from the data. This coefficient was positive for all four time-frequency measures, indicating greater change by the age-14 assessment than assumed by a strictly linear age basis. Wald z tests of the slope coefficient were significant, with the exception of theta total power, indicating a meaningful departure from linearity. Intercept and slope were significantly inversely correlated for both ITPC measures, indicating that those with higher levels of ITPC at age 11 tended to show a slower rate of change with development. (See Table 14.1 for a close approximation to these correlations from the bivariate models examined

Table 14.1 Cross-process correlations from bivariate parallel process models.

	Delta ITPC-Power		Theta ITPC-Power		Theta-Delta Power		Theta-Delta ITPC	
	r	p-value	r	p-value	r	p-value	r	p-value
Intercept1-Slope1	-.614	.001	-0.902	<.001	-0.173	.649	-0.932	<.001
Intercept2-Slope2	-.319	.422	-0.087	.846	-0.433	.355	-0.840	<.001
Intercept1-Intercept2	.551	<.001	0.240	.094	0.308	.145	-0.035	.736
Slope1-Slope2	.557	.016	0.118	.214	-0.148	.679	0.000	.957
Age-11 Residuals	NA	NA	NA	NA	0.201	.220	NA	NA
Age-14 Residuals	.438	<.001	0.246	<.001	0.274	.001	0.092	.067
Age-17 Residuals	-.193	.018	-0.034	.589	0.338	.009	0.166	.003

Tabled values are correlations between pairs of latent variables and their associated p-values for each model: ITPC and power for delta and theta, respectively, and theta and delta power or ITPC, respectively. Intercept1 and Slope1 refer to the first measure in each heading, whereas Intercept2 and Slope2 refer to the second. For instance, in the Delta ITPC-Power model, Intercept1 is the delta ITPC intercept and Intercept2 is for delta power. Age-specific residuals are the correlations between residual variance in the two measures, independent of growth trajectories. Age-11 residuals for three of the four measures were constrained to 0 to avoid negative estimates. Correlations involving these measures were therefore indeterminate.

next.) These analyses thus indicate similarities and differences in trajectories of delta and theta power and ITPC, as well as illustrate the potential of a latent time basis model in representing a nonlinear trend, which in the present situation captured nonlinear change in a particularly parsimonious model.

14.7.3 Covariation in Patterns of Change

In addition to analyses of univariate trajectories, longitudinal designs also permit one to assess how different trajectories of change are related to one another. As an illustration, we extend the univariate analyses described to the bivariate case, fitting the model to all four pairs of measures. We first examined Pearson correlation coefficients between all pairs of measures using the heatmap function in the R computing environment (R Development Team, 2019) (Figure 14.6), combined with dendrograms displaying the

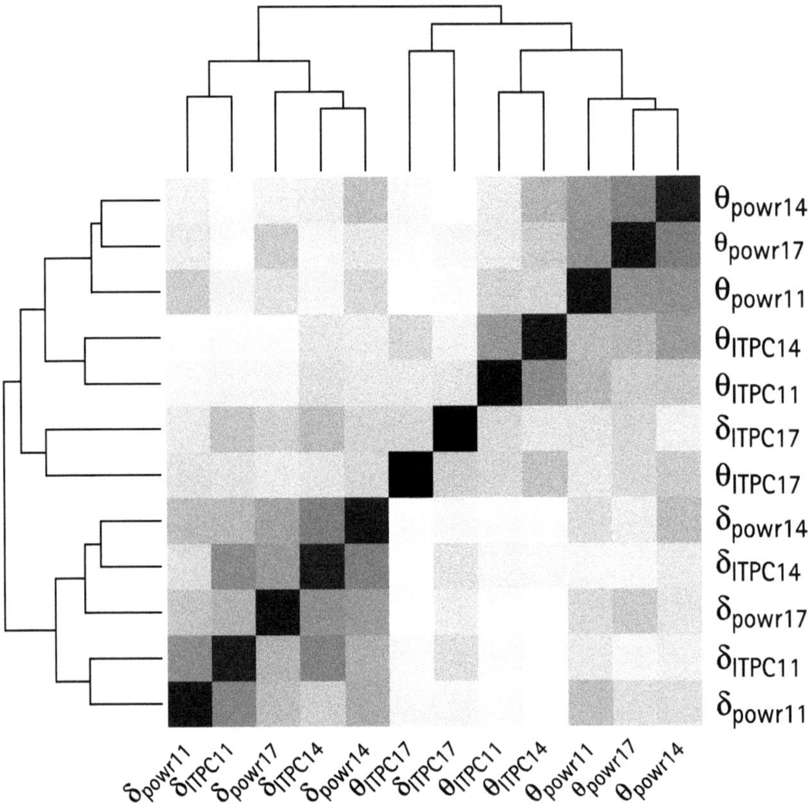

FIGURE 14.6 Matrix of correlations among time-frequency measures. The matrix of correlations among all pairs of delta and theta power and ITPC at the three waves of assessment is displayed graphically in false color (heatmap). Correlations are grouped on the basis of the results of a hierarchical clustering algorithm. Dendrograms of the clustering results are displayed on the margins. Theta and delta are indicated by their respective Greek letters and power is abbreviated as "powr." The number suffix indicates the assessment wave by its target age.

results of hierarchical clustering of the correlations. Correlations among theta and delta measures segregated into separate respective clusters, with the former in the upper right-hand corner of the correlation matrix and the latter in the lower left.[5] Within the theta cluster, measures of total power were particularly closely related, with correlations ranging from .47 to .58, whereas ITPC at 11 and 14 formed a subcluster. By contrast, delta total power and ITPC at 14 and total power at 17 tended to cluster together, and the measures that formed a subcluster were the two age-11 measures. The largest correlation coefficient was .58, between age-14 and age-17 theta power (between waves) as well as between age-14 delta power and ITPC (within a wave). Age-17 ITPC for both theta and delta were relatively independent of the other measures, with a median correlation of .06 and .10, respectively. This pattern suggests a greater degree of shared variance among delta power and ITPC than among theta power and ITPC, and relative independence of ITPC measures at age 17. In addition, a shift in the pattern of associations around the age-14 assessment is suggested.

To determine the degree to which change in one measure was associated with change in another, we fit a bivariate parallel process model. The parallel process model allows for the latent influences on each measure to be related to the latent influences on the other measure. Specifically, the model includes correlations (covariances) between the two intercepts and the two slopes as well as a regression of the slope of change in one measure on the intercept of the other measure. The latter permits an assessment of the degree to which the level of time-frequency ITPC, say, affects the rate of change in time-frequency power. This model is illustrated in Figure 14.7.

We fit four such models, one for each pairing of measures. Modeling the trajectory in each measure recapitulates the univariate analyses just described. This is evident in Table 14.1, which presents the most relevant parameter estimates from these models, in that the first two rows present correlations between the latent intercept and slope characterizing growth in a given measure. These were significant only for delta and theta ITPC. The parameters of greatest interest here are those that characterize relationships between the two measures: the correlations between latent intercepts ($\rho_{ITPIITPC}$ in Figure 14.7) and slopes ($\rho_{STPSITPC}$) and the effect of one latent intercept on slope of the other measure ($\gamma_{ITPSITPC}$ and $\gamma_{IITPCSTP}$). The former, cross-domain correlations appear in the shaded rows in Table 14.1. Only the correlations between intercepts and slopes for delta were significant, with results indicating that the greater the initial level of delta ITPC power, the greater the initial level of delta ITPC and the greater the rate of change in ITPC, the greater the rate of change in power. In addition, mean initial level in delta power was inversely associated with the rate of change in ITPC ($p = .008$). This would lead delta power and ITPC become somewhat more distinct with development, consistent with the pattern of correlations in Figure 14.6. It is also consistent with the

[5] Although PCA yields orthogonal components and varimax is an orthogonal rotation method, any type of rotation of the PCA solution results in lack of orthogonality of scores, loadings or both, depending on the form of normalization used (Jolliffe, 2002). Normalizing eigenvectors to unit length, as we did, sacrifices independence of component scores while maintaining orthogonality of the loadings.

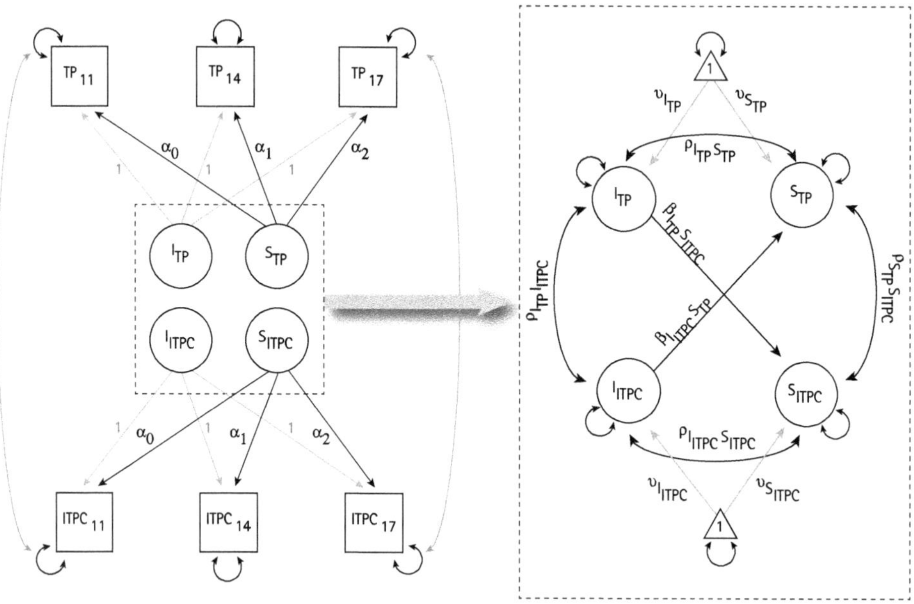

FIGURE 14.7 Partial path diagram for a parallel process bivariate growth model. Latent variables are represented by circles, whereas manifest variables are represented by squares. TP is time-frequency total power, ITPC is intertrial phase coherence. Each is measured at three different assessment waves in a longitudinal design. The left-hand panel focuses on the growth model for total power (top) and ITPC (bottom). The mean level of each is modeled as a linear function of a latent intercept (I_{TP} and I_{ITPC}) and latent slope (S_{TP} and S_{ITPC}). The loadings of each observed variable on its latent intercept are fixed at 1. The Greek letter a is used to represent the loadings on the latent slope. In the latent age basis model used in the example described in the text, ao is fixed at 0 and a2 at 1, but a1 is estimated from the data. The coefficient corresponding to a1 thus gives the proportion of total growth between 11 and 17 observed at age 14. At each assessment wave, the two regression residuals covary (are correlated), which allows for age-specific influences common to power and ITPC. These are drawn for ages 11 and 17 but not for age 14 so as to avoid an unnecessarily crowded diagram. The right-hand panel "explodes" the center of the left-hand figure to emphasize the relationships between the two latent growth processes. These consist of two correlations, between the two latent intercepts and the two latent slopes, and two regression paths predicting mean change in TP from initial level of ITPC and vice versa. The triangle containing the number 1 is used to represent the mean level of intercept and slope for TP and ITPC, respectively. In reality, these are intercepts, rather than means, because the model includes regressions of the latent variables on sex (not shown). In both panels, unlabeled circular double-headed arrows indicate residual variances.

principle of development toward progressive commitment and differentiation (Stiles & Jernigan, 2010), which implies that power and ITPC become increasingly functionally specialized. The effect was small, however; expressing the regression coefficient as a proportion of the residual variance in ITPC slope, this effect accounted for 1.6% of the variance. For the other three models—theta ITPC-power, theta-delta ITPC, and theta-delta power—neither regression of slope in one on intercept of the other nor the correlations between the two intercepts or the two slopes were significant. All p-values were greater than .360, although the ersatz effect size for the regression of delta ITPC on rate of change in theta ITPC was somewhat larger than that of the significant effect in the delta model. In addition, the correlations between paired intercepts and slopes were not significant. Thus, the two processes in each model developed independently.

In broad terms, aspects of these results were anticipated by the pattern of correlations in Figure 14.6 in that correlations between theta and delta measures were relatively small in magnitude and total power and ITPC correlations tended to segregate for theta compared to delta. However, most of the covariances between wave-specific residuals were significant and positive for both the age-14 and age-17 assessment waves. That is, in the absence of significant cross-measure regression effects for the most part, most of the covariance between measures was due to age-specific influences. This was especially true of the age-14 wave (and to a lesser extent the age-17 wave). Individual differences in pubertal status may induce correlations between time-frequency measures that are independent of developmental trajectories in these measures through hormonal effects on brain organization, an inference not readily gleaned from the correlation matrix.

14.7.4 Conclusion

We do not by any means consider these results definitive. For one thing, the parallel process model is only one possible approach, and it does not provide a direct assessment of influences of one measure on *change* in the other. This requires a more complex model with more time points (McArdle, 2009). Our results here are intended solely to illustrate some of the types of insights that longitudinal-developmental designs provide. In both investigations, we found that a common set of dimensions of time-frequency power, whether raw or dB-transformed, accounted well for time-frequency activity in adolescence and, in the first investigation, into early adulthood. The consistency of these dimensions across age in both instances was striking. Nevertheless, the particular configuration of these dimensions shifted in subtle ways with age. For instance, congruence coefficients between successive assessment waves for the component depicted in Figure 14.4 were all 0.95 or greater, indicating very close congruence between one wave and the next. Nevertheless, the coefficient for the same component between age-11 and age-24 waves was 0.77, indicating much less congruence over the course of adolescence. Thus, developmental change occurred through a gradual shift in time-frequency space of the locus of activity.

In our original investigation, we found that piecewise linear regression models accounted well for mean levels of the data, which suggests distinct phases characterize the development of time-frequency activity in the rotated heads oddball task. The timing of the transition between phases varied among components, which helps to illuminate age-related differences in the nature of the brain response. In our second, more preliminary investigation, our ability to replicate this finding was hamstrung by the fact that only three data points were available. Nevertheless, we found evidence of nonlinearity with development. For the purposes of this investigation, we averaged component scores over electrodes, but future work can benefit from the availability of a denser electrode array to further illuminate developmental aspects of time-frequency activity.

14.8 Final Thoughts

Research conducted to date has yielded several consistent general findings. The majority of studies have reported consistent associations between age and time-frequency power and phase-locking (ITPC). The association with ITPC is positive, such that phase-locking increases with age. As discussed, the nature of the association with power tends to be negative in studies using raw power or baseline subtraction but positive in studies examining modulation of power, such as by means of a dB transform, what Makeig and colleagues have called event-related spectral perturbations (ERSP; Makeig, 1993). These general trends are interpreted to mean that time-frequency responses to stimuli and events in an experimental setting become more mature in some sense. However, the consistency of these trends across paradigms presents something of an interpretive challenge for researchers. If more "efficient" responses are observed in a go/no-go task among older participants, does this reflect a change in the processes involved in withholding a response specifically or a developmental change in information processing more generally? Comparing age-related trends across a variety of tasks differing in processing and response demands will be necessary to begin answering this question.

A developmental perspective prompts a shift in the types of questions one asks. For instance, finding age-related differences in mean levels of a time-frequency phenomenon is really the first step of inquiry into the phenomenon (Rutter, 1988). This is in part because age differences involve other differences among individuals as well that are correlated with age. Observing age differences also suggests lines of follow-up research. If younger adolescents differ from older ones on a measure such as time-frequency power, how is this related to other time-frequency or performance measures (e.g., Liu et al., 2014; Papenberg et al., 2013)? At what point in development do these differences disappear? Do age-related differences in time-frequency features related to cognitive control in one task, such as increased theta power, predict similar age-related differences in those same features in another task? Failure to find such differences might call into question whether these features truly reflect cognitive-control processes. It might also indicate that the developmental course of these time-frequency features varies because of differences in the cognitive control processes required by tasks that appear similar on the surface. Control of interference in the attentional field by flanking stimuli in an Eriksen flanker tasks is likely somewhat different from control of response selection or inhibition in a go/no-go task. Yet at the same time, the similarities between these two forms of control suggest that the time-frequency activity elicited by each should be related and the relationship between them can be traced longitudinally within subjects.

Comprehensive examination of developmental aspects of time-frequency features thought to reflect specific psychological or neural processes can also shed light on those processes. For example, developmental neuroscience models differ somewhat with respect to how sensitivity to reward develops during adolescence. Some propose a rapid increase early in adolescence followed by a plateau (e.g., Casey et al., 2008; Somerville & Casey, 2010) whereas others propose that the early increase is followed by a subsequent decrease (Luciana & Collins, 2012). Time-frequency features associated with the reward positivity, for example, are thought to reflect activity of the ventral striatum and areas involved in reward processing (Bowers et al., 2018). These models make different predictions about the developmental course of reward-related time-frequency features. If, the developmental course of time-frequency features thought to reflect reward processing differs from that predicted by neuroscience models, it might suggest the models are wrong. But it might also suggest that our interpretation of the meaning of those time-frequency features is wrong.

Finally, in addition to characterizing age-related trends in a given time-frequency feature, a developmental perspective is concerned with understanding how aspects that develop early might affect those that develop later. For instance, a developmental approach to time-frequency connectivity analysis might ask how network patterns characteristic of children constrain the subsequent development of these patterns in adolescence? Are any such constraints more or less continuous with development or are they characterized by significant discontinuities? Answering questions such as these will require a longitudinal approach, perhaps in combination with cross-sectional studies. The number of studies addressing age-related differences in time-frequency activity has increased in recent years, an encouraging development that is expanding our knowledge in important ways. We suggest that expanding the types of questions being asked is likely to be fruitful as well if we are to move from description to explanation.

REFERENCES

Basar, E. (1980). *EEG-brain dynamics: Relation between EEG and brain evoked potentials.* Amsterdam.

Begleiter, H., Porjesz, B., Bihari, B., & Kissin, B. (1984). Event-related brain potentials in boys at risk for alcoholism. *Science*, *225*, 1493–1496.

Bender, S., Weisbrod, M., Bornfleth, H., Resch, F., & Oelkers-Ax, R. (2005). How do children prepare to react? Imaging maturation of motor preparation and stimulus anticipation by late contingent negative variation. *NeuroImage*, *27*, 737–752.

Bernat, E. M., Malone, S. M., Williams, W. J., Patrick, C. J., & Iacono, W. G. (2007). Decomposing delta, theta, and alpha time-frequency ERP activity from a visual oddball task using PCA. *International Journal of Psychophysiology*, *64*, 62–74.

Bernat, E. M., Williams, W. J., & Gehring, W. J. (2005). Decomposing ERP time-frequency energy using PCA. *Clinical Neurophysiology*, *116*, 1314–1334.

Bowers, M. E., Buzzell, G. A., Bernat, E. M., Fox, N. A., & Barker, T. V. (2018). Time-frequency approaches to investigating changes in feedback processing during childhood and adolescence. *Psychophysiology*, *55*, e13208.

Brandmaier, A. M., Ghisletta, P., & Oertzen, T. V. (2020). Optimal planned missing data design for linear latent growth curve models. *Behavioral Research Methods*, *52*, 1445–1458. doi:10.3758/s13428-019-01325-y

Burwell, S. J., Malone, S. M., Bernat, E. M., & Iacono, W. G. (2014). Does electroencephalogram phase variability account for reduced P3 brain potential in externalizing disorders? *Clinical Neurophysiology*, *125*, 2007–2015.

Carlson, S. R. & Iacono, W. G. (2008). Deviant P300 amplitude development in males is associated with paternal externalizing psychopathology. *Journal of Abnormal Psychology*, *117*, 910–923. https://doi.org/10.1037/a0013443

Casey, B. J., Jones, R. M., & Hare, T. A. (2008). The adolescent brain. *Annals of the New York Academy of Sciences*, *1124*, 111–126. https://doi.org/10.1196/annals.1440.010

Chorlian, D. B., Rangaswamy, M., Manz, N., Kamarajan, C., Pandey, A. K., Edenberg, H., … Porjesz, B. (2015). Gender modulates the development of theta event related oscillations in adolescents and young adults. *Behavioral Brain Research*, *292*, 342–352. doi:10.1016/j.bbr.2015.06.020

Chorlian, D. B., Rangaswamy, M., Manz, N., Meyers, J. L., Kang, S. J., Kamarajan, C., … Porjesz, B. (2017). Genetic correlates of the development of theta event related oscillations in adolescents and young adults. *International Journal of Psychophysiology*, *115*, 24–39. doi:10.1016/j.ijpsycho.2016.11.007

Cohen, M. X. (2014). *Analyzing neural time series data: theory and practice*: MIT press.

Crowley, M. J., van Noordt, S. J. R., Wu, J., Hommer, R. E., South, M., Fearon, R. M. P., & Mayes, L. C. (2014). Reward feedback processing in children and adolescents: medial frontal theta oscillations. *Brain and Cognition*, *89*, 79–89.

Curran, P. J., & Bauer, D. J. (2011). The disaggregation of within-person and between-person effects in longitudinal models of change. *Annual Review of Psychology*, *62*, 583–619. doi:10.1146/annurev.psych.093008.100356

D'Souza, D. & Karmiloff-Smith, A. (2016). Why a developmental perspective is critical for understanding human cognition. *Behavioral and Brain Sciences*, *39*, e122. doi:10.1017/S0140525X15001569

DuPuis, D., Ram, N., Willner, C. J., Karalunas, S., Segalowitz, S. J., & Gatzke-Kopp, L. M. (2015). Implications of ongoing neural development for the measurement of the error-related negativity in childhood. *Developmental Science*, *18*, 452–468.

Ehlers, J., Struber, D., & Başar-Eroğlu, C. (2016). Multistable perception in children: Prefrontal delta oscillations in the developing brain. *International Journal of Psychophysiology*, *103*, 129–134. doi:10.1016/j.ijpsycho.2015.02.013

Ferrer, E., Salthouse, T. A., Stewart, W. F., & Schwartz, B. S. (2004). Modeling age and retest processes in longitudinal studies of cognitive abilities. *Psychology and Aging*, *19*, 243–259.

Grimm, K. J., Ram, N., & Hamagami, F. (2011). Nonlinear growth curves in developmental research. *Child Development*, *82*, 1357–1371. doi:10.1111/j.1467-8624.2011.01630.x

Hallquist, M. N. & Wiley, J. F. (2018). Mplus Automation: An R package for facilitating large-scale latent variable analyses in Mplus. Structural Equation Modeling, *48*(2), 1–18. doi:10.1080/10705511.2017.1402334

Harper, J., Malone, S. M., & Iacono, W. G. (2019). Target-related parietal P3 and medial frontal theta index the genetic risk for problematic substance use. *Psychophysiology*, *56*(8), e13383. doi:10.1111/psyp.13383

Helwig, N. E. (2019). Robust nonparametric tests of general linear model coefficients: A comparison of permutation methods and test statistics. *NeuroImage*, *201*, 116030. doi:10.1016/j.neuroimage.2019.116030

Hill, S. Y. & Shen, S. (2002). Neurodevelopmental patterns of visual P3b in association with familial risk for alcohol dependence and childhood diagnosis. *Biological Psychiatry*, *51*, 621–631.

Hwang, K., Ghuman, A. S., Manoach, D. S., Jones, S. R., & Luna, B. (2016). Frontal preparatory neural oscillations associated with cognitive control: A developmental study comparing young adults and adolescents. *NeuroImage*, *136*, 139–148. doi:10.1016/j.neuroimage.2016.05.017

Iacono, W. G., Carlson, S. R., Taylor, J., Elkins, I. J., & McGue, M. (1999). Behavioral disinhibition and the development of substance use disorders: Findings from the Minnesota Twin Family Study. *Development and Psychopathology*, *11*, 869–900.

Iacono, W. G. & Malone, S. M. (2011). Developmental endophenotypes: Indexing genetic risk for substance abuse with the P300 brain event-related potential. *Child Development Perspectives*, *5*, 239–247.

Iacono, W. G., Malone, S. M., & Vrieze, S. I. (2017). Endophenotype best practices. *International Journal of Psychophysiology*, *111*, 115–144. doi:10.1016/j.ijpsycho.2016.07.516

Jolliffe, I. T. (2002). *Principal component analysis*. Springer.

Keyes, M. A., Malone, S. M., Elkins, I. J., Legrand, L. N., McGue, M., & Iacono, W. G. (2009). The enrichment study of the Minnesota twin family study: Increasing the yield of twin families at high risk for externalizing psychopathology. *Twin Research and Human Genetics*, *12*, 489.

Kolev, V. & Yordanova, J. (1997). Analysis of phase-locking is informative for studying event-related EEG activity. *Biological Cybernetics*, *76*, 229–235. doi:10.1007/s004220050335

Little, R. J. & Rubin, D. B. (2002). *Statistical analysis with missing data* (2nd ed.). Wiley.

Liu, S., Rovine, M. J., & Molenaar, P. (2012). Selecting a linear mixed model for longitudinal data: Repeated measures analysis of variance, covariance pattern model, and growth curve approaches. *Psychological Methods*, *17*, 15.

Liu, Z.-X., Woltering, S., & Lewis, M. D. (2014). Developmental change in EEG theta activity in the medial prefrontal cortex during response control. *NeuroImage*, *85*, 873–887.

Luciana, M., & Collins, P. F. (2012). Incentive motivation, cognitive control, and the adolescent brain: Is it time for a paradigm shift? *Child Development Perspectives*, *6*, 392–399. doi:10.1111/j.1750-8606.2012.00252.x

Makeig, S. (1993). Auditory event-related dynamics of the EEG spectrum and effects of exposure to tones. *Electroencephalography and Clinical Neurophysiology*, *86*, 283–293. doi:10.1016/0013-4694(93)90110-h

Malone, S. M., Harper, J., Bernat, E. M., & Iacono, W. G. (2021). Stability and change in the development of time-frequency activity in adolescence and early adulthood. Preprint [online] https://doi.org/10.31219/osf.io/8kdh

Mathes, B., Khalaidovski, K., Wienke, A. S., Schmiedt-Fehr, C., & Basar-Eroglu, C. (2016). Maturation of the P3 and concurrent oscillatory processes during adolescence. *Clinical Neurophysiology*, *127*, 2599–2609.

McArdle, J. J. (2009). Latent variable modeling of differences and changes with longitudinal data. *Annual Review of Psychology, 60*, 577–605.

Mišić, B., Mills, T., Vakorin, V. A., Taylor, M. J., & McIntosh, A. R. (2014). Developmental trajectory of face processing revealed by integrative dynamics. *Journal of Cognitive Neuroscience, 26*, 2416–2430.

Muthén, L. K., & Muthén, B. O. (1998-2012). *Mplus User's Guide* (7th ed.). Muthén & Muthén.

Papenberg, G., Hämmerer, D., Müller, V., Lindenberger, U., & Li, S.-C. (2013). Lower theta inter-trial phase coherence during performance monitoring is related to higher reaction time variability: a lifespan study. *NeuroImage, 83*, 912–920.

R Development Team. (2019). *R: A Language and Environment for Statistical Computing.* R Foundation for Statistical Computing. Retrieved from https://www.R-project.org/

Rutter, M. (1988). Epidemiological approaches to developmental psychopathology. *Archives of General Psychiatry, 45*, 486–495. doi:10.1001/archpsyc.1988.01800290106013

Sander, M. C., Werkle-Bergner, M., & Lindenberger, U. (2012). Amplitude modulations and inter-trial phase stability of alpha-oscillations differentially reflect working memory constraints across the lifespan. *NeuroImage, 59*, 646–654. doi:10.1016/j.neuroimage.2011.06.092

Schneider, J. M., Abel, A. D., Ogiela, D. A., McCord, C., & Maguire, M. J. (2018). Developmental differences in the neural oscillations underlying auditory sentence processing in children and adults. *Brain and Language, 186*, 17-25.

Schneider, J. M., Abel, A. D., Ogiela, D. A., Middleton, A. E., & Maguire, M. J. (2016). Developmental differences in beta and theta power during sentence processing. *Developmental Cognitive Neuroscience, 19*, 19--30.

Somerville, L. H. & Casey, B. J. (2010). Developmental neurobiology of cognitive control and motivational systems. *Current Opinion in Neurobiology, 20*, 236–241. doi:10.1016/j.conb.2010.01.006

Stiles, J. & Jernigan, T. L. (2010). The basics of brain development. *Neuropsychology Review, 20*, 327–348. doi:10.1007/s11065-010-9148-4

Tallon-Baudry, C., Bertrand, O., Delpuech, C., & Pernier, J. (1996). Stimulus specificity of phase-locked and non-phase-locked 40 Hz visual responses in human. *Journal of Neuroscience, 16*, 4240–4249.

Tucker, L. R. (1951). *A method for synthesis of factor analysis studies* (Vol. No. PRS-984). Educational Testing Service.

Uhlhaas, P. J., Roux, F., Singer, W., Haenschel, C., Sireteanu, R., & Rodriguez, E. (2009). The development of neural synchrony reflects late maturation and restructuring of functional networks in humans. *Proceedings of the National Academy of Sciences of the United States of America, 106*, 9866–9871.

Wienke, A. S., Basar-Eroglu, C., Schmiedt-Fehr, C., & Mathes, B. (2018). Novelty N2-P3a complex and theta oscillations reflect improving neural coordination within frontal brain networks during adolescence. *Frontiers in Behavioral Neuroscience, 12*, 218.

Williams, W. J. (2001). Reduced interference time-frequency distributions: Scaled decompositions and interpretations. In L. Debnath (Ed.), *Wavelet transforms and time-frequency signal analysis* (pp. 381–417). Birkhauser.

Wilson, S., Haroian, K., Iacono, W. G., Krueger, R. F., Lee, J. J., Luciana, M., . . . Vrieze, S. (2019). Minnesota Center for Twin and Family Research. *Twin Research and Human Genetics, 22*(6), 746–752. doi:10.1017/thg.2019.107

Yordanova, J., & Kolev, V. (1996). Brain theta response predicts P300 latency in children. *Neuroreport*, *8*, 277–280. doi:10.1097/00001756-199612200-00055

Yordanova, J. & Kolev, V. (1998). Developmental changes in the theta response system: a single-sweep analysis. *Journal of Psychophysiology*, *12*, 113–126.

Yordanova, J., & Kolev, V. (2008). Event-related brain oscillations in normal development. In L. A. Schmidt & S. J. Segalowitz (Eds.), *Developmental psychophysiology: Theory, systems, and methods* (pp. 15–68). Cambridge University Press.

CHAPTER 15

THETA-BETA POWER RATIO

An Electrophysiological Signature of Motivation, Attention and Cognitive Control

DENNIS J. L. G. SCHUTTER AND
J. LEON KENEMANS

15.1 Introduction

The preservation of frequencies across mammalian species is testament to the importance of rhythmic neural activity. Indeed, oscillatory activity at its multiple temporal scales is proposed to have guided the large volumetric expansion of the mammalian brain and its functions throughout the course of evolution (Buzsaki & Watson, 2012). The unique amplitude and frequency modulatory properties of electric signals suggest a neural syntax by which the brain can function on the different functional neuroanatomical levels (Buzsaki & Watson, 2012). Even though the localization of EEG activity in the brain is challenging, the different frequencies that make up the human EEG signal, which typically range between 1 to 50 Hz, correspond to more or less identifiable structures in the brain.

In agreement with this view, electric stimulation of the ascending reticular activating system in the brain-stem elicits 1–4 Hz (delta) cortical responses (Guyton, 1976). Delta activity is also a dominant rhythm during deep sleep and unconscious states and reflects large deactivation of the cerebral cortex and dominant activity associated with activity in the brain stem. Electric stimulation of limbic areas, on the other hand, can elicit 7-Hz (theta) activity in mammals (Gray, 1982). Activity in the theta frequency range (4–7Hz) has been observed in limbic areas, including the septo-hippocampal complex, amygdala, and anterior cingulate cortex (ACC) (Asada et al., 1999: Gray, 1982: Rahman et al., 2018).

EEG activity between 8 and 12 Hz (alpha) is the strongest frequency band produced in adult humans, and results in a wide area of cortical structures under the influence of functional decoupling with the thalamus. For example, alpha waves are generated when the visual cortex is deprived of sensory input (i.e., eyes closed), yet instantly disappear when the thalamocortical coupling is restored and the thalamus starts to transfer signals from our eyes to the visual cortex again (i.e., eyes open). This observation had led researchers to propose that alpha waves represent a so-called cortical idling rhythm that is inversely related to processing activity (Adrian & Matthews, 1934). In addition to the posterior areas, activity in the alpha band is also present in the more anterior regions and involved in motor, cognitive, and emotion-related functions (e.g., Hoptman & Davidson, 1998: Harmon-Jones & Gable, 2018).

Furthermore, EEG frequencies of 13 Hz and higher (beta) have been proposed to reflect synchronized field potentials in more local cortico-cortical circuits (Lubar, 1997). Beta activity can be registered across the cerebral cortex including the more anterior regions of the cerebral cortex (Engel & Fries, 2010). The association between beta activity and an anterior distribution concurs with the proposed posterior-anterior gradient model in which the natural peak EEG frequency increases from the posterior perceptual to the anterior action dedicated areas of the cerebral cortex (Rosanova et al., 2009). In sum, neural rhythms are preserved across the mammalian species, and they provide a signature of functionally dedicated brain regions and circuits.

The ratio of frequency bands is the relative contribution of neural rhythms to characterize the interdependency between these neural signals that can be recorded from the scalp. For example, scalp-recorded resting state frontocentral theta activity can at least in part be explained by input from signals coming from the subcortical system, whereas beta activity represents endogenous inhibitory activity that is more cortical of origin (Schutter & van Honk, 2005). Hence, the ratio between theta and beta power arguably reflects the balance between bottom-up excitatory subcortical signals and the cortical signals associated with top-down regulation and attentional control (Schutte et al., 2017). In this conceptual framework, a high theta-beta power ratio indicates a dominant bottom-up subcortical input over top-down cortical regulation, whereas a low theta-beta power ratio implies a dominant top-down cortical regulation over the bottom-up subcortical system. The general idea behind the theta-beta power ratio is that the ratio captures the interaction of more motivation-driven processes (theta) on the one hand, and attention- and cognitive-related processes (beta) on the other.

15.2 Theta-beta power ratio in attention deficit/hyperactivity disorder (ADHD): A meta-analytic perspective

Given the importance of rhythms as an organizing principle underlying brain function, EEG activity is increasingly being examined in studies on the biological underpinnings of normal and abnormal human behavior.

In the 1970s clinical-oriented scientists started to investigate possible associations between EEG rhythms and psychological functioning.

Among the first ideas that came under scientific investigation was the hypothesis that inattention and hyperactivity seen in children with ADHD result from a chronically under-aroused central nervous system (CNS) causing higher-than-normal resting state theta activity (Satterfield & Cantwell, 1974). The so-called hypo-arousal model of ADHD was postulated after observations of children with ADHD showing lower-than-normal electrodermal activity (Satterfield & Dawson, 1971). Paralleled by observations of dominant theta (4–7 Hz) and alpha activity (8–12 Hz) during rest and increased activity in the beta band (>13 Hz) during mental activity in healthy individuals (Jasper et al., 1938), Joel Lubar proposed that ADHD children may particularly have an underactive cortical system as indicated by the increased theta and reduced beta wave activity (Lubar, 1991). Further support for the EEG hypothesis was provided by a study that deployed a quantitative analysis of EEG in boys with ADHD (Mann et al., 1992). Results led to the idea that an elevated resting state theta-beta power ratio may in fact be an electrophysiological EEG marker of ADHD (Karakas & Barry, 2017). While the link between theta-beta power ratio and ADHD has proven reliable, the original idea of linking higher ratios with low CNS activity as indexed by electrodermal activity has not received much additional support ever since (Barry et al., 2004).

ADHD is a disorder marked by an ongoing pattern of inattention and/or hyperactivity-impulsivity that interferes with daily-life functioning and normal development. (https://www.nimh.nih.gov/health/topics/attention-deficit-hyperactivity-disorder-adhd/index.shtml). Features of inattention are lack of persistence, difficulties in sustaining focus, and disorganized thoughts. Importantly, these problems are not related to intelligence. Hyperactivity is the constant moving about both in appropriate and inappropriate situations. Furthermore, hyperactivity is associated with extreme restlessness and behaviors such as excessive fiddling, tapping, and talking. Persons with ADHD tend to be impulsive and engage in hasty actions without considering the possible negative consequences of their actions. As a result, they prefer immediate rewards and show an inability to delay gratification. Impulsive individuals can be socially demanding and may appear indifferent to the needs of others.

Research over the last 40 years has identified a number of EEG patterns associated with ADHD. In addition to elevated theta activity over the anterior scalp regions, lower alpha and beta activity over the posterior scalp regions have been reported. In addition, higher than normal posterior delta activity, as well as increases in the theta-alpha and theta-beta power ratio have been found in children with ADHD (Clarke et al., 2001a). As previously mentioned, increased slow wave activity (<8 Hz) and decreased fast wave activity (>10 Hz), elevations of theta activity and theta-beta power ratio are among the most consistent findings (Saad et al., 2018). Figure 15.1 shows an example of resting state theta and beta oscillations obtained with a non-invasive EEG scalp recording in a healthy young adult.

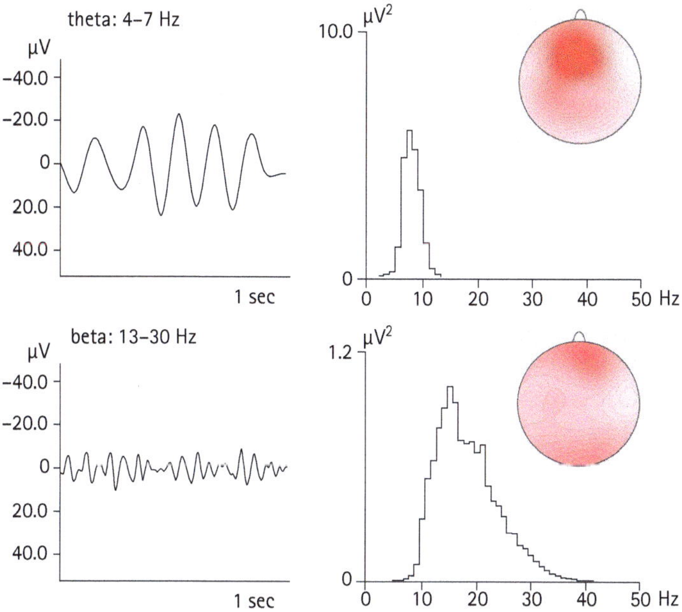

FIGURE 15.1 Example of band-pass filtered scalp recorded theta and beta oscillations and spectral power distributions.

15.3 Theta-beta power ratio in ADHD diagnostics

The replicability of these findings together with results showing that the pharmacological agent methylphenidate (Ritalin) improves ADHD symptoms paralleled by a lowering of theta activity and decrease in the theta-beta power ratio (Arns et al., 2018) has led to the suggestion that the resting state theta-beta power ratio may be a biomarker for ADHD and can be used in diagnostics. In one of the first studies, three experiments were reported that aimed to replicate previous findings, examine criterion-related validity of EEG on inattention, and determine test-retest reliability of the EEG (Monastra et al., 2001).

In the first experiment, earlier findings of significantly higher theta-beta power ratios in patients aged between 6–20 years with inattentive and hyperactive-combined ADHD as compared to controls, were replicated. The second experiment showed that the theta-beta power ratio derived attentional index could reliably differentiate between ADHD and non-ADHD participants (sensitivity: 90%, specificity: 94%). In the third experiment a high test-retest correlation coefficient (0.96) of the theta-beta power ratio was found in a sample of fifty-five volunteers that underwent an EEG session on two occasions thirty days apart. Comparable sensitivity (87%) and specificity (94%) scores in identifying ADHD within a clinical sample (n = 159) were reported by another multicenter validation study (Snyder et al., 2008). However, the theta-beta power ratio could not reliably differentiate between ADHD and comorbidities or alternative diagnoses. Based on these findings the authors advised the use of EEG as a complementary measure in the clinical evaluation of ADHD (Snyder et al., 2008).

This conclusion follows up results from the first meta-analysis by Snyder & Hall (2006), who investigated the presence of specific EEG traits in ADHD and evaluated nine high quality studies (n = 1498) in which the primary analysis of interest concerned a group comparison between ADHD patient's diagnosis according to the criteria of the Diagnostic and Statistical Manual of Mental Disorders and controls on resting state theta and beta activity. Standardized mean effect size (Glass' delta) and 95% confidence intervals (CI) were calculated. Results showed significantly higher theta (pooled effect size: 1.31; 95% CI: 1.14–1.48) and lower beta activity (pooled effect size: −0.51; 95% CI: −0.65 to −0.35) in patients compared to controls.

Consistent with these observations a strong significant effect was observed for an increase of theta-beta power ratio in ADHD patients versus controls (pooled effect size: 3.08; 95% CI: 2.90– 3.26). A second fixed effects model meta-analysis published in 2013 reported a notably lower overall effect size, Hedges' d: 0.75; 95% CI: 0.57– 0.93, favoring increased theta-beta power ratio in children and adolescents with ADHD as compared to controls (Arns et al., 2013).[1] Furthermore, considerable heterogeneity in effect sizes across studies was observed, indicating that the variance of the effect size across the included studies was greater than to be expected from sampling error. Even though the effect size was still large, subsequent analyses showed no strong support for the applicability of the theta-beta power ratio in diagnostics (Arns et al., 2013).

Despite the inconclusive scientific evidence, in July 2013, the U.S. Food and Drug Administration (FDA) approved the use of the Neuropsychiatric EEG-Based Assessment Aid Health for the assessment of ADHD based on the theta-beta power ratio (FDA, 2013). To further examine this, we updated the meta-analysis by Arns et al. (2013) and included additional studies that reported group differences between theta-beta power ratios of children and adolescent with and without an ADHD diagnosis. In total, eighteen studies that were published in peer-reviewed English scientific journals including 2425 participants (1600 patients and 825 controls) were analysed. The standardized effect size (Hedges' delta) and variance were calculated for each study (Hedges & Olkin, 1985).

Based on the assumption that, in addition to sampling error, effect sizes contain a true random component, a random effects model was performed. Results showed an estimated mean effect size of 0.92 (95% CI: 0.40–1.44) (Figure 15.2).

[1] This meta-analysis included nine studies and used different selection criteria compared to the earlier meta-analysis. As a result, only two of the nine studies used in the original meta-analysis by Snyder and colleagues (2008) were considered eligible. In addition, this study-based effect size calculation used the pooled standard deviation of the ADHD and control groups instead of exclusively using the SD of the control group in computing the Glass' delta.

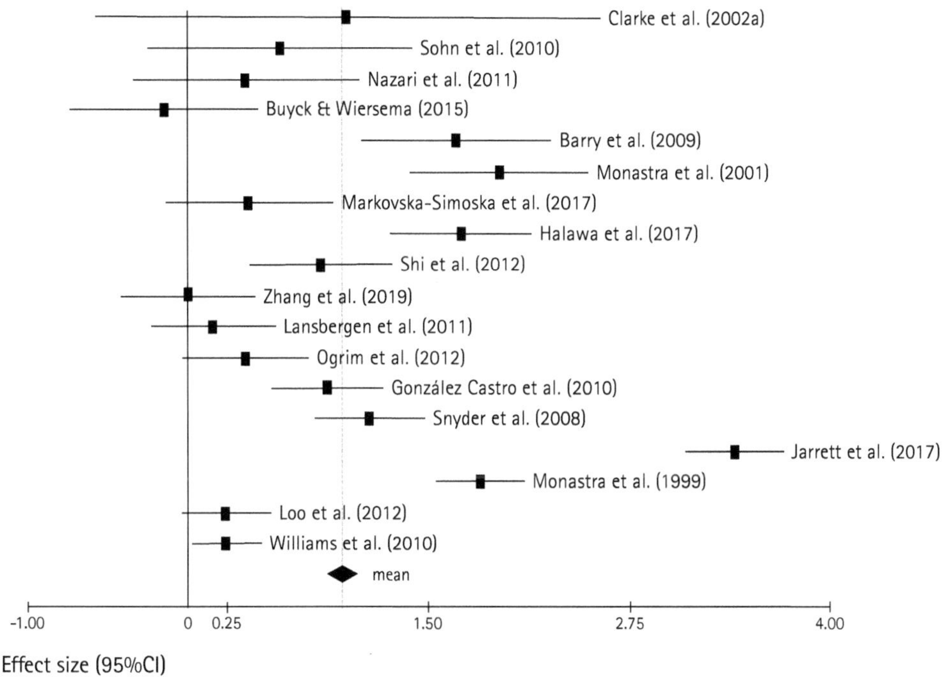

FIGURE 15.2 Forest plot of the mean effect size and 95% CI intervals of the included studies.

The test for study heterogeneity was not significant, $Q_{total} = 12.95$, $p = 0.74$. Rosenberg's fail-safe number analysis indicates that 2415 studies with null findings ($p < 0.05$) are required to change the significant meta-analytic result into a non-significant finding. In addition, Orwin's method shows that 68 studies are needed to reduce the mean effect size of 0.92 to a small effect size of 0.2. As neither method accounts for sample size or variance of the studies, these fail-safe number results should be interpreted with caution. While the mean effect size observed in the current meta-analysis is larger than the effect size reported in 2013, the overlapping 95% CIs show that the difference between the effect sizes is not statistically significant at the $p=0.05$ level. The wide 95% CI interval indicates considerable uncertainty in terms of the true value of the effect size and the use of the theta-beta power ratio for diagnosing ADHD in children and adolescents remains unclear (Arns et al., 2013). Several authors have proposed that adding cognitive, emotive and genetic information to the theta-beta power ratio may have substantial diagnostic as well as prognostic value (Williams et al., 2010).

It is suggested that determining theta activity relative to the individual alpha peak-frequency may contribute to the diagnostic value of the resting state theta-beta power ratio (Saad et al., 2018). This suggestion adds to the idea of multiple theta rhythms and that the dependency on conventional fixed frequency bands may be problematic in the classification of ADHD as well as for gaining a deeper understanding on the relation between EEG ratios and their functional meaning.

Determining the so-called transition-frequency reflects the true physiological boundary between theta and alpha activity, and may present a physiological plausible alternative as this frequency marks the intersection of resting state with active EEG (Saad et al., 2018: Klimesch, 1999). Perhaps the individualized peak alpha frequency may also be useful to indicate the transition from alpha to beta.

Personalized approaches like these may be able to control for individual differences in EEG make-up unrelated to ADHD symptoms and improve the specificity (true negative rate) and sensitivity (true positive rate) of the theta-beta power ratio in diagnostics.

Even although the concept of resting state theta-beta power ratio as a biomarker for diagnostic purposes is promising, further research is needed to further develop this concept. In addition, higher resting state theta-beta power ratio in ADHD children has been suggested to be indicative for treatment outcome (Loo et al., 1999: Clarke et al., 2002b). Support for this idea comes from results showing that good as compared to poor responders to methylphenidate have higher resting state theta-beta power ratio prior to treatment (Clarke et al., 2002a). This fits the observation that stimulant medication can increase beta activity in children with ADHD, particularly in frontal cortical regions, in addition to improving symptoms related to attentional focus and response inhibition (Loo et al., 2004).

Together, the normalization of dopaminergic activity in the subcortical circuits as indicated by a reduction in theta activity, and the increase in beta activity, may explain the lowering of the power ratio in response to methylphenidate. Indeed, oscillatory beta activity is in turn linked to dopamine levels at sites of cortical input to subcortical areas including the basal ganglia (Jenkinson & Brown, 2011). While the resting state theta-beta power ratio may be a positive response predictor, a multi-center, prospective open-label trial found that lower frontal alpha (9 Hz) peak frequency, but not the resting state theta-beta ratio in male adolescent ADHD patients, is a predictor for non-response to methylphenidate (Arns et al., 2018). This negative predictor response was found to be independent from age, medication dosage, and baseline symptom severity. The lower individual anterior alpha peak is suggestive of a maturational lag in the structural development of the frontal cortex (Whitford et al., 2007). Together these findings stress the importance of considering both functional (e.g., theta-beta power ratio) and structural aspects (e.g., alpha peak) of brain organization for understanding the efficacy of treatments. More recent work points towards the direction that EEG measures may be more useful biomarkers of ADHD outcome or treatment response rather than diagnosis (Sari Gokten et al., 2019: Loo et al., 2018). Finally, the different findings leave open the possibility that ADHD symptoms can have several neuro-etiological origins and should be evaluated within an environmental context.

15.4 Functional correlates of the theta-beta power ratio

In addition to research on the clinical features and applicability of the theta-beta power ratio in ADHD, basic neuroscientific research has also made considerable progress in identifying cognitive and affective processes associated with the theta-beta power ratio. This line of research provides insights not only into the specific symptomatology of ADHD, but also in understanding the neurocognitive processes within the general population.

Experimental human brain research has shown that the prefrontal cortex (PFC) plays a prominent role in the regulation of attention and motivational tendencies, and other executive control functions (Corbetta et al., 2008).

On the other hand, for theta specifically, based on combined fMRI and electrocorticography, significant associations have been reported between theta power and BOLD activity in "intrinsic" resting-state networks, including the default-mode network (DMN, the "idling network") and a network involved in episodic and working memory (Hacker et al., 2017). Furthermore, the CNS stimulant nicotine reduces both theta power (Knott et al., 2007), as well DMN activity (Hahn et al., 2007). Nicotine-induced reduction in DMN activity correlates positively with nicotine-induced improvements in task performance (Hahn et al., 2007). These data support an interpretation that theta power indexes idling and/or hypo-arousal.

In addition, patients responding to the CNS stimulant methylphenidate demonstrate reductions in the theta-beta power ratio (Isiten et al., 2017). Notably, the methylphenidate-related change was driven by an increase in beta, rather than a reduction in theta, power. If indeed beta power is more closely related to cerebral cortical activity, then the reductions in ADHD symptoms may be due to improved executive control functions. Raclopride positron emission tomography (PET) showed that methylphenidate (dose: 0.8 ± 0.11 mg/kg) potentiates dopaminergic activity by blocking dopamine transporters causing an increase of synaptically available DA in the human brain (especially the striatum, Volkow et al., 2001). The findings were replicated at lower dosages which are more commonly prescribed in daily practice (Gottlieb, 2001).

As methylphenidate also increases noradrenergic activity (Kuczenski & Segal, 2001), catecholaminergic dysfunctions is argued to contribute to elevated theta-beta power ratios in individuals with ADHD who respond to methylphenidate (Isiten et al., 2017).

Based on the proposed neural generators of theta and beta oscillations, in conjunction with the dopaminergic dysfunctions described earlier, a neuroanatomical model has been developed to explain the motivational, cognitive, and motoric symptoms of ADHD. The model is a midbrain ventral tegmental area (VTA)-centered framework that includes both the mesolimbic and mesocortical dopaminergic pathway. The mesolimbic pathway refers to the connection between the VTA and nucleus accumbens (ventral striatum), implicated in reward motivation and subcortical motor planning routines. In addition, the VTA extensively projects to the septo-hippocampal complex and amygdala indicative of a complex subcortical circuit dedicated to instrumental learning and memory processes (Hefco et al., 2003).

The connection between the VTA and PFC constitutes the backbone of the mesocortical pathway. In particular, the VTA projections to the medial frontal cortex and ACC provide a neural basis for monitoring and top-down regulatory functions (Di Michele et al., 2005).

Next to the septo-hippocampal complex, the ACC is another brain area that operates in the theta frequency range (Asada et al., 1999). It is suggested that together these regions make up a circuit dedicated to monitoring and goal-directed motor planning, whereas the ventral and dorsolateral parts of the FC take part in executive functions including working memory, attentional control, and top-down regulation.

The functionality of the latter anterior cortical regions is suggested to involve beta activity. From this viewpoint, the theta-beta power ratio may provide a global index for cortico-subcortical interactions wherein slow wave (theta) activity represents the subcortical bottom up and fast wave (beta) activity in cortical top-down processes (Schutter et al., 2006). In particular, the low tonic dopaminergic activity in the cortico-subcortical circuits and subsequently elevated theta-beta power ratio could perhaps provide a neural marker for attention, impulsivity, reward sensitivity and motor activity in individuals with and without ADHD (Heilman et al., 1991: Robbins, 2000). A more specific formulation of this idea holds that a reduction in dopamine level is associated with disinhibition of the ACC causing an increase in spontaneous theta oscillations and higher responsivity (e.g., the frontal-related negativity (FRN)) to external feedback (Holroyd & Coles, 2002).

Thus, high theta relative to low beta activity may reflect a state of heightened subcortical drive in conjunction with cortical hypoarousal. Furthermore, the ACC is among the prime neural generators of theta oscillations and may a take central position between the subcortical and prefrontal brain areas. This idea is further supported by research showing that the ACC and the anterior portions of the medial PFC can exert direct top-down control over the VTA via monosynaptic excitatory projections (Holroyd and Yeung, 2012: Elston et al., 2019). Increased 4 Hz (theta) oscillatory signaling between the ACC and VTA has been demonstrated during anticipatory decision making, feedback processing, and behavioral flexibility (Elston et al., 2019).

Studies on the spectral density of local field potentials in the ACC and VTA show that while the ACC shows peaks between 3–5 Hz, the VTA emits signals between 3–5 Hz and 7–9 Hz (Elston et al., 2019). The frequency specificity of theta activity across regions suggests the earlier proposed existence of multiple theta rhythms that each have a distinct functional role in behavior. Rhythmic excitability allows for temporal windows of coordinated spike timing to exchange signals between distal regions (Buzsáki & Watson, 2012: Cavanagh et al., 2012). Further evidence for frequency specificity was provided by a study showing that administering exogenous oscillatory field potentials at 4 Hz (low theta) increases working memory capacity, while at 7 Hz (high theta) working memory performance deteriorates (Wolinski et al., 2008). The latter idea also fits the proposed role of theta activity carrying information encoded in gamma oscillations (>30 Hz) from the hippocampus to the cerebral cortex in memory formation (Jensen and Lisman, 1996).

Taken together the studies imply the existence of multiple functionally independent theta activity related to (1) arousal, (2) motivation, (3) cognitive flexibility, and (4) memory-related processes.

Studying the frequency specific of theta and beta may therefore be particularly informative about whether a higher theta-beta power ratio, for example, reflects drowsiness as indicated by a broad increase in the theta EEG spectrum or heightened motivation as reflected by increased information transfer between the ACC and VTA as reflected by 4 Hz oscillations. As the conventional calculations of theta-beta power ratio are based on the averaging of power values in the 4–7 Hz (theta) and 13–30 Hz (beta) frequency range, the study of specific theta, as well as beta oscillatory, components may be relevant for understanding the meaning of the power ratio within a given neural and behavioral context.

15.5 Reward and punishment sensitivity

Several psychological studies have examined the proposed interrelations between theta-beta power ratio and cognitive-affective processes in healthy volunteers.

On the basis of the proposed relation between the theta-beta power ratio and impulsivity in ADHD, the predictive value of resting state theta-beta power ratio on risky decision making was investigated in a sample of healthy volunteers (Schutter & van Honk, 2005). Risky decision making was measured with the Iowa gambling task. In the Iowa gambling task, players are instructed to try to gain as much money as possible by drawing selections from a choice of four decks. Two of the decks are disadvantageous, producing immediate large rewards, but on the long term cause an overall loss due to even higher punishments. The other two decks are advantageous in the sense that immediate rewards are modest but more consistent and punishment rate is low. Decisions to choose from the decks become motivated by the reward and punishment schedules associated with the advantageous and disadvantageous decks.

Results showed that higher resting state theta-beta power ratio was associated with more risky decisions. These findings can be understood in terms of increased reward sensitivity and/or lower punishment sensitivity. Importantly, participants did show a cumulative increase in advantageous decision making during the course of the task. Even though volunteers acquired knowledge about the reward-punishment contingency, subjects with high theta-beta power ratio were nonetheless more inclined to take risky decisions as compared to subjects with a lower theta-beta power ratio.

The association between theta-beta power ratio and disadvantageous decision making was replicated in a follow-up experiment (Massar et al., 2014) which also explored the specific contributions of reward and punishment sensitivity to the theta-beta power ratio by testing how well subjects were able to learn associations between symbols that signal a high probability of winning money and low probability of winning nothing (reward learning), and symbols that signal a high probability of losing money and low probability of losing nothing (punishment learning). Results showed that the association between theta-beta power ratio and associative learning was driven by the positive correlation between theta oscillations and reward learning. This finding fits with the proposed role of theta oscillations in the dopaminergic mesolimbic pathway.

Results of an earlier study demonstrated that the nature of the relation between theta-beta power ratio and risk-taking involved feedback-related processing in the ACC (Massar et al., 2012). More specifically, a higher theta-beta power ratio was found to correlate with reduced feedback processing as reflected in lower FRN amplitudes. Additionally, this association varied as a function of punishment sensitivity as evaluated by the behavioral inhibition system questionnaire. The data suggest a bias toward motivationally driven behavior as a result of suboptimal action monitoring, leading to top-down executive control being subsequently less involved due to reduced ACC signaling. This effect becomes more prominent when subjects score higher on punishment sensitivity and arguably already experience higher default levels of uncertainty. In accordance, a meta-analysis has shown that midfrontal theta activity is linked to ACC activity during action-outcome uncertainty and cognitive control (Cavanagh & Shackman, 2015).

Notably, much like children with an anxiety disorder, children with ADHD display an intolerance of uncertainty (Gramszlo et al., 2018). Indeed, if the ACC is central to the integration of bottom-up feedback signals originating from midbrain structures and top-down executive functioning and control, disturbances in the ACC may subserve the lack of coordinated and goal-directed behaviors in ADHD. Furthermore, increased theta-beta power ratios may signify compromised ACC-related feedback processes that contribute to subjective states of uncertainty. Uncertainty is experienced when there is a lack of subjectively perceived contingency between action and outcome.

Thorndike's law of effect dictates that reward and punishment-related feedback signals are responsible for shaping situation appropriate action-outcome contingencies. In other words, the structural occurrence of rewards and punishment following a specific action leads to the formation of stronger action-outcome links (Seth, 2013). Experimentally, uncertainty can be manipulated by introducing unpredictable changes in reward-punishment contingencies. Earlier findings indicate that particularly in unpredictable environments, a situation-dependent balance between the exploitation of acquired knowledge and exploration of new options is the best behavioral strategy to maximize profit (Daw et al., 2006). In the event of a change in reward-punishment schemes, individuals have to relearn the contingencies to adjust decision making accordingly (Clark et al., 2004: Bechara et al., 1997). The presence of theta activity is associated with the experience of conflict and reward sensitivity, so it is reasonable to assume that high theta-beta power ratio is associated with suboptimal reversal learning when reward-associated responses suddenly are punished, and punishment-related responses become rewarded.

To test this hypothesis, healthy volunteers performed a reversal learning gambling task (Schutte et al., 2017). Unknown to the participant, the task consisted of different reward-punishment (RP) schedules for high-risk decision making. During phase 1 (trials 1–40) high-risk choices were rewarded in 80% of the trials, while in phase 2 (trials 41–80) the RP schedule was reversed, and high-risk choices were now punished in 80% of the trials. In phase 3 (trials 81–120), the RP schedule used in phase 1 (i.e., 80% reward for high risks) was reintroduced. The spontaneous resting state theta-beta power ratio recorded prior to the task was found to be inversely related to reversal learning, indicative of lower cognitive flexibility during unexpected changes when theta-beta power ratio is high. The results can be interpreted along the lines of lowered cortical cognitive involvement in the processing of subcortical reward and punishment signals.

In an additional study, transcranial alternating current stimulation (tACS) was applied to evaluate the effects of applying 5-Hz (theta) oscillatory electric fields to left and right frontal cortex on reversal learning in healthy volunteers (Wischnewski et al., 2016). Active, as compared to sham theta tACS, resulted in faster learning of newly introduced reward- and punishment contingencies during the task, but surprisingly subjects also became more hesitant to apply the learned contingencies to optimize decision making. Furthermore, a significant decrease of the left frontal resting state theta-beta power ratio was observed after tACS. While the lowering of the power ratio concurs with increased apprehension, left-sided lowering was accompanied by less-optimal decision making. This paradoxical finding can be explained by the anterior cortical asymmetry model of motivational direction. More specifically, lower left-sided power ratio may reflect reductions in approach-related motivation and increased behavioral inhibition (Schutter et al., 2008). In a follow up double-blind randomized cross-over study, 20 Hz (beta) tACS applied to the frontal cortex of healthy volunteers improved rule implementation during reversal learning (Wischnewski et al., 2020). Additionally, a global frontal reduction of the theta-beta EEG power ratio was observed (Wischnewski et al., 2020). The latter results fit the cognitive flexibility interpretation, whereas the reduction observed after theta tACS can be better explained as a change in motivation.

In another study by Lansbergen and colleagues (2007), a stop-signal task was administered that involved two types of trials: go-trials (i.e., square wave gratings presented on a computer screen) and stop-trials (i.e., 1000 Hz auditory tone). Whereas go-trials only contained go-signals, stop-trials contained both the go- and stop-signal. In contrast to expectations, results showed that participants with high resting state theta-beta power ratio demonstrated shorter stop-signal reaction times and fewer failures to respond to the go-stimulus on time. Perhaps, the opposite results indicate a curvilinear relationship between the theta-beta power ratio and executive control in which under certain (less complex?) circumstances higher ratios can have a positive effect on performance. This implies that during high cognitive load and motivational value the positive relations turn into negative correlations. These seemingly paradoxical results highlight the possibility that, depending on the operational and conceptual perspective adopted by researchers and theorists, high theta activity, and therefore a high theta-beta power ratio, can reflect cortical hypo-arousal and increased cortical drive, as well as enhanced working-memory function and cognitive control. This line of argumentation concurs with the proposed existence of different functions of theta waves. Finally, a recent study highlighted prestimulus high theta during the task as an indicator of drowsiness, but also stimulus-elicited theta specifically in high-conflict conditions, supposedly reflecting enhanced demands for cognitive control (Canales-Johnson et al., 2020; see also Cooper et al., 2019).

15.6 Central executive and attentional control function

The theta-beta power ratio is proposed to index a cortico-subcortical balance and reflects interactions between the brain's cognitive and motivational systems. In addition to the involvement of the motivational systems as outlined in the previous section, central-executive control functions have been associated with the theta-beta power ratio. The central executive serves three major control functions (Miyake et al., 2000). The first function (inhibition) is the ability to inhibit automatic or prepotent responses and requires attentional control to actively withstand intrusions from task-irrelevant stimuli or motivational tendencies. The second function (shifting) involves adaptive changes in attentional control based on a change in environmental demands and is the capacity to switch between mental sets and operations. The third function (updating) involves the monitoring and renewal of working memory representations. Attentional control processes play a key role in the different central executive functions as they involve the conscious regulation of bottom-up, stimulus-driven information processing streams. Within this dual-process system concept, the top-down strategic network resides in the anterior frontal cortical areas, whereas the stimulus-driven systems are localized in the more posterior (sensory) cortical and subcortical (motivation) regions.

Researchers have proposed that a high theta-beta power ratio reflects a maturational lag of frontal cortical development that can explain the suboptimal attentional control of posterior cortical and subcortical signals in children with ADHD (Clarke et al., 2001b). The discovery that the theta-beta power ratio is inversely correlated to self-reported attentional control in healthy young adults (Putman et al., 2010) has prompted a series of scientific studies that further investigated the latter association in the context of stress and anxiety (Putman et al., 2014). High levels of stress and anxiety have a negative impact on central executive functioning and performance. For example, cognitive performance anxiety (CPA) is the phenomenon in which anxiety/uncertainty about one's cognitive competence causes lower and even impaired performance.

The interaction between the theta-beta power ratio and CPA-related stress on performance was examined in healthy participants who performed a stress induction or control task. In addition to replicating the earlier inverse association between the theta-beta power ratio and self-reported attentional control, lower levels of the ratio were found to be protective for the negative effects of CPA. These finding can be explained by the inverse relation between theta-beta power ratio and attentional control (Putman et al., 2010).

In a subsequent study, a pictorial dot-probe paradigm was deployed to further investigate a role for the theta-beta power ratio in attention resource allocation to emotionally arousing stimuli. Healthy volunteers were instructed to fixate on the center of a screen and to locate a black dot that appeared either on the left or right side immediately following the simultaneous presentation of two stimuli in the left and right periphery of the visual field. Statistical analyses revealed that subjects with a relatively high theta-beta power ratio directed attention towards mildly arousing and avoided highly arousing negative pictures. Consequently, subjects with low theta-beta power ratios displayed more attention towards high arousing negative material. Even though correlational findings do not allow inferences on the directionality of statistical associations, the authors interpret these findings as further support for the idea that an increased theta-beta power ratio reflects an enhanced ability to regulate the impact of negative emotional information (Angelidis et al., 2018).

An independent follow-up experiment replicated this finding and also found that lower self-reported trait attentional control predicted more attention to mild as compared to high arousing negative stimuli (Van Son et al., 2018). These findings concur with earlier research that administered a go/no-go task in healthy volunteers. In this particular study subjects were instructed to respond (go) or inhibit responses (no-go) to fearful and happy facial expressions (Putman et al., 2010). In line with the ADHD literature (Metin et al., 2012), a significant correlation was found between response inhibition and the theta-beta power ratio (Putman et al., 2010). More specifically, stronger response inhibition to the fearful relative to the happy facial expressions was predicted by lower theta-beta power ratios (Putman et al., 2010).

Furthermore, evidence was found for a positive relation between the theta-beta power ratio and the drive scale of the behavioral activation system (Gray, 1985). The drive scale is related to the brain's seeking system, which promotes energized appetitive and approach-related behaviors that include exploration, foraging and future anticipation of rewards (Panksepp, 1998, 2012). Neuroanatomically, the seeking system corresponds to the mesolimbic dopaminergic pathway and provides input to the higher cortical areas and promotes reward-driven actions. Within this framework, theta activity may be a correlate of the seeking system that feeds information into the PFC. This information is then used in attentional control and regulatory processes, arguably reflected by the beta activity component of the ratio, to shape context-appropriate behaviors. More recently, temperamental shyness was found to correlate with higher levels of social anxiety in children with large baseline-to-task decreases in frontal theta-beta power ratio in the anticipation of having to give a speech (Poole et al., 2021). These findings were interpreted as increased attentional control and emotion regulation during the experience of stress in shy children (Poole et al., 2021). Observations that resting state the theta-beta power ratio is positively associated with distraction tendency and not re-appraisal strategies (Kobayashi et al., 2020) concur with the proposed link between theta-beta power ratio and attentional control (Putman et al., 2014).

Due to the low spatial resolution of scalp-recorded electric field potentials, localizing their neural generators is challenging. However, interleaving resting state functional magnetic resonance imaging with resting state EEG recording provides a way to examine the relations between different cortical and subcortical brain regions and the theta-beta power ratio. Recently, associations between the theta-beta power ratio and the brain's executive control network (ECN) were explored in a sample of healthy volunteers (Van Son et al., 2019b). The ECN consists of the dorsolateral PFC, ACC and parietal cortex, and governs top-down regulation and attentional control processes (Corbetta et al., 2008).

Reductions in attentional control and brain activity were investigated during mind wandering. Mind wandering refers to transient changes of attentional focus from a task to non-task related mental states and is accompanied by an increase in the theta-beta power ratio (Braboszcz & Delorme, 2011: Van Son et al., 2019a). As compared to an attentional focus condition, the reduction in ECN functional connectivity was correlated to an increase in theta-beta power ratio during mind wandering. Even though this finding does not imply that the ECN is the generator of theta activity, the result-concurs with the positive relation between mind wandering and ADHD symptomatology (Seli et al., 2015) as well as with aberrant functional connectivity of the executive control network in ADHD subjects (McCarthy et al., 2013). The results not only provide a functional neuro-anatomical substrate for the theta-beta power ratio, but also further strengthens construct validity of the ratio as a functional signature of the balance between bottom-up motivational and top-down cognitive aspects of normal and abnormal human behavior. Figure 15.3 depicts a functional neuro-anatomic framework of the theta-beta power ratio which represents a synthesis of the concepts and empirical work reviewed in this chapter.

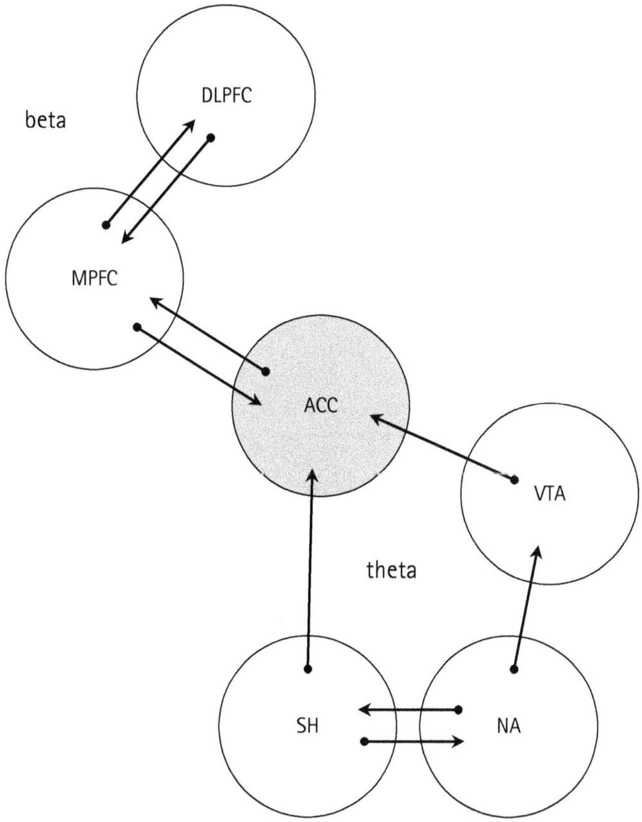

FIGURE 15.3 A functional neuro-anatomic framework of the theta-beta power ratio.

Abbreviations: ACC, Anterior cingulate cortex; DLPFC, Dorsolateral prefrontal cortex; MPFC, Medial prefrontal cortex; NA, Nucleus accumbens; SH, Septohippocampal complex; VTA, Ventral tegmental area.

15.7 Some Methodological Issues

While resting state EEG is robust and has a high test-retest reliability within a time frame of one week (Angelidis et al., 2018), longer inter-assessment intervals and methodological factors can nonetheless provide a source of error variance which can negatively affect the reliability of the theta-beta power ratio. Results can among other factors be influenced by wrong electrode placement, artifact contamination, drowsiness, changes in medication use, and comparison with suboptimal control groups. The lack of a standardized EEG protocol could make comparisons between studies and groups less reliable. So, in addition to differences already mentioned, differences in, for example, applied reference method and varying epoch and Fast Fourier Transform window lengths can yield heterogenous results. Also, age has been found to influence the relation between the theta-beta power ratio and ADHD (Clarke et al., 2001a). One line of evidence comes from work showing that the ratio is able to differentiate between ADHD in children, but not in adults (Markovska-Simoska & Jordanova, 2017).

It should furthermore be noted that whereas scientific studies are well suited to predict and demonstrate effects on the group level, specifying these effects to the individual level is more problematic. Finally, there is a difference between statistical significance and clinical relevance, as in that a reliable systematic finding is too small to have a meaningful contribution to the diagnosis, prognosis and/or treatment of a disorder. In sum, the relation between available scientific evidence and clinical usefulness of EEG in ADHD and other conditions remains a topic of research.

15.8 Beyond the Theta-Beta Power Ratio

In addition to the theta-beta power ratio, only a limited number of studies have looked at the functional relevance of other ratios. For example, an increase in parietal delta-beta power ratio has been positively associated with spontaneous emotion regulation as shown by lower ratings of discomfort after the offset of negative pictures (Tortella-Feliu et al., 2014). These results suggest that the posterior delta-beta power ratio reflects an electrophysiological dispositional feature for faster automatic recovery of unpleasant emotional responses (Tortella-Feliu et al., 2014). These findings are in agreement with a study that found a positive associations between parietal delta-beta waves and self-reported attentional control in a group of healthy volunteers (Morillas-Romero et al., 2015). In another study both the parietal delta-beta and frontal theta-beta power ratio were correlated to risky decision making (Schutter & van Honk, 2005). A study investigating ADHD subtypes in children by analyzing their EEG found that inattentive type of ADHD was associated with increased delta-beta power ratios as compared to matched controls (Aldemir et al., 2018). Although the underlying mechanism remains unclear, cognitive impairments in neurological populations go accompanied by increases in slow wave oscillations, including delta waves, in conjunction with a decrease in alpha and beta wave activity (Klass & Brenner, 1995). Perhaps, relative higher delta power as a result of lower alpha activity is linked to reduced perceptual sensitivity to external stimuli that results in less attentional resource allocation and in-depth processing. Additionally, higher delta relative to alpha oscillations has been shown to successfully discriminate between patients with acute ischemic stroke and healthy controls (Finnigan et al., 2016). Moreover, in a group of twenty patients with ischemic stroke, a significant inverse association was found between the delta-alpha EEG ratio assessed at approximately 72 hours poststroke onset and cognitive performance after 3.5 months (Schleiger et al., 2014). Similar to what was mentioned earlier, the ratio between slow delta and alpha (and beta) oscillations may be a sign of cortical (dys)function and/

or a possible attenuation of functional cortico-subcortical interactions (Schutter & Knyazev, 2012).

Future studies will continue to broaden our understanding of the structural and functional nature underlying slow-fast wave EEG ratios and will further shape the use of EEG ratios in research and the clinic. Additional questions that may be worth of investigation are the role of delta-theta and alpha-beta power ratios, and the effects of abandoning the conventional bandwidths and focus on frequency-specific oscillations.[2]

15.9 Summary and Conclusion

Following the first observation of elevated theta-beta power ratios in children with ADHD, many studies have replicated this finding. In 2013 thc FDA formally approved the use of the theta-beta power ratio as a criterion for the diagnosis of ADHD. However, recent meta-analyses suggest that the theta-beta power ratio does not reliably differentiate between individuals with and without ADHD. Recently, psychological research in healthy volunteers has established reliable associations between the theta-beta power ratio and indices of motivation and executive functioning. Consistent with observations in individuals with ADHD, higher theta-beta power ratios are associated with reduced top-down attentional control paralleled by approach-related motivation and seeking behavior. These associations are conceptualized as a cortico-limbic balance wherein beta activity stands for cortical top-down executive control and theta activity represents the bottom up limbic motivational incentives. Together with the high test-retest reliability coefficients (Angelidis et al., 2016), the theta-beta power ratio can be considered a trait marker for capturing the reciprocal relationship between cognitive control and motivational tendencies, and may have complementary value in diagnosis and treating ADHD and/or related disorders.

REFERENCES

Adrian, E. D. & Matthews, B. H. (1934). The Berger rhythm: Potential changes from the occipital lobes in man. *Brain*, *57*, 355–385.

Aldemir, R., Demirci, E., Per, H., Canpolat, M., Özmen, S., & Tokmakçi, M. (2018). Investigation of attention deficit hyperactivity disorder (ADHD) sub-types in children via EEG frequency domain analysis. *International Journal of Neuroscience*, *128*, 349–336.

Angelidis, A., van der Does, W., Schakel, L., & Putman, P. (2016). Frontal EEG theta/beta ratio as an electrophysiological marker for attentional control and its test-retest reliability. *Biological Psychology*, *121*, 49–52

[2] Strictly speaking, an oscillation refers to rhythmic electrical activity in one specific frequency (e.g., 5 Hz) and not to a range of frequencies (e.g., theta: 4–7 Hz)

Angelidis, A., Hagenaars, M., van Son, D., van der Does, W., & Putman, P. (2018). Do not look away! Spontaneous frontal EEG theta/beta ratio as a marker for cognitive control over attention to mild and high threat. *Biological Psychology*, *135*, 8–17.

Arns, M., Conners, C. K., & Kraemer, H. C. (2013). A decade of EEG theta/beta ratio research in ADHD: A meta-analysis. *Journal of Attention Disorders*, *17*, 374–383.

Arns, M., Vollebregt, M. A., Palmer, D., Spooner, C., Gordon, E., Cohn, M., ... Buitelaar, J. K. (2018). Electroencephalographic biomarkers as predictors of methylphenidate response in attention-deficit/hyperactivity disorder. *European Neuropsychopharmacology*, *28*, 881–891.

Asada, H., Fukuda, Y., Tsunoda, S., Yamaguchi, M., & Tonoike, M. (1999). Frontal midline theta rhythms reflect alternative activation of prefrontal cortex and anterior cingulate cortex in humans. *Neuroscience Letters*, *274*, 29–32.

Barry, R. J., Clarke, A. R., McCarthy, R., Selikowitz, M., Rushby, J. A., & Ploskova, E. (2004). EEG differences in children as a function of resting-state arousal level. *Clinical Neurophysiology*, *115*, 402–408.

Bechara, A., Damasio, H., Tranel, D., & Damasio, A. R. (1997). Deciding advantageously before knowing the advantageous strategy. *Science*, *275*, 1293–1295.

Braboszcz, C. & Delorme, A. (2011). Lost in thoughts: Neural markers of low alertness during mind wandering. *Neuroimage*, *54*, 3040–3047.

Buyck, I. & Wiersema, J. R. (2015). Electroencephalographic activity before and after cognitive effort in children with attention deficit/hyperactivity disorder. *Clinical EEG and Neuroscience*, *46*, 88–93.

Buzsaki, G. & Watson, B. O. (2012). Brain rhythms and neural syntax: Implications for efficient coding of cognitive content and neuropsychiatric disease. *Dialogues in Clinical Neuroscience*, *14*, 345–367.

Canales-Johnson, A., Beerendonk, L., Blain, S., Kitaoka, S., Ezquerro-Nassar, A., Nuiten, S., ... Bekinschtein, T. A. (2020). Decreased alertness reconfigures cognitive control networks. *Journal of Neuroscience*, *40*, 7142–7154.

Cavanagh, J. F., Zambrano-Vazquez, L., & Allen, J. J. (2012). Theta lingua franca: A common mid-frontal substrate for action monitoring processes. *Psychophysiology*, *49*, 220–238.

Cavanagh, J. F. & Shackman, A. J. (2015). Frontal midline theta reflects anxiety and cognitive control: Meta-analytic evidence. *Journal of Physiology*, *109*, 3–15.

Clarke, A. R., Barry, R. J., McCarthy, R., & Selikowitz, M. (2001a). EEG-defined subtypes of children with attention-deficit/hyperactivity disorder. *Clinical Neurophysiology*, *112*, 2098–2105.

Clarke, A. R., Barry, R. J., McCarthy, R., & Selikowitz, M. (2001b). Age and sex effects in the EEG: Development of the normal child. *Clinical Neurophysiology*, *112*, 806–814.

Clarke, A. R., Barry, R. J., McCarthy, R., Selikowitz, M., & Croft, R. J. (2002a). EEG differences between good and poor responders to methylphenidate in boys with the inattentive type of attention-deficit/hyperactivity disorder. *Clinical Neurophysiology*, *113*, 1191–1198.

Clarke, A. R., Barry, R. J., Bond, D., McCarthy, R., & Selikowitz, M. (2002b). Effects of stimulant medications on the EEG of children with attention-deficit/hyperactivity disorder. *Psychopharmacology*, *164*, 277–284.

Clark, L., Cools, R., & Robbins, T. W. (2004). The neuropsychology of ventral prefrontal cortex: Decision-making and reversal learning. *Brain and Cognition*, *55*, 41–53.

Cooper, P.S., Karayanidis, F., McKewen, M., McLellan-Hall, S., Wong, A.S., Skippen, P., & Cavanagh, J. F. (2019). Frontal theta predicts specific cognitive control-induced behavioural changes beyond general reaction time slowing. *NeuroImage*, *189*, 130–140.

Corbetta, M., Patel, G., & Shulman, G. L. (2008). The reorienting system of the human brain: from environment to theory of mind. *Neuron*, *58*, 306–324.

Daw, N. D., O'Doherty, D. P., Seymour, B., & Dolan R. J. (2006). Cortical substrates for exploratory decision making. *Nature*, *441*, 876–879.

Di Michele, F., Prichep, L., John, E. R., & Chabot, R. J. (2005). The neurophysiology of attention-deficit/hyperactivity disorder. *International Journal of Psychophysiology*, *58*, 81–93.

Elston, T. W., Croy, E., & Bilkey, D. K. (2019). Communication between the anterior cingulate cortex and ventral tegmental area during a cost-benefit reversal task. *Cell Reports*, *26*, 2353–2361.

Engel, A. K. & Fries, P. (2010). Beta-band oscillations: Signalling the status quo? *Current Opinion in Neurobiology*, *20*, 156–165.

Finnigan, S., Wong, A., & Read, S. (2016). Defining abnormal slow EEG activity in acute ischaemic stroke: Delta/alpha ratio as an optimal QEEG index. *Clinical Neurophysiology*, *127*, 1452–1459.

Food and Drug Administration (2013). De novo classification request for Neuropsychiatric EEG-Based Assessment Aid for ADHD (NEBA) System. Retrieved from http://www.accessdata.fda.gov/cdrh_docs/reviews/K112711.pdf

Gottlieb, S. (2001). Methylphenidate works by increasing dopamine levels. *British Medical Journal*, *322*, 259.

González Castro, P., Alvarez Pérez, L., Núñez Pérez, J. C., González-Pienda García, J. A., Alvarez García, D., & Muñiz Fernández, J. (2010). Cortical activation and attentional control in ADAH subtypes. *International Journal of Clinical and Health Psychology*, *10*, 23–39.

Gramszlo, C., Fogleman, N. D., Rosen, P. J., & Woodruff-Borden, J. (2018). Intolerance of uncertainty in children with attention-deficit/hyperactivity disorder. *Attention Deficit Hyperactivity Disorders*, *10*, 189–197.

Gray, J. A. (1982). *The neuropsychology of anxiety: An enquiry into the septo-hippocampal system*. Oxford University Press.

Gray, J. A. (1985). Emotional behaviour and the limbic system. *Advances in Psychosomatic Medicine*, *13*, 1–25.

Guyton, A. C. (1976). *Organ physiology: Structure and function of the nervous system*. W.B. Saunders.

Hacker, C. D., Snyder, A. Z., Pahwa, M., Corbetta, M., & Leuthardt, E. C. (2017). Frequency-specific electrophysiologic correlates of resting state fMRI networks. *NeuroImage*, *149*, 446–457.

Hahn, B., Ross, T. J., Yang, Y., Kim, I., Huestis, M. A., & Stein, E. A. (2007). Nicotine enhances visuospatial attention by deactivating areas of the resting brain default network. *Journal of Neuroscience*, *27*, 3477–3489.

Halawa, I. F., El Sayed, B. B., Amin, O. R., Meguid, N. A., & Abdel Kader, A. A. (2017). Frontal theta/beta ratio changes during TOVA in Egyptian ADHD children. *Neurosciences (Riyadh)*, *22*, 287–291.

Harmon-Jones, E. & Gable P. A. (2018). On the role of asymmetric frontal cortical activity in approach and withdrawal motivation: An updated review of the evidence. *Psychophysiology* [Epub], *55*. doi:10.1111/psyp.12879.

Hedges, L. V. & Olkin, I. (1985). *Statistical methods for meta-analysis*. Academic Press.

Hoptman, M. J. & Davidson, R. J. (1998). Baseline EEG asymmetries and performance on neuropsychological tasks. *Neuropsychologia*, *36*, 1343–1353.

Hefco, V., Yamada, K., Hefco, A., Hritcu, L., Tiron, A., & Nabeshima, T. (2003). Role of the mesotelencephalic dopamine system in learning and memory processes in the rat. *European Journal of Pharmacology*, *475*, 55–60.

Heilman, K. M., Voeller, K. K., & Nadeau, S. E. (1991). A possible pathophysiologic substrate of attention deficit hyperactivity disorder. *Journal of Child Neurology*, *6*, S76–81.

Holroyd, C. B. & Coles, M. G. H. (2002). The neural basis of human error processing: reinforcement learning, dopamine, and the error-related negativity. *Psychological Review*, *109*, 679–709.

Holroyd, C. B. & Yeung, N. (2012). Motivation of extended behaviors by anterior cingulate cortex. *Trends in Cognitive Sciences*, *16*, 122–128.

Isiten, H. N., Cebi, M., Sutcubasi Kaya, B., Metin, B., & Tarhan, N. (2017). Medication effects on EEG biomarkers in attention-deficit/hyperactivity disorder. *Clinical EEG and Neuroscience*, *48*, 246–250.

Jarrett, M. A., Gable, P. A., Rondon, A. T., Neal, L. B., Price, H. F., & Hilton, D. C. (2017). An EEG study of children with and without ADHD symptoms: Between-group differences and associations with sluggish cognitive tempo symptoms. *Journal of Attention Disorders*, *1*, 1087054717723986.

Jasper, H., Solomon, P., & Bradley, C. (1938) Electroencephalographic analyses of behaviour problem children. *American Journal of Psychiatry*, *95*, 641–658.

Jenkinson, N. & Brown, P. (2011). New insights into the relationship between dopamine, beta oscillations and motor function. *Trends in Neurosciences*, *34*, 611–618.

Jensen, O. & Lisman, J.E. (1996). Novel lists of 72 known items can be reliably stored in an oscillatory short-term memory network: Interaction with long-term memory. *Learning & Memory*, *3*, 257–263.

Karakaş, S. & Barry, R. J. (2017). A brief historical perspective on the advent of brain oscillations in the biological and psychological disciplines. *Neuroscience and Biobehavioral Reviews*, *75*, 335–347

Klass, D. W. & Brenner, R. P. (1995). Electroencephalography of the elderly. *Journal of Clinical Neurophysiology*, *12*, 116–131.

Klimesch, W. (1999). EEG alpha and thetaoscillations reflect cognitive and memory performance: A review and analysis. *Brain Research Reviews*, *29*, 169–195.

Knott, V. J. & Fisher, D. J. (2007). Naltrexone alteration of the nicotine-induced EEG and mood activation response in tobacco-deprived cigarette smokers. *Experimental and Clinical Psychopharmacology*, *15*, 368–381.

Kobayashi, R., Honda, T., Hashimoto, J., Kashihara, S., Iwasa, Y., Yamamoto, K., … Nakao, T. (2020). Resting-state theta/beta ratio is associated with distraction but not with reappraisal. *Biological Psychology*, *155*, 107942.

Kuczenski, R. & Segal, D.S. (2001). Locomotor effects of acute and repeated threshold doses of amphetamine and methylphenidate: relative roles of dopamine and norepinephrine. *Journal of Pharmacology and Experimental Therapeutics*, *296*, 876–83.

Lansbergen, M. M., Schutter, D. J., & Kenemans, J. L. (2007). Subjective impulsivity and baseline EEG in relation to stopping performance. *Brain Research*, 1148, 161–169.

Lansbergen, M. M., Arns, M., van Dongen-Boomsma, M., Spronk, D., & Buitelaar, J. K. (2011). The increase in theta/beta ratio on resting-state EEG in boys with attention-deficit/hyperactivity disorder is mediated by slow alpha peak frequency. *Progress in Neuropsychopharmacology & Biological Psychiatry*, *35*, 47–52.

Loo, S. K., Teale, P. D., & Reite, M. L. (1999). EEG correlates of methylphenidate response among children with ADHD: A preliminary report. *Biological Psychiatry*, *45*, 1657–1660.

Loo, S. K., Hopfer, C., Teale, P. D., & Reite, M. L. (2004). EEG correlates of methylphenidate response in ADHD: Association with cognitive and behavioral measures. *Journal of Clinical Neurophysiology, 21*, 457–464.

Loo, S. K. & Makeig, S. (2012). Clinical utility of EEG in attention-deficit/hyperactivity disorder: a research update. *Neurotherapeutics, 9*, 569–587.

Loo, S. K., McGough, J. J., McCracken, J. T., & Smalley, S. L. (2018). Parsing heterogeneity in attention-deficit hyperactivity disorder using EEG-based subgroups. *Journal of Child Psychology and Psychiatry, 59*, 223–231.

Lubar, J. F. (1991). Discourse on the development of EEG diagnostics and biofeedback for attention-deficit/hyperactivity disorders. *Biofeedback and Self-Regulation, 16*, 201–225.

Lubar, J. F. (1997). Neocortical dynamics: Implications for understanding the role of neurofeedback and related techniques for the enhancement of attention. *Applied Psychophysiology and Biofeedback, 22*, 111–126.

Mann C., Lubar, J., Zimmerman, A., Miller, C., & Muenchen, R. (1992). Quantitative analysis of EEG in boys with attention deficit hyperactivity disorder: Controlled study with clinical implications. *Pediatric Neurology, 8*, 30–36.

Markovska-Simoska, S. & Pop-Jordanova, N. (2017). Quantitative EEG in children and adults with attention deficit hyperactivity disorder: Comparison of absolute and relative power spectra and theta/beta ratio. *Clinical EEG and Neuroscience, 48*, 20–32.

Massar, S. A., Rossi, V., Schutter, D. J., Kenemans, J. L. (2012). Baseline EEG theta/beta ratio and punishment sensitivity as biomarkers for feedback-related negativity (FRN) and risk-taking. *Clinical Neurophysiology, 123*, 1958–1965.

Massar, S. A., Kenemans, J. L., & Schutter, D. J. (2014). Resting-state EEG theta activity and risk learning: sensitivity to reward or punishment? *International Journal of Psychophysiology, 91*, 172–177.

McCarthy, H., Skokauskas, N., Mulligan, A., Donohoe, G., Mullins, D., Kelly, J., ... Frold, T. (2013). Attention network hypoconnectivity with default and affective network hyperconnectivity in adults diagnosed with attention-deficit/hyperactivity disorder in childhood. *JAMA Psychiatry, 70*, 1329–1337.

Metin, B., Roeyers, H., Wiersema, J.R., van der Meere, J., & Sonuga-Barke, E. (2012). A meta-analytic study of event rate effects on go/no-go performance in attention-deficit/hyperactivity disorder. *Biological Psychiatry, 72*, 990–996.

Miyake, A., Friedman, N. P., Emerson, M. J., Witzki, A. H., Howerter, A., & Wager, T. D. (2000). The unity and diversity of executive functions and their contributions to complex "frontal lobe" tasks: A latent variable analysis. *Cognitive Psychology, 41*, 49–100.

Monastra, V. J., Lubar, J. F., Linden, M., van Deusen, P., Green, G., Wing, W., et al. (1999). Assessing attention deficit hyperactivity disorder via quantitative electroencephalography: An initial validation study. *Neuropsychology, 13*, 424–433.

Monastra, V. J., Lubar, J. F., & Linden, M. (2001). The development of a quantitative electroencephalographic scanning process for attention deficit-hyperactivity disorder: Reliability and validity studies. *Neuropsychology, 15*, 136–144.

Morillas-Romero, A., Tortella-Feliu, M., Bornas, X., & Putman, P. (2015). Spontaneous EEG theta/beta ratio and delta-beta coupling in relation to attentional network functioning and self-reported attentional control. *Cognitive Affective and Behavioral Neuroscience, 15*, 598–560.

Nazari, M. A., Wallois, F., Aarabi, A., & Berquin, P. (2011). Dynamic changes in quantitative electroencephalogram during continuous performance test in children with attention-deficit/hyperactivity disorder. *International Journal of Psychophysiology*, *81*, 230–236.

Ogrim, G., Kropotov, J., & Hestad, K. (2012). The quantitative EEG theta/beta ratio in attention deficit/hyperactivity disorder and normal controls: Sensitivity, specificity, and behavioral correlates. *Psychiatry Research*, *198*, 482–488.

Panksepp, J. (1998). *Affective neuroscience: The foundations of human and animal emotions.* Oxford University Press.

Panksepp J. & Biven, L. (2012). *The archeology of mind: Neuroevolutionary origins of human emotions.* W.W. Norton & Company.

Poole, K. L., Hassan, R., & Schmidt, L. A. (2021). Temperamental shyness, frontal EEG theta/beta ratio, and social anxiety in children. *Child Development*, *92*, 2006–2019.

Putman, P., van Peer, J., Maimari, I., & van der Werff, S. (2010). EEG theta/beta ratio in relation to fear-modulated response-inhibition, attentional control, and affective traits. *Biological Psychology*, *83*, 73–78.

Putman, P., Verkuil, B., Arias-Garcia, E., Pantazi, I., & van Schie, C. (2014). EEG theta/beta ratio as a potential biomarker for attentional control and resilience against deleterious effects of stress on attention. *Cognitive Affective and Behavioral Neuroscience*, *14*, 782–791.

Rahman, M. M., Shukla, A., & Chattarji, S. (2018). Extinction recall of fear memories formed before stress is not affected despite higher theta activity in the amygdala. *Elife*, *7*, e35450.

Robbins, T. W. (2000). Chemical neuromodulation of frontal-executive functions in humans and other animals. *Experimental Brain Research*, *133*, 130–138.

Rosanova, M., Casali, A., Bellina, V., Resta, F., Mariotti, M., & Massimini, M. (2009). Natural frequencies of human corticothalamic circuits. *Journal of Neuroscience*, *29*, 7679–7685.

Saad, J. F., Kohn, M. R., Clarke, S., Lagopoulos, J., & Hermens, D. F. (2018). Is the theta/beta EEG marker for ADHD inherently flawed? *Journal of Attention Disorders*, *22*, 815–826.

Satterfield, J. H. & Cantwell, D. P. (1974). Proceedings: CNS function and response to methylphenidate in hyperactive children. *Psychopharmacology Bulletin*, *10*, 36–38.

Satterfield, J. H. & Dawson, M. E. (1971). Electrodermal correlates of hyperactivity in children. *Psychophysiology*, *8*, 191-197.

Schleiger, E., Sheikh, N., Rowland, T., Wong, A., Read, S., & Finnigan, S. (2014). Frontal EEG delta/alpha ratio and screening for post-stroke cognitive deficits: the power of four electrodes. *International Journal of Psychophysiology*, *94*, 19–24.

Schutte, I., Kenemans, J. L., & Schutter, D. J. (2017). Resting-state theta/beta EEG ratio is associated with reward- and punishment-related reversal learning. *Cognitive Affective and Behavioral Neuroscience*, *17*, 754–763.

Schutter, D. J. & Knyazev, G. G. (2012). Cross-frequency coupling of brain oscillations in studying motivation and emotion. *Motivation and Emotion*, *36*, 46–54.

Schutter, D. J. & van Honk, J. (2005). Electrophysiological ratio markers for the balance between reward and punishment. *Cognitive Brain Research*, *24*, 685–690.

Schutter, D. J., Leitner, C., Kenemans, J. L., & van Honk, J. (2006). Electrophysiological correlates of cortico-subcortical interaction: A cross-frequency spectral EEG analysis. *Clinical Neurophysiology*, *117*, 381–387.

Schutter, D. J., de Weijer, A. D., Meuwese, J. D., Morgan, B., & van Honk, J. (2008). Interrelations between motivational stance, cortical excitability, and the frontal electroencephalogram asymmetry of emotion: A transcranial magnetic stimulation study. *Human Brain Mapping*, *29*, 574–80.

Sari Gokten, E., Tulay, E.E., Beser, B., Elagoz Yukse, M., Arikan, K., Tarhan, N., & Metin, B . (2019). Predictive value of slow and fast EEG oscillations for methylphenidate response in ADHD. *Clinical EEG and Neuroscience, 50*, 332–338.

Seli, P., Smallwood, J., Cheyne, J. A., & Smilek, D. (2015). On the relation of mind wandering and ADHD symptomatology. *Psychonomic Bulletin & Review, 22*, 629–636.

Seth, A. K. (2013). Interoceptive inference, emotion, and the embodied self. *Trends in Cognitive Sciences, 17*, 565–573.

Shi, T., Li, X., Song, J., Zhao, N., Sun, C., et al. (2012). EEG characteristics and visual cognitive function of children with attention deficit hyperactivity disorder (ADHD). *Brain & Development, 34*, 806–811.

Snyder, S. M. & Hall, J. R. (2006). A meta-analysis of quantitative EEG power associated with attention-deficit hyperactivity disorder. *Journal of Clinical Neurophysiology, 23*, 440–455.

Snyder, S. M., Quintana, H., Sexson, S. B., Knott, P., Haque, A. F. M., & Reynolds, D. A. (2008). Blinded, multi-center validation of EEG and rating scales in identifying ADHD within a clinical sample. *Psychiatry Research, 159*, 346–358.

Sohn, H., Kim, I., Lee, W., Peterson, B.S., Hong, H., et al. (2010). Linear and non-linear EEG analysis of adolescents with attention-deficit/hyperactivity disorder during a cognitive task. *Clinical Neurophysiology, 121*, 1863–1870.

Tortella-Feliu, M., Morillas-Romero, A., Balle, M., Llabrés, J., Bornas, X., & Putman, P. (2014). Spontaneous EEG activity and spontaneous emotion regulation. *International Journal of Psychophysiology, 94*, 365–372.

Van Dongen-Boomsma, M., Lansbergen, M. M., Bekker, E. M., Sandra Kooij, J. J., van der Molen, M., Kenemans, J. L., & Buitelaar, J. K. (2010). Relation between res-ting EEG to cognitive performance and clinical symptoms in adults with attention-deficit/hyperactivity disorder. *Neuroscience Letters, 469*, 102–106.

Van Son, D., Angelidis, A., Hagenaars, M. A., van der Does, W., & Putman, P. (2018). Early and late dot-probe attentional bias to mild and high threat pictures: Relations with EEG theta/beta ratio, self-reported trait attentional control, and trait anxiety. *Psychophysiology, 55*, e13274.

Van Son, D., de Blasio, F. M., Fogarty, J. S., Angelidis, A., Barry, R. J., & Putman, P. (2019a). Frontal EEG theta/beta ratio during mind wandering episodes. *Biological Psychology, 140*, 19–27.

Van Son, D., de Rover, M., De Blasio, F. M, van der Does, W., Barry, R. J., & Putman, P. (2019b). Electroencephalography theta/beta ratio covaries with mind wandering and functional connectivity in the executive control network. *Annals of the New York Academy of Sciences, 1452*, 52–64.

Volkow, N. D., Wang, G., Fowler, J. S., Logan, J., Gerasimov, M., Maynard, L., ... Franceschi, D. (2001). Therapeutic doses of oral methylphenidate significantly increase extracellular dopamine in the human brain. *Journal of Neuroscience, 21*, RC121.

Whitford, T. J., Rennie, C. J., Grieve, S. M., Clark, C. R., Gordon, E., & Williams, L. M. (2007). Brain maturation in adolescence: concurrent changes in neuroanatomy and neurophysiology. *Human Brain Mapping, 28*, 228–237.

Williams, L. M., Hermens, D. F., Thein, T., Clark, C. R., Cooper, N. J., Clarke, S. D., ... Kohn, M. R. (2010). Using brain based cognitive measures to support clinical decisions in ADHD. *Pediatric Neurology, 42*, 118–126.

Williams, L. M., Tsang, T. W., Clarke, S., & Kohn, M. (2010). An 'integrative neuroscience' perspective on ADHD: linking cognition, emotion, brain and genetic measures with implications for clinical support. *Expert Review of Neurotherapeutics, 10*, 1607–1621.

Wischnewski, M., Zerr, P., & Schutter, D. J. (2016). Effects of theta transcranial alternating current stimulation over the frontal cortex on reversal learning. *Brain Stimulation, 9*, 705–711.
Wischnewski, M., Joergensen M. L., Compen, B., & Schutter, D. J. (2020). Frontal beta transcranial alternating current stimulation improves reversal learning. Cerebral Cortex, *30*(, 3286–3295.
Wolinski, N., Cooper, N., Sauseng, P., & Romei, V. (2018). The speed of parietal theta frequency drives visuospatial working memory capacity. *PLoS Biology, 16*, e2005348.
Zhang. D. W., Li, H., Wu, Z., Zhao, Q., Song, Y., Liu, L., . . . Sun, L. (2019). Electroencephalogram Theta/Beta Ratio and Spectral Power Correlates of Executive Functions in Children and Adolescents With AD/HD. *Journal of Attention Disorders, 23*, 721–732.

CHAPTER 16

CORTICAL SOURCE LOCALIZATION IN EEG FREQUENCY ANALYSIS

WANZE XIE AND JOHN E. RICHARDS

16.1 Introduction to Source Localization

EEG signals recorded via electrodes placed on the scalp represent the postsynaptic potentials generated by mass synchronized pyramidal neurons perpendicular to the cortical surface. Source localization or source analysis is the activity that aims to identify the underlying cortical generators or sources of the EEG potentials measured on the scalp. The sources of the EEG signals may be modeled as electrical dipoles with both directionality and amplitude. The identification of the location of dipoles is often obscured because the current generated by these sources spreads in all directions in the brain and is smeared by the skull. However, recent advances made in improving the techniques for source localization have substantially increased the spatial resolution of this method and led to the tendency towards using it as a brain imaging tool (Michel & Murray, 2012).

Source localization consists of procedures creating a forward (Hallez et al., 2007) and an inverse (Grech et al., 2008) model. The forward model is created with a head model and electrodes and describes the effect of an electrical source inside the head on the EEG. The inverse model is the "inverse" of the forward model and is used with the EEG recording to compute the location and amplitude of the sources that generated the given EEG. The present chapter provides an overview of some widely adopted source analysis techniques and their applications to EEG frequency analysis. The importance of using realistic head models for source analysis is also discussed. In the last section of the chapter, we present examples for using source analysis as a neuroimaging tool in EEG frequency analysis.

16.2 Source Localization Techniques and Solutions to the Inverse Problem

There are two major approaches to source localization. Equivalent current dipole (ECD) models use a limited number of electrical dipoles that are computed with the forward model to explain the distribution of the EEG on the scalp. Distributed source models use a large set of dipole locations distributed across the brain and are calculated using the inverse model and the observed EEG to generate the amplitude of the current density (He et al., 2018; Michel et al., 2004). This section presents both models and describes how these models are used for source analysis.

16.2.1 Equivalent Current Dipole Source Localization

ECD modeling assumes that the electrical potential over the entire scalp can be explained by a small set of source dipoles (He et al., 1987). These hypothetical dipoles can vary in position, magnitude, and orientation in a 3D space. We can estimate the dipole parameters by repeated optimization of the parameters of the model so that the dipoles multiplied by the forward model fit the scalp distribution. Alternatively, a set of a priori fixed locations can be set based on theoretical specifications of the known effects, and thus only the parameters of orientation and magnitude are estimated. The ECD models are overdetermined because the number of dipole parameters is significantly less than the number of surface sensors (i.e., EEG electrodes), as the number of the estimated dipole parameters needs to be smaller than the number of electrodes on the scalp to guarantee a unique (overdetermined) solution (Michel et al., 2004).

A forward model needs to be constructed to estimate the electrical activity over the scalp in ECD modeling, as well as distributed source modeling. The forward model represents the head geometry and tissue conductivity and delineates how the activation generated by the dipoles propagates to the scalp; the so called "lead-field" matrix. The estimated electrical activity over the scalp is calculated by applying the forward model to the current dipoles with certain orientations and magnitudes (i.e., the forward solution). For ECD modeling, the output of this so-called forward solution can be compared to the actual electrical activity on the scalp, and thus the amount of variance explained by the selected current dipoles can be calculated (Richards, 2003; Scherg, 1992). The optimal solution is gained through the iteration of the forward solution with different parameters (location, orientation, and magnitude) of the dipoles until the minimal residual variance is obtained (Scherg et al., 1999).

A practical concern associated with equivalent current dipole modeling is the uncertainty about the number of dipoles and their locations that should be tested by the forward solution. One solution to this concern is to make a priori assumptions of the number and location of dipoles based on other neuroimaging data, such as functional magnetic resonance imaging (fMRI) and positron emission tomography (PET) (Agam et al., 2011; Foxe et al., 2003). For example, a meta-analysis of previous fMRI reports was conducted by Foxe and colleagues (2003) to find out the brain regions that were consistently activated in the attention task used by the authors. The location of these regions was used as a guide in their source analysis, such that the MRI coordinates of these areas were used as the seeds for the dipoles in the forward solution, with the assumption that the task-related EEG activity was generated in these areas. A priori assumptions can also be made based on theoretical rationale and findigns from previous research, such as using the fusiform and occipital face area (FFA; OFA) as the potential sources for the N170 component (Gao et al., 2019), the occipital and parietal regions for alpha oscillations (Xie et al., 2017), and the pre-central cortex for the mu rhythm (Thorpe et al., 2016). Gao and colleagues (2019) conducted a thorough literature review of the cortical source locations of the N170 ERP component found by previous EEG studies and the 3D coordinates of the FFA identified by previous fMRI research, and then used these locations as a priori defined regions of interest (ROIs) in their cortical source localization. Section 2.3 provides further information about how to define a priori ROIs for source localization.

16.2.2 Distributed Source Modeling

In distributed source models, cortical dipoles are distributed over the entire source space, and each dipole has a fixed location. The locations of the dipoles are called the "source space". The source space can be GM voxels derived from a structural MRI, the surface of the brain or "inner compartment" in boundary element methods, or the entire brain volume. A forward model is also needed for distributed source modeling. For distributed source modeling, the forward model is combined with the source space to estimate an inverse spatial filter (Grech et al., 2008). The inverse spatial filter, when multiplied by the observed scalp electrical current distribution, reconstructs the current density across the entire set of potential source locations (e.g., the current-density reconstruction [CDR] method).

The computation of the inverse spatial filter is problematic because the inverse of the lead-field matrix is "underdetermined". The relation between the scalp EEG potentials and the source activation can be simplified and illustrated by the following equation with the assumption that this relation is linear: Y (scalp potential) = K·X (source moments) + E (noise). In this equation, K stands for the transfer matrix from brain sources (the electrical "field") to the scalp potentials (the sensor "leads") and is referred to as the lead-field matrix. Mathematically, the source moments can be calculated by a transform of the previous equation: $X = K^{-1} \cdot Y$. Here, K^{-1} is the inverse of the lead-field matrix and referred to as the inverse spatial filter, which can be used to project the measured data on the scalp to the source space. The matrix inversion of K is "underdetermined" because the number of sources is substantially larger than the number of surface electrodes.

Thus, additional constraints must be imposed to obtain unique and well-posed linear inverse solutions (Grech et al., 2008; He et al., 2018).

The solution to the underdetermined construction of the inverse spatial filter is to constrain the solution by mathematical or quantitative procedures (Grech et al., 2008; Michel & He, 2012). The following paragraphs introduce a couple of widely adopted solutions in distributed source modeling: minimum norm estimation (MNE), low-resolution electromagnetic tomography (LORETA), standardized low-resolution electromagnetic tomography (sLORETA), and exact low-resolution electromagnetic tomography (eLORETA). Pascual-Marqui and colleagues have made great contribution to the development of the "LORETA family" in the past two decades (Sherlin, 2009). A number of additional solutions have been developed and utilized by the EEG research community, such as weighted MNE, shrinking LORETA FOCUSS (SLF), local autoregressive average, and beamforming techniques. Detailed description of these methods can be found elsewhere (Grech et al., 2008; Green & McDonald, 2009; Hallez et al., 2007; He et al., 2018).

The first introduced algorithm or constraint to the inverse solution in distributed source modeling was the minimum norm least-squares inverse (Hämäläinen & Ilmoniemi, 1994), often called the MNE. This algorithm minimizes the least-square error of the estimated inverse solution X in the equation described earlier, which means it results in an inverse solution with the lowest overall intensity (Hauk, 2004). The inverse solution obtained from MNE tends to be biased towards weak and superficial sources (Michel et al., 2004). To overcome the problem of this surface-restricted MNE algorithm, several methods have been developed that keep the basic mathematical relations of the MNE but alter its characteristics to resolve its weaknesses. For example, an early modification was to use a depth-weighting matrix in the formula to account for the MNE's bias toward superficial surfaces.

The LORETA algorithm was devised to add to the MNE by the additional constraint to smooth the sources with a Laplacian filter (Pascual-Marqui et al., 1994). The LORETA method was also designed to solve the "surface source" issue of MNE, such that it favors solutions with strong activation of a large number of sources and punishes solutions with small number of surface sources. The Laplacian of the sources is a measure of spatial roughness. Minimizing the Laplacian of sources leads to lower resolution in the source space, and thus the solution of LORETA algorithm is smoother than MNE and weighted MNE algorithms.

Pascual-Marqui (2002) developed the more widely used inverse algorithm sLORETA, which uses the current density estimate given by the MNE solution and weights the solution by the standardized values of the current density estimate. This partially alleviates the emphasis on superficial sources found in the MNE by enhancing the smaller deeper sources with their lower standard deviation. It was found that sLORETA results in fewer errors in reconstructed sources than MNE (Pascual-Marqui, 2002). The source localization accuracy has been compared using weighted MNE, LORETA, and sLORETA by Monte-Carlo analysis of data with different noise levels and sources with different depth in the brain (Grech et al., 2008). Grech and colleagues found that the sLORETA algorithm had the lowest level of localization error and fewest ghost sources, which are sources estimated from the inverse solution but not actually present in the simulated data. Therefore, sLORETA outperforms the other methods regarding the accuracy of source localization using simulated data with different levels of artifacts.

The eLORETA algorithm is another member of the "MNE family" that is also built upon the linear weighted MNE inverse solution and has been regarded as improvement over the previously developed LORETA and sLORETA algorithms (Pascual-Marqui, 2007). The eLORETA algorithm provides exact localization with zero error (i.e., no localization bias) to the inverse solution even in the presence of measurement and structured noise in the data. Assuming there is an activated dipole in the brain with an arbitrary orientation and known magnitude, the scalp EEG potentials generated by this dipole can be simulated. Applying the eLORETA algorithm as the inverse solution to the scalp measurement would give reconstructed current density fields. The "zero error localization" property of eLORETA means that there is zero distance between the actual point-test source and the reconstructed source with the absolute maximum amplitude. It should be noted this property has not been achieved by previously published discrete linear distributed source modeling algorithms (Pascual-Marqui, 2007).

Jatoi and colleagues (2014) compared the source localization accuracy between sLORETA and eLORETA for EEG data collected in an experiment with visual stimuli. The eLORETA algorithm was found to have enhanced ability to suppress less-significant sources in the brain compared to sLORETA. Jatoi and colleagues also found that eLORETA gave less localization error and clearer (less blurry) images as compared to sLORETA. As a member of the LORETA family, the moments in neighboring neuronal sources are also highly correlated for eLORETA.

16.2.3 The "Inverse Problem" with EEG Source Analysis

A common criticism of EEG source analysis is the "inverse problem". The complaint is that a unique solution to the underlying dipole sources is not possible, which occurs because an infinite number of dipole distributions can be constructed, combined with the forward model, and produce the same distribution of EEG potentials on the scalp.

The extent to which the inverse problem affects source analysis depends on a number of factors. The ECD modeling is the most problematic type of inverse modeling, as the resulting model can always be improved by adding one more parameter (e.g., another dipole), or a completely different set of dipole parameters (Hallez et al., 2007). There is no perfect solution to the number of dipoles necessary to find a solution with ECD modeling. Some programs using blind optimization techniques result in solutions that are physiologically unsound.

Distributed source modeling partially alleviates the major concerns with the inverse program. Distributed source models require no a priori assumption (nonparametric) of the number of dipoles (sources) in the model. This minimizes the error in source localization due to misspecification of the number and location of the dipoles (Grech et al.,

2008; Michel et al., 2004). For example, using the CDR approach, all potential source locations (e.g., brain volumes) are simultaneously estimated and the relative magnitude of the reconstructed source activity, also called the current source density (CSD), provides the putative locations of the sources. The source space of the distributed source models may be used to constrain the solution to physiologically reasonable positions. For example, since the generation of the EEG occurs in the postsynaptic potentials of the columnar pyramidal cells, the dipoles may be limited to gray matter (GM) and the dipole directions fixed as perpendicular to the cortical surface. These characteristics have led to a growing tendency in the past two decades to replace ECD optimization methods with distributed source modeling.

16.2.4 Applying a priori Information as Constraints to the EEG Inverse Solution

The linear distributed approaches devised to solve the inverse problem estimate all possible source locations simultaneously. However, there are still an infinite number of configurations of source activation that could generate the EEG scalp potentials, unless a specific set of locations are selected in advance. Anatomical and physiological information derived from other imaging modalities like fMRI, PET, and magnetoencephalography (MEG) have been be applied as constraints to the EEG inverse solution (Michel & He, 2012). In these studies, a priori anatomical and psychophysiological information is used as constraints on the inverse modeling in order to obtain a unique solution or reduce the number of potential solutions. The following two paragraphs introduce a few examples.

These external constraints can be imposed on the construction of the forward and inverse models to reduce the inverse solution space a priori (Lei et al., 2011; Phillips et al., 2002a; Phillips et al., 2002b; Yang et al., 2010) or used as a priori defined ROIs in statistical analyses to reduce the family-wise error rate (Gao et al., 2019; Xie et al., 2017). For example, Lei and colleagues (2011) used temporally coherent brain networks derived from fMRI data as the covariance priors of the EEG source reconstruction. This fMRI networks-based source imaging approach outperformed the traditional distributed source localization methods without fMRI priors and efficiently integrated the astonishing temporal resolution of EEG with the high spatial resolution of fMRI. In another study, an fMRI activation map was used as a spatial constraint to the MNE inverse solution, i.e., a fMRI-weighted MNE (Yang et al., 2010). The authors first conducted independent component analysis (ICA) of the EEG time-series. The ICs are then used to construct a regressor to fit in the fMRI signals. The output fMRI activation map was used as a spatial constraint to the estimation of the source distribution underlying its corresponding IC. Reliable source reconstructed time-frequency configuration and functional connectivity maps were obtained by using this method for both simulated and experimental data (Yang et al., 2010).

A recent study by Gao and colleagues (2019) investigated the cortical sources of the N170 face-sensitive ERP component. Individual structural MRI, fMRI, and EEG data were collected. The anatomical information of the head and brain was used to create individual realistic head models for source localization. The face-sensitive regions identified during the fMRI scan were used as the a priori defined ROIs in the analyses of the reconstructed source activity of the N170 component. This procedure reduces the family-wise error rate that would otherwise be caused by multiple comparisons or analyses across all source volumes in the brain.

16.3 Using Realistic Models and Structural MRI for Source Localization

The forward model that is used in both equivalent current dipole models and distributed source models and is created with a head model. The head model in EEG source localization describes media inside the head with their relative electrical conductivity. The head model is the link between the source volumes and the electrical potentials on the scalp in source analysis. The head models for EEG source localization have progressed from spherical models with only two or three compartments to realistic models with all different tissues of the head being segmented and assigned for their own conductivity values. This section introduces the importance of using realistic head models in EEG source analysis and its usage in empirical research. Since this chapter mostly focuses on realistic head models, an overview of other types of head models can be found elsewhere (Grech et al., 2008; Hallez et al., 2007).

16.3.1 The Effect of Distinction between Head Tissues in the Realistic Models for Source Analysis

An accurate head model that describes the materials inside the head and their relative conductivity is beneficial for source analysis (Reynolds & Richards, 2009; Vorwerk et al., 2014). The two most commonly used methods to create a realistic volume conduction head model are the boundary element method (BEM) and finite element method (FEM). The BEM model segments the head into hierarchical compartments with homogeneous conductivity profiles within a compartment. The FEM model defines the conductivity of individual voxels inside the head based on its material. For example, GM, white matter (WM), and cerebrospinal fluid (CSF) would each have its own conductivity value in a FEM model.

The creation of the FEM model used to be much more computational demanding than the BEM model (Michel et al., 2004), but the computational efficiency of computers has now been substantially improved, which makes the FEM model more applicable in source analysis. In a recent paper, Vorwerk and colleagues (2018) introduced a MATLAB-based pipeline, the "Fieldtrip-SimBio" pipeline, that aims to reduce the workload for the generation of multicompartment FEM models. This pipeline has been added to the existing package of functions to create FEM models in the Fieldtrip toolbox (Oostenveld et al., 2011), which allows for an easy construction and application of a FEM model for source analysis (Vorwerk et al., 2018).

Many studies have found that it is important to model the inhomogeneous distribution of brain and head tissues (Cho et al., 2015; Vorwerk et al., 2014). Vorwerk and colleagues (2014) comprehensively investigated the influence of modeling versus not modeling certain conductive tissues of the head on the EEG and MEG forward solution. They tested the effect of hierarchically adding new materials to a basic three-compartment head model (brain, skull, scalp) on the accuracy of source localization. Specifically, they tested the effect of the addition of realistic distribution of CSF, GM, and WM, differentiation of the skull spongiosa and compacta, and anisotropic WM conductivities on source localization. The largest increase in the accuracy of signal topography and amplitude from model to model came with the addition of CSF to the three-compartment model, followed by lesser (but still appreciable) increases when adding GM and WM and the anisotropic WM conductivities into the head models. The lack of the GM/WM distinction, or CSF, in the BEM compartment models, would result in considerable errors in the reconstructed source activities, which concurs with Cho and colleagues' (2015) study, which showed that the distinction between these brain tissues in source localization is important for analyzing EEG and MEG functional connectivity between cortical regions. Taken together, these findings highlight the importance of modeling the head conductive compartments as realistically as possible in order to increase the accuracy of source localization.

By contrast, a few other studies have compared the difference between using BEM and FEM models for source localization but found no substantial difference in the reconstructed source activation (Michel & He, 2012). For example, Birot and colleagues (2014) compared the source localization accuracy with LORETA using BEM and FEM models with 38 epileptic patients. These patients underwent scalp EEG and intracranial EEG recordings and subsequent resection surgery. The results of this study implicated that using the BEM and FEM models did not cause a significant difference in the accuracy of source localization (Birot et al., 2014). This finding suggests that the choice of a head model may not be a crucial factor for source localization using distributed localization algorithm for some applications.

16.3.2 The Uncertainty of the Skull-to-Brain Conductivity Ratio in Realistic Head Models

Another important consideration determining the accuracy of the forward and inverse solutions is to correctly model the conductivity of different tissues, especially the brain-to-skull conductivity ratio (Acar & Makeig, 2013). There is a consensus that the human skull has a much lower conductivity value than the other tissues inside the head. However, there is dispute about the correct skull conductivity value or the skull-to-brain conductivity ratio to use for creating head models (Michel & He, 2012). The ratio of 1/80 is commonly specified as the skull-to-brain conductivity ratio in head models (either spherical or realistic), with their conductivity values being 0.0042 and 0.33 S/m (Siemens per meter, reciprocal of ohm Ω), respectively. The skull-to-brain conductivity ratio of 1/80 originated from studies that measured the electrical properties of brain and head tissues (Gabriel et al., 1996; Rush & Driscoll, 1969). This ratio has been set up as the default in several source localization toolboxes, such as brain electrical source analysis (BESA; Megis Software GmbH, Gräfelfing, Germany) and Fieldtrip (Oostenveld et al., 2011), and widely used in empirical research and simulated studies (Ryynanen et al., 2006).

More recent studies have suggested that the traditional skull to brain conductivity ratio of 1/80 might be misestimated. A few studies have been conducted to estimate or measure the conductivity of human skull, and findings have shown that the correct skull-to-brain conductivity ratio may be between 1/15 to 1/30, which is much higher than the traditional ratio, and using a lower skull-to-brain conductivity ratio (e.g., 1/80) may result in shallower source locations (Acar et al., 2016; Acar & Makeig, 2013; Dannhauer et al., 2011; Lai et al., 2005; Oostendorp et al., 2000). Therefore, the traditional ratio of skull-to-brain conductivity may need to be adjusted when creating head models for cortical source localization.

Bone conductivities are age dependent, with infants and children having much higher skull conductivity values compared to adults (Odabaee et al., 2014). This means that using the adult skull conductivity values to create head models for infant and child participants could lead to inaccurate source localization results (Reynolds & Richards, 2009). To this end, some source localization software, for example, BESA, provides references for the skull-to-brain conductivity ratio for children and adolescents at different ages. Recent studies have also started to adopt age-appropriate skull conductivity values for source localization on infant EEG data (Xie, Jensen, et al., 2019).

16.3.3 The Importance of Realistic Head Models for Source Localization in Pediatric Populations

Using age-specific realistic head models for source analysis may be especially important for pediatric populations. This is because there are substantial neuroanatomical changes of the tissues inside the head over childhood, and the structure of a child brain differs greatly from an adult brain (Phan et al., 2018; Reynolds & Richards, 2009; Richards, W., 2015). For example, skull conductivity value and thickness are age dependent (Wendel, Vaisanen, Seemann, Hyttinen, & Malmivuo, 2010), such that the skull conductivity value is much higher for infants than adults (Odabaee et al., 2014). Using the adult skull conductivity values for infant EEG data may have the effect of inferring the source of the current as being shallower in the cortex than where it actually occurs.

A significant advance in cortical source analysis with pediatric participants is to use realistic head models created with individual MRIs or an age-appropriate MRI template (Guy et al., 2016; Hämäläinen et al., 2011; Ortiz-Mantilla et al., 2012; Xie et al., 2017). Although systematic estimation of skull conductivity for infants and children at different ages has not been conducted, studies have tried to use higher skull conductivity values for source localization for infants and young children data (Hämäläinen et al., 2011; Ortiz-Mantilla et al., 2012; Xie et al., 2018). There are now age-specific MRI templates available to the public for research purposes, which can be used to create realistic head models for children (e.g., the Neurodevelopment MRI Database) (Richards et al., 2016). Since the fontanels and unmended skull sutures of an infant head could allow current flow to the scalp unimpeded by the skull, a future direction is to take the fontanels and skull sutures and the conductivity and topographical differences between infants' and adults' heads into account when creating the realistic head models for infants.

16.4 Application of Source Localization in EEG Frequency Analysis

Human brain oscillatory activity is generated by neural tissue in the central nervous system spontaneously or in response to stimuli in the environment. The significance of these oscillatory signals in sensory-cognitive processes has become increasingly evident. Using source localization in EEG (time) frequency analysis has made great contributions to our understanding of the cortical mechanisms of cognitive processes associated with oscillatory activities in various frequency bands. Using source localization of EEG signals to obtain reconstructed cortical activities also allows us to investigate functional connectivity withing and between brain networks during cognitive tasks and resting-state. The current section reviews the application of source localization techniques in EEG frequency analysis and how this method has provided insights into brain functions and network organizations.

16.4.1 EEG Source Localization in Frequency Analysis for Cognitive Functions

The neural generators of brain oscillatory activity in different frequency bands during cognitive tasks are of great interests to the EEG community. The functional significance of cortical oscillations in low and high frequency bands has been specified using source localization techniques for various cognitive processes, such as spatial attention orienting and attentional control (Doesburg et al., 2009; Jones et al., 2010; Sauseng et al., 2007), sustained attention and vigilance (Kim et al., 2017), error monitoring and conflict processing (Cohen, 2011; Cohen & Ridderinkhof, 2013; Yeung et al., 2007), working memory (Maurer et al., 2015; Michels et al., 2010; Michels et al., 2008; Onton et al., 2005), motion observation and execution (Nystrom et al., 2011; Ritter et al., 2009; Thorpe et al., 2016). There is a large body of literature on the developmental origin of the relation between these cognitive processes and corresponding EEG oscillations (Bell & Cuevas, 2012). EEG source localization has also made it possible to investigate the cortical sources of EEG oscillatory activities associated with cognitive performance in infants and children (see Section 16.4.4).

The source localization techniques discussed in the previous sections open the avenue to unravel the crucial role of EEG oscillatory activity in human brain functions. For instance, Sauseng and colleagues (2007) investigated the function of the theta-band oscillations in sustained attention and executive control. They examined how EEG theta-band activity changed as a function of task difficulty and memory load. The participants were required to execute complex sequential finger movements requiring different levels of memory load, such as execution of a previously trained simple sequence, an overlearned complex sequence, or a novel complex sequence. Local frontal-midline theta activity on the scalp was found to be associated with the general level of cognitive demand, with the highest power found for the most demanding condition. This theta activity was reconstructed to the source space using the LORETA method. Specifically, the CSD was reconstructed for more than 2000 cortical voxels across the brain using distributed source modeling, separately for different cognitive conditions. The source activity was then compared voxel by voxel between the conditions, and multiple comparisons were statistically controlled. The authors found that the CSD in the anterior cingulate gyrus was significantly different between the conditions, suggesting that theta-band oscillations generated by this region may reflect the activation of an attentional system responsible for allocating cognitive resources (Sauseng et al., 2007).

Theta-band activity generated by the frontal cortex also plays an important role in error monitoring (Cohen, 2011). Cohen found that EEG theta-band activity in the frontal electrodes was associated with error-monitoring behaviors in adults. The author subsequently conducted source localization of the theta-band activity using equivalent current dipole modeling with subject-specific BEM models created with individual MRIs. One dipole that explained the most amount of variance of the scalp potentials was estimated per subject. The author found that the dipoles in the medial prefrontal and anterior cingulate cortexes best explained the variance in the frontal theta activity across subjects, suggesting that these areas are the potential sources of the distribution of the theta activity on the scalp.

EEG oscillations generated by the frontal area has been found to play an important role in maintaining vigilance (sustained attention) during cognitive tasks. Kim and colleagues (2017) investigated which brain areas are engaged in controlling the maintenance of vigilance using source analysis of EEG signals. Distributed source modeling with the sLORETA algorithm was conducted to obtain smooth source activation in different frequency bands across the entire cortex. The authors found that the activity in the left prefrontal cortex was significantly correlated with vigilance variation in the delta, beta, and gamma bands, which suggests the involvement of this brain region in maintaining vigilance. To sum up, these empirical studies demonstrated how using EEG source analysis in frequency analysis could advance our understanding of the neural mechanisms underlying different cognitive processes.

16.4.2 EEG Source Localization for Clinical Applications

Imaging the neuronal activity that generates the scalp EEG oscillations is also desirable for clinical purposes. EEG source imaging offers a useful tool to help presurgical evaluation of patients with epilepsy in a noninvasive way, as identifying the cortical epileptic zone for ictal oscillations is crucial for surgical resection. The decent accuracy (~80%) of using EEG source localization in determining the subsequently resected zone has been consistently achieved across studies on pediatric (Lu et al., 2012) and adult patients (Brodbeck et al., 2011; Yang et al., 2011). For example, Brodbeck and colleagues investigated the sensitivity and specificity of using EEG source imaging to detect the epileptic zone for over 150 patients who underwent surgery later. The authors found that using EEG source imaging resulted in high sensitivity (84%) and specificity (88%) in detecting the seizure zone. The obtained sensitivity and specificity values were even higher than some conventional neuroimaging tools (e.g., fMRI and PET). It should be noted that acceptable localization results were only obtained when a large number of electrodes (>128 channels) were used for data collection (Brodbeck et al., 2011), which suggests the importance of using sufficient electrodes to obtain a better resolution of the topographic features, that is, to avoid distortions of the distribution of scalp potentials due to large distances between electrodes (see Michel et al., 2004 for a thorough review of the effect of the number of electrodes). Spectral analysis of EEG with source localization techniques has also been applied to characterize difficulties in brain functioning for other clinical populations, such as children with ADHD (Liotti et al., 2005), people with tonic pain (Canuet et al., 2012), and Parkinsonian patients (Moazami-Goudarzi et al., 2008).

16.4.3 Applying EEG Source Analysis to Studying Functional Brain Connectivity in the Frequency Domain

One critical question in the field of cognitive neuroscience is how brain regions in large-scale networks communicate and cooperate with each other during tasks and resting-state. The application of neuroimaging tools makes it possible for researchers to investigate the dynamic interregional communications inside the brain and their development throughout the lifespan. Using fMRI allows people to obtain images of the blood-oxygen-level dependent (BOLD) signals in brain voxels. However, the temporal resolution of the BOLD response in fMRI depends on blood flow occurring several seconds following the demand for glucose via oxygenated hemoglobin. This restricts fMRI connectivity analyses to "seconds" resolution and prohibits this technique from studying dynamic changes in neural oscillations in sub-second frequency ranges. By contrast, EEG and MEG directly measure synchronized neuronal activity with msec time resolution and thus can measure dynamic changes in neural oscillations at the resolution of the neural system (e.g., sub-seconds resolution).

The examination of EEG functional connectivity between electrodes on the scalp has been used to study synchronized fluctuations in various rhythms that may reflect the dynamic coordination and network structure inside the brain (Bastos & Schoffelen, 2016; Stam, 2005). A variety of methods have been developed to evaluate the "correlation or synchronization" between the signals from two electrodes, such as the correlation between power envelope (Hipp et al., 2012), coherence analysis (Bowyer, 2016), and phase synchronization (Stam et al., 2007). In addition, previous work studying EEG functional connectivity on the scalp level with infants and children provides insights into the developmental origin of integrated brain networks (see Chu-Shore et al., 2011 for a review). For example, Cuevas and colleagues (2012) investigated the association between infants' working memory and inhibitory control performance in cognitive tasks and the coherence between frontal electrodes in the alpha band. The authors found that frontal alpha coherence was associated with inhibitory control processes in 10-month-old infants, which suggests that the cooperation between brain regions already plays an important role in cognitive performance in infancy (Cuevas et al., 2012).

The volume conduction/field spread issue of surface EEG and the effect of the reference electrode on phase synchronization and correlation values impede our understanding of the underlying neural mechanisms from the EEG functional connectivity results on the scalp (Bastos & Schoffelen, 2016; Guevara et al., 2005; Schiff, 2005). Specifically, this problem means that the activity of a cortical source can be "visible" at multiple electrodes due to the field smearing on the scalp. This will cause spurious correlation or synchronization values between electrodes. Consequently, the interpretation of the scalp-level connectivity requires considerable caution because two "connected or synchronized" electrodes may not reflect the brain regions that are functionally connected.

Applying cortical source reconstruction methods to EEG data could substantially mitigate the consequence of volume conduction. Brain functional connectivity using EEG could be done by first estimating the sources of the surface signals, deriving time-domain activity of the cortical sources via the time-domain activity of the surface activity, and then doing functional connectivity in the source-space by calculating the correlation between amplitudes (or power envelope) of times-series in cortical sources or phase synchronization in different frequency bands after Fourier transformation of the time-series in cortical sources into the frequency domain. For example, Xie and colleagues (2018) developed a pipeline to estimate functional connectivity in the cortical source space to investigate the function of different brain networks of awake infants (Figure 16.1). This pipeline includes cortical source reconstruction of EEG recordings with age-appropriate MRIs, parcellation of the source activity into brain ROIs, estimating brain functional connectivity between ROIs by calculating the weighted phase-lag index (wPLI) (Vinck et al., 2011), and applying graph theory measures to examine the overall architecture of brain networks (Figure 16.1). Using this pipeline, the authors studied how functional connectivity in infant brain networks, such as the dorsal and ventral attention, default mode, and somatosensory networks, changes across different attentional states. For instance, they showed that infant sustained attention is

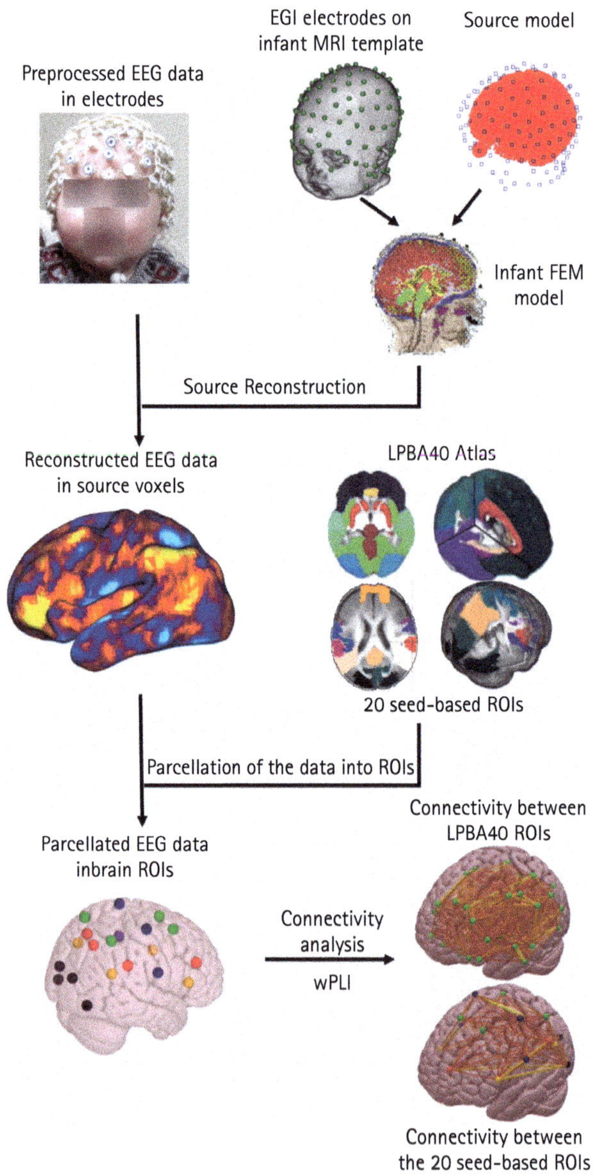

FIGURE 16.1 The pipeline for EEG functional connectivity analysis in the source space used by Xie et al. (2018).

associated with attenuated connectivity in the alpha band within the dorsal attention network and the DMN, as well as distinct network organization and efficiency indicated by graph theory measures (Xie et al., 2018). This work highlights the possibility of using source-space functional connectivity analysis to study the development of brain networks in early childhood. The following paragraphs introduce more recent advances that have been made in combining EEG source localization and functional brain connectivity analysis in the frequency domain.

EEG source localization can be used to detect brain functional connectivity during tasks. The functional connectivity between frontal and parietal cortices plays an important role in executive control. Sauseng and colleagues (2007) found a more distributed pattern of functional connectivity in the theta band between frontal and parietal regions when participants executed novel compared to memorized figure movement sequences. In a more recent study, Cohen and Ridderinkhof (2013) investigated the functional coupling between EEG signals in different frequency bands during spatial conflict processing. Inter-regional connectivity in the source space was found to differ between the congruent and incongruent conditions in a Simon task. Congruent trials induced stronger coupling between frontal theta and parietal alpha and gamma power, while incongruent trials induced stronger coupling of the theta power between the medial and lateral frontal regions. Studies also show that the inter-regional phase synchrony in low- and high-frequency bands is associated with top-down control of visual and auditory spatial attention (Doesburg et al., 2009; Doesburg et al., 2012). For example, Doesburg and colleagues (2009) found increased phase synchronization in the alpha band between the visual cortex and parietal regions contralateral to the attended visual hemifield and decreased phase synchronization between those brain regions ipsilateral to the attended visual hemifield. The Cohen and Doesburg studies used the beamforming technique for cortical source reconstruction. The beamforming techniques are not covered here and have been reviewed elsewhere (Green & McDonald, 2009).

The functional connectivity analysis in the source space of spontaneous EEG during resting-state has provided insights into the structure and functioning of brain networks, as well as the development of brain networks over childhood. Liu and colleagues (2017) detected large-scale brain networks in human adults using high-density EEG recordings along with the eLORETA technique and realistic head models for source analysis. ICA analysis of source reconstructed power envelops was conducted to extract functionally connected brain regions (networks) in different frequency bands. Their results showed that the brain networks identified with resting-state EEG data (e.g., the DMN, attention and language networks) were comparable to the brain networks obtained from prior resting-state fMRI research (Liu et al., 2017). Mantini and colleagues' (2007) simultaneous EEG-fMRI data were collected during resting-state to unravel the direct relationship between functional connectivity measured by the two metrics. Six major brain networks (e.g., the DMN, auditory, visual, and attention networks) were identified from the BOLD signals using ICA. The fluctuations in each of these fMRI-defined resting-state networks were found to be correlated with EEG power in different frequency bands (e.g., delta, theta, alpha, beta, and gamma). This study was one of the first attempts to explore the link between EEG brain rhythms and low-frequency coherent fluctuations of the BOLD signal during resting-state.

16.4.4 EEG Source Localization in Frequency Analysis for Pediatric Populations

High-density EEG offers an easy-to-use tool to measure brain functions for pediatric populations. It is often the method of choice when measuring brain activity in awake infants and older children due to its superb temporal resolution, easy application, low-cost of recordings and relatively higher tolerance to children's movements compared to other neuroimaging tools. Using EEG also makes it possible to monitor infant brain activity corresponding to infant behaviors during cognitive tasks, for example, attention (Xie et al., 2017), emotion processing (Xie et al., 2018) and recognition memory (Reynolds & Richards, 2005) tasks, which could not be done with fMRI. In addition, recent advances made in EEG source localization techniques, as reviewed earlier in the chapter, have made EEG a comprehensive and powerful brain imaging tool with reasonable spatial and superior temporal resolution (Michel & Murray, 2012). Therefore, there is growing interest in using source analysis techniques with high-density EEG recordings to examine the development of the cortical sources of neural oscillations in different frequency bands during childhood. This section gives an overview of recent progress made in using EEG source localization along with frequency analysis as a neuroimaging tool to study brain and cognitive development.

The mu rhythm ("central alpha") has been proposed to be associated with a putative human mirror neuron system (Marshall & Meltzoff, 2011). EEG source localization has been utilized to study the neurodevelopmental origin of the mirror neuron system and the functional significance of the mu rhythm during early childhood. Thorpe and colleagues (2016) investigated the structural development of the mu rhythm during motor execution. The source analysis of the mu rhythm was conducted using the sLORETA algorithm with age-appropriate head models. The results of this study showed distributed frontoparietal patterns of cortical activation in the mu rhythm. Specifically, the source locations of the mu rhythm were concentrated in the pre- and post-central gyri and the precuneus and inferior parietal regions. The source locations of the mu rhythm were found to be consistent across three age groups—the 1- and 4-year-olds and the adults. The finding of the posterior regions as part of the sources generating the mu rhythm is reminiscent of two previous studies exploring the cortical generators of the infant mu rhythm using ECD models (Nystrom, 2008; Nystrom et al., 2011). The findings of these studies shed light on the neural generators of the mu rhythm in children and its association with motion observation and execution.

Prior studies have suggested that brain rhythms can be manipulated by infant visual sustained attention. However, the cortical generators of attention associated brain rhythmic activity remained unclear. To fill this knowledge gap, a study by Xie and colleagues (2017) explored the complex patterns of EEG oscillations and their cortical sources observed during infant sustained attention. Specifically, the authors examined whether infant sustained attention, defined with the infant heart-rate attention model (Richards, 2003), was accompanied by distinct EEG rhythmic activities in the theta, alpha and beta bands in attention-related cortical areas. Cortical source reconstruction of EEG power in different frequency bands was conducted using the eLORETA method. Realistic head models were created for each participant using individual MRIs. Infant sustained attention was found to be accompanied by increased theta power in the orbitofrontal and ventral temporal areas and decreased alpha power in brain regions within the default mode network (DMN). The relation between infant sustained attention and EEG power was not found among infants aged 6 and 8 months, but the relation emerged at age 10 months and became well established by age 12 months (Figure 16.2). This study established the a connection between infant sustained attention and cortical oscillatory activity in the theta and alpha bands.

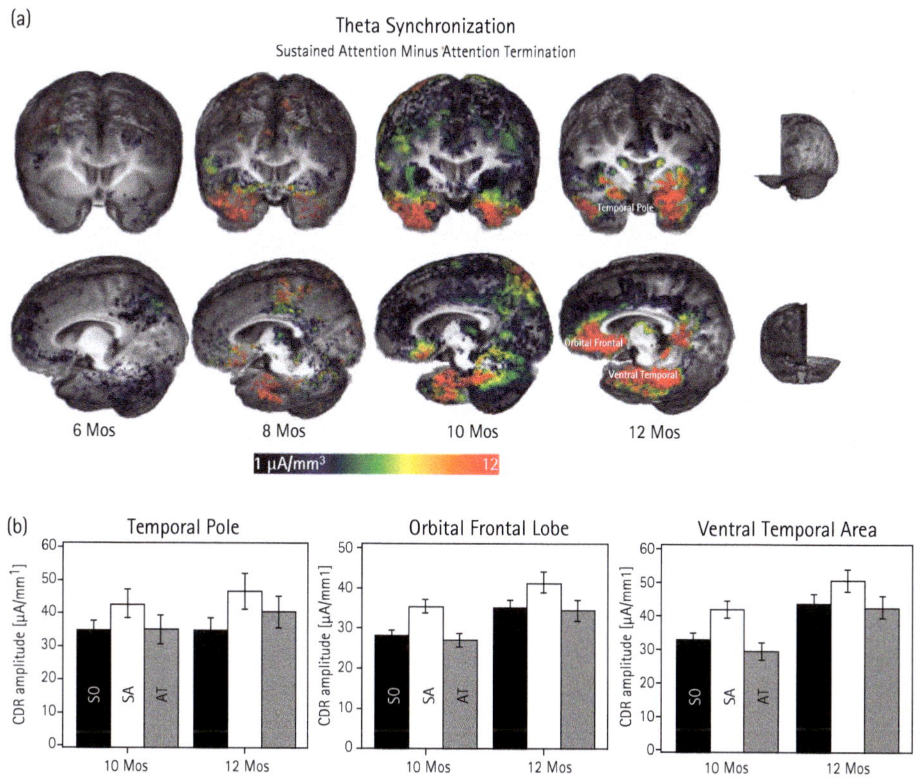

FIGURE 16.2 Development of the effect of sustained attention on infant theta source activation. (A) 3D displays for the difference in CDR amplitude between sustained attention and attention termination separately for the four ages. Age-appropriate average MRI templates were used for the display for each age. Sustained attention effect was primarily shown in the temporal pole, orbital frontal, and ventral temporal regions, especially for 10 and 12 months. (B) Bar graphs for the average CDR amplitudes in these brain networks for 10 and 12 months. Error bars represent SEMs.

Adapted from Xie et al., 2017.

Functional connectivity analysis of source reconstructed resting-state EEG signals has recently been adopted to study the development of brain networks with children. Bathelt and colleagues (2013) examined the development of brain network communities using EEG data collected when children between 2 and 6 years of age were in "resting-state" (watching a video clip of calming scenes). Cortical source reconstruction was conducted with age-appropriate head models. Graph theory measures were also employed to assess the organization and efficiency of brain networks. The authors found functional modules that resembled hub networks described for fMRI and developmental changes in graph theory measures (e.g., path length) with age (Bathelt et al., 2013). Similar methods have been used to study how deviations in early experience, such as poverty and growth faltering, shape brain networks in children living in low-income countries (Xie et al., 2019). In a recent study Xie and colleagues explored brain functional connectivity as a mediator of the relation between growth faltering and future cognitive outcomes in a longitudinal sample of impoverished Bangladeshi children. Their results revealed that whole-brain functional connectivity predicted future cognitive function and, perhaps more importantly, that brain functional connectivity at age 36 months mediated the effect of growth faltering on children's IQ assessed one year later.

16.5 Conclusion

This chapter introduces EEG source localization and its usage in (time) frequency analysis. EEG source localization offers superb temporal resolution and increasingly improved spatial resolution. Methodological progress helped to improve the accuracy of source localization, for example, techniques developed to constrain the inverse solutions in distributed source reconstruction. The importance of using realistic head models and age-appropriate conductivity values for source localization is also increasingly recognized by the field, especially for research with pediatric populations. These advances extended the application of source localization in clinical diagnosis and scientific research. Studies using EEG source localization and frequency analysis provide importance insights into the functional significance of oscillatory activities in different frequency rhythms in cognitive processes, as well as the maturation of functional brain networks over development.

REFERENCES

Acar, A. Z., Acar, C. E., & Makeig, S. (2016). Simultaneous head tissue conductivity and EEG source location estimation. *Neuroimage*, *124*(Pt A), 168–180. doi:10.1016/j.neuroimage.2015.08.032

Acar, A. Z., & Makeig, S. (2013). Effects of forward model errors on EEG source localization. *Brain Topography*, *26*(3), 378–396. doi:10.1007/s10548-012-0274-6

Agam, Y., Hamalainen, M. S., Lee, A. K., Dyckman, K. A., Friedman, J. S., Isom, M., . . . Manoach, D. S. (2011). Multimodal neuroimaging dissociates hemodynamic and electrophysiological correlates of error processing. *Proceedings of the National Academy of Sciences of the United States of America*, *108*(42), 17556–17561. doi:10.1073/pnas.1103475108

Bastos, A. M. & Schoffelen, J. M. (2016). A tutorial review of functional connectivity analysis methods and their interpretational pitfalls. *Frontiers in Systems Neuroscience*, *9*, 175. doi:10.3389/fnsys.2015.00175

Bathelt, J., O'Reilly, H., Clayden, J. D., Cross, J. H., & de Haan, M. (2013). Functional brain network organisation of children between 2 and 5 years derived from reconstructed activity of cortical sources of high-density EEG recordings. *Neuroimage*, *82*, 595-604. doi:10.1016/j.neuroimage.2013.06.003

Bell, M. A., & Cuevas, K. (2012). Using EEG to study cognitive development: issues and practices. *Journal of Cognition and Development*, *13*(3), 281–294. doi:10.1080/15248372.2012.691143

Birot, G., Spinelli, L., Vulliemoz, S., Megevand, P., Brunet, D., Seeck, M., & Michel, C. M. (2014). Head model and electrical source imaging: a study of 38 epileptic patients. *Neuroimage: Clinical*, *5*, 77–83. doi:10.1016/j.nicl.2014.06.005

Bowyer, S. M. (2016). Coherence a measure of the brain networks: past and present. *Neuropsychiatric Electrophysiology*, *2*(1), 1. doi:10.1186/s40810-015-0015-7

Brodbeck, V., Spinelli, L., Lascano, A. M., Wissmeier, M., Vargas, M. I., Vulliemoz, S., . . . Seeck, M. (2011). Electroencephalographic source imaging: a prospective study of 152 operated epileptic patients. *Brain*, *134*(Pt 10), 2887–2897. doi:10.1093/brain/awr243

Cho, J. H., Vorwerk, J., Wolters, C. H., & Knosche, T. R. (2015). Influence of the head model on EEG and MEG source connectivity analyses. *Neuroimage*, *110*, 60–77. doi:10.1016/j.neuroimage.2015.01.043

Chu-Shore, C. J., Kramer, M. A., Bianchi, M. T., Caviness, V. S., & Cash, S. S. (2011). Network analysis: Applications for the developing brain. *Journal of Child Neurology*, *26*(4), 488–500. doi:10.1177/0883073810385345

Cohen, M. X. (2011). Error-related medial frontal theta activity predicts cingulate-related structural connectivity. *Neuroimage*, *55*(3), 1373–1383. doi:10.1016/j.neuroimage.2010.12.072

Cohen, M. X. & Ridderinkhof, K. R. (2013). EEG source reconstruction reveals frontal-parietal dynamics of spatial conflict processing. *PLoS One*, *8*(2), e57293. doi:10.1371/journal.pone.0057293

Cuevas, K., Swingler, M. M., Bell, M. A., Marcovitch, S., & Calkins, S. D. (2012). Measures of frontal functioning and the emergence of inhibitory control processes at 10 months of age. *Developmental and Cognitive Neuroscience*, *2*(2), 235–243. doi:10.1016/j.dcn.2012.01.002

Dannhauer, M., Lanfer, B., Wolters, C. H., & Knosche, T. R. (2011). Modeling of the human skull in EEG source analysis. *Human Brain Mapping*, *32*(9), 1383–1399. doi:10.1002/hbm.21114

Doesburg, S. M., Green, J. J., McDonald, J. J., & Ward, L. M. (2009). From local inhibition to long-range integration: A functional dissociation of alpha-band synchronization across cortical scales in visuospatial attention. *Brain Research*, *1303*, 97–110. doi:10.1016/j.brainres.2009.09.069

Doesburg, S. M., Green, J. J., McDonald, J. J., & Ward, L. M. (2012). Theta modulation of inter-regional gamma synchronization during auditory attention control. *Brain Research*, *1431*, 77–85. doi:10.1016/j.brainres.2011.11.005

Foxe, J. J., McCourt, M. E., & Javitt, D. C. (2003). Right hemisphere control of visuospatial attention: Line-bisection judgments evaluated with high-density electrical mapping and source analysis. *Neuroimage, 19*(3), 710–726.

Gabriel, C., Gabriel, S., & Corthout, E. (1996). The dielectric properties of biological tissues. 1. Literature survey. *Physics in Medicine and Biology, 41*(11), 2231–2249. doi:10.1088/0031-9155/41/11/001

Gao, C., Conte, S., Richards, J. E., Xie, W., & Hanayik, T. (2019). The neural sources of N170: Understanding timing of activation in face-selective areas. *Psychophysiology*, 56(6), e13336. doi:10.1111/psyp.13336

Grech, R., Cassar, T., Muscat, J., Camilleri, K. P., Fabri, S. G., Zervakis, M., ... Vanrumste, B. (2008). Review on solving the inverse problem in EEG source analysis. *Journal of Neuroengineering and Rehabilitation*, 5, 25. doi:10.1186/1743-0003-5-25

Green, J. J. & McDonald, J. J. (2009). A practical guide to beamformer source reconstruction for EEG. In T. C. Handy (Ed.), *Brain signal analysis: Advances in neuroelectric and neuromagnetic methods* (pp. 76–98). MIT Press

Guevara, R., Velazquez, J. L., Nenadovic, V., Wennberg, R., Senjanovic, G., & Dominguez, L. G. (2005). Phase synchronization measurements using electroencephalographic recordings: what can we really say about neuronal synchrony? *Neuroinformatics, 3*(4), 301–314. doi:10.1385/NI:3:4:301

Guy, M. W., Zieber, N., & Richards, J. E. (2016). The cortical development of specialized face processing in infancy. *Child Development*, 87(5), 1581–1600. doi:10.1111/cdev.12543

Hallez, H., Vanrumste, B., Grech, R., Muscat, J., De Clercq, W., Vergult, A., ... Lemahieu, I. (2007). Review on solving the forward problem in EEG source analysis. *Journal of Neuroengineering and Rehabilitation, 4*, 46. doi:10.1186/1743-0003-4-46

Hämäläinen, J. A., Ortiz-Mantilla, S., & Benasich, A. A. (2011). Source localization of event-related potentials to pitch change mapped onto age-appropriate MRIs at 6 months of age. *Neuroimage, 54*(3), 1910–1918. doi:10.1016/j.neuroimage.2010.10.016

Hämäläinen, M. S. & Ilmoniemi, R. J. (1994). Interpreting magnetic fields of the brain: Minimum norm estimates. *Medical & Biological Engineering & Computing, 32*(1), 35–42.

Hauk, O. (2004). Keep it simple: a case for using classical minimum norm estimation in the analysis of EEG and MEG data. *Neuroimage, 21*(4), 1612–1621. doi:10.1016/j.neuroimage.2003.12.018

He, B., Musha, T., Okamoto, Y., Homma, S., Nakajima, Y., & Sato, T. (1987). Electric dipole tracing in the brain by means of the boundary element method and its accuracy. *IEEE Transactions on Biomedical Engineering, 34*(6), 406–414.

He, B., Sohrabpour, A., Brown, E., & Liu, Z. (2018). Electrophysiological source imaging: A noninvasive window to brain dynamics. *Annual Review of Biomedical Engineering, 20*, 171–196. doi:10.1146/annurev-bioeng-062117-120853

Hipp, J. F., Hawellek, D. J., Corbetta, M., Siegel, M., & Engel, A. K. (2012). Large-scale cortical correlation structure of spontaneous oscillatory activity. *Nature Neuroscience, 15*(6), 884–890. doi:10.1038/nn.3101

Jatoi, M. A., Kamel, N., Malik, A. S., & Faye, I. (2014). EEG-based brain source localization comparison of sLORETA and eLORETA. *Australasian Physical & Engineering Sciences in Medicine, 37*(4), 713–721. doi:10.1007/s13246-014-0308-3

Jones, S. R., Kerr, C. E., Wan, Q., Pritchett, D. L., Hamalainen, M., & Moore, C. I. (2010). Cued spatial attention drives functionally relevant modulation of the mu rhythm in primary somatosensory cortex. *Journal of Neuroscience, 30*(41), 13760–13765. doi:10.1523/JNEUROSCI.2969-10.2010

Kim, J.-H., Kim, D.-W., & Im, C.-H. (2017). Brain areas responsible for vigilance: An EEG source imaging study. *Brain Topography, 30*(3), 343–351. doi:10.1007/s10548-016-0540-0

Lai, Y., van Drongelen, W., Ding, L., Hecox, K. E., Towle, V. L., Frim, D. M., & He, B. (2005). Estimation of in vivo human brain-to-skull conductivity ratio from simultaneous extra- and intra-cranial electrical potential recordings. *Clinical Neurophysiology, 116*(2), 456–465. doi:10.1016/j.clinph.2004.08.017

Lei, X., Xu, P., Luo, C., Zhao, J., Zhou, D., & Yao, D. (2011). fMRI functional networks for EEG source imaging. *Human Brain Mapping, 32*(7), 1141–1160. doi:10.1002/hbm.21098

Liotti, M., Pliszka, S. R., Perez, R., Kothmann, D., & Woldorff, M. G. (2005). Abnormal brain activity related to performance monitoring and error detection in children with ADHD. *Cortex, 41*(3), 377–388.

Liu, Q., Farahibozorg, S., Porcaro, C., Wenderoth, N., & Mantini, D. (2017). Detecting large-scale networks in the human brain using high-density electroencephalography. *Human Brain Mapping*, 38(9), 4631–4643. doi:10.1002/hbm.23688

Lu, Y., Yang, L., Worrell, G. A., Brinkmann, B., Nelson, C., & He, B. (2012). Dynamic imaging of seizure activity in pediatric epilepsy patients. *Clinical Neurophysiology, 123*(11), 2122–2129. doi:10.1016/j.clinph.2012.04.021

Mantini, D., Perrucci, M. G., Del Gratta, C., Romani, G. L., & Corbetta, M. (2007). Electrophysiological signatures of resting state networks in the human brain. *Proceedings of the National Academy of Sciences of the United States of America, 104*(32), 13170–13175. doi:10.1073/pnas.0700668104

Marshall, P. J., & Meltzoff, A. N. (2011). Neural mirroring systems: exploring the EEG mu rhythm in human infancy. *Dev Cogn Neurosci, 1*(2), 110–123. https://doi.org/10.1016/j.dcn.2010.09.001

Maurer, U., Brem, S., Liechti, M., Maurizio, S., Michels, L., & Brandeis, D. (2015). Frontal midline theta reflects individual task performance in a working memory task. *Brain Topography, 28*(1), 127–134. doi:10.1007/s10548-014-0361-y

Michel, C. & He, B. (2012). EEG mapping and source imaging. In D. Schomer & F. H. Lopes da Silva (Eds.), *Niedermeyer's electroencephalography* (pp. 1179–1202). Lippincott Williams & Wilkins.

Michel, C. M., & Murray, M. M. (2012). Towards the utilization of EEG as a brain imaging tool. *Neuroimage, 61*(2), 371–385. doi:10.1016/j.neuroimage.2011.12.039

Michel, C. M., Murray, M. M., Lantz, G., Gonzalez, S., Spinelli, L., & Grave de Peralta, R. (2004). EEG source imaging. *Clinical Neurophysiology, 115*(10), 2195–2222. doi:10.1016/j.clinph.2004.06.001

Michels, L., Bucher, K., Luchinger, R., Klaver, P., Martin, E., Jeanmonod, D., & Brandeis, D. (2010). Simultaneous EEG-fMRI during a working memory task: Modulations in low- and high-frequency bands. *PLoS One, 5*(4), e10298. doi:10.1371/journal.pone.0010298

Michels, L., Moazami-Goudarzi, M., Jeanmonod, D., & Sarnthein, J. (2008). EEG alpha distinguishes between cuneal and precuneal activation in working memory. *Neuroimage, 40*(3), 1296–1310. doi:10.1016/j.neuroimage.2007.12.048

Moazami-Goudarzi, M., Sarnthein, J., Michels, L., Moukhtieva, R., & Jeanmonod, D. (2008). Enhanced frontal low and high frequency power and synchronization in the resting EEG of parkinsonian patients. *Neuroimage, 41*(3), 985–997. doi:10.1016/j.neuroimage.2008.03.032

Nystrom, P. (2008). The infant mirror neuron system studied with high density EEG. *Social Neuroscience, 3*(3-4), 334–347. doi:10.1080/17470910701563665

Nystrom, P., Ljunghammar, T., Rosander, K., & von Hofsten, C. (2011). Using mu rhythm desynchronization to measure mirror neuron activity in infants. *Developmental Science, 14*(2), 327–335.

Odabaee, M., Tokariev, A., Layeghy, S., Mesbah, M., Colditz, P. B., Ramon, C., & Vanhatalo, S. (2014). Neonatal EEG at scalp is focal and implies high skull conductivity in realistic neonatal head models. *Neuroimage, 96*, 73–80. doi:10.1016/j.neuroimage.2014.04.007

Onton, J., Delorme, A., & Makeig, S. (2005). Frontal midline EEG dynamics during working memory. *Neuroimage, 27*(2), 341–356. doi:10.1016/j.neuroimage.2005.04.014

Oostendorp, T. F., Delbeke, J., & Stegeman, D. F. (2000). The conductivity of the human skull: Results of in vivo and in vitro measurements. *IEEE Transactions on Biomedical Engineering, 47*(11), 1487–1492. doi:10.1109/Tbme.2000.880100

Oostenveld, R., Fries, P., Maris, E., & Schoffelen, J. M. (2011). FieldTrip: Open source software for advanced analysis of MEG, EEG, and invasive electrophysiological data. *Computational Intelligence and Neuroscience, 2011*, 156869. doi:10.1155/2011/156869

Ortiz-Mantilla, S., Hamalainen, J. A., & Benasich, A. A. (2012). Time course of ERP generators to syllables in infants: A source localization study using age-appropriate brain templates. *Neuroimage, 59*(4), 3275–3287. doi:10.1016/j.neuroimage.2011.11.048

Pascual-Marqui, R. D. (2002). Standardized low-resolution brain electromagnetic tomography (sLORETA): technical details. *Methods and Findings in Experimental and Clinical Pharmacology, 24*, 5–12.

Pascual-Marqui, R. D. (2007). Discrete, 3D distributed, linear imaging methods of electric neuronal activity. Part 1: Exact, zero error localization. *https://arxiv.org/abs/0710.3341.*

Pascual-Marqui, R. D., Michel, C. M., & Lehmann, D. (1994). Low-resolution electromagnetic tomography - A new method for localizing electrical activity in the brain. *International Journal of Psychophysiology, 18*, 49–65.

Phan, T. V., Smeets, D., Talcott, J. B., & Vandermosten, M. (2018). Processing of structural neuroimaging data in young children: Bridging the gap between current practice and state-of-the-art methods. *Developmental Cognitive Neuroscience, 33*, 206–223. doi:10.1016/j.dcn.2017.08.009

Phillips, C., Rugg, M. D., & Friston, K. J. (2002a). Anatomically informed basis functions for EEG source localization: Combining functional and anatomical constraints. *Neuroimage, 16*(3 Pt 1), 678–695.

Phillips, C., Rugg, M. D., & Fristont, K. J. (2002b). Systematic regularization of linear inverse solutions of the EEG source localization problem. *Neuroimage, 17*(1), 287–301.

Reynolds, G. D. & Richards, J. E. (2005). Familiarization, attention, and recognition memory in infancy: An event-related potential and cortical source localization study. *Developmental Psychology, 41*(4), 598–615. doi:10.1037/0012-1649.41.4.598

Reynolds, G. D. & Richards, J. E. (2009). Cortical source localization of infant cognition. *Developmental Neuropsychology, 34*(3), 312–329. doi:10.1080/87565640902801890

Richards, J. E. (2003). Cortical sources of event-related potentials in the prosaccade and antisaccade task. *Psychophysiology, 40*(6), 878–894.

Richards, J. E., Sanchez, C., Phillips-Meek, M., & Xie, W. (2016). A database of age-appropriate average MRI templates. *Neuroimage, 124*, 1254–1259. doi:10.1016/j.neuroimage.2015.04.055

Richards, J. E. X. & Xie, W. (2015). Brains for all the ages: Structural neurodevelopment in infants and children from a life-span perspective. In J. Benson (Ed.), *Advances in child development and behavior* (Vol. 48, pp. 1–52). Elsevier.

Ritter, P., Moosmann, M., & Villringer, A. (2009). Rolandic alpha and beta EEG rhythms' strengths are inversely related to fMRI-BOLD signal in primary somatosensory and motor cortex. *Human Brain Mapping, 30*(4), 1168–1187. doi:10.1002/hbm.20585

Rush, S. & Driscoll, D. A. (1969). EEG electrode sensitivity - an application of reciprocity. *IEEE Transactions on Biomedical Engineering, 16*(1), 15-&. doi:10.1109/Tbme.1969.4502598

Ryynanen, O. R., Hyttinen, J. A., & Malmivuo, J. A. (2006). Effect of measurement noise and electrode density on the spatial resolution of cortical potential distribution with different resistivity values for the skull. *IEEE Transactions on Biomedical Engineering, 53*(9), 1851–1858. doi:10.1109/TBME.2006.873744

Sauseng, P., Hoppe, J., Klimesch, W., Gerloff, C., & Hummel, F. C. (2007). Dissociation of sustained attention from central executive functions: local activity and interregional connectivity in the theta range. *European Journal of Neuroscience, 25*(2), 587–593. doi:10.1111/j.1460-9568.2006.05286.x

Scherg, M. (1992). Functional imaging and localization of electromagnetic brain activity. *Brain Topography, 5*(2), 103–111.

Scherg, M., Bast, T., & Berg, P. (1999). Multiple source analysis of interictal spikes: goals, requirements, and clinical value. *Journal of Clinical Neurophysiology, 16*(3), 214–224.

Schiff, S. J. (2005). Dangerous phase. *Neuroinformatics, 3*(4), 315–318. doi:10.1385/NI:3:4:315

Sherlin, L. (2009). Diagnosing and treating brain function through the use of low resolution electromagnetic tomography (LORETA). In T. Budzynski, H. K. Budzynski, J. Evans, & A. Abarbanel (Eds.), *Introduction to quantitative EEG and neurofeedback: Advanced theory and applications* (2nd ed.), (pp. 83–102). Academic Press.

Stam, C. J. (2005). Nonlinear dynamical analysis of EEG and MEG: Review of an emerging field. *Clinical Neurophysiology, 116*(10), 2266–2301. doi:10.1016/j.clinph.2005.06.011

Stam, C. J., Nolte, G., & Daffertshofer, A. (2007). Phase lag index: Assessment of functional connectivity from multi channel EEG and MEG with diminished bias from common sources. *Human Brain Mapping, 28*(11), 1178–1193. doi:10.1002/hbm.20346

Thorpe, S. G., Cannon, E. N., & Fox, N. A. (2016). Spectral and source structural development of mu and alpha rhythms from infancy through adulthood. *Clinical Neurophysiology, 127*(1), 254–269. doi:10.1016/j.clinph.2015.03.004

Vinck, M., Oostenveld, R., van Wingerden, M., Battaglia, F., & Pennartz, C. M. (2011). An improved index of phase-synchronization for electrophysiological data in the presence of volume-conduction, noise and sample-size bias. *Neuroimage, 55*(4), 1548–1565. doi:10.1016/j.neuroimage.2011.01.055

Vorwerk, J., Cho, J. H., Rampp, S., Hamer, H., Knosche, T. R., & Wolters, C. H. (2014). A guideline for head volume conductor modeling in EEG and MEG. *Neuroimage, 100*, 590–607. doi:10.1016/j.neuroimage.2014.06.040

Vorwerk, J., Oostenveld, R., Piastra, M. C., Magyari, L., & Wolters, C. H. (2018). The FieldTrip-SimBio pipeline for EEG forward solutions. *Biomedical Engineering Online, 17*, 37. doi:10.1186/s12938-018-0463-y

Wendel, K., Vaisanen, J., Seemann, G., Hyttinen, J., & Malmivuo, J. (2010). The influence of age and skull conductivity on surface and subdermal bipolar EEG leads. *Computational Intelligence and Neuroscience, 2010*, 397272. doi:10.1155/2010/397272

Xie, W., Jensen, K. G., Wade, M., Kumar, S., Westerlund, A., Kakon, S. H., ... Nelson, C. A. (2019). Child growth predicts brain functional connectivity and future cognitive outcomes in urban Bangladeshi children exposed to early adversities. *BMC Medicine, 17*, art. no. 199. https://doi.org/10.1186/s12916-019-1431-5

Xie, W., Mallin, B. M., & Richards, J. E. (2017). Development of infant sustained attention and its relation to EEG oscillations: An EEG and cortical source analysis study. *Developmental Science, 21*, e12562. doi:10.1111/desc.12562

Xie, W., Mallin, B. M., & Richards, J. E. (2018). Development of brain functional connectivity and its relation to infant sustained attention in the first year of life. *Developmental Science, 22*, e12703. doi:10.1111/desc.12703

Xie, W., McCormick, S. A., Westerlund, A., Bowman, L. C., & Nelson, C. A. (2018). Neural correlates of facial emotion processing in infancy. *Developmental Science, 22*, e12758. doi:10.1111/desc.12758

Xie, W., Jensen, S. K. G., Wade, M., Kumar, S., Westerlund, A., Kakon, S. H., Haque, R., Petri, W. A., & Nelson, C. A. (2019). Growth faltering is associated with altered brain functional connectivity and cognitive outcomes in urban Bangladeshi children exposed to early adversity. *BMC Med, 17*(1), 199. https://doi.org/10.1186/s12916-019-1431-5

Yang, L., Liu, Z., & He, B. (2010). EEG-fMRI reciprocal functional neuroimaging. *Clinical Neurophysiology, 121*(8), 1240–1250. doi:10.1016/j.clinph.2010.02.153

Yang, L., Wilke, C., Brinkmann, B., Worrell, G. A., & He, B. (2011). Dynamic imaging of ictal oscillations using non-invasive high-resolution EEG. *Neuroimage, 56*(4), 1908–1917. doi:10.1016/j.neuroimage.2011.03.043

Yeung, N., Bogacz, R., Holroyd, C. B., Nieuwenhuis, S., & Cohen, J. D. (2007). Theta phase resetting and the error-related negativity. *Psychophysiology, 44*(1), 39–49. doi:10.1111/j.1469-8986.2006.00482.x

CHAPTER 17

FREQUENCY CHARACTERISTICS OF SLEEP

ALPÁR S. LÁZÁR, ZSOLT I. LÁZÁR, AND RÓBERT BÓDIZS

17.1 Introduction

Our life unfolds around the 24-hour clock, structured by the periodically alternating episodes of sleep and wakefulness. This alternation is paralleled by a series of changes in the pattern of EEG frequencies and the level of consciousness. Sleep is often regarded as a state of inactivity and complete relaxation, securing a general brain recovery from prior wakefulness. Nonetheless, the magnitude of variation in the pattern of EEG frequencies in sleep are net higher compared to wakefulness, suggesting that sleep is not a homogeneous physiological state. Indeed, sleep involves multiple transiently recurrent stages characterized by specific oscillatory phenomena, often utterly absent from the waking EEG (e.g. sleep spindles, slow oscillations). The rich diversity of neural rhythms associated with total or partial lack of consciousness in non-rapid (NREM) or rapid eye movement sleep (REM), respectively, conceives sleep as a unique time window allowing for a deep insight into the functional neuro-architecture of the brain. As a result of the remarkable advances in the acquisition and quantitative analysis techniques of EEG data experienced in the last decades, as well as the continuously growing interest in sleep research across the globe, our understanding of the nature and functional significance of sleep dependent brain oscillatory phenomena has considerably evolved. There is a wealth of evidence supporting a tight link between features of sleep-dependent EEG activity and waking experience and cognitive performance. What is more, sleep EEG is a promising source of reliable biomarkers of the disease process in neurodegenerative disorders affecting the brain. Indeed, while sleep EEG is highly sensitive to extrinsic factors (e.g., pharmacological agents, physical and electric stimulation), it is remarkably modulated by intrinsic factors as well (e.g., age, sex, genetic polymorphisms, mental health, neurological disorders). Prominent sleep dependent brain oscillations—sleep slow waves and sleep spindles—are central to the cognitive and physiological function of sleep and therefore have become the target of intervention studies aimed at boosting sleep quality and cognitive performance by those sleep oscillations. This holds the promise of a new line of future treatments aiming to slow down cognitive decline associated with aging or various neurological conditions, as well as to help the rehabilitation process in acquired brain injury. The current chapter attempts to provide an overview of the most characteristic EEG rhythms and their complex coalescence/interaction in making up human sleep. The chapter aims at summarizing the state of the art without being oblivious of the historical groundwork.

17.2 Human sleep and its EEG features: The multiplicity of rhythms

The cyclic alternation of sleep and wakefulness follows a regular pattern governed by several biological timekeeping systems that modulate a series of physiological processes, including brain activity and sleep propensity, and thus govern the timing, duration, and quality of sleep.

17.2.1 Circadian Rhythms

Circadian rhythms allow organisms to adapt more effectively to environmental changes associated with the rotation of the earth (e.g., light, temperature, food availability, predation risk) and are generated by a molecular clockwork present across most cells of the body. These individual clocks are synchronized by the master circadian clock located in the suprachiasmatic nucleus (SCN) of the hypothalamus, which is entrained to the 24-h light period through the retinohypothalamic tract (Mohawk et al., 2012). The circadian clock has a substantial impact on the organism. Its effects range from the modulation of gene expressions and protein synthesis (Mohawk et al., 2012) through a variety of physiological processes, including hormonal changes, immune functions (Kizaki et al., 2015), brain arousal and neural rhythms (Cajochen et al., 2013; Dijk & Czeisler, 1995; Lazar et al., 2015), to direct effects on cognitive performance, including learning and memory (Santhi et al., 2016; Schmidt et al., 2007; Smarr et al., 2014; Wright et al., 2012). Given the diurnal nature of wake-sleep periodicity, it is not surprising that circadian rhythms are at the heart of sleep regulation.

17.2.2 Sleep Homeostasis

The other major biological time-related system involved in sleep regulation is the so-called sleep homeostasis or process H (originally called process S), which governs

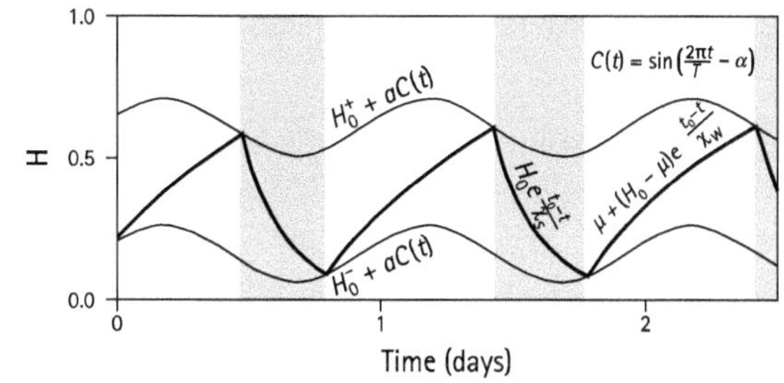

FIGURE 17.1 Sleep-wake cycles generated by the two-process model (Borbely, 1982).

Thick line represents homeostatic pressure, pressure building up during wake and decaying in sleep following an exponential dynamic; thin lines represent the circadian, C(t), modulated thresholds. White areas mark wake episodes, shaded areas mark sleep periods. Switches between wakefulness and sleep occur at intersection points between the homeostatic pressure and the circadian, C(t), modulated thresholds (thin lines).

Used parameters: T = 24 hours, α=-0.45, μ = 1, a = 0.1, χ_w = 18.2 hours, χ_s = 4.2 hours, H_0^-=0.17, H_0^+=0.6.

the balance between the duration and intensity of wakefulness and sleep. The longer one stays awake and the more stimulation the brain is exposed to, the greater sleep pressure is being accumulated, as reflected by increased slow oscillatory activity in the following sleep episode. Process H, however, does not track time awake or asleep in a linear manner, but builds up along an asymptotic function (see the thick lines within non-shaded regions in Figure 17.1), and following sleep onset it decreases exponentially (see the thick lines within shaded regions in Figure 17.1). Sleep homeostasis is closely linked to cellular metabolic-energetic processes as well as synaptic plasticity, and the most commonly used EEG marker for sleep homeostasis is delta or slow wave activity (SWA) ranging between 0.5 and 4 Hz in sleep (discussed more later). Sleep homeostasis is also under a strong age-dependent regulation, presenting a monotonically decreasing upper asymptote with aging, and is altered in depression and neurodegenerative disorders.

17.2.3 The Two-Process Model of Sleep Regulation

The effects of the two processes are synthesized in the so-called two-process model of sleep regulation (Borbely, 1982; Daan et al., 1984). The model postulates that the timing, duration, and the quality of sleep is defined by the interaction of the circadian and the homeostatic processes. That is, the switch from wake to sleep happens when the homeostatic pressure, building up exponentially towards an asymptotic value, reaches an upper threshold that consists of a mean value (defined by age for example), modulated by a circadian process (see the upper thin line in Figure 17.1.) (Borbely & Achermann, 1999; Skeldon et al., 2014). The switch from sleep to wake happens when homeostatic pressure decaying exponentially reaches a lower threshold (see the lower thin line in Figure 17.1.). Sleep duration and quality as well as numerous oscillations during sleep show both a homeostatic and a circadian modulation (Dijk & Czeisler, 1995; Lazar et al., 2015).

17.2.4 Ultradian Cycles

Besides circadian and homeostatic regulation, sleep is cyclical in itself—periods of NREM sleep alternate with REM along an ultradian rhythm in all mammals and presumably birds as well. In contrast with the largely universal period of the circadian rhythm, which is clearly determined by alternating light-dark cycles on Earth, the ultradian sleep cycles seem to be allometrically (i.e., in a body size-dependent manner) regulated and are mainly a function of the metabolism ratio of the species. The average (modal) duration of a complete NREM-REM cycle is around 90 minutes in adult human subjects repeating four to six times over a typical full night sleep, but the duration of a complete cycle in laboratory rats is around 11 minutes (Kaiser, 2013; Stupfel & Pavely, 1990; Savage & West, 2007). The time structure of the consecutive sleep cycles is complex and non-random. Cyclic variations of various forms of sleep EEG microstructural phenomena, usually associated with autonomic activations, emerge with infraslow periodicities (Halász & Bódizs, 2013). At a finer time scale, there are several EEG frequencies that are more-or-less specific for either the NREM or the REM phase of sleep. According to American Academy of Sleep Medicine (AASM) we can distinguish non-rapid eye movement sleep (N or NREM) that consists of three stages: N1 (NREM 1 or NREM stage 1), N2 (NREM 2 or NREM stage 2), and N3 (slow wave sleep, SWS), as well as rapid eye movement sleep (REM). As soon as we initiate sleep, the EEG, electromyographic (EMG), and electro-oculographic (EOG) activity goes through marked changes compared to wakefulness (see Table 17.1). The transition from lighter (N1 or NREM 1) to slow wave sleep (SWS) is associated with gradually decreasing muscle activity, heart and breathing rate, and body temperature. Moreover, EEG becomes gradually dominated by slower and slower frequency oscillations (in NREM 1 by theta, in NREM 2 by mixed frequencies including K-complexes and sleep spindles) until the entire brain will present continuous slow oscillations and delta activity (slow wave activity, SWA) in SWS. However, during REM, also called paradoxical sleep, the brain's electric and metabolic activity presents a significant increase as reflected in higher frequency, low amplitude mixed EEG frequencies resembling waking brain activity. REM is also accompanied by bursts of rapid eye movements, increased heart and breathing rates paralleled by muscle atonia. This is the stage of sleep when most dreams are reported. The scoring of sleep stages is based on conventions and, beyond characteristic EEG features, it requires other polygraphic readouts such as electrooculography (EOG) and electromyography (EMG) (Table 17.1 and Figure 17.2).

17.2.5 Infradian Rhythms

The circadian rhythm is far from being the slowest envelope of sleep rhythmicity. There are rhythms with periods significantly longer than 24 hours (infradian). Several studies suggest that circatrigintan (~30 days) periodicity in coherence with menstrual cycles

Table 17.1 EEG and polygraphic features of sleep stages according to the American Academy of Sleep Medicine (Berry et al., 2017; Iber & American Academy of Sleep, 2007).

Sleep stage	EEG (electroencephalography) features	EMG (electromyography) features	EOG (electrooculography) features
N1 (NREM 1)	Low-amplitude, mixed frequency activity in the range 4–7 Hz and attenuated alpha activity (<50% of the scoring epoch). Vertex sharp waves.	The chin EMG amplitude is variable, often lower than during wake.	Slow-rolling eye movements (conjugate, reasonably regular, sinusoidal eye movements with an initial deflection that usually lasts >500 ms).
N2 (NREM 2)	The presence of one or more K-complexes unassociated with arousal. One or more sleep spindles.	The chin EMG is of variable amplitude, but usually lower than in wake.	Usually, eye movement is not noticed during N2.
N3 (SWS)	SWS = slow wave sleep. Dominant slow wave activity: EEG waves of 0.5–2 Hz and >75 µV peak-to-peak amplitude, measured over the frontal regions, referenced to the contralateral ear or mastoid (F4–M1, F3–M2). ≥20% of a staging epoch consists of slow wave activity.	The chin EMG is of variable amplitude, often lower than in stage N2, sometimes as low as in REM sleep.	Usually, eye movements are not noticed during N3.
REM	Low-amplitude, mixed-frequency EEG activity without K-complexes or sleep spindles. Sawtooth waves may be present as well.	Low-amplitude chin EMG–baseline EMG activity, usually is at the lowest level of all recording.	Conjugated, episodic, sharply peaked, irregular eye movements.

A staging epoch is either 20 or 30 seconds long.

of females is prevalent and well measurable in human sleep EEG (Baker & Lee, 2018). Interestingly, a circatrigintan (circalunar) periodicity (~29.5 days) of sleep-dependent EEG activity has also been reported (Cajochen et al., 2013). According to this study, the lunar cycle had a significant impact on the frequencies between 0.5 and 16 Hz. More specifically, around the full moon, delta activity during NREM sleep decreased by 30%, while spindles also exhibited a marked reduction compared to the days around the new moon. This is intriguing given the role of delta waves and sleep spindles in the cognitive and physiological function of sleep (discussed later). Although there are other studies confirming an effect of lunar cycle on sleep duration and quality (Roosli et al., 2006;

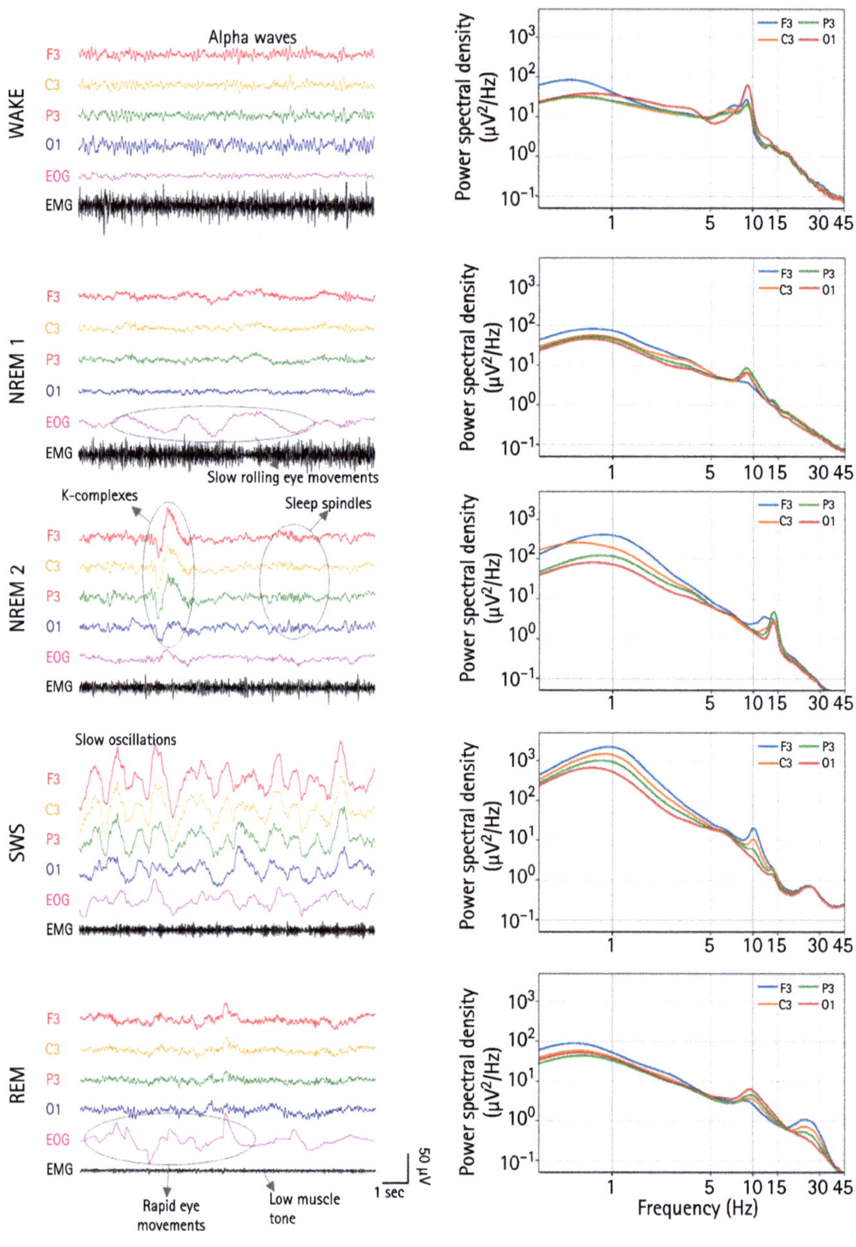

FIGURE 17.2 Polysomnographic signal characteristics of sleep stages. Characteristic electroencephalographic (EEG) and polygraphic samples (left panel) and corresponding power spectra (right panel) measured over the left frontal (F3), central (C3), parietal (P3), and occipital (O1) brain regions. EEG channels are referenced to the contralateral (A2 or M2) mastoid derivation. Artifact free segments were concatenated into a single time series. The power spectral density was calculated by employing Welch's periodogram method. (Average of the squared absolute value of the fast Fourier transform of four seconds long detrended and Hanning tapered windows with 50% overlap.) Subsequently a Gaussian smoothing was applied.

Turanyi et al., 2014), these findings are still controversial due to multiple subsequent studies failing to reproduce similar results (Cordi et al., 2014).

Details on these periodicities, rhythms, and regulatory mechanisms are discussed across the rest of the chapter. Table 17.2 provides an overview of the relevant frequencies.

Table 17.2 An overview of the frequency characteristics of human sleep

Frequency domain	Dominant period/ frequency	Principal EEG signature	Dominant method
Infradian (circatrigintan/ circalunar)*	29.5 days 0.392 μHz	Sleep architecture, oscillatory peak frequency of sleep spindles.	Repeated sleep EEG recordings analyzed by FFT/ spindle detection.
Circadian	~24 hours 11.5 μHz	Various: sleep architecture, modulation of slow wave and sleep spindle features.	Forced desynchrony protocol/free running rhythms with continuous/ repeated sleep EEG and FFT.
Ultradian–sleep cycle dynamics*	90 (80–120) min 185 μHz	Characteristic sequence of NREM and REM sleep EEG features.	FFT/wavelet/second order spectra/ modeling.
Infraslow/CAP	20–40 s 33 mHz	DC-EEG recorded fluctuations/cyclic simultaneous variation of sleep EEG patterns.	Full-band EEG/visual or automatic detection/ second order spectra.
Slow	1–10 s 1 Hz	Surface-negative large slow waves, cross-frequency coupling with higher frequency oscillations (NREM sleep).	FFT/slope and peak analysis.
Delta	0.25–1 s 2 Hz	Medium-sized, background slow waves during NREM sleep.	FFT/autoregressive spectrum/PAA
Hippocampal RSA*	0.33–0.66 s 2.25 Hz	Rhythmic slow activity in the hippocampus/ parahippocampal gyrus during REM sleep.	FFT/PAA
Alpha	83.3–125 ms 10 Hz	(Quasi-)sinusoidal waves with fluctuating amplitude and posterior dominance during eyes closed rest, microarousals during sleep and to some extent during tonic REM sleep.	FFT/wavelet/PAA
Sleep spindles/ sigma	62.5–90.9 ms 13 Hz	Waxing and waning oscillatory episodes in NREM sleep.	FFT/wavelet/PAA/ various detection methods.

(*continued*)

Table 17.2 Continued

Frequency domain	Dominant period/ frequency	Principal EEG signature	Dominant method
Beta	62.5–33.3 ms 22 Hz	Transient bouts of oscillations in REM sleep EEG ("REM beta tufts").	FFT/wavelet/PAA
Gamma	12.5–33.3 ms 40 Hz	Locally synchronized, low amplitude fast activity during SO up states and phasic REM.	FFT/wavelet
Ripples/HFO*	4–12.5 ms 100 Hz	Short (70 ms) group of waxing and waning oscillations (0.95 μV) primarily in NREM sleep.	Wavelet/visual and automatic detection.

FFT, Fast Fourier Transformation; PPA, Period Amplitude Analysis; *Evident allometry, human data provided

17.3 An overall picture of sleep EEG

Before providing a detailed analysis of specific sleep EEG oscillations, we present some core, defining features of the global picture of sleep EEG, like overall EEG amplitude, the amplitude vs. frequency relationship, power law scaling behavior of EEG spectra, H-exponent, and fractality.

EEG amplitude reaches its maxima during SWS (N3 or NREM stages 3&4 according to the Rechtschaffen and Kales criteria) in physiological circumstances (Agnew et al., 1967). Sleep EEG total power values and/or amplitudes are characterized by age-related decreases from childhood through the older ages (Astrom & Trojaborg, 1992). Moreover, women are characterized by higher overall sleep EEG power values as compared to men (Carrier et al., 2001; Bódizs et al., 2021). The age-related decrease in EEG amplitude power in children/adolescents may reflect a combination of non-neuronal factors, like the increase in skull thickness, and neural maturation (decrease in overall synaptic density). Women vs. men differences in overall sleep EEG amplitude are thought to mainly reflect differences in skull thickness (Dijk et al., 1989).

EEG rhythms with lower frequency predominate in NREM sleep Stage 2 and SWS as compared to REM sleep and wakefulness. The logarithm of sleep EEG amplitude has been shown to be a linear function (negative correlate) of the logarithm of frequency in NREM sleep of human volunteers (Feinberg et al., 1984). The steepness of the negative relationship between log amplitude and log frequency is higher in young as compared to elder subjects. The power law scaling behavior of the sleep EEG spectra is an expression of the same phenomenon. EEG spectra can in fact be well approximated by a power law decay function:

$$P_f = C \times f^{\alpha}$$

where P_f denotes power at frequency f, C is a constant expressing the overall amplitude of the EEG, and α is an exponent between −3 and −1 (Pereda et al., 1998). The power law character is demonstrated by the linear profile of the log-log representations of the spectrum in Figure 17.2. A lower absolute value of the exponent α is characteristic for wakefulness, REM sleep, and NREM Stage 2 sleep. Higher absolute α, that is, steeper spectral slope, is seen in deeper sleep (e.g., first sleep cycle, SWS, younger subjects) and more anterior recording sites. Figure 17.2 shows examples for steeper spectral slopes in deeper sleep and more anterior recording sites (characterized by locally more intense sleep). In the time domain, higher absolute α indicates a higher contribution of low-frequency components as compared to the high frequency ones. It is worth noting that the α value depends on the derivation scheme and the type of recording (surface scalp vs. electrocorticography) used for the electrophysiological recordings during sleep. Bipolar intracranial records are characterized by higher absolute values of α exponents (up to four during sleep) (Freeman & Zhai, 2009) as compared to surface scalp referential derivations (Pereda et al., 1998; Bódizs et al., 2021). Although the overall frequency composition of human sleep EEG can be approximated by power law functions through fitting a straight line to the log-log spectra and calculating its parameters, there are characteristic upward deviations from this generally descending trend, due to specific oscillatory events forming spectral peaks with specific frequencies (Bódizs et al., 2021; Chapter 23). In order to deliberately describe the power spectrum by taking into account its prominent peaks, we suggest the inclusion of a peak power function in the formula as follows (Bódizs et al., 2021):

$$P(f) = C \times f^{\alpha} \times P_{Peak}(f)$$

Peak power (P_{Peak}) at frequency f equals 1 if there is no peak and is larger than 1 if there is a spectral peak at that frequency. It has to be noted, that $P_{Peak}(f)$ is a whitened power measure, because it is characterized by roughly equal power along the frequency axis and is thus statistically independent from the spectral slope (α), which constitutes the colored noise part of the spectrum, characterized by a power-law type decrease along the frequency scale. $P_{Peak}(f)$ is also normalized in terms of amplitude differences as the term C is also partialed out from this function. By using these approaches, 191 spectral measures were effectively reduced to 4, which were efficient in characterizing known age-effects, sex-differences, and cognitive correlates of NREM sleep EEG (Figure 17.3; Bódizs et al., 2021).

Power law dependence is the hallmark of scale-free phenomena, such as fractals. Global sleep EEG features can be characterized by concepts related to self-similarity and quantified by a number of closely related measures (Ma et al., 2018; Stam, 2005). These include the correlation dimension (Grassberger & Procaccia, 1983), the scaling

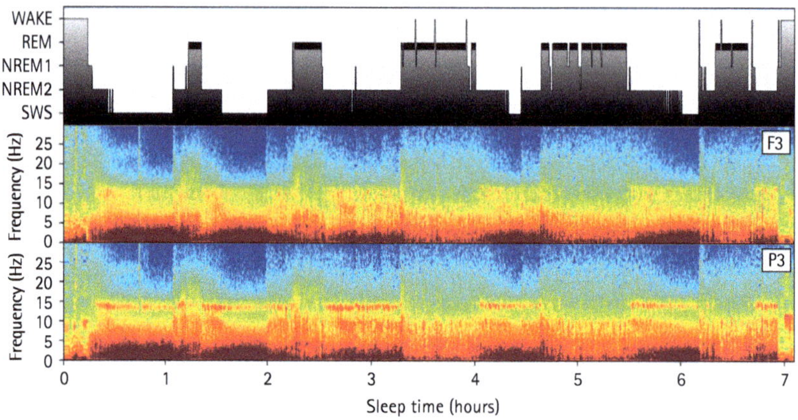

FIGURE 17.3 Evolution of EEG power spectral density during one night's sleep of a young adult measured over the left frontal (F3) and parietal (P3) brain regions. Warmer colors represent higher power (logarithmic scale). Temporal resolution is 10 seconds/pixel, each representing the average of the short-time Fourier transforms of four segments of 4-s long, Hanning tapered windows with 50% overlap.

exponent emerging from detrended fluctuation analysis (Gu & Zhou, 2006), and the Hurst exponent, H, which quantifies the smoothness of the signal (Hurst et al., 1965).

That is, a higher II, $0.5 < II < 1$, indicates that the changes in the EEG voltage tend to "go in the same direction", or maintain the direction of the deflection for longer periods (the signal is relatively smooth). In this case, there are positive correlations between the successive increments in voltage. If $0 < H < 0.5$, the overall correlation between successive increments is negative. The H-exponent is closely related to the exponent, α, of the power spectra as follows: $\alpha = 2H + 1$. Moreover, the H-exponent correlates with band-limited relative spectral power in a sleep stage- and EEG derivation-specific manner. The H-exponent of human sleep EEG is regularly larger than 0.5, which means that the sleep EEG (both NREM and REM) is characterized by long-term positive autocorrelation. As expected, the H values have been reported higher during the deepest parts of SWS compared to NREM2 and REM across all EEG derivations (Weiss et al., 2009).

In summary, sleep EEG spectra follows a power law function that is the state-specific features are best described by multiple frequency bands and their ratio, rather than single band measures. In addition, the sleep EEG has fractal structure which should be considered when characterizing different sleep-waking states in terms of EEG criteria.

17.4 Infraslow oscillations in sleep (ISO)

Infraslow oscillations in sleep (ISO) were originally defined as 0.01–0.1 Hz rhythmic or quasiperiodic changes in DC-EEG (also known as full-band EEG) recorded voltage. They can be regarded as extremely low-frequency Fourier components contributing additively to the signal. Short-circuiting of skin potentials by applying Ag/AgCl electrodes without Au coating and specific EEG gels are required in order to appropriately detect these oscillations (Vanhatalo et al., 2005). Another class of periodic phenomena in the ISO frequency range contribute multiplicatively to the signal, manifesting as modulations of the higher frequency components (Parrino et al., 2014, Lecci et al., 2017, Lazar et al.,2019). Evidence suggests that ISO modulates >0.5 Hz EEG activity, as well as interictal epileptic activity during sleep (Vanhatalo et al., 2004). Long lasting (1–5 s) occipital negative transients of 200–700 μV are seen in the DC-EEG records of preterm infants, which is probably an early form of ISO during sleep. Delta bursts have been seen to be embedded in these transients (Vanhatalo et al., 2002). The power of ISO correlates positively with the fMRI BOLD signal intensity of several subcortical regions, including the cerebellum, thalamus, and the basal ganglia. In contrast, paramedian heteromodal cortical BOLD signal intensities correlate negatively with ISO during sleep (Picchioni et al., 2011). The ISO is a global modulation of neural excitability. The generation of ISO is hypothesized to reflect the long lasting hyperpolarizations of thalamocortical cells, which might be due to the opening of some type of inwardly rectifying K^{+} channels. The latter are hypothesized to be opened by the activation of adenosine A1 receptors, while the endogenous ligand, adenosine, results from the degradation of adenosine triphosphate (ATP) after its release from glial cells (Hughes et al., 2011).

17.4.1 The Cyclic Alternating Pattern (CAP)

According to Parrino, Grassi, and Milioli (2014), "Arousals during sleep would be better conceived not as a single event but rather as part of the dynamic process, in which activating phenomena of different intensity and morphology are organized in CAP sequences". As the inter-arousal distance in CAP scoring is defined as 2–60 s, CAP can be considered roughly an EEG oscillation in the ISO frequency range, consisting in the alternation of two phases: the A phase and the B phase. The A phase is thought to be the reactive part characterized by the activation of several EEG, and in some cases autonomic, phenomena, whereas the B phase is a background activity characteristic of the actual stage of NREM sleep. CAP A phase has three subtypes. Increasing activation is a characteristic feature of these subtypes as follows: A1 contains only slow EEG components (SO), A2 is a mixture of slow and fast (alpha, beta) EEG features, while A3 is characterized by purely fast EEG components (Parrino et al., 2014). It is worth noting that A1 was recently considered a sleep defense mechanism, a so-called delta injection providing an instant sleep homeostasis in coherence with the actual sleep need and the sleeper's inner and outer environment (Halász & Bódizs, 2013). The relative time spent in CAP as compared to non-CAP sleep indicates the level of microstructural instability of sleep, with more time spent in CAP indicating more instability.

17.4.2 Infraslow Oscillations (ISO) in Slow Wave Activity and Sleep Spindles

Indirect measures (e.g. spectrum of spectra) of both ISO and CAP are reported in several studies focusing on the periodicity of sleep EEG phenomena in humans. For example, sleep EEG slow wave activity (0.5–4.5 Hz) was shown to be characterized by a 20–30 s periodicity in human volunteers (Achermann & Borbely, 1997), which might indicate a CAP-like fluctuation of slow wave amplitude (involving A1 and/or A2 phases). Sleep spindle activity (see below) was shown to be organized in rhythmic wave sequences according to multiple frequencies. One of these frequencies falls in the ISO range and implies a sleep spindle periodicity of 50–100 s (0.01–0.02 Hz) in both mice and humans (Lazar et al., 2019; Lecci et al., 2017). Using the spectrum of spectra approach (calculating the power of the multivariate time series obtained from the short-time Fourier transform for the signal), it was shown that the ISO (0.01–0.02 Hz) of the sigma (10–15 Hz) power (frequency range corresponding to sleep spindles) is coordinated with cardiac oscillations and composed of two phases: one with enhanced and one with decreased environmental alertness (Lecci et al., 2017). Moreover, the intensity of ISO in the sigma power predominate in NREM 2 and over the posterior regions in humans, and is positively associated with declarative memory recall performance following sleep (Lecci et al., 2017). Recently, ISO in sleep spindle activity were also quantified. The novel method computes the infrapower of an on/off binary square signal corresponding with individually detected sleep spindles. This infrapower is contrasted with that from a computation where the same sleep spindles and inter-spindle lapses are randomly reshuffled so that large-scale temporal structures are suppressed. The method validated using surrogate data strictly addresses the large-scale temporal structures in the sequence of spindle events, does not manifest any obvious biases and is extensible to other phasic phenomena such as slow waves. The study showed that ISO were strongest in the fast sigma band (13–15 Hz). Further, a robust infra peak of individually detected fast sleep spindles was identified. This intensity and frequency of this peak were modulated by topography, sleep stage, and sleep history. The power of ISO in fast sleep spindles is most prominent in the centro-parieto-occipital brain regions, left hemisphere, and second half of the night, independent of the number of sleep spindles. The frequency of ISO is higher over the frontal and temporal brain regions and in the first half of the night (Lazar et al., 2019).

17.5 The sleep slow oscillation (SO)

Slow EEG waves are among the most well established neurophysiological features of sleep (Figure 17.2). Their presence was apparent in the very first all-night sleep EEG records (Loomis et al., 1937). The most prominent type of slow wave activity in the sleep EEG is the slow oscillation (SO; 0.1–1 Hz). The first comprehensive description and experimental analysis of the sleep SO was given by Mircea Steriade and his colleagues (Steriade, Contreras et al., 1993; Steriade et al., 1993a, 1993b). Cycles of the SO are characterized by a more-or-less regular sequence of neuronal silence/disfacilitation, characterized by a surface negative deflection (down state) and a burst-like neuronal firing, taking the form of a surface positive component (up state). Down states are characterized by a non-selectively reduced neural oscillations, while the up states are accompanied by higher frequency rhythms (delta waves, spindles, gamma oscillations, ripples). It is worth noting that down states are more synchronized over the cortex than the up states. The preferential time for the emergence of SO events is slow wave sleep (SWS) or Stage N3, although they are present in NREM Stage 2 sleep as well (to a lower extent). Sleep SO is a travelling wave. That is, the majority of the events are generated over the prefrontal cortex. The antero-posterior direction is the most common, but not exclusive, route of travelling, with a speed of 1.2–7 m/s (Massimini et al., 2004).

SO emergence is non-monotonic during the deepening of sleep. Spontaneous or reactive (stimulation-triggered) SO down states emerge in any part of NREM sleep, except Stage 1. Some of these events are known as K-complexes, composed predominantly by a singular down state (Cash et al., 2009). K-complexes can be elicited by any kind of stimulation. Spontaneous events are thought to emerge at an average frequency of 1–1.7 events per minute. The highest K-complex emergence is seen in the ascending Stage 2 sleep periods of NREM phases (Halasz, 2005; Halász & Bódizs, 2013).

In addition to the K-complexes, NREM Stage 2 sleep is rich in SO cycles. These cycles are frequently grouped in the A phases of the CAP, which means that the gradual increase of SO during the descending slope (deepening part) of sleep cycles is periodic and burst-like (Ujma et al., 2018).

The characteristic features of the SO depend on several factors, but most of these factors (e.g., age, sleep cycles, sleep deprivation) converge as sleep pressure. High sleep pressure (early sleep, prepubertal age, rebound sleep) is characterized by higher amplitude, higher frequency (up to 1 Hz and perhaps sometimes even slightly above), and steeper SO events, which are less-frequently interrupted by secondary peaks. Aging is characterized by decreased sleep pressure, and thus decreased amplitude, frequency, and steepness of SO events. The age-related decrease in SO/delta power is attenuated in subjects with high intelligence as compared to subjects with average intelligence (Potari et al., 2017).

Sleep SO are generated in the cortex (Steriade & Amzica, 1998). Likewise, cortico-cortical connections are needed to synchronize sleep SO over cortex. However, SO spread over and are expressed in several subcortical and non-(neo)cortical structures, including the thalamus, the basal forebrain, and the hippocampus. Powerful inward transmembrane currents, mainly localized to the supragranular layers, characterize SO up states in humans. Although neuronal firing is present, it is sparse, and significantly lower, when compared to animal studies (Csercsa et al., 2010). Suppressed cell firing and hyperpolarizing currents are found in the SO down state (Csercsa et al., 2010).

17.6 Delta waves

Delta waves (activity) are medium-sized EEG waves of 1–4 Hz, prevailing in the NREM phase of sleep (maximal delta is present in SWS). The term delta activity (the spectral component corresponding to the frequency range of delta waves) is interchangeably used with slow wave activity (SWA), which normally refers to a slightly broader frequency range (0.5–4.5 Hz). Thus, delta frequency is higher than SO frequency, whereas its amplitude is lower than SO amplitude. In addition, SWA usually refers to a composite spectral measure of SO and delta activity.

Delta activity and SWA peak in the middle of consecutive sleep cycles and present a strong homeostatic regulation: their amplitude (power) declines across successive sleep cycles and increases linearly as a function of the amount of pre-sleep wakefulness (Campbell et al., 2006). Delta waves also present a strong fronto-posterior gradient with higher values over the frontal region and are associated with local sleep regulation. Those brain regions present higher homeostatic sleep pressure as reflected by higher sleep dependent delta activity, which undergoes more intense stimulation/plastic changes during the wake period prior to sleep onset (Huber et al., 2007; Huber et al., 2006; Huber et al., 2004; Kattler et al., 1994). Neural synchronization, as mediated by synaptic strength that increases with presleep stimulation and decreases with slow wave sleep, is thought to be one of the mechanisms underpinning local sleep and the homeostatic sleep drive. This theory has received significant experimental support, framing the synaptic homeostasis theory of sleep (Tononi & Cirelli, 2014).

Slow waves in sleep within the broader delta wave frequency range (0.5–4 Hz) show marked circadian regulation in a topographical manner, albeit under significant homeostatic regulation. The slope of delta waves that had been considered a surrogate marker of sleep homeostasis and shown to reflect synaptic strength (steeper slope = higher synaptic strength and higher sleep pressure) (Vyazovskiy et al., 2009) presents greater circadian than homeostatic modulation over the central brain regions (Lazar, et al., 2015). This suggests that synaptic strength and plasticity are both under a homeostatic use dependent modulation, as well as a direct circadian modulation (Frank & Cantera, 2014).

There is ongoing debate whether sleep EEG delta waves should be considered separate entities and not only reflections of the relatively sharper surface negative waves of the SO, or reflections of the faster up-down alternation caused by increased sleep pressure. Historically, delta waves have been considered separate entities (Steriade, Contreras, et al., 1993; Steriade et al., 1993a, 1993b). Two separate origins of delta activity were hypothesized: clock-like thalamocortical delta oscillations and cortical delta oscillations (Amzica & Steriade, 1998). The view of a distinct SO and delta frequency range is supported by the clear differences in delta vs. SO generation mechanisms. In contrast to SO, delta waves are generated in the thalamocortical system with the involvement of T-type Ca^{2+} channels (Crunelli et al., 2018; Lee et al., 2004). This mechanism is similar to the sleep spindle rhythmogenesis and is hypothesized to differ in the overall level of hyperpolarization of thalamocortical neurons. In the case of delta generation, the hyperpolarization of the wave is higher than in the case of sleep spindle generation.

17.7 Theta waves and rhythmic hippocampal slow activity

Perhaps one of the most persistent focuses of electrophysiological research is the phenomenon of hippocampal rhythmic slow activity (RSA), or the so-called hippocampal theta rhythms during exploratory behavior and REM sleep of different mammalian species. However, the term theta is confusing as there is evidence for clear allometry in the frequency of hippocampal RSA. Animals with larger brains are characterized by lower frequency hippocampal RSA (Blumberg, 1989), which means that human hippocampal RSA measured in REM sleep falls in the delta (1.5–3 Hz), but not in the theta, frequency range (Bodizs et al., 2001). The confusion in the literature is further increased by the fact that there are indeed clear theta frequency oscillations during the REM phases of sleep in humans, but these oscillations are of frontomedial/anterior cingular origin (frontal midline theta), and not hippocampal, as it was implicitly assumed.

17.7.1 REM Sleep Hippocampal RSA

Given the peculiarities of human neuroanatomy, direct (para)hippocampal records can only be done in clinical situations, primarily during (semi-) invasive presurgical evaluation procedures of patients suffering from pharmaco-resistant epilepsy. Such records clearly indicate that human hippocampal RSA emerges during REM sleep and its modal frequency is roughly 2 Hz (Bodizs et al., 2001) or 3 Hz (Moroni et al., 2012). These findings confirm the theory of the allometric nature of hippocampal RSA (Blumberg, 1989). There is no evidence for a phasic vs. tonic difference in REM sleep hippocampal RSA. In addition, interhemispheric coherence of REM sleep human hippocampal RSA is low (Bodizs et al., 2001; Moroni et al., 2012). Indeed, REM sleep hippocampal RSA is phase coupled with high frequency (20–240 Hz) oscillations, as has been reported in several rodent studies (Clemens et al., 2009).

17.7.2 Frontal Midline Theta Activity during NREM Stage 1 and REM Sleep

Theta rhythms of frontal origin are evidently present in the period of sleep initiation (NREM Stage 1), as well as during the REM phase of sleep. Based on the dominant topography of this activity, some authors call it frontal midline theta activity. The term refers to a train of rhythmic waves around the frequency of 6 Hz. There is evidence for the relationship between frontal midline theta and dreaming (Inanaga,

1998). There is preliminary evidence for the anterior cingulate origin of the frontally recorded theta rhythm in REM sleep (Nishida et al., 2004; Uchida et al., 2003). It has been shown that REM sleep EEG frontal theta activity is involved in emotional memory consolidation (Goldstein & Walker, 2014) and the incorporation of recent waking-life experiences in dreams (Eichenlaub et al., 2018). In addition, resilience against PTSD is indexed by higher REM sleep EEG theta activity (Cowdin et al., 2014). More recently a decrease in REM-dependent theta activity has been identified as a marker of looming neurodegeneration in pre-manifest and manifest patients carrying the mutant gene for Huntington's disease (Lazar, Panin, et al., 2015; Piano et al., 2017).

17.8 Sawtooth waves during REM sleep

Sawtooth waves are 1.5–5 Hz (mean frequency: 2.5 Hz) notched triangular waves that occur in bursts with a mean duration of 7 s (range 2–26 s) in the frontocentral region of REM sleep EEG records (Figure 17.4). The bursts have the appearance of teeth on a saw (Pearl et al., 2002). Evidence suggests the lower level of sawtooth wave activity characterizes the first REM sleep episode (Pearl et al., 2002), as well as the early emergence of sawtooth waves during the NREM-REM transition (Sato et al., 1997).

17.9 Alpha waves during sleep

When compared to presleep quiet wakefulness, the level of EEG alpha activity is attenuated during sleep. Various frequency limits for defining sleep EEG alpha activity are seen in the literature. The major issue is the differentiation of alpha from sleep spindles (Section 17.11). Realistic frequency boundaries in healthy adults are 7–10, 7–11, or 8–11 Hz. However, sleep spindles (as well as waking alpha) are slower in prepubertal children, thus the above criteria cannot be automatically transferred to all ages. Individual alpha peak frequency in the presleep period is lower than the spindle peak frequencies; thus presleep alpha peaks should be used as individual anchors for a deliberate and individualized frequency determination.

Continuous or intermittent alpha waves are most frequently considered as signs of (micro)arousals. Transient increases in alpha power are parts of the A2 and A3 types of microarousals embedded in CAP sequences, which are characterized by mixed low and high frequency and exclusively high frequency EEG components, respectively, indicating neural activation (Parrino et al., 2014). Moreover, ongoing posterior alpha activity can be considered an index of covert waking brain activity during sleep, indicating shallower/more perturbated sleep and higher reactivity to external stimuli (McKinney et al., 2011). A peculiar mix of alpha activity with sleep EEG delta waves is termed alpha-delta sleep. Alpha-delta sleep consists of epochs made up by delta wave-dominated

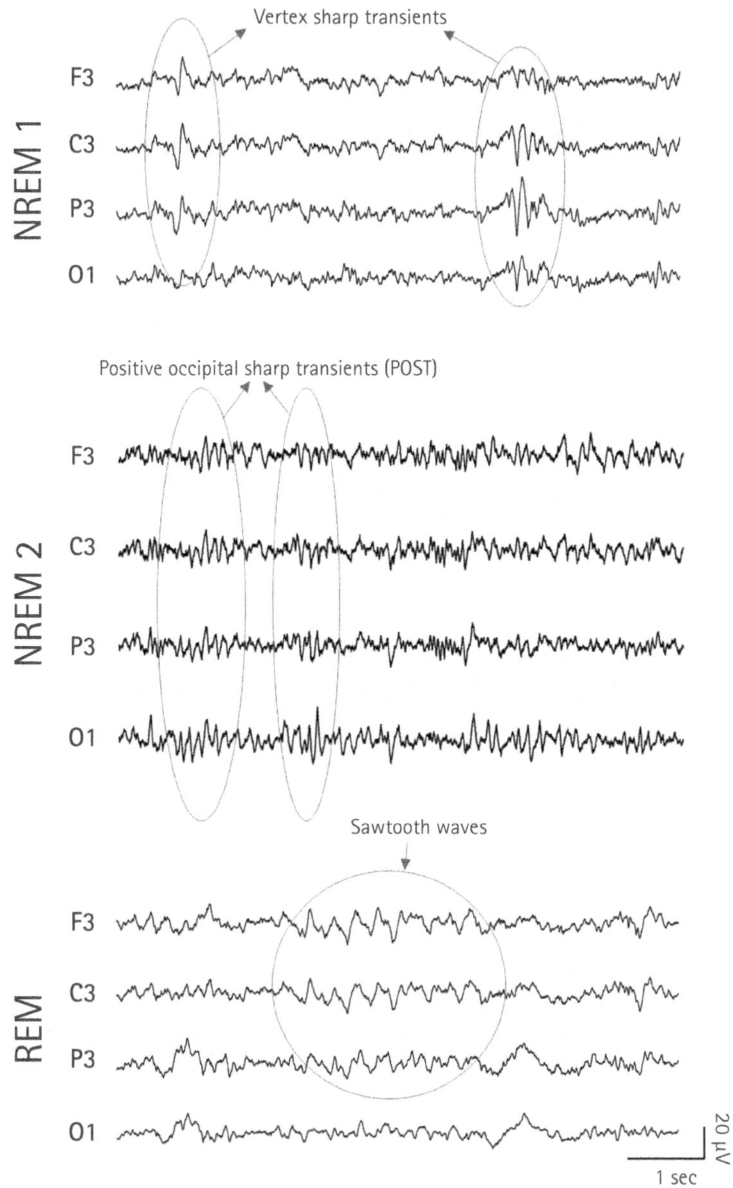

FIGURE 17.4 EEG samples with characteristic EEG oscillations including Vertex sharp transients in NREM 1, positive occipital sharp transients (POST) in NREM 2 and Sawtooth waves in REM sleep.

segments (5–20%) mixed and filled up with large amplitude alpha waves (Hauri & Hawkins, 1973). Fibromyalgia syndrome (Branco et al., 1994), major depressive disorder (Jaimchariyatam et al., 2011) and psychological features, like insecure attachment (Sloan et al., 2007), are associated with more alpha-delta waves during sleep.

The issue of alpha activity in REM sleep is somewhat more complex, however. Although, posterior alpha is a reliable measure of instantaneous sleep depth and responsiveness in REM sleep (McKinney et al., 2011), alpha activity is higher and more synchronized in tonic (no actual rapid eye movements), as compared to phasic (actual bursts of rapid eye movements) REM episodes (Simor et al., 2018; Simor et al., 2016). In addition, there is evidence of increased high alpha frequencies (10–14 Hz, which are not sleep spindles) in the REM sleep records of subjects with frequent nightmares (Simor et al., 2013).

17.10 Vertex sharp transients and positive occipital sharp transients (POSTs)

Sharp negative waves with a maximum at the vertex are frequently seen during the initiation of sleep, most frequently in NREM Stage 1 sleep (Figure 17.4). Although these transients, called vertex sharp waves, resemble some interictal epileptic discharges, these sharp waves are physiological and do not indicate the presence of any epileptic processes, even if a complete lateralization of the phenomenon is seen (Brenton & Mytinger, 2015). Vertex sharp waves and N300 components of averaged EEG responses to stimuli with different modalities have been shown to have common sources (Colrain et al., 2000).

Positive occipital sharp transients of sleep are also parts of the EEG picture of shallower phases of sleep, however, unlike vertex sharp waves, they are often seen in NREM Stage 2 sleep as well (Figure 17.4). The first thirty minutes of NREM sleep is the period of preferred emergence of positive occipital sharp transients (Egawa et al., 1983). The occipital maximum is hypothesized to be related to some activations in the visual areas of the brain.

17.11 Sleep spindles

Putative sleep spindle-like waveforms were first described by Hans Berger, the founder of human EEG, as oscillatory episodes with a 72-ms wavelength (equaling 13.89 Hz) in a female schizophrenic patient in (rectally induced) tribromoethanol narcosis (Berger, 1933). Soon thereafter, Alfred Lee Loomis described and named the sleep spindles as being very regular 14-Hz bursts, lasting 1–1.5 seconds, adding that the amplitude increases and decreases steadily before and after reaching its maximum, respectively (Loomis et al., 1935) (Figure 17.2). In the next decades, sleep spindles were mainly considered as hallmarks of Stage 2 sleep, but experimental studies clearly indicated their role in sleep protection or thalamic gating, also known as shock-absorbing function.

17.11.1 Frequency of Sleep Spindles

The frequency of sleep spindles (sigma rhythm) has been established as 14 Hz (Loomis et al., 1935; Brazier, 1961), 12–14 Hz (Rechtschaffen & Kales, 1968), 11–15 Hz (A glossary of terms most commonly used by clinical electroencephalographers, 1974; Noachtar et al., 1999), 12–16 Hz (Hori et al., 2001), and 11–16 Hz (Berry et al., 2017; Iber & American Academy of Sleep, 2007; Kane et al., 2017; Silber et al., 2007). Based on their frequency and dominant topography, sleep spindles were categorized as frontal slow (12 Hz) and centro-parietal fast (14 Hz), as well as 10 Hz global spindles (Gibbs & Gibbs, 1958). However, the 10-Hz global sleep spindles were later considered as alpha waves during sleep; thus, the existence of this type of spindle wave remained controversial (Jankel & Niedermeyer, 1985). Other studies have provided evidence that the slow spindle frequency decelerates as sleep deepens (Andrillon et al., 2011); thus, the 10-Hz spindle type of Gibbs and Gibbs (1958) might constitute a decelerated slow spindle during the deepening stages of sleep. Among the factors determining/modulating sleep spindle frequencies, genetics, age, sex, circadian phase/time of day, core body temperature, menstrual cycle phase, and sleep depth are well known (Andrillon et al., 2011; De Gennaro et al., 2008; de Zambotti et al., 2015; Dijk, 1999; Ujma et al., 2014). A common feature of sleep spindles is a slowing down of spindle frequency in successive oscillatory cycles, resulting in a frequency difference of 0.1–0.4 Hz, between the first and the last oscillatory cycle (Schonwald et al., 2011).

17.11.2 Morphology of Sleep Spindles

Besides frequency, sleep spindles are characterized by wave morphology and/or envelope. The term spindle was used in order to characterize the shape of the envelope of this wave group, characterized by an initial crescendo followed by decrescendo in amplitude reflecting the engagement of an increasing number of neurons into the rhythmic spindle frequency activity pattern followed by a gradual decrease. There is little research focusing on this aspect of sleep spindles. Some observations indicate that this spindle morphology is more an exception than the rule (Jankel & Niedermeyer, 1985). The band-pass filtered signals, as well as the so-called grouping of sleep spindles by the slow (<1 Hz) oscillation (Molle et al., 2002; Steriade & Amzica, 1998), clearly support that sleep spindle frequencies are characterized by progressively increasing amplitude followed by a gradual decrease. Regarding the polarity, sleep spindle-like waveforms induced by rhythmic thalamic stimulation in the midline areas have been shown to be surface negative (Dempsey & Morison, 1941). In the case of scalp EEG recordings, there is much controversy regarding the reference montage. The morphology of the single wave constituents of sleep spindles (single cycles) is most often tacitly considered as quasi-sinusoidal. However, there are important specifications in this regard: sleep spindles have been shown to reflect both subthreshold depolarizations of apical dendrites (smooth surface negative waves) and apical dendritic depolarization (sharp negative-positive waves) (Urakami et al., 2012). Sleep spindles with unipolar sharp peaks have been registered near the parahippocampal region in epilepsy patients undergoing presurgical examination with foramen ovale electrodes. Sharp peaks have been associated with nested (phase-coupled) 80–140 Hz ripples in these patients (Clemens et al., 2011). Sleep spindles with unusual sharp negative peaks have also been observed in infancy, up to 2 years of age (Fois, 1961). However, there is no direct evidence indicating that the above findings reflect enhanced neuronal firing during parahippocampal and cortical sleep spindles in epilepsy patients and infants, respectively.

17.11.3 Duration of Sleep Spindles

Since the publication of the scoring rules for sleep by Rechtschaffen and Kales (1968), the shortest duration of individual sleep spindles is usually defined at half-seconds. However, this is just a consensual value. In their original work describing and characterizing sleep spindles, Loomis and colleagues (1935) mention sleep spindles as short as third- or even quarter-seconds. There is no theoretical work indicating that a half-second length is a critical minimum value for a sleep spindle per se. Different experimental settings and subjects are characterized by different sleep spindle durations. The most important factor determining sleep spindle duration is age: unusually long (up to 10 seconds) sleep spindles are seen in infants, while a significant shortening progressively emerges during maturation and development, until typical 1–2-s-long spindles emerge (Spinosa & Garzon, 2007). Initial thalamic inhibition (Bartho et al., 2014) and corticothalamic feedback (Bonjean et al., 2011) are hypothesized to play a crucial role in the precise regulation of sleep spindle duration.

17.11.4 Amplitude of Sleep Spindles

The amplitude of sleep spindles depends on several specific and non-specific factors. Some trivial non-specific factors include electrode impedance, derivation (reference), and skull thickness. Specific factors include age, sleep cycle effects, sleep homeostasis, and circadian phase. In addition, several sleep spindle measures, among which amplitude is perhaps the most pertinent, depend on the method used for and criteria of defining sleep spindles. Sleep spindle amplitude is shown to be the direct function of thalamocortical entrainment of neurons to spindle frequency. There is evidence supporting the view that the more widespread a spindle is, the higher its amplitude (Andrillon et al., 2011; Dempsey & Morison, 1941).

17.11.5 Topography of Sleep Spindles

The characteristic topography of sleep spindles is twofold. The prefrontal cortices are the sites of origin of slow (~12 Hz) sleep spindles, whereas parietal cortex is characterized by the fast type (~14 Hz) sleep spindles. The central region expresses a roughly equivalent amount of slow and fast sleep spindle power.

17.11.6 The Density and Periodicity of Sleep Spindles

The ability to elicit local sleep spindles is enhanced by increasing local neural plasticity, namely, the induction of long-term potentiation (Werk et al., 2005). There are two major hypotheses on the specific mechanism of sleep spindle initiation in the thalamocortical network. The classical hypothesis was put forward by (Destexhe et al., 1998). In this view, SO up states impinge a high level of excitation in the thalamic reticular neurons by thalamoreticular (descending, thalamofugal) connections. Consequently, thalamic reticular neurons induce a high level of hyperpolarization in thalamocortical neurons, which in turn begin to fire in rebound depolarization sequences. Ascending thalamocortical pathways transfer the effects of rebound depolarizations to the cortex.

Another hypothesis implies the leading role of SO down states in the induction of a sufficient thalamic hyperpolarization by descending thalamofugal tracts. Thus, thalamic down states lag cortical ones. Thalamic sleep spindles initiate at the times of thalamic down states; they are then transferred to the cortex with a lag, which results in the coincidence of cortical sleep spindles with the forthcoming up state of the ongoing SO (Mak-McCully et al., 2017). Evidence suggests that Stage 2 sleep spindles emerge with a modal frequency of 0.25 Hz, that is the modal inter-spindle interval is 4 s (3–5 s) long (Achermann & Borbely, 1997; Evans & Richardson, 1995; Kubicki et al., 1986). According to some reports, the 4-s periodicity of sleep spindles is a characteristic feature of fast, but not of slow, sleep spindles (Spieweg et al., 1992). With the deepening of sleep, spindles become less dense, but persist during the whole NREM period. SO Up states are the preferred periods of emergence of sleep spindles in Stage 3 sleep (Fell et al., 2002; Molle et al., 2002; Staresina et al., 2015). Given the fact that the 0.5 s minimum duration criteria for sleep spindles is still in use, the spindle-like waveforms that are nested in the depolarized phases of the SO are often not considered as sleep spindles per se, but rather sleep spindle-like or spindle frequency activity. Thus, sleep spindles and/or sleep spindle-like oscillations are phase coupled with the SO; however, the frontal slow types and the parietal fast type of sleep spindles are characterized by distinct coalescence patterns. Fast sleep spindles are shown to prefer the peaks of the SO up states, whereas slow sleep spindles most frequently emerge during the transitional phase of the up and the down states of the SO (Molle et al., 2011). This is in coherence with the finding that in case of the co-occurrence of the two types of sleep spindles, the fast type precedes the slow one by several milliseconds (Zygierewicz et al., 1999).

17.11.7 Sleep Spindle Rhythmogenesis

Sleep spindles are shown to reflect the recurrent hyperpolarization rebound sequences of thalamocortical cells, the inhibition of which is caused by the NREM-dependent activation of GABAergic neurons in the reticular thalamic nucleus (Huguenard & McCormick, 2007; Steriade, 2000). The synchronization of individual thalamo-cortico-reticular loops is thought to be performed by descending inputs of the concerted up and down states of the cortical SO.

17.11.8 Sleep Spindles and Cognition

Similar to sleep slow waves, there is a wealth of evidence supporting the association between sleep spindles and cognition. On the one hand, sleep spindles have been identified as highly sensitive to individual differences often related to general and specific learning abilities. Thus, sleep spindles have been repeatedly (Ujma et al., 2018), albeit not unequivocally (Pesonen et al., 2019), shown to correlate with measures of general intelligence and memory performance (Bodizs et al., 2014; Bodizs et al., 2005; Chatburn et al., 2013; Clemens et al., 2005, 2006; Geiger et al., 2011; Hoedlmoser et al., 2014; Lustenberger et al., 2012; Schabus et al., 2006; Tessier et al., 2015; Ujma et al., 2016). On the other hand, the targeted modulation of sleep spindle activity using Transcranial Alternating Current Stimulation (tACS) and closed loop auditory stimulation has resulted in significant sleep dependent enhancement of motor learning (Lustenberger et al., 2016) and spatial navigation performance (Shimizu et al., 2018), respectively. Changes in sleep spindles (e.g. reduction in the amplitude and/or incidence of sleep spindles) have been shown to predict the development of dementia in Parkinson's disease (Latreille et al., 2015) and are associated with various neurodevelopmental disorders, such as autism spectrum disorders (Godbout et al., 2000; Limoges et al., 2005) and Williams syndrome (Bodizs et al., 2012), as well as with various mental health disorders including schizophrenia (Schilling et al., 2017), bipolar disorder (Ritter et al., 2018), and major depression (Plante et al., 2013).

Overall, there is convincing evidence that sleep spindles are generated in the thalamus, grouped by cortical slow oscillations (Steriade & Amzica, 1998). Spindles are modulated by infraslow oscillations (Lazar et al., 2019; Lecci et al., 2017) and biological timekeepers (Cajochen et al., 2013; Dijk & Czeisler, 1995; Knoblauch, Martens, Wirz-Justice, Krauchi, & Cajochen et al., 2003; Knoblauch, Martens, Wirz-Justice, & Cajochen, 2003; Knoblauch et al., 2005). They are dependent on the intrinsic features of the functional neuro-architecture of an individual's brain, and, as such, are inherently linked to cognition and mental health. However, in spite of multiple competing theories, the neural underpinnings of the associations of sleep spindles with cognition are not clear. One theory suggests memories are reinstated by spindle events and further reprocessed during subsequent spindle refractory periods (Antony et al., 2018). Another theory posits that sleep slow oscillations preferentially consolidate the stronger memory traces, which also leads to the extinction of weak memories. Sleep spindles are thought to contribute to a slow but reliable consolidation of the multiple competing memories (Wei et al., 2018).

17.12 Beta and Gamma Activity during Sleep

Historically, beta waves have been defined as EEG frequencies above 13 Hz (Kozelka & Pedley, 1990). However, it is now clear that sleep spindles constitute a specific phenomenon, thus spindles and beta oscillations have to be defined separately in spite of the clear overlap in the frequency of these phenomena. Consequently, frequencies higher than sleep spindles, but lower than gamma (16–30 Hz), can be considered as beta activity in NREM sleep, whereas a somewhat broader frequency is acceptable in states without sleep spindling, like REM sleep and wakefulness (13–30 Hz). NREM sleep beta EEG activity with frontocentral maxima is an integral part of sleep initiation, especially in children (Kellaway & Fox, 1952). Moreover, the up states of the SO are associated with an increase in beta/gamma activity (up to 80 Hz), indicating the entrainment of higher frequency oscillations by SO commonly referred to as grouping effect (Neske, 2016). SO/delta and beta activity fluctuate reciprocally across NREM and REM sleep indicating that an inverse relationship between the two activities (Uchida et al., 1992). Beta activity is evidently higher in REM as compared to NREM sleep. The anterior cingulate and the dorsolateral prefrontal cortices are characterized by prominent and coherent beta activity during REM sleep in humans (Vijayan et al., 2017).

Several reports suggest that enhanced beta/gamma EEG power (activity) is an index of cortical arousal, or disturbed sleep. Most of the reports indicate that sleep EEG (both NREM and REM) records of insomnia patients are characterized by heightened beta/gamma power as compared to records of subjects who sleep well (Perlis, Merica et al., 2001; Perlis, Smith et al., 2001).

Gamma oscillations are usually defined as 30–80 Hz waves indicating arousal and (local) activation of the cortex. If gamma oscillations are coherent, they are related to perceptual binding. As mentioned previously, there are several reports indicating similar behavior of beta and gamma oscillations during sleep (see also those indicating enhanced gamma in patients suffering from primary and secondary insomnia: Perlis, Merica, et al., 2001; Perlis, Smith, et al., 2001). The specificity of gamma waves is seen in cases in which consciousness is part of the story. Thus, frontal gamma EEG power and coherence is increased in periods of lucid dreaming, as compared to non-lucid dreaming (dream lucidity indicates the state in which the dreamer is aware of the dream-like nature of the ongoing mental activity; see Voss et al., 2009). In addition, beta and gamma activity are differentially expressed in tonic and phasic REM sleep: tonic

REM is characterized by increased beta, whereas phasic REM by increased gamma activity (Simor et al., 2016).

17.13 Ripples/HFO oscillations

Ripples are usually recorded in the invasive epilepsy monitoring setting. Slow (80–140 Hz) and fast (250–500 Hz) ripples are considered physiological and pathological (epileptic enhancement) signals, respectively. The hippocampal formation and the medial temporal lobe is the most frequent, but not exclusive origin of ripple oscillations in all wake-sleep states. It has been shown that hippocampal ripples are nested in specific phases of sleep spindle oscillations, resulting in a fine-tuned coupling of the two oscillations in human subjects (Clemens et al., 2011; Staresina et al., 2015). This hierarchical nesting (SO-sleep spindle-ripple) is thought to play a crucial role in off-line memory consolidation processes involving the reactivation of transient memory traces during sleep. Recent investigations indicate that ripples can be recorded in the scalp EEG of children as well. Mean frequency has been reported to be 102 Hz, mean duration 70 ms, and mean root mean square amplitude 0.95 μV (Mooij et al., 2017). Scalp-recorded ripples have been associated with several sleep EEG transients (Mooij et al., 2018).

17.14 Conclusion

Sleep EEG is characterized by a multitude of salient oscillatory and phasic phenomena that reflect the functional neuroarchitecture of the brain and are tightly linked with neural plasticity associated to learning and memory. The network of biological mechanisms driving sleep and interacting with sleep dependent brain oscillations are intricate, ranging from molecular mechanisms to complex behavioral and cognitive patterns. Given the small number of measurable behaviors occurring during sleep, studying the sleeping brain via EEG frequencies is a critical measure to see into the sleeping brain. Like opening a window to see and learn about the constellations in the night sky, EEG frequencies allow us to open a window to see and learn about the working brain during sleep.

REFERENCES

Achermann, P. & Borbely, A. A. (1997). Low-frequency (< 1 Hz) oscillations in the human sleep electroencephalogram. *Neuroscience, 81*(1), 213–222.

Agnew, H. W., Jr., Parker, J. C., Webb, W. B., & Williams, R. L. (1967). Amplitude measurement of the sleep electroencephalogram. *Electroencephalography and Clinical Neurophysiology*, 22(1), 84–86.

Amzica, F. & Steriade, M. (1998). Electrophysiological correlates of sleep delta waves. *Electroencephalography and Clinical Neurophysiology, 107*(2), 69–83.

Andrillon, T., Nir, Y., Staba, R. J., Ferrarelli, F., Cirelli, C., Tononi, G., & Fried, I. (2011). Sleep spindles in humans: insights from intracranial EEG and unit recordings. *Journal of Neuroscience, 31*(49), 17821–17834. doi:10.1523/jneurosci.2604-11.2011

Antony, J. W., Piloto, L., Wang, M., Pacheco, P., Norman, K. A., & Paller, K. A. (2018). Sleep Spindle Refractoriness Segregates Periods of Memory Reactivation. *Current Biology, 28*(11), 1736–1743.

Astrom, C. & Trojaborg, W. (1992). Relationship of age to power spectrum analysis of EEG during sleep. *Journal of Clinical Neurophysiology, 9*(3), 424–430.

Baker, F. C. & Lee, K. A. (2018). Menstrual cycle effects on sleep. *Sleep Medicine Clinics, 13*(3), 283–294.

Bartho, P., Slezia, A., Matyas, F., Faradzs-Zade, L., Ulbert, I., Harris, K. D., & Acsady, L. (2014). Ongoing network state controls the length of sleep spindles via inhibitory activity. *Neuron, 82*(6), 1367–1379.

Berger, H. (1933). Über das Elektrenkephalogramm des Menschen. Siebente Mitteilung. *Archiv für Psychiatrie und Nervenkrankheiten, 100*, 317–319.

Berry, R. B., Brooks, R., Gamaldo, C., Harding, S. M., Lloyd, R. M., Quan, S. F., ... Vaughn, B. V. (2017). AASM Scoring Manual Updates for 2017 (Version 2.4). *Journal of Clinical Sleep Medicine, 13*(5), 665.

Blumberg, M. S. (1989). An allometric analysis of the frequency of hippocampal theta: the significance of brain metabolic rate. *Brain Behaviour and Evolution, 34*(6), 351–356.

Bodizs, R., Gombos, F., & Kovacs, I. (2012). Sleep EEG fingerprints reveal accelerated thalamocortical oscillatory dynamics in Williams syndrome. *Research in Developmental Disabilities, 33*(1), 153–164.

Bodizs, R., Gombos, F., Ujma, P. P., & Kovacs, I. (2014). Sleep spindling and fluid intelligence across adolescent development: sex matters. *Frontiers in Human Neuroscience, 8*, 952.

Bodizs, R., Kantor, S., Szabo, G., Szucs, A., Eross, L., & Halasz, P. (2001). Rhythmic hippocampal slow oscillation characterizes REM sleep in humans. *Hippocampus, 11*(6), 747–753.

Bodizs, R., Kis, T., Lazar, A. S., Havran, L., Rigo, P., Clemens, Z., & Halasz, P. (2005). Prediction of general mental ability based on neural oscillation measures of sleep. *Journal of Sleep Research, 14*(3), 285–292.

Bodizs, R., Szalárdy, O., Horváth, C., Ujma, P. P., Gombos, F., Simor, P., ... Dresler, M. (2021). A set of composite, non-redundant EEG measures of NREM sleep based on the power law scaling of the Fourier spectrum. *Scientific Reports, 11*, 2041.

Bonjean, M., Baker, T., Lemieux, M., Timofeev, I., Sejnowski, T., & Bazhenov, M. (2011). Corticothalamic feedback controls sleep spindle duration in vivo. *Journal of Neuroscience, 31*(25), 9124–9134.

Borbely, A. A. (1982). A two process model of sleep regulation. *Human Neurobiology, 1*(3), 195–204.

Borbely, A. A. & Achermann, P. (1999). Sleep homeostasis and models of sleep regulation. *Journal of Biological Rhythms, 14*(6), 557–568.

Branco, J., Atalaia, A., & Paiva, T. (1994). Sleep cycles and alpha-delta sleep in fibromyalgia syndrome. *Journal of Rheumatology, 21*(6), 1113–1117.

Brazier, M. A. (1961). Preliminary proposal for an EEG terminology by the Terminology Committee of the International Federation for Electroencephalography and Clinical Neurophysiology. *Electroencephalography and Clinical Neurophysiology, 13*, 646–650.

Brenton, J. N. & Mytinger, J. R. (2015). Sporadic occurrence of completely lateralized vertex sharp transients of sleep is a normal phenomenon: a retrospective, blinded, case-control study. *Journal of Clinical Neurophysiology, 32*(2), 171–174.

Cajochen, C., Altanay-Ekici, S., Munch, M., Frey, S., Knoblauch, V., & Wirz-Justice, A. (2013). Evidence that the lunar cycle influences human sleep. *Current Biology, 23*(15), 1485–1488.

Campbell, I. G., Higgins, L. M., Darchia, N., & Feinberg, I. (2006). Homeostatic behavior of fast Fourier transform power in very low frequency non-rapid eye movement human electroencephalogram. *Neuroscience, 140*(4), 1395–1399.

Carrier, J., Land, S., Buysse, D. J., Kupfer, D. J., & Monk, T. H. (2001). The effects of age and gender on sleep EEG power spectral density in the middle years of life (ages 20–60 years old). *Psychophysiology, 38*(2), 232–242.

Cash, S. S., Halgren, E., Dehghani, N., Rossetti, A. O., Thesen, T., Wang, C., … Ulbert, I. (2009). The human K-complex represents an isolated cortical down-state. *Science, 324*(5930), 1084–1087.

Chatburn, A., Coussens, S., Lushington, K., Kennedy, D., Baumert, M., & Kohler, M. (2013). Sleep spindle activity and cognitive performance in healthy children. *Sleep, 36*(2), 237–243. doi:10.5665/sleep.2380

Clemens, Z., Fabo, D., & Halasz, P. (2005). Overnight verbal memory retention correlates with the number of sleep spindles. *Neuroscience, 132*(2), 529–535.

Clemens, Z., Fabo, D., & Halasz, P. (2006). Twenty-four hours retention of visuospatial memory correlates with the number of parietal sleep spindles. *Neuroscience Letters, 403*(1–2), 52–56.

Clemens, Z., Molle, M., Eross, L., Jakus, R., Rasonyi, G., Halasz, P., & Born, J. (2011). Fine-tuned coupling between human parahippocampal ripples and sleep spindles. *European Journal of Neuroscience, 33*(3), 511–520.

Clemens, Z., Weiss, B., Szucs, A., Eross, L., Rasonyi, G., & Halasz, P. (2009). Phase coupling between rhythmic slow activity and gamma characterizes mesiotemporal rapid-eye-movement sleep in humans. *Neuroscience, 163*(1), 388–396.

Colrain, I. M., Webster, K. E., Hirst, G., & Campbell, K. B. (2000). The roles of vertex sharp waves and K-complexes in the generation of N300 in auditory and respiratory-related evoked potentials during early stage 2 NREM sleep. *Sleep, 23*(1), 97–106.

Cordi, M., Ackermann, S., Bes, F. W., Hartmann, F., Konrad, B. N., Genzel, L., … Dresler, M. (2014). Lunar cycle effects on sleep and the file drawer problem. *Current Biology, 24*(12), R549–R550.

Cowdin, N., Kobayashi, I., & Mellman, T. A. (2014). Theta frequency activity during rapid eye movement (REM) sleep is greater in people with resilience versus PTSD. *Experimental Brain Research, 232*(5), 1479–1485.

Crunelli, V., Lorincz, M. L., Connelly, W. M., David, F., Hughes, S. W., Lambert, R. C., … Errington, A. C. (2018). Dual function of thalamic low-vigilance state oscillations: Rhythm-regulation and plasticity. *Nature Reviews Neuroscience, 19*(2), 107–118.

Csercsa, R., Dombovari, B., Fabo, D., Wittner, L., Eross, L., Entz, L., … Ulbert, I. (2010). Laminar analysis of slow wave activity in humans. *Brain, 133*(9), 2814–2829. doi:10.1093/brain/awq169

Daan, S., Beersma, D. G., & Borbely, A. A. (1984). Timing of human sleep: Recovery process gated by a circadian pacemaker. *American Journal of Physiology, 246*(2 Pt 2), R161–183.

Dempsey, E. W. & Morison, R. S. (1941). The production of rhythmically recurrent cortical potentials after localized thalamic stimulation. *American Journal of Physiology (Legacy Content),135*(2), 293–300.

Destexhe, A., Contreras, D., & Steriade, M. (1998). Mechanisms underlying the synchronizing action of corticothalamic feedback through inhibition of thalamic relay cells. *Journal of Neurophysiology*, *79*(2), 999–1016.

De Gennaro, L., Marzano, C., Fratello, F., Moroni, F., Pellicciari, M. C., Ferlazzo, F., ... Rossini, P. M. (2008). The electroencephalographic fingerprint of sleep is genetically determined: A twin study. *Annals of Neurology*, *64*(4), 455–460.

de Zambotti, M., Willoughby, A. R., Sassoon, S. A., Colrain, I. M., & Baker, F. C. (2015). Menstrual cycle-related variation in physiological sleep in women in the early menopausal transition. *The Journal of Clinical Endocrinology and Metabolism*, *100*(8), 2918–2926.

Dijk, D. J., Beersma, D. G., & Bloem, G. M. (1989). Sex differences in the sleep EEG of young adults: Visual scoring and spectral analysis. *Sleep*, *12*(6), 500–507.

Dijk, D. J. & Czeisler, C. A. (1995). Contribution of the circadian pacemaker and the sleep homeostat to sleep propensity, sleep structure, electroencephalographic slow waves, and sleep spindle activity in humans. *Journal of Neuroscience*, *15*(5 Pt 1), 3526–3538.

Dijk, D. J. (1999). Circadian variation of EEG power spectra in NREM and REM sleep in humans: dissociation from body temperature. *Journal of Sleep Research*, *8*(3), 189–195

Egawa, I., Yoshino, K., & Hishikawa, Y. (1983). Positive occipital sharp transients in the human sleep EEG. *Psychiatry and Clinical Neuroscience*, *37*(1), 57–65.

Eichenlaub, J. B., van Rijn, E., Gaskell, M. G., Lewis, P. A., Maby, E., Malinowski, J. E., ... Blagrove, M. (2018). Incorporation of recent waking-life experiences in dreams correlates with frontal theta activity in REM sleep. *Social Cognitive and Affective Neuroscience*, *13*(6), 637–647.

Evans, B. M. & Richardson, N. E. (1995). Demonstration pf a 3–5-s periodicity between the spindle bursts in NREM sleep in man. *Journal of Sleep Research*, *4*(3), 196–197.

Feinberg, I., March, J. D., Fein, G., & Aminoff, M. J. (1984). Log amplitude is a linear function of log frequency in NREM sleep EEG of young and elderly normal subjects. *Electroencephalography and Clinical Neurophysiology*, *58*(2), 158–160.

Fell, J., Elfadil, H., Roschke, J., Burr, W., Klaver, P., Elger, C. E., & Fernandez, G. (2002). Human scalp recorded sigma activity is modulated by slow EEG oscillations during deep sleep. *International Journal of Neuroscience*, *112*(7), 893–900.

Fois, A. (1961). *The electroencephalogram of the normal child*. Charles C Thomas.

Frank, M. G. & Cantera, R. (2014). Sleep, clocks, and synaptic plasticity. *Trends in Neuroscience*, *37*(9), 491–501.

Freeman, W. J. & Zhai, J. (2009). Simulated power spectral density (PSD) of background electrocorticogram (ECoG). *Cognitive Neurodynamics*, *3*(1), 97–103.

Geiger, A., Huber, R., Kurth, S., Ringli, M., Jenni, O. G., & Achermann, P. (2011). The sleep EEG as a marker of intellectual ability in school age children. *Sleep*, *34*(2), 181–189.

Gibbs, E. A. & Gibbs, E. L. (1958). *Atlas of electroencephalography* (2nd ed.). Addison Wesley.

A glossary of terms most commonly used by clinical electroencephalographers. (1974). *Electroencephalography and Clinical Neurophysiology*, *37*(5), 538–548.

Godbout, R., Bergeron, C., Limoges, E., Stip, E., & Mottron, L. (2000). A laboratory study of sleep in Asperger's syndrome. *Neuroreport*, *11*(1), 127–130.

Goldstein, A. N. & Walker, M. P. (2014). The role of sleep in emotional brain function. *Annual Review of Clinical Psychology*, *10*, 679–708.

Grassberger, P. & Procaccia, I. (1983). Characterization of strange attractors. *Physical Review Letters*, *50*(5), 346–349.

Gu, G.-F., & Zhou, W.-X. (2006). Detrended fluctuation analysis for fractals and multifractals in higher dimensions. *Physical Review E*, *74*(6), 061104.

Halasz, P. (2005). K-complex, a reactive EEG graphoelement of NREM sleep: An old chap in a new garment. *Sleep Med Rev*, *9*(5), 391–412.

Halász, P. & Bódizs, R. (2013). *Dynamic structure of NREM sleep*. Springer.

Hauri, P. & Hawkins, D. R. (1973). Alpha-delta sleep. *Electroencephalography and Clinical Neurophysiology*, *34*(3), 233–237.

Hoedlmoser, K., Heib, D. P., Roell, J., Peigneux, P., Sadeh, A., Gruber, G., & Schabus, M. (2014). Slow sleep spindle activity, declarative memory, and general cognitive abilities in children. *Sleep*, *37*(9), 1501–1512.

Hori, T., Sugita, Y., Koga, E., Shirakawa, S., Inoue, K., Uchida, S., ... Fukuda, N. (2001). Proposed supplements and amendments to 'A manual of standardized terminology, techniques and scoring system for sleep stages of human subjects', the Rechtschaffen & Kales (1968) standard. *Psychiatry and Clinical Neuroscience*, *55*(3), 305–310.

Huber, R., Esser, S. K., Ferrarelli, F., Massimini, M., Peterson, M. J., & Tononi, G. (2007). TMS-induced cortical potentiation during wakefulness locally increases slow wave activity during sleep. *PLoS One*, *2*(3), e276.

Huber, R., Ghilardi, M. F., Massimini, M., Ferrarelli, F., Riedner, B. A., Peterson, M. J., & Tononi, G. (2006). Arm immobilization causes cortical plastic changes and locally decreases sleep slow wave activity. *Nature Neuroscience*, *9*(9), 1169–1176.

Huber, R., Ghilardi, M. F., Massimini, M., & Tononi, G. (2004). Local sleep and learning. *Nature*, *430*(6995), 78–81.

Hughes, S. W., Lorincz, M. L., Parri, H. R., & Crunelli, V. (2011). Infraslow (<0.1 Hz) oscillations in thalamic relay nuclei basic mechanisms and significance to health and disease states. *Progress in Brain Research*, *193*, 145–162.

Huguenard, J. R. & McCormick, D. A. (2007). Thalamic synchrony and dynamic regulation of global forebrain oscillations. *Trends in Neuroscience*, *30*(7), 350–356.

Hurst, H. E., Black, R. P., & Simaika, Y. M. (1965). *Long-term storage, an experimental study*. Constable.

Iber, C. & American Academy of Sleep Medicine. (2007). *The AASM manual for the scoring of sleep and associated events: Rules, terminology and technical specifications*. American Academy of Sleep Medicine.

Inanaga, K. (1998). Frontal midline theta rhythm and mental activity. *Psychiatry and Clinical Neuroscience*, *52*(6), 555–566.

Jaimchariyatam, N., Rodriguez, C. L., & Budur, K. (2011). Prevalence and correlates of alpha-delta sleep in major depressive disorders. *Innovations in Clinical Neuroscience*, *8*(7), 35–49.

Jankel, W. R. & Niedermeyer, E. (1985). Sleep spindles. *Journal of Clinical Neurophysiology*, *2*(1), 1–35.

Kaiser, D. (2013). Infralow frequencies and ultradian rhythms. *Seminars in Pediatric Neurology*, *20*(4), 242–245.

Kane, N., Acharya, J., Benickzy, S., Caboclo, L., Finnigan, S., Kaplan, P. W., ... van Putten, M. J. A. M. (2017). A revised glossary of terms most commonly used by clinical electroencephalographers and updated proposal for the report format of the EEG findings. Revision 2017. *Clinical Neurophysiology Practice*, *2*, 170–185.

Kattler, H., Dijk, D. J., & Borbely, A. A. (1994). Effect of unilateral somatosensory stimulation prior to sleep on the sleep EEG in humans. *Journal of Sleep Research*, *3*(3), 159–164.

Kellaway, P. & Fox, B. J. (1952). Electroencephalographic diagnosis of cerebral pathology in infants during sleep: I. Rationale, technique, and the characteristics of normal sleep in infants. *The Journal of Pediatrics*, *41*(3), 262–287.

Kizaki, T., Sato, S., Shirato, K., Sakurai, T., Ogasawara, J., Izawa, T., ... Ohno, H. (2015). Effect of circadian rhythm on clinical and pathophysiological conditions and inflammation. *Critical Reviews in Immunology, 35*(4), 261–275.

Knoblauch, V., Martens, W., Wirz-Justice, A., Krauchi, K., & Cajochen, C. (2003). Regional differences in the circadian modulation of human sleep spindle characteristics. *European Journal of Neuroscience, 18*(1), 155–163.

Knoblauch, V., Martens, W. L., Wirz-Justice, A., & Cajochen, C. (2003). Human sleep spindle characteristics after sleep deprivation. *Clinical Neurophysiology, 114*(12), 2258–2267.

Knoblauch, V., Munch, M., Blatter, K., Martens, W. L., Schroder, C., Schnitzler, C., ... Cajochen, C. (2005). Age-related changes in the circadian modulation of sleep-spindle frequency during nap sleep. *Sleep, 28*(9), 1093–1101.

Kozelka, J. W. & Pedley, T. A. (1990). Beta and mu rhythms. *Journal of Clinical Neurophysiology, 7*(2), 191–207.

Kubicki, S., Meyer, C., & Rohmel, J. (1986). [The 4 second sleep spindle periodicity]. *EEG-EMG Zeitschrift fur Elektroenzephalogrphie, Elektromyographie und Verwandte Gebiete, 17*(2), 55–61.

Latreille, V., Carrier, J., Lafortune, M., Postuma, R. B., Bertrand, J. A., Panisset, M., ... Gagnon, J. F. (2015). Sleep spindles in Parkinson's disease may predict the development of dementia. *Neurobiology of Aging, 36*(2), 1083–1090.

Lazar, A. S., Lazar, Z. I., & Dijk, D. J. (2015). Circadian regulation of slow waves in human sleep: Topographical aspects. *Neuroimage, 116*, 123–134.

Lazar, A. S., Panin, F., Goodman, A. O., Lazic, S. E., Lazar, Z. I., Mason, S. L., ... Barker, R. A. (2015). Sleep deficits but no metabolic deficits in premanifest Huntington's disease. *Annals of Neurology, 78*(4), 630–648.

Lázár, Z. I., Dijk, D. J., & Lázár, A. S. (2019). Infraslow oscillations in human sleep spindle activity. *Journal of Neuroscience Methods, 316*, 22–34. .

Lecci, S., Fernandez, L. M., Weber, F. D., Cardis, R., Chatton, J. Y., Born, J., & Luthi, A. (2017). Coordinated infraslow neural and cardiac oscillations mark fragility and offline periods in mammalian sleep. *Science Advances, 3*(2), e1602026.

Lee, J., Kim, D., & Shin, H.-S. (2004). Lack of delta waves and sleep disturbances during non-rapid eye movement sleep in mice lacking alpha1G-subunit of T-type calcium channels. *Proceedings of the National Academy of Sciences of the United States of America, 101*(52), 18195–18199.

Limoges, E., Mottron, L., Bolduc, C., Berthiaume, C., & Godbout, R. (2005). Atypical sleep architecture and the autism phenotype. *Brain, 128*(Pt 5), 1049–1061.

Loomis, A. L., Harvey, E. N., & Hobart, G. (1935). Potential rhythms of the cerebral cortex during sleep. *Science, 81*(2111), 597–598.

Loomis, A. L., Harvey, E. N., & Hobart, G. A. (1937). Cerebral states during sleep, as studied by human brain potentials. *Journal of Experimental Psychology, 21*(2), 127–144.

Lustenberger, C., Boyle, M. R., Alagapan, S., Mellin, J. M., Vaughn, B. V., & Frohlich, F. (2016). Feedback-controlled transcranial alternating current stimulation reveals a functional role of sleep spindles in motor memory consolidation. *Current Biology, 26*(16), 2127–2136.

Lustenberger, C., Maric, A., Durr, R., Achermann, P., & Huber, R. (2012). Triangular relationship between sleep spindle activity, general cognitive ability and the efficiency of declarative learning. *PLoS One, 7*(11), e49561.

Ma, Y., Shi, W., Peng, C. K., & Yang, A. C. (2018). Nonlinear dynamical analysis of sleep electroencephalography using fractal and entropy approaches. *Sleep Medicine Review, 37*, 85–93.

Mak-McCully, R. A., Rolland, M., Sargsyan, A., Gonzalez, C., Magnin, M., Chauvel, P., ... Halgren, E. (2017). Coordination of cortical and thalamic activity during non-REM sleep in humans. *Nature Communications*, *8*, 15499.

Massimini, M., Huber, R., Ferrarelli, F., Hill, S., & Tononi, G. (2004). The sleep slow oscillation as a traveling wave. *Journal of Neuroscience*, *24*(31), 6862–6870.

McKinney, S. M., Dang-Vu, T. T., Buxton, O. M., Solet, J. M., & Ellenbogen, J. M. (2011). Covert waking brain activity reveals instantaneous sleep depth. *PLoS One*, *6*(3), e17351.

Mohawk, J. A., Green, C. B., & Takahashi, J. S. (2012). Central and peripheral circadian clocks in mammals. *Annual Review of Neuroscience*, 35, 445–462.

Molle, M., Bergmann, T. O., Marshall, L., & Born, J. (2011). Fast and slow spindles during the sleep slow oscillation: Disparate coalescence and engagement in memory processing. *Sleep*, *34*(10), 1411–1421.

Molle, M., Marshall, L., Gais, S., & Born, J. (2002). Grouping of spindle activity during slow oscillations in human non-rapid eye movement sleep. *Journal of Neuroscience*, *22*(24), 10941–10947.

Mooij, A. H., Frauscher, B., Goemans, S. A. M., Huiskamp, G. J. M., Braun, K. P. J., & Zijlmans, M. (2018). Ripples in scalp EEGs of children: Co-occurrence with sleep-specific transients and occurrence across sleep stages. *Sleep*, *41*(11), zsy169-zsy169.

Mooij, A. H., Raijmann, R. C. M. A., Jansen, F. E., Braun, K. P. J., & Zijlmans, M. (2017). Physiological ripples (± 100 Hz) in spike-free scalp EEGs of children with and without epilepsy. *Brain Topography*, *30*(6), 739–746.

Moroni, F., Nobili, L., De Carli, F., Massimini, M., Francione, S., Marzano, C., ... Ferrara, M. (2012). Slow EEG rhythms and inter-hemispheric synchronization across sleep and wakefulness in the human hippocampus. *Neuroimage*, *60*(1), 497–504.

Neske, G. T. (2016). The slow oscillation in cortical and thalamic networks: Mechanisms and functions. *Frontiers in Neural Circuits*, *9*, 88).

Nishida, M., Hirai, N., Miwakeichi, F., Maehara, T., Kawai, K., Shimizu, H., & Uchida, S. (2004). Theta oscillation in the human anterior cingulate cortex during all-night sleep: An electrocorticographic study. *Neuroscience Research*, *50*(3), 331–341.

Noachtar, S., Binnie, C., Ebersole, J., Mauguiere, F., Sakamoto, A., & Westmoreland, B. (1999). A glossary of terms most commonly used by clinical electroencephalographers and proposal for the report form for the EEG findings. The International Federation of Clinical Neurophysiology. *Electroencephalography and Clinical Neurophysiology Supplement*, 52, 21–41.

Parrino, L., Grassi, A., & Milioli, G. (2014). Cyclic alternating pattern in polysomnography: What is it and what does it mean? *Current Opinion in Pulmonary Medicine*, *20*(6), 533–541.

Pearl, P. L., LaFleur, B. J., Reigle, S. C., Rich, A. S., Freeman, A. A., McCutchen, C., & Sato, S. (2002). Sawtooth wave density analysis during REM sleep in normal volunteers. *Sleep Medicine*, *3*(3), 255–258.

Pereda, E., Gamundi, A., Rial, R., & Gonzalez, J. (1998). Non-linear behaviour of human EEG: Fractal exponent versus correlation dimension in awake and sleep stages. *Neuroscience Letters*, *250*(2), 91–94.

Perlis, M. L., Merica, H., Smith, M. T., & Giles, D. E. (2001). Beta EEG activity and insomnia. *Sleep Medicine Review*, *5*(5), 363–374.

Perlis, M. L., Smith, M. T., Andrews, P. J., Orff, H., & Giles, D. E. (2001). Beta/gamma EEG activity in patients with primary and secondary insomnia and good sleeper controls. *Sleep*, *24*(1), 110–117.

Pesonen, A.-K., Ujma, P., Halonen, R., Räikkönen, K., & Kuula, L. (2019). The associations between spindle characteristics and cognitive ability in a large adolescent birth cohort. *Intelligence*, *72*, 13–19.

Piano, C., Mazzucchi, E., Bentivoglio, A. R., Losurdo, A., Calandra Buonaura, G., Imperatori, C., ... Della Marca, G. (2017). Wake and sleep EEG in patients with Huntington disease: An eLORETA study and review of the literature. *Clinical EEG and Neuroscience*, *48*(1), 60–71.

Picchioni, D., Horovitz, S. G., Fukunaga, M., Carr, W. S., Meltzer, J. A., Balkin, T. J., ... Braun, A. R. (2011). Infraslow EEG oscillations organize large-scale cortical-subcortical interactions during sleep: a combined EEG/fMRI study. *Brain Research*, *1374*, 63–72.

Plante, D. T., Goldstein, M. R., Landsness, E. C., Peterson, M. J., Riedner, B. A., Ferrarelli, F., ... Benca, R. M. (2013). Topographic and sex-related differences in sleep spindles in major depressive disorder: A high-density EEG investigation. *Journal of Affective Disorders*, *146*(1), 120–125.

Potari, A., Ujma, P. P., Konrad, B. N., Genzel, L., Simor, P., Kormendi, J., ... Bodizs, R. (2017). Age-related changes in sleep EEG are attenuated in highly intelligent individuals. *Neuroimage*, *146*, 554–560.

Rechtschaffen, A. & Kales, A. (1968). *A manual of standardized terminology, techniques and scoring system for sleep stages of human subjects*. University of California Los Angeles Brain Information Service.

NINDB Neurological Information Network (US), Bethesda, Md., U. S. National Institute of Neurological Diseases and Blindness, Neurological Information Network.

Ritter, P. S., Schwabedal, J., Brandt, M., Schrempf, W., Brezan, F., Krupka, A., ... Nikitin, E. (2018). Sleep spindles in bipolar disorder - a comparison to healthy control subjects. *Acta Psychiatrica Scandinavica*, *138*(2), 163–172.

Roosli, M., Juni, P., Braun-Fahrlander, C., Brinkhof, M. W., Low, N., & Egger, M. (2006). Sleepless night, the moon is bright: Longitudinal study of lunar phase and sleep. *Journal of Sleep Research*, *15*(2), 149–153.

Santhi, N., Lazar, A. S., McCabe, P. J., Lo, J. C., Groeger, J. A., & Dijk, D. J. (2016). Sex differences in the circadian regulation of sleep and waking cognition in humans. *Proceedings of the National Academy of Sciences of the United States of America*, *113*(19), E2730–2739.

Sato, S., McCutchen, C., Graham, B., Freeman, A., von Albertini-Carletti, I., & Alling, D. W. (1997). Relationship between muscle tone changes, sawtooth waves and rapid eye movements during sleep. *Electroencephalography and Clinical Neurophysiology*, *103*(6), 627–632.

Savage, V. M. & West, G. B. (2007). A quantitative, theoretical framework for understanding mammalian sleep. *Proceedings of the National Academy of Sciences of the United States of America*, *104*(3), 1051.

Schabus, M., Hodlmoser, K., Gruber, G., Sauter, C., Anderer, P., Klosch, G., ... Zeitlhofer, J. (2006). Sleep spindle-related activity in the human EEG and its relation to general cognitive and learning abilities. *European Journal of Neuroscience*, *23*(7), 1738–1746.

Schilling, C., Schlipf, M., Spietzack, S., Rausch, F., Eisenacher, S., Englisch, S., ... Schredl, M. (2017). Fast sleep spindle reduction in schizophrenia and healthy first-degree relatives: Association with impaired cognitive function and potential intermediate phenotype. *European Archives of Psychiatry and Clinical Neuroscience*, *267*(3), 213–224.

Schmidt, C., Collette, F., Cajochen, C., & Peigneux, P. (2007). A time to think: Circadian rhythms in human cognition. *Cognitive Neuropsychology*, *24*(7), 755–789. doi:10.1080/02643290701754158

Schonwald, S. V., Carvalho, D. Z., Dellagustin, G., de Santa-Helena, E. L., & Gerhardt, G. J. (2011). Quantifying chirp in sleep spindles. *Journal of Neuroscience Methods*, *197*(1), 158–164.

Shimizu, R. E., Connolly, P. M., Cellini, N., Armstrong, D. M., Hernandez, L. T., Estrada, R., ... Simons, S. B. (2018). Closed-loop targeted memory reactivation during sleep improves spatial navigation. *Frontiers in Human Neuroscience*, *12*, 28.

Silber, M. H., Ancoli-Israel, S., Bonnet, M. H., Chokroverty, S., Grigg-Damberger, M. M., Hirshkowitz, M., ... Iber, C. (2007). The visual scoring of sleep in adults. *Journal of Clinical Sleep Medicine*, *3*(2), 121–131.

Simor, P., Gombos, F., Blaskovich, B., & Bodizs, R. (2018). Long-range alpha and beta and short-range gamma EEG synchronization distinguishes phasic and tonic REM periods. *Sleep*, *41*(3), zsx210. https://doi.org/10.1093/sleep/zsx210

Simor, P., Gombos, F., Szakadat, S., Sandor, P., & Bodizs, R. (2016). EEG spectral power in phasic and tonic REM sleep: different patterns in young adults and children. *Journal of Sleep Research*, *25*(3), 269–277.

Simor, P., Horváth, K., Ujma, P. P., Gombos, F., & Bódizs, R. (2013). Fluctuations between sleep and wakefulness: Wake-like features indicated by increased EEG alpha power during different sleep stages in nightmare disorder. *Biological Psychology*, *94*(3), 592–600.

Skeldon, A. C., Dijk, D. J., & Derks, G. (2014). Mathematical models for sleep-wake dynamics: Comparison of the two-process model and a mutual inhibition neuronal model. *PLoS One*, *9*(8), e103877.

Sloan, E. P., Maunder, R. G., Hunter, J. J., & Moldofsky, H. (2007). Insecure attachment is associated with the alpha-EEG anomaly during sleep. *Biopsychosocial Medicine*, *1*, 20.

Smarr, B. L., Jennings, K. J., Driscoll, J. R., & Kriegsfeld, L. J. (2014). A time to remember: The role of circadian clocks in learning and memory. *Behavioral Neuroscience*, *128*(3), 283–303.

Spieweg, I., Sanders, S., & Kubicki, S. (1992). Periodische Entladungen von Schlafspindeln unter Plazebo und Zopiclone. [Periodical discharges of sleep spindles under placebo and Zopiclone]. *Klinische Neurophysiologie*, *23*(04), 215–220.

Spinosa, M. J. & Garzon, E. (2007). Sleep spindles: Validated concepts and breakthroughs. *Journal of Epilepsy and Clinical Neurophysiology*, *13*, 179–182.

Stam, C. J. (2005). Nonlinear dynamical analysis of EEG and MEG: Review of an emerging field. *Clinical Neurophysiology*, *116*(10), 2266–2301.

Staresina, B. P., Bergmann, T. O., Bonnefond, M., van der Meij, R., Jensen, O., Deuker, L., ... Fell, J. (2015). Hierarchical nesting of slow oscillations, spindles and ripples in the human hippocampus during sleep. *Nature Neuroscience*, *18*(11), 1679–1686.

Steriade, M. (2000). Corticothalamic resonance, states of vigilance and mentation. *Neuroscience*, *101*(2), 243–276.

Steriade, M., & Amzica, F. (1998). Coalescence of sleep rhythms and their chronology in corticothalamic networks. *Sleep Research Online*, *1*(1), 1–10.

Steriade, M., Contreras, D., Curro Dossi, R., & Nunez, A. (1993). The slow (< 1 Hz) oscillation in reticular thalamic and thalamocortical neurons: Scenario of sleep rhythm generation in interacting thalamic and neocortical networks. *Journal of Neuroscience*, *13*(8), 3284–3299.

Steriade, M., Nunez, A., & Amzica, F. (1993a). Intracellular analysis of relations between the slow (< 1 Hz) neocortical oscillation and other sleep rhythms of the electroencephalogram. *Journal of Neuroscience*, *13*(8), 3266–3283.

Steriade, M., Nunez, A., & Amzica, F. (1993b). A novel slow (< 1 Hz) oscillation of neocortical neurons in vivo: Depolarizing and hyperpolarizing components. *Journal of Neuroscience*, *13*(8), 3252–3265.

Stupfel, M., & Pavely, A. (1990). Ultradian, circahoral and circadian structures in endothermic vertebrates and humans. *Comparative Biochemistry & Physiology A: Physiology*, *96*(1), 1–11.

Tessier, S., Lambert, A., Chicoine, M., Scherzer, P., Soulieres, I., & Godbout, R. (2015). Intelligence measures and stage 2 sleep in typically-developing and autistic children. *International Journal of Psychophysiology*, *97*(1), 58–65.

Tononi, G., & Cirelli, C. (2014). Sleep and the price of plasticity: From synaptic and cellular homeostasis to memory consolidation and integration. *Neuron*, *81*(1), 12–34.

Turanyi, C. Z., Ronai, K. Z., Zoller, R., Veber, O., Czira, M. E., Ujszaszi, A., . . . Novak, M. (2014). Association between lunar phase and sleep characteristics. *Sleep Medicine*, *15*(11), 1411–1416.

Uchida, S., Maehara, T., Hirai, N., Kawai, K., & Shimizu, H. (2003). Theta oscillation in the anterior cingulate and beta-1 oscillation in the medial temporal cortices: A human case report. *Journal of Clinical Neuroscience*, *10*(3), 371–374.

Uchida, S., Maloney, T., & Feinberg, I. (1992). Beta (20–28 Hz) and delta (0.3–3 Hz) EEGs oscillate reciprocally across NREM and REM sleep. *Sleep*, *15*(4), 352–358.

Ujma, P. P., Halasz, P., Simor, P., Fabo, D., & Ferri, R. (2018). Increased cortical involvement and synchronization during CAP A1 slow waves. *Brain Structure & Function*, *223*(8), 3531–3542.

Ujma, P. P., Sandor, P., Szakadat, S., Gombos, F., & Bodizs, R. (2016). Sleep spindles and intelligence in early childhood-developmental and trait-dependent aspects. *Developmental Psychology*, *52*(12), 2118–2129.

Urakami, Y., Ioannides, A. A., & Kostopoulos, G. K. (2012). Sleep spindles – as a Biomarker of brain function and plasticity. In I. Ajeena (Ed.), *Advances in clinical neurophysiology* (pp. 73–108). IntechOpen.

Vanhatalo, S., Palva, J. M., Holmes, M. D., Miller, J. W., Voipio, J., & Kaila, K. (2004). Infraslow oscillations modulate excitability and interictal epileptic activity in the human cortex during sleep. *Proceedings of the National Academy of Sciences of the United States of America*, *101*(14), 5053–5057.

Vanhatalo, S., Tallgren, P., Andersson, S., Sainio, K., Voipio, J., & Kaila, K. (2002). DC-EEG discloses prominent, very slow activity patterns during sleep in preterm infants. *Clinical Neurophysiology*, *113*(11), 1822–1825.

Vanhatalo, S., Voipio, J., & Kaila, K. (2005). Full-band EEG (FbEEG): An emerging standard in electroencephalography. *Clinical Neurophysiology*, *116*(1), 1–8.

Vijayan, S., Lepage, K. Q., Kopell, N. J., & Cash, S. S. (2017). Frontal beta-theta network during REM sleep. *eLife*, *6*, e18894.

Voss, U., Holzmann, R., Tuin, I., & Hobson, J. A. (2009). Lucid dreaming: a state of consciousness with features of both waking and non-lucid dreaming. *Sleep*, *32*(9), 1191–1200.

Vyazovskiy, V. V., Olcese, U., Lazimy, Y. M., Faraguna, U., Esser, S. K., Williams, J. C., . . . Tononi, G. (2009). Cortical firing and sleep homeostasis. *Neuron*, *63*(6), 865–878.

Wei, Y., Krishnan, G. P., Komarov, M., & Bazhenov, M. (2018). Differential roles of sleep spindles and sleep slow oscillations in memory consolidation. *PLoS Computational Biology*, *14*(7), e1006322

Weiss, B., Clemens, Z., Bodizs, R., Vago, Z., & Halasz, P. (2009). Spatio-temporal analysis of monofractal and multifractal properties of the human sleep EEG. *Journal of Neuroscience Methods*, *185*(1), 116–124.

Werk, C. M., Harbour, V. L., & Chapman, C. A. (2005). Induction of long-term potentiation leads to increased reliability of evoked neocortical spindles in vivo. *Neuroscience*, *131*(4), 793–800.

Wright, K. P., Lowry, C. A., & Lebourgeois, M. K. (2012). Circadian and wakefulness-sleep modulation of cognition in humans. *Frontiers in Molecular Neuroscience*, *5*, 50.

Zygierewicz, J., Blinowska, K. J., Durka, P. J., Szelenberger, W., Niemcewicz, S., & Androsiuk, W. (1999). High resolution study of sleep spindles. *Clinical Neurophysiology*, *110*(12), 2136–2147.

CHAPTER 18

A REVIEW OF OSCILLATORY BRAIN DYNAMICS IN SCHIZOPHRENIA

KEVIN M. SPENCER

18.1 Introduction

Schizophrenia is a serious neuropsychiatric disorder that affects about one percent of the world's population (Perälä et al., 2007), and imposes high social and economic costs due to its severity (Davidson et al., 2016). Schizophrenia is characterized primarily by psychotic symptoms such as auditory hallucinations and delusions, which are commonly accompanied by negative symptoms such as social withdrawal and reduced motivation, and cognitive deficits in domains such as working memory and sustained attention. This disorder has a complex etiology comprised of a number of genetic and environmental factors, and its pathophysiology includes abnormalities at the neuronal, local circuit, region-wide, and whole-brain levels (Owen et al., 2016). Schizophrenia is considered to be a neurodevelopmental disorder (Murray & Lewis, 1987; Weinberger, 1987), with a typical progression beginning with a prodromal phase in the teenage years in which subclinical symptoms become apparent, to the first psychotic episode that leads to hospitalization and diagnosis, to the chronic phase that lasts the rest of the individual's life, which may be characterized by periods of psychosis and possibly remittance (Rapoport et al., 2012). Schizophrenia is also conceptualized as part of a spectrum (Siever & Davis, 2004) that includes disorders which feature subclinical manifestations of symptoms in individuals with schizotypal personality disorder, individuals in the prodrome or at high risk for developing schizophrenia, and first-degree relatives of persons with schizophrenia. In addition, schizophrenia may occur atypically early in development, in what is termed early-onset schizophrenia, and may be mixed with symptoms of affective disorder, in schizoaffective disorder. Bipolar I disorder shares similarities with schizophrenia, including psychosis, and may be a related disorder.

Electroencephalography (EEG) has been applied to the study of schizophrenia for over 80 years (Lemere, 1936). With the advent of digital computerized spectral analysis in the 1960s, "quantitative" EEG analysis came to supplant the traditional analytical approaches based on visual inspection and analog methods (reviewed in Itil, 1977). The modern era of research into neural oscillations in schizophrenia dates from the seminal studies of the Singer and Eckhorn labs (e.g., Eckhorn et al., 1988; Gray et al., 1989), which impacted neuroscience in the 1990s. The results of these studies suggested that highly precise synchronization of the spike trains of neurons in visual cortex might serve as a mechanism whereby individual stimulus features could be bound together into coherent representations of complex objects (Singer & Gray, 1995). This neuronal synchronization was mediated by oscillations in the gamma band (30–100 Hz) of the EEG, which were generated in the local neuronal population (Gray & Singer, 1989). More generally, information might be represented throughout the brain not just by the firing rates of neurons, but by precise correlations in the timing of their firing that were organized by oscillations (Singer, 1999), defining cell assemblies as predicted by Hebb (1949). Subsequent research has led to a new view of brain function in which oscillations across the frequency spectrum mediate perceptual, cognitive, motor, and emotional functions through neuronal synchronization across a multitude of hierarchically organized time scales (Buzsáki, 2006). This view of brain function based on oscillatory brain dynamics has had a large impact on the study of the neural bases of schizophrenia.

This chapter aims to summarize the major findings to date of studies of neural oscillations in schizophrenia within the context of the development of the field of oscillatory brain dynamics in general and in schizophrenia research more particularly. It begins with research into sensory gamma oscillations, which have been the main focus of research. This is followed by research into oscillations involved in working memory processes, TMS-evoked oscillations, and spontaneous oscillatory activity. It concludes with a discussion of emerging future directions in the field.

18.2 Sensory and Perceptual Gamma Oscillations

The idea that synchronous neural activity in the gamma band mediated information processing in the brain arrived at a time in which schizophrenia was being conceptualized as a disorder not just of malfunctioning of particular brain regions, but of connectivity in both local microcircuits (McGlashan & Hoffman, 2000), and large-scale, distributed networks (e.g., Friston & Frith, 1995). Gamma oscillations had been proposed to be the primary means by which brain networks were functionally connected (e.g., Singer, 1999). When taken together with the idea that schizophrenia was a "splitting of the mind", characterized by a "disintegration of thought and personality" (Bleuler, 1911/

1950), it followed that schizophrenia might be closely associated with dysfunctional gamma oscillations.

Clementz and colleagues (1997) conducted the earliest study examining gamma activity in schizophrenia in the modern era of oscillation research, and suggested that the P50 event-related potential (ERP) sensory gating abnormality might be due to an underlying sensory gating abnormality in the early auditory-evoked gamma band response (EAGBR), one of the first gamma responses identified in humans (Pantev et al., 1991). Kwon and colleagues (1999) were the first to hypothesize that schizophrenia might be associated with a specific deficit in the generation of oscillations in the gamma band, and used the auditory steady-state response (ASSR) to examine the ability of the auditory system to be driven at frequencies in the beta (13–30 Hz) and gamma ranges. Kwon and colleagues (1999) presented click trains at frequencies of 20, 30, and 40 Hz to chronic schizophrenia patients and healthy controls. The principal finding of the study was the patients had reduced ASSR power to 40 Hz stimulation, but not to stimulation at 20 or 30 Hz, suggesting that schizophrenia was associated with a specific deficit in gamma-frequency synchronization. This finding spurred the development of a new area of research in the neuroscience of schizophrenia, the study of oscillatory brain dynamics in schizophrenia.

18.2.1 Auditory Gamma Oscillations

Two types of auditory oscillations have been studied in schizophrenia: the ASSR (Galambos et al., 1981; Picton et al., 2003), and the EAGBR (Pantev et al., 1991). ASSRs are evoked by trains of simple stimuli such as clicks presented at a constant rate, or by amplitude-modulated tones, and the scalp-recorded potentials/magnetic fields are primarily generated in the primary auditory cortex (e.g., Draganova et al., 2008; Herdman et al., 2003), with subcortical sources also being detectable (e.g., Herdman et al., 2002; Farahani et al., 2017). In humans, ASSRs show maximum amplitude/power for stimulation in the gamma band at ~40 Hz, which likely reflects the resonant frequency of its generating circuits in the auditory system (e.g., Pastor et al., 2002). The EAGBR is evoked by sounds in general. It has a peak latency of ~50 ms, coincident with the P50 ERP component, and a peak frequency of ~40 Hz. The EAGBR and ASSR frequencies have been shown to be correlated across individuals, suggesting that they share a common neural generator (Zaehle et al., 2010).

18.2.1.1 *ASSR Studies*

The ASSR is the most commonly studied gamma oscillation in schizophrenia, and deficits in the power and/or phase locking factor (PLF; Tallon-Baudry et al., 1996) aspects of this oscillation in individuals with schizophrenia have been consistently replicated (reviewed in O'Donnell et al., 2013; for meta-analysis see Thuné et al., 2016). Following Kwon and colleagues' (1999) seminal report, others have reported gamma (40 Hz, sometimes also 30 Hz) ASSR power and/or PLF deficits in chronic patients

(Brenner et al., 2003; Hong et al., 2004; Light et al., 2006; Teale et al., 2008; Krishnan et al., 2009: Spencer et al., 2009; Hamm et al., 2011; Tsuchimoto et al., 2011: Edgar et al., 2014; Hirano et al., 2015; Zhou et al., 2018).

However, an important factor to consider in schizophrenia research is that the results of studies conducted with chronic patients, that is, patients who have had schizophrenia for some time, could be confounded to some degree by the duration of the patients' illness and their exposure to medications, especially antipsychotic drugs (e.g., Leung et al., 2011). Therefore, it is important to test whether patients who have never been medicated, first-episode patients (who were recently diagnosed with schizophrenia), and unmedicated individuals within the schizophrenia spectrum show the same abnormalities as chronic patients. In general, these studies have found the same pattern of gamma-frequency ASSR deficits as in chronic schizophrenia. The gamma ASSR deficit has been reported in ultra-high-risk individuals (Tada et al., 2016), first-degree relatives (Hong et al., 2004; Puvvada et al., 2018; Rass et al., 2012), schizotypal personality disorder (Brenner et al., 2003; but not observed in Rass et al., 2012), first-episode schizophrenia (Spencer et al., 2008a; Tada et al., 2016), early-onset psychosis (Wilson et al., 2008), schizoaffective disorder (Zhou et al., 2018), and bipolar I disorder (Isomura et al., 2016; Spencer et al., 2008a; Zhou et al., 2018). One study also found the gamma ASSR deficit in bipolar I disorder without psychosis (Parker et al., 2019), which raises the question of whether this deficit is particular to psychosis. Additional studies of the ASSR in non-psychotic bipolar disorder need to be done to confirm this result.

Thus, the deficit in gamma ASSR generation appears to affect individuals with psychosis in general, soon after the first episode of psychosis. The gamma ASSR deficit is even present in first degree relatives of individuals with schizophrenia, who share genetic (and possibly environmental) risk factors with persons with schizophrenia, even though they are not psychotic and have not received antipsychotic medication. A limited amount of evidence suggests that the gamma ASSR deficit is also present in individuals at high risk of developing schizophrenia. As more studies take place with this population, it is important to ask whether the degree of the gamma ASSR deficit is predictive of the probability of their conversion to psychosis.

While the gamma ASSR is clearly sensitive to neural circuit abnormalities in psychosis, it is not so clearly associated with particular psychotic symptoms. Given the nature of the ASSR as an auditory sensory response, and the hypothesized relationships between gamma oscillations and perception, hallucination symptoms may be expected to be correlated with gamma ASSR measures. Indeed, some studies have found correlations between gamma ASSR measures and hallucination symptoms (Mulert et al., 2011; Spencer et al., 2008a,2009; Zhou et al., 2018), although the direction of these correlations is not consistent across studies.

The ASSR has become the most widely-studied gamma oscillation in schizophrenia research because of several factors:

1. It has a relatively high signal-to-noise ratio and is readily obtained.

2. It is easy to analyze, as the frequency is known a priori and the power of the ASSR can be measured with basic spectral analysis methods that are commonly found in EEG analysis software.
3. The gamma ASSR deficit in schizophrenia appears to be independent of whether or not the stimuli are attended (e.g., Hamm et al., 2015).
4. The basic set of neural generators of the ASSR are known (e.g., Herdman et al., 2002, 2003).
5. The ASSR exhibits good test-retest reliability (Legget et al., 2017; McFadden et al., 2014; Tan et al., 2015).

For these reasons, the ASSR is being extensively used in translational neuroscience studies of animal models relevant to schizophrenia. While rodents do not appear to have a clearly defined resonance frequency as humans, they do exhibit gamma ASSR deficits when administered drugs that are commonly used to model aspects of psychosis, such as *N*-methyl-D-aspartate receptor (NMDAR) antagonists (e.g., Leishman et al., 2015; Sivarao et al., 2016; Sullivan et al., 2015). Therefore, the ASSR is a promising tool for translational neuroscience research.

18.2.1.2 *EAGBR Studies*

The next most-studied gamma oscillation in schizophrenia is the EAGBR, owing to the extensive use of the auditory oddball and sensory gating paradigms in schizophrenia ERP research, from which the EAGBR can be readily measured (typically in the responses to the standard stimuli). In general, the evoked power and PLF aspects of the EAGBR are decreased in the schizophrenia spectrum. EAGBR deficits in chronic schizophrenia patients have been reported in auditory oddball (e.g., Hall et al., 2011a; Lenz et al., 2011; Oribe et al., 2019; Roach & Mathalon, 2008), sensory gating (e.g., Hall et al., 2011b; Popov et al., 2011), and tone discrimination (e.g., Leicht et al., 2010) tasks, although some studies have failed to find deficits (Brenner et al., 2009; Spencer et al., 2008b). In first-episode schizophrenia patients, various studies all reported reduced evoked power and PLF (Leicht et al., 2015; Oribe et al., 2019; Taylor et al., 2013), although Gallinat and colleagues (2004) did not find EAGBR deficits in a sample of unmedicated patients. Studies of clinical high-risk individuals are mixed: Perez and colleagues (2013) found that EAGBR evoked power was reduced but PLF did not differ compared with healthy individuals, while Oribe and colleagues (2019) did not find any reductions in clinical high-risk individuals.

Three studies of the EAGBR have provided some insight into the genetic basis of auditory gamma oscillation deficits in schizophrenia. Leicht and colleagues (2011) found EAGBR power and PLF deficits in both schizophrenia patients and their first-degree relatives in a tone discrimination task, pointing to the EAGBR deficit being an inheritable trait. Hall and colleagues (2011a) examined the EAGBR from an oddball task in a sample of schizophrenia patients and their monozygotic twins who were concordant or discordant for schizophrenia, and compared them to a sample of healthy twin pairs. Hall and colleagues (2011a) found EAGBR power and PLF deficits in the patients, with deficits to lesser degrees in the well co-twins of discordant pairs. These results were clearly indicative of a strong genetic factor in the EAGBR deficits. But in contrast to those studies, Hall and colleagues (2011b) did not find an EAGBR power deficit in the unaffected first-degree relatives of schizophrenia patients in data from a sensory gating task, which suggests that there is no heritable component of the EAGBR deficit in schizophrenia. One possible explanation for these discrepant findings is that attention plays a role in producing the EAGBR deficit, as the standard sensory gating task used by Hall and colleagues (2011b) was passive (unlike the active tasks in Hall et al., 2011a and Leicht et al., 2011, which found deficits). Thus, the EAGBR deficit in the schizophrenia spectrum and its genetic basis could involve a major contribution from reduced top-down attentional modulation of the EAGBR, rather than a dysfunctional EAGBR per se. Further work is needed to test this hypothesis.

18.2.1.3 *Summary*

The ASSR and EAGBR studies provide strong evidence for impaired generation of stimulus-evoked gamma oscillations in the auditory cortex in schizophrenia, although the ASSR results as a whole are stronger than the EAGBR findings. A natural question to ask is whether the gamma ASSR and EAGBR deficits are correlated. Roach and colleagues (2013) found that 40-Hz ASSR PLF and EAGBR PLF were indeed correlated within both patient and controls groups. Thus, auditory-evoked gamma oscillations may provide biomarkers of cortical circuit abnormalities that are heritable and present across the schizophrenia spectrum.

18.2.2 Visual Gamma Oscillations

18.2.2.1 *Gestalt Perception*

The first studies that investigated the role of gamma synchronization in perception had found evidence that gamma oscillations mediated the "binding" of visual features into coherent objects (Singer & Gray, 1995). Therefore, if schizophrenia was the result of dysfunctional binding of information in the brain via gamma oscillations, it made sense to study visual feature binding in schizophrenia. Psychophysical studies had reported abnormalities in visual perception in schizophrenia that were consistent with dysfunctional Gestalt formation (reviewed in Phillips & Silverstein, 2003). Inspired by Rodriguez and colleagues (1999), who demonstrated local and distributed gamma oscillatory activity supporting Gestalt perception (Mooney faces; Mooney & Ferguson, 1951), Spencer and colleagues (2003) had participants with schizophrenia and healthy controls perform an illusory contour detection task. In the controls, the illusory contour pattern (a Kanisza square) evoked an early visual gamma oscillation, but the non-contour pattern did not evoke this oscillation. In the schizophrenia patients, neither pattern evoked the visual gamma oscillation. Furthermore, the patients showed an abnormal pattern of inter-electrode phase synchrony (cf. Lachaux et al., 1999) in the gamma band.

In a follow-up study (Spencer et al., 2004), the oscillations phase-locked to the manual response in the same task were investigated, as these oscillations were hypothesized to be more closely related to the perception of the Gestalt than stimulus-evoked oscillations. Spencer and colleagues (2004) found that a response-locked oscillation in the gamma band occurred in controls for the illusory contour but not the non-contour pattern. In the schizophrenia patients, there was a similar response-locked oscillation, but it occurred in the beta band (13–30 Hz). The PLF aspect of this beta oscillation was correlated with some of the patients' psychotic symptom ratings, including visual hallucinations and disorganization. In a replication study with briefly presented stimuli, Spencer and Ghorashi (2014) used a dense electrode array with statistical non-parametric mapping. The illusory contour effect on the early visual evoked gamma oscillation was not found, although an enhancement of PLF for the illusory contour was found for a later evoked oscillation in controls that appeared to be an offset response. In the response-locked oscillations there was a high gamma (74–99 Hz) oscillation at fronto-central electrodes just prior to the button press that showed greater PLF for the illusory contour than the non-contour pattern in controls, and the opposite pattern in patients. The illusory contour effect on this oscillation in patients was correlated with their formal thought disorder symptoms. Further, in a study using Kanisza squares as well as diamonds, Wynn and colleagues (2015) found that in controls the early visual-evoked gamma oscillation showed enhanced PLF for illusory contours compared to the non-contour pattern, but this effect was absent in the patients. The illusory contour effect in patients was negatively correlated with their psychotic symptoms.

In another series of studies, Uhlhaas and colleagues (2006a) provided further important evidence for links between abnormal oscillatory synchronization in Gestalt perception and schizophrenia using a task in which schizophrenia patients were impaired in the perception of a Gestalt, the discrimination of upright from inverted Mooney faces (as in Rodriguez et al., 1999). Uhlhaas and colleagues (2006a) found that inter-electrode phase synchrony in the beta band elicited by upright face stimuli was delayed and reduced in schizophrenia patients compared to controls. Furthermore, this beta phase synchrony effect was correlated with the patients' psychotic symptom scores, particularly delusions and hallucinations. Thus, Uhlhaas and colleagues' (2006a) main findings were broadly consistent with those of Spencer and colleagues (2004). However, no differences were found between controls and patients in induced gamma oscillation power or gamma phase synchrony.

Magnetoencephalography (MEG) has been shown to be more sensitive than EEG to oscillations in the "high gamma" band (approximately 60–100 Hz) (Muthukumaraswamy & Singh, 2013), likely owing to its relative insensitivity to scalp muscle activity (e.g., Pope et al., 2009). Induced gamma oscillations elicited by visual stimuli that are found in local field potential recordings in animals are also readily detectable with MEG (Hall et al., 2005; Hoogenboom et al., 2006). In a follow-up study to Uhlhaas and colleagues (2006a) using MEG, Grützner and colleagues (2013) found that induced high gamma responses to upright Mooney faces were reduced in schizophrenia patients compared to controls. This deficit was inversely correlated with the patients' disorganization symptoms, consistent with psychophysical studies showing associations between disorganization and impaired Gestalt perception (e.g., Uhlhaas et al., 2006b). And in a study examining first-episode schizophrenia patients who had never been treated with antipsychotic medication, Sun and colleagues (2013) replicated this high gamma finding, which was localized to ventral visual areas, and also found that that low gamma oscillations (30–60 Hz) were intact while beta power was increased in the patients.

In the years after the publication of studies by Spencer and colleagues (2003) and Uhlhaas and colleagues (2006a), it became apparent that the calculation of EEG inter-electrode phase synchrony to measure oscillatory synchronization between brain regions was confounded by several factors, such as common reference and source mixing (e.g., Guevara et al., 2005; Palva & Palva, 2012). Revisiting the issue of long-distance synchronization between brain regions, Hirvonen and colleagues (2017) analyzed the MEG data of Grützner and colleagues (2013) using a source-based phase synchronization method that avoided the common reference problem and minimized the contribution of source mixing, and found reduced gamma phase synchronization between visual and prefrontal cortical areas in individuals with schizophrenia compared to healthy controls, and the reduction in inter-areal synchronization was correlated with the patients' total clinical symptoms.

These studies investigating Gestalt perception in schizophrenia generally found associations between oscillation abnormalities and psychotic symptoms, particularly disorganization, and are in line with the results of psychophysical studies of Gestalt perception in schizophrenia. Taken together, the results of these studies are consistent with the proposed role of gamma synchronization in visual feature binding, and support the hypothesis that schizophrenia is associated with abnormal oscillatory synchronization that is functionally related to impaired information processing and clinically relevant symptoms.

18.2.2.2 *Visual Evoked and Induced Gamma*

Taking a step back from the domain of Gestalt perception, it can be asked whether visual gamma activity in general is abnormal in schizophrenia. Visual stimulation elicits two kinds of high-frequency oscillatory activity in visual cortex (e.g., Muthukumaraswamy et al., 2010):

1. an early (~100 ms) evoked, transient oscillation that is phase-locked to stimulus onset, with a frequency that varies across the upper beta and gamma bands; and
2. an induced oscillation in the high gamma band with a later onset, that lasts for the duration of stimulation.

However, for long-duration stimuli, a transient evoked offset oscillation is also found with similar characteristics as the early evoked oscillation.

These three visual oscillations have been investigated in schizophrenia. Spencer and colleagues (2003, 2004, 2008b) found reduced PLF of the early visual evoked gamma oscillation in chronic schizophrenia patients, but in other studies this oscillation did not differ from healthy controls (Ghorashi & Spencer, 2015; Spencer & Ghorashi, 2014). Grent-'t-Jong and colleagues (2020) reported reduced PLF of the early evoked gamma oscillation in first-episode patients and clinical high-risk individuals. Grent-'t-Jong and colleagues (2016, 2020) showed that the power of the visual induced oscillation was reduced in schizophrenia patients during visual stimulation with a moving grating stimulus, and during stimulation with upright and inverted Mooney face stimuli (Grützner et al., 2013; Sun et al., 2013). And, the power of the offset response was also reduced overall in schizophrenia patients compared to controls in Grützner and colleagues (2013) and Sun and colleagues (2013). Thus, several studies have found some impairments in gamma oscillations elicited by visual stimuli in schizophrenia, but the causes for the discrepancies among studies are not clear.

As in the auditory system, steady-state stimulation has been used in the visual system to probe the integrity of its circuitry in schizophrenia, although with little consistency in the stimulation frequencies tested among studies. Using 1-Hz stimulation, Jin and colleagues (2000) found that schizophrenia patients had reduced visual steady-state response (VSSR) power at harmonics in the alpha band (10–12 Hz) compared to healthy controls. Clementz and colleagues (2004) found that the temporal dynamics of the VSSR to stimulation at 6.4 Hz were abnormal in schizophrenia, showing a delayed buildup and a longer decay time. Similarly, using 12.5-Hz stimulation, Ethridge and colleagues (2011) found a decline in VSSR power during the stimulation period in persons with schizophrenia compared to controls. Krishnan and colleagues (2005) investigated VSSRs at stimulation frequencies from 4–40 Hz and found reduced VSSR power in patients for beta and gamma frequency stimulation. But in contrast, Riečanský and colleagues (2010) found increased PLF of the initial transient phase of the VSSR for 40 Hz stimulation in patients.

In sum, there is evidence for deficits in visual gamma generation in schizophrenia, although these deficits may not be as general as with auditory gamma oscillations. There is a substantial degree of inconsistency in the early visual-evoked oscillation and gamma VSSR findings, while visual induced gamma power seems to be consistently decreased in schizophrenia (although relatively few studies have investigated this oscillation). More experiments assessing different types of stimuli, tasks, and multiple VSSR stimulation frequencies are needed to better understand visual gamma generation in schizophrenia.

18.3 Oscillations Related to Working Memory Processes

Schizophrenia is characterized by impairments in prefrontal cortex (PFC) function (e.g., Minzenberg et al., 2009), microcircuitry (e.g., Hoftman et al., 2018), and cognitive domains associated with PFC circuits: working memory and executive control (Barch,

2005). Since most of the evidence for neural circuit abnormalities in schizophrenia comes from studies of PFC microcircuitry, deficits should be found gamma oscillations elicited in tasks that engage PFC circuits. Several studies have indeed reported gamma oscillations deficits in working memory and executive control tasks in schizophrenia.

One notable series of studies employed the "preparing to overcome prepotency" (POP) task, in which a rule governing the spatial correspondence between a response cue and the manual response is maintained in working memory. On low cognitive control trials, the response cue matches the response hand, while on high cognitive control trials, the response cue signals the opposite response hand. In the first study reporting strong evidence for PFC gamma deficits in working memory in schizophrenia, Cho and colleagues (2006) examined gamma power in chronic schizophrenia patients and controls performing the POP task, finding that increased cognitive control requirements elicited higher PFC gamma power during the delay period at electrodes over the PFC in controls but not in patients, who made more errors in the high control condition. Gamma power during the late part of the delay was correlated with accuracy in the controls, and negatively correlated with disorganization symptoms in the patients. In a study comparing medicated and unmedicated first-episode patients with controls in the POP task, Minzenberg and colleagues (2010) found a similar pattern of cognitive control-related gamma deficits at frontal electrodes in both groups of patients, indicating that the gamma deficit was not related to antipsychotic medication, and was present early in the course of the disorder. In healthy individuals, cognitive-control-related gamma in the POP task was correlated with gamma-amino-butyric acid (GABA) neurotransmission, but not in schizophrenia patients (Frankle et al., 2015). And Minzenberg and colleagues (2015) showed that the administration of modafinil, a drug that improves cognition through its actions on the dopamine and norepinephrine systems, enhanced performance on the POP task and the associated cognitive control-related gamma power in schizophrenia patients.

Using a version of the Sternberg paradigm, Haenschel and colleagues (2009) provided further support for the findings of impaired gamma oscillations associated with working memory maintenance in schizophrenia. Participants had to encode from one to three objects in working memory and then maintain them for 12 seconds, until a probe item was presented. Hence, this task enabled the dissociation of processes involved in working memory encoding, rehearsal, and retrieval. The study found that their early-onset schizophrenia patients had worse performance than controls as memory load (number of objects) increased and showed deficits in stimulus-evoked oscillations in several bands in the encoding phase. In the late delay period, induced gamma power at frontal sites in the controls peaked in the three-item load condition, whereas induced gamma power peaked for the patients in the two-item condition. This kind of abnormal memory load/response function has been observed in functional neuroimaging studies (e.g., Jansma et al., 2004). The patients also had reduced induced theta and gamma power during the retrieval phase compared to controls.

These studies demonstrate that gamma activity likely originating in the PFC involved in working memory and executive control processes is impaired in schizophrenia, and point to avenues for restoring cognitive function to normal levels. This area of research makes the best link between studies of neural microcircuits and oscillatory activity related to particular cognitive functions in schizophrenia.

18.4 Transcranial Magnetic Stimulation (TMS)-Evoked Oscillations

Single TMS pulses evoke EEG oscillations in the cortex underneath the stimulated scalp site. Evidence suggests that the frequency of these TMS-evoked EEG oscillations varies for different cortical areas (Ferrarelli et al., 2012; Rosanova et al., 2009): ~10 Hz for visual cortex, ~20 Hz for parietal cortex, and ~30 Hz for prefrontal cortex. It has been proposed that the frequency of these oscillations reflects the resonant frequency of the stimulated cortical area (Rosanova et al., 2009), which is likely to be determined by specific characteristics of the circuitry in that area. Therefore, TMS-evoked EEG oscillations could be used to directly probe the integrity of cortical circuits in neuropsychiatric disorders.

To date, only a handful of studies have been conducted in this area. Ferrarelli and colleagues (2008) examined the oscillations evoked by TMS to the frontal cortex in healthy persons and schizophrenia patients and found that the oscillations were reduced in total power and PLF in the patients. In a follow-up study, Ferrarelli and colleagues (2012) obtained TMS-evoked oscillations to stimulation over prefrontal, premotor, motor, and parietal cortex. They found that the frequency, total power, and PLF of the TMS-evoked oscillation were lower in patients than controls for stimulation at the prefrontal and premotor sites. Canali and colleagues (2015) stimulated over premotor cortex in patients with schizophrenia, bipolar disorder, and depression, and found a reduction in the TMS-evoked oscillation frequency in all the patient groups compared to controls. And in a study of first-episode psychosis patients, Ferrarelli and colleagues (2019) found reduced beta/low gamma oscillations evoked by TMS of the motor cortex. The results of these studies are consistent with the hypothesized gamma oscillation deficits in frontal cortical areas in schizophrenia (see Section 18.3). TMS-evoked oscillations could thus provide a powerful approach to investigate cortical circuits directly without needing to use particular sensory stimuli or tasks to engage the brain region of interest.

18.5 Spontaneous Oscillations

In modern neuroscience, brain activity is measured with methods at different spatial and temporal scales: single-unit spiking, local field potentials, intracranial and scalp EEG, and functional neuroimaging, among others. The patterns of activity observed with all of these methods can be classified into two categories: *evoked* activity, which results directly from sensory or other stimulation; and *intrinsic* or *spontaneous* activity, which is the ongoing activity does not directly result from stimulation. Due to the highly recurrent architecture of connections in the brain, spontaneous activity makes up most of neural activity, while evoked activity makes up only a small portion (Raichle, 2010). Classical views of brain function have considered spontaneous activity to be simply "noise", with evoked activity simply being added to the random ongoing activity. However, more recent evidence has demonstrated that spontaneous activity is structured in space and time (Tsodyks et al., 1999), and much of the variance in evoked responses can be accounted for by spontaneous activity (Arieli et al., 1996). Spontaneous activity appears to reflect the spatial and temporal connectivity patterns in local and large-scale networks, and in part represents "predictions" of expected sensory input based on past experience, which interact with and modulate incoming stimulus evoked responses (Ringach, 2009).

EEG is a powerful method with which to study spontaneous and evoked activity in the brain as it is sensitive to neural activity across most of the neural frequency spectrum. However, the neural generators and functional significance of many types of spontaneous activity in the EEG are largely unknown. Infraslow (0.01–0.1 Hz) and slow (0.1–1 Hz) oscillations in the EEG may correspond to fluctuations in the blood oxygenation level dependent signal measured with functional magnetic resonance imaging, as these kinds of activity occur in the same frequency bands and may share the same neural generators in large-scale networks of brain regions (Hiltunen et al., 2014). At the opposite ends of the frequency and spatial spectra, broadband (as opposed to narrowband) gamma activity (30–100 Hz) in the EEG may reflect asynchronous neuronal spiking in local circuits (e.g., Burke et al., 2015; Yizhar et al., 2011). Recent studies suggest that oscillatory activity across the frequency spectrum is hierarchically organized (Buzsáki & Draguhn, 2004), with the phase of low-frequency oscillations modulating the power of higher-frequency oscillations (reviewed in Canolty & Knight, 2010).

EEG studies have repeatedly demonstrated that various types of sensory- and task-evoked responses tend to be decreased in individuals with schizophrenia compared to healthy control subjects. ERPs ranging from early sensory evoked components like the P50, to purely cognitive components like the P300, generally show decreased, rather than increased amplitudes in schizophrenia (Javitt et al., 2008). And as reviewed earlier, event-related oscillations typically show decreased evoked power and/or phase locking. But while schizophrenia is generally associated with *decreases* in evoked brain activity, spontaneous oscillatory activity during waking appears to be generally *increased* in particular frequency bands. (In fact, findings of increased spontaneous oscillation power in schizophrenia go back several decades; see Itil, 1977.) One major question is whether evoked activity deficits are caused to some degree by increased spontaneous activity.

18.5.1 Resting State Oscillations

There have been many reports of increased low-frequency power in the resting state EEG of individuals with schizophrenia, typically in the delta (1–4 Hz) and/or theta (4–8 Hz) bands, although alpha (8–13 Hz) power does not seem to show a consistent pattern in the literature (e.g., Boutros et al., 2008; Clementz et al., 1994; Narayanan et al., 2014; Schulman et al., 2011; Sponheim et al., 2000). Increased delta/theta power in individuals with psychotic bipolar disorder has been reported as well (e.g., Clementz et al., 1994; Narayanan et al., 2014; Sponheim et al., 2000). Increased delta/theta power has been found in first-episode psychosis patients (Clementz et al., 1994; Sponheim et al., 2000), although Ranlund and colleagues (2014) found increased delta/theta power in chronic but not first-episode psychosis patients. Increased delta/theta power has not generally been found in relatives of psychosis patients (e.g., Clementz et al., 1994; Ranlund et al., 2014), although Narayanan and colleagues (2014) reported an increase in delta power in SZ relatives. Finally, van Tricht and colleagues (2014) observed increased delta and theta power in individuals at clinical high risk for psychosis who later transitioned to psychosis, in comparison to controls and to clinical high-risk individuals who did not transition to psychosis. Thus, the sum of the evidence suggests that increased delta and/or theta power is associated with psychosis and may even occur before the onset of psychosis in at-risk individuals. Furthermore, increased delta/theta power is found in unmedicated schizophrenia patients (Boutros et al., 2008), so it does not appear to be an effect of antipsychotic treatment.

Reports of increased spontaneous high frequency (beta [13–30 Hz] and gamma bands) power during the resting state also go back decades (Itil, 1977), but the contribution of artifacts to those reports has been debated, as scalp muscle (e.g., Pope et al., 2009) and saccadic spike potential (e.g., Keren et al., 2010) artifacts can confound measurements of high frequency power. Recent studies have used independent component analysis (Jung et al., 2000) to remove such artifacts. For example, Narayanan and colleagues (2014) analyzed the resting EEG in a large sample of individuals with schizophrenia and psychotic bipolar disorder and their relatives. They found increased resting power in participants with schizophrenia and psychotic bipolar disorder in the low beta range (13–20 Hz), but not in the upper beta and gamma bands. Using MEG (which is less sensitive to scalp muscular artifacts than EEG) in conjunction with ICA artifact correction, Grent-‘t-Jong and colleagues (2018) found widespread reductions of resting gamma power in chronic schizophrenia patients. However, different gamma patterns were observed in other illness stages: first-episode patients showed increased gamma in visual areas with gamma decreases elsewhere, and clinical high-risk individuals demonstrated widespread increases in high gamma power. Consistent with the latter finding, Ramyead and colleagues (2015) also found increased resting gamma power in at-risk individuals who subsequently converted to psychosis. Thus, the evidence so far suggests that different stages of schizophrenia may be associated with different patterns of spontaneous gamma activity in the resting state. Further studies of clinical high-risk and first-episode patients are necessary to confirm this hypothesis, particularly longitudinal studies.

18.5.2 Spontaneous Oscillations During Tasks

If schizophrenia is associated with increased EEG power in particular frequency bands during the resting state, what about during sensory stimulation or task performance? Winterer and colleagues (2000, 2004, 2006) found increased induced or "noise" power with decreased "signal" (ERP) amplitude in schizophrenia patients in the delta/theta range at frontal electrodes during auditory oddball tasks. This pattern was also found in unaffected siblings (Winterer et al., 2004, 2006), and was associated with the COMT genotype, which affects dopamine signaling in the cortex (Winterer et al., 2006). Dopamine has been hypothesized to affect the signal-to-noise ratio of cortical responses (e.g., Rolls et al., 2008).

In ASSR tasks, Spencer (2012) and Hirano and colleagues (2015) examined induced gamma power during the baseline and stimulus periods, using dipole source localization to construct a spatial filter to focus on auditory cortex activity. They found that broadband induced gamma power was increased in the auditory cortex of individuals with schizophrenia compared to healthy individuals during both the baseline and stimulus periods. Induced gamma power was inversely correlated with ASSR PLF for 40-Hz stimulation, demonstrating an interaction between evoked and spontaneous EEG activity. In addition, auditory hallucination symptoms in the patients were correlated with broadband gamma power in the left auditory cortex only during 40-Hz stimulation (Hirano et al., 2015).

These studies suggest that spontaneous EEG activity is increased in schizophrenia during sensory stimulation and task performance, although the affected frequency bands may depend on stimulus and task factors. An important question is whether increased spontaneous activity in the baseline period of a task is responsible for the apparent deficit in subsequent evoked activity in individuals with schizophrenia. An increase in baseline power in schizophrenia patients could cause the power of an evoked response, typically measured against the baseline, to appear to be reduced in comparison to healthy individuals, even if the true magnitude of the evoked response was unaffected. Winterer and colleagues (2000) proposed such an effect of decreased signal-to-noise ratio and evidence of this effect appears in a schizophrenia-related animal model (Lazarewicz et al., 2010) and in a VSSR task (Ethridge et al., 2011). The negative correlation between 40-Hz ASSR PLF and spontaneous gamma power in Hirano and colleagues (2015) may reflect a different effect: the interference of noise on the PLF measure, as PLF is reduced as overlapping noise increases (e.g., Ding & Simon, 2013; Muthukumaraswamy & Singh, 2011). These signal-to-noise ratio effects could be considered artifactual, in that they produce apparent effects due to the imperfect nature of the measures. Another possibility is that increased noise in a neural circuit actually disrupts oscillatory synchronization in that circuit. Distinguishing that phenomenon from artifactual effects may require measurement of the affected oscillation at the local circuit level.

18.5.3 Sleep Oscillations

Investigations of spontaneous oscillatory activity during sleep in schizophrenia have focused on two oscillations: slow (delta) waves and sleep spindles (7–15 Hz). Slow waves are generated in the cortex, whereas spindles originate in the thalamic reticular nucleus (TRN) and are transmitted to the cortex (Steriade, 2006). Spindle generation in the thalamus involves interactions among inhibitory interneurons and excitatory thalamocortical projection neurons in the TRN (e.g., Jacobsen et al., 2001). Corticothalamic projections initiate spindles through glutamatergic inputs to TRN cells at NMDARs. The oscillations tend to be linked through phase-amplitude coupling, such that sleep spindles are more likely to occur during the up-state phase of the slow wave (e.g., Staresina et al., 2015). Both oscillations are hypothesized to support the consolidation of long-term memories during sleep (Diekelmann & Born, 2010).

Compared to gamma oscillations, there have been relatively few studies of sleep oscillations in schizophrenia, despite the detailed knowledge of the circuitry that generates these oscillations. While spontaneous EEG activity during the resting state or task performance tends to be increased in schizophrenia, spontaneous oscillatory patterns during sleep tend to be reduced. The findings on slow waves are mixed, with some reports of deficits (e.g., D'Agostino et al., 2018; Sarkar et al., 2010; Sekimoto et al., 2011) but other studies not finding slow wave abnormalities (e.g., Ferrarelli et al., 2007, 2010). In contrast, there is strong evidence that spindle generation is impaired in schizophrenia (e.g., Ferrarelli et al., 2007, 2010; Manoach et al., 2014; Wamsley et al., 2012). Spindle deficits have been reported in medication-naïve first-episode patients (Manoach et al., 2014) and unaffected first-degree relatives of schizophrenia patients (D'Agostino et al., 2018; Manoach et al., 2014), indicating that spindle impairment is a trait marker of schizophrenia and is not due to antipsychotic medication. While spindle impairment is associated with reduced thalamic volume (Buchmann et al., 2014), it has also been correlated with abnormally increased thalamocortical functional connectivity (Baran et al., 2019). With regards to symptomatology, spindle deficits have been associated with psychotic symptom ratings (Ferrarelli et al., 2010; Wamsley et al., 2012). Notably, there is substantial evidence that sleep spindle deficits are associated with impairments in memory consolidation in schizophrenia (reviewed in Manoach et al., 2016), pointing to spindle-generating circuitry as a potential target for treatment of both psychotic symptoms and impaired cognition in schizophrenia.

18.6 Circuit Mechanisms Underlying Oscillation Abnormalities in Schizophrenia

Neural oscillation research has made an important impact on our understanding of schizophrenia due to the cross-fertilization of this field with basic research into oscillation mechanisms and neuropathological studies of neural circuit abnormalities in schizophrenia. Unlike ERPs, for which the generating mechanisms are unknown, the basic mechanisms underlying some oscillations are known to a fair degree. For example, cortical gamma oscillations are generated by recurrent interactions between excitatory pyramidal cells and fast-spiking, parvalbumin-expressing (PV+) inhibitory interneurons (Buzsáki & Wang, 2012). Neuropathological research has identified deficits in pyramidal cells and PV+ interneurons (e.g., Lewis et al., 2012) that can be used to inform models of gamma generation, enabling researchers to test hypotheses about gamma abnormalities in schizophrenia with methods such as neuropharmacology (e.g., Sivarao et al., 2016), optogenetics (e.g., Carlén et al., 2012; Yizhar et al., 2011), and computational modeling (e.g., Jadi et al., 2016; Kömek et al., 2012; Spencer, 2009). This section reviews some of the integrative research into the neural circuit mechanisms of oscillation abnormalities in schizophrenia and focuses on the areas of NMDAR hypofunction and reduced synaptic connectivity in which a major proportion of research has been done.

18.6.1 NMDAR Hypofunction

The glutamate hypothesis of schizophrenia (Moghaddam & Javitt, 2012) offers clues to the neural circuit abnormalities that may be responsible for both evoked gamma oscillation deficits and increased spontaneous gamma activity in this disorder. In healthy people, acute administration of NMDAR antagonists produces a constellation of positive and negative symptoms, cognitive deficits, and neurophysiological abnormalities that resemble those found in schizophrenia (Krystal et al., 2003). In animals, the behavioral and neurophysiological abnormalities resulting from acute administration of NMDAR antagonists (reviewed in Amann et al., 2010) and genetic knockouts of NMDARs (e.g., Belforte et al., 2010; Korotkova et al., 2010) resemble those found in schizophrenia. In particular, acute NMDAR antagonism reduces the amplitude of some ERPs (Amann et al., 2010) and gamma oscillations (e.g., Lazarewicz et al., 2010; Leishman et al., 2015; Sivarao et al., 2016; Sullivan et al., 2015), in addition to increasing broadband spontaneous gamma power (reviewed in Hunt & Kasicki, 2013). Genetic ablation of NMDARs also increases spontaneous gamma power in adult animals (e.g., Carlén et al., 2012; Korotkova et al., 2010; Tatard-Leitman et al., 2015). Increased spontaneous gamma as a result of NMDAR hypofunction results from increased excitability of pyramidal cells, either through disinhibition from reduced excitatory drive to fast-spiking interneurons (Carlén et al., 2012; Homayoun & Moghaddam, 2007), and/or by alterations in cell membrane properties of the pyramidal cells themselves (Tatard-Leitman et al., 2015). With regard to increased spontaneous delta/theta power, in the thalamus, NMDAR antagonism increases delta band activity, which is transmitted to the hippocampus (Zhang et al., 2012) and the cortex (Hunt & Kasicki, 2013).

In healthy humans, besides the behavioral effects that resemble psychotic symptoms, NMDAR antagonism via acute ketamine administration decreases the amplitude of some ERPs (e.g., Gunduz-Bruce et al., 2012) and increases broadband gamma power during rest (Muthukumaraswamy et al., 2015; Rivolta et al., 2015). This increase in spontaneous gamma power is consistent with increased cortical excitability, which has been shown with TMS to result from NMDAR antagonist administration (Di Lazzaro et al., 2003). But in contrast to the animal model studies, ketamine in humans decreases resting state cortical delta activity (Muthukumaraswamy et al., 2015).

18.6.2 Reduced Synaptic Connectivity

The scalp-recorded EEG and MEG mainly reflect the spatially and temporally summated post-synaptic potentials in the dendritic trees of cortical pyramidal cells (Nunez, 1981). Dendritic spines are the main sources of excitatory input to pyramidal cells, and reductions in dendritic spine density have been consistently reported in the prefrontal cortex in schizophrenia (e.g., Glantz & Lewis, 2000; Sweet et al., 2009). Decreased dendritic spine density is thought to reflect a general reduction in synaptic connectivity (both excitatory and inhibitory) in local cortical circuits, and is manifested as reduced cortical volume and thickness at the macroscopic level of structural magnetic resonance imaging studies of schizophrenia (Selemon & Goldman-Rakic, 1999). Thus, deficits of oscillation power in schizophrenia could be caused by reduced synaptic connectivity in the generating cortex. In support of this hypothesis, a few studies have found correlations between the structure of the superior temporal gyrus (STG) (which contains auditory sensory and association cortex) and ASSR measures. Edgar and colleagues (2014) found a correlation between left STG cortical thickness and 40-Hz ASSR total power in the left hemisphere, although in controls and not schizophrenia patients. Similarly, Kim and colleagues (2019) reported a correlation between right STG gray volume and 40-Hz ASSR evoked power in controls but not patients. Hirano and colleagues (2020) found that the gray matter volume of left Heschl's gyrus (located in the anterior portion of the STG and containing primary auditory cortex, the main ASSR generator) was correlated with 40-Hz ASSR PLF and inversely correlated with spontaneous gamma power in patients. While these three sets of findings involve different oscillation measures and diagnostic groups, they do provide support for the idea that cortical structure can influence gamma oscillations.

Using a computational model, we simulated the reduction of excitatory and inhibitory connections in a cortical region and found that pyramidal cell excitability increased as excitatory and inhibitory connections were eliminated (Spencer, 2009). A recent experimental study showed that reduced density of dendritic spines was associated with increased pyramidal cell excitability and schizophrenia-like behaviors in rodents, which could be reversed with antipsychotic medication (Kim et al., 2015). Thus, increased spontaneous gamma power in schizophrenia might also be associated with decreased dendritic spine density and cortical volume/thickness deficits (Hirano et al., 2020). As subchronic administration of phencyclidine, an NMDAR antagonist, can result in decreased dendritic spine density (Hajszan et al., 2006), NMDAR hypofunction could be involved in synaptic connectivity reductions in schizophrenia as well.

18.7 Conclusions and Future Directions

Within the last two decades, research into neural oscillations has generated a body of findings that has made a substantial impact on our conceptualization of schizophrenia, understanding of its pathophysiology, and methods of testing new treatments. The study of gamma oscillations has had the most success, due to the cross-fertilization of clinical studies with neuropathological, basic, and computational research. As details of the specific cellular elements that are disturbed in schizophrenia are revealed, it is possible to construct hypotheses about the consequences of these circuit abnormalities in animal and computational models and try to relate them to the human clinical data.

The field has matured in step with the broader field of oscillatory brain dynamics. In most studies the potential sources of EEG artifacts are now understood and corrected for, and the importance of having adequate numbers of epochs is appreciated. Studies have better statistical power, and mass univariate statistics are being used with the appropriate corrections for multiple tests. The influence of confounds like common reference and volume conduction is more widely appreciated for connectivity studies. Thus, the potential for false positive findings seems to be less than before.

And yet, to this researcher, the promise of oscillation research for ultimately helping develop new treatments for schizophrenia (and other neuropsychiatric disorders) seems far from being fulfilled. The most complete body of research has been on the gamma ASSR, which is being used as a biomarker in translational research, yet this oscillation has limited relevance for cognition or symptoms in schizophrenia (except possibly for auditory hallucinations). The areas of research into perceptual organization and working memory, while being the best motivated by psychological and mechanistic theories, have not yet yielded reliable measures that can be used in translational research. So, much more work remains to be done. I still believe that as research progresses into the neural mechanisms underlying different oscillations, and the nature of neural circuit abnormalities in schizophrenia is understood in greater detail, it will become possible to construct and test mechanistic hypotheses with greater precision, both in vivo and *in silico*. Ultimately this research effort may come full circle, leading to the use of neurostimulation methods to modulate and "correct" aberrant oscillations in individuals with schizophrenia, and in this way, re-integrate the broken mind.

ACKNOWLEDGMENTS

This work was supported by grants I01 CX001443 from the US Department of Veterans Affairs, R01 MH093450 from the National Institutes of Health, and an Independent

Investigator Award from the Brain and Behavior Research Foundation. The author reports no conflicts of interest.

REFERENCES

Amann, L. C., Gandal, M. J., Halene, T. B., Ehrlichman, R. S., White, S. L., McCarren, H. S., & Siegel, S. J. (2010). Mouse behavioral endophenotypes for schizophrenia. *Brain Research Bulletin, 83*(3–4), 147–161.

Arieli, A., Sterkin, A., Grinvald, A., & Aertsen, A. (1996). Dynamics of ongoing activity: Explanation of the large variability in evoked cortical responses. *Science, 273*, 1868–1871.

Baran, B., Karahanoğlu, F. I., Mylonas, D., Demanuele, C., Vangel, M., Stickgold, R., Anticevic, A., & Manoach, D. S. (2019). Increased thalamocortical connectivity in schizophrenia correlates with sleep spindle deficits: evidence for a common pathophysiology. *Biological Psychiatry: Cognitive Neuroscience and Neuroimaging, 4*, 706–714.

Barch, D. M. (2005). The cognitive neuroscience of schizophrenia. *Annual Review of Clinical Psychology, 1*, 321–353.

Belforte, J. E., Zsiros, V., Sklar, E. R., Jiang, Z., Yu, G., Li, Y., ... Nakazawa, K. (2010). Postnatal NMDA receptor ablation in corticolimbic interneurons confers schizophrenia-like phenotypes. *Nature Neuroscience, 13*(1), 76–83.

Bleuler, E. (1911/1950). *Dementia Praecox or the Group of Schizophrenias*. International Universities Press.

Boutros, N. N., Arfken, C., Galderisi, S., Warrick, J., Pratt, G., & Iacono, W. (2008). The status of spectral EEG abnormality as a diagnostic test for schizophrenia. *Schizophrenia Research, 99*(1–3), 225–237.

Brenner, C. A., Kieffaber, P. D., Clementz, B. A., Johannesen, J. K., Shekhar, A., O'Donnell, B. F., & Hetrick, W. P. (2009). Event-related potential abnormalities in schizophrenia: A failure to "gate in" salient information? *Schizophrenia Research, 113*(2–3), 332–338.

Brenner, C., Sporns, O., Lysaker, P. H., & O'Donnell, B. F. (2003). EEG synchronization to modulated auditory tones in schizophrenia, schizoaffective disorder, and schizotypal personality disorder. *American Journal of Psychiatry, 160*(December), 2238–2240.

Buchmann, A., Dentico, D., Peterson, M. J., Riedner, B. A., Sarasso, S., Massimini, M., Tononi, G., & Ferrarelli, F. (2014). Reduced mediodorsal thalamic volume and prefrontal cortical spindle activity in schizophrenia. *NeuroImage, 102*, 540–547.

Burke, J. F., Ramayya, A. G., & Kahana, M. J. (2015). Human intracranial high-frequency activity during memory processing: neural oscillations or stochastic volatility? *Current Opinion in Neurobiology, 31*, 104–110.

Buzsáki, G. (2006). *Rhythms of the brain*. Oxford University Press.

Buzsáki, G. & Draguhn, A. (2004). Neuronal oscillations in cortical networks. *Science, 304*, 1926–1929.

Buzsáki, G. & Wang, X.-J. (2012). Mechanisms of gamma oscillations. *Annual Review of Neuroscience, 35*, 203–225.

Canali, P., Sarasso, S., Rosanova, M., Casarotto, S., Sferrazza-Papa, G., Gosseries, O., ... Benedetti, F. (2015). Shared reduction of oscillatory natural frequencies in bipolar disorder, major depressive disorder and schizophrenia. *Journal of Affective Disorders, 184*, 111–115.

Canolty, R. T. & Knight, R. T. (2010). The functional role of cross-frequency coupling. *Trends in Cognitive Sciences, 14*, 506–515.

Carlén, M., Meletis, K., Siegle, J. H., Cardin, J. A, Futai, K., Vierling-Claassen, D., ... Tsai, L.-H. (2012). A critical role for NMDA receptors in parvalbumin interneurons for gamma rhythm induction and behavior. *Molecular Psychiatry, 17*(5), 537–548.

Cho, R. Y., Konecky, R. O., & Carter, C. S. (2006). Impairments in frontal cortical gamma synchrony and cognitive control in schizophrenia. *Proceedings of the National Academy of Sciences of the United States of America, 103*(52), 19878–19883.

Clementz, B. A., Blumenfeld, L. D., & Cobb, S. (1997). The gamma band response may account for poor P50 suppression in schizophrenia. *Neuroreport, 8*(18), 3889–3893.

Clementz, B. A., Keil, A., & Kissler, J. (2004). Aberrant brain dynamics in schizophrenia: Delayed buildup and prolonged decay of the visual steady-state response. *Cognitive Brain Research, 18*(2), 121–129. doi:10.1016/j.cogbrainres.2003.09.007

Clementz, B., Sponheim, S., Iacono, W. G., & Beiser, M. (1994). Resting EEG in first-episode schizophrenia patients, bipolar psychosis patients, and their first-degree relatives. *Psychophysiology, 31*, 486–494.

D'Agostino, A., Castelnovo, A., Cavallotti, S., Casetta, C., Marcatili, M., Gambini, O., Canevini, M., Tononi, G., Riedner, B., Ferrarelli, F., & Sarasso, S. (2018). Sleep endophenotypes of schizophrenia: slow waves and sleep spindles in unaffected first-degree relatives. *NPJ Schizophrenia, 4*(1), 2.

Davidson, M., Kapara, O., Goldberg, S., Yoffe, R., Noy, S., & Weiser, M. (2016). A nation-wide study on the percentage of schizophrenia and bipolar disorder patients who earn minimum wage or above. *Schizophrenia Bulletin, 42*(2), 443–447.

Diekelmann, S. & Born, J. (2010). The memory function of sleep. *Nature Reviews Neuroscience, 11*, 114–126.

Di Lazzaro, V., Oliviero, A., Profice, P., Pennisi, M. A., Pilato, F., Zito, G., ... Tonali, P. A. (2003). Ketamine increases human motor cortex excitability to transcranial magnetic stimulation. *The Journal of Physiology, 547*(Pt 2), 485–496.

Ding, N. & Simon, J. Z. (2013). Power and phase properties of oscillatory neural responses in the presence of background activity. *Journal of Computational Neuroscience, 34*(2), 337–343.

Draganova, R., Ross, B., Wollbrink, A., & Pantev, C. (2008). Cortical steady-state responses to central and peripheral auditory beats. *Cerebral Cortex, 18*(5), 1193–1200.

Eckhorn, R., Bauer, R., Jordan, W., Brosch, M., Kruse, W., Munk, M., & Reitboeck, H. (1988). Coherent oscillations: a mechanism of feature linking in the visual cortex? Multiple electrode and correlation analyses in the cat. *Biological Cybernetics, 60*, 121–130.

Edgar, J. C., Chen, Y.-H., Lanza, M., Howell, B., Chow, V. Y., Heiken, K., ... Cañive, J. M. (2014). Cortical thickness as a contributor to abnormal oscillations in schizophrenia? *NeuroImage: Clinical, 4*, 122–129.

Ethridge, L., Moratti, S., Gao, Y., Keil, A., & Clementz, B. A. (2011). Sustained versus transient brain responses in schizophrenia: The role of intrinsic neural activity. *Schizophrenia Research, 133*(1–3), 106–111.

Farahani, E. D., Goossens, T., Wouters, J., & van Wieringen, A. (2017). Spatiotemporal reconstruction of auditory steady-state responses to acoustic amplitude modulations: Potential sources beyond the auditory pathway. *NeuroImage, 148*, 240–253.

Ferrarelli, F., Huber, R., Peterson, M. J., Massimini, M., Murphy, M., Riedner, B. A., ... Tononi, G. (2007). Reduced sleep spindle activity in schizophrenia patients. *American Journal of Psychiatry, 164*(3), 483–492.

Ferrarelli, F., Kaskie, R. E., Graziano, B., Reis, C. C., & Casali, A. G. (2019). Abnormalities in the evoked frontal oscillatory activity of first-episode psychosis: A TMS/EEG study. *Schizophrenia Research*, *206*, 436–439.

Ferrarelli, F., Massimini, M., Peterson, M. J., Riedner, B. A., Lazar, M., Murphy, M. J., . . . Tononi, G. (2008). Reduced evoked gamma oscillations in the frontal cortex in schizophrenia patients: A TMS/EEG study. *American Journal of Psychiatry*, *165*(8), 996–1005.

Ferrarelli, F., Peterson, M., Sarasso, S., Riedner, B. A., Murphy, M. J., Benca, R. M., . . . Tononi, G. (2010). Thalamic dysfunction in schizophrenia suggested by whole-night deficits in slow and fast spindles. *American Journal of Psychiatry*, *167*(November), 1339–1348.

Ferrarelli, F., Sarasso, S., Guller, Y., Riedner, B. A, Peterson, M. J., Bellesi, M., ... Tononi, G. (2012). Reduced natural oscillatory frequency of frontal thalamocortical circuits in schizophrenia. *Archives of General Psychiatry*, *69*, 766–774.

Frankle, W. G., Cho, R. Y., Prasad, K. M., Mason, N. S., Paris, J., Himes, M. L., . . . Narendran, R. (2015). In vivo measurement of GABA transmission in healthy subjects and schizophrenia patients. *American Journal of Psychiatry*, *172*, 1148–1159.

Friston, K. J. & Frith, C. D. (1995). Schizophrenia: A disconnection syndrome? *Clinical Neuroscience*, *3*, 89–97.

Galambos, R., Makei, S., & Talmachoff, P. J. (1981). A 40-Hz auditory potential recorded from the human scalp. *Proceedings of the National Academy of Sciences of the United States of America*, 78(4), 2643–2647.

Gallinat, J., Winterer, G., Herrmann, C. S., & Senkowski, D. (2004). Reduced oscillatory gamma-band responses in unmedicated schizophrenic patients indicate impaired frontal network processing. *Clinical Neurophysiology*, *115*(8), 1863–1874.

Ghorashi, S. & Spencer, K. M. (2015). Attentional load effects on beta oscillations in healthy and schizophrenic individuals. *Frontiers in Psychiatry*, *6*(October), 149.

Glantz, L. A. & Lewis, D. A. (2000). Decreased dendritic spine density on prefrontal cortical pyramidal neurons in schizophrenia. *Archives of General Psychiatry*, *57*(1), 65–73.

Gray, C. M., Konig, P., Engel, A. K., & Singer, W. (1989). Oscillatory responses in cat visual cortex exhibit inter-columnar synchronization which reflects global stimulus properties. *Nature*, *338*, 334–337.

Gray, C. M. & Singer, W. (1989). Stimulus-specific neuronal oscillations in orientation columns of cat visual cortex. *Proceedings of the National Academy of Sciences of the United States of America*, *86*, 1698–1702.

Grent-'t-Jong, T., Gajwani, R., Gross, J., Gumley, A. I., Krishnadas, R., Lawrie, S. M., . . . Uhlhaas, P. J. (2020). Association of magnetoencephalographically measured high-frequency oscillations in visual cortex with circuit dysfunctions in local and large-scale networks during emerging psychosis. *JAMA Psychiatry*, *77*(8), 852–862.

Grent-'t-Jong, T., Gross, J., Goense, J., Wibral, M., Gajwani, R., Gumley, A. I., ... Uhlhaas, P. J. (2018). Resting-state gamma-band power alterations in schizophrenia reveal E/I-balance abnormalities across illness-stages. *eLife*, *7*, e37799.

Grent-'t-Jong, T., Rivolta, D., Sauer, A., Grube, M., Singer, W., Wibral, M., & Uhlhaas, P. J. (2016). MEG-measured visually induced gamma-band oscillations in chronic schizophrenia: Evidence for impaired generation of rhythmic activity in ventral stream regions. *Schizophrenia Research*, *176*(2–3), 177–185.

Grützner, C., Wibral, M., Sun, L., Rivolta, D., Singer, W., Maurer, K., & Uhlhaas, P. J. (2013). Deficits in high- (>60 Hz) gamma-band oscillations during visual processing in schizophrenia. *Frontiers in Human Neuroscience*, 7(March), 88.

Guevara, R., Luis Pérez Velazquez, J., Nenadovic, V., Wennberg, R., Senjanovic, G., & Garcia Dominguez, L. (2005). Phase synchronization measurements using electroencephalographic recordings: What can we really say about neuronal synchrony? *Neuroinformatics*, *3*, 301–313.

Gunduz-Bruce, H., Reinhart, R. M. G., Roach, B. J., Gueorguieva, R., Oliver, S., D'Souza, D. C., ... Mathalon, D. H. (2012). Glutamatergic modulation of auditory information processing in the human brain. *Biological Psychiatry*, *71*(11), 969–977.

Haenschel, C., Bittner, R. a, Waltz, J., Haertling, F., Wibral, M., Singer, W., ... Rodriguez, E. (2009). Cortical oscillatory activity is critical for working memory as revealed by deficits in early-onset schizophrenia. *Journal of Neuroscience*, *29*(30), 9481–9489.

Hajszan, T., Leranth, C., & Roth, R. H. (2006). Subchronic phencyclidine treatment decreases the number of dendritic spine synapses in the rat prefrontal cortex. *Biological Psychiatry*, *60*(6), 639–644.

Hall, M., Taylor, G., Salisbury, D. F., & Levy, D. L. (2011a). Sensory gating event-related potentials and oscillations in schizophrenia patients and their unaffected relatives. *Schizophrenia Bulletin*, *37*(6), 1187–1199.

Hall, M., Taylor, G., Sham, P., Schulze, K., Rijsdijk, F., Picchioni, M., ... Salisbury, D. F. (2011). The early auditory gamma-band response is heritable and a putative endophenotype of schizophrenia. *Schizophrenia Bulletin*, *37*(4), 778–787.

Hall, S. D., Holliday, I. E., Hillebrand, A., Singh, K. D., Furlong, P. L., Hadjipapas, A., & Barnes, G. R. (2005). The missing link: Analogous human and primate cortical gamma oscillations. *NeuroImage*, *26*, 13–17.

Hamm, J. P., Bobilev, A. M., Hayrynen, L. K., Hudgens-Haney, M. E., Oliver, W. T., Parker, D. A., ... Clementz, B. A. (2015). Stimulus train duration but not attention moderates γ-band entrainment abnormalities in schizophrenia. *Schizophrenia Research*, *165*(1), 97–102.

Hamm, J. P., Gilmore, C. S., Picchetti, N. A. M., Sponheim, S. R., & Clementz, B. A. (2011). Abnormalities of neuronal oscillations and temporal integration to low- and high-frequency auditory stimulation in schizophrenia. *Biological Psychiatry*, *69*(10), 989–996.

Hebb, D. O. (1949). *The organization of behavior*. Wiley.

Herdman, A. T., Lins, O., Roon, P. Van, Stapells, D. R., Scherg, M., & Picton, T. W. (2002). Intracerebral sources of human auditory steady-state responses. *Brain Topography*, *15*, 69–86.

Herdman, A. T., Wollbrink, A., Chau, W., Ishii, R., Ross, B., & Pantev, C. (2003). Determination of activation areas in the human auditory cortex by means of synthetic aperture magnetometry. *NeuroImage*, *20*(2), 995–1005.

Hiltunen, T., Kantola, J., Abou Elseoud, A., Lepola, P., Suominen, K., Starck, T., ... Palva JM (2014). Infra-slow EEG fluctuations are correlated with resting-state network dynamics in fMRI. *Journal of Neuroscience*, *34*, 356–362.

Hirano, Y., Oribe, N., Kanba, S., Onitsuka, T., Nestor, P. G., & Spencer, K. M. (2015). Spontaneous gamma activity in schizophrenia. *JAMA Psychiatry*, *72*(8), 813–821.

Hirano, Y., Oribe, N., Onitsuka, T., Kanba, S., Nestor, P. G., Hosokawa, T., ... Spencer, K. M. (2020). Auditory cortex volume and gamma oscillation abnormalities in schizophrenia. *Clinical EEG and Neuroscience*, *51*, 244–251.

Hirvonen, J., Wibral, M., Palva, J. M., Singer, W., Uhlhaas, P., & Palva, S. (2017). Whole-brain source-reconstructed MEG-data reveal reduced long-range synchronization in chronic schizophrenia. *eNeuro*, *4*(5), e0338–17.2017.

Hoftman, G. D., Dienel, S. J., Bazmi, H. H., Zhang, Y., Chen, K., & Lewis, D. A. (2018). Altered gradients of glutamate and gamma-aminobutyric acid transcripts in the cortical visuospatial working memory network in schizophrenia. *Biological Psychiatry*, *83*(8), 670–679.

Homayoun, H. & Moghaddam, B. (2007). NMDA receptor hypofunction produces opposite effects on prefrontal cortex interneurons and pyramidal neurons. *Journal of Neuroscience, 27*(43), 11496–11500.

Hong, L. E., Summerfelt, A., McMahon, R., Adami, H., Francis, G., Elliott, A., ... Thaker, G. K. (2004). Evoked gamma band synchronization and the liability for schizophrenia. *Schizophrenia Research, 70*(2–3), 293–302.

Hoogenboom, N., Schoffelen, J., Oostenveld, R., Parkes, L. M., & Fries, P. (2006). Localizing human visual gamma-band activity in frequency, time and space. *NeuroImage, 29*(3), 764–773.

Hunt, M. J. & Kasicki, S. (2013). A systematic review of the effects of NMDA receptor antagonists on oscillatory activity recorded in vivo. *Journal of Psychopharmacology, 27*(11), 972–986.

Isomura, S., Onitsuka, T., Tsuchimoto, R., Nakamura, I., Hirano, S., Oda, Y., ... Kanba, S. (2016). Differentiation between major depressive disorder and bipolar disorder by auditory steady-state responses. *Journal of Affective Disorders, 190*, 800–806.

Itil, T. M. (1977). Qualitative and quantitative EEG findings in schizophrenia. *Schizophrenia Bulletin, 3*(1), 61–79.

Jacobsen, R. B., Ulrich, D., & Huguenard, J. R. (2001). GABA(B) and NMDA receptors contribute to spindle-like oscillations in rat thalamus in vitro. *Journal of Neurophysiology, 86*, 1365–1375.

Jadi, M. P., Behrens, M. M., & Sejnowski, T. J. (2016). Abnormal gamma oscillations in N-methyl-D-aspartate receptor hypofunction models of schizophrenia. *Biological Psychiatry, 79*(9), 716–726.

Jansma, J. M., Ramsey, N. F., van der Wee, N. J. A., & Kahn, R. S. (2004). Working memory capacity in schizophrenia: A parametric fMRI study. *Schizophrenia Research, 68*(2–3), 159–171.

Javitt, D. C., Spencer, K. M., Thaker, G. K., Winterer, G., & Hajós, M. (2008). Neurophysiological biomarkers for drug development in schizophrenia. *Nature Reviews Drug Discovery, 7*(1), 68–83.

Jin, Y., Castellanos Jr, A., Solis Jr, E. R., & Potkin, S. G. (2000). EEG resonant responses in schizophrenia: A photic driving study with improved harmonic resolution. *Schizophrenia Research, 44*, 213–220.

Jung, T. P., Makeig, S., Humphries, C., Lee, T. W., McKeown, M. J., Iragui, V., & Sejnowski, T. J. (2000). Removing electroencephalographic artifacts by blind source separation. *Psychophysiology, 37*, 163–178.

Keren, A. S., Yuval-Greenberg, S., & Deouell, L. Y. (2010). Saccadic spike potentials in gamma-band EEG: characterization, detection and suppression. *NeuroImage, 49*(3), 2248–2263.

Kim, S., Jang, S.-K., Kim, D.-W., Shim, M., Kim, Y.-W., Im, C.-H., & Lee, S.-H. (2019). Cortical volume and 40-Hz auditory-steady-state responses in patients with schizophrenia and healthy controls. *NeuroImage: Clinical, 22*, 101732.

Kim, I. H., Rossi, M. A., Aryal, D. K., Racz, B., Kim, N., Uezu, A., ... Soderling, S. H. (2015). Spine pruning drives antipsychotic-sensitive locomotion via circuit control of striatal dopamine. *Nature Neuroscience, 18*(6), 883–891.

Kömek, K., Bard Ermentrout, G., Walker, C. P., & Cho, R. Y. (2012). Dopamine and gamma band synchrony in schizophrenia—insights from computational and empirical studies. *European Journal of Neuroscience, 36*(2), 2146–2155.

Korotkova, T., Fuchs, E. C., Ponomarenko, A., von Engelhardt, J., & Monyer, H. (2010). NMDA receptor ablation on parvalbumin-positive interneurons impairs hippocampal synchrony, spatial representations, and working memory. *Neuron, 68*(3), 557–569.

Krishnan, G. P., Hetrick, W. P., Brenner, C. A., Shekhar, A., Steffen, A. N., & O'Donnell, B. F. (2009). Steady state and induced auditory gamma deficits in schizophrenia. *NeuroImage*, *47*(4), 1711–1719.

Krishnan, G. P., Vohs, J. L., Hetrick, W. P., Carroll, C. A., Shekhar, A., Bockbrader, M. A., & O'Donnell, B. F. (2005). Steady state visual evoked potential abnormalities in schizophrenia. *Clinical Neurophysiology*, *116*(3), 614–624.

Krystal, J. H., D'Souza, D. C., Mathalon, D., Perry, E., Belger, A., & Hoffman, R. (2003). NMDA receptor antagonist effects, cortical glutamatergic function, and schizophrenia: Toward a paradigm shift in medication development. *Psychopharmacology*, *169*(3–4), 215–233.

Kwon, J. S., O'Donnell, B. F., Wallenstein, G. V, Greene, R. W., Hirayasu, Y., Nestor, P. G., . . . McCarley, R. W. (1999). Gamma frequency-range abnormalities to auditory stimulation in schizophrenia. *Archives of General Psychiatry*, *56*(11), 1001–1005.

Lachaux, J. P., Rodriguez, E., Martinerie, J., & Varela, F. J. (1999). Measuring phase synchrony in brain signals. *Human Brain Mapping*, *8*(4), 194–208.

Lazarewicz, M. T., Ehrlichman, R. S., Maxwell, C. R., Gandal, M. J., Finkel, L. H., & Siegel, S. J. (2010). Ketamine modulates theta and gamma oscillations. *Journal of Cognitive Neuroscience*, *22*(7), 1452–1464.

Legget, K. T., Hild, A. K., Steinmetz, S. E., Simon, S. T., & Rojas, D. C. (2017). MEG and EEG demonstrate similar test-retest reliability of the 40 Hz auditory steady-state response. *International Journal of Psychophysiology*, *114*, 16–23.

Leicht, G., Andreou, C., Polomac, N., Lanig, C., Schöttle, D., Lambert, M., & Mulert, C. (2015). Reduced auditory evoked gamma band response and cognitive processing deficits in first episode schizophrenia. *The World Journal of Biological Psychiatry*, *16*(6), 387–397.

Leicht, G., Karch, S., Karamatskos, E., Giegling, I., Möller, H.-J., Hegerl, U., . . . Mulert, C. (2011). Alterations of the early auditory evoked gamma-band response in first-degree relatives of patients with schizophrenia: Hints to a new intermediate phenotype. *Journal of Psychiatric Research*, *45*(5), 699–705.

Leicht, G., Kirsch, V., Giegling, I., Karch, S., Hantschk, I., Möller, H.-J., . . . Mulert, C. (2010). Reduced early auditory evoked gamma-band response in patients with schizophrenia. *Biological Psychiatry*, *67*(3), 224–231.

Leishman, E., O'Donnell, B. F., Millward, J. B., Vohs, J. L., Rass, O., Krishnan, G. P., . . . Morzorati, S. L. (2015). Phencyclidine disrupts the auditory steady state response in rats. *PLoS One*, *10*(8), e0134979.

Lemere, F. (1936). The significance of individual differences in the Berger rhythm. *Brain*, *59*, 366–375.

Lenz, D., Fischer, S., Schadow, J., Bogerts, B., & Herrmann, C. S. (2011). Altered evoked γ-band responses as a neurophysiological marker of schizophrenia? *International Journal of Psychophysiology*, *79*(1), 25–31.

Leung, M., Cheung, C., Yu, K., Yip, B., Sham, P., Li, Q., . . . McAlonan, G. (2011). Gray matter in first-episode schizophrenia before and after antipsychotic drug treatment. Anatomical likelihood estimation meta-analyses with sample size weighting. *Schizophrenia Bulletin*, *37*(1), 199–211.

Lewis, D. A., Curley, A. A., Glausier, J. R., & Volk, D. W. (2012). Cortical parvalbumin interneurons and cognitive dysfunction in schizophrenia. *Trends in Neurosciences*, *35*(1), 57–67.

Light, G. A., Hsu, J. L., Hsieh, M. H., Meyer-Gomes, K., Sprock, J., Swerdlow, N. R., & Braff, D. L. (2006). Gamma band oscillations reveal neural network cortical coherence dysfunction in schizophrenia patients. *Biological Psychiatry*, *60*(11), 1231–1240.

Manoach, D. S., Demanuele, C., Wamsley, E. J., Vangel, M., Montrose, D. M., Miewald, J., ... Keshavan, M. S. (2014). Sleep spindle deficits in antipsychotic-naive early course schizophrenia and in non-psychotic first-degree relatives. *Frontiers in Human Neuroscience*, *8*(October), 762.

Manoach, D. S., Pan, J. Q., Purcell, S. M., & Stickgold, R. (2016). Reduced sleep spindles in schizophrenia: A treatable endophenotype that links risk genes to impaired cognition? *Biological Psychiatry*, *80*(8), 599–608.

McFadden, K. L., Steinmetz, S. E., Carroll, A. M., Simon, S. T., Wallace, A., & Rojas, D. C. (2014). Test-retest reliability of the 40 Hz EEG auditory steady-state response. *PLoS One*, *9*(1), e85748.

McGlashan, T. H. & Hoffman, R. E. (2000). Schizophrenia as a disorder of developmentally reduced synaptic connectivity. *Archives of General Psychiatry*, *57*(7), 637–648.

Minzenberg, M. J., Firl, A. J., Yoon, J. H., Gomes, G. C., Reinking, C., & Carter, C. S. (2010). Gamma oscillatory power is impaired during cognitive control independent of medication status in first-episode schizophrenia. *Neuropsychopharmacology*, *35*(13), 2590–2599.

Minzenberg, M. J., Laird, A. R., Thelen, S., Carter, C. S., & Glahn, D. C. (2009). Meta-analysis of 41 functional neuroimaging studies of executive function in schizophrenia. *Archives of General Psychiatry*, *66*(8), 811–822.

Minzenberg, M. J., Yoon, J. H., Cheng, Y., & Carter, C. S. (2015). Sustained modafinil treatment effects on control-related gamma oscillatory power in schizophrenia. *Neuropsychopharmacology*, *41*(5), 1231–1240.

Moghaddam, B. & Javitt, D. (2012). From revolution to evolution: the glutamate hypothesis of schizophrenia and its implication for treatment. *Neuropsychopharmacology*, *37*(1), 4–15.

Mooney, C. M. & Ferguson, G. A. (1951). A new closure test. *Canadian Journal of Psychology*, *5*, 129–133.

Mulert, C., Kirsch, V., Pascual-Marqui, R., McCarley, R. W., & Spencer, K. M. (2011). Long-range synchrony of γ oscillations and auditory hallucination symptoms in schizophrenia. *International Journal of Psychophysiology*, *79*(1), 55–63.

Murray, R. M. & Lewis, S. W. (1987). Is schizophrenia a neurodevelopmental disorder? *British Medical Journal*, *295*(6600), 681–682.

Muthukumaraswamy, S. D., Shaw, A. D., Jackson, L. E., Hall, J., Moran, R., & Saxena, N. (2015). Evidence that subanesthetic doses of ketamine cause sustained disruptions of NMDA and AMPA-mediated frontoparietal connectivity in humans. *Journal of Neuroscience*, *35*(33), 11694–11706.

Muthukumaraswamy, S. D. & Singh, K. D. (2011). A cautionary note on the interpretation of phase-locking estimates with concurrent changes in power. *Clinical Neurophysiology*, *122*(11), 2324–2325.

Muthukumaraswamy, S. D. & Singh, K. D. (2013). Visual gamma oscillations: The effects of stimulus type, visual field coverage and stimulus motion on MEG and EEG recordings. *NeuroImage*, *69*, 223–230.

Muthukumaraswamy, S. D., Singh, K. D., Swettenham, J. B., & Jones, D. K. (2010). Visual gamma oscillations and evoked responses: variability, repeatability and structural MRI correlates. *NeuroImage*, *49*(4), 3349–3357.

Narayanan, B., O'Neil, K., Berwise, C., Stevens, M. C., Calhoun, V. D., Clementz, B. A., ... Pearlson, G. D. (2014). Resting state electroencephalogram oscillatory abnormalities in schizophrenia and psychotic bipolar patients and their relatives from the bipolar and schizophrenia network on intermediate phenotypes study. *Biological Psychiatry*, *76*(6), 456–465.

Nunez, P. L. (1981). *Electric Fields of the Brain*. Oxford University Press.

O'Donnell, B. F., Vohs, J. L., Krishnan, G. P., Rass, O., Hetrick, W. P., & Morzorati, S. L. (2013). The auditory steady-state response (ASSR): A translational biomarker for schizophrenia. Supplements to Clinical Neurophysiology, *62*, 101–112.

Oribe, N., Hirano, Y., del Re, E., Seidman, L. J., Mesholam-Gately, R. I., Woodberry, K. A., ... Spencer, K. M. (2019). Progressive reduction of auditory evoked gamma in first episode schizophrenia but not clinical high risk individuals. *Schizophrenia Research*, *208*, 145–152.

Owen, M. J., Sawa, A., & Mortensen, P. B. (2016). Schizophrenia. *The Lancet*, *388*(10039), 86–97.

Parker, D. A., Hamm, J. P., McDowell, J. E., Keedy, S. K., Gershon, E. S., Ivleva, E. I., ... Clementz, B. A. (2019). Auditory steady-state EEG response across the schizo-bipolar spectrum. *Schizophrenia Research*, *209*, 218–226.

Palva, S. & Palva, J. M. (2012). Discovering oscillatory interaction networks with M/EEG: challenges and breakthroughs. *Trends in Cognitive Sciences*, *16*(4), 219–230.

Pantev, C., Makeig, S., Hoke, M., Galambos, R., Hampson, S., & Gallen, C. (1991). Human auditory evoked gamma-band magnetic fields. *Proceedings of the National Academy of Sciences of the United States of America*, *88*, 8996–9000.

Perälä, J., Suvisaari, J., Saarni, S. I., Kuoppasalmi, K., Isometsä, E., Pirkola, S., ... Lönnqvist, J. (2007). Lifetime prevalence of psychotic and bipolar I disorders in a general population. *Archives of General Psychiatry*, *64*(1), 19–28.

Perez, V. B., Roach, B. J., Woods, S. W., Srihari, V. H., McGlashan, T. H., Ford, J. M., & Mathalon, D. H. (2013). Early auditory gamma-band responses in patients at clinical high risk for schizophrenia. *Supplements to Clinical Neurophysiology*, *62*(415), 147–162.

Pastor, M. A., Artieda, J., Arbizu, J., Marti-climent, J. M., Pen, I., & Masdeu, J. C. (2002). Activation of Human Cerebral and Cerebellar Cortex by Auditory Stimulation at 40 Hz. *Journal of Neuroscience*, *22*(23), 10501–10506.

Phillips, W. A. & Silverstein, S. M. (2003). Convergence of biological and psychological perspectives on cognitive coordination in schizophrenia. *Behavioral and Brain Sciences*, *26*, 65–82.

Picton, T. W., John, M. S., Dimitrijevic, A., & Purcell, D. (2003). Human auditory steady-state responses. *International Journal of Audiology*, *42*, 177–219.

Pope, K. J., Fitzgibbon, S. P., Lewis, T. W., Whitham, E. M., & Willoughby, J. O. (2009). Relation of gamma oscillations in scalp recordings to muscular activity. *Brain Topography*, *22*(1), 13–17.

Popov, T., Jordanov, T., Weisz, N., Elbert, T., Rockstroh, B., & Miller, G. A. (2011). Evoked and induced oscillatory activity contributes to abnormal auditory sensory gating in schizophrenia. *NeuroImage*, *56*(1), 307–314.

Puvvada, K. C., Summerfelt, A., Du, X., Krishna, N., Kochunov, P., Rowland, L. M., ... & Hong, L. E. (2018). Delta vs gamma auditory steady state synchrony in schizophrenia. *Schizophrenia Bulletin*, *44*(2), 378–387.

Raichle, M. E. (2010). Two views of brain function. *Trends in Cognitive Sciences*, *14*, 180–190.

Ramyead, A., Kometer, M., Studerus, E., Koranyi, S., Ittig, S., Gschwandtner, U., ... Riecher-Rossler, A. (2015). Aberrant current source-density and lagged phase synchronization of neural oscillations as markers for emerging psychosis. *Schizophrenia Bulletin*, *41*(4), 919–929.

Ranlund, S., Nottage, J., Shaikh, M., Dutt, A., Constante, M., Walshe, M., ... Bramon, E. (2014). Resting EEG in psychosis and at-risk populations - A possible endophenotype? *Schizophrenia Research*, *153*(1–3), 96–102.

Rapoport, J. L., Giedd, J. N., & Gogtay, N. (2012). Neurodevelopmental model of schizophrenia: Update 2012. *Molecular Psychiatry*, *17*(12), 1228–1238.

Rass, O., Forsyth, J. K., Krishnan, G. P., Hetrick, W. P., Klaunig, M. J., Breier, A., ... Brenner, C. A. (2012). Auditory steady state response in the schizophrenia, first-degree relatives, and schizotypal personality disorder. *Schizophrenia Research*, *136*(1–3), 143–149.

Riečanský, I., Kašpárek, T., Rehulová, J., Katina, S., & Přikryl, R. (2010). Aberrant EEG responses to gamma-frequency visual stimulation in schizophrenia. *Schizophrenia Research*, *124*(1–3), 101–109.

Ringach, D. L. (2009). Spontaneous and driven cortical activity: Implications for computation. *Current Opinion in Neurobiology*, *19*, 439–444.

Rivolta, D., Heidegger, T., Scheller, B., Sauer, A., Schaum, M., Birkner, K., ... Uhlhaas, P. J. (2015). Ketamine dysregulates the amplitude and connectivity of high-frequency oscillations in cortical-subcortical networks in humans: evidence from resting-state magnetoencephalography-recordings. *Schizophrenia Bulletin*, *41*(5), 1105–1114.

Roach, B. J., Ford, J. M., Hoffman, R. E., & Mathalon, D. H. (2013). Converging evidence for gamma synchrony deficits in schizophrenia. *Supplements to Clinical Neurophysiology*, *62*, 163–180.

Roach, B. J. & Mathalon, D. H. (2008). Event-related EEG time-frequency analysis: An overview of measures and an analysis of early gamma band phase locking in schizophrenia. *Schizophrenia Bulletin*, *34*(5), 907–926.

Rodriguez, E., George, N., Lachaux, J.-P., Martinerie, J., Renault, B., & Varela, F. J. (1999). Perception's shadow: Long-distance synchronization of human brain activity. *Nature*, *397*, 430–433.

Rolls, E. T., Loh, M., Deco, G., & Winterer, G. (2008). Computational models of schizophrenia and dopamine modulation in the prefrontal cortex. *Nature Reviews Neuroscience*, *9*(9), 696–709.

Rosanova, M., Casali, A., Bellina, V., Resta, F., Mariotti, M., & Massimini, M. (2009). Natural frequencies of human corticothalamic circuits. *Journal of Neuroscience*, *29*(24), 7679–7685.

Sarkar, S., Katshu, M. Z. U. H., Nizamie, S. H., & Praharaj, S. K. (2010). Slow wave sleep deficits as a trait marker in patients with schizophrenia. *Schizophrenia Research*, *124*(1–3), 127–133. https://doi.org/10.1016/j.schres.2010.08.013

Schulman, J. J., Cancro, R., Lowe, S., Lu, F., Walton, K. D., & Llinás, R. R. (2011). Imaging of thalamocortical dysrhythmia in neuropsychiatry. *Frontiers in Human Neuroscience*, *5*(July), 69.

Selemon, L. D., & Goldman-Rakic, P. S. (1999). The reduced neuropil hypothesis: a circuit based model of schizophrenia. *Biological Psychiatry*, *45*, 17–25.

Sekimoto, M., Kato, M., Watanabe, T., Kajimura, N., & Takahashi, K. (2011). Cortical regional differences of delta waves during all-night sleep in schizophrenia. *Schizophrenia Research*, *126*(1–3), 284–290.

Siever, L. J. & Davis, K. L. (2004). The pathophysiology of schizophrenia disorders: perspectives from the spectrum. *American Journal of Psychiatry*, *161*(March), 398–413.

Singer, W. (1999). Neuronal synchrony: A versatile code for the definition of relations? *Neuron*, *24*, 49–65.

Singer, W. & Gray, C. M. (1995). Visual feature integration and the temporal correlation hypothesis. *Annual Review of Neuroscience*, *18*, 555–586.

Sivarao, D. V, Chen, P., Senapati, A., Yang, Y., Fernandes, A., Benitex, Y., ... Ahlijanian, M. K. (2016). 40 Hz auditory steady-state response is a pharmacodynamic biomarker for cortical NMDA receptors. *Neuropsychopharmacology, 41*(9), 2232–2240.

Spencer, K. M. (2009). The functional consequences of cortical circuit abnormalities on gamma oscillations in schizophrenia: insights from computational modeling. *Frontiers in Human Neuroscience, 3*, 33.

Spencer, K. M. (2012). Baseline gamma power during auditory steady-state stimulation in schizophrenia. *Frontiers in Human Neuroscience, 5*(January), 190.

Spencer, K. M., & Ghorashi, S. (2014). Oscillatory dynamics of Gestalt perception in schizophrenia revisited. *Frontiers in Psychology, 5*(February), 68.

Spencer, K., Nestor, P., Niznikiewicz, M. A., Salisbury, D. F., Shenton, M. E., & McCarley, R. W. (2003). Abnormal neural synchrony in schizophrenia. *Journal of Neuroscience, 23*(19), 7407–7411.

Spencer, K. M., Nestor, P. G., Perlmutter, R., Niznikiewicz, M. A, Klump, M. C., Frumin, M., ... McCarley, R. W. (2004). Neural synchrony indexes disordered perception and cognition in schizophrenia. *Proceedings of the National Academy of Sciences of the United States of America, 101*(49), 17288–17293.

Spencer, K. M., Niznikiewicz, M. A, Nestor, P. G., Shenton, M. E., & McCarley, R. W. (2009). Left auditory cortex gamma synchronization and auditory hallucination symptoms in schizophrenia. *BMC Neuroscience, 10*, 85.

Spencer, K. M., Niznikiewicz, M. A., Shenton, M. E., & McCarley, R. W. (2008b). Sensory-evoked gamma oscillations in chronic schizophrenia. *Biological Psychiatry, 63*(8), 744–747.

Spencer, K. M., Salisbury, D. F., Shenton, M. E., & McCarley, R. W. (2008a). Gamma-band auditory steady-state responses are impaired in first episode psychosis. *Biological Psychiatry, 64*(5), 369–375.

Sponheim, S. R., Clementz, B. a, Iacono, W. G., & Beiser, M. (2000). Clinical and biological concomitants of resting state EEG power abnormalities in schizophrenia. *Biological Psychiatry, 48*(11), 1088–1097.

Staresina, B. P., Bergmann, T. O., Bonnefond, M., van der Meij, R., Jensen, O., Deuker, L., ... Fell, J. (2015). Hierarchical nesting of slow oscillations, spindles and ripples in the human hippocampus during sleep. *Nature Neuroscience, 18*(11), 1679–1686.

Steriade, M. (2006). Grouping of brain rhythms in corticothalamic systems. *Neuroscience, 137*(4), 1087–1106.

Sullivan, E. M., Timi, P., Hong, L. E., & O'Donnell, P. (2015). Effects of NMDA and GABA-A receptor antagonism on auditory steady-state synchronization in awake behaving rats. *International Journal of Neuropsychopharmacology, 18*(7), pyu118.

Sun, L., Castellanos, N., Grützner, C., Koethe, D., Rivolta, D., Wibral, M., ... Uhlhaas, P. J. (2013). Evidence for dysregulated high-frequency oscillations during sensory processing in medication-naïve, first episode schizophrenia. *Schizophrenia Research, 150*(2–3), 519–525.

Sweet, R. A., Henteleff, R. A., Zhang, W., Sampson, A. R., & Lewis, D. A. (2009). Reduced dendritic spine density in auditory cortex of subjects with schizophrenia. *Neuropsychopharmacology, 34*(2), 374–389.

Tada, M., Nagai, T., Kirihara, K., Koike, S., Suga, M., Araki, T., ... Kasai, K. (2016). Differential alterations of auditory gamma oscillatory responses between pre-onset high-risk individuals and first-episode schizophrenia. *Cerebral Cortex, 26*(3), 1027–1035.

Tallon-Baudry, C., Bertrand, O., Delpuech, C., & Pernier, J. (1996). Stimulus specificity of phase-locked and non-phase-locked 40 Hz visual responses in human. *Journal of Neuroscience, 16*(13), 4240–4249.

Tan, H.-R. M., Gross, J., & Uhlhaas, P. J. (2015). MEG—measured auditory steady-state oscillations show high test–retest reliability: A sensor and source-space analysis. *NeuroImage, 122*, 417–426.

Tatard-Leitman, V. M., Jutzeler, C. R., Suh, J., Saunders, J. A., Billingslea, E. N., Morita, S., … Siegel, S. J. (2015). Pyramidal cell selective ablation of N-methyl-D-aspartate receptor 1 causes increase in cellular and network excitability. *Biological Psychiatry, 77*(6), 556–568.

Taylor, G. W., McCarley, R. W., & Salisbury, D. F. (2013). Early auditory gamma band response abnormalities in first hospitalized schizophrenia. *Supplements to Clinical Neurophysiology, 62*, 131–145.

Teale, P., Collins, D., Maharajh, K., Rojas, D. C., Kronberg, E., & Reite, M. (2008). Cortical source estimates of gamma band amplitude and phase are different in schizophrenia. *NeuroImage, 42*(4), 1481–1489.

Thuné, H., Recasens, M., & Uhlhaas, P. J. (2016). The 40-Hz auditory steady-state response in patients with schizophrenia. *JAMA Psychiatry, 73*(11), 1145–1153.

Tsodyks, M., Kenet, T., Grinvald, A., & Arieli, A. (1999). Linking spontaneous activity of single cortical neurons and the underlying functional architecture. *Science, 286*, 1943–1946.

Tsuchimoto, R., Kanba, S., Hirano, S., Oribe, N., Ueno, T., Hirano, Y., … Onitsuka, T. (2011). Reduced high and low frequency gamma synchronization in patients with chronic schizophrenia. *Schizophrenia Research, 133*(1–3), 99–105.

Uhlhaas, P. J., Linden, D. E. J., Singer, W., Haenschel, C., Lindner, M., Maurer, K., & Rodriguez, E. (2006a). Dysfunctional long-range coordination of neural activity during Gestalt perception in schizophrenia. *Journal of Neuroscience, 26*(31), 8168–8175.

Uhlhaas, P. J., Phillips, W. a, Mitchell, G., & Silverstein, S. M. (2006b). Perceptual grouping in disorganized schizophrenia. *Psychiatry Research, 145*(2–3), 105–117.

van Tricht, M. J., Ruhrmann, S., Arns, M., Müller, R., Bodatsch, M., Velthorst, E., … Nieman, D. H. (2014). Can quantitative EEG measures predict clinical outcome in subjects at clinical high risk for psychosis? A prospective multicenter study. *Schizophrenia Research, 153*(1–3), 42–47.

Wamsley, E. J., Tucker, M. A., Shinn, A. K., Ono, K. E., McKinley, S. K., Ely, A. V, … Manoach, D. S. (2012). Reduced sleep spindles and spindle coherence in schizophrenia: Mechanisms of impaired memory consolidation? *Biological Psychiatry, 71*(2), 154–161.

Weinberger, D. R. (1987). Implications of normal brain development for the pathogenesis of schizophrenia. *Archives of General Psychiatry, 44*(7), 660–669.

Wilson, T. W., Hernandez, O. O., Asherin, R. M., Teale, P. D., Reite, M. L., & Rojas, D. C. (2008). Cortical gamma generators suggest abnormal auditory circuitry in early-onset psychosis. *Cerebral Cortex, 18*(2), 371–378.

Winterer, G., Coppola, R., Goldberg, T. E., Egan, M. F., Jones, D. W., Sanchez, C. E., & Weinberger, D. R. (2004). Prefrontal broadband noise, working memory, and genetic risk for schizophrenia. *American Journal of Psychiatry, 161*(March), 490–500.

Winterer, G., Egan, M. F., Kolachana, B. S., Goldberg, T. E., Coppola, R., & Weinberger, D. R. (2006). Prefrontal electrophysiologic "noise" and catechol-O-methyltransferase genotype in schizophrenia. *Biological Psychiatry, 60*(6), 578–584.

Winterer, G., Ziller, M., Dorn, H., Frick, K., Mulert, C., Wuebben, Y., ... Coppola, R. (2000). Schizophrenia: Reduced signal-to-noise ratio and impaired phase-locking during information processing. *Clinical Neurophysiology, 111*(5), 837–849.

Wynn, J. K., Roach, B. J., Lee, J., Horan, W. P., Ford, J. M., Jimenez, A. M., & Green, M. F. (2015). EEG findings of reduced neural synchronization during visual integration in schizophrenia. *PLoS One, 10*(3), e0119849.

Yizhar, O., Fenno, L. E., Prigge, M., Schneider, F., Davidson, T. J., O'Shea, D. J., ... Deisseroth, K. (2011). Neocortical excitation/inhibition balance in information processing and social dysfunction. *Nature, 477*(7363), 171–178.

Zaehle, T., Lenz, D., Ohl, F. W., & Herrmann, C. S. (2010). Resonance phenomena in the human auditory cortex: Individual resonance frequencies of the cerebral cortex determine electrophysiological responses. *Experimental Brain Research, 203*(3), 629–635.

Zhang, Y., Yoshida, T., Katz, D. B., & Lisman, J. E. (2012). NMDAR antagonist action in thalamus imposes δ oscillations on the hippocampus. *Journal of Neurophysiology, 107*(11), 3181–3189.

Zhou, T.-H., Mueller, N. E., Spencer, K. M., Mallya, S. G., Lewandowski, K. E., Norris, L. A., ... Hall, M.-H. (2018). Auditory steady state response deficits are associated with symptom severity and poor functioning in patients with psychotic disorder. *Schizophrenia Research, 201*, 278–286.

CHAPTER 19

EEG FREQUENCY TECHNIQUES FOR IMAGING CONTROL FUNCTIONS IN ANXIETY

JASON S. MOSER, COURTNEY LOUIS, LILIANNE GLOE, STEFANIE RUSSMAN BLOCK, AND SPENCER FIX

19.1 Overview and Introduction

This chapter reviews and discusses the application of EEG frequency techniques to imaging and understanding motivational/emotional and cognitive control functions in anxiety. The first three sections present a historical account of how EEG frequency techniques have been used to understand emotional, motivational, and cognitive components of anxiety. The final section considers these different EEG frequency metrics vis-à-vis theoretical models of anxiety and how they are being, or could be, used to advance both science and treatment. Particularly, this chapter focuses on how anxiety relates to motivational/emotional and cognitive control dynamics involved in balancing bottom-up and top-down influences on behavior. As previous chapters in this volume primarily review the measurement and nature of EEG frequency signals, here we instead focus on the applications of these methods to the study of anxiety.

Anxiety is a common human experience that often involves a mix of cognitive, physiological, and behavioral manifestations (Barlow, 2002). It lies on a continuum from mild levels that can motivate individuals to increase vigilance and respond adaptively to threat or challenge (Barlow, 2002), but at excessive levels represents the most common mental health problem worldwide (U.S. Burden of Disease Collaborators, 2018). It is a multi-dimensional construct that many have separated into anxious apprehension and anxious arousal subtypes, with the former dominated by verbal worries and ruminations focused on future ambiguous threat and the latter by somatic tension and physiological hyperactivity centered on clear immediate threats (Heller et al., 1997, Heller & Nitschke, 1998). Given its multifaceted nature, anxiety and its associated impairments have been tackled from motivational/emotional, cognitive and neuropsychological perspectives (Eysenck et al., 2007; Gray & McNaughton, 2000; Heller et al., 1997, Heller & Nitschke, 1998). In the review that follows, we examine anxiety across the spectrum of severity and, thus, how EEG frequency metrics reflect the dynamic interplay of motivation, emotion, and cognition in adaptive and maladaptive forms of anxiety.

19.2 The First Wave: Alpha and Anxiety

19.2.1 Early Foundations and the Right-Sided Bias

Historically, the largest body of research examining anxiety and EEG frequency band activity focused on alpha frontal asymmetry, or the difference in alpha power at lateral frontal electrode sites between the right and left hemispheres. This is examined by subtracting the natural log of alpha power in the left hemisphere from that of the right hemisphere (i.e., F4–F3, F6–F5, and F8–F7), resulting in an asymmetry score (Reznik & Allen, 2018). Because alpha power is thought to reflect reduced cortical activity, greater relative right alpha asymmetry is interpreted as decreased cortical activity in the right hemisphere (Allen et al., 2004).

Such research grew out of an emerging interest in understanding emotion in the 1970s (e.g., Ekman & Friesen, 1971; Ledoux, 1978) and simultaneous observations of the hemispheric specialization of emotions in patients with brain lesions (Gainotti, 1972; Sackeim et al., 1982; Silberman & Weingartner, 1986) as well as patients undergoing neurological procedures (see Davidson & Fox, 1982). Specifically, it was proposed that the left hemisphere is involved the generation of positive affect, whereas the right hemisphere is involved in the generation of negative affect, including anxiety.

Davidson and colleagues (Davidson, 1979, 1984, 1988; Davidson & Fox, 1982) were the first to apply EEG methodology to this framework. In a series of novel papers, they introduced an approach-withdrawal model of behavior to explain frontal laterality and the neural basis of emotion. According to this model, greater relative left-frontal activity (i.e., reduced alpha power in the left hemisphere) is related to approach behaviors and positive affect, whereas greater relative right-frontal activity is related to withdrawal behavior and negative affect. Here, anxiety is positioned as a manifestation of negative affect and withdrawal (Fox, 1991).

In support of this model, several studies across a variety of populations—infants, children, non-human primates, and adults—have implicated right alpha frontal asymmetry in anxiety. In several papers, Davidson and Fox (Davidson & Fox, 1989; Fox et al., 1992;

Fox & Davidson, 1987) reported that in 10-month-old infants, distress, gaze aversion, and crying in response to maternal separation was associated with greater relative right-frontal brain activity. This was found to be both a trait and state marker of behavior, as resting-state asymmetry was predictive of later anxious behaviors during the separation (Davidson & Fox, 1989; Fox et al., 1992), and asymmetry scores increased during the separation (Fox & Davidson, 1987). Anxious behaviors, internalizing problems, and diagnoses of anxiety disorders (based on direct observation and parent report) in preschool (Fox et al., 1995; Fox et al., 1996) and school-age children (Baving et al., 2002; Schmidt et al., 1999) also demonstrated a relationship to relative right-frontal activation both at rest and during a stress induction. Similarly, child anxious temperament (e.g., high reactivity and behavioral inhibition), which is predictive of the development of clinically significant social anxiety (Hirshfeld-Becker et al., 2007), has been reported to be associated with right anterior activity (McManis et al., 2002). The same patterns of alpha frontal asymmetry can be observed in non-human primates (Kalin et al., 1998), and can be decreased with the administration of anxiolytic benzodiazepines (Davidson et al., 1992). In adults, self-reported trait-anxiety on the State-Trait-Anxiety Inventory (STAI) (Spielberger, 1983) has been reported to correlate with relative right-frontal asymmetry in healthy young adults (Adolph & Margraf, 2017; Blackhart et al., 2006). In clinical populations, those with obsessive compulsive disorder (OCD) were found to have greater relative right-frontal asymmetry (Ischebeck et al., 2014; Kuskowski et al., 1993) and cognitive behavioral therapy has been found to shift asymmetry from relative right to left activity in patients with social anxiety disorder (Moscovitch et al., 2011). In summary, relative right-frontal alpha asymmetry seems to reflect tendencies to withdraw from stress or novelty and a propensity towards anxious temperament.

19.2.2 Conflicting Findings and Anxiety Subtypes

An equally large body of work, however, has not supported the hypothesis that anxiety relates to greater right compared to left-frontal activation. Indeed, several studies have reported no frontal asymmetry differences in anxious populations (Kemp et al., 2010; Metzger et al., 2004; Nitschke et al., 1999; Shankman et al., 2008; Smith et al., 2018), whereas others have reported greater relative left-frontal activation in anxiety (Heller et al., 1997). These mixed findings could be due to the fact that anxiety is not a unitary construct and led to the proposal that anxiety subtypes may show different patterns of alpha asymmetry (Heller et al., 1997). Anxious apprehension, characterized by worry and rumination typically observed in generalized anxiety disorder (GAD) (Barlow, 1991), has been posited to be associated with greater left-frontal activity due to the lateralization of language to the left hemisphere (Heller et al., 1997). On the other hand, anxious arousal, characterized by somatic symptoms of anxiety that are typically observed in panic disorder and post-traumatic stress disorder (PTSD), has been posited to be associated with greater relative right-frontal activation (Heller et al., 1997).

The few EEG studies of alpha frontal asymmetry in patient populations (Buchsbaum et al., 1985; Smith et al., 2016) have supported the hypothesis that anxious apprehension is predominately related to greater left-hemisphere activity, though findings in non-patient populations have been mixed, possibly due to methodological differences. For example, Heller and colleagues (1997) found that anxious apprehension, defined as scores above the 90th percentile on the STAI, was associated with left-frontal asymmetry, though as noted earlier, other studies have reported the opposite findings using this measure (Adolph & Margraf, 2017; Blackhart et al., 2006) and one study found no asymmetry differences (Smith et al., 2018). Studies that have used the Penn State Worry Questionnaire (Meyer et al., 1990) as a measure of anxious apprehension have also been mixed, with one study reporting greater left-frontal asymmetry (Smith et al., 2016), and two reporting no asymmetry differences (Nitschke et al., 1999; Smith et al., 2018). Hofmann et al. (2005) reported that a worry induction in healthy college students resulted in increased left-frontal alpha asymmetry, whereas, Carter, Johnson, and Borkovec (1986) reported no association between a worry induction and alpha asymmetry, though greater left asymmetry was reported in the beta band. Recently, Nusslock and colleagues (2018) reported that the typical pattern of relative right-frontal activity in patients with depression was no longer present when considering those with comorbid anxiety disorders (e.g., separation, GAD, and OCD). Together, these findings highlight the importance of considering anxiety measurement and population in the interpretation of alpha asymmetry findings.

With regards to anxious arousal and relative right-frontal asymmetry, all investigations, with the exception of those in PTSD, have supported the hypothesis that anxious arousal can be distinguished from anxious apprehension by relative right-frontal activation. Self-reported anxious arousal (Mathersul et al., 2008; Nitschke et al., 1999; Smith et al., 2016), anxiety inductions (Avram et al., 2010; Davidson et al., 2000; Isotani et al., 2001) and clinical levels of anxious arousal (e.g., panic disorder) (Wiedemann et al., 1999) have all been shown to correlate with relative right-frontal activation.

Reports in participants with PTSD have failed to find frontal asymmetry differences in either direction (Kemp et al., 2010; Metzger et al., 2004; Rabe et al., 2006; Shankman et al., 2008), though Rabe and colleagues (2006) did find right-frontal activity to be associated with greater symptom severity when participants viewed trauma-related pictures. Kemp and colleagues (2010) did not find differences between healthy controls and participants with PTSD; however, they did find an association between PTSD severity and right-frontal asymmetry within the PTSD group. These mixed findings in PTSD could be due to the heterogeneous presentation of the disorder, age differences in the populations studied, and the possibility that PTSD may be neurobiologically distinct from other anxiety disorders (Lobo et al., 2015; Shankman et al., 2008).

In addition to being associated with frontal alpha asymmetry, arousal has also been related to alpha power in posterior regions (i.e., parietal and temporal) (Heller, 1993). Indeed, PTSD patients demonstrate greater relative right-parietal activity (Kemp et al., 2010; Metzger et al., 2004; Rabe, Beauducel, et al., 2006). Although some have reported similar findings in other anxious populations (Bruder et al., 1997; Heller et al., 1997; Mathersul et al., 2008; Smith et al., 2016), there have been insufficient studies focusing on posterior alpha activity to draw definitive conclusions.

19.2.3 The Devil is in the Details: Methodological Heterogeneity and Ambiguity in Neural Sources

Several other methodological issues may limit the interpretation of alpha asymmetry findings. First, there are inconsistencies regarding electrode site selection, with some studies examining only lateral frontal sites and others examining only midfrontal sites—although there is no current theory that predicts why alpha at these sites would be different (Reznik & Allen, 2018). Second, past studies have differed in EEG processing and referencing procedures. For instance, referencing procedures have included the use of an average scalp reference, the vertex (i.e., CZ), the linked mastoids, or current-source-density (CSD) transform. The majority of alpha activity in the brain is found over the parietal and occipital cortices (Smith et al., 2018), and referencing procedures that do not use CSD are susceptible to contaminating activity from more central and posterior sites (Reznik & Allen, 2018). Third, frontal alpha asymmetry is not specific to anxiety. It has also been linked to other types of psychopathology including major depressive disorder (Allen & Reznik, 2015), attention deficit hyperactivity disorder (Keune et al., 2011), and bipolar disorder (Nusslock et al., 2015). Thus, reported differences in anxiety and asymmetry could (a) either be due to comorbidities if these are not accounted for and/or (b) reflect cognitive, behavioral, or emotional processes affected across multiple disorders in a dimensional manner. Beyond psychopathology, alpha frontal asymmetry has been linked with a multitude of constructs such as guilt (Amodio et al., 2007), anger (Harmon-Jones, 2007), racial bias (Amodio, 2010), day and time of year (Peterson & Harmon-Jones, 2009; Velo et al., 2012), liking of dessert (Gable & Harmon-Jones, 2008), homesickness in college freshman (Steiner & Coan, 2011), feelings towards exercise (Hall et al., 2010), listening to rock music (Field et al., 1998), handedness (Brookshire & Casasanto, 2012), and attractiveness of the experimenter (Wacker et al., 2013). These findings suggest that there could be unaccounted moderators in previous studies on anxiety, thus making it difficult to determine the functional significance of frontal asymmetry.

Finally, a significant question remaining in EEG frontal asymmetry research is the underlying biological mechanism producing hemispheric differences. Occipital and thalamic brain regions have been described as alpha generators (Smith et al., 2018), and combined EEG-functional magnetic resonance imaging (fMRI) research has demonstrated that alpha power correlates positively with functional connectivity between occipital and thalamic areas (Scheeringa et al., 2012). Additionally, alpha power is positively correlated with resting-state functional connectivity of the default-mode network, but negatively correlated with functional connectivity and activity of the dorsal attention network (Laufs et al., 2003; Mantini et al., 2007). Studies that have directly examined correlates of alpha asymmetry have implicated the anterior cingulate cortex (ACC; Gorka et al., 2015; Nash et al., 2012; Smith et al., 2018), the dorsal lateral prefrontal cortex (Smith et al., 2018), the pre and postcentral gyri (Lubar et al., 2003) and the amygdala (Zotev et al., 2016). Collectively, this suggests that alpha asymmetry could reflect attention to motivationally relevant information and the execution of intended actions.

In sum, there has been great interest over the past 50 years in understanding how alpha asymmetry relates to anxiety. Models of emotional valence, motivation, physiological arousal, and verbal processing have been proposed to explain the reported findings. However, because anxiety is a heterogeneous construct that may involve and/or interact in complex ways with cognitive processes such as attention and motivation, further research is needed to fully understand the functional significance of alpha asymmetry in anxiety.

19.3 The Second Wave: Frontal Midline Theta and Anxiety

A more detailed treatment of frontal-midline theta (FMT), itself, is provided in Chapter 9; therefore, the focus of our review centers on how FMT has been used to understand cognitive and related processes in anxiety. As such, research on FMT and anxiety is relatively new.

19.3.1 Early Days and a Primer on FMT as an Index of Medial Frontal Cortex Function

Prior to landmark observations by Cavanaugh & Schackman (2015), research on FMT and anxiety was scarce and methodologically limited. An early series of studies conducted by Mizuki and colleagues examined how anxiolytics, state and trait anxiety assessed using the STAI and Generalized Anxiety Disorder were related to FMT during a continuous addition task (Mizuki et al., 1983; Mizuki et al., 1989; Mizuki et al., 1992; Suetsugi et al., 2000). Findings showed that lower levels of trait anxiety were related to greater theta activity during the arithmetic test; however, they also revealed that reductions in state anxiety in those with high trait anxiety were related to decreases in theta activity (Mizuki et al., 1992). However, these results are difficult to interpret in light of a number of limitations. First, all but one study utilized only male participants. Second, all of the studies were significantly hampered by small samples. Finally, these studies involved tasks that are quite different from the typical two-choice reaction time tasks commonly used today. Thus, it is difficult to conclude much from these early attempts to understand relations between anxiety and FMT. Nonetheless, they demonstrated that FMT was modulated in the context of anxiety.

A large number of studies has examined the role of medial frontal cortical activity in cognitive control efforts, particularly when these efforts are aimed at reducing anxiety or the circumstances which contribute to anxiety. The cingulate, especially the anterior and midcingulate cortices (ACC and MCC, respectively), is a cortical area which is highly interconnected with other brain systems, including emotional centers in the limbic system, cognitive association areas in the prefrontal cortex, sensory and motor processing areas in the parietal and occipital lobes, and deeper structures within the brain (Van Den Heuvel et al., 2009). As such, the cingulate is well positioned to integrate information from a variety of inputs and coordinate action-oriented outputs to maximize behavioral adaptability (Laurienti et al., 2003). Furthermore, the MCC's involvement is critical whenever new or conflicting information must be evaluated for its relevance to the current behavioral set (Botvinick et al., 2004). Indeed, researchers have suggested the MCC has domain-general functionality that is engaged whenever there is uncertainty about behavior and its outcomes (Cavanagh et al., 2012). Because uncertainty is a central aspect to feelings of anxiety and to anxiety disorders, the MCC has been a target for research efforts aimed at understanding the underpinnings of anxiety-related brain dynamics (Cavanagh & Shackman, 2015; Shackman et al., 2011).

A growing body of work suggests a primary function of the MCC is to integrate cognitive, affective, and sensory information to enable behavioral adaptation (Shackman et al., 2011). MCC activity has been implicated when incoming information is novel, conflicting, punishing, or difficult to assimilate into the current action plan. In particular, FMT oscillations in the MCC have been observed across a variety of tasks whenever cognitive control is needed to alter behavior (Cavanagh & Frank, 2014). Because of its ability to facilitate long-range communication between disparate brain regions, theta oscillations are a candidate mechanism for how the MCC helps organize and coordinate neural processes related to action monitoring and response selection. Indeed, FMT has been posited to represent an alarm signal that is integral for the recruitment of areas in the prefrontal cortex that mediate cognitive control (Cavanagh & Frank, 2014). A wide range of studies demonstrate that individuals high in anxiety generate larger FMT power during cognitive tasks, suggesting anxiety and FMT hyperactivity are intimately related. Moreover, larger FMT control signals are associated with more cautious or inhibited responding, providing strong evidence that FMT is a biomarker that can be used to study the intersection of anxiety and cognitive control (Cavanagh & Shackman, 2015).

19.3.2 Relationships Between Anxiety and FMT-Related ERP Components

Among individuals high in anxiety, heightened FMT responses to novel, conflicting, or punishing stimuli have been observed indirectly for several decades within typical event-related potential (ERP) paradigms. The wide array of tasks that elicit FMT is evidence of the domain-general role that FMT plays in dealing with uncertainty through the recruitment of cognitive control resources. Time-domain ERP components that have been consistently related to FMT include the error-related negativity (ERN; Luu et al., 2004; Trujillo & Allen, 2007), feedback negativity (FN; Foti et al., 2014; Watts et al., 2018), and the N2 (Harper et al., 2014), all of which tend to be increased in anxious individuals.

The ERN is a response-locked ERP component that is thought to reflect action-monitoring that is involved in the recruitment of cognitive control during post-error behavioral adjustment (Gehring et al., 2018). Studies investigating the time-frequency dynamics have demonstrated that the ERN primarily reflects theta-band activity that is generated in the MCC (Luu et al., 2004; Trujillo & Allen, 2007). ERN amplitude has been positively associated with anxiety, worry, and behavioral inhibition in flanker (Moser et al., 2012; Riesel et al., 2017), go/no-go (Moadab et al., 2010), and trial and error learning (Judah et al., 2016) tasks (for reviews see (Cavanagh & Shackman, 2015; Moser et al., 2013). The functional significance of the larger ERN in anxiety is hotly debated, with some suggesting it reflects the degree to which errors are seen as aversive (Weinberg et al., 2016), whereas others suggest it reflects compensatory deployment of cognitive control to counteract the distracting effects of worry (Moser et al., 2013; Schroder et al., 2017).

Another FMT-related ERP component thought to be generated by the MCC and associated with anxiety is the feedback-locked FN (Foti et al., 2014; Watts et al., 2018). Like other FMT-related ERP components, the FN can be understood as a signal from the salience network indicating new information (e.g., feedback) is highly relevant for adapting to changing circumstances. As such, the future-oriented emotional state of anxiety can prime the salience network to be hypersensitive to performance-related feedback, making individuals high in trait anxiety more likely to consistently produce an elevated FN (Grupe & Nitschke, 2013). Congruent with this framework, FN amplitude has shown positive associations with negative affect during a monetary incentive delay task (Santesso et al., 2011) and behavioral inhibition in go/no-go (De Pascalis et al., 2010) and spatial matching (Balconi & Crivelli, 2010) tasks.

Parallel to the ERN and FN, the N2 ERP component has been shown to be highly related to FMT (Harper et al., 2014; Van Noordt et al., 2016) and is also sensitive to trait and state anxiety differences. Although there is some debate as to which process(es) the N2 most likely reflects, it is associated with performance monitoring, conflict resolution, and response inhibition. Individuals high in anxiety have been shown to produce larger N2s in a variety of contexts, including during go/no-go (Righi et al., 2009; Sehlmeyer et al., 2010) and flanker (Schmid et al., 2015) tasks. Larger N2 amplitude to conflict trials in anxious individuals might reflect the recruitment of compensatory effort to overcome distraction by anxious thoughts or a regulation strategy aimed at reducing current and future negative affect (Dreisbach & Fischer, 2012; Kool et al., 2010; Lindström et al., 2013; Schouppe et al., 2012).

19.3.3 Anxiety and Time-Frequency Representations of MCC Activity

Recent work has utilized time-frequency analyses to directly examine the relationship between FMT and anxiety during a variety of tasks. Congruent with studies using time-domain metrics, time-frequency representations of MCC activity consistently show a positive association between anxiety and FMT (Cavanagh & Shackman, 2015).

A growing body of work demonstrates that time-frequency representations of the ERN show a similar relationship to state and trait anxiety as time-domain measures. During a flanker task, individuals diagnosed with GAD were found to produce larger ERN, error-related FMT, and correlations between single-trial FMT and reaction time during a flanker task than did control participants (Cavanagh et al., 2017). Furthermore, each of these three variables uniquely predicted group membership, suggesting FMT may not exclusively be an evoked signal (Cohen & Donner, 2013). In another study,

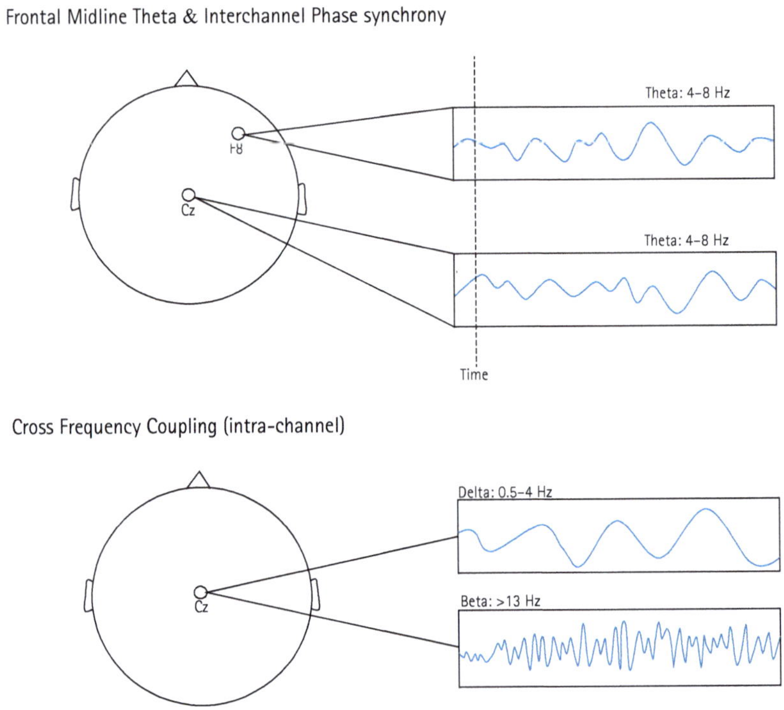

FIGURE 19.1 Simplistic graphic illustrating frontal midline theta and interchannel phase synchrony cross frequency coupling, described in the text. Frontal midline theta metrics are calculated at midline sites as theta power and phase synchrony between sites. Cross frequency coupling metrics include correlations between amplitudes of slower (i.e., delta) and faster (i.e., beta) frequencies as well as correlations between phase and amplitude of these frequencies or ratios of frequency power (i.e., slow wave/fast wave).

adolescent participants engaged in a flanker task while in a social anxiety-producing situation and displayed greater error-related FMT and medial-to-lateral inter-channel phase synchrony (ICPS; Buzzell et al., 2018; see Figure 19.1). This finding provides some evidence that state anxiety can also lead to greater performance monitoring and cognitive control. There are mixed findings, however, concerning the relationship between chronic anxiety—in this case chronic worry—and medial-to-lateral ICPS, with one study showing reduced ICPS (Moran et al., 2015) and another reporting the opposite (Cavanagh et al., 2017).

Studies utilizing time-frequency analyses have also revealed associations between anxiety and feedback-related FMT. One study found that FMT was positively correlated with general worry during a gambling task for both gains and losses in a community sample with clinically elevated levels of anxiety (Ellis et al., 2018). Moreover, FMT displayed a stronger relationship to loss feedback in this study further supporting the hypothesis that individuals high in anxiety over-utilize cognitive control to regulate behavior in the face of aversive outcomes (Cavanagh & Shackman, 2015). Cavanagh and colleagues (2019) found a similar positive relationship between FMT and loss feedback among individuals high in anxiety, suggesting heightened FMT may be used to boost avoidance and cautious behavior. Another study using path analyses revealed pre-decision FMT fully mediated the relationship between trait anxiety and risk behavior during a decision-making task, with individuals high in anxiety making less risky decisions, as a function of pre-decision FMT, than those with low trait anxiety (Schmidt et al., 2018). Similarly, increased FMT has been found to predict less risky gambling decisions in the Balloon Analogue Risk Task (Zhang & Gu, 2018). Although bidirectional theta information flow was detected between medial and lateral frontal areas during decision making, trait anxiety was only correlated with theta power generated in the lateral PFC. This provides further evidence that anxiety's effect on theta dynamics may not be limited to FMT power and likely includes lateral PFC power and connectivity between MCC and lateral PFC.

More socially-oriented studies have found relationships between anxiety and FMT that run parallel to those found with more traditional cognitive tasks. For example, one study found that FMT responses during International Affective Picture System image viewing depended on an individual's level of anxiety (Aftanas et al., 2003). Individuals high in trait anxiety produced the largest FMT increase after threatening images, while those low in trait anxiety displayed greater FMT increases to pleasant images. This interaction is congruent with the gambling literature described earlier, which supports an association between high anxiety and punishment-sensitive enhanced FMT. Greater FMT has also been found to associate with unexpected social rejection feedback among participants with social anxiety disorder (Harrewijn et al., 2018a).

In sum, there is ample evidence indicating that anxiety is associated with a range of FMT metrics. Anxiety appears to be associated with larger FMT signals, which may reflect a greater need for cognitive control of behavior in chronically anxious individuals or consequential anxious states. There is some inconsistency regarding the direction of the association between anxiety and network-related FMT metrics (i.e., ICPS) that will require further investigation to disentangle potentially confounding or moderating effects.

19.4 The Third Wave: Cross Frequency Coupling and Anxiety

19.4.1 A Primer on Cross Frequency Coupling

Cross frequency coupling (CFC) metrics are also emerging as tools for understanding cognition and emotion in anxiety. CFC is often computed between slow waves (SW), such as delta (1–3 Hz) or theta (4–7 Hz), and fast waves (FW), such as alpha (8–12), beta (13–30 Hz), or gamma (>30 Hz; see Figure 19.1 for an illustration). Work in this area has generally considered delta and theta to reflect reward seeking and emotional arousal, respectively, generated by subcortical brain regions (Knayzev, 2003). Alpha, beta, and gamma are considered to reflect heightened attentional vigilance (Knayazev, 2003), inhibitory control (Engel & Fried, 2010; Jenkinson & Brown, 2011), and sustained attention/working memory (Jenson et al., 2007), respectively, in the neocortex. As such, SW-FW coupling is thought to index increased communication between subcortical and cortical regions of the brain.

CFC metrics in the anxiety literature mostly include amplitude-amplitude coupling during resting or anticipatory states, although some have also investigated phase-amplitude coupling (i.e., the power of a higher frequency along specific phases of a lower frequency; Poppelaars et al., 2018). On the other hand, SW/FW ratios of frequency power are computed to examine the relative strength of frequencies, with higher ratios indicating a predominance of slow waves (i.e., less control over emotional processes). Therefore, coupling provides an index of the associated strength of SW and FW power or phase, whereas ratios provide information on the relative strength of SW to FW power.

An early study by Knyazev and Slobodskaya (2003) investigated the relationship between the behavioral inhibition system (BIS), a motivational system highly implicated in anxiety-related symptoms and behaviors (Gray & McNaughton, 2000), and CFC anticorrelations (i.e., negative relationships between frequencies). BIS, as conceived by Gray and McNaughton, is similar in many ways to behaviorally inhibited temperament in toddlers and young children described earlier yet developed separately in parallel as a way to understand motivational processes involved in personality and psychopathology. They found that BIS was related to a higher alpha-delta anticorrelation, concluding that behaviorally inhibited adults experience heightened activation in cortical regions combined with less activation in limbic regions of the brain. Although the interpretation of alpha-delta anticorrelations remains obscure, it provided a framework to investigate SW-FW relations as cortical "cross-talk", which has gone on to focus on delta or theta and beta.

19.4.2 Delta-Beta Coupling and Anxiety

Two early studies of particular relevance to the topic of this chapter demonstrated an association between higher delta-beta coupling and anxiety-related individual differences. Specifically, Schutter and Van Honk (2005) found increased delta-beta coupling in individuals with high baseline cortisol levels, relative to those with low cortisol levels. Van Peer and colleagues (2008) subsequently reported that males characterized by high BIS evinced high delta-beta coupling in frontal sites that was further increased by the administration of cortisol (vs. placebo). These findings led to further inquiry as to whether delta-beta coupling is related to trait individual differences in anxiety-related constructs, anxious states, or both.

Adopting Gray and McNaughton's (2000) theory of anxiety, Knyazev, Savostyanov, and Levin (2005) posited that apprehensive states are distinct from dispositional anxiety (i.e., BIS), because they involve coactivation of both BIS and the behavioral activation system (BAS). Theoretically, this occurs in uncertain contexts when the probability of winning and losing are equal. As such, these contexts lead to the activation of both BIS (i.e., cortical activation) and BAS (i.e., subcortical activation) with no particular predominance of one over the other. In a series of studies with predominantly female samples, high- and low-stress groups were given negative feedback on a behavioral task and were told that this could potentially cause them to lose money. Results revealed that those in the high-stress groups showed greater delta-beta and delta-gamma[1] coupling than those in the control group (Knyazev et al., 2005; 2006, 2011). Notably, increases in delta-beta coupling were comparable for both high and low trait anxious individuals, although high trait anxious individuals showed an even greater increase in coupling. Thus, anxious states and traits may be additive with respect to increases in delta-beta coupling. Source analysis studies further suggest that increases in delta-beta coupling as a function of apprehensive states involves activity in ACC and orbitofrontal cortex (OFC; Knyazev et al., 2011). How exactly delta-beta coupling relates to ACC and OFC activity remains unknown; however, delta may be the primary driver of this coupling and coordination of brain regions involved in preparing an adaptive behavioral response (Knyazev et al., 2011).

CFC studies of clinical anxiety have primarily focused on social anxiety. Miskovic and colleagues (2010), for instance, observed increases in delta-beta coupling during speech anticipation (vs. resting) in socially anxious individuals. Research from the same group also found that 12 weeks of cognitive behavioral therapy served to decrease delta-beta

[1] Only one study has examined delta-gamma coupling (Knyazev et al., 2005), and the interpretation of this result is consistent with general SW-FW coupling interpretations.

correlations over treatment during speech anticipation periods (Miskovic et al., 2011), and that children of socially anxious parents showed increased delta-beta coupling (Miskovic et al., 2011). However, negative associations between delta and beta have also been reported in social anxiety (Harrewijn et al., 2016). Moreover, Harrewjin and colleagues (2018b) recently found that this negative association between delta and beta is heritable and speculated that the differences in the direction of coupling could be due to different stress manipulations utilized by research groups (anticipatory stress vs. task-induced stressor), or different ways of computing frequency power densities (i.e., absolute vs. relative power). If the delta-beta correlation, however, is generally understood as reflecting increased neural communication, then the mere presence of a relationship between their frequencies could still indicate important cross-talk (regardless of direction), although the exact nature of this cross-talk remains vague. Further complicating the relationship between anxiety and delta-beta coupling, one study showed that OCD is characterized by delta-beta decoupling (Velikova et al., 2010). Clearly more research will be needed to clarify how clinical levels of anxiety-related problems associate with delta-beta coupling.

19.4.3 Frontal Theta/Beta Ratio and Anxiety

Beyond delta-beta coupling, anxiety has also been examined in relation to frontal theta/beta ratio insomuch as frontal theta/beta ratio is an index of attentional control (AC), a core component of executive function involved in balancing bottom-up selection and top-down control influences on behavior. As previously mentioned, the examination of ratios provides a measure of relative dominance of SW over FW or vice versa. For instance, Putman and colleagues (2010) found that higher self-reported trait anxiety and attentional control were related to lower theta/beta ratio (i.e., more control) in an all-female sample. Subsequent studies replicated these associations (Angelidis et al., 2018; Putman et al., 2014). Putman and colleagues (2014) further demonstrated that individuals with a lower theta/beta ratio showed lesser effects of a stress manipulation on change in self-reported AC. Theta/beta ratio is also lower during on-task periods relative to mind wandering episodes (Van Son et al., 2019). Together, these findings have been interpreted as indicating that theta/beta ratio reflects the degree of top-down control over bottom-up processes, with higher theta/beta ratio indicting less control and lower theta/beta ratio indicating greater control (Putman et al., 2014). It is interesting, then, that trait anxiety shows a negative association with theta/beta ratio rather than a positive one, as anxiety has been related to poorer attentional control (Eysenck et al., 2007; Shi et al., 2019). We return to this seeming contradiction in the next section.

In all, SW-FW coupling and ratios appear to reflect a number of anxiety-related processes, including AC. Specifically, BIS, apprehensive states and clinical levels of anxiety appear to relate to greater SW-FW coupling, although there are quite mixed findings regarding clinical anxiety. Higher trait anxiety and self-reported AC associate with lower SW/FW ratios. Both CFC metrics appear to reflect interactions between cognition and emotion—that is, top-down versus bottom-up control of processing resources—that may serve compensatory or coping functions in anxious contexts and for anxious individuals.

19.5 The New Wave: Theoretical Considerations, Future Directions and Conclusions

The previous section review indicates the exciting promise of EEG frequency techniques to illuminate anxiety-related processes, including motivation and cognitive control. In this final section, we first consider relevant cognitive, motivational, and emotional theories that might help synthesize the reviewed literature. We then turn to potential avenues for future work and finish with some brief concluding remarks.

19.5.1 Theoretical Considerations

Throughout this review, much of the literature seems to point to the presence of a dynamic interplay between cognition, motivation, and emotion in the links between EEG frequency metrics and anxious states and traits. In particular, the interplay between emotional processes involved in the generation and detection of internal or external threats, motivational processes dedicated to approaching and/or avoiding these threats, and cognitive processes involved in the monitoring and control of these forces seems to be positioned front and center in anxiety-related situations and symptoms. Figure 19.2 shows a hypothetical model of the interplay between motivational conflicts and cognitive control in anxiety, which is seen as intimately related to the dynamic process of co-activation of approach and avoidance motivational systems that produce conflicts

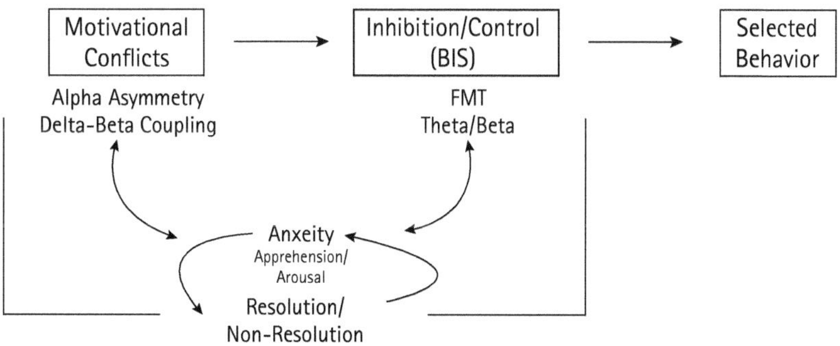

FIGURE 19.2 A hypothetical model of the interplay between motivational conflicts and cognitive control in anxiety.

detected and weighed by the BIS during which ongoing behavior is inhibited and controlled until there is a resolution of the conflict and a behavior is selected. Anxiety is generally related to enhanced conflicts between motivational systems, inhibition, and control. Alpha asymmetry and metrics and theta/beta ratio may index BIS-related inhibitory and related control functions. Apprehension and arousal subcomponents of anxiety are proposed to have different relations to these dynamics. The model more likely captures the role of apprehension whereas arousal is proposed to have more direct links to fear and avoidance motivational systems that may be less related to control systems.

There is much overlap between the concepts of emotion, motivation, and cognition, and, thus, we do not intend to communicate that they are completely separable constructs (cf. Miller, 1996, 2010). Rather, we assert that consideration of these processes is critical to better understanding what EEG frequency metrics might reflect with respect to anxiety, as well as to evaluate the utility of EEG frequency metrics in detecting risk, maintenance factors, and treatment targets for anxiety-related interference and disorders. In the next section, we consider merging the theoretical contributions of Gray and McNaughton (2000), Eysenck and colleagues (2007), and Heller and colleagues (Heller et al., 1997, Heller & Nitschke, 1998) with findings from the cognitive neuroscience of inhibitory control (over emotion) as offering a unique window into advancing our knowledge of how EEG frequency metrics might reflect the dynamic interplay of emotion, motivation, and cognition in anxiety.

Gray and McNaughton's (2000) neuropsychological model of anxiety draws on earlier work by Gray (1975, 1982) regarding motivation and personality (see also Corr et al., 2013 for a more recent review). This work largely delineated the BAS and BIS as two motivational systems relevant to personality and anxiety. The updated version of this model, the Reinforcement Sensitivity Theory (RST), further decomposed the BIS into the BIS and the fight-flight-freeze system (FFFS), which is involved in fear-related behaviors aimed at defending an organism from active threat. The BIS, on the other hand, is posited to be involved in the detection and resolution of approach-avoidance conflicts—that is, conflicts driven by coactivation of the BAS and FFFS. The BIS therefore inhibits ongoing behavior to allow for risk assessment and scanning so as to resolve the approach-avoid conflict by enacting the appropriate behavior in context.

The attentional control theory (ACT) of anxiety (Eysenck et al., 2007) more specifically aims to describe the nature of the relationship between anxiety and cognition. Eysenck and colleagues conclude that anxiety disrupts the balance of the bottom-up salience system and the top-down control system such that the bottom-up system is generally more active than the top-down control system. To resolve this imbalance in the presence of task-related goals, anxious individuals are hypothesized to deploy compensatory top-down control to meet the demands of the task at hand. We suggest that the interplay of the bottom-up and top-down control systems in anxiety proposed by Eysenck and colleagues bears striking resemblance to the approach-avoid conflicts detected and resolved by the BIS in Gray's RST. That is, anxiety often involves conflicts between threat-driven, avoidance, or distraction processes and goal-driven, approach processes involved in producing appropriate behaviors to meet the demands of the current context.

To further add nuance to this interplay, however, consider the seminal work by Heller and colleagues (Heller et al., 1997; Heller & Nitschke, 1998; Sharp et al., 2015) disentangling different dimensions of anxiety and depression. This work makes important distinctions between anxious apprehension (worry), anxious arousal (fear and panic), and anhedonic depression across psychometric and neural measures. Neurally, anxious apprehension has been associated with left-frontal activity and anxious arousal with right-parietal activity (reviewed by Sharp et al., 2015). The association between anxious apprehension and left-frontal activity has been interpreted as reflecting the involvement of language production and verbal working memory centers in worry whereas anxious arousal may not only involve right-parietal activity but rather several parietal and frontal regions involved in threat detection and environmental scanning (Sharp et al., 2015). Nonetheless, findings have generally distinguished between anxious apprehension and anxious arousal, pointing to heterogeneity in the anxiety construct that must be considered when attempting to understand the involvement of neural measures like EEG frequency metrics.

Together, we suggest that when interpreting the function of the various EEG frequency metrics reviewed above, one must consider the important interplay of emotion, motivation, and cognition as well as unique differences between dimensions of anxiety—e.g., anxious apprehension and arousal. Moreover, considering the context in which associations between EEG frequency metrics and anxiety emerge is critical—i.e., at rest, in the presence of threat, and/or during completion of an affective neutral task.

One common theme that seems to arise from the review is the link between anxiety and metrics of control. We see this most strikingly in the frontal midline theta (FMT) and cross-frequency coupling (CFC) literatures. We suggest that this link might reflect the important role of the BIS in anxiety. That is, FMT and CFC metrics might reflect the dynamic detection and attempted resolution of conflicts between approach and avoidance motivations that arise when anxious individuals or individuals subjected to anxious states attempt to complete a task at hand on which they have been given instructions to perform well. Studies linking FMT and other CFC metrics to activity of medial prefrontal cortices (most importantly, the ACC or MCC) seem to further support this interpretation given the wealth of data indicating the ACC/MCC's role in inhibition, conflict processing, and integration of negative affect and cognitive control (Bartholomew et al., 2019; Popov et al., 2018; Shackman et al., 2011; Spielberg et al., 2015). A seeming "hub" for emotion-cognition interactions, the ACC/MCC might be critically involved in the BIS via FMT oscillations and their associations with other frequencies (e.g., beta; see also Popov et al., 2018). As such, ACC/MCC-mediated processes may reflect this "dance" between approach and avoidance motivations in contexts in which fear-related avoidance and goal-directed behaviors stand in conflict with one another. This process, as Gray and McNaughton (2000) propose, captures the very nature of the BIS and associated anxiety symptoms. EEG frequency approaches may offer the ideal metrics to index these important dynamics given their millisecond precision and inherent ability to reflect activity of systems/networks.

19.5.2 Future Directions

There is a strong foundation of knowledge produced by the extant studies on associations between anxiety and EEG frequency metrics, although a number of areas are ripe for further investigation and theory development. One particularly curious contradiction is the functional significance of theta activity. In the literature on FMT, theta is hypothesized to signal the need for control whereby a larger theta signal leads to greater control and adaptive behavioral adjustments. On the other hand, in the CFC literature, theta—i.e., a slow wave—is hypothesized to reflect emotional arousal and related subcortical processes such that its preponderance or dominance over faster wave activity indicates lesser control. It will be important to reconcile these two literatures, as currently they appear to be evolving in parallel. This reconciliation may be challenging at present given the differences in paradigms and calculations of theta used in these two literatures. Both the FMT and CFC literatures focus on frontal theta, and, therefore, it follows that there should be some relation between these different metrics.

It also seems important to gain new insights into the exact relationship between anxiety constructs and alpha asymmetry. Indeed, it represents the historically largest literature on the association between anxiety and EEG frequency metrics, yet much is still left unknown. The various competing models of alpha asymmetry likely occlude clarity with respect to its association with anxiety, however, perhaps Gray and McNaughton's (2000) model of motivation and personality could shed some light. For example, given the important role of the BIS in anxiety and other related EEG frequency metrics, it could be difficult to isolate associations between alpha asymmetry and anxiety because of the likely co-activation of approach (left-sided activity) and avoidance (right-sided activity) motivational systems. The work of Heller and colleagues (Sharp et al., 2015) further demonstrates the importance of considering the heterogeneity of anxiety constructs when examining associations between anxiety and alpha asymmetry, among other neural measures of motivation and cognition. Thus, future work will need to be more precise with setting up paradigms that test clearer predictions involving specific motivational contexts and anxiety constructs.

In general, the literature on associations between anxiety and EEG frequency metrics will have to grapple with different motivational contexts and anxiety constructs. Enhanced precision in theory will drive more sophisticated methods for testing related predictions concerning approach-avoidance conflicts and separable dimensions of anxiety. For example, whereas some have shown associations between "lumped" anxiety constructs and FMT (Cavanagh & Schackman, 2015), others have shown that "splitting" anxiety constructs reveals important distinctions (e.g., Moser et al., 2013).

Beyond the EEG frequency metrics and anxiety constructs themselves, this literature must also incorporate the roles of development and gender, both of which show important influences on anxiety and its associations with neural measures of motivation/emotion and cognitive control. For instance, there is growing evidence demonstrating important changes in the association between anxiety and the ERN—a theta-based ERP—across development (Moser et al., 2017). Gender also moderates this association, such that females show a larger association (Moser et al., 2016) and a greater change in association across development (Ip et al., 2019). Furthermore, racial, ethnic, and cultural differences will have to be considered to fully understand the generalizability of these findings. For example, related research shows higher heart rate variability (HRV) in Black individuals, which may reflect compensatory control mechanisms deployed to manage the sustained and cumulative adverse effects of discrimination (Kemp et al., 2016). Such a compensatory control response may also be reflected in the EEG frequency metrics reviewed herein.

Additional methodological innovations will also further advance this area of clinical neuroscience. For instance, the application of graph theoretic (Bolanos et al., 2013) and directed network (Liu et al., 2014; Popov et al., 2018) methods to such EEG frequency techniques is just beginning and will likely provide unique insights into associations between anxiety and motivational and cognitive control systems. This work will also further benefit from integrating data across the lab and the "real-world" through combining EEG imaging techniques with experience-sampling methods and deploying wearable technology (cf. Hur et al., 2019).

Finally, we would be remiss if we did not acknowledge the importance of considering the psychological and philosophical machinations involved in attempting to make meaning of associations between various psychological and biological measures. It is important for future work in this area to avoid terminology such as "underlying" in describing the involvement of neural oscillations in self-reported experiences of anxiety as they incorrectly insinuate a causal mechanism and further the fallacy that more micro-level data (e.g., neural activity) are superior to more macro-level data (e.g., self-report of anxiety; Sharp & Miller, 2019). Rather, such neural markers are likely involved in the instantiation of experiences and resolution of motivational conflicts and anxiety symptoms. The ways in which these neural, self-report, and behavioral measures are combined and weighed should involve careful consideration of the goals to be achieved (e.g., intervention development; See Taschereau-Dumouchel, Michel, Lau, Hofmann & LeDoux, 2022 for a recent review of this approach).

19.5.3 Conclusions

The literature considering associations between anxiety and EEG frequency measures has a long history and generally follows the historical trends present in other areas of psychological science. More recently, early motivational and emotional models have been overtaken by or integrated into cognitive models of anxiety and related constructs. A long-view perspective on the literature suggests strong associations between anxiety-related constructs and EEG frequency markers of motivational and cognitive control processes. A central theme in this literature is the involvement of the BIS that can be seen reflected in a variety of EEG frequency metrics and generally reveals the dynamic interplay between the co-activation of approach and avoidance motivations in anxious individuals and states as the execution of adaptive behavioral responses is attempted. This area of clinical neuroscience is ripe for future work and should involve advances in theory, generalizability, and methodology.

REFERENCES

Adolph, D. & Margraf, J. (2017). The differential relationship between trait anxiety, depression, and resting frontal α-asymmetry. *Journal of Neural Transmission*, *124*(3), 379–386. https://doi.org/10.1007/s00702-016-1664-9

Aftanas, L. I., Pavlov, S. V., Reva, N. V., & Varlamov, A. A. (2003). Trait anxiety impact on the EEG theta band power changes during appraisal of threatening and pleasant visual stimuli. *International Journal of Psychophysiology*, *50*(3), 205–212.

Allen, J. J. B., Coan, J. A., & Nazarian, M. (2004). Issues and assumptions on the road from raw signals to metrics of frontal EEG asymmetry in emotion. *Biological Psychology*, *67*(1–2), 183–218. https://doi.org/10.1016/j.biopsycho.2004.03.007

Allen, J. J. B. & Reznik, S. J. (2015). Frontal EEG asymmetry as a promising marker of depression vulnerability: Summary and methodological considerations. *Current Opinion in Psychology*, *4*, 93–97. https://doi.org/10.1016/j.copsyc.2014.12.017

Amodio, D. M. (2010). Coordinated roles of motivation and perception in the regulation of intergroup responses: Frontal cortical asymmetry effects on the P2 event-related potential and behavior. *Journal of Cognitive Neuroscience*, *22*(11), 2609–2617. https://doi.org/10.1162/jocn.2009.21395

Amodio, D. M., Devine, P. G., & Harmon-Jones, E. (2007). A dynamic model of guilt: Implications for motivation and self-regulation in the context of prejudice: Research article. *Psychological Science*, *8*(6), 524–530. https://doi.org/10.1111/j.1467-9280.2007.01933.x

Angelidis, A., Hagenaars, M., van Son, D., van der Does, W., & Putman, P. (2018). Do not look away! Spontaneous frontal EEG theta/beta ratio as a marker for cognitive control over attention to mild and high threat. *Biological Psychology*, *135*, 8–17.

Avram, J., Balteş, F. R., Miclea, M., & Miu, A. C. (2010). Frontal EEG activation asymmetry reflects cognitive biases in anxiety: Evidence from an emotional face Stroop task. *Applied Psychophysiology Biofeedback*, *35*, 285–292. https://doi.org/10.1007/s10484-010-9138-6

Balconi, M. & Crivelli, D. (2010). FRN and P300 ERP effect modulation in response to feedback sensitivity: The contribution of punishment-reward system (BIS/BAS) and behaviour identification of action. *Neuroscience Research*, *66*(2), 162–172.

Barlow, D. H. (1991). Disorders of emotion. *Psychological Inquiry*, *2*(1), 58–71. https://doi.org/10.1207/s15327965pli0201_15

Barlow, D. H. (2002). *Anxiety and its disorders* (2nd ed.). Guilford Press.

Bartholomew, M. E., Yee, C. M., Heller, W., Miller, G. A., & Spielberg, J. M. (2019). Reconfiguration of brain networks supporting inhibition of emotional challenge. *NeuroImage*, *186*, 350–357.

Baving, L., Laucht, M., & Schmidt, M. H. (2002). Frontal brain activation in anxious school children. *Journal of Child Psychology and Psychiatry and Allied Disciplines*, *43*(2), 265–274. https://doi.org/10.1111/1469-7610.00019

Blackhart, G. C., Minnix, J. A., & Kline, J. P. (2006). Can EEG asymmetry patterns predict future development of anxiety and depression? A preliminary study. *Biological Psychology*, *72*(1), 46–50. https://doi.org/10.1016/j.biopsycho.2005.06.010

Bolanos, M., Bernat, E.M., He, B., & Aviyente, S. (2013). A weighted small world network measure for assessing functional con- nectivity. *Journal of Neuroscience Methods*, *212*, 133–42.

Botvinick, M. M., Cohen, J. D., & Carter, C. S. (2004). Conflict monitoring and anterior cingulate cortex: an update. *Trends in Cognitive Sciences*, *8*(12), 539–546.

Brookshire, G., & Casasanto, D. (2012). Motivation and motor control: Hemispheric specialization for approach motivation reverses with handedness. *PLoS One*, *7*(4), e36036. https://doi.org/10.1371/journal.pone.0036036

Bruder, G. E., Fong, R., Tenke, C. E., Leite, P., Towey, J. P., Stewart, J. E., . . . Quitkin, F. M. (1997). Regional brain asymmetries in major depression with or without an anxiety disorder: A quantitative electroencephalographic study. *Biological Psychiatry*, *41*, 939–948. https://doi.org/10.1016/S0006-3223(96)00260-0

Buchsbaum, M. S., Hazlett, E., Sicotte, N., Stein, M., Wu, J., & Zetin, M. (1985). Topographic EEG changes with benzodiazepine administration in generalized anxiety disorder. *Biological Psychiatry*, *20*(8), P837–842. https://doi.org/10.1016/0006-3223(85)90208-2

Buzzell, G. A., Barker, T. V., Troller-Renfree, S. V., Bernat, E. M., Bowers, M. E., Morales, S., . . . Fox, N. A. (2018). Adolescent cognitive control, theta oscillations, and social motivation. *bioRxiv* [online], 366831.

Cavanagh, J. F., Bismark, A. W., Frank, M. J., & Allen, J. J. (2019). Multiple dissociations between comorbid depression and anxiety on reward and punishment processing: Evidence from computationally informed EEG. *Computational Psychiatry*, 3, 1–17.

Cavanagh, J. F. & Frank, M. J. (2014). Frontal theta as a mechanism for cognitive control. *Trends in Cognitive Sciences*, *18*(8), 414–421.

Cavanagh, J. F., Meyer, A., & Hajcak, G. (2017). Error-specific cognitive control alterations in generalized anxiety disorder. *Biological Psychiatry: Cognitive Neuroscience and Neuroimaging*, 2(5), 413–420.

Cavanagh, J. F. & Shackman, A. J. (2015). Frontal midline theta reflects anxiety and cognitive control: Meta-analytic evidence. *Journal of Physiology-Paris*, *109*(1-3), 3–15.

Cavanagh, J. F., Zambrano-Vazquez, L., & Allen, J. J. (2012). Theta lingua franca: A common mid-frontal substrate for action monitoring processes. *Psychophysiology*, *49*(2), 220–238.

Carter, W. R., Johnson, M. C., & Borkovec, T. D. (1986). Worry: An electrocortical analysis. *Advances in* Behaviour Research and Therapy, *8*(4), 193–204. https://doi.org/10.1016/0146-6402(86)90004-4

Cohen, M. X. & Donner, T. H. (2013). Midfrontal conflict-related theta-band power reflects neural oscillations that predict behavior. *Journal of Neuro*physiology, *110*(12), 2752–2763.

Corr, P. J., DeYoung, C. G., & McNaughton, N. (2013). Motivation and personality: A neuropsychological perspective. *Social and Personality Psychology Compass*, *7*(3), 158–175.

Davidson, R. J. (1979). Frontal versus perietal EEG asymmetry during positive and negative affect. *Psychophysiology*, *16*(2), 202–203.

Davidson, R. J. (1984). Affect, Cognition, and Hemispheric Specialization. In C. E. Izard, J. Kagan, & R. B. Zajonc (Eds.), *Emotions, cognition, and behavior* (pp. 320–365). Cambridge University Press.

Davidson, R. J. (1988). EEG measures of cerebral asymmetry: Conceptual and methodological issues. International Journal of Neuroscience, *39*(1–2), 71–89. https://doi.org/10.3109/00207458808985694

Davidson, R. J. & Fox, N. A. (1982). Asymmetrical brain activity discriminates between positive and negative affective stimuli in human infants. Science, 218(4578), 1235–1237. https://doi.org/10.1126/science.7146906

Davidson, R. J. & Fox, N. A. (1989). Frontal brain asymmetry predicts infants' response to maternal separation. Journal of Abnormal Psychology, 98(2), 127–131. https://doi.org/10.1037/0021-843X.98.2.127

Davidson, R. J., Kalin, N. H., & Shelton, S. E. (1992). Lateralized effects of diazepam on frontal brain electrical asymmetries in rhesus monkeys. Biological Psychiatry, 32(5), 438–451. https://doi.org/10.1016/0006-3223(92)90131-I

Davidson, R. J., Marshall, J. R., Tomarken, A. J., & Henriques, J. B. (2000). While a phobic waits: Regional brain electrical and autonomic activity in social phobics during anticipation of public speaking. *Biological Psychiatry*, *47*(2), 85–95. https://doi.org/10.1016/S0006-3223(99)00222-X

De Pascalis, V., Varriale, V., & D'Antuono, L. (2010). Event-related components of the punishment and reward sensitivity. *Clinical Neurophysiology*, *121*(1), 60–76.

Dreisbach, G. & Fischer, R. (2012). Conflicts as aversive signals. *Brain and Cognition*, *78*(2), 94–98.

Ekman, P. & Friesen, W. V. (1971). Constants across cultures in the face and emotion. *Journal of Personality and Social Psychology*, *17*(2), 124–129. https://doi.org/10.1037/h0030377

Ellis, J. S., Watts, A. T., Schmidt, N., & Bernat, E. M. (2018). Anxiety and feedback processing in a gambling task: Contributions of time-frequency theta and delta. *Biological Psychology*, *136*, 1–12.

Engel, A. K. & Fries, P. (2010). Beta-band oscillations—signaling the status quo? *Current Opinion in Neurobiology*, *20*(2), 156–165.

Eysenck, M. W., Derakshan, N., Santos, R., & Calvo, M. G. (2007). Anxiety and cognitive performance: Attentional control theory. *Emotion*, *7*(2), 336.

Field, T., Martinez, A., Nawrocki, T., Pickens, J., Fox, N. A., & Schanberg, S. (1998). Music shifts frontal EEG in depressed adolescents. *Adolescence*, *33*(129), 109–116.

Foti, D., Weinberg, A., Bernat, E. M., & Proudfit, G. H. (2014). Anterior cingulate activity to monetary loss and basal ganglia activity to monetary gain uniquely contribute to the feedback negativity. *Clinical Neurophysiology*, *126*(7), 1338–1347.

Fox, N. A. (1991). If it's not left, it's right. *American Psychologist*, *46*(8), 863–872. https://doi.org/10.1037/0003-066X.46.8.863

Fox, N. A., Bell, M. A., & Jones, N. A. (1992). Individual differences in response to stress and cerebral asymmetry. *Developmental Neuropsychology*, *8*(2–3), 161–184. https://doi.org/10.1080/87565649209540523

Fox, N. A., & Davidson, R. J. (1987). Electroencephalogram asymmetry in response to the approach of a stranger and maternal separation in 10-month-old infants. Developmental Psychology, 23(2), 233–240. https://doi.org/10.1037/0012-1649.23.2.233

Fox, N. A., Rubin, K. H., Calkins, S. D., Marshall, T. R., Coplan, R. J., Porges, S. W., … Stewart, S. (1995). Frontal activation asymmetry and social competence at four years of age. *Child Development*, *66*(6), 1770-1784. https://doi.org/10.1111/j.1467-8624.1995.tb00964.x

Fox, N. A., Schmidt, L. A., Calkins, S. D., Rubin, K. H., & Coplan, R. J. (1996). The role of frontal activation in the regulation and dysregulation of social behavior during the preschool years. *Development and Psychopathology*, *8*(1), 89–102. https://doi.org/10.1017/S0954579400006982

Gable, P., & Harmon-Jones, E. (2008). Relative left frontal activation to appetitive stimuli: Considering the role of individual differences. *Psychophysiology*, *45*(2), 275–278. https://doi.org/10.1111/j.1469-8986.2007.00627.x

Gainotti, G. (1972). Emotional Behavior and Hemispheric Side of the Lesion. *Cortex*, *8*(1), 41–55. https://doi.org/10.1016/S0010-9452(72)80026-1

Gehring, W. J., Goss, B., Coles, M. G. H., Meyer, D. E., & Donchin, E. (2018). The error-related negativity. *Perspectives in Psychological Science*, *13*(2), 200–204. doi:10.1177/1745691617715310

Gorka, S. M., Phan, K. L., & Shankman, S. A. (2015). Convergence of EEG and fMRI measures of reward anticipation. *Biological Psychology*, *112*, 12–19. https://doi.org/10.1016/j.biopsycho.2015.09.007

Gray, J. A. (1975). *Elements of a two-process theory of learning*. Academic Press.

Gray, J. A. (1982). T*he neuropsychology of anxiety: An enquiry into the functions of the septo-hippocampal system*. Oxford University Press.

Gray, J. A. & McNaughton, N. (2000). *The neuropsychology of anxiety: An enquiry into the functions of the septo-hippocampal system* (2nd ed.). Oxford University Press.

Grupe, D. W. & Nitschke, J. B. (2013). Uncertainty and anticipation in anxiety: an integrated neurobiological and psychological perspective. *Nature reviews Neuroscience*, *14*(7), 488.

Hall, E. E., Ekkekakis, P., & Petruzzello, S. J. (2010). Predicting affective responses to exercise using resting EEG frontal asymmetry: Does intensity matter? *Biological Psychology*, *83*(3), 201–206. https://doi.org/10.1016/j.biopsycho.2010.01.001

Harmon-Jones, E. (2007). Trait anger predicts relative left frontal cortical activation to anger-inducing stimuli. *International Journal of Psychophysiology*, *66*(2), 154–160. https://doi.org/10.1016/j.ijpsycho.2007.03.020

Harper, J., Malone, S. M., & Bernat, E. M. (2014). Theta and delta band activity explain N2 and P3 ERP component activity in a go/no-go task. *Clinical Neurophysiology*, *125*(1), 124–132.

Harrewijn, A., Van der Molen, M. J. W., & Westenberg, P. M. (2016). Putative EEG measures of social anxiety: Comparing frontal alpha asymmetry and delta–beta cross-frequency correlation. *Cognitive, Affective, & Behavioral Neuroscience*, *16*(6), 1086–1098.

Harrewijn, A., van der Molen, M. J., van Vliet, I. M., Houwing-Duistermaat, J. J., & Westenberg, P. M. (2018a). Delta-beta correlation as a candidate endophenotype of social anxiety: A two-generation family study. *Journal of Affective Disorders*, *227*, 398–405.

Harrewijn, A., van der Molen, M. J., van Vliet, I. M., Tissier, R. L., & Westenberg, P. M. (2018b). Behavioral and EEG responses to social evaluation: A two-generation family study on social anxiety. *NeuroImage: Clinical*, *17*, 549–562.

Heller, W. (1993). Neuropsychological mechanisms of individual differences in emotion, personality, and arousal. Neuropsychology, 7(4), 476–489. https://doi.org/10.1037/0894-4105.7.4.476

Heller, W. & Nitschke, J. B. (1998). The puzzle of regional brain activity in and anxiety: The importance of subtypes and comorbidity. *Cognition & Emotion*, *12*(3), 421–447.

Heller, W., Nitschke, J. B., Etienne, M. A., & Miller, G. A. (1997). Patterns of regional brain activity differentiate types of anxiety. *Journal of Abnormal Psychology*, *106*(3), 376–385. https://doi.org/10.1037/0021-843X.106.3.376

Hirshfeld-Becker, D. R., Biederman, J., Henin, A., Faraone, S. V., Davis, S., Harrington, K., & Rosenbaum, J. F. (2007). Behavioral inhibition in preschool children at risk is a specific predictor of middle childhood social anxiety: A five-year follow-up. *Journal of Developmental and Behavioral Pediatrics*, 28(3), 225–233. https://doi.org/10.1097/01.DBP.0000268559.34463.do

Hofmann, S. G., Moscovitch, D. A., Litz, B. T., Kim, H. J., Davis, L. L., & Pizzagalli, D. A. (2005). The worried mind: Autonomic and prefrontal activation during worrying. *Emotion*, *5*(4), 464–475. https://doi.org/10.1037/1528-3542.5.4.464

Hur, J., Stockbridge, M. D., Fox, A. S. & Shackman, A. J. (2019). Dispositional negativity, cognition, and anxiety disorders: An integrative translational neuroscience framework. *Progress in Brain Research*, *247*, 375–436.

Ip, K. I., Liu, Y., Moser, J., Mannella, K., Hruschak, J., Bilek, E., … Fitzgerald, K. (2019). Moderation of the relationship between the error-related negativity and anxiety by age and gender in young children: A preliminary investigation. *Developmental Cognitive Neuroscience*, *39*, 100702.

Ischebeck, M., Endrass, T., Simon, D., & Kathmann, N. (2014). Altered frontal EEG asymmetry in obsessive-compulsive disorder. *Psychophysiology*, *51*(7), 596–601. https://doi.org/10.1111/psyp.12214

Isotani, T., Tanaka, H., Lehmann, D., Pascual-Marqui, R. D., Kochi, K., Saito, N., … Sasada, K. (2001). Source localization of EEG activity during hypnotically induced anxiety and relaxation. *International Journal of Psychophysiology*, *41*(2), 143–153. https://doi.org/10.1016/S0167-8760(00)00197-5

Jenkinson, N. & Brown, P. (2011). New insights into the relationship between dopamine, beta oscillations and motor function. *Trends in Neurosciences*, *34*(12), 611–618.

Jensen, O., Kaiser, J., & Lachaux, J. P. (2007). Human gamma-frequency oscillations associated with attention and memory. *Trends in Neurosciences*, *30*(7), 317–324.

Judah, M. R., Grant, D. M., Frosio, K. E., White, E. J., Taylor, D. L., & Mills, A. C. (2016). Electrocortical evidence of enhanced performance monitoring in social anxiety. *Behavior Therapy*, *47*(2), 274–285.

Kalin, N. H., Larson, C., Shelton, S. E., & Davidson, R. J. (1998). Asymmetric frontal brain activity, cortisol, and behavior associated with fearful temperament in rhesus monkeys. *Behavioral Neuroscience*, *112*(2), 286–292. https://doi.org/10.1037/0735-7044.112.2.286

Kemp, A. H., Griffiths, K., Felmingham, K. L., Shankman, S. A., Drinkenburg, W., Arns, M., … Bryant, R. A. (2010). Disorder specificity despite comorbidity: Resting EEG alpha asymmetry in major depressive disorder and post-traumatic stress disorder. *Biological Psychology*, *85*(2), 350–354. https://doi.org/10.1016/j.biopsycho.2010.08.001

Kemp, A. H., Koenig, J., Thayer, J. F., Bittencourt, M. S., Pereira, A. C., Santos, I. S., ... & Benseñor, I. M. (2016). Race and resting-state heart rate variability in Brazilian civil servants and the mediating effects of discrimination: An ELSA-Brasil cohort study. *Psychosomatic Medicine*, *78*(8), 950–958.

Keune, P. M., Schönenberg, M., Wyckoff, S., Mayer, K., Riemann, S., Hautzinger, M., & Strehl, U. (2011). Frontal alpha-asymmetry in adults with attention deficit hyperactivity disorder: Replication and specification. *Biological Psychology*, *87*(2), 306–310. https://doi.org/10.1016/j.biopsycho.2011.02.023

Knyazev, G. G., Savostyanov, A. N., & Levin, E. A. (2005). Uncertainty, anxiety and brain oscillations. *Neuroscience Letters*, *387*, 121–125.

Knyazev, G. G. & Slobodskaya, H. R. (2003). Personality trait of behavioral inhibition is associated with oscillatory systems reciprocal relationships. *International Journal of Psychophysiology*, *48*(3), 247–261.

Knyazev, G. G., Schutter, D. J., & van Honk, J. (2006). Anxious apprehension increases coupling of delta and beta oscillations. *International Journal of Psychophysiology*, *61*(2), 283–287

Knyazev, G. G. (2007). Motivation, emotion, and their inhibitory control mirrored in brain oscillations. *Neuroscience & Biobehavioral Reviews, 31*(3), 377–395.

Knyazev, G. G. (2011). Cross-frequency coupling of brain oscillations: an impact of state anxiety. *International Journal of Psychophysiology, 80*(3), 236–245.

Kool, W., McGuire, J. T., Rosen, Z. B., & Botvinick, M. M. (2010). Decision making and the avoidance of cognitive demand. *Journal of Experimental Psychology, 139*(4), 665.

Kuskowski, M. A., Malone, S. M., Kim, S. W., Dysken, M. W., Okaya, A. J., & Christensen, K. J. (1993). Quantitative EEG in obsessive-compulsive disorder. *Biological Psychiatry, 33*(6), 423–430. https://doi.org/10.1016/0006-3223(93)90170-I

Laufs, H., Kleinschmidt, A., Beyerle, A., Eger, E., Salek-Haddadi, A., Preibisch, C., & Krakow, K. (2003). EEG-correlated fMRI of human alpha activity. *NeuroImage, 19*(4), 1463-1476. https://doi.org/10.1016/S1053-8119(03)00286-6

Laurienti, P. J., Wallace, M. T., Maldjian, J. A., Susi, C. M., Stein, B. E., & Burdette, J. H. (2003). Cross-modal sensory processing in the anterior cingulate and medial prefrontal cortices. *Human Brain Mapping, 19*(4), 213–223.

Ledoux, J. (1978). *The neurology of emotion*. Casa Editrice.

Liu, Y., Moser, J. S., Aviyente, S. (2014). Network community structure detection for directional neural networks inferred from multichannel multisubject EEG data. *IEEE Transactions on Biomedical Engineering, 61*, 1919–1930.

Lindström, B. R., Mattsson-Mårn, I. B., Golkar, A., & Olsson, A. (2013). In your face: Risk of punishment enhances cognitive control and error-related activity in the corrugator supercilii muscle. *PloS One, 8*(6), e65692.

Lobo, I., Portugal, L. C., Figueira, I., Volchan, E., David, I., Garcia Pereira, M., & De Oliveira, L. (2015). EEG correlates of the severity of posttraumatic stress symptoms: A systematic review of the dimensional PTSD literature. *Journal of Affective Disorders, 183*, 210–220. https://doi.org/10.1016/j.jad.2015.05.015

LoBue, V., Coan, J. A., Thrasher, C., & DeLoache, J. S. (2011). Prefrontal asymmetry and Parent-Rated temperament in infants. *PLoS One, 6*(7): e22694. https://doi.org/10.1371/journal.pone.0022694

Lubar, J. F., Congedo, M., & Askew, J. H. (2003). Low-resolution electromagnetic tomography (LORETA) of cerebral activity in chronic depressive disorder. *International Journal of Psychophysiology, 49*(3), 175–185. https://doi.org/10.1016/S0167-8760(03)00115-6

Luu, P., Tucker, D. M., & Makeig, S. (2004). Frontal midline theta and the error-related negativity: neurophysiological mechanisms of action regulation. *Clinical Neurophysiology, 115*(8), 1821–1835.

Mantini, D., Perrucci, M. G., Del Gratta, C., Romani, G. L., & Corbetta, M. (2007). Electrophysiological signatures of resting state networks in the human brain. *Proceedings of the National Academy of Sciences of the United States of America, 104*(32), 13170–13175. doi:10.1073/pnas.0700668104

Mathersul, D., Williams, L. M., Hopkinson, P. J., & Kemp, A. H. (2008). Investigating models of affect: Relationships among EEG alpha asymmetry, depression, and anxiety. *Emotion, 8*(4), 560–572. https://doi.org/10.1037/a0012811

McManis, M. H., Kagan, J., Snidman, N. C., & Woodward, S. A. (2002). EEG asymmetry, power, and temperament in children. *Developmental Psychobiology, 41*(2), 169–177. https://doi.org/10.1002/dev.10053

Metzger, L. J., Carson, M. A., Paulus, L. A., Paige, S. R., Lasko, N. B., Pitman, R. K., & Orr, S. P. (2004). PTSD arousal and depression symptoms associated with increased right-sided parietal EEG asymmetry. *Journal of Abnormal Psychology, 113*(2), 324–329. https://doi.org/10.1037/0021-843X.113.2.324

Meyer, T. J., Miller, M. L., Metzger, R. L., & Borkovec, T. D. (1990). Development and validation of the Penn State Worry Questionnaire. *Behaviour Research and Therapy, 28*(6), 487–495. https://doi.org/10.1016/0005-7967(90)90135-6

Miller, G. A. (1996). How we think about cognition, emotion, and biology in psychopathology. *Psychophysiology, 33*, 615–628.

Miller, G. A. (2010). Mistreating psychology in the decades of the brain. *Perspectives on Psychological Science, 5*,716–743.

Miskovic, V., Ashbaugh, A. R., Santesso, D. L., McCabe, R. E., Antony, M. M., & Schmidt, L. A. (2010). Frontal brain oscillations and social anxiety: A cross-frequency spectral analysis during baseline and speech anticipation. *Biological Psychology, 83*(2), 125–132.

Miskovic, V., Campbell, M. J., Santesso, D. L., Van Ameringen, M., Mancini, C. L., & Schmidt, L. A. (2011). Frontal brain oscillatory coupling in children of parents with social phobia: A pilot study. *The Journal of Neuropsychiatry and Clinical Neurosciences, 23*(1), 111–114.

Miskovic, V., Moscovitch, D. A., Santesso, D. L., McCabe, R. E., Antony, M. M., & Schmidt, L. A. (2011). Changes in EEG cross-frequency coupling during cognitive behavioral therapy for social anxiety disorder. *Psychological Science, 22*(4), 507–516.

Mizuki, Y., Hashimoto, M., Tanaka, T., Inanaga, K., & Tanaka, M. (1983). A new physiological tool for assessing anxiolytic effects in humans: frontal midline theta activity. *Psychopharmacology, 80*(4), 311–314.

Mizuki, Y., Suetsugi, M., Imai, T., Kai, S., Kajimura, N., & Yamada, M. (1989). A physiological marker for assessing anxiety level in humans: frontal midline theta activity. *Psychiatry and Clinical Neurosciences, 43*(4), 619–626.

Mizuki, Y., Kajimura, N., Kai, S., Suetsugi, M., Ushijima, I., & Yamada, M. (1992). Differential responses to mental stress in high and low anxious normal humans assessed by frontal midline theta activity. *International Journal of Psychophysiology, 12*(2), 169–178.

Moadab, I., Gilbert, T., Dishion, T. J., & Tucker, D. M. (2010). Frontolimbic activity in a frustrating task: Covariation between patterns of coping and individual differences in externalizing and internalizing symptoms. *Development and Psychopathology, 22*(2), 391–404.

Moran, T. P., Bernat, E. M., Aviyente, S., Schroder, H. S., & Moser, J. S. (2015). Sending mixed signals: Worry is associated with enhanced initial error processing but reduced call for subsequent cognitive control. Social *Cognitive and Affective Neuroscience, 10*(11), 1548–1556.

Morillas-Romero, A., Tortella-Feliu, M., Bornas, X., & Putman, P. (2015). Spontaneous EEG theta/beta ratio and delta–beta coupling in relation to attentional network functioning and self-reported attentional control. *Cognitive, Affective, & Behavioral Neuroscience, 15*(3), 598–606.

Moscovitch, D. A., Santesso, D. L., Miskovic, V., McCabe, R. E., Antony, M. M., & Schmidt, L. A. (2011). Frontal EEG asymmetry and symptom response to cognitive behavioral therapy in patients with social anxiety disorder. *Biological Psychology*. https://doi.org/10.1016/j.biopsycho.2011.04.009

Moser, J. S., Moran, T. P., & Jendrusina, A. A. (2012). Parsing relationships between dimensions of anxiety and action monitoring brain potentials in female undergraduates. *Psychophysiology, 49*(1), 3–10.

Moser, J. S., Moran, T. P., Schroder, H. S., Donnellan, M. B., & Yeung, N. (2013). On the relationship between anxiety and error monitoring: A meta-analysis and conceptual framework. *Frontiers in Human Neuroscience*, *7*, 466. doi:10.3389/fnhum.2013.00466

Moser, J. S., Moran, T. P., Kneip, C., Schroder, H. S., & Larson, M. J. (2016). Sex moderates the association between symptoms of anxiety, but not obsessive compulsive disorder, and error-monitoring brain activity: A meta-analytic review. Psychophysiology, 53, 21–29.

Moser, J.S. (2017). The nature of the relationship between anxiety and the error-related negativity across development. Current Behavior Neuroscience Reports, 4(4), 309–321.

Nash, K., Inzlicht, M., & McGregor, I. (2012). Approach-related left prefrontal EEG asymmetry predicts muted error-related negativity. *Biological Psychology*, *91*(1), 96–102. https://doi.org/10.1016/j.biopsycho.2012.05.005

Nitschke, J. B., Heller, W., Palmieri, P. A., & Miller, G. A. (1999). Contrasting patterns of brain activity in anxious apprehension and anxious arousal. *Psychophysiology*, *36*(5), 628–637. https://doi.org/10.1017/S0048577299972013

Nusslock, R., Shackman, A. J., McMenamin, B. W., Greischar, L. L., Davidson, R. J., & Kovacs, M. (2018). Comorbid anxiety moderates the relationship between depression history and prefrontal EEG asymmetry. *Psychophysiology*, 55(1), e12953. https://doi.org/10.1111/psyp.12953

Nusslock, R., Walden, K., & Harmon-Jones, E. (2015). Asymmetrical frontal cortical activity associated with differential risk for mood and anxiety disorder symptoms: An RDoC perspective. *International Journal of Psychophysiology*, *98*(2, Pt.2), 249–261. https://doi.org/10.1016/j.ijpsycho.2015.06.004

Peterson, C. K. & Harmon-Jones, E. (2009). Circadian and seasonal variability of resting frontal EEG asymmetry. *Biological Psychology*, *80*(3), 315–320. https://doi.org/10.1016/j.biopsycho.2008.11.002

Poppelaars, E. S., Harrewijn, A., Westenberg, P. M., & van der Molen, M. J. W. (2018). Frontal delta-beta cross-frequency coupling in high and low social anxiety: An index of stress regulation? Cognitive, Affective & Behavioral Neuroscience, 18(4), 764–777.

Popov, T., Westner, B. U., Silton, R. L., Sass, S. M., Spielberg, J. M., Rockstroh, B., Heller, W., & Miller, G. A. (2018). Time course of brain network reconfiguration supporting inhibitory control. *Journal of Neuroscience*, *38*, 4348–4356.

Putman, P., Verkuil, B., Arias-Garcia, E., Pantazi, I., & van Schie, C. (2014). EEG theta/beta ratio as a potential biomarker for attentional control and resilience against deleterious effects of stress on attention. *Cognitive, Affective, & Behavioral Neuroscience*, *14*(2), 782–791.

Putman, P., Arias-Garcia, E., Pantazi, I., & van Schie, C. (2012). Emotional Stroop interference for threatening words is related to reduced EEG delta–beta coupling and low attentional control. *International Journal of Psychophysiology*, *84*, 194–200.

Putman, P., van Peer, J., Maimari, I., & van der Werff, S. (2010). EEG theta/beta ratio in relation to fear-modulated response-inhibition, attentional control, and affective traits. *Biological Psychology*, *83*(2), 73–78.

Rabe, S., Beauducel, A., Zöllner, T., Maercker, A., & Karl, A. (2006). Regional brain electrical activity in posttraumatic stress disorder after motor vehicle accident. *Journal of Abnormal Psychology*, *115*(4), 687–698. https://doi.org/10.1037/0021-843X.115.4.687

Rabe, S., Zöllner, T., Maercker, A., & Karl, A. (2006). Neural correlates of posttraumatic growth after severe motor vehicle accidents. *Journal of Consulting and Clinical Psychology*, *74*(5), 880–886. https://doi.org/10.1037/0022-006X.74.5.880

Reznik, S. J., & Allen, J. J. B. (2018). Frontal asymmetry as a mediator and moderator of emotion: An updated review. *Psychophysiology, 55*(1), e12965. https://doi.org/10.1111/psyp.12965

Riesel, A., Goldhahn, S., & Kathmann, N. (2017). Hyperactive performance monitoring as a transdiagnostic marker: Results from health anxiety in comparison to obsessive–compulsive disorder. *Neuropsychologia, 96*, 1–8.

Righi, S., Mecacci, L., & Viggiano, M. P. (2009). Anxiety, cognitive self-evaluation and performance: ERP correlates. *Journal of Anxiety Disorders, 23*(8), 1132–1138.

Sackeim, H. A., Greenberg, M. S., Weiman, A. L., Gur, R. C., Hungerbuhler, J. P., & Geschwind, N. (1982). Hemispheric asymmetry in the expression of positive and negative emotions. Neurologic evidence. *Archives of Neurology, 39*(4), 210–218.

Santesso, D. L., Bogdan, R., Birk, J. L., Goetz, E. L., Holmes, A. J., & Pizzagalli, D. A. (2011). Neural responses to negative feedback are related to negative emotionality in healthy adults. Social *Cognitive and Affective Neuroscience, 7*(7), 794–803.

Scheeringa, R., Petersson, K. M., Kleinschmidt, A., Jensen, O., & Bastiaansen, M. C. M. (2012). EEG alpha power modulation of fMRI resting-state connectivity. *Brain Connectivity, 2*(5), 254–264. https://doi.org/10.1089/brain.2012.0088

Schmid, P. C., Kleiman, T., & Amodio, D. M. (2015). Neural mechanisms of proactive and reactive cognitive control in social anxiety. *Cortex, 70*, 137–145.

Schmidt, L. A., Fox, N. A., Schulkin, J., & Gold, P. W. (1999). Behavioral and psychophysiological correlates of self-presentation in temperamentally shy children. *Developmental Psychobiology, 35*(2), 119–135. https://doi.org/10.1002/(SICI)1098-2302(199909)35:2<119::AID-DEV5>3.0.CO;2-G

Schmidt, B., Kanis, H., Holroyd, C. B., Miltner, W. H., & Hewig, J. (2018). Anxious gambling: Anxiety is associated with higher frontal midline theta predicting less risky decisions. *Psychophysiology, 55*(10), e13210.

Schouppe, N., De Houwer, J., Richard Ridderinkhof, K., & Notebaert, W. (2012). Conflict: Run! Reduced Stroop interference with avoidance responses. *The Quarterly Journal of Experimental Psychology, 65*(6), 1052–1058.

Schroder, H. S., Glazer, J. E., Bennett, K. P., Moran, T. P., & Moser, J. S. (2017). Suppression of error-preceding brain activity explains exaggerated error monitoring in females with worry. *Biological Psychology, 122*, 33–41. doi:10.1016/j.biopsycho.2016.03.013

Schutter, D. J. & Honk, J. V. (2004). Decoupling of midfrontal delta–beta oscillations after testosterone administration. *International Journal of Psychophysiology, 53*, 71–73.

Schutter, D. J. & Van Honk, E. J. (2005). Salivary cortisol levels and the coupling of midfrontal delta-beta oscillations. *International Journal of Psychophysiology, 55*, 127–129.

Sehlmeyer, C., Konrad, C., Zwitserlood, P., Arolt, V., Falkenstein, M., & Beste, C. (2010). ERP indices for response inhibition are related to anxiety-related personality traits. *Neuropsychologia, 48*(9), 2488–2495.

Shackman, A. J., Salomons, T. V., Slagter, H. A., Fox, A. S., Winter, J. J., & Davidson, R. J. (2011). The integration of negative affect, pain and cognitive control in the cingulate cortex. *Nature Reviews Neuroscience, 12*(3), 154–167. doi:10.1038/nrn2994

Shankman, S. A., Silverstein, S. M., Williams, L. M., Hopkinson, P. J., Kemp, A. H., Felmingham, K. L., … Clark, C. R. (2008). Resting electroencephalogram asymmetry and posttraumatic stress disorder. *Journal of Traumatic Stress, 21*(2), 190–198. https://doi.org/10.1002/jts.20319

Sharp, P. B., Miller, G. A., & Heller, W. (2015). Transdiagnostic dimensions of anxiety: neural mechanisms, executive functions, and new directions. *International Journal of Psychophysiology, 98*(2), 365–377.

Sharp, P. B. & Miller, G. A. (2019). Reduction and autonomy in psychology and neuroscience: A call for pragmatism. *Journal of Theoretical and Philosophical Psychology, 39*(1), 18.

Shi, R., Sharpe, L., & Abbott, M. (2019). A meta-analysis of the relationship between anxiety and attentional control. *Clinical Psychology Review, 72*, 101754.

Silberman, E. K. & Weingartner, H. (1986). Hemispheric lateralization of functions related to emotion. *Brain and Cognition, 5*(3), 322–353. https://doi.org/10.1016/0278-2626(86)90035-7

Smith, E. E., Cavanagh, J. F., & Allen, J. J. B. (2018). Intracranial source activity (eLORETA) related to scalp-level asymmetry scores and depression status. *Psychophysiology, 55*(1), e13019. https://doi.org/10.1111/psyp.13019

Smith, E. E., Zambrano-Vazquez, L., & Allen, J. J. B. (2016). Patterns of alpha asymmetry in those with elevated worry, trait anxiety, and obsessive-compulsive symptoms: A test of the worry and avoidance models of alpha asymmetry. *Neuropsychologia, 85*, 118–126. https://doi.org/10.1016/j.neuropsychologia.2016.03.010

Spielberg, J. M., Miller, G. A., Heller, W., & Banich, M. T. (2015). Flexible brain network reconfiguration supporting inhibitory control. *Proceedings of the National Academy of Sciences of the United States of America, 112*(32), 10020–10025.

Spielberger, C. D. (1983). State-Trait Anxiety Inventory (STAI). Mind *Garden, 94061*(650), 261–3500. https://doi.org/10.1002/9780470479216.corpsy0943

Steiner, A. R. W., & Coan, J. A. (2011). Prefrontal asymmetry predicts affect, but not beliefs about affect. *Biological Psychology, 88*, 65–71. https://doi.org/10.1016/j.biopsycho.2011.06.010

Suetsugi, M., Mizuki, Y., Ushijima, I., Kobayashi, T., Tsuchiya, K., Aoki, T., & Watanabe, Y. (2000). Appearance of frontal midline theta activity in patients with generalized anxiety disorder. *Neuropsychobiology, 41*(2), 108–112.

Taschereau-Dumouchel, V., Michel, M., Lau, H., Hofmann, S. G., & LeDoux, J. E. (2022). Putting the "mental" back in "mental disorders": a perspective from research on fear and anxiety. Molecular Psychiatry. https://doi.org/10.1038/s41380-021-01395-5

Trujillo, L. T. & Allen, J. J. (2007). Theta EEG dynamics of the error-related negativity. *Clinical Neurophysiology, 118*(3), 645–668.

U.S. Burden of Disease Collaborators. (2018). The state of US health, 1990-2016. Burden of diseases, injuries, and risk factors among US states. *Journal of the American Medical Association, 319*, 1444–1472.

Van Den Heuvel, M. P., Mandl, R. C., Kahn, R. S., & Hulshoff Pol, H. E. (2009). Functionally linked resting-state networks reflect the underlying structural connectivity architecture of the human brain. *Human Brain Mapping, 30*(10), 3127–3141.

van der Molen, M. J., Harrewijn, A., & Westenberg, P. M. (2018). Will they like me? Neural and behavioral responses to social-evaluative peer feedback in socially and non-socially anxious females. *Biological Psychology, 135*, 18–28.

Van Noordt, S. J., Campopiano, A., & Segalowitz, S. J. (2016). A functional classification of medial frontal negativity ERPs: Theta oscillations and single subject effects. *Psychophysiology, 53*(9), 1317–1334.

Van Peer, J. M., Roelofs, K., & Spinhoven, P. (2008). Cortisol administration enhances the coupling of midfrontal delta and beta oscillations. *International Journal of Psychophysiology, 67*, 144–150.

van Son, D., De Blasio, F. M., Fogarty, J. S., Angelidis, A., Barry, R. J., & Putman, P. (2019). Frontal EEG theta/beta ratio during mind wandering episodes. *Biological Psychology*, *140*, 19–27.

Velikova, S., Locatelli, M., Insacco, C., Smeraldi, E., Comi, G., & Leocani, L. (2010). Dysfunctional brain circuitry in obsessive–compulsive disorder: source and coherence analysis of EEG rhythms. *NeuroImage*, *49*(1), 977–983.

Velo, J. R., Stewart, J. L., Hasler, B. P., Towers, D. N., & Allen, J. J. B. (2012). Should it matter when we record? Time of year and time of day as factors influencing frontal EEG asymmetry. *Biological Psychology*, *91*(2), 283–291. https://doi.org/10.1016/j.biopsycho.2012.06.010

Wacker, J., Mueller, E. M., Pizzagalli, D. A., Hennig, J., & Stemmler, G. (2013). Dopamine-D2-receptor blockade reverses the association between trait approach motivation and frontal asymmetry in an approach-motivation context. *Psychological Science*, *24*(4), 489–497. https://doi.org/10.1177/0956797612458935

Watts, A. T., Tootell, A. V., Fix, S. T., Aviyente, S., & Bernat, E. M. (2018). Utilizing time-frequency amplitude and phase synchrony measure to assess feedback processing in a gambling task. *International Journal of Psychophysiology*, *132*, 203–212.

Weinberg, A., Meyer, A., Hale-Rude, E., Perlman, G., Kotov, R., Klein, D. N., & Hajcak, G. (2016). Error-related negativity (ERN) and sustained threat: Conceptual framework and empirical evaluation in an adolescent sample. *Psychophysiology*, *53*(3), 372–385. doi:10.1111/psyp.12538

Wiedemann, G., Pauli, P., Dengler, W., Lutzenberger, W., Birbaumer, N., & Buchkremer, G. (1999). Frontal brain asymmetry as a biological substrate of emotions in patients with panic disorders. *Archives of General Psychiatry*, *56*(1), 78–84. https://doi.org/10.1001/archpsyc.56.1.78

Zhang, D. & Gu, R. (2018). Behavioral preference in sequential decision-making and its association with anxiety. *Human Brain Mapping*, *39*(6), 2482–2499.

Zotev, V., Yuan, H., Misaki, M., Phillips, R., Young, K. D., Feldner, M. T., & Bodurka, J. (2016). Correlation between amygdala BOLD activity and frontal EEG asymmetry during real-time fMRI neurofeedback training in patients with depression. *NeuroImage: Clinical*, *11*, 224–238. https://doi.org/10.1016/j.nicl.2016.02.003

PART IV

CHAPTER 20

BIVARIATE FUNCTIONAL CONNECTIVITY MEASURES FOR WITHIN- AND CROSS-FREQUENCY COUPLING OF NEURONAL OSCILLATIONS

J. MATIAS PALVA AND SATU PALVA

20.1 Introduction

Oscillations are ubiquitous in electrophysiological recordings of spontaneous brain activity at all measurable scales, from single neurons and micro-electrode recordings to large neuronal masses observable with EEG and MEG from scalp. Numerous synaptic mechanisms and neuronal microcircuit patterns serve the production of oscillatory behavior in neuronal populations ranging in size from some tens of neurons to millions of neurons in macroscopic cortical areas. Functional correlations among neuronal populations arise primarily through direct structural connections. However, in neuronal systems operating near a critical phase transition, these correlations may exceed the spatial and temporal range of the actual underlying structural interactions. Several lines of evidence indicate that human brain activity exhibits such critical-like dynamics, which implies that correlations of neuronal activities are likely to exceed the connections defined by direct axonal, or "structural", connections and temporally the time scales of the associated synaptic time constants.

In studies of brain dynamics, correlations between the activity time series of neurons or populations of neurons are termed "functional" connectivity (FC). In this chapter, we outline the key methods for measuring bivariate (pairwise) FC among electrophysiological time series. As a preface, we provide an overview of the essential features of the underlying (univariate) signals and the corresponding analysis methods. We aim to dissect these methods both by the information they yield about the underlying physiological interactions as well as by limitations posed by the typical confounders in MEG and EEG data.

20.2 Dissecting Broadband, Quasi-Periodic, and Periodic Signals for Functional Connectivity Analysis

Before decisions about time-series analysis methods are made, it is important to understand the to-be-studied system and grasp whether the system operates with broadband or narrowband processes. This is because the analyses of broad- and narrowband signals involve distinct concepts and implications. In the context of brain data and EEG in particular, a distinction has been drawn between supposedly aperiodic forms of activity, such as the "1/f-component" of the EEG power spectrum, and activities that are clearly periodic, such as the alpha oscillations that exhibit a power spectral peak (Cole & Voytek, 2017). However, in between genuine aperiodic and periodic signals, that is, between genuine noise and oscillations, there is likely to be a wide range of electrophysiological phenomena that are "quasi-periodic" (Palva & Palva, 2012; Palva & Palva 2018). Quasi-periodic neuronal oscillations are genuine oscillations that arise via oscillatory mechanisms and embody the functional implications and consequences of oscillations, but that are short-lived and have such variance in frequency that they do not show up as a clear peak in power-spectral analyses (Palva & Palva, 2012; Palva & Palva, 2018). For example, infra-slow (0.01–0.1 Hz) and slow (0.1–1 Hz) fluctuations, despite their 1/f-like power spectrum, have been suggested to be quasi-periodic oscillations in both EEG (Monto et al., 2008; Palva & Palva, 2012; Palva & Palva, 2018) and fMRI (Abbas et al., 2019; Belloy et al., 2019; Zhang et al., 2020) data.

Functional connectivity in genuine broadband activity can be meaningfully analyzed only with methods such as the classical time-domain *cross correlation* measures (Pearson and Spearman correlation coefficients, canonical correlation, and many others) or information theoretical measures such as *mutual information* and its many derivatives. These time-domain measures can be applied also to quasi-periodic or periodic signals such as neuronal oscillations, but they are suboptimal because they ignore the frequency-limited nature of the periodicity and include "noise" and signals in other frequency bands. Moreover, broadband coupling measures, by definition, ignore the most important parameter of neuronal oscillations—their phase. As this chapter focuses on neuronal oscillations, we will not delve deeper into the analysis of aperiodic signals.

Measures intended for analyses of oscillations (discussed later) can also be applied to broadband signals and will converge to the correct result. However, they are suboptimal by involving narrow-band filtering as well as phase-estimation that is irrelevant in broadband signals because in true aperiodic correlations, only time rather than phase lags may exist. These measures also exploit only limited frequency ranges for correlation estimation, which distributes the statistical power in the signal to multiple tests (Figure 20.1).

The vast majority of brain signals are (i) quasi-periodic or periodic, (ii) arise via frequency-limited, time-scale specific neuronal mechanisms, (iii) are phenomenologically confined to a limited frequency band, and finally, (iv) involve the notion of phase as the central mechanistic concept that spans from neuronal firing to behavioral correlations. Thus, initiation of any analysis with an analysis approach intended for frequency-resolved narrow-band oscillations is justifiable and gives an appropriate representation of the signal features.

20.3 Phase and Amplitude Correlations of Neuronal Oscillations

Phase and amplitude time series of neuronal activity in electrophysiological signals can be extracted by filtering. When estimated for two or more such time series, bivariate correlation metrics for pairs of such phase and amplitude estimates can be applied to uncover the patterns of correlations in these data, that is, to quantify the phase and amplitude FC.

Phase is a quantitative description of the state of a given process/component and its temporal evolution within a period (or a cycle) of the oscillation. From micro-electrode recordings to macroscopic EEG signals, observations of genuine phase correlations imply correlations of neuronal excitability fluctuations at the cellular level and the potential for consistent spike-timing relationships.

The physiological and mechanistic implications of amplitude, on the other hand, are always dependent on the experimental setting because the amplitude measurement is dependent on biological and physical signal generation and conduction mechanisms. The population-level summation of electric fields leading to the measured signal amplitude is largely dependent on the coherence of the underlying synaptic inputs. This is because cancellation of electric currents effectively suppresses signals from incoherent sources. Thus, the amplitude of oscillations is primarily an approximate index of local synchronization and only secondarily dependent on the net magnitude of source currents.

Phase and amplitude of neuronal oscillations thus reflect partially distinct physiological phenomena and, correspondingly, inter-areal phase and amplitude correlations tap into physiologically and functionally distinct forms of large-scale coordination of neuronal activities.

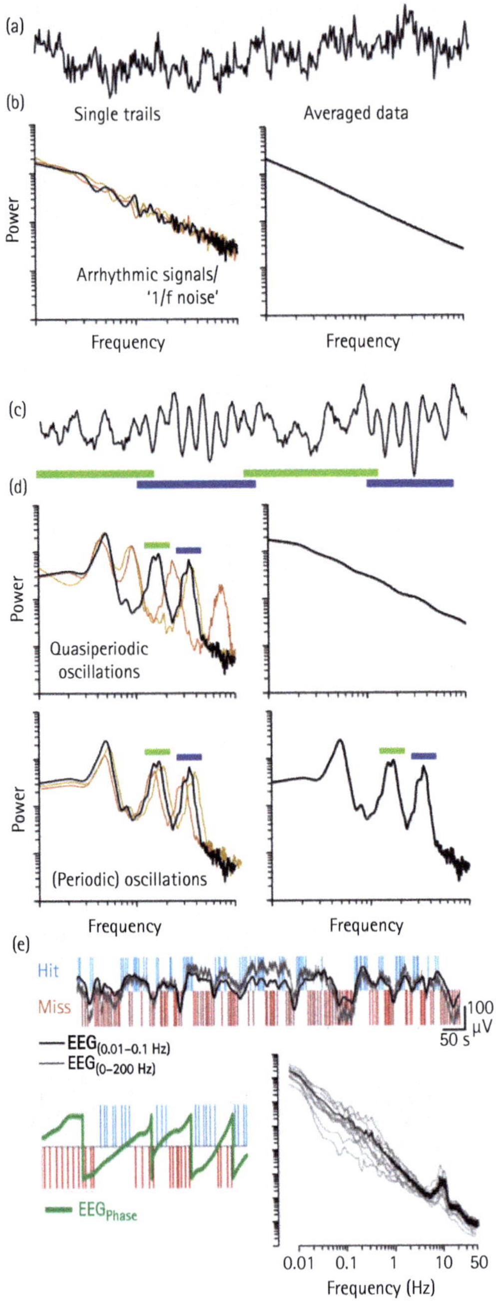

FIGURE 20.1 Conceptual illustration of aperiodic, quasiperiodic, and periodic signals. (A) The time series of 1/f noise yield both (B) in a single realization (left) and in averaged data (right) 1/f power spectra. (C) Signals composed of mixtures of transient periodicities from multiple sources or quasiperiodic oscillators with a large variability in oscillation frequencies give rise to power spectra (D) that may exhibit peaks at the corresponding frequencies (green and blue bars) in single trials and yet the averaged spectra may not exhibit peaks. Only oscillations with a stable period retain peaks in averaged power spectra. (E) Human behavioral performance is characterized by power-law scale-free dynamics where the detected (Hits) and undetected (Misses) stimuli in a threshold-stimulus detection task are clustered with power- law and long-range temporal correlations. Adapted from Palva and Palva 2018.

20.4 Phase- and Amplitude-Time Series Estimation for Periodic and Quasi-Periodic Signals

This chapter discusses the principal approaches for measuring phase and amplitude correlations from neuronal data. Further, it aims to illustrate the adaptation of phase-locking methods to the assessment of frequency-resolved amplitude correlations, as well as to quantification of cross-frequency correlations, such as phase-amplitude and cross-frequency phase correlations, among signals in different frequency bands.

All these interaction metrics are based on estimates of phase and amplitude. To explicitly represent frequency-band-limited phase and amplitude dynamics, most real-valued continuous signals can be transformed by filtering to a complex form. A popular choice is to convolve the signal with a complex wavelet, such as the Gabor or Morlet wavelet (Sinkkonen et al., 1995; Tallon-Baudry et al., 1996). Such Gaussian wavelet filtering yields optimal time-frequency-resolution compromise and is superior to other filtering approaches in most cases (Sinkkonen et al., 1995). The Morlet wavelet w is defined by:

$$W(t, f_0) = A \exp\left(-t^2 / 2\sigma_t^2\right) \exp\left(2i\pi f_0 t\right) \quad (1)$$

where t denotes time, i the imaginary unit, f_0 the center frequency of the wavelet, and σ_t the standard deviation of the wavelet's Gaussian envelope in time domain. A is a scaling parameter and for conventional filtering that aims to yield a filtered signal with the same amplitude as the original signal, A should be set to $A = 2/\sum(w)$ where $\sum(w)$ denotes the integral of wavelet's amplitude envelope (simply the sum of wavelet values for a discrete realization). Notably, this yields a wavelet spectrum that is flat for 1/f noise. The wavelet family should be designed with a constant ratio m

$$m = f_0 / \sigma f \quad (2)$$

where σ_f the standard deviation of the wavelet's gaussian envelope in frequency domain and given by $\sigma_f = 1/2\,\pi\sigma_t$. This ratio, or "the m parameter" thus defines the time-frequency compromise so that small values (such as $m = 5$) favors time resolution whereas larger values (such as $m = 7.5$) yields improved frequency resolution at the expense of time resolution.

In applications where other forms of filtering are used, and the filter yields real valued output, the complex form can be obtained by finding the imaginary part with the Hilbert transform. As an important non-linear filtering approach, empirical mode decomposition (Huang & Wu, 2008) is an adaptive and data-driven method that identifies major signal components iteratively from the fastest to slowest frequencies and yields time series that may follow frequency fluctuations better than conventional fixed-frequency filtering.

For segments of stationary signals, the phase estimation can be achieved with spectral measures as well, such as with the Fourier transform. These three approaches are formally equivalent (Bruns, 2004). However, spectral methods yield constant-sized frequency bins while wavelets can and should be scaled with frequency so that they have constant time-frequency resolution. This is often overlooked but is as a fundamental disadvantage for the spectral methods. For any given fixed time-window size, different frequencies will have different amounts of cycles in the window where they are presumed stationary. Because neuronal oscillations are "stationary" for just a few cycles, spectral measures will be optimal for a very limited frequency range, and for most other frequencies will include too few or too many cycles of the oscillations in the time window.

In the filtering approach, let us denote the filtering operation with a center frequency f generically with the operator $\mathbf{T}_f$ so that the filtered signal $x(t,f)$ is obtained from the original signal $x_0(t)$ by

$$x(t,f) = \mathbf{T}_{f,complex}\,[\mathrm{x}_0(\mathrm{t})] = \mathrm{a}_x(t,f)\,\exp[i\theta_x(\mathrm{t,f})] \tag{2}$$

where i is the imaginary unit, $a_x(t,f)$ denotes the amplitude and $\theta_x(t,f)$ the phase of $x(t,f)$, and $\mathrm{T}f_{,complex}$ a complex filter such as the Morlet wavelet transform. If the filtering is performed with a real-valued filter, $\mathbf{T}_{f,real}$, the imaginary signal is obtained by the Hilbert transform, so that

$$x(t,f) = \mathbf{T}_{f,real}\,[\mathrm{x}_0(\mathrm{t})] + i\mathbf{H}\left\{\mathbf{T}_{f,real}\,[\mathrm{x}_0(\mathrm{t})]\right\} \tag{3}$$

where H denotes the Hilbert transform.

From either approach, let us denote an array of such time series by $X(t,f) = [x_j(t,f)] = A_x(t,f)\exp[i\Theta_x(t,f)]$, where $j = 1 \ldots n_s$ and n_s is the number of sensors or sources. Comparable phase time series can also be produced from point process time series (Rosenblum et al., 2004).

20.5 Definition of Phase Difference as the Basis for a Range of Interaction Metrics

The time series of phase and amplitude enable the estimation of bivariate phase and amplitude correlations between any pair of signals. In addition to phase synchronization per se, a range of phase and amplitude interactions can be quantified with the phase-locking estimation formalism and will be described here first. Phase locking is statistically indicated by a non-random phase or phase-difference distribution.

The generic n:m-phase difference ω_{nm} of signals x and y (Tass et al., 1998) is given by

$$\omega_{n:m} = n\theta_x - m\theta_y \tag{4}$$

where the small integers n and m determine the frequency ratio $nf_x = mf_y$ of the center frequencies of filters T_{fx} and T_{fy}. For assessing the classical within-frequency phase synchrony, $n = m = 1$. This phase difference is conveniently obtained in complex form from the filtered x and y simply by

$$\exp[i\omega_{n:m}] = \left(x/|x|\right)^n \; (y/|y|)^{*m} \tag{5}$$

where the asterisk* denotes the complex conjugate. Hence, for frequencies f_x and f_y, the $n_S \times n_S$-sized pair-wise complex phase difference matrix $\Phi_{n:m}$ of a single sample with each signal against each other signal is given by the complex outer product, ⊗, so that

$$\Phi_{n:m}(t, f_x f_y) = (X/|X|)^n \otimes (Y/|Y|)^{*m} \tag{6}$$

This expression thus provides the phase difference with which one may quantify classical within-frequency phase synchronization but also cross-frequency phase synchronization (CFS). CFS is a cross-frequency coupling (CFC) mechanism that indicates a non-random phase difference between different low- (LF) and high-frequency (HF) frequency pairs at small-integer ratios n:m (LF:HF). CFS enables temporally precise coordination of neuronal processing by establishing the possibility for systematic spike-timing relationships among oscillatory assemblies at different frequencies and can thus serve functional integration and coordination across within-frequency synchronized large-scale networks (Fell & Axmacher, 2011; Palva et al., 2005; Palva & Palva, 2017; Siebenhühner et al., 2016; Siebenhühner et al., 2020).

The phase-difference approach can also be used to represent more complex bivariate relationships such as phase-amplitude coupling (PAC), or "nested oscillations", that are characterized by the phase-locking of the amplitude envelope of a faster oscillation with the phase of a slower oscillation. PAC has been suggested to reflect the regulation of sensory information processing in faster frequencies by excitability fluctuations imposed by slower oscillations (Canolty & Knight, 2010; Fell & Axmacher, 2011; Jensen & Colgin, 2007; Jensen & Lisman, 2013, Schroeder & Lakatos, 2009). The complex phase difference matrix, Φ_{PAC}, is given by filtering the amplitude envelopes A_x of the faster oscillation at f_x with the filter **T*fy*** that was used to obtain the slower oscillation at $f_y < f_x$ (Vanhatalo et al., 2004):

$$\Phi_{PAC} = \left[\mathbf{T}_{fy}(A_x)/\mathbf{T}_{fy}(A_x)|\right] \otimes (Y/|Y|)^* \tag{7}$$

It is important to note that this quantification of the non-uniformity of Φ_{PAC} with a phase-locking value (Section 20.6) assumes a unimodal distribution of the fast oscillation amplitudes along the slow oscillation's phase. An interaction in which the amplitude of the fast oscillation peaks twice, or more formally m times during the slow cycle, can be explicitly assessed as a *1:m*-amplitude-phase interaction. Here the amplitude envelopes A_x are filtered with a filter $\mathbf{T}_{mfy}$ at frequency mf_y and then the corresponding metric, Φ_{PACm}, is given simply by

$$\Phi_{PACm} = \left[\mathbf{T}_{2fy}(A_x)/|\mathbf{T}_{2fy}(A_x)|\right] \otimes (Y/|Y|)^{*m} \tag{8}$$

Φ_{PACm} is quantifiable with the phase-locking value. Alternatively, *1:m*-amplitude-phase coupling can be implicitly assessed with Φ_{PAC} and a histogram-based statistic (Section 20.7).

Characterization of amplitude dynamics with phase differences and circular statistics can be further extended to amplitude-amplitude interactions by filtering the amplitude envelopes A_x and A_y to a slower frequency band $f_z < f_x, f_y$ with $\mathbf{T}_{fz}$. Now the complex phase difference matrix, Φ_{AA}, is

$$\Phi_{AA} = \left[\mathbf{T}_{fz}(A_x)/|\mathbf{T}_{fz}(A_x)|\right] \otimes \left[\mathbf{T}_{fz}(A_y)/|\mathbf{T}_{fz}(A_y)|\right]^* \tag{9}$$

The advantage of characterizing amplitude interactions (Figure 20.2) with phase estimators is that they yield frequency-resolved estimates of amplitude correlations and that phase dynamics of amplitude fluctuations are independent of variability in the absolute amplitudes and hence large amplitude events do not exert a strong bias like they do in cross-correlation analyses for example. For broadband amplitude correlations, however, conventional correlation-coefficient based analysis, that is, simply obtaining the correlation coefficient between two amplitude time series, is likely to be more efficient.

20.6 Quantification of Phase Locking: The Phase-Locking Value and Alternative Measures

With the phase or phase difference data, the presence of statistically significant phase locking can be assessed with a number of circular statistics. The most common of them is the phase locking value/factor (PLV, PLF) that is given by an average of complex phases (Sinkkonen et al., 1995). Let us first define the complex PLV (cPLV) as simply the average across the complex phase differences

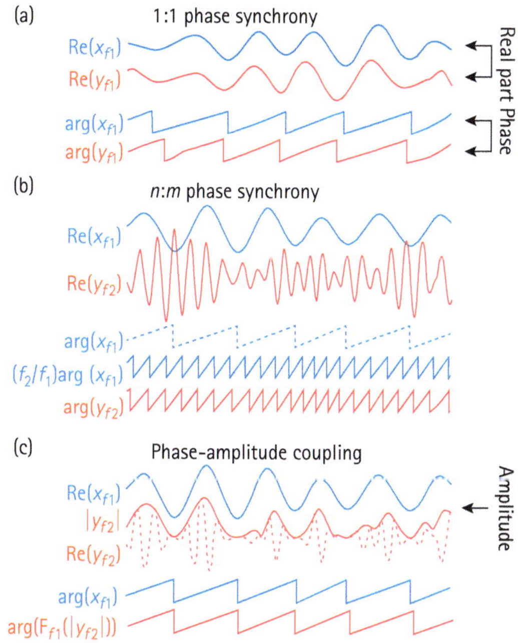

FIGURE 20.2 Schematic illustration of within-frequency and cross-frequency phase and amplitude interactions. (A) Within-frequency synchronization is a case of generic n:m-phase synchrony with $n = m = 1$. Re(.) denotes the real part of a complex filtered signal and arg(.) its phase. (B) A simulated example of 1:4 phase synchronization. (C) An example of phase-amplitude coupling where the phase of the slow and the amplitude of the fast oscillations are correlated.

Adapted from Palva & Palva, 2012.

$$\text{cPLV} = \text{n}_t^{-1} \sum \Phi \tag{10}$$

where n_t is the number of complex phase difference Φ samples pooled across trials and/or time. PLV is the absolute value of cPLV (PLV = |cPLV|) and is 1 for perfect coupling (delta-function phase distribution) and approaches 0 for a uniform phase distribution when $n_t \to \infty$. If the phase-difference samples Φ are independent and the marginal phase distributions are uniform, the no-interaction null hypothesis is characterized by uniformly distributed Φ and can be tested with the Rayleigh test.

When Φ are pooled across time, that is, when the independence condition is not met because of redundancy, and/or when the underlying process is not sinusoidal (see Nikulin et al., 2007), surrogate data are needed for statistical testing. For spontaneous data, the surrogates may be constructed by random rotation of the time series before estimation of the phase differences in order to preserve endogenous and filter-induced autocorrelations and avoid the underestimation of surrogate distribution (Palva et al, 2005; Siebenhühner et al., 2020). For event-related data, the surrogates may be constructed by trial shuffling. If the interaction estimates involve components that are time-locked to the events, such as evoked responses, the artificial interactions that they cause must be accounted for with forward-and-inversed-modeled surrogate data in trial shuffling (Hirvonen et al., 2018).

20.7 Considerations for using PLV for measuring phase synchrony and CFC

While PLV is a conceptually straightforward and computationally extremely efficient "first choice" for interaction estimation in many applications, several characteristics of PLV warrant attention. First, PLV is biased, that is, positively dependent on the number of samples (Vinck et al., 2010; Aydore et al., 2013). Hence, if two conditions are to be compared, the number of samples must be equalized. Alternatively, a similar but unbiased metric, pairwise-phase consistency (PPC, Vinck et al., 2010) that is equivalent to squared PLV (Aydore et al., 2013) can be used but its computational cost scales with $n_s^2 n_t^2$ whereas the cost of PLV is proportional only to $n_s^2 n_t$. PPC yields sensitivity similar to PLV. Second, PLV is appropriate only for unimodal phase distributions. The non-uniformity of arbitrary distributions can be estimated with Shannon-entropy and conditional-probability based indices for phase (difference) histograms (Tass et al., 1998). Third, PLV is a measure of correlation and does not disclose causal relationships or the lack thereof between the signals. Directional phase-specific interactions can be estimated with histogram-based phase transfer entropy without sample-size or linear mixing bias (Lobier et al., 2014). Directional interactions can be estimated with mixed phase and amplitude information with a range of metrics such as Granger causality, partial directed coherence, transfer entropy, and many others. Fourth, PLV among all other interaction metrics is crucially dependent on the signal-to-noise ratio of the phase estimate. Only at signal-to-noise ratios greater than 2–3, the phase estimates reach relative independence from amplitude. Hence, albeit phase is mathematically independent of amplitude, the accuracy of phase estimates in noisy signals with variable amplitudes will be dynamically dependent on the amplitude. When making inferences about differences in phase locking, one needs to consider the putative contribution of concurrent differences in signal-to-noise ratio (Palva et al., 2010).

20.8 Linear mixing inflates PLV and other correlation metrics and causes artificial false positives

One-to-one-phase and amplitude correlations, but not CFS or PAC, estimated with PLV are directly influenced by linear mixing such as volume conduction in electrophysiological recordings, which leads to inflated PLV values and false positive phase correlation observations in the presence of no true correlations. For amplitude correlations, the same applies also to correlation coefficients. This problem is pervasive across all forms of electrophysiological recordings: from micro-electrodes to intra-cranial stereo-EEG and electrocorticography, and to scalp EEG, volume conduction causes linear mixing of electric potentials. While volume conduction does little to influence MEG, signal mixing between cortical sources and scalp sensors leads to comparable linear mixing (Figure 20.3). Finally, demixing the sensor signals by source modeling leaves residual linear mixing, known as source leakage (Palva et al., 2018).

This problem can be partially solved by noting that linear mixing always has a zero-time and zero-phase lag influence, which is fully captured by the real part of PLV or that of coherence. On the other hand, as true neuronal interactions supposedly often involve conduction-delay-related phase lags, excluding the zero-lag contribution becomes an

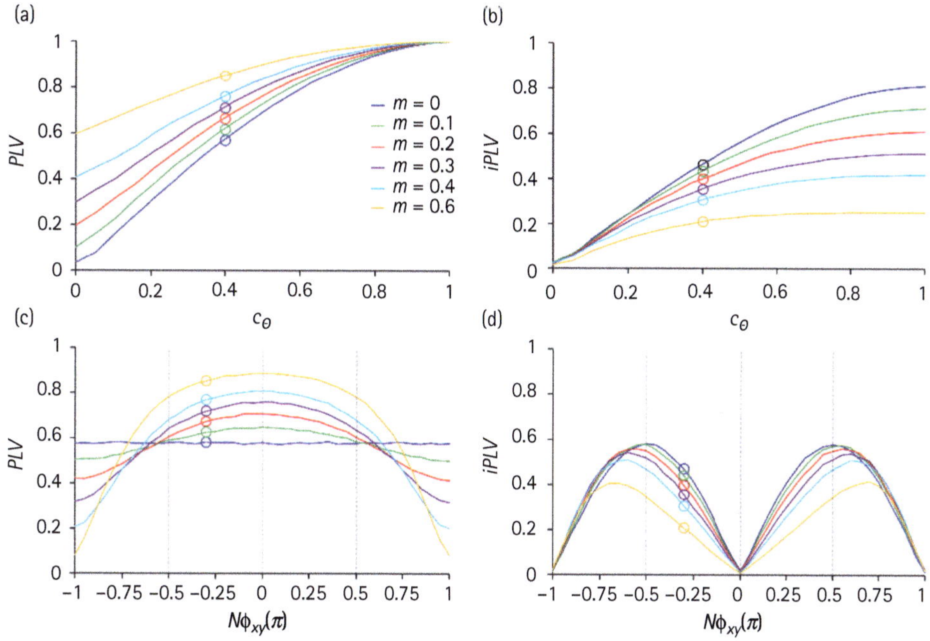

FIGURE 20.3. Phase coupling measures *PLV* and *iPLV* are influenced not only by the phase coupling strength c_Θ, but also by the phase difference $n\phi_{xy}$ and linear mixing m between the studied signals. Here, phase coupling and linear signal mixing (m = 0 (blue), 0.1 (green), 0.2 (red), 0.3 (violet), 0.4 (cyan), and 0.6 (orange)) were simulated between two signals for a range of coupling and phase lag values. (A) *PLV* between the signals as a function of c_Θ and m for $n\phi_{xy} = -0.3$. Open circles at $c_\Theta = 0.4$ indicate the coupling strength used in C and D. (B) *iPLV* between the signals as a function of c_Θ and m. (C) *PLV* as a function of $n\phi_{xy}$, when c_Θ was set to 0.4. Notably, PLV is greatly influenced by the phase difference when signal mixing is strong even though without mixing, the magnitude of PLV is independent of the phase lag. Open circles at $n\phi_{xy} = -0.3$ indicate the $n\phi_{xy}$ used in A and B. (D) The strength of *iPLV* depends on the phase difference and is biased towards large phase difference so that *iPLV* is abolished when $n\phi_{xy} = 0$ or $n\phi_{xy} = \pm\pi$.

Adapted from Palva et al., 2018.

approach for obtaining a measure that focuses on neuronal phase coupling and is insensitive to linear mixing. Measures such as imaginary coherence (iCoh, Nolte et al., 2004), the imaginary part of PLV (iPLV) (Palva & Palva, 2012), and weighted phase-lag index (wPLI, Vinck et al., 2011) ignore the real part of the complex phase differences. These measures are hence insensitive to linear mixing while revealing the true phase-lagged interactions. Of these, wPLI is superior to iPLV and iCoh in not being influenced by the magnitude of linear mixing. The obvious shortcoming of each these approaches is, nonetheless, their insensitivity to true near-zero-phase lag interactions and the dependence of the correlation value on the phase difference per se in addition to the strength of the interaction.

The equivalent for iPLV and wPLI for amplitude correlations is orthogonalized cross correlation (oCC) where the two signals are orthogonalized prior to the estimation of the correlation coefficient (Brookes et al., 2012, Hipp et al., 2012). It is important to note that the orthogonalized amplitude correlation coefficients are not independent of concurrent phase coupling. In fact, they are non-trivially affected by the presence of true phase coupling and linear mixing in a phase-difference dependent manner and may yield both false positive and negative findings (Palva et al., 2018b).

20.9 Ghost interactions constitute false positives in all bivariate interaction estimates

The exclusion of artificial interactions with coupling measures that are not inflated by linear mixing, such as iPLV, wPLI, and oCC, has been often claimed to categorically "account" for the problem of linear mixing causing false positive detections. Although these measures are de facto immune to artificial false positives and can be overly conservative by missing true near-zero-phase interactions, they still yield abundant "spurious" or "ghost" false positive interactions due to signal spread. This is because field spread in the vicinity of a true nonzero phase interaction mirrors the true interactions into spurious ghost interactions, that appear as false positives with any bivariate interaction measure in a manner that can never be accounted for in bivariate fashion (Figure 20.4).

There are two principal approaches to addressing the problem of spurious "ghost" interactions. One is to perform symmetric multivariate orthogonalization where the cortical-parcel-source signals are orthogonalized in a multivariate fashion (rather than bivariate as for oCC) (Colclough et al., 2015). Symmetric orthogonalization overcomes the problem of spurious ghost interactions by simultaneously removing zero phase-lag components from all source time series using Löwdin orthogonalization for gradient descent. All-to-all amplitude correlations are estimated with partial correlation of amplitude envelopes to keep direct and remove indirect interactions. Because the partial correlation matrix is expected to be sparse, a graphical lasso regularization of the inverse

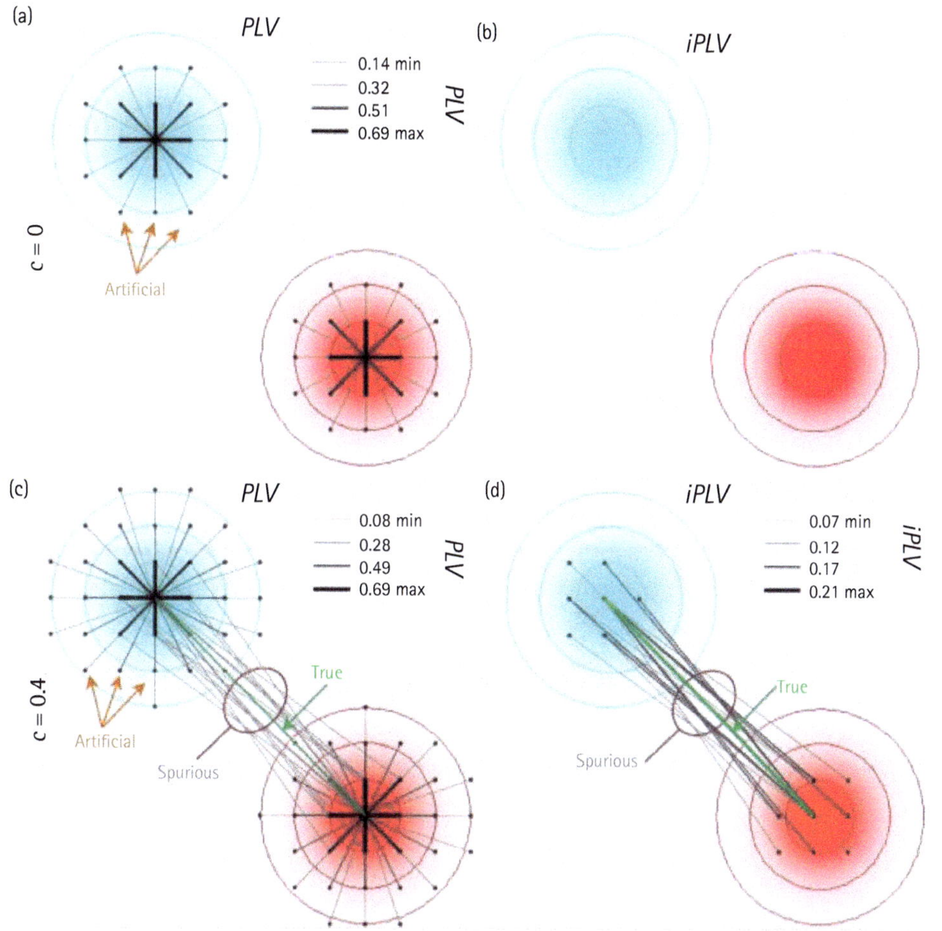

FIGURE 20.4 (A) Linear mixing leads to observations of false positive *artificial PLV* interactions, *within* the mixing region even in the absence of true correlations. Activity of 169 uncoupled ($c = 0$) sources (black dots) in a 13×13 grid was simulated and the 20 strongest *PLV* edges of the two sources-of-interest (centers of the cyan and red regions) are visualized. The cyan and red color gradients indicate the Gaussian mixing strength. (B) *iPLV* analysis of the same data as in A shows that *iPLV* does not lead to observations of artificial interactions in the absence of true interactions. (C) True phase correlations are mirrored into false positive *spurious* correlations, between the two mixing regions in the condition where there is a true interaction ($c = 0.9$) between two sources-of-interest (here the centers of the mixing regions). (D) Even though *iPLV* does not yield *artificial* false positives, it detects the spurious "ghost" interactions similarly to *PLV*. Spurious correlations arise because any two sources in separate mixing regions partially retain the nonzero phase difference of their center sources.

covariance matrix is applied to penalize near-zero elements, which reduces noise in the partial correlation graph. This leads symmetric orthogonalization to attenuate both signal-leakage caused and true indirect couplings. The main limitations of this method are that it is principally applicable only to the estimation of amplitude correlations and that it is limited by the rank of the data due to its dependence on singular value decomposition. For MEG/EEG data that are preprocessed with signal space separation (SSS) and temporal SSS methods, the rank of the data (~degrees of freedom) is often limited to 60–70 (Haumann et al., 2016). Thus, symmetric orthogonalization is applicable only to cortical networks with less than ~60 parcels in total, such as the 19 regions per hemisphere used in Colclough et al. (2015).

Another approach to controlling spurious connections is "hyperedge bundling", which uses any bivariate connectivity mapping as the basis and bundles observed connections on the basis of their proximity in source geometry. Hyperedge bundling reduces the false positive rate by a factor of 10-100 with moderate to little decrease on the true positive rate. The main advantages of this approach are that it does not involve complex mathematical transformations prior to interaction estimation and may be used with any bivariate metric in a manner that retains the metric's sensitivity. Moreover, hyperedge bundling enables high-resolution analyses with hundreds of cortical parcels, that is, redundant oversampling of the source space, which safeguards the analysis outcome from the possibility that the actual neuronal source constellations or degrees of freedom in the data are different from those of the used parcellation scheme. In the worst-case scenario, coarse parcellations can misrepresent or miss source areas that fall in between the parcels or are much smaller than the parcels.

Finally, the overall effects of signal mixing on estimates of inter-areal phase- and amplitude coupling can be mitigated by optimizing the MEG/EEG inverse modelling procedure. The inverse transformed source dipole time series can be collapsed into parcel time series in a manner that maximizes the source reconstruction accuracy and minimizes the effects of source leakage (Korhonen et al., 2014). The fidelity-weighted inverse modeling has higher reconstruction accuracy than a regular inverse operator for the given parcellation, because it gives greater weight to sources with better reconstruction accuracy for the signals from the parcels they belong into.

20.10 Data-driven approach for mapping oscillatory connectomes

Initial MEG and EEG source connectivity studies used region-of-interest-based approaches for phase and amplitude correlation analyses but this approach has been all but replaced by data-driven all-to-all connectivity mapping, or "connectomics", studies. Data-driven analyses approaches reveal the most robust effects in data regardless of the original hypothesis and as such provide more rigorous hypothesis testing compared to hypothesis-driven analyses that may be confounded by several forms of biases and explicit errors such as circularity in analysis targeting (Kriegeskorte, 2009). Data-driven analysis of both local and especially large-scale oscillatory interactions provide a comprehensive view on brain dynamics that is not biased by selection criteria. In addition, several approaches for network analysis as described below can be applied to compensate for the presence of artefactual and spurious connections.

Graph theory and complex network theory enable the analysis and prediction of the behaviors of a wide variety of complex systems. In human brain networks, these have included graph theoretical network analysis (Bullmore & Sporns, 2009; Rubinov & Sporns, 2010) that can also be used to characterize the properties of oscillatory networks or connectomes that have been obtained with data-driven analysis approaches (Palva et al., 2010a; Palva et al., 2010b; Siebenhühner et al., 2016; Siebenhühner et al., 2020). This approach describes oscillatory networks at three levels: at the entire graph level (i.e., the connectome), level of edges (i.e., connections), and vertices (i.e., brain regions or parcels). Vertex centrality estimates can be used to identify those vertices (brain regions) that are hubs in networks of interareal oscillatory interactions and likely to play an important role in neuronal communication. Several metrics can, in addition, be used to characterize specific topological attributes such as clustering, path lengths, and modularity of oscillatory networks (Palva et al., 2010b; Zhigalov et al., 2017).

20.11 Confounders in local and inter-areal CFC

In addition to source mixing described, estimates of CFC may be inflated by false positive couplings arising from non-sinusoidal and nonzero mean signals. False positives are caused by the artificial higher-frequency components produced when non-sinusoidal signals are filtered into narrow bands and by artificial lower-frequency components arising from filtering of nonzero-mean waveforms. Because these are ubiquitous in electrophysiological signals, their filter artefacts constitute a significant confounder to CFS and PAC estimation. Approaches based on waveform analysis (Kramer et al., 2008; Aru et al., 2015, Van Driel et al., 2015) have been proposed to reduce the artefactual connections arising from non-sinusoidal signals. Nevertheless, filter-artefact-caused spurious CFC, in particular CFS, is difficult to dissociate from genuine CFC by inspection of the waveform shape of any single signal in isolation. Local CFC estimates are thus prone to ambiguous results. CFC is necessarily genuine when there is evidence for two distinct coupled processes while spurious CFC arises from a single process with signal components distributed to distinct frequency bands because of filter artefacts (Figure 20.5). These two distinct processes can be identified by connection-by-connection testing of whether CFC can unambiguously be attributed to two separable processes using graph-theoretical approaches (Figure 20.5, Siebenhühner et al., 2020). If CFC is spurious between areas A and B, these regions are necessarily also connected by 1:1 phase synchrony and local CFC that together may lead to a spurious observation of inter-areal CFC. The absence of either PS or local CFC, in contrast, indicates that the observed inter-areal CFC cannot be attributable to a single source and is thus genuine.

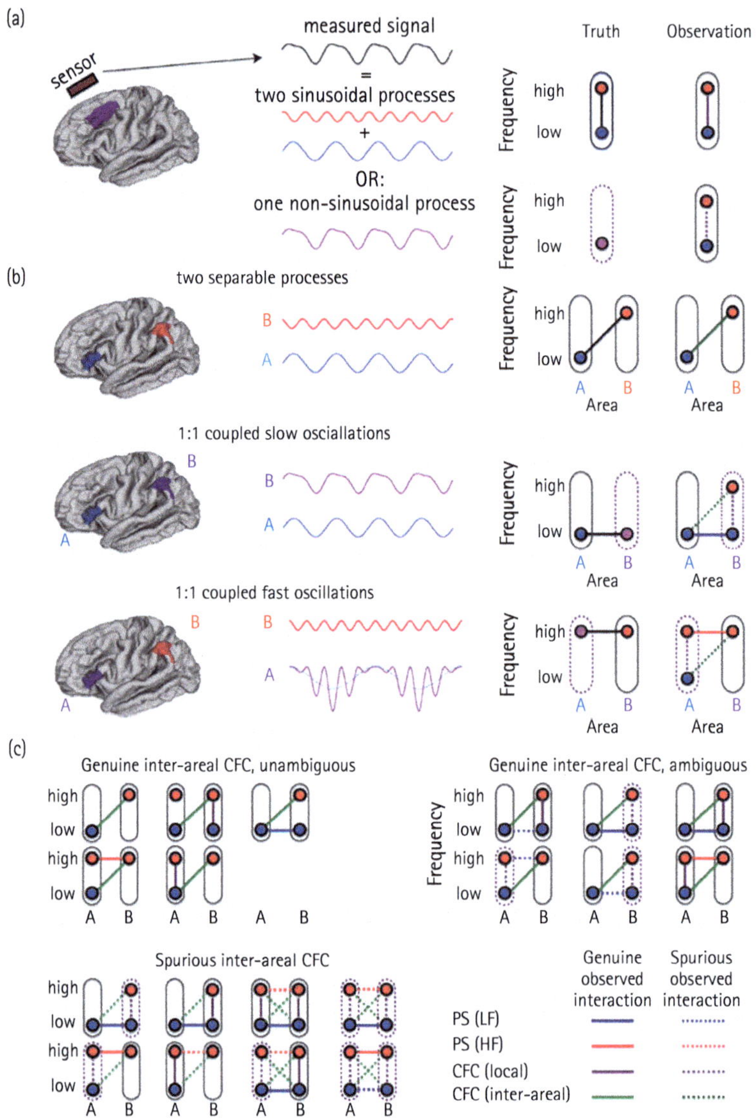

FIGURE 20.5 Schemata for dissociating the genuine from putatively spurious CFC. (A) Spurious observations of *local* are seen when a non-sinusoidal signal is filtered. (B) Inter-areal CFC can be proven to be genuine if it unambiguously originates from separable neuronal signals. Spurious inter-areal CFC is always accompanied by spurious local CFC at one or both locations that are also inter-areally coupled via 1:1 phase synchrony either at low frequency or high frequency so that a "triangle motif" with the observed (spurious) inter-areal CFC is formed. (C) Graph-theory based approach can identify all triangle motifs that might contain spurious inter-areal CFC. This approach only identifies those inter-areal CFC observations as genuine, which are not part of a full triangle motif, whereas all others are discarded. Since this may include also genuine connections, this approach provides a lower bound for the number of genuine connections.

Modified from Siebenhühner et al., 2020.

REFERENCES

Abbas, A., Belloy, M., Kashyap, A., Billings, J., Nezafati, M., Schumacher, E., & Keilholz, S. (2019). Quasi-periodic patterns contribute to functional connectivity in the brain. *NeuroImage, 191*, 193–204. doi:10.1016/j.neuroimage.2019.01.07

Aru, J., Aru, J., Priesemann, V., Wibral, M., Lana, L., Pipa, G., Singer, W., & Vicente, R. (2015). Untangling cross-frequency coupling in neuroscience. *Current Opinion in Neurobiology, 31*, 51–61. doi:10.1016/j.conb.2014.08.002

Aydore, S., Pantazis, D., & Leahy, R. M. (2013). A note on the phase locking value and its properties. *NeuroImage, 74*, 231–244.

Belloy, M. E., Naeyaert, M., Abbas, A., Shah, D., Vanreusel, V., van Audekerke, J., ... Verhoye M (2018). Dynamic resting state fMRI analysis in mice reveals a set of quasi-periodic patterns and illustrates their relationship with the global signal *NeuroImage, 180*(Pt B), 463–484. doi:10.1016/j.neuroimage.2018.01.075.

Brookes, M. J., Woolrich, M. W., & Barnes, G. R. (2012). Measuring functional connectivity in MEG: A multivariate approach insensitive to linear source leakage. *NeuroImage, 63*(2), 910–920.

Bruns, A. (2004). Fourier-, Hilbert- and wavelet-based signal analysis: Are they really different approaches? *Journal of Neuroscience Methods, 137*, 321–332.

Bullmore, E. & Sporns, O. (2009). Complex brain networks: Graph theoretical analysis of structural and functional systems. *Nature Reviews Neuroscience, 10*, 186–198.

Canolty, R. T. & Knight, R. T. (2010). The functional role of cross-frequency coupling. *Trends in Cognitive Science, 14*(11), 506–515. doi:10.1016/j.tics.2010.09.001

Colclough, G. L., Brookes, M. J., Smith, S. M., & Woolrich, M.W. (2015). A symmetric multivariate leakage correction for MEG connectomes. *NeuroImage, 117*, 439–448. doi:10.1016/j.neuroimage.2015.03.071.

Fell, J. & Axmacher, N. (2011). The role of phase synchronization in memory processes. *Nature Reviews Neuroscience, 12*(2), 105–118. doi:10.1038/nrn2979

Hipp, J. F., Hawellek, D. J., Corbetta, M., Siegel, M., & Engel, A. K. (2012). BOLD fMRI correlation reflects frequency-specific neuronal correlation. *Nature Neuroscience, 15*, 884–889.

Hirvonen, J., Monto, S., Wang, S. H., Palva, J. M., & Palva, S. (2018). Dynamic large-scale network synchronization from perception to action. *Network Neuroscience*, 2(4), 442–463. https://doi.org/10.1162/netn_a_0003

Huang, N. E. & Wu, Z. (2008). A review on Hilbert-Huang transform: Method and its applications to geophysical studies. *Reviews of Geophysics* [online], *46*, 1–23.

Kramer, M. A., Tort, A. B. L., & Kopell, N. J. (2008). Sharp edge artifacts and spurious coupling in EEG frequency comodulation measures. *Journal of Neuroscience Methods, 170*(2), 352–357. doi:10.1016/j.jneumeth.2008.01.020

Kriegeskorte, N., Simmons, W. K., Bellgowan, P. S., & Baker, C. I. (2009). Circular analysis in systems neuroscience: The dangers of double dipping. *Nature Neuroscience, 12*(5), 535–40. doi:10.1038/nn.2303

Jensen, O. & Colgin, L. L. (2007). Cross-frequency coupling between neuronal oscillations. *Trends in Cognitive Sciences, 11*, 267–269. doi:10.1016/j.tics.2007.05.003

Lisman, J. E. & Jensen, O. (2013). The theta-gamma neural code. *Neuron, 77*, 1002–1016. doi:10.1016/j.neuron.2013.03.007

Haumann NT, Parkkonen L Kliuchko M, Vuust P , Brattico E Comparing the Performance of Popular MEG/EEG Artifact Correction Methods in an Evoked-Response Study Comput Intell Neurosci 2016;2016:7489108.

Hyafil, A., Giraud, A. L., Fontolan, L., & Gutkin, B. (2015). Neural cross-frequency coupling: Connecting architectures, mechanisms, and functions. *Trends in Neuroscience, 38*, 725–740. doi:10.1016/j.tins.2015.09.001

Lobier, M., Siebenhühner, F., Palva, S., & Palva, J. M. (2014). Phase transfer entropy: A novel phase-based measure for directed connectivity in networks coupled by oscillatory interactions. *NeuroImage, 85*(Pt 2), 853–872.

Monto S, Palva S, Voipio J, Palva JM (2008) Very slow EEG fluctuations predict the dynamics of stimulus detection and oscillation amplitudes in humans. *Journal of Neuroscience* 28: 8268–8272.

Nolte, G., Bai, O., Wheaton, L., Mari, Z., Vorbach, S., & Hallett, M. (2004). Identifying true brain interaction from EEG data using the imaginary part of coherency. *Clinical Neurophysiology, 115*, 2292–2307.

Nikulin VV, Linkenkaer-Hansen K, Nolte G, Lemm S, Müller KR, Ilmoniemi RJ, Curio G. Eur J Neurosci. 2007 A novel mechanism for evoked responses in the human brain. Eur J Neurosci. 2007 May;25(10):3146–54.

Palva, J. M. & Palva, S. (2012). Discovering oscillatory interaction networks with MEG/EEG: Challenges and breakthroughs. *Trends in Cognitive Sciences, 16*, 219–230.

Palva, J. M. & Palva, S. (2017). Functional integration across oscillatory frequencies by cross-frequency phase synchronization. *European Journal of Neuroscience, 48*(7), 2399–2406. doi:10.1111/ejn.13767

Palva, J. M., Palva, S., & Kaila, K. (2005). Phase synchrony among neuronal oscillations in the human cortex. *Journal of Neuroscience, 25*, 3962–3397.

Palva, J. M., Monto, S., Kulashekhar, S., & Palva, S. (2010a). Neuronal synchrony reveals working memory networks and predicts individual memory capacity. *Proceedings of the National Academy of Science of the United States of America, 107*, 7580–7585.

Palva, J. M., Wang, S. H., Palva, S., Zhigalov, A., Monto, S., Brookes, M. J., Schoffelen, J. M., & Jerbi, K. (2018) Ghost interactions in MEG/EEG source space: A note of caution on inter-areal coupling measures. *NeuroImage, 173*, 632–643. doi:10.1016/j.neuroimage.2018.02.032

Palva, S., Monto, S., & Palva, J. M. (2010b). Graph properties of synchronized cortical networks during working memory maintenance. *NeuroImage, 49*, 3257–3268.

Palva, S. & Palva, J. M. (2018). Roles of brain criticality and multi-scale oscillations in temporal predictions for sensorimotor processing. *Trends in Neuroscience, 41*(10), 729–743. https://doi.org/10.1016/j.tins.2018.08.008

Schroeder, C. E. & Lakatos, P. (2009). Low-frequency neuronal oscillations as instruments of sensory selection. *Trends in Neuroscience, 32*, 9–18.

Rosenblum, M. G., Pikovsky, A. S., & Kurths, J. (2004). Synchronization approach to analysis of biological systems. *Fluctuation and Noise Letters, 4*, 52–63.

Rubinov, M. & Sporns, O. (2010). Complex network measures of brain connectivity: Uses and interpretations. *NeuroImage*, 52, 1059–1069

Sinkkonen, J., Tiitinen, H., & Näätänen, R. (1995). Gabor filters: An informative way for analysing event-related brain activity. *Journal of Neuroscience Methods, 56*, 99–104.

Tallon-Baudry, C., Bertrand, O., Delpuech, C., & Pernier, J. (1996). Stimulus specificity of phase-locked and non-phase-locked 40 Hz visual responses in human. *Journal of Neuroscience, 16*, 4240–4249.

Tass, P., Rosenblum, M. G., Weule, J., Kurths, J., Pikovsky, A., Volkmann, J., Schnitzler, A., & Freund, H.-J. (1998). Detection of n:m phase locking from noisy data: Application to magnetoencephalography. *Physics Review Letters, 81*, 3291.

Siebenhühner, F., Wang, S. H., Arnulfo, G., Lampinen, A., Nobili, L., Palva, J. M., & Palva, S. (2020). Resting-state cross-frequency coupling networks in human electrophysiological recordings. *PLoS Biology, 18*(5), e3000685. https://doi.org/10.1371/journal.pbio.3000685.

Siebenhühner, F., Wang, S. H., Palva, J. M., & Palva, S. (2016). Cross-frequency synchronization connects networks of fast and slow oscillations during visual working memory maintenance. *Elife*, 5, e13451. doi:10.7554/eLife.13451

van Driel, J., Cox, R., & Cohen, M. X. (2015). Phase-clustering bias in phase–amplitude cross-frequency coupling and its removal. *Journal of Neuroscience Methods, 254*, 60–72. doi:10.1016/j.jneumeth.2015.07.014

Vanhatalo, S., Palva, J. M., Holmes, M. D., Miller, J. W., Voipio, J., & Kaila, K. (2004). Infraslow oscillations modulate excitability and interictal epileptic activity in the human cortex during sleep. *Proceedings of the National Academy of Science of the United States of America, 101*, 5053–5057.

Vinck, M., Oostenveld, R., van Wingerden, M., Battaglia, F., & Pennartz, C. M. (2011). An improved index of phase-synchronization for electrophysiological data in the presence of volume-conduction, noise and sample-size bias. *NeuroImage*, 55, 1548–1565.

Vinck, M., van Wingerden, M., Womelsdorf, T., Fries, P., & Pennartz, C. M. (2010). The pairwise phase consistency: A bias-free measure of rhythmic neuronal synchronization. *NeuroImage*, 51, 112–122.

Zhang, X., Pan, W. J., & Keilholz, S. D. (2020). The relationship between BOLD and neural activity arises from temporally sparse events. *NeuroImage, 207*, 116390. doi:10.1016/j.neuroimage.2019

Zhigalov, A., Arnulfo, G., Nobili, L., Palva, S., & Palva, J. M. (2017). Modular co-organization of functional connectivity and scale-free dynamics in the human brain. *Network Neuroscience, 1*(2), 143–165.

CHAPTER 21

MULTIVARIATE METHODS FOR FUNCTIONAL CONNECTIVITY ANALYSIS

SELIN AVIYENTE

21.1 Introduction

Cognition and perception are founded on the coordinated activity of neural populations communicating among different specialized brain regions (Uhlhaas & Singer, 2006). Neurons that synchronously oscillate across different frequency bands provide the fundamental mechanism for information transfer (Bastos & Schoffelen, 2016), allowing coordinated activity in the normally functioning brain (Uhlhaas & Singer, 2016; Buzsáki & Draguhn, 2004; Fries, 2009). This neural coordination is spatio-temporally dynamic (Lakatos et al., 2008), and the oscillatory synchronization among different regions is dynamically adjusted based on the cognitive task (Uhlhaas & Singer, 2016). Tononi and colleagues (1998, 1994) argue that brain functionality is based on functional segregation and integration; the former establishes that specialized activity occurs due to segregated neuronal populations within dedicated brain regions (Rubinov & Sporns, 2010), while the latter consists of the combination of multiple distributed regions and serves as the basis for coherent cognition and behavior (Rubinov & Sporns, 2010).

Methods for quantifying functional connectivity include linear correlation, mutual information, coherence, synchronization likelihood, phase-locking value (PLV) (Lachaux et al., 1999; Aviyente et al., 2011), and pairwise phase consistency (Pereda et al., 2005; Quiroga et al., 2002). In order to quantify both linear and nonlinear relationships in the brain signals, phase synchronization, as defined in the context of two chaotic oscillators, has emerged as an alternative method for the assessment of functional connectivity and is quantified through PLV (Aviyente et al., 2011; Aviyente

& Yener Mutlu, 2011; Dimitriadis et al., 2013). As a measure for functional connectivity, PLV was introduced by Lachaux and colleagues (1999) and it estimates the synchrony between two signals by looking at the circular variance of their phase difference across trials. In comparison to other linear and nonlinear methods, PLV is more sensitive to nonlinear effects (Jalili et al., 2013). In addition, this metric contributes to the assessment of brain rhythms and their related cognitive processes, for example, alpha, beta, delta, and theta in the low-frequency bands and gamma bands in the higher frequencies (Pereda et al., 2005; Lachaux et al., 1999; Aydore et al., 2013; Aviyente et al., 2011).

Although PLV is a promising measure for quantifying functional connectivity, it is still limited due to its bivariate nature. Specifically, it does not provide information regarding the integration across multiple regions in the brain. In addition, functional connectivity results from bivariate measures are difficult to interpret and computationally expensive for systems with large number of regions. In order to overcome these drawbacks, researchers propose multivariate phase synchrony measures (Stam & Van Dijk, 2002; Carmeli et al., 2005; Mutlu & Aviyente, 2012; Al-Khassaweneh et al., 2016). The two main approaches to quantifying multivariate synchronization are spectral and graph theoretic methods.

The application of bivariate measures to multivariate data sets with N time-series results in an $N \times N$ matrix of bivariate indices, which leads to a large amount of mostly redundant information. Therefore, it is necessary to reduce the complexity of the data set in such a way to reveal the relevant underlying structures using multivariate analysis methods. The basic approach used for multivariate phase synchronization is to trace the observed pairwise correspondences back to a smaller set of direct interactions using approaches such as partial coherence adapted to phase synchronization (Schelter et al., 2006). Another complementary way to achieve such a reduction is cluster analysis—a separation of the parts of the system into different groups such that the signal interdependencies within each group tend to be stronger than in between groups (Newman, 2006a, 2006b). Allefeld and colleagues have proposed two complementary approaches to identify synchronization clusters and applied their methods to EEG data (Allefeld & Kurths, 2004; Allefeld et al., 2007; Allefeld et al., 2005; Allefeld & Kurths, 2003). Allefeld and Kurths (2004) present a mean-field approach that assumes the existence of a single synchronization cluster that all oscillators contribute to a different extent. The authors define the to-cluster synchronization strength of individual oscillators to identify multivariate synchronization. This method has the disadvantage of assuming a single cluster and thus cannot identify the underlying clustering structure. Allefeld and colleagues (2007) introduce an approach that addresses the limitation of the single cluster approach using methods from random matrix theory. This method is based on the eigenvalue decomposition of the pairwise bivariate synchronization matrix and appears to allow identification of multiple clusters. Each eigenvalue greater than 1 is associated with a synchronization cluster and quantifies its strength within the data set. The internal structure of each cluster is described by the corresponding eigenvector.

Combining the eigenvalues and the eigenvectors, one can define a participation index for each oscillator and its contribution to different clusters. This method assumes that the synchrony between systems belonging to different clusters (i.e. between-cluster synchronization) is equal to zero and requires an adjustment for proper computation of the participation indices in the case that there is between-cluster synchronization. Despite the usefulness of eigenvalue decomposition for the purposes of cluster identification, Allefeld and Bialonski (2007) demonstrate that there are important special cases—clusters of similar strength that are slightly synchronized to each other—where the assumed one-to-one correspondence of eigenvectors and clusters is completely lost. Other alternative measures that quantify multivariate relationships include the directed transfer function and Granger causality defined for an arbitrary number of channels (Granger, 1969; Baccala & Sameshima, 2001). Both of these methods have been applied to study interdependencies and causal relationships, however, are limited to stationary processes and linear dependencies.

On the other hand, graph theory provides the means for characterizing the functional connections in the brain using a complex network model (Bullmore & Sporns, 2009). Functional connectivity networks are constructed by considering the different brain regions or electrodes/sensors as nodes and the relationships between different nodes, quantified by bivariate functional connectivity measures such as PLV and correlation, as edges. In this manner, functional connectivity networks can take advantage of the widely available set of techniques for characterizing complex networks. In terms of brain networks, these measures have been grouped as measures of functional segregation and functional integration (Rubinov & Sporns, 2010). Measures of functional segregation include the clustering coefficient, transitivity, and modularity (Rubinov & Sporns, 2010; de Vico Fallani et al., 2014). On the other hand, measures of functional integration include the characteristic path length and the global efficiency. By computing measures that characterize network structure, such as the small-world measure and the degree distribution, it has been shown that functional connectivity networks exhibit features of complex networks, including the small-world network (Bullmore & Sporns, 2009; Bassett & Bullmore, 2017; Bassett & Bullmore, 2006), and both small-world and scale-free networks (van den Heuvel et al., 2008). Although graph theoretic measures have contributed greatly to the advancements in the study of functional connectivity networks, these measures present some drawbacks. Measures employed in the characterization of network structure such as the mean clustering coefficient, the characteristic path length and the global efficiency may be affected by certain characteristics of the network. Examples include how nodes with low degree affect the clustering coefficient, and the dependence of the characteristic path length and the global efficiency in the shortest path between nodes, when networks may rely on other mechanisms than the shortest path for communication.

This chapter reviews spectral, direct multivariate, and graph theoretic measures for quantifying multivariate synchrony within a group of oscillators.

21.2 Spectral Methods for Multivariate Synchronization

The most common approach to quantifying multivariate connectivity is to utilize the $N \times N$ bivariate connectivity matrix, A. Since this matrix usually contains redundant information, tools from dimensionality reduction and random matrix theory are utilized to extract information from this matrix such as the spectrum of the connectivity matrix, that is, the distribution of the eigenvalues. There are a variety of methods used to quantify multivariate synchrony based on the distribution of these eigenvalues.

21.2.1 S-Estimator

S-estimator is one of the most commonly used multivariate synchronization metrics. It quantifies the amount of synchronization within a group of oscillators using the eigenvalue spectrum of the correlation, covariance or functional connectivity matrix:

$$S = 1 + \frac{\sum_{i=1}^{N} \lambda_i \log(\lambda_i)}{\log(N)}, \tag{21.1}$$

where λ_is are the N normalized eigenvalues. This measure is complementary to the entropy of the normalized eigenvalues of the correlation matrix. The more dispersed the eigenspectrum is the higher the entropy would be. If all of the oscillations in a group are completely synchronized, that is, the entries of the pairwise functional connectivity matrix are all equal to 1, then all of the eigenvalues except one will be equal to zero, and the value of S will be equal to 1 indicating perfect multivariate synchrony. This measure can quantify the amount of synchronization within a group of signals and thus is useful as a global complexity measure.

21.2.2 Omega Complexity

A variation of S-estimator is Ω-complexity. Using the eigenvalues of the correlation matrix as λ_i, the omega complexity can be computed as:

$$\Omega = \exp\left(-\sum_{i=1}^{N} \frac{\lambda_i}{N} \log \frac{\lambda_i}{N} \right). \tag{21.2}$$

Ω-complexity varies between 1 (maximum synchronization) and N (minimum synchrony, i.e., maximum de-synchronization). In order to scale the above measure between a value close to 0 (for minimum synchrony) and 1 (for maximum synchrony), one can compute Omega as $1/\Omega$.

21.2.3 Global Field Synchronization

A multivariate synchronization can be calculated in frequency domain by means of global field synchronization (GFS) measure. First, the time-series are converted to frequency-domain using fast Fourier transform. This results in the sine and cosine coefficients. At a given frequency *f*, the multivariate signals can be visualized in two-dimensional sine-cosine maps. As the entries are getting more scattered, the phase synchrony between them worsens. Let $\lambda_1(f)$ and $\lambda_2(f)$ be the eigenvalues of the covariance matrix along these two vectors (coefficients of sine and cosine). GFS at a given frequency is calculated as

$$GFS(f)=\frac{|\lambda_1(f)-\lambda_2(f)|}{\lambda_1(f)+\lambda_2(f)}. \tag{21.3}$$

If the sine-cosine clouds lie on a straight line, one of the eigenvalues equals to 0 and the covariance is completely explained by a single principal component; the GFS takes a value of 1 for such cases. This corresponds to complete phase synchrony at a given frequency. In a non phase-synchronized case, the two eigenvalues are close to each other, leading to GFS values close to 0. In order to obtain GFS values in a certain frequency range, similar to cross-coherence, one can get the average of GFS values over that frequency range.

21.3 Direct Measures of Multivariate Synchronization

Omega complexity and S-estimator are all based on the computation of the covariance or bivariate connectivity matrix. Therefore, the metrics are not direct measures quantifying multivariate synchrony and will be influenced by the bias affecting bivariate measures. A solution to this problem is to use multivariate phase synchrony (MPS) measure, which is indeed an extension to PLV. Having extracted the instantaneous phases from the individual time-series (using the Hilbert transform, for instance), the MPS is computed as

$$MPS=\frac{1}{LN}\sum_{t=1}^{L}\cdot\left|\sum_{i=1}^{N}e^{j\Phi_i(t)}\right| \tag{21.4}$$

MPS measures the mean phase coherence between the time-series, averaged over the observation samples. It ranges from 0 for completely unsynchronized systems to 1 for completely synchronized systems. However, this measure does not directly quantify the relationship of phases across oscillators as it relies on the absolute phase rather than the phase differences. For this reason, in this section we discuss a more recent method to quantify multivariate phase synchrony directly from the phase differences of the observed oscillators (Al-Khassaweneh et al., 2016; Mutlu & Aviyente, 2012).

21.3.1 Hyper-Spherical Phase Synchrony

Bivariate phase synchrony is based on the circular variance of the two-dimensional direction vectors on a unit circle (1-sphere), obtained by mapping the phase differences between two oscillators, $\{\Phi_{1,2}^k(t) = \Phi_1^k(t) - \Phi_2^k(t)\}_{k=1,2,\ldots,K}$ where K is the total number of trials, between the two time-series onto a Cartesian coordinate system. If the circular variance of these direction vectors is low, the time-series are said to be locked to each other. This idea can be extended to the multivariate case by considering the phase differences across M oscillators. In order to capture global phase information, the phase difference between the phase of each oscillator and the phase of the resultant vector of the remaining oscillators is defined, that is, $\theta_i^k(t) = \Phi_i^k(t) - \arg\left\{\sum_{m=1,m\neq i}^{M} \exp\left(j\Phi_m^k(t)\right)\right\}$. Using the $(M-1)$ angular coordinates, a direction vector $\Gamma^k(t) = \left[\gamma_1^k(t),\ldots,\gamma_M^k(t)\right]$ can be formed by mapping the angular coordinates

$(\theta_1,\ldots,\theta_{M-1})$ on a unit $(M-1)$-sphere as:

$$
\begin{aligned}
\gamma_1^k(t) &= \cos\left(\theta_1^k(t)\right),\\
\gamma_2^k(t) &= \sin\left(\theta_1^k(t)\right)\times\cos\left(\theta_2^k(t)\right),\\
\gamma_3^k(t) &= \sin\left(\theta_1^k(t)\right)\times\sin\left(\theta_2^k(t)\right)\times\cos\left(\theta_3^k(t)\right),\\
&\vdots\\
\gamma_{M-1}^k(t) &= \sin\left(\theta_1^k(t)\right)\times\ldots\times\sin\left(\theta_{M-2}^k(t)\right)\times\cos\left(\theta_{M-1}^k(t)\right),\\
\textit{and}\ \gamma_M^k(t) &= \sin\left(\theta_1^k(t)\right)\times\ldots\times\sin\left(\theta_{M-2}^k(t)\right)\times\sin\left(\theta_{M-1}^k(t)\right).
\end{aligned}
\tag{21.5}
$$

For example, $\gamma_1^k(t)$ is the x coordinate of a vector on the unit circle at angular position $\theta_1^k(t)$, while $\gamma_2^k(t)$ is the x coordinate of a vector on a circle with radius $\sin\left(\theta_1^k(t)\right)$ at angular position $\theta_2^k(t)$. Similar analysis applies to the remaining $\theta_i^k(t)$ s. Thus, every $\gamma_i^k(t)$ is just the x coordinate of a vector on a circle with radius $r_i^k(t) = \prod_{j=1}^{i-1} sin\left(\theta_j^k(t)\right)$, for $i = 2,3,\ldots,M$ and with a phase $\theta_i^k(t)$. The equation for $r_i(t)$ shows that as i increases, $\gamma_i^k(t)$ will have less impact on the overall synchrony. This means that the choice of the first phase difference, $\theta_1^k(t)$, will have a high impact on the measured synchrony.

Equation (21.5) may also be interpreted as follows. Every $\gamma_i^k(t)$ is the x projection of the y coordinate of the previous $\gamma_{i-1}^k(t)$ on the x-axis with a phase $\theta_i^k(t)$, i.e. define x and y coordinates of the rotating vector for each trial k as

$$
\begin{aligned}
\gamma_{x_1}^k(t) &= \cos\left(\theta_1^k(t)\right),\\
\gamma_{y_1}^k(t) &= \sin\left(\theta_1^k(t)\right),\\
\gamma_{x_2}^k(t) &= \sin\left(\theta_1^k(t)\right)\times\cos\left(\theta_2^k(t)\right),\\
\gamma_{y_2}^k(t) &= \sin\left(\theta_1^k(t)\right)\times\sin\left(\theta_2^k(t)\right),\\
\gamma_{x_3}^k(t) &= \sin\left(\theta_1^k(t)\right)\times\sin\left(\theta_2^k(t)\right)\times\cos\left(\theta_3^k(t)\right),\\
\gamma_{y_3}^k(t) &= \sin\left(\theta_1^k(t)\right)\times\sin\left(\theta_2^k(t)\right)\times\sin\left(\theta_3^k(t)\right),\\
&\vdots\\
\gamma_{x_M}^k(t) &= \sin\left(\theta_1^k(t)\right)\times\ldots\times\sin\left(\theta_{M-1}^k(t)\right)\times\cos\left(\theta_M^k(t)\right),\\
\gamma_{y_M}^k(t) &= \sin\left(\theta_1^k(t)\right)\times\ldots\times\sin\left(\theta_{M-1}^k(t)\right)\times\sin\left(\theta_M^k(t)\right),
\end{aligned}
\tag{21.6}
$$

where the superscripts x and y refer to the projection coordinates.

This recursive structure in the definition of the x- and y-axis for each oscillator causes a dependence of the synchrony metric on the order of the phases. To solve this problem, we consider both the x and y coordinates for all oscillators and normalize these coordinates by $d_i^k(t) = r_i^k(t) = \prod_{j=1}^{i-1} sin\left(\theta_j^k(t)\right)$ for $i = 2,3,\ldots,M$. This will result in unit radius for all i. Therefore, the multivariate phase synchrony measure, hyper-torus phase synchrony (HTS), is given by

$$HTS(t) = \frac{1}{N\times\sqrt{M}}\left\|\sum_{k=1}^{N} D^k(t)\right\|_2 \tag{21.7}$$

where $D^k(t) = \left[\frac{\gamma_{x_1}^k(t)}{d_1^k(t)}, \frac{\gamma_{y_1}^k(t)}{d_1^k(t)}, \ldots, \frac{\gamma_{x_M}^k(t)}{d_M^k(t)}, \frac{\gamma_{y_M}^k(t)}{d_M^k(t)}\right]$.

This HTS metric can also be simplified as

$$HTS(t) = \frac{1}{\sqrt{M}}\sqrt{PLV_1^2(t) + \ldots + PLV_M^2(t)}$$

$$= \sqrt{\frac{1}{M}\sum_{i=1}^{M} PLV_i^2(t)}, \tag{21.8}$$

where PLV_i quantifies the synchronization of each oscillator with respect to a common reference angle with $\theta_i^k(t,\omega)$ as described above and PLV_i is given by

$$PLV_i(t) = \frac{1}{N}\left|\sum_{k=1}^{N} \exp\left(j\theta_i^k(t)\right)\right|$$

$$= \sqrt{\langle\cos\theta_i^k(t)\rangle^2 + \langle\sin\theta_i^k(t)\rangle^2}. \tag{21.9}$$

The maximum value of HTS is 1, when there is complete phase synchronization among oscillators. On the other hand, HTS is theoretically 0 when the oscillators are independent.

Some advantages of this formulation with respect to the spectral methods discussed in the previous section is its computational efficiency and flexibility. First, for N oscillators this metric requires the computation of N PLV values, whereas the spectral methods require the computation of $N(N-1)$ pairs of PLV values. Second, this metric is directly applicable for quantifying the synchrony within a group of oscillators, whereas the spectral methods are usually applied to the whole connectivity network.

21.4 Graph Theoretic Metrics

In the last decade, connectivity-based methods have had a prominent role in characterizing normal brain organization as well as alterations due to various brain disorders. Functional connectivity networks (FCNs) are obtained by recording physiological signals from different brain regions and then computing the pairwise similarity (Stam & Reijneveld, 2007; Bullmore & Bassett, 2011). In this context, the nodes of the network correspond to the brain regions and the edges correspond to their functional connectivity. In the context of synchronization, the $N \times N$ bivariate connectivity matrix, A can be treated as the adjacency matrix of the graph.

As with other real-world connected systems and relational data, studying the topology of interactions in the brain has profound implications in the comprehension of complex phenomena, such as the emergence of coherent behaviour and cognition or the capability to functionally reorganize after brain lesions (i.e. brain plasticity) (Sporns, 2018). In practice, graph metrics (or indices) such as clustering coefficient, path length and efficiency measures are often used to characterize the "small-world" properties of brain networks. Centrality metrics such as degree, betweenness, closeness, and eigenvector centrality are used to identify the crucial areas within the network. Community structure analysis, which detects the groups of regions more densely connected between themselves than expected by chance, is also essential for understanding brain network organization and topology.

The definition of the nodes (or vertices) for brain graphs is modality specific. In sensor-based modalities, such as EEG, brain nodes are commonly assigned directly to sensors or to electrodes. However, volume conduction in EEG causes the signal at each sensor to be a mixture of blurred activity from different inner cortical sources. This effect can either be ignored, in which case brain nodes will suffer from a biased non-neural dependence, or it can be addressed in several ways, such as using spatial filters (Kayser & Tenke, 2006), choosing functional connectivity metrics that attenuate volume conduction (Stam et al., 2007), or using cortical source reconstruction. After defining brain nodes, assigning links between them is the subsequent crucial modelling step. In functional neuroimaging, the links of a brain graph are given by evaluating the similarity between two brain signals, through functional connectivity (FC) measures such as phase synchrony, Pearson's correlation or coherence. FC methods fall into two broad categories: those measuring symmetric mutual interaction (undirected weighted links) and those measuring asymmetric information propagation (directed weighted links). FC across N recording sites can be described by an $N \times N$ adjacency matrix A containing all the pairwise FC measures a_{ij} corresponding to the weighted links of the brain graph. Since most of the graph theoretic metrics are defined on binary (unweighted) graphs, initial efforts on the analysis of brain network topologies were implemented via the binarization of the weights using some arbitrarily chosen thresholds with some good success. A simple way of building a graph from a weighted FC matrix is to apply a threshold τ to each element of the matrix, such that if $a_{ij} \geq \tau$, then an edge is drawn between the corresponding nodes, but if $a_{ij} < \tau$, no edge is drawn (Achard et al., 2006). This thresholding operation thus binarizes the weight matrix and converts the continuously variable edge weights to either 1 or 0. By varying the threshold τ used to construct a binary graph from a continuous weight matrix, the connection density of the network is made denser or less dense. If the threshold is low and many weak weights are added to the graph as edges then the connection density will increase; if the threshold is high and only the strongest weights are represented as edges, then the connection density will decrease. Since the choice of τ may be arbitrary, a lot of the binarization techniques rely on choosing a threshold to achieve a certain graph density. Moreover, in recent work, instead of choosing a single τ, a range of τ values have been chosen to generate multiple binary graphs from a single FC network. In this manner, the topology of the connectivity network can be analyzed across different scales. However, thresholding poses the problem of over-simplifying FCNs and, more importantly, there is no generally accepted criterion to select the threshold (Bassett & Bullmore, 2017; Lee et al., 2012). Moreover, the size and density of the thresholded network vary based on the chosen threshold value (Rubinov & Sporns, 2010). Recent studies show that the significance of the difference between groups is strongly dependent on the threshold parameter, that is, the power of the statistical analysis varies with the threshold (Langer et al., 2013). Recently, extensions of graph theoretic measures have been proposed for weighted networks to address some of these issues (Bolanos et al., 2013; Bassett et al., 2011). It has also been shown that graph theoretic measures, such as the clustering measure and the small-world parameter, are very sensitive to the size of the network, that is, the number of nodes, and the density of the connections. Thus, comparing two networks with different edge density may lead to wrong conclusions, making it difficult to disentangle experimental effects from those introduced by differences in the average degree (Bassett & Bullmore, 2017; Muldoon et al., 2016). For these reasons, in this chapter we focus on metrics suitable for undirected and weighted links.

Graphs can be investigated at different levels of scale, and specific measures capture graph attributes at local (nodal) and global (network-wide) scales (Kruschwitz et al., 2015; de Vico Fallani et al., 2014). Nodal measures include simple statistics such as node degree or strength, while global measures express networkwide attributes such as the path length or the efficiency. Intermediate scales can be accessed via hierarchical neighborhoods around single graph elements, or by considering subgraphs or motifs. Motifs are defined as subsets of network nodes and their mutual edges whose patterns of connectivity can be classified into distinct motif categories. In empirical networks, these categories often occur in characteristic frequencies that can be compared with distributions from appropriate (random) null models. Even though a multitude of graph theoretics have been defined to analyze complex networks, here we focus on metrics that are particularly useful for characterizing functional connectivity networks.

21.4.1 Small-World

Watts and Strogatz (1998) showed that graphs with many local connections and a few random long distance connections are characterized by a high cluster coefficient (like ordered networks) and a short path length (like random networks). Such near-optimal networks, which are intermediate between ordered and random networks, are designated as small-world networks. Many neuroimaging studies have shown that both structural and functional brain networks shared similar small-world properties of short path length and high clustering (Bassett & Bullmore, 2006; Bassett & Bullmore, 2017). Small-world networks are simultaneously strongly clustered and integrated. This phenomenon of small-worldness is captured by the small-world parameter, which is the ratio of the normalized clustering coefficient to the normalized path length. For a weighted network, the small-world parameter is given as (Rubinov & Sporns, 2010; Humphries & Gurney, 2008):

$$\sigma^w = \frac{C^w / C^w_{rand}}{L^w / L^w_{rand}} \tag{21.10}$$

where C and C_{rand} are the clustering coefficients of the network and a random network with the same degree distribution, respectively, and L and L_{rand} are the characteristic path lengths of the network and a random network with the same degree distribution, respectively. In this definition, the clustering coefficient is a measure of segregation and reflects mainly the fraction of clustered connectivity available around individual nodes.

The clustering coefficient for a weighted network is defined as (Onnela et al., 2005):

$$C^w = \frac{1}{N}\sum\nolimits_{i\in N} \frac{2t_i^w}{k_i\left(k_i - 1\right)}, \tag{21.11}$$

where t_i^w is the weighted geometric mean of the triangles around a node i.

Similarly, the characteristic path length of the network is the average shortest path length between all pairs of nodes in the network. For a weighted network it is calculated as (Rubinov & Sporns, 2010):

$$L^w = \frac{1}{N}\sum\nolimits_{i\in N} \frac{\sum\nolimits_{j\in N, j\neq i} d_{ij}^w}{\left(n-1\right)}, \tag{21.12}$$

where d_{ij}^w is the shortest weighted path length between node i and j.

Even though the small-world parameter has been widely used to characterize functional brain networks, it has some shortcomings. First, it is not clear in most cases how the small-worldness of the brain network relates to biological properties. Second, a single parameter that summarizes the network topology is not sufficient and does not tell the whole story about the network structure. Third, the small-world scalar σ can be greater than 1 even in cases when the normalized path length is much greater than one; because it is defined as a ratio, if $C >> 1$ and $L > 1$, the scalar $\sigma > 1$. This means that a small-world network will always have $\sigma > 1$, but not all networks with $\sigma > 1$ will be small-world (some of them may have greater path length than random graphs). Finally, the measure is strongly driven by the density of the graph, and denser networks will naturally have smaller values of σ even if they are in fact generated from an identical small-world model.

21.4.2 Modularity

Among the most widely encountered and biologically meaningful aspects of brain networks is their organization into distinct network communities or modules (Meunier,

Lambiotte et al., 2009; Laumann et al., 2015; Chavez et al., 2010; Fair et al., 2009; Ferrarini et al., 2009; Meunier, Archard et al., 2009; Power et al., 2011). Modules are useful to partition larger networks into basic "building blocks," that is, internally densely connected clusters that are more weakly interconnected among each other. Modular partitions have neurobiological significance as their boundaries separate functionally related neural elements, define critical bridges and hubs that join communities, channel and restrict the flow of neural signals and information, and limit the uncontrolled spread of perturbations.

There are numerous computational techniques for extracting communities and modules from complex networks. One of the most widely used approaches in network neuroscience is modularity maximization, which aims to divide a given network into a set of nonoverlapping communities by maximizing a global objective function, the modularity metric (Newman, 2006a). Originally, this metric was formulated to detect communities whose internal density of connections is maximal, relative to a degree-preserving null model. A good partitioning of a network is expected to have high modularity Q with $Q = (\textit{fraction of edges within communities}) - (\textit{expected fraction of such edges})$, where the expected fraction of edges is evaluated for a random graph.

While modularity characterizes the brain network at a finer scale compared to the small-world parameter, it still has some shortcomings. First, modularity maximization has some shortcomings such as coming up with degenerate solutions, that is, numerous partitions may result in the same value of the modularity metric. Second, modularity optimization cannot identify modules below a certain size. One way to address these issues and represent the full multi-scale structure of brain networks is to perform consensus clustering across multiple spatial resolutions—an approach that combines sampling the entire range of possible spatial resolutions with a hierarchical consensus clustering procedure.

21.4.3 Centrality

Numerous measures quantify the potential of individual nodes and edges to influence the global state of the network. Many of them allow the identification of network hubs generally defined as highly central parts of the network (Sporns et al., 2007). The number of connections maintained by a node (its degree) or the combined weight of these connections (its strength) often provides a strong indicator of influence or centrality. Other measures of centrality take advantage of the layout of the shortest paths within the network and record the number of such paths that pass through a given node or edge—a measure called the betweenness. Another way to approach centrality is by referencing the relation of nodes and edges to a network's community structure. The participation coefficient quantifies the diversity of a given node's connections across multiple modules—high participation indicates that many of these connections are made across modules, thus linking structurally and functionally the distinct communities (Guimera & Nunes Amaral, 2005). This measure is particularly useful in brain networks as it can be applied to both structural and functional network data.

21.5 Illustrative Example

This section illustrates the differences between the various multivariate metrics introduced in this chapter. Moran and colleagues (2015) used an EEG dataset from a previously published cognitive control-related error processing study. The study was designed following the experimental protocol approved by the Institutional Review Board (IRB) of the Michigan State University. The data collection was performed in accordance with the guidelines and regulation established by this protocol. Written and informed consent was collected from each participant before data collection.

The experiment consisted of a speeded-reaction Flanker task (Eriksen & Eriksen, 1974), in which subjects identified the middle letter on a five-letter string, being congruent (e.g. MMMMM) or incongruent (e.g. MMNMM) with respect to the Flanker letters. Flanker letters (e.g. MM MM) were shown during the first 35 ms of each trial, and during the following 100 ms the Flanker and target letters were shown on the screen. This was followed by an inter-trial interval of variable duration ranging from 1200 ms to 1700 ms. A total of six blocks consisting of 80 trials composed the experiment, and letters were changed between blocks. EEG responses were recorded by the 64 electrode ActiveTwo system (BioSemi, Amsterdam, The Netherlands). The sampling frequency was 512 Hz. Trials containing artifacts were rejected and volume conduction was reduced through the Current Source Density (CSD) Toolbox (Kayser & Tenke, 2006). A total of 18 subjects and 58 channels were considered for the analysis, for which the total number of error trials ranged from 20 to 61. The same number of correct responses was chosen randomly. In this example, we explore the effectiveness of the different multivariate measures by applying them to the $N \times N$ bivariate functional connectivity matrices corresponding to error-related negativity (ERN) and the correct-related negativity (CRN). Previous studies have shown that the ERN is associated with increased synchronization in the theta band (4–8 Hz) between electrodes in the central and lateral frontal regions (Aviyente et al., 2011). For this reason, an FCN was constructed for each subject by averaging the PLV over the time window 25–75 ms and the frequency bins corresponding to the theta band per subject and response type. This results in two FCNs of size 58×58 per subject, one corresponding to error responses and the other to correct responses.

Table 21.1 summarizes the results of applying four of the metrics discussed in this chapter. For the S-estimator and Omega complexity, in both cases the results indicate that the FCNs for the correct response have higher levels of synchronization compared to the error response. However, these measures do not necessarily quantify how this increased synchronization is distributed across the network and does not reflect the organization of the network. The graph theoretic metrics, on the other hand, address this issue. We note that the small-world measure is slightly higher for error networks compared to correct ones. In particular, the FCNs constructed from error responses are small-world while the ones from correct response are not. This difference between

Table 21.1 Mean ± Standard deviation of the different multivariate metrics on example FCNs

Metric	ERN	CRN
S-estimator	0.2147 ± 0.0282	0.3216 ± 0.01
Omega Complexity	0.8825 ± 0.0017	0.8891 ± 0.0006
Small-World	1.0078 ± 0.0086	0.9938 ± 0.001
Modularity	0.0329 ± 0.0154	0.0041 ± 0.0019

small-world parameters is statistically significant using a two-tailed t-test ($p < 0.001$). This implies that the FCN constructed from the error response is more clustered and has smaller average path length indicating a better integreated system. Similarly, the modularity values are higher for error network compared to correct FCNs with the difference being significant ($p < 0.001$). This difference in modularity shows that the network has a more modular structure, that is, groups of nodes that are highly integrated, for the error response.

Next, we illustrate the use of the direct multivariate phase synchrony measure, HTS, on the same data. Unlike the metrics discussed earlier, HTS does not require the $N \times N$ connectivity matrix and can be directly applied to a group of oscillators or electrodes. A preliminary analysis shows that HTS yields increased synchronization for the central and lateral frontal regions for the error response, whereas there is no topographical differentiation of synchrony for the correct response. For this reason, we focus on ERN networks and assess whether the medial and lateral electrodes formed a stronger network than that between medial and occipital electrodes by computing HTS among electrode FCz and the left lateral (F1, F3, F5), right lateral (F2, F4, F6), and occipital (Oz, O1, O2) electrodes. Table 21.2 shows the multivariate synchrony among these regions, and as expected, the medial and lateral regions exhibit higher synchrony when compared to the medial and occipital regions. Moreover, we examined the statistical significance of the difference in multivariate synchrony of these regions. The multivariate synchrony among the medial and left and right lateral regions is not significantly different. On the

Table 21.2 Mean ± Standard deviation of HTS for error-related negativity networks over different brain regions

Electrodes	HTS
FCz,F1, F3, F5	0.2649 ± 0.049
FCz, F2, F4, F6	0.2845 ± 0.042
FCz, Oz, O1, O2	0.2294 ± 0.036

other hand, lateral-medial synchrony is significantly different than that within the occipital regions.

21.6 Conclusions

This chapter reviewed different metrics that have been introduced to quantify multivariate synchronization in the brain. The first group of methods focus on spectral properties of the bivariate phase synchrony matrix and as such quantify the spread of the eigenvalues of this $N \times N$ matrix. Even though these measures are successful at capturing the global synchronization within the brain, they do not pinpoint to the actual mechanisms underlying the observed global synchrony values. Graph theoretic metrics, on the other hand, can characterize the network topology across multiple scales ranging from the micro to the macro-scale. Thus, they can provide explanations to the observed synchrony values in terms of different network models such as small-world and modular structures. In order to better characterize brain networks, these analyses need to be performed at the subnetwork level. The recently proposed hyper-torus phase synchrony measure addresses this issue by computing the synchronization within a group of electrodes or a subnetwork. This type of analysis offers the best trade-off between bivariate metrics that focus on pairs of electrodes and multivariate metrics that focus on the whole brain.

References

Achard, S., Salvador, R., Whitcher, B., Suckling, J., & Bullmore, E. T. (2006). A resilient, low-frequency, small-world human brain functional network with highly connected association cortical hubs. *Journal of Neuroscience*, *26*(1), 63–72.

Al-Khassaweneh, M., Villafane-Delgado, M., Yener Mutlu, A., & Aviyente, S. (2016). A measure of multivariate phase synchrony using hyperdimensional geometry. *IEEE Transactions on Signal Processing*, *64*(11), 2774–2787.

Allefeld, C. & Bialonski, S. (2007). Detecting synchronization clusters in multivariate time series via coarse-graining of Markov chains. *Physical Review* E, *76*(6), 66207–66215.

Allefeld, C., Frisch, S., & Schlesewsky, M. (2005). Detection of early cognitive processing by event-related phase synchronization analysis. *Neuroreport*, *16*(1), 13–16.

Allefeld, C. & Kurths, J. (2003). Multivariate phase synchronization analysis of EEG data. *IEICE Transactions on Fundamentals of Electronics, Communications and Computer Sciences*, *86*(9), 2218–2221.

Allefeld, C. & Kurths, J. (2004). An approach to multivariate phase synchronization analysis and its application to event-related potentials. *International Journal of Bifurcation and Chaos*, *14*(2), 417–426.

Allefeld C., Müller, M., & Kurths, J. (2007). Eigenvalue decomposition as a generalized synchronization cluster analysis. *International Journal of Bifurcation Chaos*, *17*, 3493–3497.

Aviyente, S., Bernat, E. M., Evans, W. S., & Sponheim, S. R. (2011). A phase synchrony measure for quantifying dynamic functional integration in the brain. *Human Brain Mapping, 32*(1), 80–93.

Aviyente, S. & Yener Mutlu, A. (2011). A time-frequency-based approach to phase and phase synchrony estimation. *IEEE Transactions on Signal Processing, 59*(7), 3086–3098.

Aydore, S., Pantazis, D., & Leahy, R. M. (2013). A note on the phase locking value and its properties. *Neuroimage, 74*, 231–244.

Baccala, L. A. & Sameshima, K. (2001). Partial directed coherence: A new concept in neural structure determination. *Biological Cybernetics, 84*(6), 463–474.

Bassett D. S. & Bullmore, E. (2006). Small-world brain networks. *The Neuroscientist, 12*(6), 512–523.

Bassett D. S. & Bullmore, E. T. (2017). Small-world brain networks revisited. *The Neuroscientist, 23*(5), 499–516.

Bassett, D. S., Wymbs, N. F., Porter, M. A., Mucha, P. J., Carlson, J. M., & Grafton, S. T. (2011). Dynamic reconfiguration of human brain networks during learning. *Proceedings of the National Academy of Sciences, 108*(18) 7641–7646.

Bastos, A. M. & Schoffelen, J. M. (2016). A tutorial review of functional connectivity analysis methods and their interpretational pitfalls. *Frontiers in Systems Neuroscience* [online], *9*, 175.

Bolanos, M., Bernat, E. M., He, B., & Aviyente, S. (2013). A weighted small world network measure for assessing functional connectivity. *Journal of Neuroscience Methods, 212*(1), 133–142.

Bullmore E. T. & Bassett, D. S. (2011). Brain graphs: Graphical models of the human brain connectome. *Annual Review of Clinical Psychology, 7*, 113–140.

Bullmore, E. & Sporns, O. (2009). Complex brain networks: Graph theoretical analysis of structural and functional systems. *Nature Reviews Neuroscience, 10*(3), 186–198.

Buzsáki, G. & Draguhn, A. (2004). Neuronal oscillations in cortical networks. *Science, 304*(5679), 1926–1929.

Carmeli, C., Knyazeva, M. G., Innocenti, G. M., & De Feo, O. (2005). Assessment of EEG synchronization based on state-space analysis. *Neuroimage, 25*(2), 339–354.

Chavez, M., Valencia, M., Navarro, V., Latora, V., & Martinerie, J. (2010). Functional modularity of background activities in normal and epileptic brain networks. *Physical Review Letters, 104*(11), 118701.

Dimitriadis, S. I., Laskaris, N. A., & Tzelepi, A. (2013). On the quantization of time-varying phase synchrony patterns into distinct functional connectivity microstates (fcμstates) in a multi-trial visual ERP paradigm. *Brain Topography, 26*(3), 397–409.

de Vico Fallani, F., Richiardi, J., Chavez, M., & Achard, S. (2014). Graph analysis of functional brain networks: Practical issues in translational neuroscience. *Philosophical Transactions of the Royal Society B: Biological Sciences, 369*(1653), 20130521.

Eriksen, B. A. & Eriksen, C. W. (1974). Effects of noise letters upon the identification of a target letter in a nonsearch task. *Perception & Psychophysics, 16*(1), 143–149.

Fair, D. A., Cohen, A. L., Power, J. D., Dosenbach, N. U. F., Church, J. A., Miezin, F. M., Schlaggar, B. L., & Petersen, S. E. (2009). Functional brain networks develop from a "local to distributed" organization. *PLoS Computational Biology, 5*(5), e1000381.

Ferrarini, L., Veer, I. M., Baerends, E., van Tol, M-J., Renken, R. J., van der Wee, N. J. A., ... Milles, J. (2009). Hierarchical functional modularity in the resting-state human brain. *Human Brain Mapping, 30*(7), 2220–2231.

Fries, P. (2009). Neuronal gamma-band synchronization as a fundamental process in cortical computation. *Annual Review of Neuroscience*, *32*, 209–224.

Granger, C. W. J. (1969). Investigating causal relations by econometric models and cross-spectral methods. *Econometrica*, *37*(3), 424–438.

Giulio, T., McIntosh, A. R., Russell, D. P., & Edelman, G. M. (1998). Functional clustering: Identifying strongly interactive brain regions in neuroimaging data. *NeuroImage*, *7*(2), 133–149.

Guimera, R. & Nunes Amaral, L. A. (2005). Functional cartography of complex metabolic networks. *Nature*, *433*(7028), 895–900.

Humphries, M. D. & Gurney, K. (2008). Network "small-world-ness": A quantitative method for determining canonical network equivalence. *PloS One*, *3*(4), e0002051.

Jalili, M., Barzegaran, E., & Knyazeva, M. G. (2013). Synchronization of EEG: Bivariate and multivariate measures. *IEEE Transactions on Neural Systems and Rehabilitation Engineering*, *22*(2), 212–221.

Kayser J. & Tenke, C. E. (2006). Principal components analysis of Laplacian waveforms as a generic method for identifying ERP generator patterns: I. evaluation with auditory oddball tasks. *Clinical Neurophysiology*, *117*(2), 348–368.

Kruschwitz, J. D., List, D., Waller, L., Rubinov, M., & Walter, H. (2015). Graphvar: A user-friendly toolbox for comprehensive graph analyses of functional brain connectivity. *Journal of Neuroscience Methods*, *245*, 107–115.

Lachaux, J-P., Rodriguez, E., Martinerie, J., & Varela, F. J. (1999). Measuring phase synchrony in brain signals. *Human Brain Mapping*, *8*, 194–208.

Lakatos, P., Karmos, G., Mehta, A. D., Ulbert, I., & Schroeder, C. E. (2008). Entrainment of neuronal oscillations as a mechanism of attentional selection. *Science*, *320*(5872), 110–113.

Langer, N., Pedroni, A., & Jäncke, L. (2013). The problem of thresholding in small-world network analysis. *PloS One*, *8*(1), e53199.

Laumann, T. O., Gordon, E. M., Adeyemo, B., Snyder, A. Z., Joo, S. J., Chen, M-Y., . . . Petersen, S. E. (2015). Functional system and areal organization of a highly sampled individual human brain. *Neuron*, *87*(3), 657–670.

Lee, H., Kang, H., Chung, M. K., Kim, B-N., & Lee, D. S. (2012). Persistent brain network homology from the perspective of dendrogram. *IEEE Transactions on Medical Imaging*, *31*(12), 2267–2277.

Meunier, D., Achard, S., Morcom, A., & Bullmore, E. (2009). Age-related changes in modular organization of human brain functional networks. *NeuroImage*, *44*(3), 715–723.

Meunier, D., Lambiotte, R., Fornito, A., Ersche, K., & Bullmore, E. T. (2009). Hierarchical modularity in human brain functional networks. *Frontiers in Neuroinformatics*, *3*, 37.

Moran, T. P., Bernat, E. M., Aviyente, S., Schroder, H. S., & Moser, J. S., (2015). Sending mixed signals: Worry is associated with enhanced initial error processing but reduced call for subsequent cognitive control. *Social Cognitive and Affective Neuroscience*, *10*(11), 1548–1556.

Muldoon, S. F., Bridgeford, E. W., & Bassett, D. S. (2016). Small-world propensity and weighted brain networks. *Scientific Reports*, *6*(1) 1–13.

Newman, M. E. J. (2006a). Finding community structure in networks using the eigenvectors of matrices. *Physical Review E*, *74*(3), 036104.

Newman, M. E. J. (2006b). Modularity and community structure in networks. *Proceedings of the National Academy of Sciences of the United States of America*, *103*(23), 8577–8582.

Onnela, J-P., Saramäki, J., Kertész, J., & Kaski, K. (2005). Intensity and coherence of motifs in weighted complex networks. *Physical Review E*, *71*(6), 065103.

Pereda, E., Quiroga, R. Q., & Bhattacharya, J. (2005). Nonlinear multivariate analysis of neurophysiological signals. *Progress in Neurobiology*, *77*(1–2), 1–37.

Power, J. D., Cohen, A. L., Nelson, S. M., Wig, G. S., Barnes, K. A., Church, J. A., ... Petersen, S. E. (2011). Functional network organization of the human brain. *Neuron, 72*(4), 665–678.

Quiroga, R. Q., Kraskov, A., Kreuz, T., & Grassberger, P. (2002). Performance of different synchronization measures in real data: A case study on electroencephalographic signals. *Physical Review E, 65*(4), 041903.

Rubinov, M. & Sporns, O. (2010). Complex network measures of brain connectivity: Uses and interpretations. *NeuroImage, 52*(3), 1059–1069.

Schelter, B., Winterhalder, M., Dahlhaus, R., Kurths, J., & Timmer, J. (2006). Partial phase synchronization for multivariate synchronizing systems. *Physical Review Letters, 96*(20), 208103.

Sporns, O. (2018). Graph theory methods: Applications in brain networks. *Dialogues in Clinical Neuroscience, 20*(2), 111.

Sporns, O., Honey, C. J., & Kötter, R. (2007). Identification and classification of hubs in brain networks. *PloS One, 2*(10), e1049.

Stam, C. J., Nolte, G., & Daffertshofer, A. (2007). Phase lag index: Assessment of functional connectivity from multi-channel EEG and MEG with diminished bias from common sources. *Human Brain Mapping, 28*(11), 1178–1193.

Stam C. J. & Reijneveld, J. C. (2007). Graph theoretical analysis of complex networks in the brain. *Nonlinear Biomedical Physics, 1*(1), 1–19.

Stam, C. J. & Van Dijk, B. W. (2002). Synchronization likelihood: An unbiased measure of generalized synchronization in multivariate data sets. *Physica D: Nonlinear Phenomena, 163*(3–4), 236–251.

Tononi, G., Sporns, O., & Edelman, G. M. (1994). A measure for brain complexity: relating functional segregation and integration in the nervous system. *Proceedings of the National Academy of Sciences of the United States of America, 91*(11), 5033–5037.

Uhlhaas, P. J. & Singer, W. (2006). Neural synchrony in brain disorders: Relevance for cognitive dysfunctions and pathophysiology. *Neuron, 52*(1), 155–168.

Uhlhaas P. J. & Singer, W. (2010). Abnormal neural oscillations and synchrony in schizophrenia. *Nature Reviews Neuroscience, 11*(2), 100–113.

van den Heuvel, M. P., Stam, C. J., Boersma, M., & Hulshoff Pol, H. E. (2008). Small-world and scale-free organization of voxel-based resting-state functional connectivity in the human brain. *NeuroImage, 43*(3), 528–539.

Watts, D. J. & Strogatz, S. H. (1998). Collective dynamics of "small-world" networks. *Nature, 393*(6684), 440–442.

Yener Mutlu, A. & Aviyente, S. (2012). Hyperspherical phase synchrony for quantifying multivariate phase synchronization. *2012 IEEE Statistical Signal Processing Workshop (SSP)*, 888–891.

CHAPTER 22

BRAIN STIMULATION APPROACHES TO INVESTIGATE EEG OSCILLATIONS

FLORIAN H. KASTEN AND
CHRISTOPH S. HERRMANN

22.1 Introduction

Several of the previous chapters in this book laid out the numerous associations between oscillatory brain activity and various domains of cognitive functioning that have been discovered over the course of the last century. At the same time, altered patterns of this oscillatory brain activity have been observed in many neurological and psychiatric diseases (Herrmann & Demiralp, 2005; Uhlhaas & Singer, 2006, 2012).

The recording of brain signals using electro- or magnetoencephalography (EEG/MEG) or by means of invasive techniques (e.g., intracortical EEG or electrocorticography), has strongly contributed to our knowledge in this area, and will continue to do so. However, as these methods can only provide observational data, inference about the functional role of brain oscillations for cognitive functions remains correlational. That is, oscillatory activity is observed as the dependent variable while participants are engaged in different cognitive tasks or task conditions, which are experimentally manipulated (independent variable, Figure 22.1A). Whether the observed oscillatory activity has a direct causal influence on the investigated function, or whether it reflects byproducts of the underlying neural processing, cannot be resolved by this type of experimental design. In order to demonstrate such causal relationships, one needs to revert the design and experimentally manipulate the oscillatory activity in question (independent variable) and measure the resulting

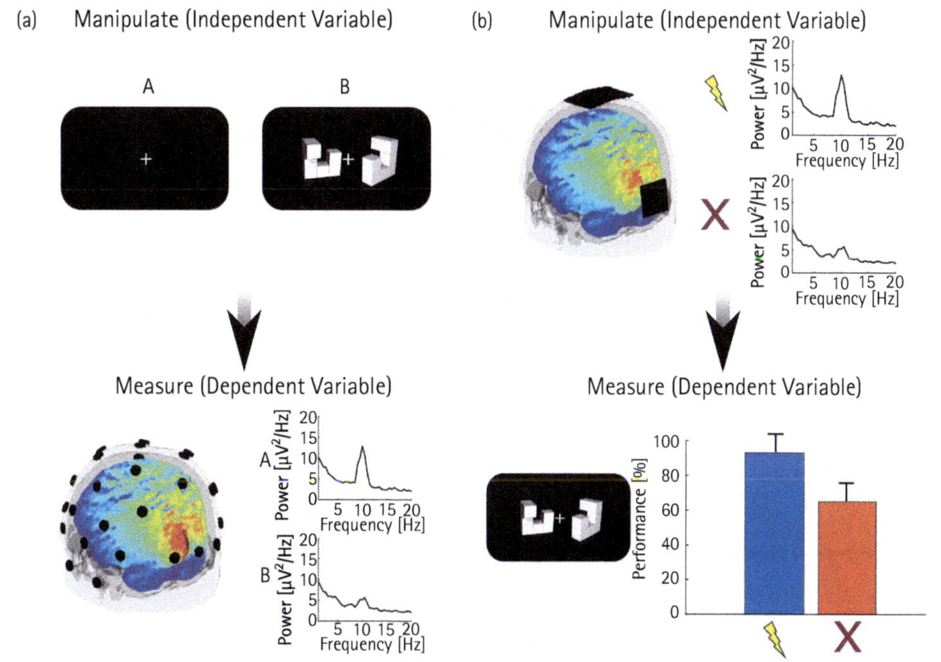

FIGURE 22.1 Experimental designs to investigate brain oscillations. (A) In traditional EEG experiments participants are exposed to different experimental conditions (e.g. a task free fixation interval and a stimulus). An EEG is recorded and the frequency content of the signal is compared between task conditions and correlated with task performance. (B) In order to establish causal relationships between the brain oscillation and the cognitive process, one needs to revert the design and manipulate a feature of the oscillation of interest (here power around 10 Hz) by an intervention. If behavioral changes are observed in response to the intervention, this can be taken as evidence for a causal involvement of the oscillations in the investigated process.

behavioral changes as the dependent variable (Bergmann et al., 2016; Herrmann et al., 2016b; Figure 22.1B).

In principle, a variety of methods allow researchers to modulate brain oscillations. However, many of these approaches, such as pharmacological interventions (e.g., Kopp et al., 2004), optogenetics (Sohal, 2012), or intracranial electrical stimulation (Alagapan et al., 2016; Fröhlich & McCormick, 2010), are highly invasive and their application mostly restricted to animal models and small groups of patients. For example, light-driven activation of fast-spiking interneurons at 40 Hz has been shown to cause gamma band increase of local field potentials in mice (Cardin et al., 2009) and low-frequency direct cortical stimulation in the alpha range has been shown induce state-dependent modulation of oscillatory brain activity in the alpha and theta band in epilepsy patients (Alagapan et al., 2016). While invasive stimulation offers potential for more reliable effects due to stronger and more focal perturbation of brain activity, they require opening of the scalp and skull or even penetration of brain tissue making them unsuitable for application in healthy human subjects. Other methods like rhythmic stimulation with sensory stimuli (e.g., rhythmic visual stimulation with flickering light causes steady state visual evoked responses) or neurofeedback training have been shown to modulate oscillatory activity in the brain and can be applied in healthy human subjects. However, steady-state responses exhibit their effects on brain oscillations indirectly via the sensory systems and it is debated whether the observed oscillatory activity reflects a modulation of brain oscillations or merely a superposition of event-related potentials (but see Notbohm et al., 2016). Neurofeedback training can be used in many different contexts, including studying higher cognitive processes. However, the intervention is rather time consuming as the training usually requires several sessions on separate days before its effects can be measured.

During the past decades, a branch of techniques has evolved to noninvasively modulate brain activity by either applying magnetic pulses via transcranial magnetic stimulation (Barker et al., 1985; Hallett, 2000), or weak electric currents by means of transcranial electrical stimulation (tES). The latter is an umbrella term covering several electrical stimulation methods including transcranial direct current stimulation (tDCS), transcranial alternating current stimulation (tACS), and transcranial random noise stimulation (Woods et al., 2016). Among these methods, the rhythmic, repetitive application of TMS (rTMS) and tACS are considered particularly promising to study the functional role of brain oscillations in cognition. The latter works via the application of an alternating current usually of sinusoidal shape. Although, depending on the specific limitations of the hardware used for stimulation, almost any type of waveform can be created (Figure 22.2). For example, recent work applied tACS using the envelope of

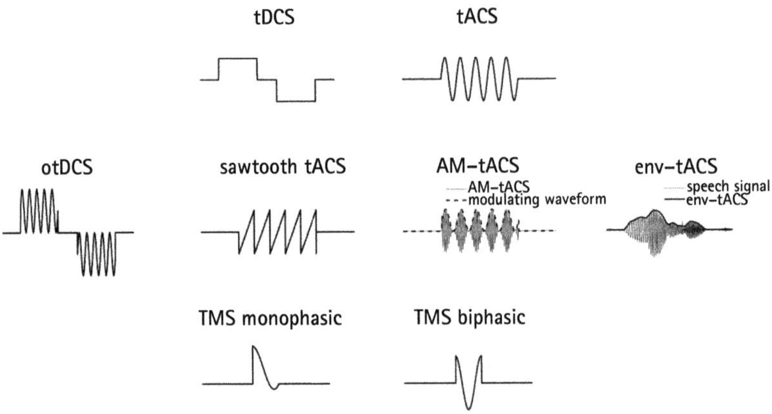

FIGURE 22.2 Stimulation waveforms used in tES and TMS. Top: Waveforms conventionally used for tDCS and tACS. Middle: Alternative waveforms used to stimulate oscillatory activity in the brain. For otDCS, a sinusoidal tACS waveform is combined with a DC offset. TACS using sawtooth waves cause large power in the spectrum at harmonic frequencies, which makes residual artifacts easier to detect. In amplitude modulated-tACS a high-frequency carrier waveform is modulated in amplitude by a low-frequency waveform at the frequency of the target brain oscillation. For envelope-tACS the stimulation waveform is extracted from the envelope of a speech signal. Bottom: Waveform shapes of mono- and biphasic TMS pulses.

human speech signals to improve speech intelligibility (Riecke et al., 2018; Wilsch et al., 2018). Others administered tACS with sawtooth waves or amplitude modulated sine waves that are supposed to allow easier removal of stimulation artifacts from concurrent electrophysiological recordings (Dowsett & Herrmann, 2016; Kasten et al., 2018b; Witkowski et al., 2016). Some researchers also applied alternating currents together with a DC offset, which has been referred to as oscillating transcranial direct current stimulation (otDCS; Marshall et al., 2006; Neuling et al., 2012a). RTMS and tACS are believed to synchronize/entrain endogenous brain oscillations to the externally applied driving force, thus specifically targeting the brain oscillation of interest and probing its causal role during a particular task (Herrmann et al., 2013; Reato et al., 2013; Thut et al., 2011b, 2011a). In addition, effects outlasting the duration of stimulation by several minutes or even hours have been observed in many rTMS and tACS studies (Veniero et al., 2015). These findings give rise to the hope that in the future these methods may offer new treatment options for psychiatric and neurological conditions by restoring dysfunctional oscillatory activity.

22.2 Mechanisms Underlying rTMS and tES

TMS exploits the principles of electromagnetic induction of electric fields in the brain (Barker et al., 1985; Wagner et al., 2009). To this end, strong, transient currents are fed through a coil of wire placed above the scalp in the proximity of the targeted brain area. The high-intensity current creates a magnetic field with magnetic flux passing perpendicularly to the plane of the coil. The rapidly changing magnetic field in turn passes through the skull and induces current in the brain tissues underneath the scalp, flowing in loops parallel to the coil plane (Figure 22.3A; Hallett, 2000; Wagner et al., 2009). The resulting electric fields cause changes in membrane polarization sufficiently strong to modulate neural excitability by changing membrane polarization and even trigger the firing of action potentials (Wagner et al., 2009). Different coil shapes, like circular, double cone, or figure of eight, are available that vary with respect to their intensity, focality, and the depth of brain areas that can be reached with the stimulation (Hallett, 2000; Klomjai et al., 2015). TMS can be administered using different stimulation protocols. The application of a single TMS pulse can already elicit effects and impair visual perception (Amassian et al., 1989). Traditionally, low-frequency rTMS (<1 Hz) is considered to suppress cortical excitability while high-frequency rTMS (>5Hz) is considered facilitatory (Klomjai et al., 2015). Variants of rTMS include theta burst stimulation, where short bursts of high-frequency stimulation (40 Hz) are repeated at theta frequency (5 Hz). Besides the overall excitatory and inhibitory effects of rTMS, the method can also be used to modulate and investigate brain oscillations (Hanslmayr et al., 2014; Thut et al., 2011b).

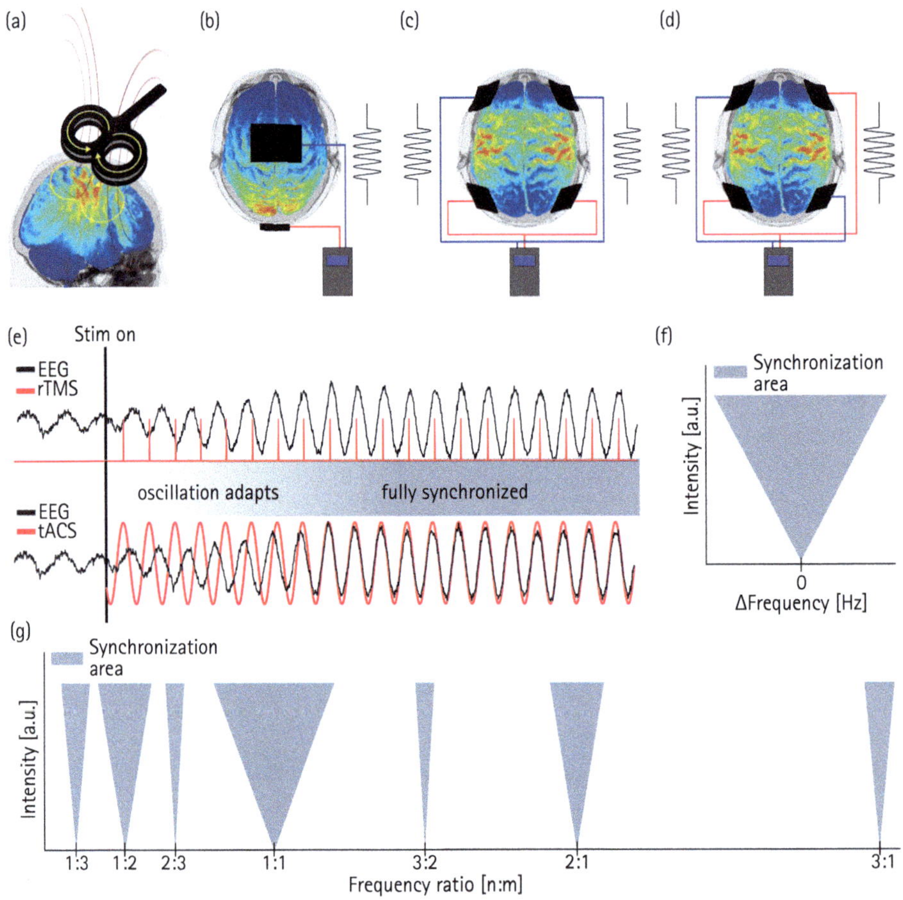

FIGURE 22.3 Modes and mechanisms of stimulation. (A) Figure-eight coil TMS and induced magnetic/electric field. (B) Simple tACS montage with electrodes placed above Cz and Oz of the international 10–20 system to target occipital-parietal regions. (C & D) Electrode montages to modulate inter-hemispheric synchronization. Stimulation between electrode pairs is applied either in-phase (C) or anti-phase (D). (E) Synchronization of endogenous brain oscillations to external stimulation via rTMS (top) and tACS (bottom). The external driving force causes the brain oscillation to adapt in frequency and phase to the stimulation. Due to the increased synchronous neuronal activity, an amplitude increase is expected. Please note that real EEG recordings are strongly corrupted by artifacts arising from stimulation. (F) The "Arnold tongue" illustrates the relationship between the intensity of a driving force coupled to an oscillator and their frequency differences on the resulting synchronization. The farther apart the frequency of the driving force and the oscillator, the higher the intensity needed to cause the oscillator to synchronize. If the frequencies of the oscillator and driving force match, very small intensities are needed. (G) Arnold tongues at different n:m (n perturbations within m oscillatory cycles) frequency ratios.

In contrast to TMS, tES methods like tDCS and tACS have much more subtle effects on neuronal activity. Here, a direct or alternating current is passed through the scalp by two or more stimulation electrodes (Figure 22.3B–D). Commonly, saline soaked sponge electrodes, fixed to the scalp via rubber bands, or electrically conductive rubber electrodes attached using a conductive, adhesive paste are used for this purpose. Some systems also make use of Ag/AgCl electrodes similar to those used in standard EEG systems. Results from simulations of the current flow through the head, as well as findings from invasive recordings in animals and human cadavers suggest that a large proportion of the current is directly shunted through the skin (Miranda et al., 2006; Neuling et al., 2012b; Opitz et al., 2016; Vöröslakos et al., 2018). However, at the same time, small amounts of the injected current still reach the target regions in the brain. In case of tDCS, these currents shift the resting potentials of the stimulated neurons, resulting in slight depolarization of the cell membrane under the anode and hyperpolarization under the cathode. While these changes in membrane potentials are too small to directly elicit neural firing, early work on animal models in vivo and in vitro suggests that they can alter neuronal excitability by shifting the membrane potential such that more or less incoming excitatory postsynaptic potentials are required to reach the cells' firing threshold (Bikson et al., 2004; Bindman et al., 1964; Creutzfeldt et al., 1962; Jefferys, 1981).

When an alternating current is applied (e.g., via tACS), the electric field rhythmically alternates between hyperpolarizing and depolarizing the cell membrane during the negative and positive half cycles of the sinusoidal stimulation waveform. Computer simulations as well as in vivo and in vitro recordings demonstrated that these phasic modulations of membrane polarization temporally align neural firing to the frequency of the externally applied stimulation (Deans et al., 2007; Fröhlich & McCormick, 2010; Krause et al., 2019; Ozen et al., 2010; Reato et al., 2010). The electric field strengths necessary to achieve this effect are comparable to the electric fields reaching cortical pyramidal cells during tES (Antal & Herrmann, 2016; Opitz et al., 2016).

Although the mechanisms of tACS and rTMS differ in terms of their effects on the cellular level, both techniques are assumed to cause synchronization/entrainment of the endogenous oscillatory brain activity to the external driving force (Thut et al., 2011a). Synchronization phenomena can be observed everywhere in nature, from chirping of crickets to the beating of the heart, as well as in human-made technical systems like pendulum clocks or electric circuits. The basic principles of these phenomena are universal and can be described using the framework of synchronization theory (Pikovsky et al., 2003). Synchronization requires the presence of a so-called self-sustained oscillator. Such oscillators are characterized by an internal source of energy, allowing the oscillator to exhibit some sort of rhythmic activity until the energy source is consumed (Pikovsky et al., 2003). If two or more oscillators with similar eigenfrequencies (i.e., the frequency at which a system tends to oscillate in the absence of any external driving force) are weakly coupled, their rhythmic activity synchronizes; that is, they align their rhythmic activity in frequency and phase (Pikovsky et al., 2003). Similarly, if coupled to an external driving force, the oscillator synchronizes in frequency and phase to the external signal. Importantly, the further apart the frequency of the external force is to the eigenfrequency of the oscillator, the more energy is needed to synchronize it (Pikovsky et al., 2003). Driving forces at frequencies too far away from the oscillators' eigenfrequency and with too little strength cannot achieve synchronization. In contrast, driving forces that exactly match the frequency of the oscillator can achieve synchronization even at very small intensities. When the relationship between frequency difference and driving force strength on the synchronization strength is visualized in a diagram with synchronization strength coded in color, a triangular shape can be observed, which has been referred to as the Arnold Tongue (Figure 3G; Pikovsky et al., 2003). Besides 1:1 frequency matching between oscillator and driving force, synchronization can also occur if the driving force perturbs the oscillator at harmonic or subharmonic frequencies. Such harmonic entrainment can occur for any $n:m$ (n perturbations within m oscillatory cycles) frequency relationship (where $n, m \in \mathbb{N}$). For example, an oscillator with a frequency of 10 Hz synchronizes to driving forces with frequencies around 10 Hz, but also to those with 2.5 Hz, 3.33 Hz, 6.66 Hz, 15 Hz, 20 Hz, and 30 Hz, corresponding to the 1:3, 1:2, 2:3, 3:2, 2:1, and 3:1 Arnold tongues (Figure 22.3G). The strength of synchronization decays the higher the harmonic frequency of the external driving force is (Pikovsky et al., 2003).

In the human brain, synchronization of oscillatory activity to external driving forces has been observed in response to flickering light stimulation (Herrmann, 2001; Notbohm et al., 2016) and rTMS (Herrmann et al., 2016a; Thut et al., 2011b). Further, the effect of sinusoidal alternating currents on local field potentials and multi-unit activity follows the rules of synchronization theory in computer simulations as well as in in vivo and in vitro recordings (Fröhlich & McCormick, 2010; Krause et al., 2019; Negahbani et al., 2018; Ozen et al., 2010; Reato et al., 2010). However, as recordings of M/EEG activity during stimulation are contaminated by a strong, electromagnetic artifact, direct evidence for these mechanisms of actions for tACS effects in humans is largely missing so far (an overview of approaches for tACS artifact removal in M/EEG and their associated problems is reviewed in Kasten & Herrmann, 2019).

These mechanisms describe the general framework currently assumed to underlie effects of rTMS and tACS during stimulation (also called online effects). A common observation in brain stimulation experiments is that behavioral and physiological effects persist after the stimulation is switched off (reviewed in Veniero et al., 2015). These aftereffects (or offline effects) can last for several minutes or even hours after both rTMS (Schindler et al., 2008; Schutter et al., 2001) and tACS (Kasten et al., 2016; Neuling et al., 2013; Wischnewski et al., 2018). After tACS, these outlasting effects tend to be rather frequency specific. The application of rTMS, in contrast, seems to elicit more broadband aftereffects in all frequency bands irrespective of the stimulation frequency (Veniero et al., 2015). While the pattern of effects observed during this sustained period may appear similar to those observed during stimulation, it is important to emphasize that the underlying mechanisms responsible for those effects may be different. This phenomenon has been well documented for effects of tDCS on motor evoked potentials (MEPs). Here, the selective, pharmacological blockage of NMDA receptors abolished the induction of long-lasting aftereffects, while not affecting online effects of the stimulation (Nitsche et al., 2003). In contrast, blockage of calcium and sodium channels abolished or reduced the enhancing effect of anodal tDCS on MEP size during and after stimulation (Nitsche et al., 2003). These results indicate that the online effect of tDCS depends on membrane polarization, modulating the conductance of sodium and calcium channels. Offline effects appear to be caused by NMDA receptor mediated plasticity, although online effects seem to be necessary to induce the offline effect (Nitsche et al., 2003).

While the mechanism underlying online effects of rTMS and tACS is assumed to be entrainment, offline effects have been suggested to be caused by entrainment echoes, a state of sustained synchronization after stimulation is switched off (Hanslmayr et al., 2014), or processes of spike-timing dependent plasticity (STDP; Vossen et al., 2015; Zaehle et al., 2010). STDP relies on the temporal relation of pre- and postsynaptic potentials. If a presynaptic potential repeatedly precedes a postsynaptic potential, the synaptic connection is strengthened, which is referred to as long-term potentiation (LTP). If the presynaptic potential repeatedly follows the postsynaptic potential, long-term depression (LTD) occurs and the synaptic connection is weakened (Markram et al., 1997). In a neural circuit that generates an oscillation, pathways exist by which the activity of a neuron is fed back to its own synaptic input connections via other neurons. Depending on the length and speed of these connections, each spike requires a certain time to travel through these circuits, which determines its intrinsic frequency (Zaehle et al., 2010). Following the principles of STDP, repetitive stimulation (e.g., with rTMS or tACS) with frequencies slightly below the circuits' intrinsic frequency should lead to a strengthening of synaptic connections because stimulation can systematically precede the circulating action potentials. The application of other frequencies in turn, cannot cause such effects, as the stimulation does not match the circuits' intrinsic frequency, and membrane polarization does not coincide with incoming neural spiking (Vossen et al., 2015; Zaehle et al., 2010).

Although entrainment echoes may exist during the first few seconds after stimulation, the STDP model seems more suited to explain the long-lasting changes elicited by tACS and rTMS, which have been observed for up to an hour (Kasten et al., 2016; Veniero et al., 2015; Vossen et al., 2015). Wischnewski and colleagues (2018) provided direct evidence for such involvement of synaptic plasticity in the generation of tACS aftereffects. In that study, administration of an NMDA receptor antagonist was observed to abolish aftereffects induced by tACS in the beta range over the primary motor cortex. This suggests a role of NMDA receptor-mediated plasticity in the generation of tACS aftereffects similar to those observed after tDCS (Nitsche et al., 2003).

22.3 Targets for Brain Stimulation

During the design of a brain stimulation experiment a couple of choices have to be made that can severely impact both the stimulation success and the interpretability of results.

This section provides an overview of these choices and examples of how different features of brain oscillations are targeted to investigate their causal role for different domains of cognitive functioning.

Usually, the design process starts with identifying a brain oscillation that correlates with a cognitive function, or ideally an output measure of a concrete task (e.g., reaction time or performance). Depending on the kind of association, brain stimulation can be used to target different features of the oscillation, such as its amplitude, frequency, or phase. Some relationships may also depend on cross-frequency interactions (Palva et al., 2005) or synchronization between distinct brain areas (Siegel et al., 2012). Another important aspect is related to the stimulation montage. One needs to decide about the positioning and orientation of the stimulation electrodes or the TMS coil. Usually, the observed correlations between a cognitive function and a brain oscillation are limited to specific areas of the brain. Oscillations in the same frequency range may serve different functions within distinct areas of the brain. Based on neuroimaging results, for example, from source localization of oscillatory activity, a target region in the brain is derived and a montage is chosen that maximizes the electric field strength in this area. A couple of software tools are available that simulate the expected current flow caused by transcranial magnetic and electrical stimulation (e.g., ROAST: Huang et al., 2017a; Simnibs: Thielscher et al., 2015). Recent validation studies using invasive electrophysiology in animals and humans indicate that these models are able to predict the spatial extent of the electric field in the brain quite accurately but tend to overestimate the strength of the induced electric fields (Huang et al., 2017b; Opitz et al., 2016). Even more advanced tools are able to automatically optimize electrode montages to target specific brain areas with specific directions of current flow (Baltus et al., 2018b; Huang et al., 2018; Saturnino et al., 2019; Wagner et al., 2016).

As soon as a feature of an oscillation and a brain region are selected for stimulation, appropriate control conditions for the experiment must be established. The choice of adequate control conditions is not a trivial problem. Stimulation using TMS or tACS can induce visual and somatosensory perceptions that may confound results if compared to a stimulation-free control condition (Schutter, 2016; Turi et al., 2013, 2014). In addition, one would ideally compare the selected stimulation condition with stimulation of all other possible stimulation frequencies at all possible brain regions in order to demonstrate frequency and region-specific effects. However, due to a huge parameter space this is practically impossible. Let us consider a very simple case of tACS with only two electrodes that can be placed on any of the 32 locations of the international 10–20 system for EEG electrodes. In addition, stimulation frequencies are limited to only one frequency within each of the five major EEG frequency bands. This parameter space already amounts to almost 5,000 possible control conditions. The inclusion of more fine-grained electrode locations and stimulation frequencies or stimulating at different intensities easily increases the number of possible control conditions to several million. In practice, stimulation of the target oscillations is thus usually compared to one or two control frequencies applied to the same region, or the same stimulation frequency is applied to one or two control regions. Other researchers also compare stimulation effects against a sham stimulation. During sham stimulation, stimulation is usually applied in a way that mimics visual and somatosensory sensations, but that is considered ineffective in modulating brain activity. The aim is to blind participants towards their experimental condition and to avoid confounds of behavioral results by participants' expectations. The exact choice of appropriate control conditions heavily depends on the research question and should allow to rule out alternative explanations how stimulation may have affected the task.

A straightforward target for stimulation, and probably the most popular, is the amplitude of a brain oscillation. Several studies demonstrated that tACS and rTMS can elevate the power of brain oscillations in the range of the stimulated frequency band (Kasten et al., 2016; Vossen et al., 2015; Wischnewski et al., 2018; Zaehle et al., 2010). Moreover, both techniques have been shown to also modulate event-related changes in oscillatory power (Kasten et al., 2018a; Kasten & Herrmann, 2017; Klimesch et al., 2003; Wischnewski & Schutter, 2017). While rTMS is frequently used to stimulate oscillatory activity during specific periods of a task (Hanslmayr et al., 2014; Klimesch et al., 2003; Romei et al., 2010), similar protocols using tACS show less-reliable effects (Braun et al., 2017; Stonkus et al., 2016), which might be explained by insufficient stimulation durations or state-dependent effects (Kasten et al., 2018a; Kasten & Herrmann, 2017; Strüber et al., 2015). More commonly, tACS is applied in a continuous manner while participants are engaged in a cognitive task. Such continuous stimulation during tasks that involve induced event-related oscillations appears to foster the pre-existing patterns of event-related power change (Kasten et al., 2018a; Kasten & Herrmann, 2017).

Boosting oscillatory activity with rTMS has been used to study the role of posterior pre-stimulus alpha oscillations for basic visual perception (Romei et al., 2010), as well as for higher visual-spatial processing (Klimesch et al., 2003) and short-term memory (Sauseng et al., 2009). Enhancing beta oscillations with rTMS disrupted the encoding of items in a verbal memory task (Hanslmayr et al., 2014). In these experiments, authors applied short trains of rTMS in the frequency band of interest to specifically alter the amplitude of the oscillation during certain time intervals (e.g., before or after stimulus presentation). To demonstrate that the observed behavioral changes are specific to one frequency band, the authors also applied stimulation at control frequencies where no effects were expected (and indeed not found).

As outlined, tACS (and its derivative otDCS) are mostly applied in a continuous manner. One of the first applications was to increase slow-wave oscillations during sleep to probe their role for memory consolidation (Marshall et al., 2006). Here, participants' memory performance significantly increased after stimulation during overnight sleep as compared to a control condition. In the awake brain, tACS has been used to study the role of alpha amplitude in creativity (Lustenberger et al., 2015), visual-spatial cognitive performance (Kasten & Herrmann, 2017), attention (Wöstmann et al., 2018, Kasten et al, 2020), and many more cognitive domains (for a recent overview see Klink et al. 2020). In these experiments, tACS was applied continuously over a brain region during task execution to increase participants' alpha amplitude.

If the frequency of a brain oscillation is of interest, one would try to speed up or slow the target oscillation by applying stimulation at frequencies slightly above or below its intrinsic frequency. According to synchronization theory, such stimulation should shift the intrinsic frequency in the brain towards the frequency of the external driving force (Pikovsky et al., 2003). This approach has been used, for example, to investigate the causal role of occipital alpha oscillations in the generation of the sound-induced double-flash illusion (Cecere et al., 2015). The illusion occurs when two sound beeps are presented within a time window of ~100 ms together with a single visual flash. Participants perceive a second illusory flash in this kind of experiment (Shams et al., 2002). The authors were able to alter the length of the temporal window during which the illusion occurs by stimulating participants ± 2 Hz above or below their individual alpha frequency (Cecere et al., 2015). In another experiment, the perceived frequency of an illusory jitter was modulated in a similar manner by applying amplitude modulated tACS ± 1 Hz above or below participants' individual alpha frequency (Minami & Amano, 2017). In the auditory domain, tACS has been used to test whether the frequency of gamma oscillations determines the temporal resolution of the auditory system. In that study, participants were stimulated above or below their individual gamma frequency (Baltus et al., 2018a, 2018b). By accelerating the individual gamma frequency, participants were able to detect smaller gaps in continuous sound streams of noise. However, when the individual gamma frequency was decelerated, no modulation of gap detection performance was observed (Baltus et al., 2018a, 2018b). Beyond investigations into basic sensory processing, the frequency of brain oscillations had also been modulated in the context of working memory performance. It is well known that humans can uphold 7 ± 2 items in their working/short term memory (Miller, 1956). Later, this capacity has been associated with individual gamma and theta frequencies. Specifically, it is argued that the number of gamma cycles that fit into the positive half wave of a theta oscillation determines the capacity (Lisman & Idiart, 1995; Lisman & Jensen, 2013). Vosskuhl and colleagues (2015) tested this relationship and demonstrated that slowing down participants' theta frequency with tACS such that more gamma cycles fit into one theta oscillation increased working memory capacity in a digit span task.

Investigations into the role of oscillatory phase or phase relationships for cognition are comparatively challenging as they require precise stimulus timing or complex stimulation protocols. Neuling and colleagues (2012a) presented brief auditory targets at specific phases of a 10 Hz otDCS waveform applied to the auditory cortex. In line with previous correlational evidence (Mathewson et al., 2009), participants' detection performance systematically varied depending on the phase of the 10-Hz stimulation at which the target was presented. Similar modulations of detection performance with tACS phase was found for 4-Hz stimulation in an auditory detection task (Riecke et al., 2015) and for stimulation of somatosensory cortex at mu frequency in a somatosensory detection task (Gundlach et al., 2016). In a different type of experiment, authors applied tACS to two distinct brain regions with the same stimulation waveform applied either in or out of phase (with 0- or 180-degree phase shift, Figure 22.3C,D). This way, stimulation is thought to either increase (in phase) or decrease (out of phase) the synchrony of the targeted brain regions within a specific frequency band. Such modulation of inter-regional synchrony, when applied in the gamma frequency range, modulates perception of ambiguous movements (Helfrich et al., 2014a; Strüber et al., 2014). Increasing/decreasing fronto-parietal synchronization in the theta frequency range in the left hemisphere with in-phase vs. anti-phase tACS has been shown to increase/decrease reaction times in a visual memory-matching task (Polanía et al., 2012). Other authors extended this concept to even target interpersonal synchrony (Novembre et al., 2017; Szymanski et al., 2017). In one study, tACS in the beta range was simultaneously applied to the motor cortex of two participants (Novembre et al., 2017). Here, in-phase stimulation in the beta range increased synchrony in a joint finger tapping task as compared to anti-phase stimulation or sham. No such effect was found for stimulation in other frequency bands. In another study, simultaneous tACS with same-phase same-frequency vs. different-phase different-frequency in the theta range over fronto-parietal sites did not show effects on a synchronous drumming task (Szymanski et al., 2017). An important concern about in-phase vs. anti-phase stimulation protocols has been raised recently. In order to directly infer a phase dependent relationship of a behavioral measure and a brain oscillation, it is crucial that the only difference between stimulation conditions is in the phase difference between the stimulated areas. Saturnino and colleagues (2017) simulated the electric fields created by different stimulation montages targeting inter-regional synchrony with in-phase and anti-phase stimulation. Their results demonstrated that the electric field patterns differed during in- and anti-phase stimulation for many of the employed montages, giving rise to the concern that effects of the stimulation could also originate from differences in field strength or even differences in the brain regions that have been (co-)stimulated during the in- and anti-phase stimulation (Saturnino et al., 2017). However, the authors also suggest stimulation montages that can avoid such confounds. Especially, montages using a stimulation electrode in the center, surrounded by several return electrodes or a large ring electrode, are recommended to avoid such confounds (Saturnino et al., 2017).

Besides direct targeting of a brain oscillation with stimulation of the same frequency band, sometimes cross-frequency interactions are of interest or used to indirectly modulate an oscillation via stimulation of a different frequency band. For example, oscillations in the alpha and gamma range are well known to show an antagonistic relationship. When alpha oscillations increase, gamma oscillations are suppressed, and vice versa (Jensen & Mazaheri, 2010). Some authors therefore applied stimulation in the gamma frequency range to suppress alpha oscillations (Boyle & Frohlich, 2013). As increased activity in the alpha band is negatively correlated with vigilance. This antagonistic stimulation approach has recently been used to counteract increasing reaction times in a sustained attention task (Loffler et al., 2018). Along the same lines, rTMS in the alpha range has been observed to alter alpha-gamma cross-frequency coupling during a visual working memory task (Hamidi et al., 2009). An exceptionally sophisticated tACS cross-frequency protocol has been used to study theta-gamma coupling in a working memory task. As discussed earlier, short-term or working memory capacity is thought to depend on the number of gamma oscillations that fit into the positive half of a theta cycle (Lisman and Idiart, 1995; Lisman and Jensen, 2013), and can be increased by slowing the frequency of fronto-parietal theta oscillations (Vosskuhl et al., 2015). To further study the relationship of theta and gamma oscillations in working memory, Alekseichuk and colleagues (2016) applied tACS using short trains of a high-frequency stimulation at different gamma frequencies superimposed on either the positive or the negative half wave of a slow stimulation waveform in the theta range. In that study, only gamma stimulation applied during the positive but not during the negative half of the theta cycle fostered participants' working memory performance.

The aforementioned studies exemplify how brain stimulation can be used to study the role of various features of oscillatory brain activity for cognition. Especially tACS can be used very flexibly. As it allows high control over timing and waveform shapes, it can be used to study complex phase or cross-frequency relationships. In contrast, rTMS protocols allow to stimulate during specific, transient time periods of a cognitive process as it induces stronger/faster effects.

22.4 Combining Brain Stimulation and Neuroimaging

A major challenge in the context of brain stimulation is the measurement of stimulation effects. Ideally one would like to monitor the changes in brain activity elicited by stimulation that cause the behavioral effects. This would not only allow validation of whether a technique induced its effects on brain activity in the expected direction but also strengthen our knowledge about the general mechanisms of stimulation. Although there are elaborate theories about the mechanisms underlying stimulation effects backed up by evidence from animal models and computer simulations, direct observations into these mechanisms in humans are sparse and difficult to obtain. The application of strong magnetic fields or electric currents introduces massive distortions to electrophysiological recordings (e.g., M/EEG). Naturally, these distortions are several orders of magnitude larger than the concurrently recorded brain signals and in most cases spectrally overlap with the brain oscillation of interest (Kasten et al., 2018a).

In recent years, different strategies have been employed to measure effects of stimulation on human brain activity. A common approach is to discard the artifact-contaminated recordings obtained during stimulation and to focus on aftereffects in M/EEG. Such aftereffects are commonly observed after rTMS and tACS at different frequencies (Veniero et al., 2015) and have been related to sustained alterations of behavioral measures (Kasten and Herrmann, 2017). On the one hand, these after effects offer important evidence for the ability of brain stimulation techniques to modulate brain oscillations. Such sustained changes are of major interest, especially in the context of potential clinical applications. Only if the stimulation can elicit long-lasting changes of dysfunctional oscillatory activity can it offer real potential for the treatment of neurological or psychiatric disorders. On the other hand, the mechanisms of action underlying online effects and aftereffects may be different. Online effects of rTMS and tACS are usually assumed to result from entrainment of the endogenous brain oscillations, whereas offline effects are likely to result from mechanisms of synaptic plasticity (Vossen et al., 2015; Wischnewski et al., 2018; Zaehle et al., 2010). While there might be some relation between online effects and aftereffects (Helfrich et al., 2014b), inference about online effects from aftereffects is inherently difficult and cannot substitute direct measurements.

Some researchers have recorded brain signals during stimulation using measurement modalities that are less susceptible to distortions of the stimulation. For example, fMRI can be recorded during tACS with little to no distortions introduced to the signal (Antal et al., 2014; Vosskuhl et al., 2016). Initial studies were able to show that stimulation with tACS at different frequencies can modulate the blood-oxygen-level-dependent (BOLD) response (Violante et al., 2017; Vosskuhl et al., 2016) and connectivity profiles (Cabral-Calderin et al., 2016; Violante et al., 2017; Weinrich et al., 2017). This way, the authors were able to causally test relationships between brain oscillations and specific patterns of BOLD-signal activation previously observed in concurrent EEG-fMRI (e.g., Goldman et al., 2002; Laufs et al., 2003). A major drawback of this approach is, however, that the hemodynamic response measured in fMRI is a rather indirect measure to draw conclusions about effects of stimulation on oscillatory activity in the brain. In order to obtain direct insights to the effects occurring during stimulation, simultaneous recordings of electrophysiological signals with EEG and MEG and are ultimately required. In recent years different strategies have been developed with the ambitious goal to clean these recordings from the massive stimulation artifacts.

The magnetic fields generated by TMS are too strong to be measured in the MEG, as the pulses will harm the sensitive electronics of the systems. Concurrent recording of EEG signals during TMS application is generally feasible, albeit technically challenging. Each TMS pulse is accompanied by a couple different artifacts that need to be considered during the conductance of the measurements and data analysis (Ilmoniemi et al., 2015; Ilmoniemi & Kičić, 2010; Veniero et al., 2009). Although the TMS pulses themselves only last for less than a millisecond, the strong electromagnetic artifact can induce eddy currents to EEG leads that can saturate EEG amplifiers for several hundreds of milliseconds and polarize electrode contacts. Specialized TMS compatible EEG hardware has been developed to handle this strong distortion without saturation. With such specialized hardware, the duration of intervals contaminated with the TMS artifact can be reduced to ~5 ms (Veniero et al., 2009). In addition to the TMS pulse itself, recharging of capacitors in the TMS electronics can induce a second artifact in the EEG signal. Depending on the stimulation intensity, this artifact can follow the pulse by some tens of milliseconds (Veniero et al., 2009). Apart from electromagnetic artifacts, the strong stimulation has the potential to trigger different physiological responses like eye movements/blinks and muscle artifacts. Further, the strong magnetic force can cause vibration of the coil and click sounds which can elicit unwanted somatosensory or auditory evoked potentials (Ilmoniemi & Kičić, 2010; Rogasch et al., 2017). In order to obtain meaningful results from concurrent TMS-EEG, one needs to account for these artifacts—ideally by avoiding/reducing them when performing the experiment, for example, by carefully arranging the EEG cables relative to the coil or by slowing down the recharging times of the TMS device (Veniero et al., 2009). The remaining distortions in the signal have to be removed by advanced data processing procedures.

The data segment containing the massive, sharp artifact of the TMS pulse (lasting 10–25 ms after the pulse) is commonly removed from the data and subsequently replaced by a linear or cubic interpolation (Bergmann et al., 2012; Thut et al., 2011b). The same approach can be used to get rid of the recharger artifact following the pulse (Thut et al., 2011b). However, here the latency of the artifact has to be carefully identified as the artifact strongly depends on the TMS intensity (Veniero et al., 2009). To remove physiological artifacts from eye movements or muscle activity, blind-source separation methods such as principle component analysis (PCA), independent component analysis (ICA), or signal space projection are widely used (Ilmoniemi et al., 2015; Rogasch et al., 2017).

Transcranial alternating current stimulation can be measured using both EEG and MEG. Again, the stimulation contaminates the recorded signals with a strong electromagnetic artifact and recording hardware with sufficient dynamic range is needed to avoid saturation of the recordings. When combined with EEG, direct connection of stimulation and recording electrodes via bridges of gel or saline solution should be avoided. When used in the MEG, stimulation parameters should be carefully tested on a phantom head to rule out that the electromagnetic artifact imposes harm to the sensor array. In contrast to the transient stimulation pulses used in TMS, tACS applies a continuous alternating current. Consequently, recordings will not contain artifact-free segments as long as the stimulation is switched on, precluding the use of interpolation methods as with TMS. During the last couple of years different strategies have been employed aiming to suppress the tACS artifact from M/EEG recordings (Kasten & Herrmann, 2019).

In concurrent tACS-EEG, some authors created a template of the artifact waveform and subtracted it from the EEG recording (Dowsett & Herrmann, 2016; Helfrich et al., 2014b; Kohli & Casson, 2015; Voss et al., 2014). The method assumes that the stimulation artifact has a stable size and shape over time, while signals originating from the brain fluctuate randomly. To create the template, multiple EEG segments, time-locked to the same phase of the artifact waveform (e.g., the zero-crossing of the signal), are averaged. This way, brain activity represented in the segments averages out, while the shape and size of the artifact is retained. The subtraction of the template should in turn remove the artifact from the EEG signal, while leaving the superimposed brain activity intact. As the approach achieved non-optimal results, some authors subsequently applied PCA to remove residual artifacts from the data (Helfrich et al., 2014b).

In the MEG, spatial filtering approaches, such as synthetic aperture magnetometry (Soekadar et al., 2013) and linearly constrained minimum variance (LCMV) beamforming (Neuling et al., 2015), have been suggested to suppress artifacts from tACS and tDCS. Beamformers have been designed to separate signals originating from different directions and have widespread applications for example in radar and sonar technologies (Van Veen & Buckley, 1988). In neuroscience, beamformers are used to localize sources of brain activity seen in M/EEG signals and can work in both the time (Van Veen et al., 1997) and the frequency domains (Gross et al., 2001). The filters are designed to pass signals from a specific location in the brain, while attenuating signals from all other sources. To obtain a spatial map of brain activation, multiple filters with different spatial pass-bands are constructed on a predefined grid of possible source locations (Van Veen et al., 1997). The crucial feature of the LCMV beamformer in the context of concurrent tACS-MEG is its insensitivity to highly correlating sources (Neuling et al., 2015). The spatial filters of the LCMV beamformer are designed to minimize the variance of the output signal at each source location (hence the name). If two or more spatially distinct, highly correlating sources are present in the data, the common variance of the signals cancels to minimize the filter output (Van Veen et al., 1997). Such high correlations between distinct sources are unlikely to naturally occur in the brain and the LCMV beamformer is relatively robust to moderate correlations between sources (Van Veen et al., 1997). During concurrent tACS-MEG, the massive stimulation artifact propagates to virtually all sensors with high consistency. Thus, the beamformer can cancel out large proportions of the artifact waveform (Neuling et al., 2015). As a consequence, however, the artifact suppression capabilities of the beamformer are naturally limited by the degree to which the artifact signals are correlated (or uncorrelated) over the sensor array (Mäkelä et al., 2017).

Both methods, the template subtraction and the beamformer approach, assume a stationary, invariant artifact waveform, but this assumption has recently been challenged. Physiological processes such as heartbeat and respiration can lead to small changes in body impedance and elicit small head movements that can modify the size of the recorded artifact waveform compromising the artifact suppression capabilities of these methods (Noury et al., 2016; Noury & Siegel, 2017). These systematic changes in artifact size manifest in an amplitude modulation of the tACS waveform that causes side-bands around the main stimulation frequency, which survive artifact suppression attempts (Noury et al., 2016). Additionally, these processes may also affect artifact suppression performance directly at the stimulation frequency. Variations in artifact strength cannot be incorporated when constructing a template of the artifact to be subtracted from EEG data. As a consequence, any deviation of the artifact strength from the averaged template remains in the signal as a residual artifact. In a similar manner, artifact suppression capabilities of LCMV beamforming might be corrupted if the systematic changes reduce the signal correlation of the artifact waveform over sensors (Mäkelä et al., 2017).

While the presence of residual tACS artifacts hinders the proper analysis of stimulation effects on spontaneous brain oscillations, some authors argue that the methods attenuate the stimulation artifact sufficiently to analyze tACS effects on event-related oscillations (Kasten et al., 2018a; Neuling et al., 2017; Noury & Siegel, 2018), because the residual artifact cancels out if two intervals (e.g., a pre- and a post-stimulus interval) that contain a similar residual artifact are contrasted. Importantly, this approach can only work if the absolute difference of the conditions is computed. Relative measures that involve a division of one condition by the other, as done for example to compute event-related (de-) synchronization (Pfurtscheller & Lopes Da Silva, 1999), are vulnerable to systematic bias by the residual tACS artifact (Kasten et al., 2018a; Kasten & Herrmann, 2019). Further, the subtraction requires that the artifact size is uncorrelated with the task. Such systematic modulations of the artifact size may arise if a task elicits head movements or changes in body impedance (e.g., emotional stimuli) and should be ruled out by appropriate control analyses (Kasten et al., 2018a; Kasten & Herrmann, 2019).

In order to improve the performance of artifact cleaning approaches or to directly avoid artifact contamination within the frequency range of interest, the use of alternative stimulation waveforms such as sawtooth waves (Dowsett & Herrmann, 2016) or stimulation with an amplitude modulated, high frequency waveform (Witkowski et al., 2016) has been proposed (Figure 22.2 shows examples of both waveforms). Sawtooth waves contain strong harmonic peaks in the frequency spectrum that allow for easier detection and rejection of data segments contaminated by residual artifacts (Dowsett & Herrmann, 2016). Amplitude modulated waveforms consist of a high-frequency carrier signal, which is modulated in amplitude by the lower-frequency stimulation waveform. In principle, such a signal only contains power at the carrier frequency and two sidebands, but not at the frequency of the modulating waveform itself, thus shifting the stimulation artifact into higher frequencies and avoiding the spectral overlap between the brain signal of interest and the artifact (Witkowski et al., 2016). Recently, a computer simulation demonstrated that a cortical network, oscillating in the alpha frequency range, can be entrained to the modulation frequency of such a signal, although the effect was substantially weaker as compared to stimulation with a pure sine wave (Negahbani et al., 2018). Further, there are challenges to the assumption that M/EEG recordings of AM-tACS are completely artifact free in the range of the modulation frequency as nonlinear behavior of the involved hardware has been shown to reintroduce such artifacts (Kasten et al., 2018a; Minami & Amano, 2017).

22.5 Safety Aspects

In general, the application of rTMS and tACS is considered safe if applied within normal dosage ranges (intensities and durations) and by trained personnel. Expert congresses have yielded publications that detail guidelines for TMS and tES safety (Antal et al., 2017;Rossi et al. 2020). This section provides a brief introduction to safety aspects associated with rTMS and tACS. For details, the reader is referred to the aforementioned publications.

The most severe adverse effect associated with rTMS is the potential to induce seizures. Such events are reported extremely rarely and in most cases stimulation parameters were chosen outside the recommended dosage ranges. However, in some patient groups or under specific medication the threshold for seizures may be lowered (Rossi et al., 2020). Dosage limits for rTMS depend on the number of applied pulses, stimulation intensity, frequency, and inter-train intervals. Contraindication for the use of TMS is the presence of hardware containing metallic parts in the proximity of the TMS coil (e.g., cochlear implants, pulse generators, or medical pumps). The strong magnetic force may cause malfunction of the devices. When combined with EEG, rTMS can cause heating of the electrodes and an increase in the risk of skin burns. The use of smaller electrodes (pallet or annulus shaped electrodes), can greatly reduce the amount of heating (Ilmoniemi & Kičić, 2010).

Transcranial electrical stimulation is generally considered safe within the range of 0–4 mA stimulation intensity for <60 minutes over a single session (Antal et al., 2017). For tACS, stimulation durations of up to 45 minutes at 1.5 mA have been used without severe adverse effects (Laczó et al., 2012). The application of electric currents via scalp electrodes bears the risk of skin burns, if electrodes are incorrectly applied (e.g., drying of sponge electrodes or use of tap water instead of saline solution). So far, there are no confirmed incidence of seizures reported in the context of tES (Antal et al., 2017). In addition to adverse effects that impose serious safety issues, it should be acknowledged that both methods can cause discomfort for participants. The application of electric currents can lead to the sensation of tingling and itching or heating under the electrodes. Further, parts of the current may polarize cells of the retina and induce a flickering sensation (so-called phosphenes). The amount of visual and somatosensory sensation depends on stimulation intensities, frequencies, and electrode montages (Turi et al., 2013, 2014). Comparatively high intensities may even be painful for some participants. TMS can induce phosphenes and contractions of scalp muscles in the proximity of the stimulated brain area. In order to increase participant safety and comfort, standardized questionnaires are available to screen participants for risk factors (diseases, medication, etc.) and exclusion criteria (implants) and to evaluate commonly reported adverse effects (Antal et al., 2017; Brunoni et al., 2011; Rossi et al., 2009).

22.6 Challenges and Future Directions

Probing causal relationships between brain oscillations and cognitive processes by means of non-invasive brain stimulation is a rapidly growing and developing field. Methods to non-invasively modulate oscillatory activity in the brain offer promising pathways to advance basic scientific research as well as clinical practice. However, many of the fundamental mechanisms of the stimulation are still not fully understood and require further investigation. The development and improvement of measurement and analysis techniques for electrophysiological signals acquired during stimulation can strongly contribute to this knowledge.

An important issue concerning all non-invasive brain stimulation techniques is the variability of stimulation effects and its potential sources. A recent meta-analysis found that effects of tACS on cognitive measures to be in the small to moderate range (Schutter & Wischnewski, 2016) and several studies failed to induce effects in behavioral or EEG measures (e.g., Fekete et al., 2018; Stecher & Herrmann, 2018; Veniero et al., 2017). There is a wide range of potential factors that may affect stimulation success or even the direction of effects. For example, there is accumulating evidence that brain stimulation effects are state dependent (Feurra et al., 2013; Neuling et al., 2013; Ruhnau et al., 2016; Silvanto et al., 2008), with some brain states being more susceptible to stimulation than others. For example, when participants close their eyes, alpha oscillations tend to increase strongly. During such a state of strong activity, brain stimulation failed to further increase power in the alpha band (Neuling et al., 2013; Ruhnau et al., 2016). Nevertheless, some involvement of the stimulated oscillation in the a given task or stimulation setting seems necessary to see effects (Feurra et al., 2013). Such results indicate that the overall context of the stimulation may play a crucial role. Subtle differences in the context, such as the lightning conditions of the room, may modulate or mask stimulation effects (Stecher et al., 2017).

Apart from contextual influences, differences between individuals could potentially explain large proportions of effect variability. Of increasing interest are differences in individual anatomy. Humans differ with respect to skull thickness and folding of the cortex. Applying stimulation at standard locations (e.g., according to EEG electrode positions) can result in substantially different electric field patterns in the brain (Laakso et al., 2015). Recent work has shown that these e-field differences can account for a large amount of variability of brain stimulation effects (Antonenko et al., 2019; Kasten et al., 2019). Utilizing improved modelling software to individualize stimulation montages may increase the reliability of stimulation effects (Huang et al., 2018; Saturnino et al., 2019; Wagner et al., 2016). In the context of aftereffects, a long list of factors is known to have the potential to modulate the induction of cortical plasticity by means of non-invasive brain stimulation. Those factors include participants' sex, age, genetics, and use of medication or psychoactive substances (Ridding & Ziemann, 2010). Unfortunately, most of the work has been done in the context of tDCS and TMS protocols not targeting brain oscillations. Thus far, direct investigations into the role of these factors for modulatory effects of rTMS and tACS on brain oscillations is largely missing and requires more investigation. Individual factors may not only alter the induction of plasticity dependent aftereffects, but also influence the brain oscillation of interest. For example, nicotine has been shown to diminish or abolish the induction of aftereffects in the context of tDCS and TMS protocols (Grundey et al., 2012; Thirugnanasambandam et al., 2011). At the same time, nicotine can alter the frequency and amplitudes of brain oscillations in the alpha and theta range (Domino et al., 2009; Herning et al., 1983). Tobacco smoking initially reduces power of alpha and theta oscillations, while tobacco withdrawal is associated with a power increase (Herning et al., 1983). In the context of stimulation protocols targeting alpha or theta oscillations, such effects could severely confound experimental results. Strengthening the understanding of determinants of brain stimulation effects has the potential to reduce the variability of stimulation outcomes, for example by identifying participant (or patient) groups not responding to brain stimulation.

Dysfunctional patterns of oscillatory activity in the brain have been associated with a wide range of psychiatric and neurological disease (Herrmann & Demiralp, 2005; Uhlhaas & Singer, 2006, 2012). TACS is seen as an especially promising technique to restore such dysfunctional oscillations and reduce related symptoms as devices are small, portable, comparably cheap and relatively easy to apply after short training (Antal et al., 2017; Bikson et al., 2018). Further, the devices allow high levels of control over stimulation waveforms, enabling the design of stimulation protocols for very specific applications. For example, recently two studies applied tACS using the envelope of human speech to improve speech intelligibility, offering potential to combine brain stimulation with hearing aids (Riecke et al., 2018; Wilsch et al., 2018). There are also recent results from a first clinical trial, applying tACS to restore reduced power in the alpha band in order to reduce auditory hallucinations in patients with schizophrenia (Ahn et al., 2019; Mellin et al., 2018). While the authors find some indication that tACS may be capable of inducing long-term changes to dysfunctional oscillatory activity and lead to symptom improvement, a lot more research is needed to demonstrate its clinical value.

A disadvantage of rTMS and tACS is that both methods can only target superficial regions in the brain, but cannot reach areas in the depth of the brain without co-stimulating the overlaying cortex. Recently, stimulation with interfering electric fields (so-called temporal interference stimulation, TIS) has been proposed as a possible solution for this (Grossmann et al., 2017). TIS is based on the idea that alternating currents with slightly different high frequencies are injected to the brain via two pairs of electrodes. While superficial areas of the brain are stimulated at high frequencies outside of the physiological range, stimulation gives rise to an amplitude modulation or beat-frequency in regions where the two electric fields overlap (von Conta et al. 2021). Such amplitude-modulated stimulation waveforms can in principle interact with neural oscillations, however, substantially higher stimulation intensities may be required in comparison to conventional tACS (Negahbani et al. 2018, Esmaipour et al. 2020). While there is first evidence from computational modelling and from animal models, the effects of TIS in humans have so far only been demonstrated in one study applying TIS to superficial areas of the motor cortex (Ma et al. 2022). Overall, TIS has great potential to offer new applications and stimulation targets for non-invasive brain stimulation. However, more research is needed to establish the efficacy of TIS in humans, especially in targeting deep brain regions, and its underlying mechanisms.

22.7 Conclusions

Non-invasive stimulation to modulate oscillations in the human brain is increasingly popular in human neuroscience. This chapter presents approaches that have the potential to strengthen our knowledge about brain oscillations and their functional role for cognition, as they allow to probe the of causal relationships between different features of oscillatory activity and human cognition. Further, they may offer new pathways to treat neurological and psychiatric disease. However, more research is needed to explore the underlying mechanisms of the stimulation and understand the processes and factors determining stimulation success and the direction of effects. In particular, the improvement of analysis methods for concurrent recordings of brain activity during stimulation can strongly contribute to our knowledge in this area.

REFERENCES

Ahn, S., Mellin, J. M., Alagapan, S., Alexander, M. L., Gilmore, J. H., Jarskog, L. F., & Fröhlich, F. (2019). Targeting reduced neural oscillations in patients with schizophrenia by transcranial alternating current stimulation. *NeuroImage*, *186*, 126–136. doi:10.1016/j.neuroimage.2018.10.056

Alagapan, S., Schmidt, S. L., Lefebvre, J., Hadar, E., Shin, H. W., & Fröhlich, F. (2016). Modulation of cortical oscillations by low-frequency direct cortical stimulation is state-dependent. *PLOS Biology*, *14*, e1002424. doi:10.1371/journal.pbio.1002424

Alekseichuk, I., Turi, Z., Amador de Lara, G., Antal, A., & Paulus, W. (2016). Spatial working memory in humans depends on theta and high gamma synchronization in the prefrontal cortex. *Current Biology*, *26*, 1513–1521. doi:10.1016/j.cub.2016.04.035

Amassian, V. E., Cracco, R. Q., Maccabee, P. J., Cracco, J. B., Rudell, A., & Eberle, L. (1989). Suppression of visual perception by magnetic coil stimulation of human occipital cortex. *Electroencephalography and Clinical Neurophysiology*, *74*, 458–462. doi:10.1016/0168-5597(89)90036-1

Antal, A., Alekseichuk, I., Bikson, M., Brockmöller, J., Brunoni, A. R., Chen, R., ... Paulus, W. (2017). Low intensity transcranial electric stimulation: Safety, ethical, legal regulatory and application guidelines. *Clinical Neurophysiology*, *128*(9), 1774–1718. doi:10.1016/j.clinph.2017.06.001

Antal, A., Bikson, M., Datta, A., Lafon, B., Dechent, P., Parra, L. C., & Paulus, W. (2014). Imaging artifacts induced by electrical stimulation during conventional fMRI of the brain. *NeuroImage*, *85*, 1040–1047. doi:10.1016/j.neuroimage.2012.10.026

Antal, A. & Herrmann, C. S. (2016). Transcranial alternating current and random noise stimulation: Possible mechanisms. *Neural Plasticity*, *2016*, 3616807. doi:10.1155/2016/3616807

Antonenko, D., Thielscher, A., Saturnino, G. B., Aydin, S., Ittermann, B., Grittner, U., et al. (2019). Towards precise brain stimulation: Is electric field simulation related to neuromodulation? *Brain Stimulation*, *12*(5), 1159–1168. doi:10.1016/j.brs.2019.03.072

Baltus, A., Vosskuhl, J., Boetzel, C., & Herrmann, C. S. (2018a). Transcranial alternating current stimulation modulates auditory temporal resolution in elderly people. *European Journal of Neuroscience*, *51*(5), 1328–1338. doi:10.1111/ejn.13940

Baltus, A., Wagner, S., Wolters, C. H., & Herrmann, C. S. (2018b). Optimized auditory transcranial alternating current stimulation improves individual auditory temporal resolution. *Brain Stimulation*, *11*, 118–124. doi:10.1016/j.brs.2017.10.008

Barker, A. T., Jalinous, R., & Freeston, I. L. (1985). Non-invasive magnetic stimulation of human motor cortex. *The Lancet*, *1*(8437), 1106–1107. doi:10.1016/s0140-6736(85)92413-4

Bergmann, T. O., Karabanov, A., Hartwigsen, G., Thielscher, A., & Siebner, H. R. (2016). Combining non-invasive transcranial brain stimulation with neuroimaging and electrophysiology: Current approaches and future perspectives. *NeuroImage, 140*, 4–19. doi:10.1016/j.neuroimage.2016.02.012

Bergmann, T. O., Molle, M., Schmidt, M. A., Lindner, C., Marshall, L., Born, J., & Siebner, H. R. (2012). EEG-guided transcranial magnetic stimulation reveals rapid shifts in motor cortical excitability during the human sleep slow oscillation. *Journal of Neuroscience, 32*, 243–253. doi:10.1523/JNEUROSCI.4792-11.2012

Bikson, M., Brunoni, A. R., Charvet, L. E., Clark, V. P., Cohen, L. G., Deng, Z-D, ... Lisanby, S. H. (2018). Rigor and reproducibility in research with transcranial electrical stimulation: An NIMH-sponsored workshop. *Brain Stimulation, 11*, 465–480. doi:10.1016/j.brs.2017.12.008

Bikson, M., Inoue, M., Akiyama, H., Deans, J. K., Fox, J. E., Miyakawa, H., & Jefferys, J. G. R. (2004). Effects of uniform extracellular DC electric fields on excitability in rat hippocampal slices in vitro. *Journal of Physiology, 557*, 175–190. doi:10.1113/jphysiol.2003.055772

Bindman, L. J., Lippold, O. C. J., & Redfearn, J. W. T. (1964). The action of brief polarizing currents on the cerebral cortex of the rat (1) during current flow and (2) in the production of long-lasting after-effects. *Journal of Physiology, 172*, 369–382. doi:10.1113/jphysiol.1964.sp007425

Boyle, M. R. & Frohlich, F. (2013). EEG feedback-controlled transcranial alternating current stimulation. In L. S. Colzato (Ed.), *2013 6th International IEEE/EMBS Conference on Neural Engineering (NER)* (pp. 140–143). IEEE. doi:10.1109/NER.2013.6695891

Braun, V., Sokoliuk, R., & Hanslmayr, S. (2017). On the effectiveness of event-related beta tACS on episodic memory formation and motor cortex excitability. *Brain Stimulation, 10*, 910–918. doi:10.1016/j.brs.2017.04.129

Brunoni, A. R., Amadera, J., Berbel, B., Volz, M. S., Rizzerio, B. G., & Fregni, F. (2011). A systematic review on reporting and assessment of adverse effects associated with transcranial direct current stimulation. *International Journal of Neuropsychopharmacology, 14*, 1133–1145. doi:10.1017/S1461145710001690

Cabral-Calderin, Y., Williams, K. A., Opitz, A., Dechent, P., & Wilke, M. (2016). Transcranial alternating current stimulation modulates spontaneous low frequency fluctuations as measured with fMRI. *NeuroImage, 141*, 88–107. doi:10.1016/j.neuroimage.2016.07.005

Cardin, J. A., Carlén, M., Meletis, K., Knoblich, U., Zhang, F., Deisseroth, K., Tsai, L-H., & Moore, C. I. (2009). Driving fast-spiking cells induces gamma rhythm and controls sensory responses. *Nature, 459*, 663–667. doi:10.1038/nature08002

Cecere, R., Rees, G., and Romei, V. (2015). Individual differences in alpha frequency drive crossmodal illusory perception. *Current Biology, 25*, 231–235. doi:10.1016/j.cub.2014.11.034

Creutzfeldt, O. D., Fromm, G. H., & Kapp, H. (1962). Influence of transcortical d-c currents on cortical neuronal activity. *Experimental Neurology, 5*, 436–452. doi:10.1016/0014-4886(62)90056-0

Deans, J. K., Powell, A. D., & Jefferys, J. G. R. (2007). Sensitivity of coherent oscillations in rat hippocampus to AC electric fields. *Journal of Physiology, 583*, 555–565. doi:10.1113/jphysiol.2007.137711

Domino, E. F., Ni, L., Thompson, M., Zhang, H., Shikata, H., Fukai, H., Sakaki, T., & Ohya, I. (2009). Tobacco smoking produces widespread dominant brain wave alpha frequency increases. *International Journal of Psychophysiology, 74*, 192–198. doi:10.1016/j.ijpsycho.2009.08.011

Dowsett, J. & Herrmann, C. S. (2016). Transcranial alternating current stimulation with sawtooth waves: Simultaneous stimulation and EEG recording. *Frontiers in Human Neuroscience, 10*, 1–10. doi:10.3389/fnhum.2016.00135

Esmaeilpour, Zeinab, Greg Kronberg, Davide Reato, Lucas C. Parra, and Marom Bikson. "Temporal Interference Stimulation Targets Deep Brain Regions by Modulating Neural Oscillations." *Brain Stimulation*, 14, no. 1 (January 1, 2021): 55–65. https://doi.org/10.1016/j.brs.2020.11.007.

Fekete, T., Nikolaev, A. R., Knijf, F. De, Zharikova, A., & van Leeuwen, C. (2018). Multi-electrode alpha tACS during varying background tasks fails to modulate subsequent alpha power. *Frontiers in Neuroscience*, *12*, 428. doi:10.3389/fnins.2018.00428

Feurra, M., Pasqualetti, P., Bianco, G., Santarnecchi, E., Rossi, A., & Rossi, S. (2013). State-dependent effects of transcranial oscillatory currents on the motor system: What you think matters. *Journal of Neuroscience*, *33*, 17483–9. doi:10.1523/JNEUROSCI.1414-13.2013

Fröhlich, F. & McCormick, D. A. (2010). Endogenous electric fields may guide neocortical network activity. *Neuron*, *67*, 129–143. doi:10.1016/j.neuron.2010.06.005

Goldman, R. I., Stern, J. M., Engel, J., & Cohen, M. S. (2002). Simultaneous EEG and fMRI of the alpha rhythm. *Neuroreport*, *13*, 2487–2492. doi:10.1097/00001756-200212200-00022

Gross, J., Kujala, J., Hamalainen, M., Timmermann, L., Schnitzler, A., & Salmelin, R. (2001). Dynamic imaging of coherent sources: Studying neural interactions in the human brain. *Proceedings of the National Academy of Sciences of the United States of America*, *98*, 694–699. doi:10.1073/pnas.98.2.694

Grossman, Nir, David Bono, Nina Dedic, Suhasa B. Kodandaramaiah, Andrii Rudenko, Ho-Jun Suk, Antonino M. Cassara, et al. "Noninvasive Deep Brain Stimulation via Temporally Interfering Electric Fields." *Cell*, 169, 6 (2017): 1029–1041.e16. https://doi.org/10.1016/j.cell.2017.05.024.

Grundey, J., Thirugnanasambandam, N., Kaminsky, K., Drees, A., Skwirba, A. C., Lang, N., ... Nitsche, M. A. (2012). Rapid effect of nicotine intake on neuroplasticity in non-smoking humans. *Frontiers in Pharmacology*, *3*, 1–9. doi:10.3389/fphar.2012.00186

Gundlach, C., Müller, M. M., Nierhaus, T., Villringer, A., & Sehm, B. (2016). Phasic modulation of human somatosensory perception by transcranially applied oscillating currents. *Brain Stimulation*, *9*, 712–719. doi:10.1016/j.brs.2016.04.014

Hallett, M. (2000). Transcranial magnetic stimulation and the human brain. *Nature*, *406*, 147–150. doi:10.1038/35018000

Hamidi, M., Slagter, H. A., Tononi, G., & Postle, B. R. (2009). Repetitive transcranial magnetic stimulation affects behavior by biasing endogenous cortical oscillations. *Frontiers in Integrative Neuroscience*, *3*, 14. doi:10.3389/neuro.07.014.2009

Hanslmayr, S., Matuschek, J., & Fellner, M. C. (2014). Entrainment of prefrontal beta oscillations induces an endogenous echo and impairs memory formation. *Current Biology*, *24*, 904–909. doi:10.1016/j.cub.2014.03.007

Helfrich, R. F., Knepper, H., Nolte, G., Strüber, D., Rach, S., Herrmann, C. S., et al. (2014a). Selective modulation of interhemispheric functional connectivity by HD-tACS shapes perception. *PLoS Biology*, *12*(12), e1002031. doi:10.1371/journal.pbio.1002031

Helfrich, R. F., Schneider, T. R., Rach, S., Trautmann-Lengsfeld, S. A., Engel, A. K., & Herrmann, C. S. (2014b). Entrainment of brain oscillations by transcranial alternating current stimulation. *Current Biology*, *24*, 333–339. doi:10.1016/j.cub.2013.12.041

Herning, R. I., Jones, R. T., & Bachman, J. (1983). EEG changes during tobacco withdrawal. *Psychophysiology*, *20*, 507–512. doi:10.1111/j.1469-8986.1983.tb03004.x

Herrmann, C. S. (2001). Human EEG responses to 1–100 Hz flicker: Resonance phenomena in visual cortex and their potential correlation to cognitive phenomena. *Experimental Brain Research*, *137*, 346–53. doi:10.1007/s002210100682

Herrmann, C. S. & Demiralp, T. (2005). Human EEG gamma oscillations in neuropsychiatric disorders. *Clinical Neurophysiology, 116*, 2719–2733. doi:10.1016/j.clinph.2005.07.007

Herrmann, C. S., Murray, M. M., Ionta, S., Hutt, A., & Lefebvre, J. (2016a). Shaping intrinsic neural oscillations with periodic stimulation. *Journal of Neuroscience. 36*, 5328–5337. doi:10.1523/JNEUROSCI.0236-16.2016

Herrmann, C. S., Rach, S., Neuling, T., and Strüber, D. (2013). Transcranial alternating current stimulation: A review of the underlying mechanisms and modulation of cognitive processes. *Frontiers in Human Neuroscience, 7*, 1–13. doi:10.3389/fnhum.2013.00279

Herrmann, C. S., Strüber, D., Helfrich, R. F., & Engel, A. K. (2016b). EEG oscillations: From correlation to causality. *International Journal of Psychophysiology, 103*, 12–21. doi:10.1016/j.ijpsycho.2015.02.003

Huang, Y., Datta, A., Bikson, M., & Parra, L. C. (2017a). Realistic vOlumetric-Approach to Simulate Transcranial Electric Stimulation – ROAST – a fully automated open-source pipeline. *bioRxiv* [online]. doi:10.1101/217331

Huang, Y., Liu, A. A., Lafon, B., Friedman, D., Dayan, M., Wang, X., ... Parra, L. C. (2017b). Measurements and models of electric fields in the in vivo human brain during transcranial electric stimulation. *Elife, 6*, e18834. doi:10.7554/eLife.18834

Huang, Y., Thomas, C., Datta, A., & Parra, L. C. (2018). Optimized tDCS for targeting multiple brain regions: An integrated implementation. *2018 40th Annual International Conference of the IEEE Engineering in Medicine and Biology Society (EMBC) (IEEE), 2018*, 3545–3548. doi:10.1109/EMBC.2018.8513034

Ilmoniemi, R. J., Hernandez-Pavon, J. C., Makela, N. N., Metsomaa, J., Mutanen, T. P., Stenroos, M., & Sarvas, J. (2015). Dealing with artifacts in TMS-evoked EEG. *2015 37th Annual International Conference of the IEEE Engineering in Medicine and Biology Society (EMBC), 2015*, 230–233. doi:10.1109/EMBC.2015.7318342

Ilmoniemi, R. J. & Kičić, D. (2010). Methodology for combined TMS and EEG. *Brain Topography, 22*, 233–248. doi:10.1007/s10548-009-0123-4

Jefferys, J. G. (1981). Influence of electric fields on the excitability of granule cells in guinea-pig hippocampal slices. *Journal of Physiology, 319*, 143–52. doi:10.1113/jphysiol.1981.sp013897

Jensen, O. & Mazaheri, A. (2010). Shaping functional architecture by oscillatory alpha activity: gating by inhibition. *Frontiers in Human Neuroscience, 4*, 186. doi:10.3389/fnhum.2010.00186

Kasten, F. H., Dowsett, J., & Herrmann, C. S. (2016). Sustained aftereffect of α-tACS lasts up to 70 min after stimulation. *Frontiers in Human Neuroscience, 10*, 245. doi:10.3389/fnhum.2016.00245

Kasten, F. H., Duecker, K., Maack, M. C., Meiser, A., & Herrmann, C. S. (2019). Integrating electric field modeling and neuroimaging to explain inter-individual variability of tACS effects. *Nature Communications, 10*, art. no. 5427. doi:10.1038/s41467-019-13417-6

Kasten, F. H. & Herrmann, C. S. (2017). Transcranial alternating current stimulation (tACS) enhances mental rotation performance during and after stimulation. *Frontiers in Human Neuroscience, 11*, 1–16. doi:10.3389/fnhum.2017.00002

Kasten, F. H. & Herrmann, C. S. (2019). Recovering brain dynamics during concurrent tACS-M/EEG: An overview of analysis approaches and their methodological and interpretational pitfalls. *Brain Topography*. doi:10.1007/s10548-019-00727-7

Kasten, Florian H., Tea Wendeln, Heiko I. Stecher, and Christoph S. Herrmann. "Hemisphere-Specific, Differential Effects of Lateralized, Occipital–Parietal α- versus γ-TACS on

Endogenous but Not Exogenous Visual-Spatial Attention." *Scientific Reports,* 10, no. 1 (December 2020): 12270. https://doi.org/10.1038/s41598-020-68992-2.

Kasten, F. H., Maess, B., & Herrmann, C. S. (2018a). Facilitated event-related power modulations during transcranial alternating current stimulation (tACS) revealed by concurrent tACS-MEG. *eNeuro* [online] 5(3). doi:10.1523/ENEURO.0069-18.2018

Kasten, F. H., Negahbani, E., Fröhlich, F., & Herrmann, C. S. (2018b). Non-linear transfer characteristics of stimulation and recording hardware account for spurious low-frequency artifacts during amplitude modulated transcranial alternating current stimulation (AM-tACS). *NeuroImage, 179,* 134–143. doi:10.1016/j.neuroimage.2018.05.068

Klimesch, W., Sauseng, P., & Gerloff, C. (2003). Enhancing cognitive performance with repetitive transcranial magnetic stimulation at human individual alpha frequency. *European Journal of Neuroscience, 17,* 1129–1133. doi:10.1046/j.1460-9568.2003.02517.x

Klink, Katharina, Sven Paßmann, Florian H. Kasten, and Jessica Peter. "The Modulation of Cognitive Performance with Transcranial Alternating Current Stimulation: A Systematic Review of Frequency-Specific Effects." *Brain Sciences,* 10, no. 12 (December 2020): 932. https://doi.org/10.3390/brainsci10120932.

Klomjai, W., Katz, R., & Lackmy-Vallée, A. (2015). Basic principles of transcranial magnetic stimulation (TMS) and repetitive TMS (rTMS). *Annals of Physical and Rehabilitation Medicine, 58,* 208–213. doi:10.1016/j.rehab.2015.05.005

Kohli, S. & Casson, A. J. (2015). Removal of transcranial a.c. current stimulation artifact from simultaneous EEG recordings by superposition of moving averages. *2015 37th Annual International Conference of the IEEE Engineering in Medicine and Biology Society (EMBC), 2015,* 3436–3439. doi:10.1109/EMBC.2015.7319131

Kopp, C., Rudolph, U., Low, K., & Tobler, I. (2004). Modulation of rhythmic brain activity by diazepam: $GABA_A$receptor subtype and state specificity. *Proceedings of the National Academy of Sciences of the United States of America, 101,* 3674–3679. doi:10.1073/pnas.0306975101

Krause, M. R., Vieira, P. G., Csorba, B. A., Pilly, P. K., & Pack, C. C. (2019). Transcranial alternating current stimulation entrains single-neuron activity in the primate brain. *Proceedings of the National Academy of Sciences of the United States of America, 116,* 5747–5755. doi:10.1073/pnas.1815958116

Laakso, I., Tanaka, S., Koyama, S., De Santis, V., & Hirata, A. (2015). Inter-subject variability in electric fields of motor cortical tDCS. *Brain Stimulation, 8,* 906–913. doi:10.1016/j.brs.2015.05.002

Laczó, B., Antal, A., Niebergall, R., Treue, S., & Paulus, W. (2012). Transcranial alternating stimulation in a high gamma frequency range applied over V1 improves contrast perception but does not modulate spatial attention. *Brain Stimulation,* 5, 484–491. doi:10.1016/j.brs.2011.08.008

Laufs, H., Kleinschmidt, A., Beyerle, A., Eger, E., Salek-Haddadi, A., Preibisch, C., & Krakow, K. (2003). EEG-correlated fMRI of human alpha activity. *NeuroImage, 19,* 1463–1476. doi:10.1016/S1053-8119(03)00286-6

Lisman, J. E. & Idiart, M. A. (1995). Storage of 7+/-2 short-term memories in oscillatory subcycles. *Science, 267*(5203), 1512–1515. doi:10.1126/science.7878473

Lisman, J. E. & Jensen, O. (2013). The theta-gamma neural code. *Neuron, 77,* 1002–1016. doi:10.1016/j.neuron.2013.03.007

Loffler, B. S., Stecher, H. I., Fudickar, S., de Sordi, D., Otto-Sobotka, F., Hein, A., & Herrmann, C. S. (2018). Counteracting the slowdown of reaction times in a vigilance experiment with

40-Hz Transcranial alternating current stimulation. *EEE Transactions on Neural Systems and Rehabilitation* Engineering, *26*(10), 2053–2061. doi:10.1109/TNSRE.2018.2869471

Lustenberger, C., Boyle, M. R., Foulser, A. A., Mellin, J. M., & Fröhlich, F. (2015). Functional role of frontal alpha oscillations in creativity. *Cortex*, *67*, 74–82. doi:10.1016/j.cortex.2015.03.012

Ma, R., Xia, X., Zhang, W., Lu, Z., Wu, Q., Cui, J., Song, H., Fan, C., Chen, X., Zha, R., Wei, J., Ji, G.-J., Wang, X., Qiu, B., & Zhang, X. (2022). High Gamma and Beta Temporal Interference Stimulation in the Human Motor Cortex Improves Motor Functions. *Frontiers in Neuroscience*, 15. https://www.frontiersin.org/article/10.3389/fnins.2021.800436

Mäkelä, N., Sarvas, J., & Ilmoniemi, R. J. (2017). Proceedings #17. A simple reason why beamformer may (not) remove the tACS-induced artifact in MEG. *Brain Stimulation*, *10*, e66–e67. doi:10.1016/j.brs.2017.04.110

Markram, H., Lübke, J., Frotscher, M., & Sakmann, B. (1997). Regulation of synaptic efficacy by coincidence of postsynaptic APs and EPSPs. *Science*, *275*, 213–5. doi:10.1126/science.275.5297.213

Marshall, L., Helgadóttir, H., Mölle, M., & Born, J. (2006). Boosting slow oscillations during sleep potentiates memory. *Nature*, *444*, 610–613. doi:10.1038/nature05278

Mathewson, K. E., Gratton, G., Fabiani, M., Beck, D. M., & Ro, T. (2009). To see or not to see: Prestimulus phase predicts visual awareness. *Journal of Neuroscience*, *29*, 2725–2732. doi:10.1523/JNEUROSCI.3963-08.2009

Mellin, J. M., Alagapan, S., Lustenberger, C., Lugo, C. E., Alexander, M. L., Gilmore, J. H., Jarskog, L. F., & Fröhlich, F. (2018). Randomized trial of transcranial alternating current stimulation for treatment of auditory hallucinations in schizophrenia. *European Psychiatry*, *51*, 25–33. doi:10.1016/j.eurpsy.2018.01.004

Miller, G. A. (1956). The magical number seven, plus or minus two: Some limits on our capacity for processing information. *Psychology Review*, *63*, 81–97. doi:10.1037/h0043158

Minami, S. & Amano, K. (2017). Illusory jitter perceived at the frequency of alpha oscillations. *Current Biology*, *27*, 2344-2351.e4. doi:10.1016/j.cub.2017.06.033

Miranda, P. C., Lomarev, M., & Hallett, M. (2006). Modeling the current distribution during transcranial direct current stimulation. *Clinical Neurophysiology*, *117*, 1623–1629. doi:10.1016/j.clinph.2006.04.009

Negahbani, E., Kasten, F. H., Herrmann, C. S., & Frohlich, F. (2018). Targeting alpha-band oscillations in a cortical model with amplitude-modulated high-frequency transcranial electric stimulation. *NeuroImage*, *173*, 3–12. doi:10.1016/j.neuroimage.2018.02.005

Neuling, T., Rach, S., & Herrmann, C. S. (2013). Orchestrating neuronal networks: Sustained after-effects of transcranial alternating current stimulation depend upon brain states. *Frontiers in Human Neuroscience*, *7*, 161. doi:10.3389/fnhum.2013.00161

Neuling, T., Rach, S., Wagner, S., Wolters, C. H., & Herrmann, C. S. (2012a). Good vibrations: Oscillatory phase shapes perception. *NeuroImage*, *63*, 771–778. doi:10.1016/j.neuroimage.2012.07.024

Neuling, T., Ruhnau, P., Fuscà, M., Demarchi, G., Herrmann, C. S., & Weisz, N. (2015). Friends, not foes: Magnetoencephalography as a tool to uncover brain dynamics during transcranial alternating current stimulation. *NeuroImage*, *118*, 406–413. doi:10.1016/j.neuroimage.2015.06.026

Neuling, T., Ruhnau, P., Weisz, N., Herrmann, C. S., & Demarchi, G. (2017). Faith and oscillations recovered: On analyzing EEG/MEG signals during tACS. *NeuroImage*, *147*, 960–963. doi:10.1016/j.neuroimage.2016.11.022

Neuling, T., Wagner, S., Wolters, C. H., Zaehle, T., & Herrmann, C. S. (2012b). Finite-element model predicts current density distribution for clinical applications of tDCS and tACS. *Frontiers in Psychiatry*, *3*, 1–10. doi:10.3389/fpsyt.2012.00083

Nitsche, M. A, Fricke, K., Henschke, U., Schlitterlau, A., Liebetanz, D., Lang, N., et al. (2003). Pharmacological modulation of cortical excitability shifts induced by transcranial direct current stimulation in humans. *Journal of Physiology*, *553*, 293–301. doi:10.1113/jphysiol.2003.049916

Notbohm, A., Kurths, J., & Herrmann, C. S. (2016). Modification of brain oscillations via rhythmic light stimulation provides evidence for entrainment but not for superposition of event-related responses. *Frontiers in Human Neuroscience*, *10*, 10. doi:10.3389/fnhum.2016.00010

Noury, N., Hipp, J. F., & Siegel, M. (2016). Physiological processes non-linearly affect electrophysiological recordings during transcranial electric stimulation. *NeuroImage*, *140*, 99–109. doi:10.1016/j.neuroimage.2016.03.065

Noury, N. & Siegel, M. (2017). Phase properties of transcranial electrical stimulation artifacts in electrophysiological recordings. *NeuroImage*, *158*, 406–416. doi:10.1016/j.neuroimage.2017.07.010

Noury, N. & Siegel, M. (2018). Analyzing EEG and MEG signals recorded during tES, a reply. *NeuroImage*, *167*, 53–61. doi:10.1016/j.neuroimage.2017.11.023

Novembre, G., Knoblich, G., Dunne, L., & Keller, P. E. (2017). Interpersonal synchrony enhanced through 20 Hz phase-coupled dual brain stimulation. *Social Cognitive and Affective Neuroscience*, *12*(4), 662–670. doi:10.1093/scan/nsw172.

Opitz, A., Falchier, A., Yan, C. G., Yeagle, E. M., Linn, G. S., Megevand, P., et al. (2016). Spatiotemporal structure of intracranial electric fields induced by transcranial electric stimulation in humans and nonhuman primates. *Scientific Reports*, *6*, 1–11. doi:10.1038/srep31236

Ozen, S., Sirota, A., Belluscio, M. A., Anastassiou, C. A., Stark, E., Koch, C., et al. (2010). Transcranial electric stimulation entrains cortical neuronal populations in rats. *Journal of Neuroscience*, *30*, 11476–11485. doi:10.1523/JNEUROSCI.5252-09.2010

Palva, J. M., Palva, S., & Kaila, K. (2005). Phase synchrony among neuronal oscillations in the human cortex. *Journal of Neuroscience*, *25*, 3962–72. doi:10.1523/JNEUROSCI.4250-04.2005

Pfurtscheller, G., & Lopes Da Silva, F. H. (1999). Event-related EEG/MEG synchronization and desynchronization: Basic principles. *Clinical Neurophysiology*, *110*, 1842–1857. doi:10.1016/S1388-2457(99)00141-8

Pikovsky, A., Rosenblum, M., & Kurths, J. (2003). *Synchronization: A Universal concept in nonlinear sciences*. Cambridge University Press. doi:10.1063/1.1554136

Polanía, R., Nitsche, M. A., Korman, C., Batsikadze, G., & Paulus, W. (2012). The importance of timing in segregated theta phase-coupling for cognitive performance. *Current Biology*, *22*, 1314–1318. doi:10.1016/j.cub.2012.05.021

Reato, D., Rahman, A., Bikson, M., & Parra, L. C. (2010). Low-intensity electrical stimulation affects network dynamics by modulating population rate and spike timing. *Journal of Neuroscience*, *30*, 15067–15079. doi:10.1523/JNEUROSCI.2059-10.2010.

Reato, D., Rahman, A., Bikson, M., & Parra, L. C. (2013). Effects of weak transcranial alternating current stimulation on brain activity-a review of known mechanisms from animal studies. *Frontiers in Human Neuroscience*, *7*, 687. doi:10.3389/fnhum.2013.00687

Ridding, M. C. & Ziemann, U. (2010). Determinants of the induction of cortical plasticity by non-invasive brain stimulation in healthy subjects. *Journal of Physiology, 588*, 2291–304. doi:10.1113/jphysiol.2010.190314

Riecke, L., Formisano, E., Herrmann, C. S., & Sack, A. T. (2015). 4-Hz transcranial alternating current stimulation phase modulates hearing. *Brain Stimulation, 8*, 777–783. doi:10.1016/j.brs.2015.04.004

Riecke, L., Formisano, E., Sorger, B., Başkent, D., & Gaudrain, E. (2018). Neural entrainment to speech modulates speech intelligibility. *Current Biology, 28*, 161-169.e5. doi:10.1016/j.cub.2017.11.033

Rogasch, N. C., Sullivan, C., Thomson, R. H., Rose, N. S., Bailey, N. W., Fitzgerald, P. B., et al. (2017). Analysing concurrent transcranial magnetic stimulation and electroencephalographic data: A review and introduction to the open-source TESA software. *NeuroImage, 147*, 934–951. doi:10.1016/j.neuroimage.2016.10.031

Romei, V., Gross, J., & Thut, G. (2010). On the role of prestimulus alpha rhythms over occipito-parietal areas in visual input regulation: Correlation or causation? *Journal of Neuroscience, 30*, 8692–8697. doi:10.1523/JNEUROSCI.0160-10.2010

Rossi, Simone, Andrea Antal, Sven Bestmann, Marom Bikson, Carmen Brewer, Jürgen Brockmöller, Linda L. Carpenter, et al. "Safety and Recommendations for TMS Use in Healthy Subjects and Patient Populations, with Updates on Training, Ethical and Regulatory Issues: Expert Guidelines." *Clinical Neurophysiology*, 132, no. 1 (January 1, 2021): 269–306. https://doi.org/10.1016/j.clinph.2020.10.003.

Ruhnau, P., Neuling, T., Fuscá, M., Herrmann, C. S., Demarchi, G., & Weisz, N. (2016). Eyes wide shut: Transcranial alternating current stimulation drives alpha rhythm in a state dependent manner. *Scientific Reports, 6*, 27138. doi:10.1038/srep27138

Saturnino, G. B., Madsen, K. H., Siebner, H. R., & Thielscher, A. (2017). How to target inter-regional phase synchronization with dual-site transcranial alternating current stimulation. *NeuroImage, 163*, 68–80. doi:10.1016/j.neuroimage.2017.09.024

Saturnino, G. B., Siebner, H. R., Thielscher, A., & Madsen, K. H. (2019). Accessibility of cortical regions to focal TES: Dependence on spatial position, safety, and practical constraints. *NeuroImage, 203*, 116–183. doi:10.1016/j.neuroimage.2019.116183

Sauseng, P., Klimesch, W., Heise, K. F., Gruber, W. R., Holz, E., Karim, A. A., et al. (2009). Brain oscillatory substrates of visual short-term memory capacity. *Current Biology, 19*, 1846–1852. doi:10.1016/j.cub.2009.08.062

Schindler, K., Nyffeler, T., Wiest, R., Hauf, M., Mathis, J., Hess, C. W., et al. (2008). Theta burst transcranial magnetic stimulation is associated with increased EEG synchronization in the stimulated relative to unstimulated cerebral hemisphere. *Neuroscience Letters, 436*, 31–34. doi:10.1016/j.neulet.2008.02.052

Schutter, D. J. L. G. (2016). Cutaneous retinal activation and neural entrainment in transcranial alternating current stimulation: A systematic review. *NeuroImage, 140*, 83–88. doi:10.1016/j.neuroimage.2015.09.067

Schutter, D. J. L. G. & Wischnewski, M. (2016). A meta-analytic study of exogenous oscillatory electric potentials in neuroenhancement. *Neuropsychologia, 86*, 110–118. doi:10.1016/j.neuropsychologia.2016.04.011

Schutter, D. J., van Honk, J., D'Alfonso, A. A., Postma, A., & de Haan, E. H. (2001). Effects of slow rTMS at the right dorsolateral prefrontal cortex on EEG asymmetry and mood. *Neuroreport, 12*, 445–7. doi:10.1097/00001756-200103050-00005

Shams, L., Kamitani, Y., & Shimojo, S. (2002). Visual illusion induced by sound. *Cognitive Brain Research*, *14*, 147–152. doi:10.1016/S0926-6410(02)00069-1

Siegel, M., Donner, T. H., & Engel, A. K. (2012). Spectral fingerprints of large-scale neuronal interactions. *Nature Reviews Neuroscience*, *13*, 121–134. doi:10.1038/nrn3137

Silvanto, J., Muggleton, N., & Walsh, V. (2008). State-dependency in brain stimulation studies of perception and cognition. *Trends in Cognitive Sciences*, *12*, 447–454. doi:10.1016/j.tics.2008.09.004

Soekadar, S. R., Witkowski, M., Cossio, E. G., Birbaumer, N., Robinson, S. E., & Cohen, L. G. (2013). In vivo assessment of human brain oscillations during application of transcranial electric currents. *Nature Communications*, *4*, 2032. doi:10.1038/ncomms3032

Sohal, V. S. (2012). Insights into cortical oscillations arising from optogenetic studies. *Biol. Psychiatry* *71*, 1039–1045. doi:10.1016/j.biopsych.2012.01.024

Stecher, H. I. & Herrmann, C. S. (2018). Absence of alpha-tACS aftereffects in darkness reveals importance of taking derivations of stimulation frequency and individual alpha variability into account. *Frontiers in Psychology*, *9*, 1–9. doi:10.3389/fpsyg.2018.00984

Stecher, H. I., Pollok, T. M., Strüber, D., Sobotka, F., & Christoph, S. (2017). Ten minutes of α-tACS and ambient illumination independently modulate EEG α-power. *Frontiers in Human Neuroscience*, *11*. doi:10.3389/fnhum.2017.00257

Stonkus, R., Braun, V., Kerlin, J. R., Volberg, G., & Hanslmayr, S. (2016). Probing the causal role of prestimulus interregional synchrony for perceptual integration via tACS. *Scientific Reports*, *6*, 32065. doi:10.1038/srep32065

Strüber, D., Rach, S., Neuling, T., Herrmann, C. S., & Kar, K. (2015). On the possible role of stimulation duration for after-effects of transcranial alternating current stimulation. *Frontiers in Cellular Neuroscience*, *9*, 311. doi:10.3389/fnsys.2015.00148

Strüber, D., Rach, S., Trautmann-Lengsfeld, S. A., Engel, A. K., & Herrmann, C. S. (2014). Antiphasic 40 Hz oscillatory current stimulation affects bistable motion perception. *Brain Topography*, *27*, 158–171. doi:10.1007/s10548-013-0294-x

Szymanski, C., Müller, V., Brick, T. R., von Oertzen, T., & Lindenberger, U. (2017). Hyper-transcranial alternating current stimulation: Experimental manipulation of inter-brain synchrony. *Frontiers in Human Neuroscience*, *11*, 1–15. doi:10.3389/fnhum.2017.00539

Thielscher, A., Antunes, A., & Saturnino, G. B. (2015). Field modeling for transcranial magnetic stimulation: A useful tool to understand the physiological effects of TMS? *2015 37th Annual International Conference of the IEEE Engineering in Medicine and Biology Society (EMBC)*, *2015*, 222–225. doi:10.1109/EMBC.2015.7318340

Thirugnanasambandam, N., Grundey, J., Adam, K., Drees, A., Skwirba, A. C., Lang, N., … Nitsche, M. A. (2011). Nicotinergic impact on focal and non-focal neuroplasticity induced by non-invasive brain stimulation in non-smoking humans. *Neuropsychopharmacology*, *36*, 879–86. doi:10.1038/npp.2010.227

Thut, G., Schyns, P. G., & Gross, J. (2011a). Entrainment of perceptually relevant brain oscillations by non-invasive rhythmic stimulation of the human brain. *Frontiers in Psychology*, *2*, 1–10. doi:10.3389/fpsyg.2011.00170

Thut, G., Veniero, D., Romei, V., Miniussi, C., Schyns, P., & Gross, J. (2011b). Rhythmic TMS causes local entrainment of natural oscillatory signatures. *Current Biology*, *21*, 1176–1185. doi:10.1016/j.cub.2011.05.049

Turi, Z., Ambrus, G. G., Ho, K.-A., Sengupta, T., Paulus, W., & Antal, A. (2014). When size matters: Large electrodes induce greater stimulation-related cutaneous discomfort than smaller electrodes at equivalent current density. *Brain Stimulation*, *7*, 460–467. doi:10.1016/j.brs.2014.01.059

Turi, Z., Ambrus, G. G., Janacsek, K., Emmert, K., Hahn, L., Paulus, W., et al. (2013). Both the cutaneous sensation and phosphene perception are modulated in a frequency-specific manner during transcranial alternating current stimulation. *Restorative Neurology and Neuroscience*, *31*, 275–85. doi:10.3233/RNN-120297

Uhlhaas, P. J. & Singer, W. (2006). Neural synchrony in brain disorders: Relevance for cognitive dysfunctions and pathophysiology. *Neuron*, *52*, 155–168. doi:10.1016/j.neuron.2006.09.020

Uhlhaas, P. J. & Singer, W. (2012). Neuronal dynamics and neuropsychiatric disorders: Toward a translational paradigm for dysfunctional large-scale networks. *Neuron*, *75*, 963–980. doi:10.1016/j.neuron.2012.09.004

Van Veen, B. D. & Buckley, K. M. (1988). Beamforming: A versatile approach to spatial filtering. *IEEE ASSP Magazine*, *5*, 4–24. doi:10.1109/53.665

Van Veen, B. D., Van Drongelen, W., Yuchtman, M., & Suzuki, A. (1997). Localization of brain electrical activity via linearly constrained minimum variance spatial filtering. *IEEE Transactions in Biomedical Engineering*, *44*, 867–880. doi:10.1109/10.623056

Veniero, D., Benwell, C. S. Y., Ahrens, M. M., & Thut, G. (2017). Inconsistent effects of parietal α-tACS on pseudoneglect across two experiments: A failed internal replication. *Frontiers in Psychology*, *8*, 1–14. doi:10.3389/fpsyg.2017.00952

Veniero, D., Bortoletto, M., & Miniussi, C. (2009). TMS-EEG co-registration: On TMS-induced artifact. *Clinical Neurophysiology*, *120*, 1392–1399. doi:10.1016/j.clinph.2009.04.023

Veniero, D., Vossen, A., Gross, J., & Thut, G. (2015). Lasting EEG/MEG aftereffects of rhythmic transcranial brain stimulation: Level of control over oscillatory network activity. *Frontiers in Cellular Neuroscience*, *9*, 477. doi:10.3389/fncel.2015.00477

Violante, I. R., Li, L. M., Carmichael, D. W., Lorenz, R., Leech, R., Hampshire, A., Rothwell, J. C., & Sharp, D. J. (2017). Externally induced frontoparietal synchronization modulates network dynamics and enhances working memory performance. *Elife*, *6*. doi:10.7554/eLife.22001

Conta, Jill von, Florian H. Kasten, Branislava Ćurčić-Blake, André Aleman, Axel Thielscher, and Christoph S. Herrmann. "Interindividual Variability of Electric Fields during Transcranial Temporal Interference Stimulation (TTIS)." *Scientific Reports*, 11, no. 1 (October 13, 2021): 20357. https://doi.org/10.1038/s41598-021-99749-0.

Vöröslakos, M., Takeuchi, Y., Brinyiczki, K., Zombori, T., Oliva, A., Fernández-Ruiz, A., . . . Berényi, A. (2018). Direct effects of transcranial electric stimulation on brain circuits in rats and humans. *Nature Communications*, *9*, 483. doi:10.1038/s41467-018-02928-3

Voss, U., Holzmann, R., Hobson, A., Paulus, W., Koppehele-Gossel, J., Klimke, A., & Nitsche, M. A. (2014). Induction of self-awareness in dreams through frontal low current stimulation of gamma activity. *Nature Neuroscience*, *17*, 810–812. doi:10.1038/nn.3719

Vossen, A., Gross, J., & Thut, G. (2015). Alpha power increase after transcranial alternating current stimulation at alpha frequency (α-tACS) reflects plastic changes rather than entrainment. *Brain Stimulation*, *8*, 499–508. doi:10.1016/j.brs.2014.12.004

Vosskuhl, J., Huster, R. J., & Herrmann, C. S. (2015). Increase in short-term memory capacity induced by down-regulating individual theta frequency via transcranial alternating current stimulation. *Frontiers in Human Neuroscience*, *9*, 257. doi:10.3389/fnhum.2015.00257

Vosskuhl, J., Huster, R. J., & Herrmann, C. S. (2016). BOLD signal effects of transcranial alternating current stimulation (tACS) in the alpha range: A concurrent tACS–fMRI study. *NeuroImage*, *140*, 118–125. doi:10.1016/j.neuroimage.2015.10.003

Wagner, S., Burger, M., & Wolters, C. H. (2016). An optimization approach for well-targeted transcranial direct current stimulation. *SIAM Journal of Applied Mathematics*, *76*, 2154–2174. doi:10.1137/15M1026481

Wagner, T., Rushmore, J., Eden, U., & Valero-Cabre, A. (2009). Biophysical foundations underlying TMS: Setting the stage for an effective use of neurostimulation in the cognitive neurosciences. *Cortex*, *45*, 1025–1034. doi:10.1016/j.cortex.2008.10.002

Weinrich, C. A., Brittain, J. S., Nowak, M., Salimi-Khorshidi, R., Brown, P., & Stagg, C. J. (2017). Modulation of long-range connectivity patterns via frequency-specific stimulation of human cortex. *Current Biology*, *27*, 3061-3068.e3. doi:10.1016/j.cub.2017.08.075

Wilsch, A., Neuling, T., Obleser, J., & Herrmann, C. S. (2018). Transcranial alternating current stimulation with speech envelopes modulates speech comprehension. *NeuroImage*, *172*, 766–774. doi:10.1016/j.neuroimage.2018.01.038

Wischnewski, M., Engelhardt, M., Salehinejad, M. A., Schutter, D. J. L. G., Kuo, M.-F., & Nitsche, M. A. (2018). NMDA receptor-mediated motor cortex plasticity after 20 Hz transcranial alternating current stimulation. *Cerebral Cortex*, 29(7), 2924–2931. doi:10.1093/cercor/bhy160

Wischnewski, M. & Schutter, D. J. L. G. (2017). After-effects of transcranial alternating current stimulation on evoked delta and theta power. *Clinical Neurophysiology*, *128*, 2227–2232. doi:10.1016/j.clinph.2017.08.029

Witkowski, M., Garcia-Cossio, E., Chander, B. S., Braun, C., Birbaumer, N., Robinson, S. E., & Soekadar, S. R. (2016). Mapping entrained brain oscillations during transcranial alternating current stimulation (tACS). *NeuroImage*, *140*, 89–98. doi:10.1016/j.neuroimage.2015.10.024

Woods, A. J., Antal, A., Bikson, M., Boggio, P. S., Brunoni, A. R., Celnik, P., ... Nitsche, M. A.. (2016). A technical guide to tDCS, and related non-invasive brain stimulation tools. *Clinical Neurophysiology*. *127*, 1031–1048. doi:10.1016/j.clinph.2015.11.012

Wöstmann, M., Vosskuhl, J., Obleser, J., & Herrmann, C. S. (2018). Opposite effects of lateralised transcranial alpha versus gamma stimulation on auditory spatial attention. *Brain Stimulation*, *11*, 752–758. doi:10.1016/j.brs.2018.04.006

Zaehle, T., Rach, S., & Herrmann, C. S. (2010). Transcranial alternating current stimulation enhances individual alpha activity in human EEG. *PLoS One* 5, 13766. doi:10.1371/journal.pone.0013766

CHAPTER 23

PARAMETERIZING NEURAL FIELD POTENTIAL DATA

BRADLEY VOYTEK

23.1 Introduction

The proliferation of large-scale, single-unit electrophysiological recording has been a boon for modern neuroscience, permitting us to record greater numbers of neurons in more parts of the brain simultaneously (Buzsáki, 2004). These studies provide tremendous insight regarding how neurons translate input into action, perception, and cognition, and how disease disrupts the normal function of neuronal circuits. While these forms of recording are direct measures of neuronal activity, they are invasive and penetrating—irreparably damaging brain tissue—and therefore they are performed almost exclusively in animals. As opposed to microscale single-unit recordings, meso- and macroscale recordings are aggregated from neuronal populations—hereafter collectively referred to as field potentials. Field potential methods include invasive local field potentials (LFP) and electrocorticography (ECoG), as well as non-invasive magneto- and electroencephalography (M/EEG) (Buzsáki et al., 2012). M/EEG is especially important because it can be performed non-invasively in humans, with EEG being the only functional approach that is used clinically.

Field potential recordings are a less direct measure of neural activity than single-unit approaches: instead of expressly capturing neuronal spiking of the underlying local population, field potentials are largely composed of the integrated postsynaptic currents of the millions of inputs to the region (Buzsáki et al., 2012; Pesaran et al., 2018). These field potential signals exhibit aperiodic and periodic properties (Miller et al., 2012; Gao et al., 2017; Donoghue et al., 2020). The periodic signal consists of both tonic and bursting neural oscillations (Feingold et al., 2015; Jones, 2016; Peterson & Voytek, 2017); oscillations are widely studied, and are linked to numerous physiological, cognitive, behavioral, and disease states (Engel et al., 2001; Buzsáki & Draguhn, 2004; Kopell et al., 2014; Voytek & Knight, 2015).

There is a growing literature focusing on the fact that oscillations are nonsinusoidal (Sherman et al., 2016; Cole & Voytek, 2017, 2018, 2019; Cole et al., 2017; Jackson et al., 2019); specific time-domain characteristics of their nonsinusoidality potentially capture dynamics of the underlying neural circuit (Sherman et al., 2016; Cole & Voytek, 2018). Most analyses of oscillations are conducted on canonically defined frequency bands. This is often done without consideration of the aperiodic component, which has a 1/f-like characteristic in the power spectrum that likely arises from the double-exponential time domain shape of postsynaptic currents (Freeman & Zhai, 2009; Gao et al., 2017). Because power at any given frequency is a mixture of both periodic and aperiodic activity, measuring band-limited power with the presumption that the resulting numerical power value captures *only oscillatory power* is problematic (Haller et al., 2020).

First and foremost, this chapter highlights problems regarding the traditional interpretation of field potential data, and then presents recent approaches for minimizing the damage that our presumptions cause (as related to physiological interpretations of field potential analyses). For example, because field potentials include a mixture of features—aperiodic, periodic, and transients—one cannot be certain that band-limited power values extracted from traditional analyses includes an oscillation at all. Therefore, if we wish to talk about oscillations and their various functional correlates, we must first verify—to the best of our ability—that an oscillation is present above and beyond the aperiodic signal.

As an additional caveat, we must also ensure that spectral power above the aperiodic signal reflects an oscillation at that frequency, rather than a harmonic of a slower, nonsinusoidal rhythm (Cole & Voytek, 2017). Finally, because the aperiodic signal of neural power spectra has received less attention than oscillations, we discuss the origins of this signal, its possible physiological interpretations, and emerging research regarding its cognitive and behavioral relevance. I argue that these three features—oscillations, nonsinusoidal waveforms, and the aperiodic signal—simultaneously exist in the same field potential data, but likely have different physiological interpretations. Without careful parameterization *of all of these features simultaneously*, it is easy to misinterpret their physiological relevance. The goal of this chapter is to help researchers mitigate the propagation of potential physiological (mis)interpretations of their oscillation results by encouraging explicit parameterization of field potential data.

23.2 What is an oscillation?

Oscillations are one of the most widely used analytic approaches in human neuroscience. Part of the excitement about neural oscillations is their ubiquity: they are conserved across species (Bullock, 1981), correlate with numerous cognitive and behavioral processes (Engel et al., 2001; Varela et al., 2001), are disrupted in a variety of neurological and psychiatric disorders (Voytek & Knight, 2015), and have been causally linked to neural information routing and the shaping of spiking networks (Varela et al.,

2001, 2001; Fries, 2005). Mechanistically, oscillatory networks are said to coordinate distributed neural ensembles via synchronizing spike-timing via phase coordination of spiking (Fries, 2005); that is, oscillations have been argued to be the mechanism by which brain regions form functional, dynamic communication networks (Fries, 2005; Voytek et al., 2015a; Helfrich & Knight, 2016).

The vast majority of studies examining neural oscillations assume classic, canonical bands of interest. These are approximately defined as: delta (1–4 Hz), theta (4–8 Hz), alpha (8–12 Hz), beta (16–30 Hz), and low gamma (30–60 Hz). However, these frequency bands are only loosely related to the underlying physiology; there exists a great deal of variability across species (Bullock, 1981), age (Obrist, 1954), and cognitive/behavioral state (Klimesch, 1999; Haegens et al., 2014; Samaha et al., 2015). One concern with such a priori, band-limited analyses is that they are often performed without first examining the power spectrum to account for this intra- and interindividual variability in, for example, center frequency. This means that predetermined frequency bands may include nonoscillatory activity from outside the true physiological oscillatory band—whose center frequency and bandwidth may not fall exactly within a canonical band—thus masking crucial behaviorally and physiologically relevant information.

While there are several methods for identifying individual differences in oscillations, most are restricted to identifying the frequency at which the power spectrum peaks within a specific sub-band (Haegens et al., 2014); additionally, these methods are limited to finding only one oscillation within a specified band while ignoring other potentially physiologically relevant oscillations and their inter-relationships (Donoghue et al., 2020). Importantly, all of these methods measure total band power, not band power *relative to* the aperiodic signal, thus conflating the two processes (Donoghue et al., 2020).

To further complicate matters, oscillations are often not tonic and sustained, but rather appear in brief, intermittent bursts (Feingold et al., 2015; Jones, 2016; Peterson & Voytek, 2017). The functional significance of bursting is supported by a rapidly growing literature, primarily based on cortical LFP recordings from rodents and nonhuman primates (Feingold et al., 2015; Lundqvist et al., 2016; Sherman et al., 2016). Example works have shown that oscillatory burst probability increases during a variety of cognitive and behavioral tasks such as working memory (Lundqvist et al., 2016) or motor control and action completion (Feingold et al., 2015; Sherman et al., 2016). Historically, the separation between the functional role of bursting and sustained oscillations can be highlighted by research into the visual alpha rhythm. The earliest EEG studies of alpha treated it as an idling rhythm (Jasper & Penfield, 1949; Klimesch, 2012), one that is sustained while at rest and becomes intermittent when awake and attentive. The idling perspective was complicated by studies revealing that alpha power in posterior (visual) regions increases with working memory load across many paradigms (Jensen, 2002). The fact that alpha power increases during active working memory conflicts with higher alpha power during eyes closed restful states. Thus, the idling hypothesis was updated to an account where alpha acts to rhythmically inhibit neural gain (Jensen & Mazaheri, 2010). In this updated view, regions not receiving sensory information, or otherwise engaging in active behavioral responses, were said to be suppressed to avoid introducing noise into the larger network. When stimulated, task-relevant regions are disinhibited, restoring cortical processing. (For an in-depth perspective on the alpha rhythm, see Chapter 10.)

Recent computational work from our lab has challenged this classic view of alpha as a suppression rhythm (Peterson & Voytek, 2017). When alpha oscillations are modeled specifically as balanced gain modulation, alpha can suppress visual detection. When suppression acts over long periods (e.g., >1 sec) this strongly reduces the overall gain, per the standard view. That is, this low-gain state reflects an "idle" state of activity, serving to inhibit firing output for any given input. Surprisingly, however, the model shows that when alpha is in a short, bursting mode (<0.5 sec, or 5 cycles), alpha actually *enhances* neuronal firing. That is, modeling and physiological work uncovered an enticing result: that oscillatory bursts may play a different functional role compared to sustained oscillations *of the same frequency*. This result—that oscillatory bursts may have a different physiological effect than sustained rhythms—highlights the importance of parameterizing not just the power of an oscillation, but its short-time temporal dynamics.

23.3 Nonsinusoidal oscillations

Neural oscillations are not smoothly varying sinusoids. Rather, they manifest many different nonsinusoidal characteristics. Across brain regions, species, and frequencies, there do exist a variety of stereotyped nonsinusoidal shapes such as human motor cortical beta oscillations (Cole et al., 2017; Jackson et al., 2019), sleep slow oscillations (Steriade et al., 1993; Amzica & Steriade, 1998), the sensorimotor mu rhythm (Kuhlman, 1978; Arroyo et al., 1993), and especially the rodent hippocampal theta rhythm (Belluscio et al., 2012; Cole & Voytek, 2018). The nonsinusoidal shape of oscillation waveforms is an exciting, re-emerging candidate for potentially indexing circuit-level physiology, such as synaptic input synchrony (Sherman et al., 2016). It is becoming increasingly apparent that nonsinusoidal oscillatory waveform shape carries physiological information (Sherman et al., 2016; Cole & Voytek, 2017, 2018). Because there does not seem to be a theoretical reason why brain oscillations should be sinusoidal, it may be that the diverse set of neuronal activation and synaptic current dynamics present in different oscillations determine the waveform shape in the nearby field potential.

Several nonsinusoidal features of an oscillation can be parameterized, such as a waveform's symmetry (the ratio between the rise period and the decay period), the sharpness of each oscillation peak, the rise and decay times for each cycle's voltage, and so on. To quantify these features, an oscillation can explicitly be broken up into individual cycles by identifying the locations of peaks and troughs and then computing features on the raw voltage time series (Figure 23.1) (Cole & Voytek, 2019). We recently reviewed studies that have related the waveform shape of brain oscillations to neurophysiology (Cole & Voytek, 2017). One of the most well-supported of these relationships shows that

(a)

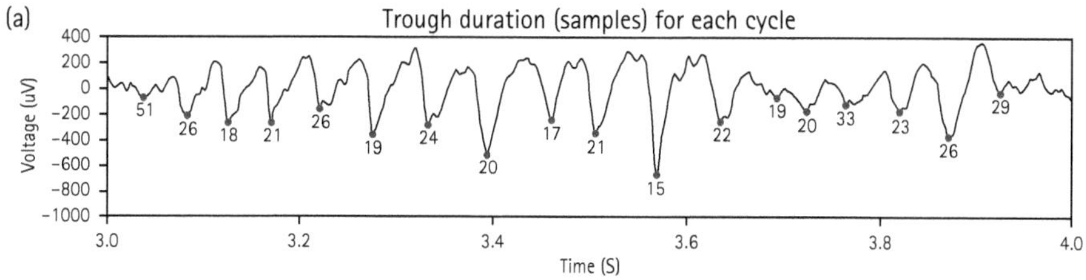

	amplitude	period	time_rise	time_trough	time_decay	time_peak	volt_rise	volt_trough	volt_decay	volt_peak
0	27.442521	46	26	18	20	22	77.542938	-18.474257	127.051091	108.576834
1	34.980215	48	37	19	11	21	123.623268	-84.545266	143.613947	59.068681
2	19.290058	40	31	24	9	29	161.573095	-87.419348	126.497350	39.078002
3	18.898692	53	7	31	46	30	113.252293	15.490505	58.663243	74.153747
4	20.070945	29	16	10	13	13	149.175399	31.155748	97.587049	128.742797

(b)

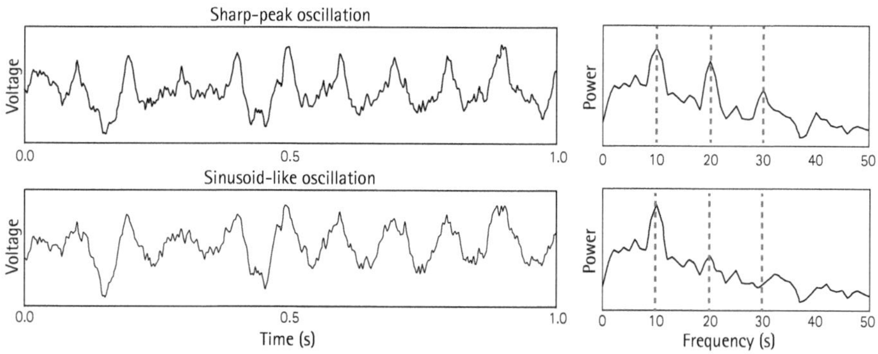

FIGURE 23.1 Nonsinusoidal oscillations: (A, Top) Example of a time-domain parameterization approach for quantifying nonsinusoidal features of neural oscillations, on a cycle-by-cycle basis. Here, the duration of the trough is annotated directly on each cycle's trough. (A, Bottom) For each cycle, features such as amplitude, period, rise- and decay times, asymmetries, and so on, are extracted and stored. (B, Left) Two simulated neural time series where the main difference between the two is the "sharpness" of each waveform peak. The top rhythm is sharper and less sinusoidal than the bottom. (B, Right) The power spectral representations of these two time series are markedly different, such that the sharpness of the waveforms in the top trace are accompanied by significant harmonic peaks that are absent in the smoother rhythm (horizontal dashed blue lines at 10, 20, and 30 Hz).

sharper extrema of oscillations relate to higher synchrony of a neuronal population. This has been shown, for example, for motor cortical beta oscillations (Sherman et al., 2016) and for alpha oscillations in the rat gustatory cortex (Tort et al., 2010).

While narrowband power above the aperiodic signal is the definition of an oscillation, even this definition can be misleading due to a natural consequence of the mathematics of how nonsinusoidal waveforms are captured by the Fourier transform. This is critically important, because we showed recently how the harmonic effects that result from nonsinusoidal waveform shape can also lead to the appearance of strong cross-frequency coupling (Cole et al., 2017)—the statistical interaction between multiple different oscillatory processes (Canolty et al., 2006; Canolty & Knight, 2010)—when in fact no such multiple-oscillation interaction need exist. Given the role that cross-frequency coupling has been argued to play in organizing neural activity across regions (Canolty & Knight, 2010; Voytek et al., 2013, 2015a), it is imperative that the physiological nature of any statistical interaction between multiple oscillations is clear.

Such cross-frequency coupling can take many forms, including phase-phase coupling and phase-amplitude coupling (PAC) (Bruns et al., 2000; Varela et al., 2001; Canolty & Knight, 2010). The results of experiments leveraging such approaches are often interpreted as implying mechanism: for example, phase-phase coupling may suggest multiplexing of information across different frequencies while PAC may represent the biasing of population spiking by an oscillation electric field. It has been recently shown that patients with Parkinson's disease have significantly higher beta-to-high gamma (70–150 Hz) PAC in their motor cortices. With the application of subthalamic deep brain stimulation (DBS), this PAC significantly decreases in a manner that relates to improved clinical outcomes (de Hemptinne et al., 2015). We recently analyzed those same data to show that, rather than abolishing pathological multiple-oscillation cross-frequency coupling, DBS effects could more parsimoniously be described by a reduction in the waveform sharpness of a single beta oscillator (Cole et al., 2017).

The DSB example highlights the practical, clinical importance of considering whether *measured outputs from algorithmic approaches* to oscillations analysis, such as PAC, derive multiple rhythmic processes vs. a single non-sinusoidal oscillator. This point was elegantly demonstrated in a recent paper that leveraged a large population of intracranial human recordings across the neocortex (2,750 contacts from 33 patients) (Vaz et al., 2017). This paper showed that in some cases—often in anterior frontal and ventral temporal regions—significant PAC arose from coupled oscillations, where two independent rhythms are truly coupled. In other cases—mostly in somatosensory regions and lateral temporal cortex—significant PAC arose from the nonsinusoidality of a *single rhythm*. That is, the PAC *metric* only informs about the statistical strength of the relationship, not the physiological or signal-processing origin that gives rise to the PAC.

23.4 The aperiodic signal

Until recently, the aperiodic signal has been largely referred to as electrophysiological "noise" or the "background." There is mounting evidence, however, that this signal carries physiological information (Gao et al., 2017) and is dynamically altered by cognitive and perceptual states (Waschke et al., 2017; Dave et al., 2018), as well as in aging (Voytek et al., 2015b; Waschke et al., 2017; Dave et al., 2018) and disease (Peterson et al., 2018; Robertson et al., 2019; Veerakumar et al., 2019).

This aperiodic signal has previously been referred to as the 1/f background and/or neural noise, because power at any given frequency is inversely related to the frequency itself (1/f) and looks like colored (pink or brown) "noise" seen in digital signal processing. However, this is likely not *just* "noise", because, as noted, field potentials are dominated by postsynaptic currents across relatively large populations (Buzsáki et al., 2012; Pesaran et al., 2018). These currents are driven by input from neurons that are integrated across a large number of neurons and thus can appear noise-like. However, we have shown that the 1/f-like nature of the signal can arise simply as a consequence of the nonlinear nature of post-synaptic currents, and that changes in the exponent—or degree—of the 1/f drop-off may reflect the relative balance of excitatory and inhibitory (EI) currents coming in to the region (Gao et al., 2017). For this reason, we refer to this feature as the aperiodic signal, rather than referring to it as "noise" (Donoghue et al., 2020).

The prospect that the aperiodic signal may index EI balance is exciting, but much more work is required to assess the strength and accuracy of this relationship. Unfortunately, because field potential spectra are a combination of periodic and aperiodic components, and because the large-power periodic oscillatory bumps can easily lead to mismeasurements of the aperiodic component, a method for carefully and accurately separating and parametrizing those components is critical. This is especially important given the mounting computational and animal single-unit evidence that suggests that EI balance may be the physiological basis for top-down gain modulation (Chance et al., 2002) and persistent delay period activity required for working memory maintenance (Lim & Goldman, 2013).

23.5 Parameterizing Neural Field Potentials

In order for researchers to state that oscillatory power in frequency *X* has altered as a function of task, condition, behavior, or group, they must first demonstrate that such an oscillation exists in their data, and that the apparent change in oscillatory power is due to a true reduction in that oscillation as opposed to a change in its center frequency moving it outside the analyzed band, a shift in its bandwidth, or from a shift in the aperiodic signal. Additionally, these oscillations need to be demonstrated to be true rhythms, and not harmonics caused by nonsinusoidal features of a rhythm at another frequency.

Problematically, standard analytic approaches conflate periodic parameters (center frequency, power, bandwidth) with aperiodic ones (offset, exponent). This compromises physiological interpretations. Additionally, separating bursting from sustained oscillations can be difficult, if not impossible, if just looking at the spectral representation of the data (Jones, 2016). To further compound the problem, a single nonsinusoidal, rhythmic process that has "sharp" extrema will manifest, in the power spectrum, as an oscillatory bump at the frequency of the rhythm *as well as* exhibiting smaller bumps at integer harmonics of the rhythm's primary frequency. This means that a sharp 10-Hz oscillation will also appear as a bump at 20 Hz, and possibly at 30 Hz and beyond. In such a scenario—one that is common due to the ubiquity of nonsinusoidal brain rhythms—a narrowband analysis at, say, 20 Hz, would give the researcher the impression that a 20-Hz beta oscillation exists in their data in conjunction with the 10-Hz alpha, when in fact all that is present is a nonsinusoidal 10-Hz rhythm. Further, should this 10-Hz sharpness change in relation to behavioral state, such as a task-related reduction in sharpness resulting in a "smoother" 10-Hz shape, power in the 20-Hz harmonic would concomitantly be reduced. This could easily lead to the results being interpreted as a task-related reduction in 20-Hz beta power, when in fact no such beta oscillation truly existed in the first place.

There are currently several algorithms for identifying oscillations in specific ways that have attempted to address some of these concerns individually, but never conjointly. In particular, an approach called BOSC (Better OSCillation Detector) (Hughes et al., 2012), begins by fitting a linear regression to the *log-log* PSD to estimate the aperiodic signal. This is used to determine a power threshold, which is then used in combination with a duration threshold to define oscillations in wavelet-based decompositions of the time series data. However, a significant limitation of this and other similar methods is that a simple linear fit of the background spectrum can be significantly skewed by the presence of oscillations—especially large oscillations—and therefore mischaracterizes the aperiodic signal.

Another similar approach is the irregular-resampling auto-spectral analysis (IRASA) method, which seeks to explicitly separate the periodic and aperiodic components through a resampling procedure (Wen & Liu, 2016). Though conceptually similar, this resampling method is computationally much more expensive, and may have trouble separating large amplitude oscillations from the aperiodic signal. Other methods, such as principle component variants fail to separate periodic and aperiodic features, and require manual component selection (Miller et al., 2012). Consistent with previous work, we show that the aperiodic signal is of significant physiological and behavioral interest, although all of these methods treat it as a nuisance variable, such as correcting for it via spectral whitening rather than a feature to be explicitly modeled.

To overcome the limitations of traditional narrowband analyses, to reduce the errors caused by conflating periodic and aperiodic features, and to address the nonsinusoidal nature of field potential signals, we recently developed several open-source algorithms for parameterizing field potential data. The first is a semi-automated parameterization of neural power spectra (Donoghue et al., 2020). Spectra are parameterized as a linear combination of the aperiodic component and putative periodic oscillations (Figure 23.2). This method thus operates upon frequency representations of time-series field potential data. The algorithm considers the PSD as the linear sum of an aperiodic signal upon which there oscillatory "bumps," or frequency regions of power over and above this aperiodic signal, can exist. These bumps are considered to be putative oscillations and modeled individually as Gaussians. Each Gaussian then characterizes the frequency definition of the oscillation, whereby the mean, amplitude, and standard deviation can be interpreted as the center frequency, power, and bandwidth, respectively,

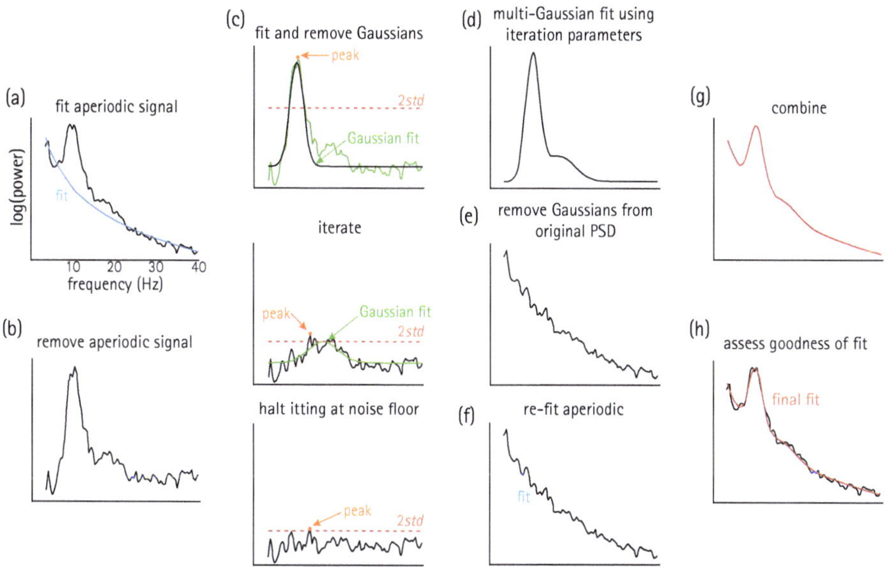

FIGURE 23.2 Parametrization algorithm applied to field potential data. (A) The power spectral density (PSD) is first fit with an estimated aperiodic signal (blue), defined by two parameters: a slope and an offset. (B) The estimated aperiodic portion of the signal is subtracted from the raw PSD, the residuals of which are assumed to be a mix of periodic oscillatory peaks and noise. (C) The maximum (peak) of the residuals is found. If this peak is above the noise floor (2*std*; red dashed line) then a Gaussian (green) is fit around this peak based on the peak's frequency, amplitude, and estimated bandwidth. The fitted Gaussian is then subtracted, and the process is iterated until the noise floor is reached (bottom). These values are used as seeds for the multi-Gaussian fitting in D. (D) Having identified the number of putative oscillations, based on the number of peaks above the noise floor, multi-Gaussian fitting is then performed on the aperiodic-adjusted signal from B to account for the joint power contributed by all the putative oscillations, together. (E) This multi-Gaussian model is then subtracted from the original PSD from A. (F) This is done to give a better estimate of the aperiodic signal—one that is less corrupted by the large oscillations present in the original PSD. (G) This re-fit aperiodic signal is combined with the multi-Gaussian model to give the final fit. (H) The final fit—here parameterized as a line (aperiodic signal) and two Gaussians (putative oscillations)—captures >99% of the variance of the original PSD. In this example, the extracted parameters for the aperiodic signal are: broadband offset = -21.4 au; slope = -1.12 au/Hz. Two Gaussians were found, with the parameters: (1) frequency = 10.0 Hz amplitude = 0.69 au, bandwidth = 3.18 Hz; (2) frequency = 16.3 Hz, amplitude = 0.14 au, bandwidth = 7.03 Hz.

This figure appears in https://doi.org/10.1038/s41593-020-00744-x

of the oscillation. The final outputs of the algorithm are the definition of the fit aperiodic signal, and the definitions for *N* Gaussians, where *N* is the number of oscillations found in the PSD. Notably, this algorithm extracts all these parameters together in a manner that accounts for potentially overlapping oscillations; it also minimizes the degree to which they are confounded and requires no specification of canonical oscillation bands.

As for waveform shape methods, multiple approaches have been developed. In particular, instantaneous phase estimation methods for hippocampal theta have been modified to consider the peak and trough locations in a broadband signal to account for nonsinusoidal features. Another approach, empirical mode decomposition, has the theoretical capability of extracting a nonsinusoidal oscillation in a single component (Sweeney-Reed et al., 2018), although in practice on neural signals the rhythms of interest spread across multiple components, meaning the decomposition does not reliably separate components of interest.

Though there are still many open questions regarding approaches to analyzing waveform shape, we have developed a set of available methods that cover critical basics (Figure 23.1). Our current toolbox for parameterizing cycle-by-cycle waveform features (*bycycle*) breaks an oscillation up into individual cycles by identifying the locations of peaks and troughs and then computing features on the raw voltage time-series. After the signal is segmented into cycles, each cycle is characterized by a set of parameters. The amplitude of the cycle is computed as the average voltage difference between the trough and the two adjacent peaks. The period is defined as the time between the two peaks. Rise-decay symmetry is the fraction of the period that was composed of the rise time. Peak-trough symmetry is the fraction of the period, encompassing the previous peak and current trough, that was composed of the peak. The distributions of these features can be computed across all cycles in a signal in order to compare oscillation properties in different neural signals.

However, in order to analyze cycle-by-cycle features, we must first identify, in the time domain, whether a signal has an oscillatory cycle or not. After computing features for each cycle, following the waveform shape algorithm given, an additional step is done to determine whether each cycle is part of an oscillatory burst. Traditional burst-detection methods are relatively simple: bursts are detected by comparing instantaneous, band-limited power at each time point to the median power (usually three times median) in the channel of interest (Feingold et al., 2015). When the amplitude rises above this threshold, the signal is said to have entered a bursting state; when it falls below another threshold (e.g., 1.5 times median) it is said to exit the burst state. While this is acceptable in some cases, for very stable oscillations and very intermittent oscillations, this criterion becomes unreliable because it will often fail to separate a stable oscillation from brief nonoscillatory segments. Further, this median-threshold procedure has never been explicitly tested against ground-truth simulated neural data. Oscillatory bursts are identified as time periods in which consecutive cycles in the time-series had similar amplitudes, similar periods, and rise and decay flanks that are predominantly monotonic (Cole & Voytek, 2019).

Unlike our approaches, these other methods are not generalizable across the numerous different forms of field potential data, nor are they designed and optimized to be parallelizable or to run in a cloud environment. Additionally, none are designed to examine all of the features in concert, which we show is critical for teasing apart the physiological interpretations of changes to oscillations, the aperiodic signal, and waveform shape.

23.6 Discussion

The fact that both periodic oscillations and the aperiodic signal are seen in field potential data hints at the promise for those signals to bridge across those spatial scales. In studying neural oscillations, the analyses we apply are often very complicated, mathematically intensive, and full of assumptions, both explicit and implicit. Therefore, careful consideration of the methods applied to our data is paramount, as seemingly arbitrary choices in the hyperparameters of our analysis, such as the minimum number of spikes required for including a neuron in an analysis or the length and precise cutoff frequencies of a filter, can have large impacts on the results and ultimate conclusions. Often, in-depth knowledge of the techniques is required in order to appropriately choose hyperparameters and assure the validity of our conclusions. Because considerable effort is required to obtain this knowledge, this means that we will often make honest mistakes in our analysis and as peer reviewers, because we often miss the statistical confounds that may underlie highly impactful results.

Therefore, we first advocate paying deliberate and careful attention to the raw data to help gain a maximal understanding of broad features; complicated methods can often transform our data in ways we do not expect. If we do not understand how we see the ultimate effect by looking at the raw data, there is reason for concern, or at least further investigation. It can similarly be useful to apply multiple methods to our data. When it comes to analyzing neural oscillations, there are several analytic options to choose from (e.g., in the frequency or time-domains), and so this choice should be made consciously.

The choice for the analytic method applied can strongly impact the ultimate conclusion. As we showed, phase-amplitude coupling analysis and sharpness analysis were capturing the same phenomena in the data. However, the former favored a conclusion of coupled oscillations, whereas we favored the latter conclusion concerning synchrony of transmembrane currents. The choice of developing a time-domain approach to analyzing neural oscillations was very deliberate: as physical processes, including neural dynamics, happen over time (as opposed to being generated in the frequency domain), there are advantages in analyzing signals in this natural domain and directly measuring and accounting for non-stationarities in the dynamics.

While it is certainly a biased perspective, we believe that neural oscillations research would be better positioned if analysis of these rhythms combined time-domain and frequency-domain approaches, using careful parameterization. Time-domain cycle-by-cycle parameterization offers several advantages over frequency-domain approaches. First, it directly quantifies waveform asymmetry, which is only indirectly and ambiguously captured in spectral analysis (i.e., similar harmonic patterns can be generated by different oscillations that produce diverse waveforms). Second, cycle-by-cycle parameterization inherently runs an oscillation detection algorithm, so oscillatory features are only analyzed on appropriate portions of the signal (i.e., where the oscillation is observable). Third, cycle-by-cycle parameterization offers time-resolved estimates of oscillatory features with an appropriate degree of temporal resolution: not only is symmetry measured for a single cycle, but the estimates of amplitude and frequency are intuitively measured in terms of peak-to-trough voltage and trough-to-trough time, respectively. However, "instantaneous" estimates of amplitude and frequency that are comparatively used offer amplitude and frequency estimates at every point in time. We have previously demonstrated that the cycle-by-cycle approach can provide more sensitive measures of amplitude and frequency, compared to more traditional filter-and-Hilbert transform approaches, by better differentiating simulated experimental conditions (Cole & Voytek, 2019).

One downside of the time-domain approach outlined is that it can miss relatively weaker oscillations, and that fine-tuning of oscillation-detection hyperparameters can influence results. For this reason, we advocate including a frequency-domain parameterization approach to help identify which frequencies have power above the aperiodic signal, during which times. In this way, both parameterization approaches can be used synergistically.

To date, tens of thousands of studies have been published regarding neural oscillations, their function, and their behavioral, cognitive, and disease correlates. Most of these studies have been conducted using traditional approaches that assume canonical frequency bands—delta, theta, alpha/mu, beta, and gamma. Often these oscillation bands are described as having roles themselves, such as theta being equated with memory, and alpha with attention and wakefulness. This approach arises due to letting predefined bands guide analyses, rather than allowing the data to guide the analyses. That is, canonical band analyses commit us to tacit acceptance of predefined oscillatory bands having a functional role, rather than considering the underlying physiological mechanisms that may generate different spectral features and addressing inter-individual differences. With proper parameterization, it is possible to broaden our perspective, allowing us to take full advantage of the rich variance present in oscillatory data. This increases analytical power and can potentially provide greater insight into physiological mechanisms underlying oscillations and the role that oscillatory variability may play in explaining individual differences in cognitive functioning in health, aging, and disease.

ACKNOWLEDGEMENTS

I sincerely thank Scott Cole, Thomas Donoghue, Richard Gao, Matar Haller, Erik Peterson, Natalie Schaworonkow, Avgusta Shestyuk, Tammy Tran, and Roemer van der Meij for all the years of hard work, careful thought, and patience with these ideas. My support for this work came from a Sloan Research Fellowship (FG-2015-66057), the Whitehall Foundation (2017-12-73), the National Science Foundation under grant BCS-1736028, and the National Institute of General Medical Sciences grant 1R01GM134363-01. The author declares no competing financial interests. Portions of this chapter are based on/informed by Donoghue et al., 2020; Cole & Voytek, 2018; Peterson & Voytek, 2018.

References

Amzica, F. & Steriade, M. (1998). Electrophysiological correlates of sleep delta waves. *Electroencephalography and Clinical Neurophysiology, 107*, 69–83.

Arroyo, S., Lesser, R. P., Gordon, B. , Uematsu, S., Jackson, D., & Webber, R. (1993). Functional significance of the mu rhythm of human cortex: An electrophysiologic study with subdural electrodes. *Electroencephalography and Clinical Neurophysiology, 87*, 76–87.

Belluscio, M. A., Mizuseki, K., Schmidt, R., Kempter, R., & Buzsáki, G. (2012). Cross-frequency phase-phase coupling between theta and gamma oscillations in the hippocampus. *Journal of Neuroscience, 32*, 423–435.

Bruns, A., Eckhorn, R., Jokeit, H., & Ebner, A. (2000). Amplitude envelope correlation detects coupling among incoherent brain signals: *Neuroreport, 11*, 1509–1514.

Bullock, T. H. (1981). Neuroethology deserves more study of evoked responses. *Neuroscience, 6*(7), 1203–1215.

Buzsáki, G. (2004). Large-scale recording of neuronal ensembles. *Nature Neuroscience, 7*, 446–451.

Buzsáki, G., Anastassiou, C. A., & Koch, C. (2012). The origin of extracellular fields and currents—EEG, ECoG, LFP and spikes. *Nature Reviews Neuroscience, 13*, 407–420.

Buzsáki, G. & Draguhn, A. (2004). Neuronal oscillations in cortical networks. *Science, 304*, 1926–1929.

Canolty, R. T., Edwards, E., Dalal, S. S., Soltani, M., Nagarajan, S. S., Kirsch, H. E., ... Knight, R. T. (2006). High gamma power is phase-locked to theta oscillations in human neocortex. *Science, 313*, 1626–1628.

Canolty, R. T. & Knight, R. T. (2010). The functional role of cross-frequency coupling. *Trends in Cognitive Sciences, 14*, 506–515.

Chance, F. S., Abbott, L. F., & Reyes, A. D. (2002). Gain modulation from background synaptic input. *Neuron, 35*, 773–782.

Cole, S. & Voytek, B. (2019). Cycle-by-cycle analysis of neural oscillations. *Journal of Neurophysiology, 122*(2), 849–861.

Cole, S. R., van der Meij, R., Peterson, E. J., de Hemptinne, C., Starr, P. A., & Voytek, B. (2017). Nonsinusoidal beta oscillations reflect cortical pathophysiology in Parkinson's disease. *Journal of Neuroscience, 37*, 4830–4840.

Cole, S. R. & Voytek, B. (2017). Brain oscillations and the importance of waveform shape. *Trends in Cognitive Sciences, 21*, 137–149.

Cole, S. R. & Voytek, B. (2018). Hippocampal theta bursting and waveform shape reflect CA1 spiking patterns. *bioRxiv* [online]. http://biorxiv.org/lookup/doi/10.1101/452987

Dave, S., Brothers, T. A., & Swaab, T. Y. (2018). 1/ f neural noise and electrophysiological indices of contextual prediction in aging. *Brain Research, 1691*, 34–43.

de Hemptinne, C., Swann, N. C., Ostrem, J. L., Ryapolova-Webb, E. S., San Luciano, M., Galifianakis, N. B., & Starr, P. A. (2015). Therapeutic deep brain stimulation reduces cortical phase-amplitude coupling in Parkinson's disease. *Nature Neuroscience, 18*, 779–786.

Engel, A. K., Fries, P., & Singer, W. (2001). Dynamic predictions: Oscillations and synchrony in top–down processing. *Nature Reviews Neuroscience, 2*, 704–716.

Feingold, J., Gibson, D. J., DePasquale, B., & Graybiel, A. M. (2015). Bursts of beta oscillation differentiate postperformance activity in the striatum and motor cortex of monkeys performing movement tasks. *Proceedings of the National Academy of Sciences of the United States of America, 112*, 13687–13692.

Freeman, W. J. & Zhai, J. (2009). Simulated power spectral density (PSD) of background electrocorticogram (ECoG). *Cognitive Neurodynamics*, *3*, 97–103.

Fries, P. (2005). A mechanism for cognitive dynamics: neuronal communication through neuronal coherence. *Trends in Cognitive Sciences*, *9*, 474–480.

Gao, R., Peterson, E. J., & Voytek, B. (2017) Inferring synaptic excitation/inhibition balance from field potentials. *NeuroImage*, *158*, 70–78.

Haegens, S., Cousijn, H., Wallis, G., Harrison, P. J., & Nobre, A. C. (2014). Inter- and intra-individual variability in alpha peak frequency. *NeuroImage*, *92*, 46–55.

Donoghue, T., Haller, M., Peterson, E., Varma, P., Sebastian, P., Gao, R., ... Voytek, B. (2020). Parameterizing neural power spectra into periodic and aperiodic components. *Nature Neuroscience*, *23*,1655–1665.

Helfrich, R. F. & Knight, R. T. (2016). Oscillatory dynamics of prefrontal cognitive control. *Trends in Cognitive* Sciences, *20*, 916–930.

Hughes, A. M., Whitten, T. A., Caplan, J. B., & Dickson C. T. (2012). BOSC: A better oscillation detection method, extracts both sustained and transient rhythms from rat hippocampal recordings. *Hippocampus*, *22*, 1417–1428.

Jackson, N., Cole, S. R., Voytek, B., & Swann, N. C. (2019). Characteristics of waveform shape in Parkinson's disease detected with scalp electroencephalography. *eNeuro*, *6*(3), ENEURO.0151-19.2019.

Jasper, H. & Penfield, W. (1949). Electrocorticograms in man: Effect of voluntary movement upon the electrical activity of the precentral gyrus. *Archiv für Psychiatrie & Nervenkrankh*, *183*, 163–174.

Jensen, O. (2002). Oscillations in the alpha band (9-12 Hz) increase with memory load during retention in a short-term memory task. *Cerebral Cortex*, *12*, 877–882.

Jensen, O. & Mazaheri, A. (2010). Shaping functional architecture by oscillatory alpha activity: Gating by inhibition. *Frontiers in Human Neuroscience*, *4*, 186. doi:10.3389/fnhum.2010.00186

Jones, S. R. (2016). When brain rhythms aren't 'rhythmic': Implication for their mechanisms and meaning. *Current Opinion in Neurobiology*, *40*, 72–80.

Klimesch, W. (1999). EEG alpha and theta oscillations reflect cognitive and memory performance: A review and analysis. *Brain Research Reviews*, *29*, 169–195.

Klimesch, W. (2012). Alpha-band oscillations, attention, and controlled access to stored information. *Trends in Cognitive Sciences*, *16*, 606–617.

Kopell, N. J., Gritton, H. J., Whittington, M. A., & Kramer, M. A. (2014). Beyond the connectome: The dynome. *Neuron*, *83*, 1319–1328.

Kuhlman, W. N. (1978). Functional topography of the human mu rhythm. *Electroencephalography and Clinical Neurophysiology*, *44*, 83–93.

Lim, S. & Goldman, M. S. (2013). Balanced cortical microcircuitry for maintaining information in working memory. *Nature Neuroscience*, *16*,1306–1314.

Lundqvist, M., Rose, J., Herman, P., Brincat, S. L., Buschman, T. J., & Miller, E. K. (2016). Gamma and beta bursts underlie working memory. *Neuron*, *90*, 152–164.

Miller, K. J., Hermes, D., Honey, C. J., Hebb, A. O., Ramsey, N. F., Knight, R. T., Ojemann, J. G., & Fetz, E. E. (2012). Human motor cortical activity is selectively phase-entrained on underlying rhythms. *PLoS Computational Biology*, *8*, e1002655.

Obrist, W. D. (1954). The electroencephalogram of normal aged adults. *Electroencephalography and Clinical Neurophysiology*, *6*, 235–244.

Pesaran, B., Vinck, M., Einevoll, G. T., Sirota, A., Fries, P., Siegel, M., … Srinivasan, R. (2018). Investigating large-scale brain dynamics using field potential recordings: analysis and interpretation. *Nature Neuroscience*, http://www.nature.com/articles/s41593-018-0171-8

Peterson, E. J., Rosen, B. Q., Campbell, A. M., Belger, A., & Voytek, B. (2018). 1/f neural noise is a better predictor of schizophrenia than neural oscillations. *bioRxiv* [online]. http://biorxiv.org/lookup/doi/10.1101/113449

Peterson, E. J. & Voytek, B. (2017). Alpha oscillations control cortical gain by modulating excitatory-inhibitory background activity. *bioRxiv* [online]. http://biorxiv.org/lookup/doi/10.1101/185074

Robertson, M. M., Furlong, S., Voytek, B., Donoghue, T., Boettiger, C. A., & Sheridan, M. A. (2019). EEG power spectral slope differs by ADHD status and stimulant medication exposure in early childhood. *Journal of Neurophysiology, 122*(6), 2427–2437. doi:10.1152/jn.00388.2019.

Samaha, J., Bauer, P., Cimaroli, S., & Postle, B. R. (2015). Top-down control of the phase of alpha-band oscillations as a mechanism for temporal prediction. *Proceedings of the National Academy of Sciences of the United States of America, 112*, 8439–8444.

Sherman, M. A., Lee, S., Law, R., Haegens, S., Thorn, C. A., Hämäläinen, M. S., Moore, C. I., & Jones, S. R. (2016). Neural mechanisms of transient neocortical beta rhythms: Converging evidence from humans, computational modeling, monkeys, and mice. *Proceedings of the National Academy of Sciences of the United States of America, 113*, E4885–E4894.

Steriade, M., Contreras, D., & Curr, F. (1993). The slow (4 Hz) oscillation in reticular thalamic and thalamocortical neurons: Scenario of sleep rhythm generation in interacting thalamic and neocortical networks. *Journal of Neuroscience, 13*(8), 3284–3299.

Sweeney-Reed, C. M., Nasuto, S. J., Vieira, M. F., & Andrade, A. O. (2018). Empirical mode decomposition and its extensions applied to EEG analysis: A review. *Advances in Data Science and Adaptive Analysis, 10*(2), 1840001.

Tort, A. B. L., Fontanini, A., Kramer, M. A., Jones-Lush, L. M., Kopell, N. J., & Katz, D. B. (2010). Cortical networks produce three distinct 7-12 Hz rhythms during single sensory responses in the awake rat. *Journal of Neuroscience, 30*, 4315–4324.

Varela, F., Lachaux, J-P., Rodriguez, E., & Martinerie, J. (2001). The brainweb: Phase synchronization and large-scale integration. *Nature Reviews Neuroscience, 2*, 229–239.

Vaz, A. P., Yaffe, R. B., Wittig, J. H., Inati, S. K., & Zaghloul, K. A. (2017). Dual origins of measured phase-amplitude coupling reveal distinct neural mechanisms underlying episodic memory in the human cortex. *NeuroImage, 148*, 148–159.

Veerakumar, A., Tiruvadi, V., Howell, B., Waters, A. C., Crowell, A. L., Voytek, B., … Mayberg, H. S. (2019). Field potential 1/f activity in the subcallosal cingulate region as a candidate signal for monitoring deep brain stimulation for treatment-resistant depression. *Journal of Neurophysiology, 122*(3), 1023–1035.

Voytek, B., D'Esposito, M., Crone, N., & Knight, R. T. (2013). A method for event-related phase/amplitude coupling. *NeuroImage, 64*, 416–424.

Voytek, B., Kayser, A. S., Badre, D., Fegen, D., Chang, E. F., Crone, N. E., … D'Esposito, M. (2015a). Oscillatory dynamics coordinating human frontal networks in support of goal maintenance. *Nature Neuroscience, 18*, 1318–1324.

Voytek, B. & Knight, R. T. (2015). Dynamic network communication as a unifying neural basis for cognition, development, aging, and disease. *Biological Psychiatry, 77*, 1089–1097.

Voytek, B., Kramer, M. A., Case, J., Lepage, K. Q., Tempesta, Z. R., Knight, R. T., & Gazzaley, A. (2015b). Age-related changes in 1/f neural electrophysiological noise. *Journal of Neuroscience*, *35*, 13257–13265.

Waschke, L., Wöstmann, M., & Obleser, J. (2017). States and traits of neural irregularity in the age-varying human brain. *Scientific Reports*, *7*, art no. 17381. https://doi.org/10.1038/s41598-017-17766-4

Wen, H. & Liu, Z. (2016). Separating fractal and oscillatory components in the power spectrum of neurophysiological signal. *Brain Topography*, *29*, 13–26.

INDEX

For the benefit of digital users, indexed terms that span two pages (e.g., 52–53) may, on occasion, appear on only one of those pages.

Tables and figures are indicated by *t* and *f* following the page number

C

G

H

I

S